Machine Tools and Machining Practices

Volume 2 Machine Tools and Machining Practices

Formerly of Foothill and DeAnza Colleges **Warren T. White**

Machine Technology Lane Community College **John E. Neely**

Richard R. Kibbe

Machine Technology Lane Community College **Roland O. Meyer**

JOHN WILEY & SONS
New York · Chichester · Brisbane · Toronto · Singapore

This book was set in Melior by Graphic Arts Composition. The designer was A Good Thing, Inc. in coordination with the Wiley Design Department. New drawings were executed by the Wiley Illustration Department. Joan Tobin supervised production.

Copyright © 1977, by John Wiley & Sons, Inc.

All rights reserved. Published simultaneously in Canada.

Reproduction or translation of any part of this work beyond that permitted by Sections 107 or 108 of the 1976 United States Copyright Act without the permission of the copyright owner is unlawful. Requests for permission or further information should be addressed to the Permissions Department, John Wiley & Sons, Inc.

Library of Congress Cataloging in Publication Data:

Main entry under title:

Machine tools and machining practices.

   Includes indexes.
   1. Machine-tools.   2. Machine-shop practice.
I. White, Warren T.

TJ1185.M224   621.8   76-27863
ISBN 0-471-94036-4(v.2)

Printed in the United States of America

20 19 18 17 16 15 14 13 12 11

preface

The two volumes in this series are intended for students who are seriously thinking of becoming machinists, either through apprenticeship training, vocational schools, or community college programs. The content deals with topics usually presented in a combined lecture and laboratory program of 1200 to 1600 hours extending over an appropriate number of school terms. In writing these textbooks, we have attempted to overcome some of the limitations of conventional books in the field of machine tools and machining practices.

The structure of this series gives the instructor maximum flexibility in shop and classroom application. Although the order of topic presentation represents traditional sequencing derived nationally from numerous course outlines, the internal structure of the book allows the material to be taken up in an order best suited to the individual teacher. The traditional structure of chapters has been eliminated and replaced by lettered sections representing major divisions of the topic. Each section is prefaced by an introductory component that provides an overview of the section topic. To stimulate interest, some historical information is included, followed by a general description of the types and applications of that specific group of tools or machines.

The detailed treatments of the topics are contained in the instructional units following each section introduction. Each unit follows a standardized format. It begins with a statement of purpose that explains why the information is important to a machinist. This statement is followed by a list of specific objectives to be accomplished by the student. It has been our experience that learning is reinforced if the student has the opportunity to test himself. Therefore, each instructional unit is followed by a self-test; answers appear in Appendix II. When the student is confident that he knows the material in the unit, he may then take the unit post-test. Many of the units also contain worksheets or exercises that allow the student to apply what he has learned.

The work of machining is a highly visual one. Thus, for the benefit of the students using this series, we have created a visually intensive format. This arrangement hopefully offers some of the advantages of other external visual media with the added advantage that the reader may study an illustration as long as necessary to glean the maximum amount of information.

Furthermore, photographs have been taken with the student in mind. Whenever possible, the camera sees the tool or machine as the student would see them. Many photographs showing tools and machines as they would be used in everyday applications have been preferred to static views of them. The use of line drawings has been minimized and overlaid photographs substituted in order to add a further degree of realism.

We have attempted to keep the series as up to date as possible. Archaic machining practices have been deemphasized. Furthermore, considerable space has been devoted to the treatment of related technologies. This has not been done to present complete courses in the related areas, but to emphasize the importance of these related topics to the prospective machinist.

The series is accompanied by an Instructor's Manual containing the unit post-tests, answers to post-tests, additional project drawings, and suggested sources for audio-visual aids. The Manual also contains suggestions regarding the organization and management for those instructors who might be interested in using this series in an individualized instructional setting.

Warren T. White
John E. Neely
Richard R. Kibbe
Roland O. Meyer

acknowledgments

We greatly appreciate the generous efforts of Doris Neely and Janet White in taking care of the correspondence, in preparing the manuscript, and in assisting in the organization of the entire project.

We want to acknowledge our appreciation to Richard L. McKee of Naperville, Illinois and Henry Kapell of Antioch, Illinois for their contribution to the writing of the grinding section of Volume II.

We thank all individuals who served as models for photographs. We also thank the following associations, societies, educational institutions, and individuals for providing service, facilities, assistance, and review of the manuscript.

Ackerman, William T., Orange Coast College, Costa Mesa, California
Alan Hancock College, Santa Maria, California
Allan, John Jr., Engineering and Technology Division, DeAnza College, Cupertino, California
American Society for Metals, Metals Park, Ohio
American Society of Mechanical Engineers, New York, New York
Bayard, John A., Santa Ana College, Santa Ana, California
Boulton, Franklin, Utah State Technical College, Salt Lake City, Utah
Bratt, George, Citrus College, Azusa, California
Brooker, Floyd E.
Busch, Ted, Minneapolis, Minnesota
California State University at Fresno, Department of Industrial Arts and Technology, Fresno, California
DeAnza College, Engineering and Technology, Cupertino, California
Dixon, Richard, Alan Hancock College, Santa Maria, California
Epsilon Pi Tau Fraternity, Alpha Lambda Chapter, California State University at Fresno, Fresno, California
Ferris, Sherman, Larrimer County Voc-Tech, Fort Collins, Colorado
Foothill College District, Office of Technical Education, Individualized Machinist's Curriculum Project, Los Altos Hills, California
Garcia, Manuel R., Department of Industrial Arts and Technology, California State University at Fresno, Fresno, California
Griffiths, John, Linn-Benton Community College, Albany, Oregon
Hall, Charles W., Paul C. Hays Technical School, Grove City, Ohio
Herreman, Glenn, Hewlett-Packard Company, Palo Alto, California
Keen, Harvey, K & M Tool, Inc., Eugene, Oregon
Keuhn, Karl, Jr., Muskingun Area Joint Vocational School, Zanesville, Ohio
Larrimer County Voc-Tech, Fort Collins, Colorado
Lane Community College, Mechanics Department, Eugene, Oregon
Lewis, William, North Kentucky State Voc-Tech School, Covington, Kentucky

Link, Robert S., Triton College, River Grove, Illinois
Linn-Benton Community College, Business and Industrial Division, Albany, Oregon
Moss, Farris S., North County Technical School, Florrisant, Missouri
Musser, Jonathan A., Saratoga, California
National Machine Tool Builders Association, McLean, Virginia
National Screw Machine Products Association, Cleveland, Ohio
Neff, John, Engineering and Technology, DeAnza College, Cupertino, California
Owens, Aubrey F., Spartanburg Technical College, Spartanburg, South Carolina
Randol, Stan M., Engineering and Technology Division, DeAnza College, Cupertino, California
Starr, Edmund A., Chairman of Applied Arts Department, Yuba College, Marysville, California
Utah State Technical College, Salt Lake City, Utah
Valmassoi, Nino, Pasadena City College, Pasadena, California
West, Jack, Community College of Denver, Denver, Colorado
Wilmouth, Henry, North County Technical School, St Louis County, Missouri
Yeley, Max, Ferris State College, Big Rapids, Michigan
Yuba College, Applied Arts Department, Marysville, California

Our special thanks go to the following firms and their employees with whom we corresponded and visited and who supplied us with invaluable technical information and illustrations.

Accurate Diamond Tool Corporation, Hackensack, New Jersey
Aloris Tool Company, Inc., Clifton, New Jersey
American Bechler, Norwalk, Connecticut
American Chain & Cable Company, Inc., Wilson Instrument Division, Bridgeport, Connecticut
American Iron & Steel Institute, Washington, D.C.
American Machinist, New York, New York
American SIP Corporation, Elmsford, New York
Ameropean Industries, Inc., Hamden, Connecticut
Ames Research Center, National Aeronautics and Space Administration (NASA), Mountain View, California
Armco Steel Corporation, Middletown, Ohio
Barber-Colman Company, Rockford, Illinois
Barnes Drill Company, Rockford, Illinois
Barret Centrifugals, Worcester, Massachusetts
Bay State Abrasives, Division of Dresser Industries, Westborough, Massachusetts
Bendix Corporation, South Bend, Indiana
Bethlehem Steel Corporation, Bethlehem, Pennsylvania
Boyer-Schultz Corporation, Broadview, Illinois
Bridgeport Milling Machine Division, Textron, Inc., Bridgeport, Connecticut
Brown & Sharpe Manufacturing Company, North Kingstown, Rhode Island
Bryant Grinder Corporation, Springfield, Vermont
Buck Tool, Kalamazoo, Michigan

Acknowledgments

Burgmaster Division of Houdaille, Inc., Los Angeles, California
The Carborundum Company, Niagara Falls, New York
Cincinnati Incorporated, Cincinnati Ohio
Cincinnati Milacron, Inc., Cincinnati, Ohio
Clausing Corporation, Kalamazoo, Michigan
Cleveland Twist Drill Company, Cleveland, Ohio
Cone-Blanchard Machine Company, Windsor, Vermont
Dake Corporation, Grand Haven, Michigan
Desmond-Stephan Manufacturing Company, Urbana, Ohio
Diamond Abrasive Corporation, New York, New York
W. C. Dillon & Company, Inc., Van Nuys, California
DoAll Company, Des Plaines, Illinois
Dover Publications Inc., New York, New York
The du Mont Corporation, Greenfield, Massachusetts
El-Jay, Inc., Eugene, Oregon
Elm Systems, Inc., Arlington Heights, Illinois
Enco Manufacturing Company, Chicago, Illinois
Engis Corporation, Morton Grove, Illinois
Ex-Cell-O Corporation, Troy, Michigan
Exolon Company, Tonawanda, New York
Federal Products Corporation, Providence, Rhode Island
Fellows Corporation, Springfield, Vermont
Floturn, Inc., Division of Lodge & Shipley Company, Cincinnati, Ohio
France Studios, Eugene, Oregon
Gaertner Scientific Corporation, Chicago, Illinois
General Electric Company, Detroit, Michigan, and Specialty Materials Department, Worthington, Ohio
Geometric Tool, New Haven, Connecticut
Giddings & Lewis, Fond Du Lac, Wisconsin
Gleason Works, Rochester, New York
Great Lakes Screw, Chicago, Illinois
Greenfield Tap & Die, Division of TRW, Inc., Greenfield, Massachusetts
Greenlee Brothers, Rockford, Illinois
Hammond Machinery Builders, Kalamazoo, Michigan
Hardinge Brothers, Inc., Elmira, New York
Harig Products, Inc., Elgin, Illinois
Harry M. Smith & Associates, Santa Clara, California
Heald Machine Division, Cincinnati Milacron Company, Worcester, Massachusetts
Hewlett-Packard Company, Palo Alto and Santa Clara, California
Hitachi Magna-Lock Corporation, Big Rapids, Michigan
Hitchiner Manufacturing Company, Inc., Milford, New Hampshire
Houston Instrument, Austin, Texas
Illinois/Eclipse, Division of Illinois Tool Works, Inc., Chicago, Illinois
Industrial Plastics Products, Inc., Forest Grove, Oregon
Industrial Press, New York, New York
Ingersoll Milling Machine Company, Rockford, Illinois
Investment Casting Corporation, Springfield, New Jersey
Jarvis Products Corporation, Middletown, Connecticut
J. K. Gill Company, Eugene, Oregon

Acknowledgments

K & M Tool, Inc., Eugene, Oregon
Kasto-Racine, Inc., Monroeville, Pennsylvania
Kennametal, Latrobe, Pennsylvania
Koering, HPM Division, Mount Gilead, Ohio
Landis Tool Company, Division of Litton Industries, Waynesboro, Pennsylvania
Lapmaster Division, Crane Packing Company, Morton Grove, Illinois
LeBlond, Inc., Cincinnati, Ohio
Lees Bradner, Cleveland, Ohio
Litronix, Inc., Cupertino, California
Lodge & Shipley Company, Cincinnati, Ohio
Louis Levin & Son, Inc., Culver City, California
M & M Tool Manufacturing Company, Dayton, Ohio
Madison Industries, Division of Amtel, Inc., Providence, Rhode Island
Magnaflux Corporation, Chicago, Illinois
Mahr Gage Company, New York, New York
Mattison Machine Works, Rockford, Illinois
Megadiamond Industries, New York, New York
Minnesota Mining and Manufacturing Company (3M), St. Paul, Minnesota
The MIT Press, Cambridge, Massachusetts
Mohawk Tools, Inc., Montpelier, Ohio
Monarch Machine Tool Company, Sidney, Ohio
Moog, Inc., Hydra Point Division, Buffalo, New York
Moore Special Tool Company, Bridgeport, Connecticut
MTI Corporation, New York, New York
National Broach & Machine Division, Lear Siegler, Inc., Detroit Michigan
National Twist Drill & Tool Division, Lear Siegler, Inc., Rochester, Michigan
New Britain Machine Division, Litton Industries, New Britain, Connecticut
Newcomer Products, Inc., Latrobe, Pennsylvania
Norton Company, Worcester, Massachusetts
Pacific Machinery & Tool Steel Company, Portland, Oregon
PMC Industries, Wickliffe, Ohio
Pratt & Whitney Machine Tool Division of The Colt Industries Operating Corporation, West Hartford, Connecticut
Precision Diamond Tool Company, Elgin, Illinois
Production Components, Inc., Ranchita, California
Ralmike's Tool-A-Rama, South Plainfield, New Jersey
Rank Precision Industries, Inc., Des Plaines, Illinois
Republic Steel Corporation, Cleveland, Ohio
Roberts Consolidated Industries, Inc., City of Industry, California
Rockford Machine Tool Company, Rockford, Illinois
Sellstrom Manufacturing Company, Palatine, Illinois
Shore Instrument & Manufacturing Company, Inc., Jamaica, New York
Sipco Machine Company, Marion, Massachusetts
Snap-On Tools Corporation, Kenosha, Wisconsin
Southwestern Industries, Inc., Los Angeles, California

Acknowledgments

Speedfam Corporation, Des Plaines, Illinois
Standard Gage Company, Poughkeepsie, New York
L. S. Starret Company, Athol, Massachusetts
Sunnen Products Company, St. Louis, Missouri
Superior Electric Company, Bristol, Connecticut
Surface Finishes, Inc., Addison, Illinois
Taft-Peirce Manufacturing Company, Woonsocket, Rhode Island
Taper Micrometer Corporation, Worcester, Massachusetts
Thompson Vacuum Company, Sarasota, Florida
Tinius Olsen Testing Machine Company, Inc., Willow Grove, Pennsylvania
Ultramatic Equipment Company, Addison, Illinois
Unison Corporation, Madison Heights, Michigan
United States Steel Corporation, Pittsburgh, Pennsylvania
Volstro Manufacturing Company, Inc., Philadelphia, Pennsylvania
Waldes Kohinoor, Inc., Long Island City, New York
O. S. Walker Company, Inc., Worcester, Massachusetts
Walton Company, Hartford, Connecticut
Warner & Swasey Company, Cleveland, Ohio, and King of Prussia, Pennsylvania
Weldon Tool Company, Cleveland, Ohio
Whitman & Barnes Division, TRW, Inc., Plymouth, Michigan
Whitnon Spindle Division, Mite Corporation, Farmington, Connecticut
J. H. Williams Division, TRW, Inc., Buffalo, New York
Wilton Tool Division, Wilton Corporation, Des Plaines, Illinois

W.T.W.
J.E.N.
R.R.K.
R.O.M.

contents

volume II

section a
vertical band machines

unit 1
Vertical Band Machine Safety — 11

unit 2
Preparing to Use the Vertical Band Machine — 15

unit 3
Using the Vertical Band Machine — 27

section b
vertical milling machines

unit 1
Vertical Milling Machine Safety — 39

unit 2
The Vertical Spindle Milling Machine — 41

unit 3
Vertical Milling Machine Operations — 45

unit 4
Cutting Tools for the Vertical Milling Machine — 52

unit 5
Setups on the Vertical Milling Machine — 58

unit 6
Using End Mills — 65

unit 7
Using the Offset Boring Head — 69

section c
horizontal milling machines

unit 1
Horizontal Milling Machine Safety — 87

unit 2
Plain and Universal Horizontal Milling Machines — 90

unit 3
Types of Spindles, Arbors, and Adaptors — 94

unit 4
Arbor-Driven Milling Cutters — 99

unit 5
Setting Speeds and Feeds on the Horizontal Milling Machine — 106

unit 6
Workholding and Locating Devices on the Milling Machine — 110

unit 7
Plain Milling on the Horizontal Milling Machine — 116

unit 8
Using Side Milling Cutters on the Horizontal Milling Machine — 127

unit 9
Using Face Milling Cutters on the Horizontal Milling Machine — 136

section d
dividing the circle

unit 1
Indexing Devices — 153

unit 2
Direct and Simple Indexing — 157

unit 3
Angular Indexing — 159

section e
gears

unit 1
Introduction to Gears — 171

unit 2
Spur Gear Terms and Calculations — 176

unit 3
Cutting a Spur Gear — 181

unit 4
Gear Inspection and Measurement — 185

section f
shapers and planers

unit 1
Features and Tooling of the Horizontal Shaper — 209

unit 2
Cutting Factors on Shapers and Planers — 224

unit 3
Shaper Safety and Using the Shaper — 229

unit 4
Features and Tooling of the Planer — 242

unit 5
Planer Safety and Using the Planer — 248

section g
physical properties and the heat treating of metals

unit 1
Mechanical and Physical Properties of Metals — 265

unit 2
The Crystal Structure of Metals — 282

unit 3
Phase Diagrams for Steels — 292

unit 4
I-T Diagrams and Cooling Curves — 303

unit 5
Hardenability of Steels and Tempered Martensite — 310

unit 6
Heat Treating Steels — 319

section h
grinding machines

unit 1
Selection and Identification of Grinding Wheels — 357

unit 2
Grinding Wheel Safety — 369

unit 3
Care of Abrasive Wheels: Trueing, Dressing, and Balancing — 379

unit 4
Grinding Fluids — 387

unit 5
Horizontal Spindle, Reciprocating Table Surface Grinders — 396

unit 6
Workholding on the Surface Grinder — 404

unit 7
Form Grinding and Wheel Dressing — 411

unit 8
Using the Surface Grinder — 418

unit 9
Problems and Solutions in Surface Grinding — 428

unit 10
Cylindrical Grinders — 435

unit 11
Using the Cylindrical Grinder — 445

unit 12
Problems and Solutions in Cylindrical Grinding — 451

unit 13
Cutter and Tool Grinder Features and Components — 457

unit 14
Cutter and Tool Grinder Safety and General Setup Procedures — 464

unit 15
Sharpening Plain Milling Cutters — 470

unit 16
Sharpening Slitting Saws and Stagger Tooth Milling Cutters — 476

unit 17
Sharpening of End Mills — 484

unit 18
Sharpening Form Relieved Cutters — 492

unit 19
Sharpening Machine Reamers — 499

unit 20
Miscellaneous Cutter and Tool Grinding Operations — 506

section i
numerical control of machine tools

unit 1
Numerical Control Dimensioning — 535

unit 2
Numerical Control Tape, Tape Preparation, Tape Code Systems, and Tape Readers ... 542

unit 3
Basic Format Numerical Control Programs ... 548

unit 4
Operating the N/C Machine Tool ... 573

appendix I
Self-test answers ... 579

appendix II
precision vise
Drawings I, II, III, and IV ... 608

v-block
Drawing I ... 612

Index ... 613

volume I

section a
technical foundations

unit 1
Review of Shop Mathematics ... 5

unit 2
Using Electronic Calculators ... 16

unit 3
Reviewing Plane Geometry ... 20

unit 4
Reviewing Right Triangle Trigonometry ... 29

unit 5
Reading Shop Drawings ... 38

unit 6
Introduction to Mechanical Hardware ... 48

unit 7
Using a Machinist's Handbook ... 64

unit 8
Shop Safety ... 69

section b
hand tools

unit 1
Work Holding for Hand Operations ... 79

unit 2
Arbor and Shop Presses ... 82

unit 3
Noncutting Hand Tools ... 91

unit 4
Cutting Hand Tools: Hacksaws ... 99

unit 5
Cutting Hand Tools: Files ... 104

section c
dimensional measurement

unit 1
Systems of Measurement ... 141

unit 2
Using Steel Rules ... 146

unit 3
Using Vernier Calipers and Vernier Depth Gages ... 157

unit 4
Using Surface Plates ... 170

unit 5
Using Micrometer Instruments ... 175

unit 6
Mechanical Dial and Electronic Digital Measuring Instruments ... 201

unit 7
Using Comparison Measuring Instruments ... 211

unit 8
Using Gage Blocks ... 236

unit 9
Using Angular Measuring Instruments ... 247

section d
materials group

unit 1
Safety in Material Handling ... 261

unit 2
Pig Iron and Steel Making — 267

unit 3
Steel Finishing Processes — 275

unit 4
Selection and Identification of Steels — 282

unit 5
Selection and Identification of Nonferrous Metals — 290

unit 6
Hardening, Tempering, and Case Hardening — 296

unit 7
Annealing, Normalizing, and Stress Relieving — 305

unit 8
Other Metals Used in Machining — 308

unit 9
Casting Processes — 312

unit 10
Rockwell and Brinell Hardness Testers — 323

unit 11
Nondestructive Testing — 336

unit 12
Nonmetallic Materials — 344

section e
cutoff machines

unit 1
Cutoff Machine Safety — 353

unit 2
Using Reciprocating and Horizontal Band Cutoff Machines — 358

unit 3
Abrasive and Cold Saws — 370

section f
layout

unit 1
Basic Semiprecision Layout Practices — 391

unit 2
Basic Precision Layout Practices — 397

section g
drilling machine group

unit 1
Drilling Machine Safety — 415

unit 2
The Drill Press — 418

unit 3
Drilling Tools — 423

unit 4
Hand Grinding of Drills on the Pedestal Grinder — 434

unit 5
Sharpening Drills by Machine — 440

unit 6
Work Locating and Holding Devices on Drilling Machines — 447

unit 7
Operating Drilling Machines — 456

unit 8
Countersinking and Counterboring — 466

unit 9
Reaming in the Drill Press — 471

section h
taps and dies

unit 1
Hand Reamers — 483

unit 2
Identification and Application of Types of Taps — 488

unit 3
Tapping Procedures — 495

unit 4
Thread Cutting Dies and Their Uses — 503

section i
turning machine group

unit 1
Turning Machine Safety — 517

unit 2
The Engine Lathe — 523

unit 3
Engine Lathe Maintenance and Adjustments — 529

unit 4
Toolholders and Tool Holding for the Lathe ... 535

unit 5
Cutting Tools for the Lathe ... 542

unit 6
Lathe Spindle Tooling ... 551

unit 7
Operating the Machine Controls ... 559

unit 8
Facing and Center Drilling ... 565

unit 9
Turning between Centers ... 575

unit 10
Alignment of the Lathe Centers ... 590

unit 11
Drilling, Boring, Reaming, Knurling, Recessing, Parting, and Tapping in the Lathe ... 595

unit 12
Sixty Degree Thread Information and Calculations ... 614

unit 13
Cutting Unified External Threads ... 624

unit 14
Basic and Advanced Thread Measurement ... 636

unit 15
Cutting of Unified Internal Threads ... 642

unit 16
Taper Turning, Taper Boring, and Forming ... 647

unit 17
Using Steady and Follower Rests in the Lathe ... 666

unit 18
Additional Thread Turning ... 672

unit 19
Additional Thread Forms ... 676

unit 20
Cutting Acme Threads on the Lathe ... 680

unit 21
Cutting Metric Threads with English Measure Lathes ... 685

unit 22
Using Carbides and Other Tool Materials on the Lathe ... 690

unit 23
Infrequently Used Lathe Applications ... 707

appendix I

table 1
Natural Trigonometric Functions ... 714

table 2
Sine-Bar Constants ... 726

table 3
Inch-Metric Conversion Table ... 729

appendix II
Answers to Self-Tests ... 731

appendix III
instructions for completing the c-clamps, the parallel clamp, and the paper punch ... 769

c-clamp:
Drawings I and II ... 769

parallel clamp:
Drawing III ... 772

paper punch:
Drawing IV ... 773

index ... 776

Machine Tools and Machining Practices

section a
vertical band machines

The vertical band machine (Figure 1) is often called the handiest machine tool in the machine shop. Perhaps the reason for this is the wide variety of work that can be accomplished on this versatile machine tool. The vertical band machine or vertical band saw is similar in general construction to its horizontal counterpart. Basically, it consists of an endless band blade or other band tool that runs on a driven and idler wheel. The band tool runs vertically at the point of the cut where it passes through a worktable on which the workpiece rests. The workpiece is pushed into the blade and the direction of the cut is guided by hand or mechanical means.

ADVANTAGES OF BAND MACHINES

Shaping of material with the use of a saw blade or other band tool is often called **band machining**. The reason for this is that the band machine can perform other machining tasks aside from simple sawing. These include band friction sawing, band filing, and band polishing.

In any machining operation, a piece of stock material is cut by various processes to form the final shape and size of the part desired. In most machining operations, all of the unwanted material must be reduced to chips in order to uncover the final shape and size of the workpiece. With a band saw, only a small portion of the unwanted material must be reduced to chips in order to uncover the final workpiece shape and size (Figure 2). A piece of stock material can often be shaped to final size by one or two saw cuts. A further advantage is gained in that the band saw cuts a very narrow kerf. A minimum amount of material is wasted. Other machining operations may require that a large amount of material be wasted as chips in order to uncover the final size and shape of the workpiece.

A second important advantage in band sawing machines is **contouring ability.** Contour band sawing is the ability of the saw to cut intricate curved shapes that would be nearly impossible to machine by other methods (Figure 3). The sawing of intricate shapes can be accomplished by a combination of hand and power feeds. On vertical band machines so equipped, the workpiece is steered by manual operation of the handwheel. The hydraulic table feed varies according to the saw pressure on the workpiece. This greatly facilitates contouring sawing operations.

Figure 1. Leighton A. Wilkie bandsaw of 1933 was the last basic machine tool to be developed (Courtesy of the DoAll Company).

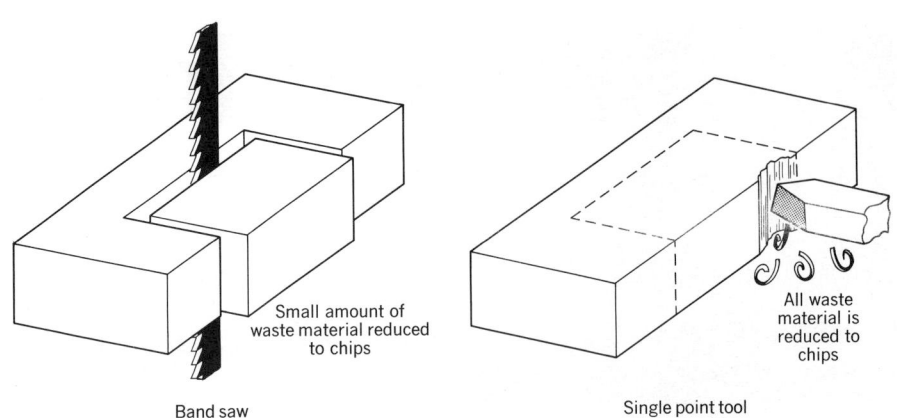

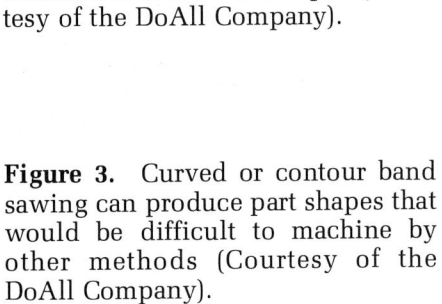

Figure 2. Sawing can uncover the workpiece shape in a minimum number of cuts.

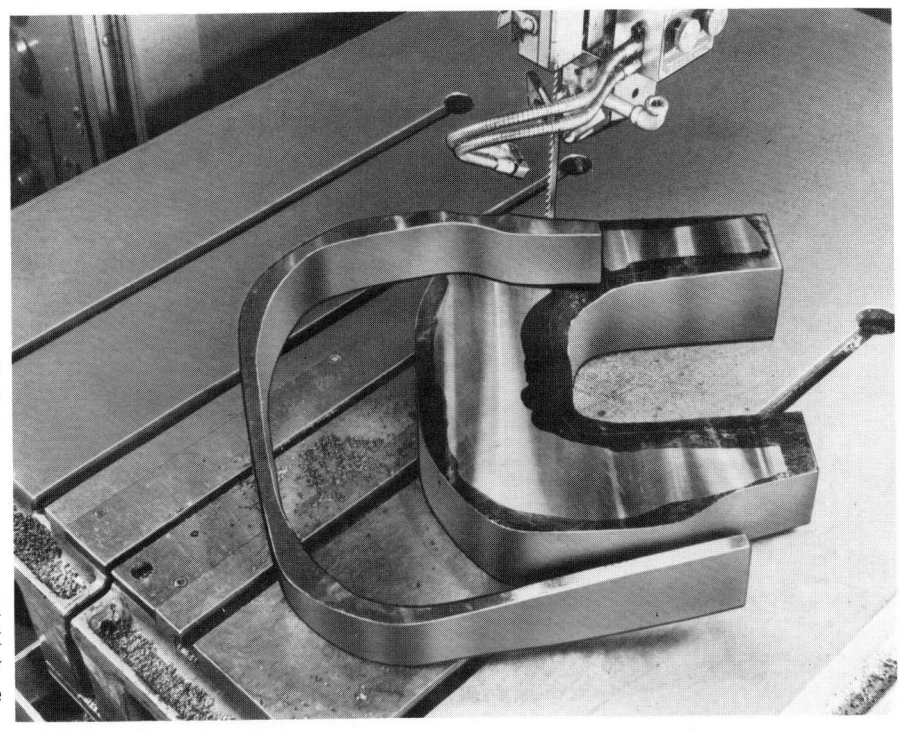

Figure 3. Curved or contour band sawing can produce part shapes that would be difficult to machine by other methods (Courtesy of the DoAll Company).

Band sawing and band machining have several other advantages. There is no limit to the length, angle, or direction of the cut (Figure 4). Workpieces larger than the band machine can be cut (Figure 5). Since the band tool is fed continuously past the work, cutting efficiency is high. A band tool, whether it be a saw blade, band file, or grinding band, has a large number of cutting points passing the work. In most other machining operations, only one or a fairly low number of cutting points pass the work. With the band tool, wear is distributed over these many cutting points. Tool life is prolonged.

TYPES OF BAND MACHINES
General Purpose Band Machine with Fixed Worktable

The general purpose band machine is found in most machine shops (Figure 6). This machine tool has a nonpower-fed worktable that can be tilted in order to make angle cuts. The table may be tilted 10

Figure 4. Splitting a large diameter ring on the vertical band machine (Courtesy of the DoAll Company).

Figure 5. Workpieces larger than the machine tool can be cut (Courtesy of the DoAll Company).

Figure 6. General purpose vertical band machine (Courtesy of the DoAll Company).

Figure 7. Vertical band machine worktable can be tilted 10 degrees left (Courtesy of the California Community Colleges — Individualized Machinist's Curriculum Project).

degrees left (Figure 7). Tilt on this side is limited by the saw frame. The table may be tilted 45 degrees right (Figure 8). On large machines, table tilt left may be limited to 5 degrees.

The workpiece may be pushed into the blade by hand. Mechanical (Figure 9) or mechanical-hydraulic feeding mechanisms are also used. A band machine may be equipped with a hydraulic tracing attachment. This accessory uses a stylus contacting a template or

Figure 8. Vertical band machine worktable tilted 45 degrees right (Courtesy of the California Community Colleges — Individualized Machinist's Curriculum Project).

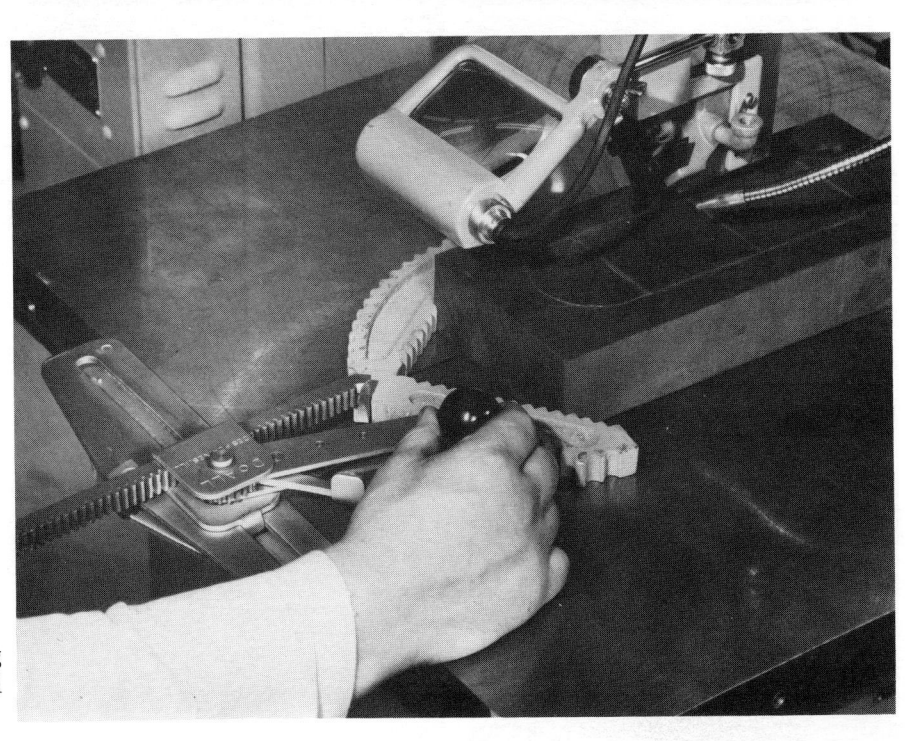

Figure 9. Mechanical work feeding mechanism (Courtesy of the DoAll Company).

pattern. The tracing accessory guides the workpiece during the cut (Figure 10).

Band Machines with Power-Fed Worktables Heavier construction is used on these machine tools. The worktable is moved hydraulically. The operator is relieved of the need to push the workpiece into the cutting blade. The direction of the cut can be guided by a steering mechanism (Figure 11). A roller chain wraps around the workpiece and passes over a sprocket at the back of the worktable. The sprocket is connected to a steering wheel at the front of the worktable. The operator can then guide the workpiece and keep the saw cutting along the proper lines. The workpiece rests on

Figure 10. Hydraulic tracing accessory (Courtesy of the DoAll Company).

roller bearing stands. These permit the workpiece to turn freely while it is being steered.

High Tool Velocity Band Machines On the high tool velocity band machine (Figure 12) band speeds can range as high as 10 to 15,000 feet per minute (FPM). These machine

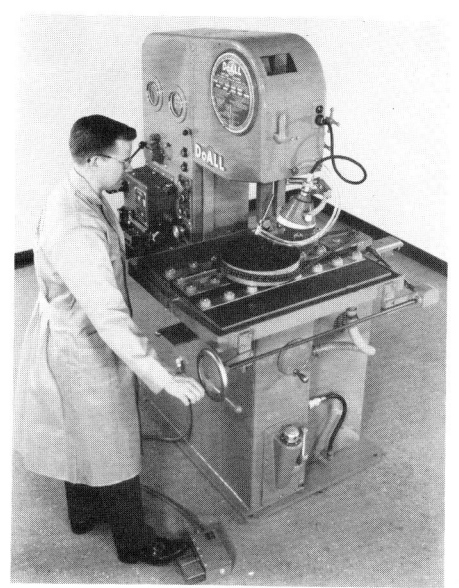

Figure 11. Heavy duty vertical band machine with power-fed worktable (Courtesy of the DoAll Company).

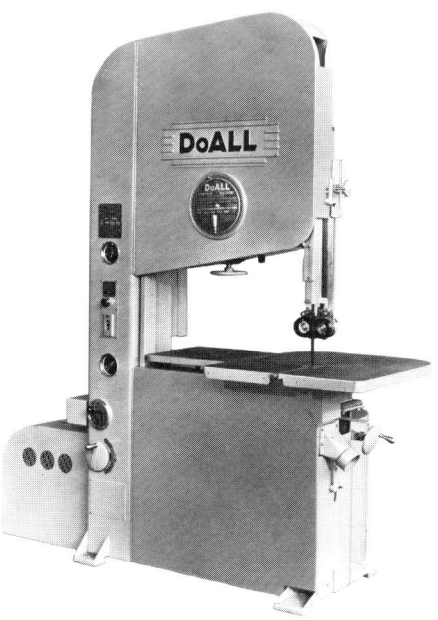

Figure 12. High tool velocity vertical band machine (Courtesy of the DoAll Company).

tools are used in many band machining applications. They are frequently found cutting nonmetal products. These include applications such as trimming plastic laminates (Figure 13) and cutting fiber materials (Figure 14).

Figure 13. Trimming plastic laminates on the high tool velocity band machine (Courtesy of the DoAll Company).

Figure 14. Cutting fiber material on the high tool velocity band machine (Courtesy of the DoAll Company).

Section A Vertical Band Machines

Large Capacity Band Machines This type of band machine is used on large workpieces. The entire saw is attached to a swinging column. The workpiece remains stationary and the saw is moved about to accomplish the desired cuts (Figure 15).

APPLICATIONS OF THE VERTICAL BAND MACHINE
Conventional and Contour Sawing Vertical band machines are used in many conventional sawing applications. They are found in the foundry trimming sprues and risers from castings. The band machine can accommodate a large casting and make widely spaced cuts (Figure 16). Production trimming of castings is easily accomplished with the high tool velocity band machine (Figure 17). Band saws are also useful in ripping operations (Figure 18). In the machine shop, the vertical band machine is used in general purpose, straight line, and contour cutting mainly in sheet and plate stock.

Friction Sawing Friction sawing can be used to cut materials that would be impossible or very difficult to cut by other means. In friction sawing, the workpiece is heated by friction created between it and the cutting blade. The blade melts its way through the work. Friction sawing can be used to cut hard materials such as files. Tough materials such as stainless steel wire brushes can be trimmed by friction sawing (Figure 19).

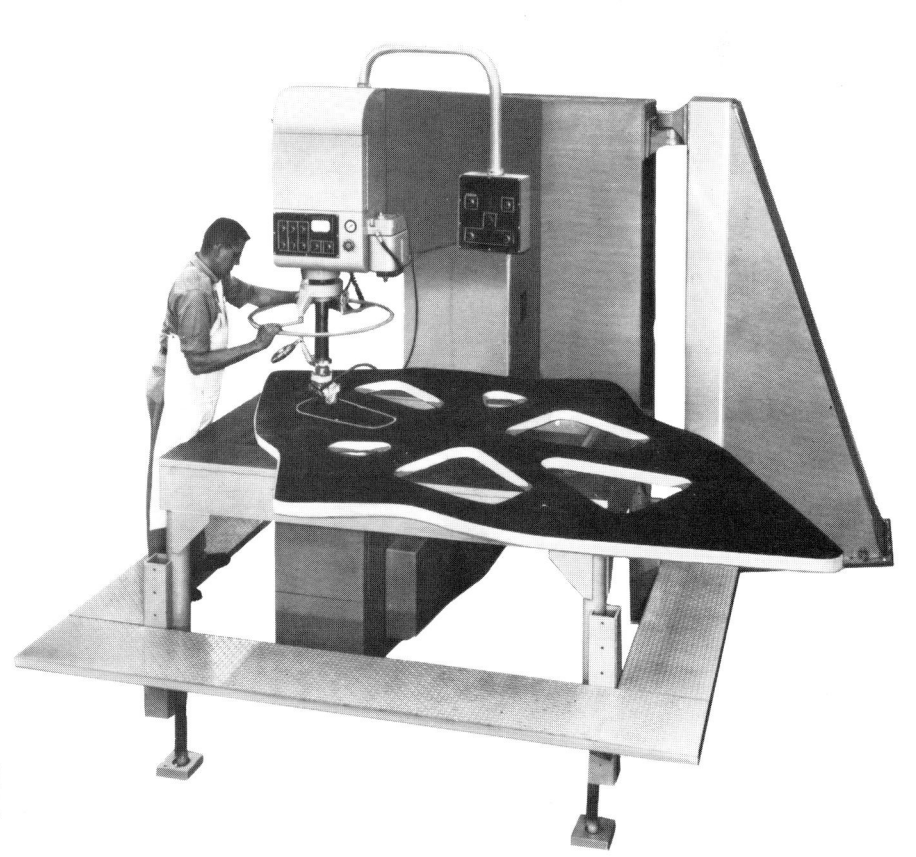

Figure 15. Large capacity vertical band machine (Courtesy of the DoAll Company).

Figure 16. Trimming casting sprues and risers on the vertical band machine (Courtesy of the DoAll Company).

Figure 17. Production trimming of castings on the vertical band machine (Courtesy of the DoAll Company).

Figure 18. Ripping on the vertical band machine (Courtesy of the DoAll Company).

Figure 19. Trimming stainless steel wire brushes by friction sawing (Courtesy of the DoAll Company).

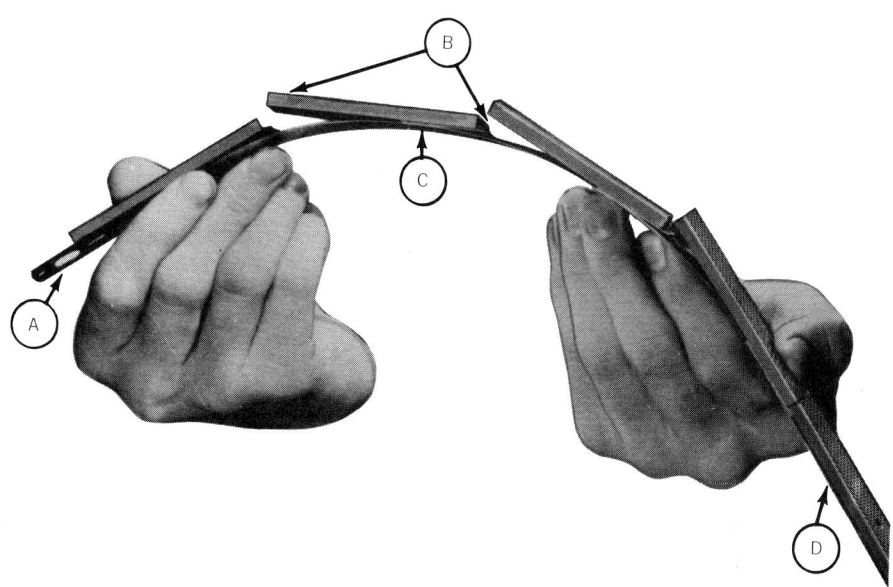

Figure 20. Band file (Courtesy of the DoAll Company).

Figure 21. Internal band filing (Courtesy of the DoAll Company).

Band Filing and Band Polishing The band file consists of file segments attached to a spring steel band (D) (Figure 20). As the band file passes through the work an interlock closes and keeps the file segment tight (B). The interlock then releases, permitting the file segment to roll around the band wheel. A space is provided for chip clearance between the band and file segment (C). The band file has a locking slot so that the ends can be joined to form a continuous loop (A). Special guides are required for both file and polishing bands. Band files can be used in both internal (Figure 21) and external (Figure 22) filing applications. They are also used in applications such as filing large gear teeth to shape and size (Figure 23).

Figure 22. External band filing (Courtesy of the DoAll Company).

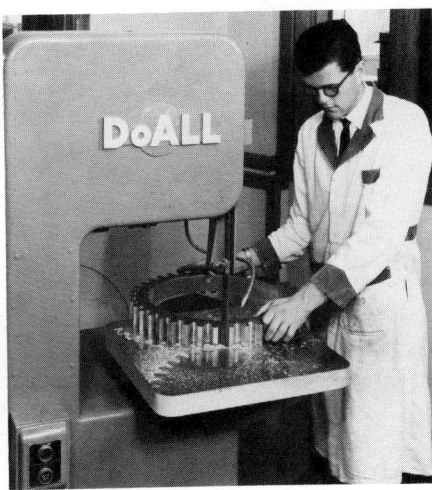

Figure 23. Band filing a large spur gear (Courtesy of the DoAll Company).

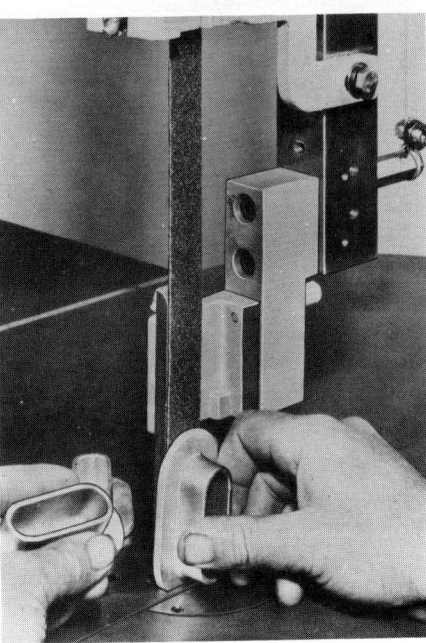

Figure 24. Band polishing (Courtesy of the DoAll Company).

In band polishing, a continuous abrasive strip is used (Figure 24). The grit of the abrasive can be varied depending on the surface finish desired.

review questions

1. Name four types of band machines.
2. Which type would be particularly useful in cutting nonmetal soft materials like plastics?
3. What other machining operations aside from sawing can be accomplished on the band machine?
4. What is contour sawing?
5. Name two advantages of band machines.

unit 1
vertical band machine safety

The vertical band machine, like any other machine tool, is a motor-driven device with a number of moving parts. The machine is designed with sufficient power to perform its required tasks. Anytime that mechanical parts with sufficient power are put into motion, potential hazards exist for the operator. You must remember that any machine tool is only an extension of the operator's intelligence. A machine has no intelligence of its own. Once turned on, it cannot distinguish between cutting material and cutting fingers. The potential hazards of a machine tool must be kept in mind at all times. A lax safety attitude on the part of the operator can only lead to an accident. Many accidents are caused by doing something that you should not do. Short-cutting proper operating procedure may result in cutting a finger, hand, or arm, or losing an eye. This kind of accident can end a machinist's career.

objectives After completing this unit, you will be able to:
1. Identify potential hazards on and around vertical band machines.
2. Describe safe vertical band machine cutting procedure.

information

Machine tools, the vertical band machine included, are designed to multiply power. Only through this can they be expected to cut through tough metallic materials and shape them to precise sizes. The power of a machine is multiplied many times beyond the strength of the operator. If you become entangled in a powerful moving mechanism, chances are that the machine will continue to operate and cause serious injury to fingers, hands, or arms.

The machine tool industry is well aware of safety considerations. State and federal authorities are also concerned with machine safety and industrial safety in general. Both industry and government are continuously striving to build safer machines and establish safe machine operating procedures. However, no machine tool can be built with total safety. It is up to you as the operator to bridge the gap between safety built into a machine tool and total safe operation.

SAFETY CONSIDERATIONS ON VERTICAL BAND MACHINES

Personal Protection
Eye protection is a primary consideration on any machine tool. Vertical band machines use a hardened steel blade, file, or abrasive band. There is always the chance that metal particles may be ejected from the machine during the cutting operation. **Always wear** an approved **face shield** (Figure 1), **goggles**, or **safety glasses.** Safety glasses with **side shields** or a face shield

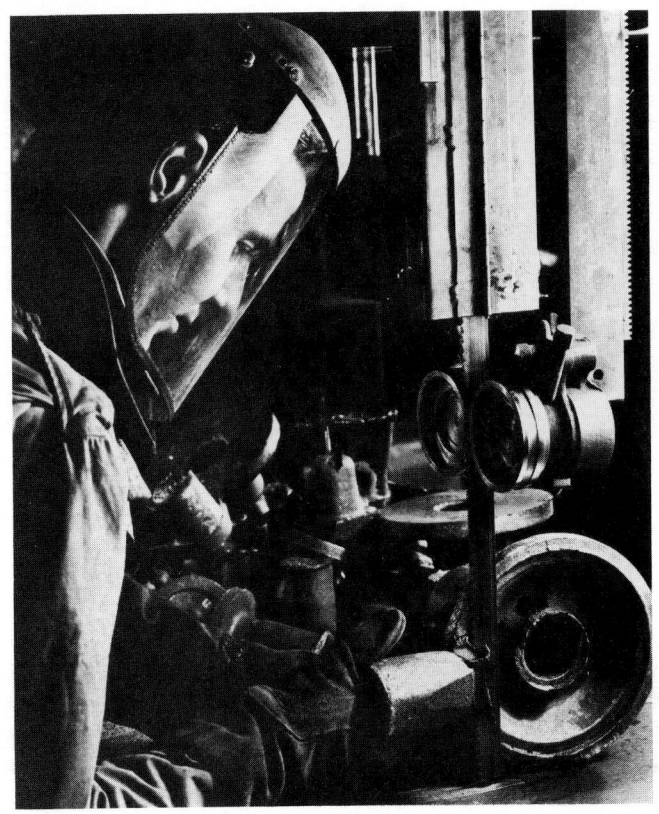

Figure 1. The operator is wearing a full face shield while sawing (Courtesy of the DoAll Company).

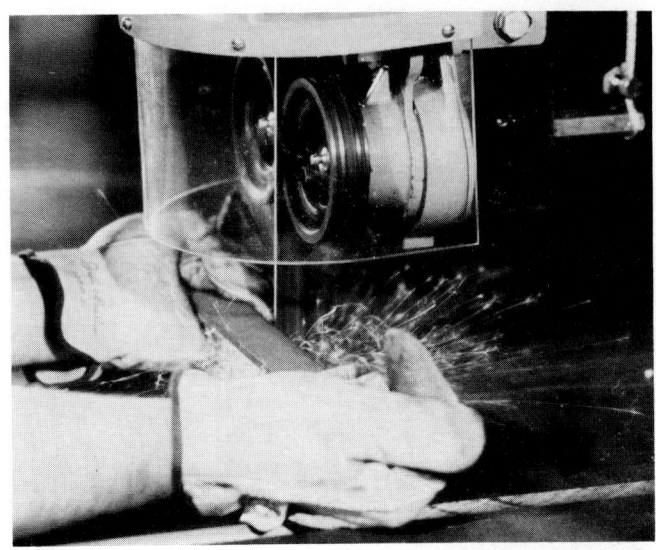

Figure 2. Gloves may be necessary in high speed or friction sawing (Courtesy of the California Community Colleges — Individualized Machinist's Curriculum Project).

must be worn when **friction** or **high speed sawing.**

Gloves should not be worn around any machine tool. An exception to this is for friction or high speed sawing or when handling band blades. Gloves will protect hands from the sharp saw teeth. If you wear gloves during friction or high speed sawing (Figure 2), be extra careful that they do not become entangled in the blade or other moving parts.

Remove your rings and wristwatch before operating any machine tool. These can be extremely hazardous if they become caught in a moving part. Rings are especially dangerous. If they should become caught, a finger can be lost very easily.

Chips produced by sawing operations should be **removed with** a suitable **brush. Do not use your hands and do not use a rag.** The sharp metal can cut your hands and a rag may be caught in a moving part of the machine. Chips also imbed themselves in rags, creating an additional hazard.

A machine tool cannot distinguish between rags and clothing. **Roll up your sleeves,** and keep your shirt tucked in. Remove or tuck in your necktie if you are wearing one. If you have long hair, keep it properly secured out of danger.

Machine Guarding on Vertical Band Machines
The entire blade must be guarded except at the point of the cut (Figure 3). This is effectively accomplished by enclosing the wheels and blade behind guards that are easily opened for adjustments to the machine. Wheel and blade guard must be closed at all times during machine operation. The guidepost guard moves up and down with the guidepost (Figure 4). The operator is protected from an exposed blade at this point. For maximum safety, set the guidepost $\frac{1}{8}$ to $\frac{1}{4}$ in. above the workpiece.

Band machines may have one or two idler wheels. On machines with two idler wheels, a short blade running over only one wheel may be used. Under this condition, an additional blade guard at the left side of the wheel is required (Figure 5). This guard is removed when operating over two idler wheels as the blade is then behind the wheel guard.

Roller blade guides are used in friction and high speed sawing. A roller guide shield is used to provide protection for the operator (Figure 6). Depending on the material being cut, the entire cutting area may be enclosed (Figure 7). This would apply to the cutting of hard, brittle materials such as granite and glass. Diamond blades

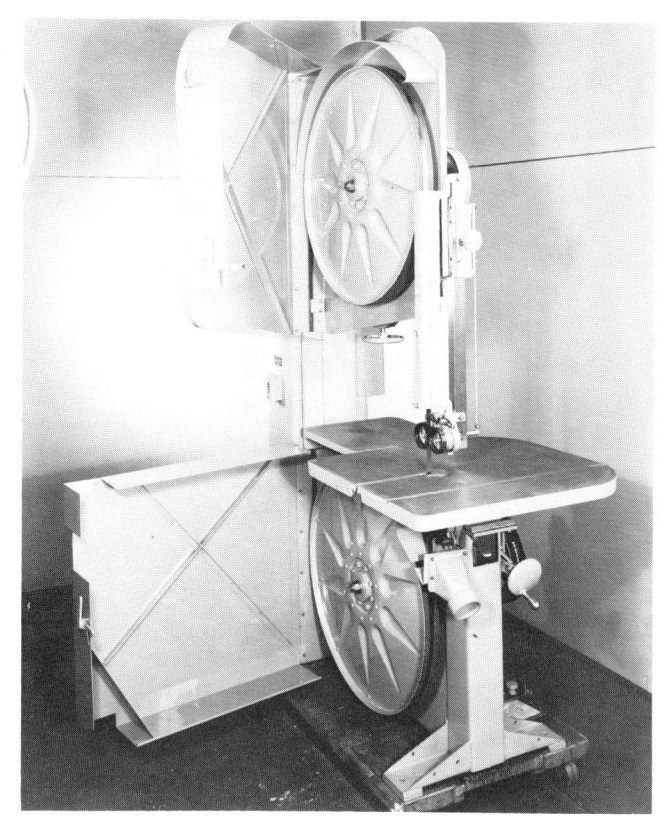

Figure 3. Wheel and blade guards on the vertical band machine (Courtesy of the DoAll Company).

Figure 4. Guidepost guard (California State University at Fresno).

Figure 5. Left side blade guard when using a short blade over one idler wheel (California State University at Fresno).

Figure 6. Roller blade guide shield (Courtesy of the California Community Colleges — Individualized Machinist's Curriculum Project).

Figure 7. When cutting brittle materials, the entire work area may be guarded (Courtesy of the DoAll Company).

are frequently used in cutting these materials. The clear shield protects the operator while permitting him to view the operation. Cutting fluids are also prevented from spilling on the floor. In any sawing operation making use of cutting fluids, see that they do not spill on the floor around the machine. This creates an extremely dangerous situation, not only for you but for others in the shop as well.

Any time that a guard is removed or opened, exposing the operating mechanism of the band machine, the electrical service to the tool should be turned off. Some band machines may be unplugged from the electrical outlet. The circuit breaker supplying the machine circuit can be switched off on machines that are permanently wired. If the machine is to be out of service for an extended period, the circuit breaker should be switched off and tagged with an appropriate warning.

Normally you will not disconnect the electrical service for a routine adjustment such as changing a blade. For total safety, it is good practice to take the time to secure the power to the machine. If a routine adjustment such as changing a belt is being made, you must insure that no other person is likely to turn on the machine while your hands are in contact with belts and pulleys.

Safe Cutting Procedure on the Vertical Band Machine
The **primary danger** in operating the vertical band machine is **accidental contact with the cutting blade.** Workpieces are often hand guided. One advantage in sawing machines is that the pressure of the cut tends to hold the workpiece against the saw table. However, hands are often in close proximity to the blade. If you should contact the blade accidentally, an injury is almost sure to occur. You will not have time even to think about withdrawing your fingers before they are cut. Keep this in mind at all times when operating a band saw.

Always use a **pusher** against the workpiece whenever possible (Figure 8). This will keep your fingers away from the blade. Be careful as you are about to complete a cut. As the blade clears through the work, the pressure that you are applying is suddenly released and your hand or finger could be carried into the blade. As you approach the end of the cut, **reduce the feeding pressure** as the blade cuts through.

The vertical band machine is generally not

Figure 8. Using a pusher (Courtesy of the California Community Colleges — Individualized Machinist's Curriculum Project).

used to cut round stock. This can be extremely hazardous and should be done on the horizontal band machine, where round stock can be secured in a vice. Hand-held round stock will turn if it is cut on the vertical band machine. This can cause an injury and may damage the blade as well. If round stock must be cut on the vertical band saw, it must be clamped securely in a vise, vee block, or other suitable workholding fixture.

Be sure to select the proper blade for the sawing requirements. Install it properly and apply the correct blade tension. Band tension should be rechecked after a few cuts. New blades will tend to stretch to some degree during their break-in period. Band tension may have to be readjusted.

Safety extends to the machine as well as to the operator. Never abuse any machine tool. They cost a great deal of money and in many cases are purchased with your tax dollars. When cutting on the vertical band machine, bring the workpiece gently into the blade. Use correct speeds, feeds, and blade types. Shift gears only at slow speeds or with the machine stopped. Treat this machine tool with the same respect as you would any expensive precision machine.

SAFETY CHECKLIST
FOR VERTICAL BAND MACHINES

1. Have you removed your rings and wristwatch?
2. Are you using the correct blade for the sawing task and is the tension set properly?

3. Are feeds and speeds properly set?
4. Are all guards in place?
5. Is cutting fluid likely to spill on the floor?
6. Do you have eye protection?
7. Is the guidepost with upper blade guide set as close to the work as possible?
8. Are you using a pusher when fingers are close to the blade?

self-evaluation

SELF-TEST

1. What is the biggest danger on the vertical band machine?

2. Describe conditions under which gloves are acceptable?

3. What is a pusher and how is it used?

4. Why should round stock not be cut on the vertical band machine?

5. What is the primary piece of safety equipment for the band machine operator?

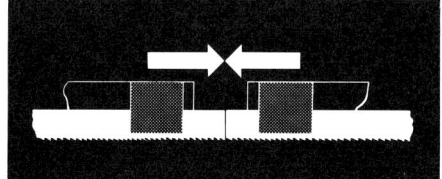

unit 2
preparing to use the vertical band machine

A machine tool can perform at maximum efficiency only if it has been properly maintained, adjusted, and set up. Before the vertical band machine can be used for a sawing or other band machining operation, several important preparations must be made. These include welding saw blades into bands and making several adjustments on the machine tool.

objectives After completing this unit, you will be able to:
1. Weld band saw blades.
2. Prepare the vertical band machine for operation.

information

Welding Band Saw Blades

Band saw blade is frequently supplied in rolls. The required length is measured and cut and the ends are welded together to form an endless band. Most band machines are equipped with a band welding attachment. These are frequently attached to the machine tool. They may also be separate pieces of equipment (Figure 1).

The **band welder** is a resistance-type butt welder. They are often called **flash** welders because of the bright flash and shower of sparks created during the welding operation. The metal in the blade material has a certain resistance to the flow of an electric current. This resistance causes the blade metal to heat as the electric current flows during the welding operation. The blade metal is heated to a temperature that permits the ends to be forged together under pressure. When the forging temperature is reached, the ends of the blade are pushed together by mechanical pressure. They fuse, forming a resistance weld. The band weld is then annealed or softened and dressed to the correct thickness by grinding.

Welding band saw blades is a fairly simple operation and you should master it as soon as possible. Blade welding is frequently done in the machine shop. New blades are always being prepared. Sawing operations, where totally enclosed workpiece features must be cut, require that the blade be inserted through a starting hole in the workpiece and then welded into a band. After the enclosed cut is made, the blade is broken apart and removed.

Preparing the Blade for Welding.

The first step is to cut the required length of blade stock for the band machine that you are using. Blade stock can be cut with snips or with the band shear (Figure 2). Many band machines have a blade shear near the welder. The required length of blade will usually be marked on the saw frame. Blade length, B_L, for two wheel sawing machines can be calculated by the formula

$$B_L = \pi D + 2L$$

where D is the diameter of the band wheel and L is the distance between band wheel centers. Set the tension adjustment on the idler wheel about midrange so that the blade will fit after welding. Most machine shops will have a permanent reference mark, probably on the floor, that can be used for gaging blade length.

After cutting the required length of stock, the ends of the blade must be ground so that they are square when positioned in the welder. Place the ends of the blade together so that the teeth are

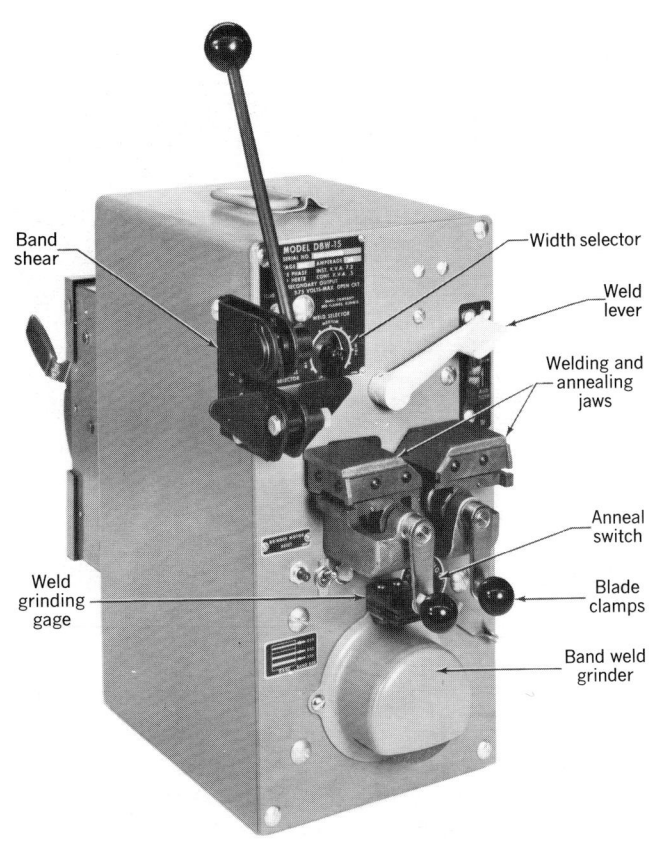

Figure 1. Band blade welder (Courtesy of the DoAll Company).

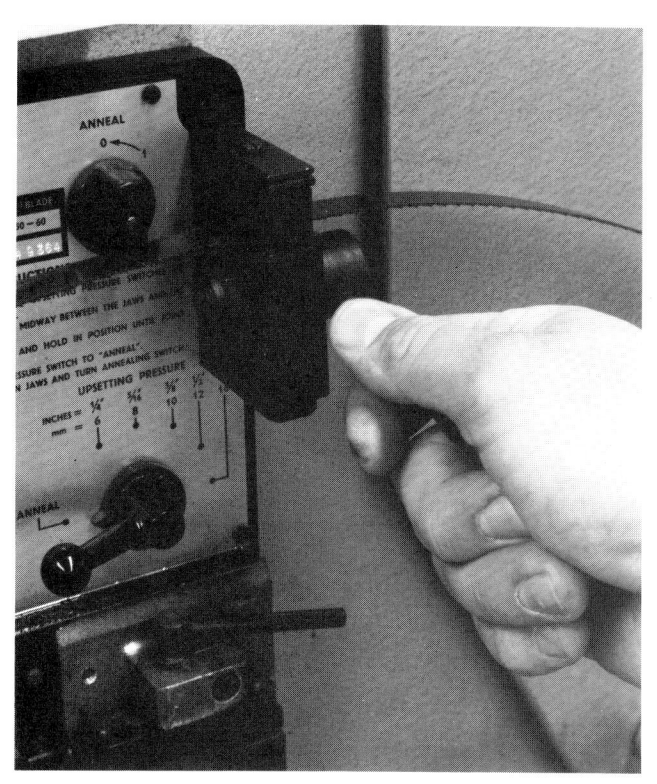

Figure 2. Blade shear (California State University at Fresno).

Figure 3. Placing the blade ends together with the teeth opposed (California State University at Fresno).

Figure 4. End grinding the blade on the pedestal grinder (California State University at Fresno).

opposed (Figure 3). Grind the blade ends in this position. The grinding wheel on the blade welder may be used for this operation. Blade ends may also be ground on the pedestal grinder (Figure 4). Grinding the blade ends with the teeth opposed will insure that the ends of the blade are square when the blade is positioned in the welder. Any small error in grinding will be canceled when the teeth are placed in their normal position.

Proper grinding of the blade ends permits correct tooth spacing to be maintained. After the blade has been welded, the tooth spacing across the weld should be the **same** as any other place on the band. Tooth set should be aligned as well. A certain amount of blade material is consumed in the welding process. Therefore, the blade must be ground correctly if tooth spacing is to be maintained. The amount consumed by the welding process may vary with different blade welders. You will have to determine this by experimentation. For example, if 1/4 in. of blade length is consumed in welding, this would amount to about one tooth on a four pitch blade. Therefore, one tooth should be ground from the blade. This represents the amount lost in welding (Figure 5). Be sure to grind only the tooth and not the end of the blade. The number of teeth to grind from a blade will vary according to the pitch and amount of material consumed by a specific welder. The weld should occur at the bottom of the tooth **gullet**. Exact tooth spacing can be somewhat difficult to obtain. You may have to practice end grinding and welding several pieces of scrap blades until you are familiar with the proper welding and tooth grinding procedure.

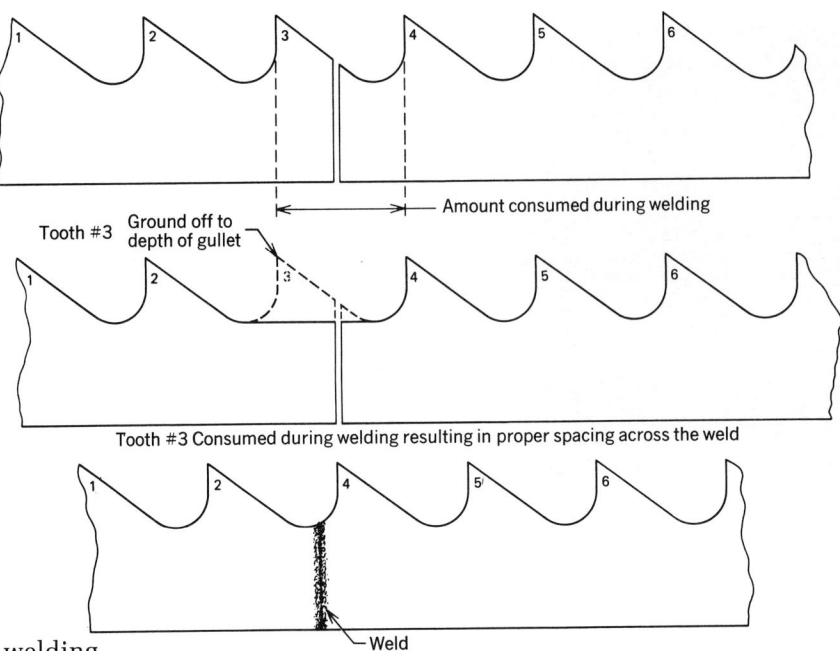

Figure 5. The amount of blade lost in welding.

The jaws of the blade welder should be clean before attempting any welding. Postion the blade ends in the welder jaws (Figure 6). The saw teeth should point toward the back. This prevents scoring of the jaws when welding blades of different widths. A uniform amount of blade should extend from each jaw. The blade ends must contact squarely in the center of the gap between the welder jaws. Be sure that the blade ends are not offset or overlapped. Tighten the blade clamps.

Welding the Blade into an Endless Band
Adjust the welder for the proper width of blade to be welded. **Wear eye protection** and stand to one side of the welder during the welding operation. Depress the weld lever. A flash with a shower of sparks will occur (Figure 7). In this brief operation, the movable jaw of the welder moved toward the stationary jaw. The blade ends were heated to forging temperature by a flow of electric current, and the molten ends of the blades were pushed together, forming a solid joint.

The blade clamps should be loosened before releasing the weld lever. This prevents scoring of the welder jaws by the now welded band. A correctly welded band will have the weld **flash** evenly distributed across the weld zone (Figure 8). Tooth spacing across the weld should be the same as the rest of the band.

Figure 6. Placing the blade in the welder (Courtesy of the California Community Colleges — Individualized Machinist's Curriculum Project).

Annealing the Weld
The metal in the weld zone is hard and brittle immediately after welding. For the band to function, the weld must be **annealed** or **softened.** This improves strength qualities of the weld. Place the band in the annealing jaws with the teeth pointed out (Figure 9). This will concentrate annealing heat away from the saw teeth. A small amount of compression should be placed on the movable welder jaw prior to clamping the band. This permits the jaw to move as the annealing heat expands the band.

Figure 7. Welding the blade into a band (Courtesy of the DoAll Company).

Figure 8. Weld flash should be evenly distributed after welding (California State University at Fresno.)

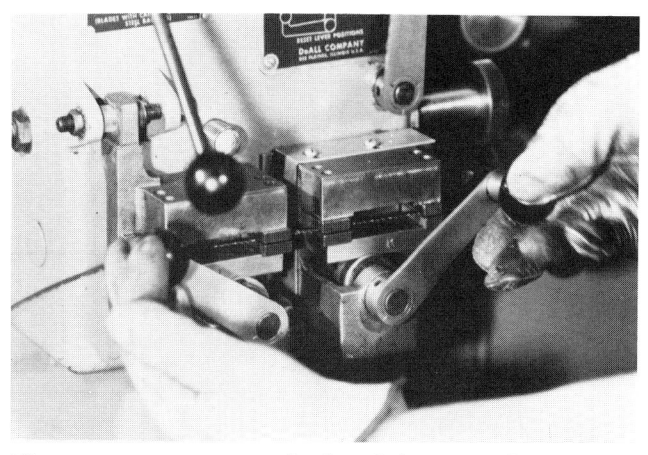

Figure 9. Positioning the band for annealing (Courtesy of the California Community Colleges — Individualized Machinist's Curriculum Project).

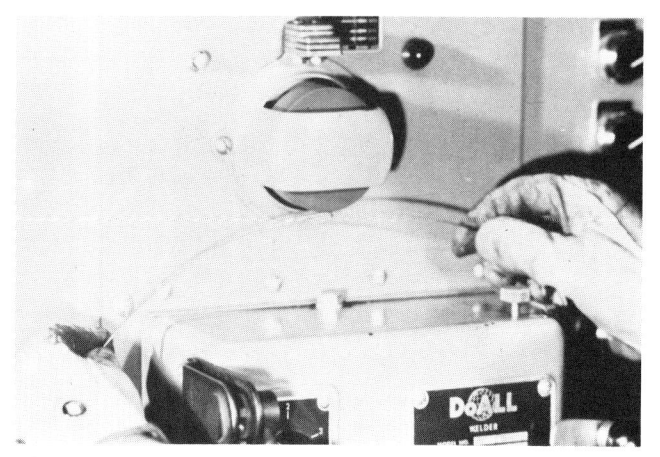

Figure 10. Grinding the band weld (Courtesy of the California Community Colleges — Individualized Machinist's Curriculum Project).

It is most important not to overheat the weld during the annealing process. Overheating can destroy an otherwise good weld, causing it to become brittle. The correct annealing temperature is determined by the color of the weld zone during annealing. This should be a dull red color. Depress the anneal switch and watch the band heat. When the dull red color appears, release the anneal switch immediately and let the band begin to cool. As the weld cools, depress the anneal switch briefly several times to slow the cooling rate. Too rapid cooling can result in a band weld that is not properly annealed.

Grinding the Weld

Some machinists prefer to grind the band weld prior to annealing. This permits the annealing color to be seen more easily. More often, the weld is ground after the annealing process. However, it is good practice to anneal the blade weld further after grinding. This will eliminate any hardness induced during the grinding operation. The grinding wheel on the band welder is designed for this operation. The top and bottom of the grinding wheel are exposed so that both sides of the weld can be ground (Figure 10). **Be careful not to grind the teeth** when grinding a band weld. This will destroy the tooth set. Grind the band weld evenly on both sides (Figure 11). The weld should be ground to the same thickness as the rest of the band. If the weld area is ground thinner, the band will be weakened at that point. As you grind, check the band thickness in the gage (Figure 12) to determine proper thickness (Figure 13).

PROBLEMS IN BAND WELDING

Several problems may be encountered in band welding (Figure 14). These include misaligned pitch, blade misalignment, insufficient welding heat, or too much welding heat. You should learn to recognize and avoid these problems. The best way to do this is to obtain some scrap blades and practice the welding and grinding operations.

INSTALLING AND ADJUSTING BAND GUIDES ON THE VERTICAL BAND MACHINE

Band guides must be properly installed if the band machine is to cut accurately and if damage to the band is to be prevented. Be sure to use the **correct width guides** for the band (Figure 15). The band must be fully supported except for the teeth. Using wide band guides with a narrow band will destroy tooth set as soon as the machine is started.

Band guides are set with a **guide setting gage.** Install the right-hand band guide and tighten the lock screw just enough to hold the guide insert in place. Place the setting gage in the left guide slot and adjust the position of the right

Figure 12. Band weld thickness gage (Courtesy of the California Community Colleges — Individualized Machinist's Curriculum Project).

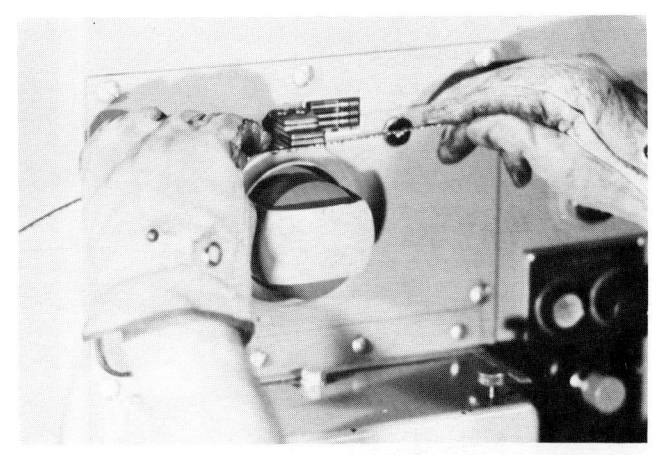

Figure 13. Gaging the weld thickness in the grinding gage (Courtesy of the California Community Colleges — Individualized Machinist's Curriculum Project).

guide insert so that it is in contact with both the vertical and diagonal edges of the gage (Figure 16). Check the **backup bearing** at this time. Clear any chips that might prevent it from turning freely. If the backup bearing cannot turn freely, it will be scored by the band and damaged permanently.

Install the right-hand guide insert and make the adjustment for band thickness using the same setting gage (Figure 17). The thickness of the band will be marked on the tool. Be sure that this is the same as the band that will be used. The lower band guide is adjusted in a like manner. Use the setting gage on the same side as it was used when adjusting the top guides (Figure 18).

Figure 11. The saw teeth must not be ground while grinding the band weld (California State University at Fresno).

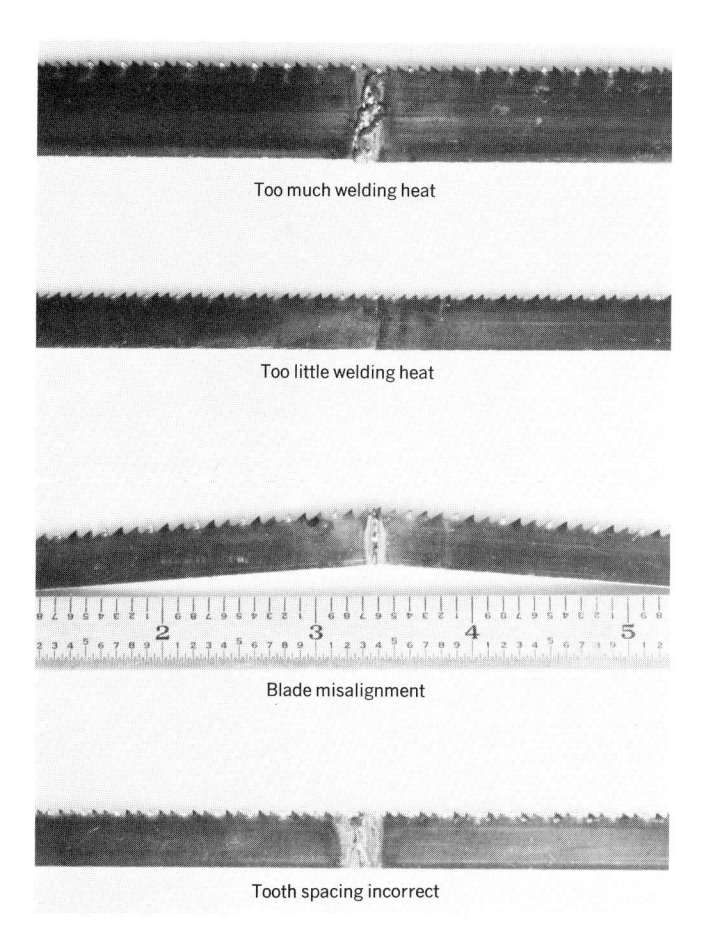

Figure 14. Problems in band welding.

Figure 15. Band guides must fully support the band but must not extend over the saw teeth (Courtesy of the California Community Colleges — Individualized Machinist's Curriculum Project).

Roller band guides are used in high speed sawing applications where band velocities exceed 2000 FPM. They are also used in friction

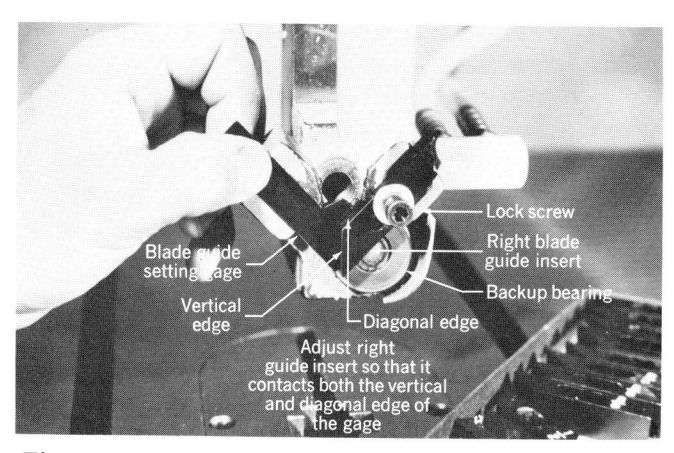

Figure 16. Using the saw guide setting gage (Courtesy of the California Community Colleges — Individualized Machinist's Curriculum Project).

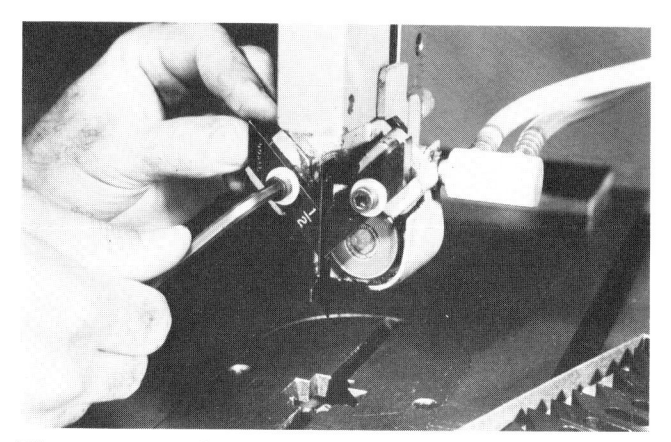

Figure 17. Adjusting the band guides for band thickness (Courtesy of the California Community Colleges — Individualized Machinist's Curriculum Project).

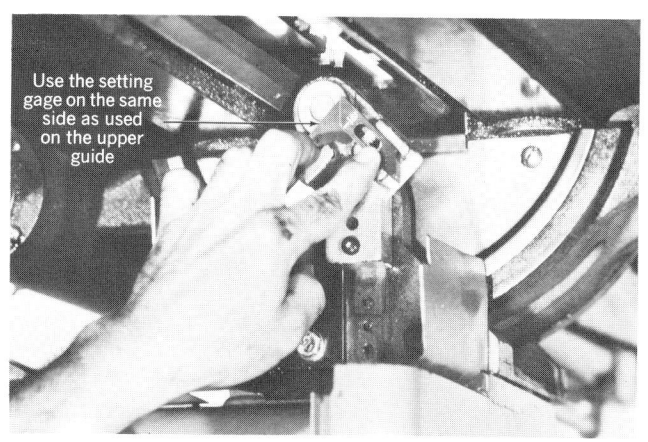

Figure 18. When adjusting the lower guide, use the setting gage on the same side as on the upper guide (Courtesy of the California Community Colleges — Individualized Machinist's Curriculum Project).

sawing operations. The roller guide should be adjusted so that it has .001 to .002 in. clearance with the band (Figure 19).

ADJUSTING THE COOLANT NOZZLE

A band machine may be equipped with flood or mist coolant. Mist coolant is liquid coolant mixed with air. Certain sawing operations may require only small amounts of coolant. With the mist system, liquid coolant is conserved and is less likely to spill on the floor. When cutting with flood coolant, be sure that the runoff returns to the reservoir and does not spill on the floor. Flood coolant may be introduced directly ahead of the band (Figure 20).

Flood or mist coolant may be introduced through a nozzle in the upper guidepost assembly (Figure 21). Air and liquid are supplied to the inlet side of the nozzle by two hoses (Figure 22). The coolant nozzle must be installed (Figure 23) and preset (Figure 24) prior to installing the band. For mist coolant set the nozzle end ½ in. from the face of the band guide. The setting for flood is ⅜ in.

INSTALLING THE BAND ON THE VERTICAL BAND MACHINE

Open the upper and lower wheel covers and remove the filler plate for the worktable. It is **safer to handle the band with gloves to protect your hands** from the sharp saw teeth. The hand tension crank is attached to the upper idler wheel

Figure 20. Coolant may be introduced directly ahead of the band (Courtesy of the DoAll Company).

Figure 21. Mist and flood coolant nozzles (Courtesy of the California Community Colleges — Individualized Machinist's Curriculum Project).

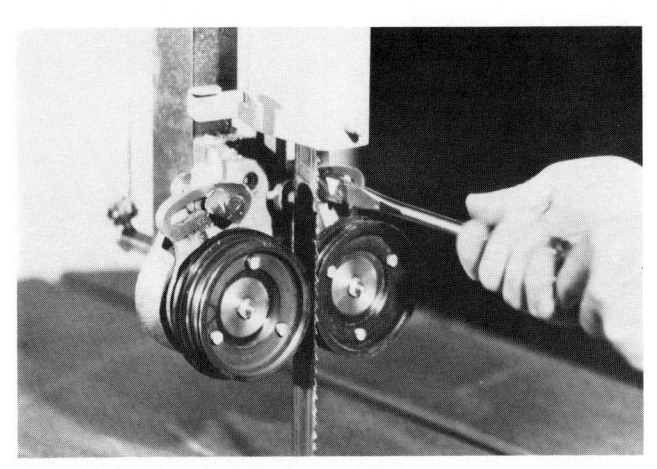

Figure 19. Adjusting roller band guides (Courtesy of the California Community Colleges — Individualized Machinist's Curriculum Project).

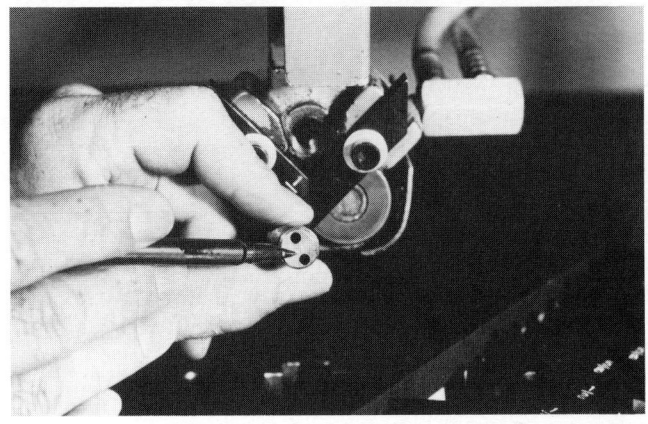

Figure 22. Inlet side of the coolant nozzle (Courtesy of the California Community Colleges — Individualized Machinist's Curriculum Project).

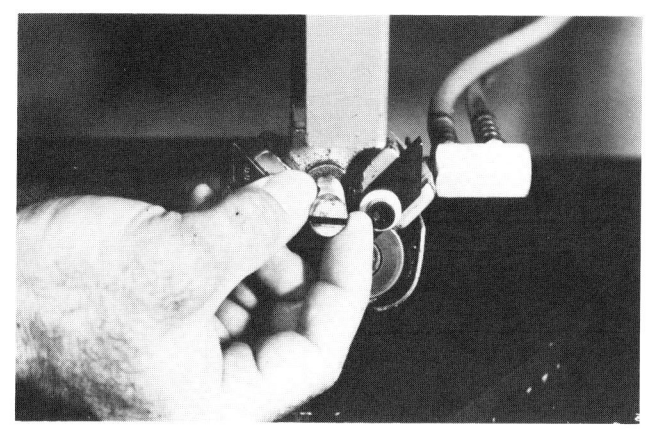

Figure 23. Installing the coolant nozzle (Courtesy of the California Community Colleges — Individualized Machinist's Curriculum Project).

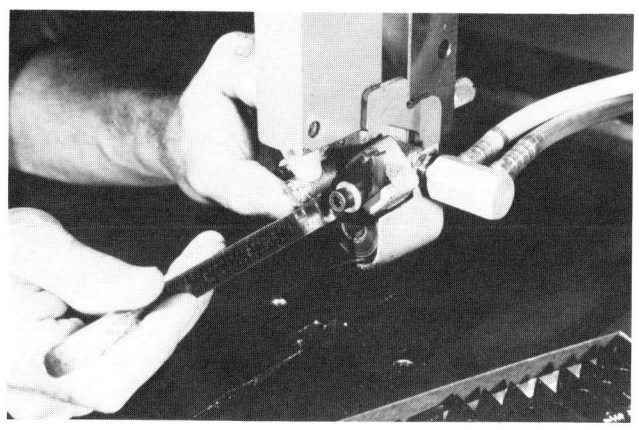

Figure 24. Presetting the coolant nozzle position before installing the band guides (Courtesy of the California Community Colleges — Individualized Machinist's Curriculum Project).

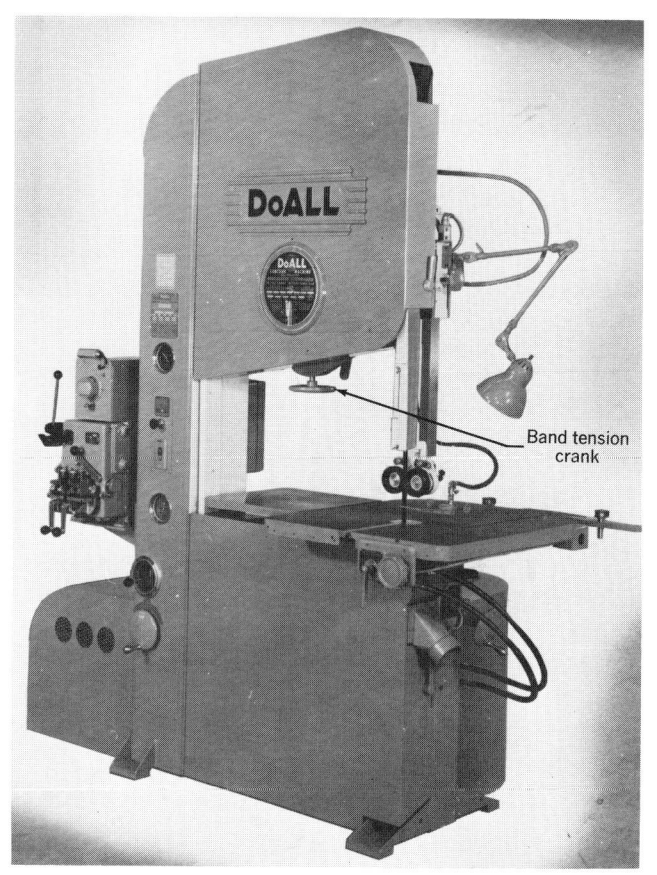

Figure 25. Band tension crank (Courtesy of the DoAll Company).

(Figure 25). Turn the crank to lower the wheel to a point where the band can be placed around the drive and idler wheels. Be sure to install the band so the teeth point in the direction of the cut. This is always in a **down direction toward the worktable.** If the saw teeth seem to be pointed in the wrong direction, the band may have to be turned inside out. This can be done easily. Place the band around the drive and idler wheels and turn the tension crank so that tension is placed on the band. Be sure that the band slips into the upper and lower guides properly. Replace the filler plate in the worktable.

Adjusting Band Tension
Proper **band tension** is important to accurate cutting. A high tensile strength band should be used whenever possible. Tensile strength refers to the strength of the band to withstand stretch. The correct band tension is indicated on the **band tension dial** (Figure 26). Adjust the tension for the width of band that you are using. After a new band has been run for a short time, recheck the tension. New bands tend to stretch during their initial running period.

Adjusting Band Tracking
Band tracking refers to the position of the band as it runs on the idler wheel tires. On the vertical band machine, the idler wheel can be tilted to adjust tracking position. The band tracking position should be set so that the back of the band just touches the backup bearing in the guide assembly. Generally, you will not have to adjust band tracking very often. After you have installed a blade, check the tracking position. If it is incorrect, consult your instructor for help in adjusting the tracking position.

The tracking adjustment is made with the

motor off and the speed range transmission in neutral. This permits the band to be rolled by hand. Two knobs are located on the idler wheel hub. The outer knob (Figure 27) tilts the wheel. The inner knob is the tilt lock. Loosen the lock knob and adjust the tilt of the idler wheel while rolling the band by hand. When the correct tracking position is reached, lock the inner knob. If the band machine has three idler wheels, adjust band tracking on the top wheel first. Then adjust tracking position on the back wheel.

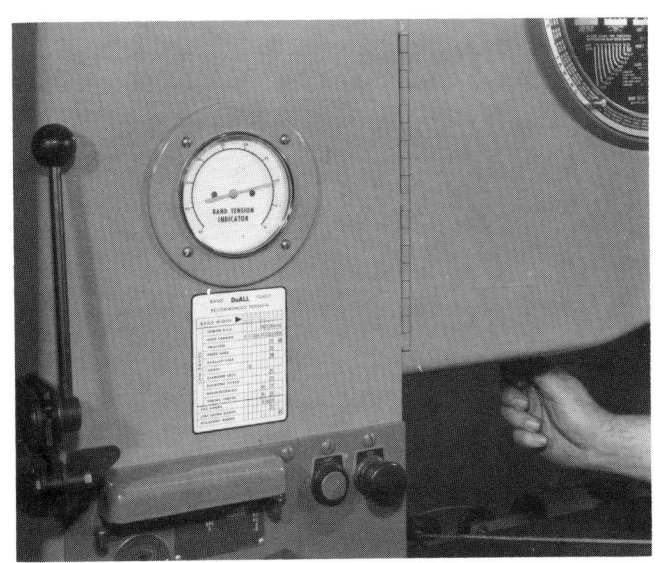

Figure 26. Band tension dial (Courtesy of the DoAll Company).

Figure 27. Adjusting the band tracking position by tilting the idler wheel (Courtesy of the California Community Colleges — Individualized Machinist's Curriculum Project).

OTHER ADJUSTMENTS ON THE VERTICAL BAND MACHINE

The hub of the variable speed pulley should be lubricated weekly (Figure 28). While the drive

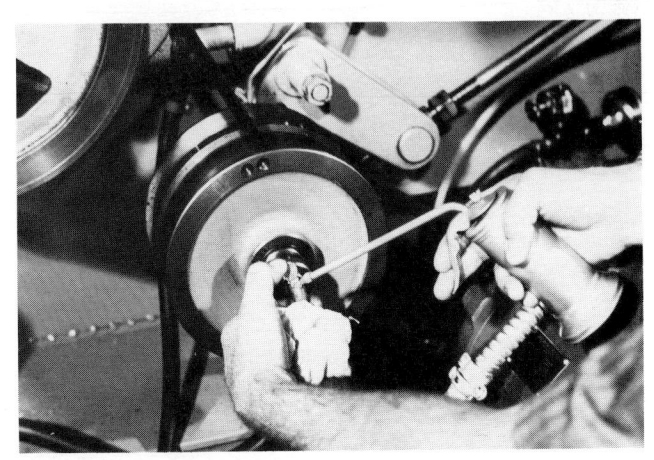

Figure 28. Lubricating the variable speed pulley hub (Courtesy of the California Community Colleges — Individualized Machinist's Curriculum Project).

Figure 29. Band wheel chip brush (Courtesy of the California Community Colleges — Individualized Machinist's Curriculum Project).

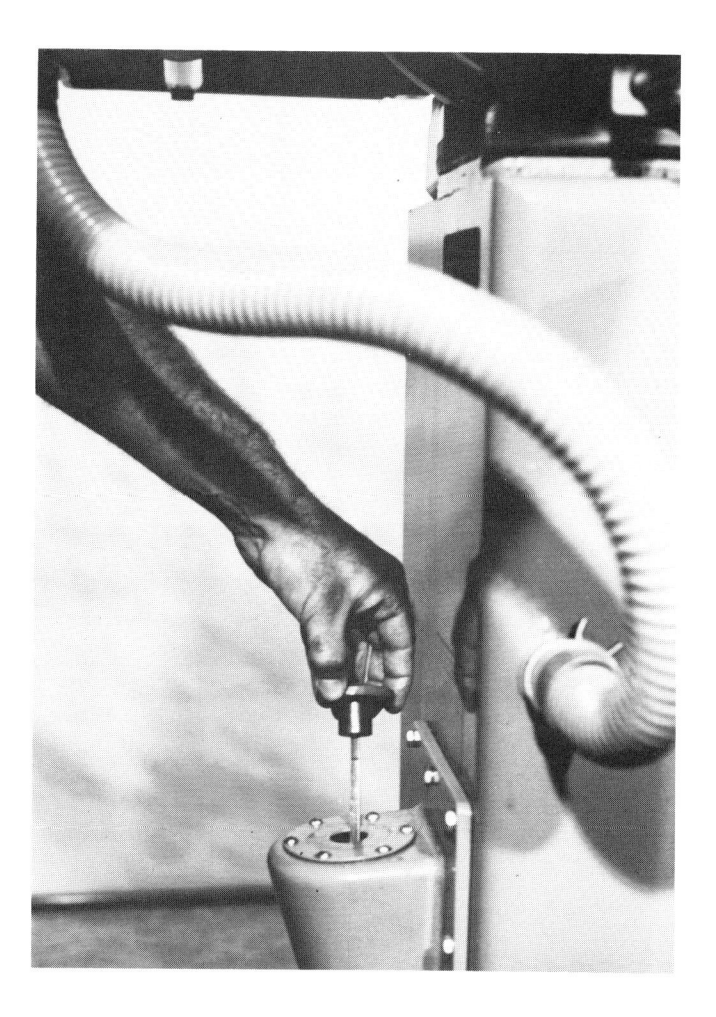

mechanism guard is open, check the oil level in the speed range transmission. The band machine may be equipped with a chip brush on the band wheel (Figure 29). This should be adjusted frequently. Chips that are transported through the band guides can score the band and make it brittle. The hydraulic oil level should be checked daily on band machines with hydraulic table feeds (Figure 30).

Figure 30. Checking the hydraulic oil level on the vertical band machine (Courtesy of the California Community Colleges — Individualized Machinist's Curriculum Project).

self-evaluation

SELF-TEST

1. Describe the blade end grinding procedure.

2. Describe the band welding procedure.

3. Describe the weld grinding procedure.

4. What is the purpose of the band blade guide?

5. Why is it important to use a band guide of the correct width?

6. What tool can be used to adjust band guides?

7. What is the function of annealing the band weld?

8. Describe the annealing process.

9. What is band tracking?

10. How is band tracking adjusted?

EXERCISES
1. Obtain a worn out band blade and cut several suitable lengths. Practice end grinding, welding, annealing, and weld grinding procedures.
2. Practice folding band saw blades into small coils suitable for storage.

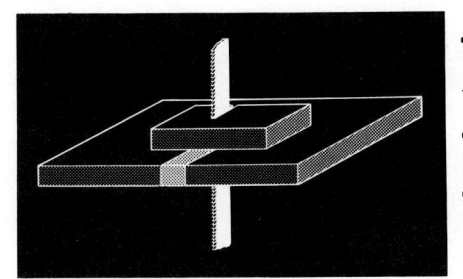

unit 3
using the vertical band machine

After a machine tool has been properly adjusted and set up, it can be used to accomplish a machining task. In the preceding unit, you had an opportunity to prepare the vertical band machine for use. In this unit, you will be able to operate this versatile machine tool.

objectives After completing this unit, you will be able to:
1. Use the vertical band machine job selector.
2. Operate the band machine controls.
3. Perform typical sawing operations on the vertical band machine.

information

SELECTING A BLADE FOR THE VERTICAL BAND MACHINE

Blade materials include standard carbon steel where the saw teeth are fully hardened but the back of the blade remains soft. The standard carbon steel blade is available in the greatest combination of width, set, pitch, and gage.

The carbon alloy steel blade has hardened teeth and also a hardened back. The harder back permits sufficient flexibility of the blade but, because of increased tensile strength, a higher band tension may be used. Because of this, cutting accuracy is greatly improved. The carbon alloy blade material is well suited to contour sawing.

High speed steel and bimetallic high speed steel blade materials are used in high production and severe sawing applications where blades must have long wearing characteristics. The high speed steel blade can withstand much more heat than the carbon or carbon alloy materials. On the bimetallic blade, the cutting edge is made from one type of high speed steel, while the back is made from another type of high speed steel that has been selected for high flexibility and high tensile strength. High speed and bimetallic high speed blades can cut longer, faster, and more accurately.

Band blade selection will depend on the sawing task. You should review saw blade terminology discussed in the cutoff machine section (Volume I). The first consideration is blade pitch. The pitch of the blade should be such that at least two teeth are in contact with the workpiece. This generally means that fine pitch blades with more teeth per inch will be used in thin materials. Thick material requires coarse pitch blades so that chips will be more effectively cleared from the kerf.

Remember that there are three tooth **sets** that can be used (Figure 1). **Raker** and **wave** set are the most common in the metalworking industries. **Straight** set may be used for cutting thin materials. Wave set is best for accurate cuts through materials with variable cross sections. Raker set may be used for general purpose sawing.

You also have a choice of **tooth forms** (Figure 2). **Precision** or **regular** tooth form is best for accurate cuts where a good finish may be required. **Hook** form is fast cutting but leaves a rougher finish. **Skip** tooth is useful on deep cuts where additional chip clearance is required.

Several special bands are also used. **Straight, scalloped,** and **wavy edges** are used for cutting nonmetallic substances where saw teeth would tear the material (Figure 3). **Continuous** (Figure 4) and **segmented** (Figure 5) **diamond edged band** is used for cutting very hard nonmetallic materials.

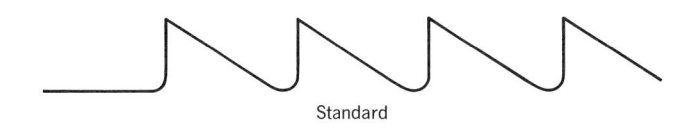

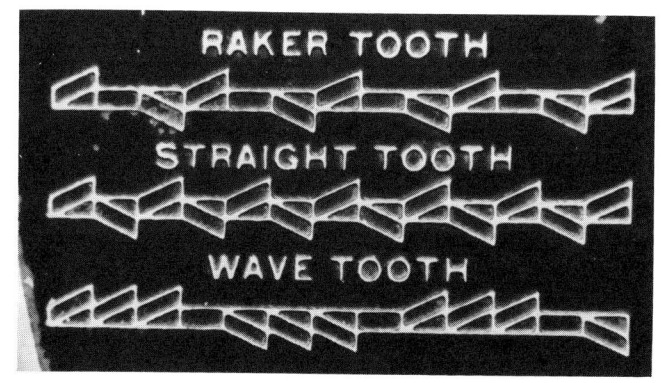

Figure 1 Blade set patterns (California State University at Fresno).

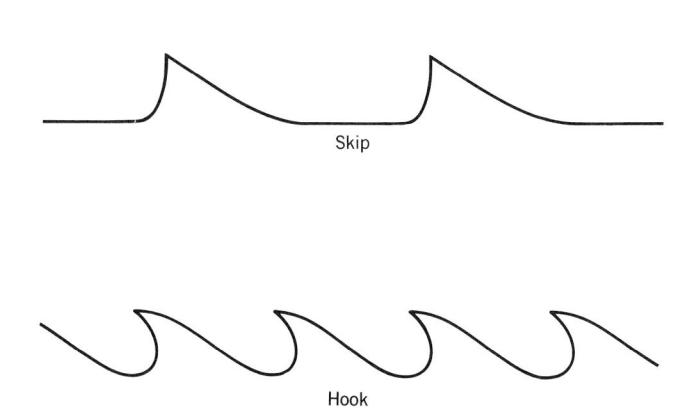

Figure 2. Saw tooth forms.

Figure 3. Straight, scalloped, and wavy edge bands (Courtesy of the DoAll Company).

Figure 4. *(left)* Continuous edge diamond band (Courtesy of the DoAll Company).

Figure 5. *(right)* Segmented diamond edge band (Courtesy of the DoAll Company.)

USING THE JOB SELECTOR ON THE VERTICAL BAND MACHINE

Most vertical band machines are equipped with a **job selector.** This device will be of great aid to you in accomplishing a sawing task. Job selectors are usually attached to the machine tool. They are frequently arranged by material. The material to be cut is located on the rim of the selector. The selector disk is then turned until the sawing data for the material can be read (Figure 6).

The job selector yields much valuable information. Sawing velocity in feet per minute is the most important. The band must be operated at the correct cutting speed for the material. If it is not, the band may be damaged or productivity will be low. Saw velocity is read at the top of the column and is dependent on material thickness. The job selector also indicates recommended pitch, set, feed, and temper. The job selector will provide information on sawing of nonmetallic materials (Figure 7). Information on band filing can also be determined from the job selector.

SETTING SAW VELOCITY ON THE VERTICAL BAND MACHINE

Most vertical band machines are equipped with a **variable speed drive** that permits a wide selection of band velocities. This is one of the factors that make the band machine such a versatile machine tool. Saw velocities can be selected that permit successful cutting of many materials.

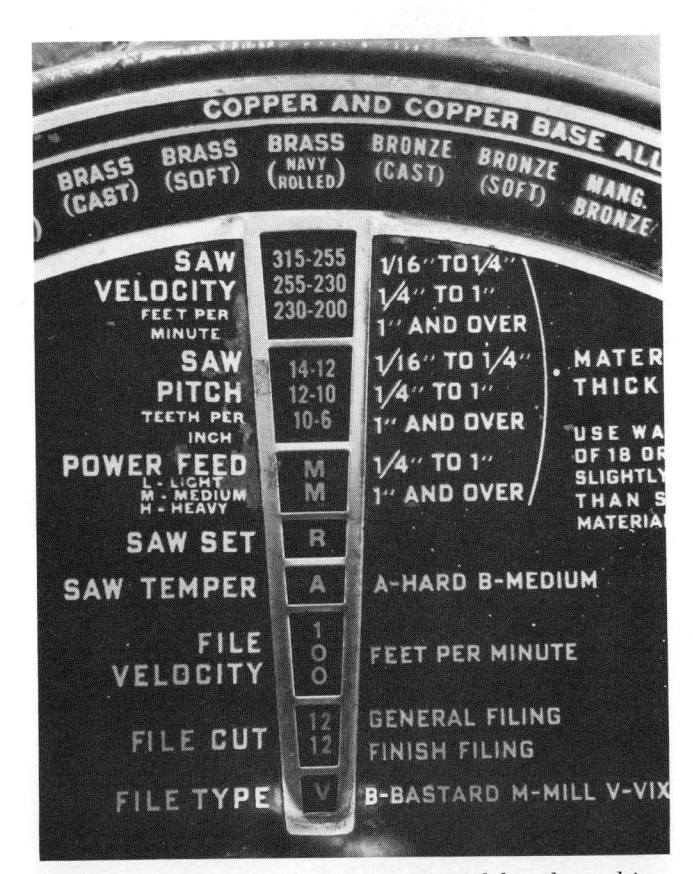

Figure 6. Job selector on the vertical band machine (California State University at Fresno).

The typical variable speed drive uses a split flange pulley to vary the speed of the drive wheel (Figure 8). As the flanges of the pulley are spread apart by adjusting the speed control, the belt runs deeper in the pulley groove. This is the same as running the drive belt on a smaller diameter pulley. Slower speeds are obtained. As the flanges of

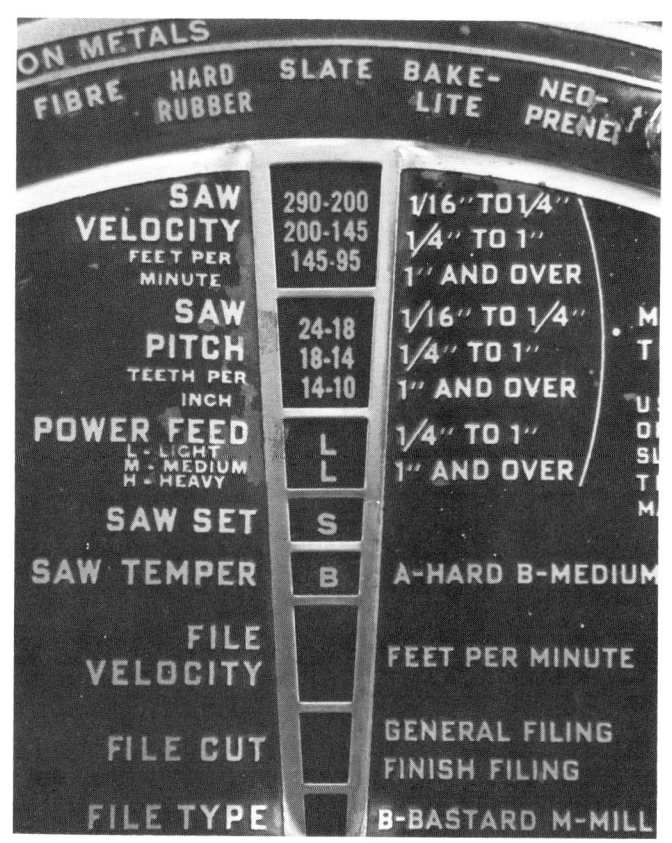

Figure 7. The job selector set for a nonferrous material (California State University at Fresno).

Figure 8. Vertical band machine variable speed drive (Courtesy of the DoAll Company).

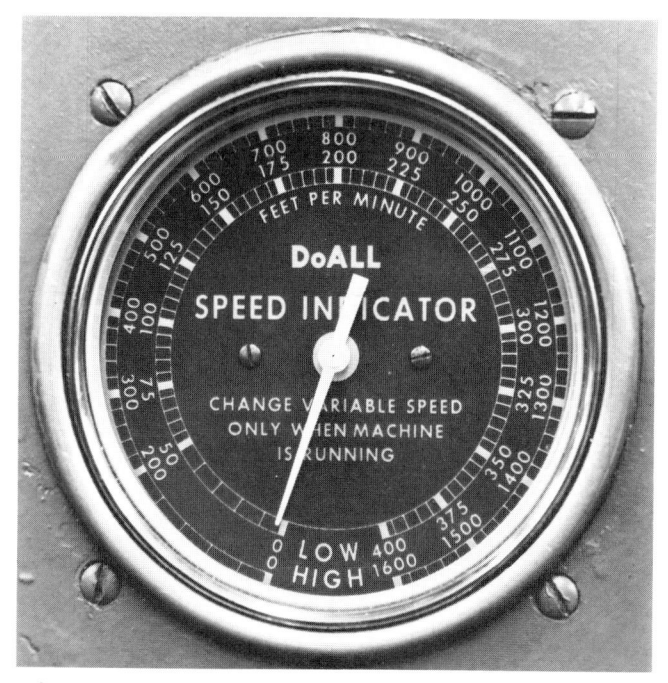

Figure 9. Band velocity indicator (California State University at Fresno).

the pulley flanges are adjusted for less spread, the belt runs toward the outside. This is equivalent to running the belt on a larger diameter pulley. Faster speeds are obtained.

Setting Band Velocity
Band velocity is indicated on the **band velocity indicator** (Figure 9). Remember that band velocity is measured in **feet per minute.** The inner scale indicates band velocity in the low speed range. The outer scale indicates velocity in the high speed range. Band velocity is regulated by adjusting the speed control (Figure 10). Adjust this control only while the motor is running, as this adjustment moves the flanges of the variable speed pulley.

Setting Speed Ranges
Most band machines with a variable speed drive have a **high** and **low speed range.** High or low speed range is selected by operating the **speed range shift lever** (Figure 10). This setting must be made while the band is stopped or is running at the lowest speed in the range. If the machine is set in high range and it is desired to go to low range, turn the band velocity control wheel until the band has slowed to the lowest speed possible. The speed range shift may now be changed to low speed. If the machine is in low speed and it is desired to shift to high range, slow the band to the lowest speed before shifting speed ranges. A speed range shift made while the band is running at a fast speed may damage the speed

Figure 10. Speed range and band velocity controls (California State University at Fresno).

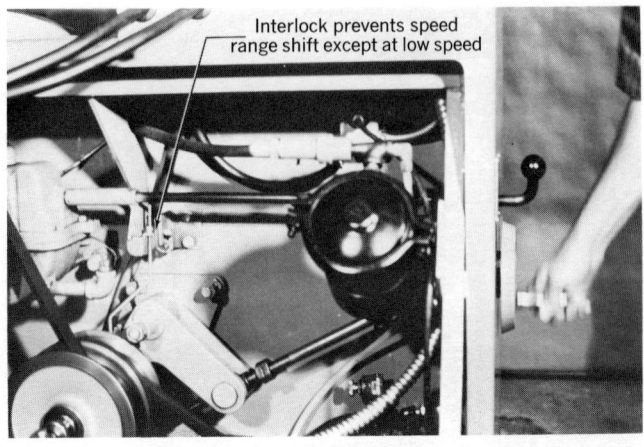

Figure 11. An interlock prevents shifting speed ranges except at low speed (Courtesy of the California Community Colleges—Individualized Machinist's Curriculum Project).

range transmission gears. Some band machines are equipped with an interlock to prevent speed range shifts except at low band velocity (Figure 11).

STRAIGHT CUTTING
ON THE VERTICAL BAND MACHINE

Adjust the upper guidepost so that it is as close to the workpiece as possible (Figure 12). This will maximize safety by properly supporting and guarding the band. Accuracy of the cut will also be aided. The guidepost is adjusted by loosening the clamping knob and moving the post up or down according to the workpiece thickness.

Be sure to use a band of the **proper pitch** for the thickness of the material to be cut. If the band pitch is too fine, the teeth will clog (Figure 13). This can result in stripping and breakage of the saw teeth due to overloading (Figure 14). Cutting productivity will also be reduced. Slow cutting will result from using a fine pitch band on thick material (Figure 15). The correct pitch for thick material (Figure 16) results in much more effi-

Figure 12. Adjusting the upper guide post (Courtesy of the DoAll Company).

cient cutting in the same amount of time and at the same feeding pressure.

As you begin a cut, feed the workpiece **gently** into the band. A sudden shock will cause the saw teeth to chip or fracture (Figure 17). This will reduce band life quickly. See that chips are cleared from the band guides. These can score

Figure 13. Using too fine a pitch blade results in clogged teeth (Courtesy of the California Community Colleges — Individualized Machinist's Curriculum Project).

Figure 14. Stripped and broken teeth resulting from overloading the saw (Courtesy of the DoAll Company).

Figure 15. Saw cut with a fine pitch blade in thick material (Courtesy of the California Community Colleges — Individualized Machinist's Curriculum Project).

Figure 16. Saw cut with correct pitch band for thick material (Courtesy of the California Community Colleges — Individualized Machinist's Project).

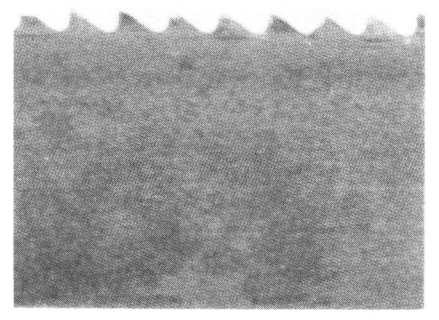

Figure 17. Chipped and fractured teeth resulting from shock and vibration (Courtesy of the DoAll Company).

the band (Figure 18), making it brittle and subject to breakage.

Cutting Fluids

Cutting fluids are an important aid to sawing many materials. They **cool** and **lubricate** the band and **remove chips** from the kerf. Many band machines are equipped with a mist coolant system. Liquid cutting fluids are mixed with air to form a mist. With mist, the advantages of the coolant are realized without the need to collect

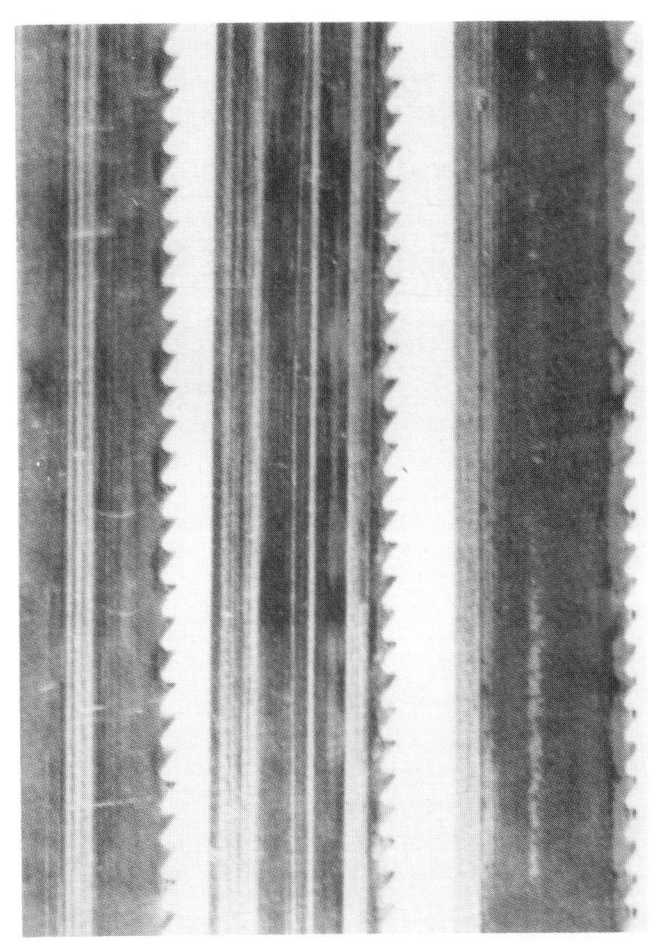

Figure 18. Scored bands can become brittle and loose flexibility (Courtesy of the DoAll Company).

Figure 19. Adjusting the coolant and air mix on the vertical band machine (Courtesy of the California Community Colleges — Individualized Machinist's Curriculum Project).

and return large amounts of liquids to a reservoir. If your band machine uses mist coolant, set the liquid flow first and then add air to create a mist (Figure 19). Do not use more coolant than is necessary. Overuse of air may cause a mist fog around the machine. This is both unpleasant and hazardous, as coolant mist should not be inhaled.

CONTOUR CUTTING
ON THE VERTICAL BAND MACHINE

Contour cutting is the ability of the band machine to cut around corners and produce intricate shapes. The ability of the saw to cut a specific radius depends on its **width**. The job selector will provide information on the minimum radius that can be cut with a blade of a given width (Figure 20). As you can see, a narrow band can cut a smaller radius than a wide band.

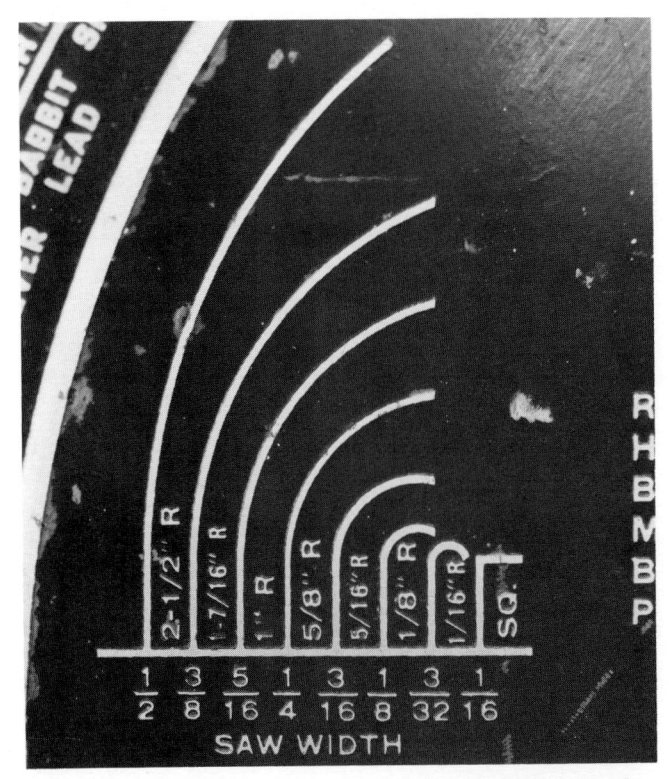

Figure 20. Minimum radius per saw width chart on the job selector (California State University at Fresno).

Unit 3 Using the Vertical Band Machine

The set of the saw needs to be adequate for the corresponding band width. It is a good idea to make a test contour cut in a piece of scrap material. This will permit you to determine if the **saw set** is adequate to cut the desired radius. If the saw set is not adequate, you may not be able to keep the saw on the layout line as you complete the radius cut (Figure 21).

If you are cutting a totally enclosed feature, be sure to insert the saw blade through the starting hole in the workpiece before welding it into a band. Also, be sure that the teeth are pointed in the right direction.

Figure 21. Band set must be adequate for the band width if the layed out radius is to be cut (Courtesy of the California Community Colleges — Individualized Machinist's Curriculum Project).

self-evaluation

SELF-TEST

1. Name three saw blade sets and describe the applications of each.

   a. _____

   b. _____

   c. _____

2. When might scalloped or wavy edged bands be used?

3. What information is found on the job selector?

4. In what units of measure are band velocities measured?

5. Explain the operating principle of the variable speed pulley.

6. Explain the selection of band speed ranges.

7. What machine safety precaution must be observed when selecting speed ranges?

8. Describe the upper guidepost adjustment.

9. What does band pitch have to do with sawing efficiency?

10. What does band set have to do with contouring?

EXERCISES
1. Use the job selector and list all of the pertinent sawing data for the following materials.
   a. Mild steel.
   b. Aluminum.
   c. Cast iron.
2. Familiarize yourself with the band machine controls.
3. Obtain some scrap material. Consult the job selector and set proper band velocity. Make several straight and contouring cuts to familiarize yourself with band machine operation.
4. Use the band machine on your shop projects whenever it is appropriate.

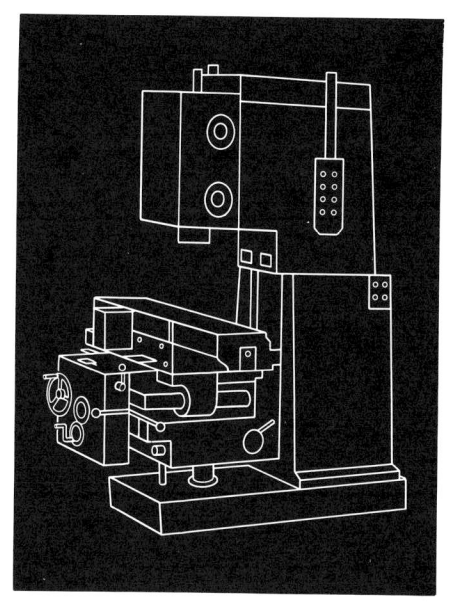

section b
vertical milling machines

The vertical milling machine is a relatively new development in comparison to the horizontal milling machine. The first vertical milling machines appeared in the 1860s (Figure 1). The vertical milling machine was a development more closely related to the drill press than to the horizontal spindle milling machine. The basic difference between drill presses and the earliest vertical milling machines was that the entire spindle assembly, pulleys and all, was moved vertically. This arrangement meant that the bearings that supported the cutting tool were always reasonably close to the tool, which permitted side thrust to be taken more readily. On the other hand, a drill press with a quill that extends some distance from the support bearing is not suitable for the side loading usually encountered in vertical milling.

The next significant step came in the mid-1880s (Figure 2) with the adaptation of the "knee and column" from the horizontal milling machine, which allowed the milling table to be raised and lowered in relation to the spindle. Also, the spindle heads on some of these machines could be tilted to an angular relationship to the table.

Just after the turn of the twentieth century, vertical milling machines began to appear with power feeds on the spindle, housed

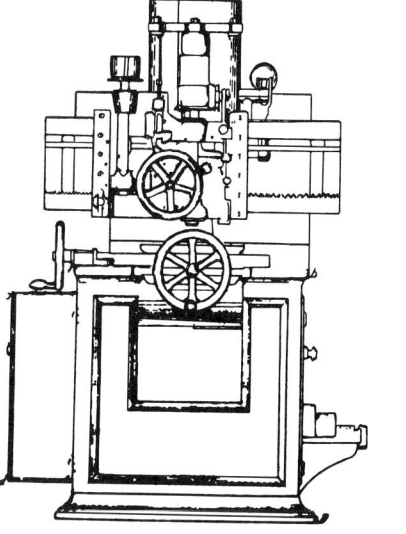

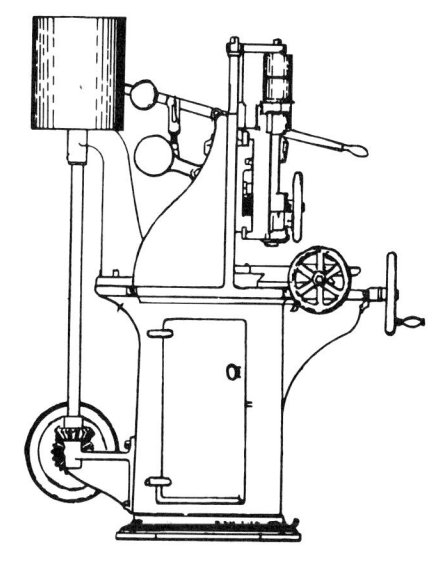

Figure 1. Vertical milling machines of 1862 (U.S. Patent) (Courtesy of The MIT Press, Massachusetts Institute of Technology).

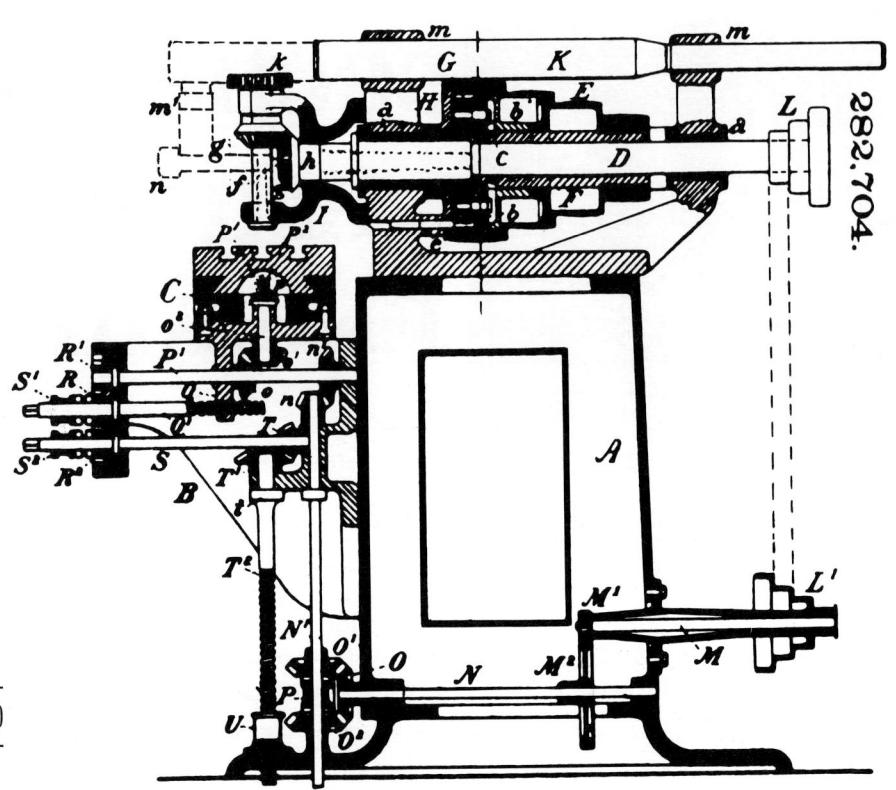

Figure 2. Cosgrove's vertical milling machine of 1883 (U.S. Patent) (Courtesy of The MIT Press, Massachusetts Institute of Technology).

Figure 3. Hydraulic tracing controls on a vertical milling machine especially developed for "die sinking" (Courtesy of Cincinnati Milacron).

in a heavy duty quill. By 1906 the structural development of the vertical milling machine was essentially complete. There were various types of toolroom and production vertical milling machines, with either fixed or swiveling heads, having either fixed position or knee and column tables and either hand or power feed for longitudinal and lateral table movement. Some had vertical power feed as well.

In the first decade of the twentieth century, micrometers and vernier scales had been applied to vertical milling machines to make them suitable for precise hole locating, known as jig boring. The LeBlond Company produced a machine especially for that purpose in 1910.

Developments relative to vertical milling machines after 1910 related mostly to drive and control issues. Machines began to appear with automatic table cycles and, by 1920, electrical servomechanisms were applied to the vertical milling machine for operations like die sinking. By 1927, hydraulic tracing controls were developed and applied to vertical milling machines (Figure 3).

Other types of sensing for machine control have also been used. Optical means have been used to activate servomechanisms for milling to a line on a drawing, called line tracing (Figure 4).

Control systems, not limited to vertical milling machines, have appeared that activate machine control movements from information stored on punched or magnetic tape (Figure 5), called numerical control (NC), or from computer numerical control (CNC).

The manually operated vertical milling machine (Figure 6) is certainly one of the most important basic machine tools. It is

Section B Vertical Milling Machines

Figure 4. Line tracing vertical milling machine (Photo courtesy, Bridgeport Milling Machines, a Division of Textron, Inc.)

Figure 5. *(right)* Numerically controlled vertical milling machine (Ames Research Center — NASA).

Figure 6. *(left)* A popular type of manually operated vertical milling machine with accessory slotting attachment. This is called a ram-type turret mill. (Photo courtesy, Bridgeport Milling Machines, a Division of Textron, Inc.).

convenient to use for various operations such as drilling, boring, and slotting as well as milling.

In some respects, it is even more versatile than the lathe, because in its various forms and with adaptations, it comes very close to being a machine tool that can reproduce itself.

In this section you will be called on to identify safe working practices on the vertical milling machine and to be able to name components and functions. You will be shown a variety of machine setups and operations. The cutting tools of the vertical milling machine come in a remarkable variety of forms and adaptations for specialized work. You will be shown a variety of ways to hold and locate workpieces relative to the cutting tools and how to determine reasonable speeds and feeds rates for vertical milling. Finally, you will apply the offset boring head for accurately sizing nonstandard holes.

The vertical milling machine is an interesting and challenging machine tool, capable of a large variety of work. It is important for you to be able to use this machine competently.

review questions

1. How does the vertical milling machine differ basically from the drill press?
2. What does a "knee and column" do on a vertical milling machine?
3. What is jig boring?
4. After the vertical milling machine had reached its basic design maturity about 1910, what refinements took place?
5. Why is the vertical milling machine considered to be even more versatile than the lathe?

unit 1
vertical milling machine safety

Safety is partly common sense and partly learning safe working practices. This unit points out the factors that make for safe vertical milling machine operation.

objectives After completing this unit, you will be able to identify safe vertical milling machine practices.

information

Safe operation of any machine tool requires that the operator learn about any hazards that exist so that he can protect himself from them. It is very important that the operator is alert when working on a machine. Being tired, sick, or emotionally upset is dangerous when operating equipment. Many prescription drugs should not be taken before driving a car because they affect one's reflexes. The same is true when one is operating machinery. Playing tricks on someone or indulging in horseplay is intended to provide some laughs, but it is not very funny when someone gets hurt or badly injured by it. The common excuse is: "I didn't intend for this to happen." Don't be part of such games. A safe machine operator thinks before doing. He questions what will happen if he turns this lever or operates this control. If in doubt, he will find out before doing it.

Proper dress is important for a safe machine operation. Short sleeves or tightly fitting sleeves are protection against being caught in a revolving spindle. Rings, bracelets, earrings, and even wristwatches can become dangerous when worn around machinery, and they should be removed. Eye protection in the form of safety goggles and face shields should always be worn in a machine shop. Flying particles from your machine or someone else's machine can blind you in a split second. Long hair should be covered under a cap. The feet should be protected with shoes that prevent chips from cutting through the soles and that prevent serious foot injuries from falling objects. Gloves should not be worn while operating equipment because of the danger of being caught by the machine.

All machine guards should be in place prior to starting a machine. Observe other machines in operation around you to make sure they are guarded properly. Report any unsafe or missing guards to your supervisor. If you observe any unsafe practices by other machine operators, report them to the person in charge. A safe workplace involves everyone.

Safety also involves keeping a clean machine and keeping the area surrounding it clean. Any oil or coolant spills on the floor should be wiped up immediately to avoid slipping and falling. Chips should be swept up with a brush or broom and deposited in chip or trash containers. Do not handle chips with your bare hands or you will get cut. Dirty and oily rags should be kept in closed containers and should not accumulate in piles on the floor. Do not use an air hose to clean a machine. Flying chips can hurt you or those around you. When you lift heavy workpieces or machine attachments on or off the machine, ask someone to help you. When you are lifting anything, use proper lifting methods. The right way to lift things is with your back straight and vertical and your legs doing the lifting.

Be careful when handling tools and sharp edged workpieces to avoid getting cut. Use a rag to protect your hand. Workpieces should be rigidly supported and tightly clamped to withstand the usually high cutting forces encountered in machining. When a workpiece comes loose while machining, it is usually ruined and, often, so is the cutter. The operator can also be hurt by flying particles from the cutter or workpiece.

The cutting tools should be securely fastened in the machine spindle to prevent any movement during the cutting operation. Cutting tools need to be operated at the correct revolutions per minute (RPM) and feedrate for any given material. Excessive speeds and feeds can break the cutting tools. On vertical milling machines, care has to be exercised when swivelling the workhead to make angular cuts. After loosening the clamping bolts that hold the workhead to the overarm, retighten them lightly to create a slight drag. There should be enough friction between the workhead and the overarm that the head only swivels when pressure is applied to it. If the clamping bolts are completely loosened, the weight of the heavy spindle motor will flip the workhead upside down or until it hits the table, possibly smashing the operator's hand or a workpiece.

Measurements are frequently made during machining operations. Do not make any measurements until the spindle has come to a complete stop or standstill.

self-evaluation

SELF-TEST

1. Why is it dangerous to operate machinery when tired, sick, or upset?

2. Why is safe dress for machine work important?

3. How are chips handled?

4. When is eye protection worn?

5. Why do machine guards need to be in place before operating a machine?

6. Why is a clean work area important?

7. How are heavy objects lifted?

8. What are danger points when handling tools?

9. How is a vertical milling machine head swiveled safely?

10. How are safe measurements made on a machine?

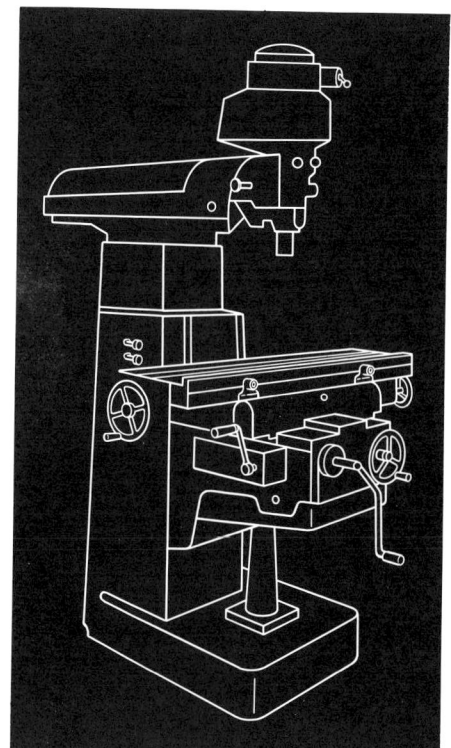

unit 2
the vertical spindle milling machine

Milling machine parts and components have names to make their identification easier. Knowing these names is helpful in locating trouble spots or in operating the machine controls.

objectives After completing this unit, you will be able to:
1. Identify the important parts of a vertical milling machine.
2. Perform routine maintenance.

information

The vertical milling machine is a very versatile tool in a machining operation. Figure 1 identifies many of its important parts. The *column* is the backbone of the machine; it rests on the base. The front or face of the column is accurately machined to provide a guide for the vertical travel of the knee. The upper part of the *column* is machined to provide a swivel capability to the ram. The *knee* supports and guides the saddle. The *saddle* provides cross travel for the machine and is the support and guide for the table. The *table* gives longitudinal travel to the machine and it holds the workpiece or workholding devices. The ram can be adjusted toward or away from the column to increase the working capacity of the milling machine. The *toolhead* is attached to the end of the ram. The toolhead can be swiveled on some milling machines in one or two planes. The six listed assemblies are the major components of the vertical milling machine. Most of these components have controls or parts that are important to know. The

Section B Vertical Milling Machines

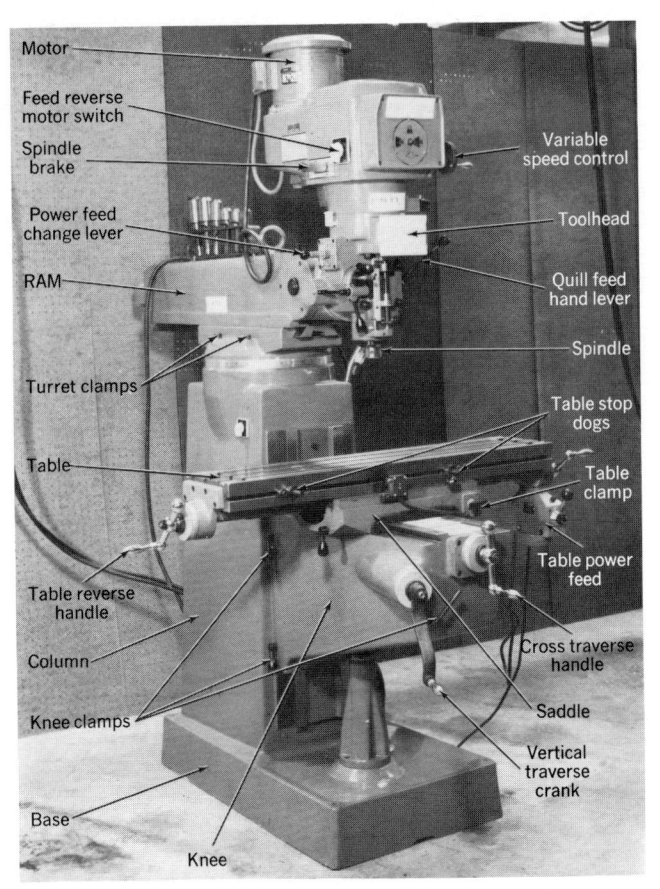

Figure 1. The important parts of a vertical milling machine (Lane Community College).

upward, the power feed disengages when the quill reaches its upper limit. The micrometer dial allows depth stop adjustments in .001 in. increments. The quill clamp is used to lock the quill in the head to get maximum rigidity when milling. The spindle brake or spindle lock is needed to keep the spindle from rotating when installing or removing tools from it. The toolhead is swiveled on the ram by loosening the clamping nuts on the toolhead and then turning the swivel adjustment until the desired angle is obtained.

The ram is adjusted toward or away from the column by the ram positioning pinion. The ram also swivels on the column after the turret clamps are loosened.

The table is moved manually with the table traverse hand wheel. Table movement toward or away from the column is accomplished with the cross traverse hand wheel. Raising and lowering the knee is done with the vertical traverse crank. Each of these three axes of travel can be adjusted in .001 in. increments with micrometer dials. The table, saddle, and knee can be locked securely in position with the clamping levers

toolhead (Figure 2) contains the motor, which powers the spindle. Speed changes are made with V-belts, gears, or variable speed drives. When changing speeds into the high or low speed range, the spindle has to be stopped. The same is true for V-belt or gear-driven speed changes. On variable drives the spindle has to be revolving while speed changes are made. The spindle is contained in a quill. The quill can be extended and retracted into the toolhead by a quill feed hand lever or hand wheel. The quill feed hand lever is used to position rapidly the quill or to drill holes. The quill feed hand wheels give a controlled slow manual feed as needed when boring holes.

Power feed to the quill is obtained by engaging the feed control lever. Different power feeds are available through the power feed change lever. The power feed is automatically disengaged when the quill dog contacts the adjustable micrometer depth stop (Figure 3). When feeding

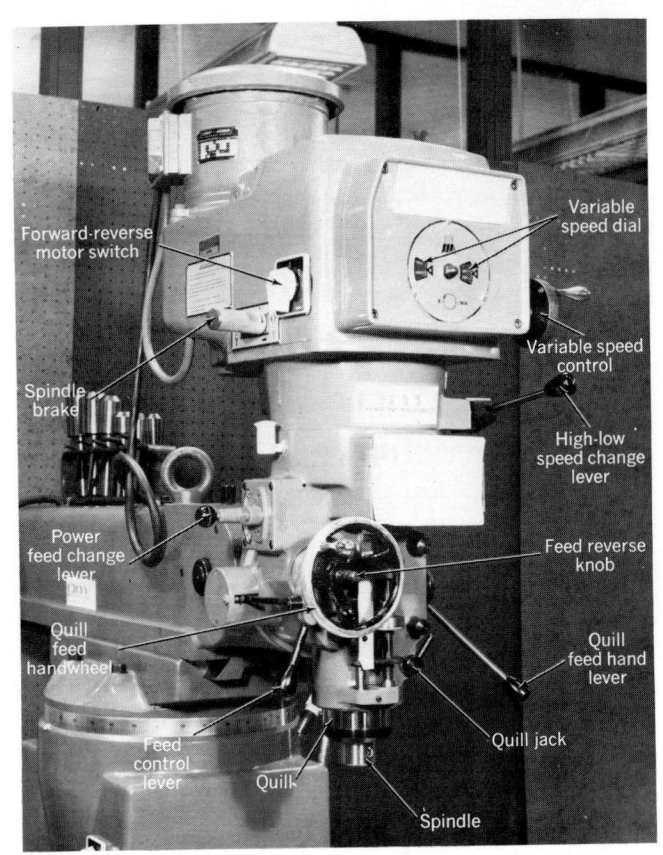

Figure 2. The toolhead (Lane Community College).

Figure 3. Quill stop (Lane Community College).

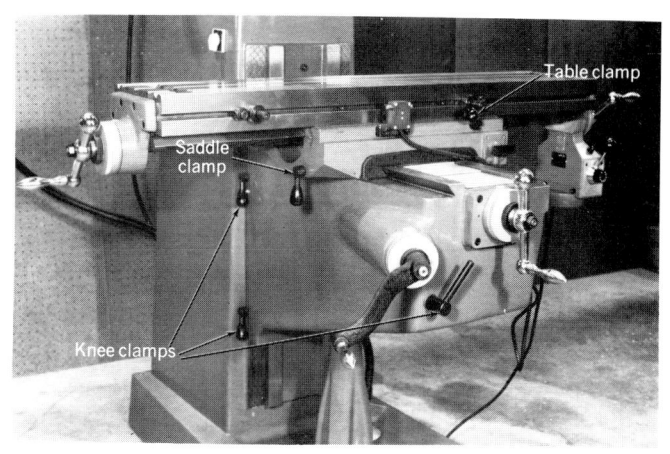

Figure 4. Clamping devices (Lane Community College).

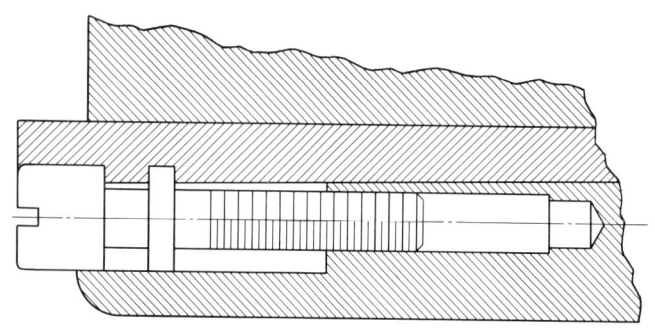

Figure 5. Gib adjusting screw (Courtesy of Cincinnati Milacron).

shown in Figure 4. During machining, all axes except the moving one should be locked. This will increase the rigidity of the setup. Do not use these clamping devices to compensate for wear on the machine slides. If the machine slides become loose, make adjustments with the gib adjustment screws (Figure 5). Turning this screw in will tighten a tapered gib. Make a partial turn on the screw, then try moving the unit with the hand wheel. Repeat this operation until a free but not loose movement is obtained. Too tight an adjustment squeezes the lubricant from the slides, resulting in rapid wear.

Before operating any of the machine controls, the machine should be lubricated. The lubrication follows the cleaning of the machine. Do not use air to blow off the machine; use a rag. Lubrication is needed on all moving parts. Follow the machine manufacturer's recommendation as to the kind of lubricant required.

self-evaluation

SELF-TEST 1. Name the six major components of a vertical milling machine.

a. _____

b. _____

c. _____

d. _____

e. _____

f. _____

2. Which parts are used to move the table longitudinally?

_____

3. Which parts are used to move the saddle?

_____

4. What moves the quill manually?

_____

5. What is the purpose of the table clamp?

_____

6. What is the purpose of the spindle brake?

_____

7. What is important when changing the spindle speed range from high to low?

_____

8. Why is the toolhead fastened to a ram?

_____

9. How is a loose table movement adjusted?

_____

10. What is the purpose of the quill clamp?

_____

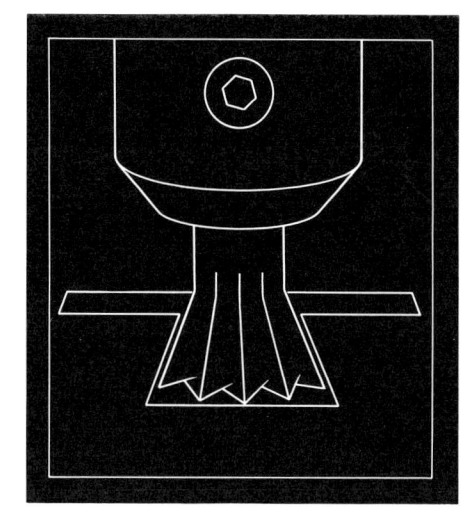

unit 3
vertical milling machine operations

The vertical milling machine is one of the most versatile machines found in a machine shop. This unit will illustrate some of the many possible vertical milling machine operations.

objective After completing this unit, you will be able to identify and select vertical milling machine setups for a variety of different machining operations.

information

Many vertical milling machine operations such as the milling of steps are performed with end mills (Figure 1). Two surfaces can be machined at one time, both square to each other. The ends of workpieces can be machined square and to a given length by using the peripheral teeth of an end mill (Figure 2). Center cutting end mills make their own starting hole when used to mill a pocket or cavity (Figure 3).

Prior to making any milling cuts, the outline of the cavity should be accurately laid out on the workpiece for a guide or reference line. Only when finish cuts are made should these layout lines disappear.

Good milling practice is to rough out the cavity to within .030 in. of finish size before making any finishing cuts.

When you are milling a cavity, the direction of the feed should be against the rotation of the cutter (Figure 4). This assures positive control over the distance that the cutter travels and prevents the workpiece from being pulled into the cutter because of backlash. When you reverse the direction of table travel, you will have to compensate for the backlash in the table feed mechanism. During any milling operation, all table movements should be locked except the one that is moving to obtain the most rigid setup possible. Spiral fluted end mills may work their way out of a split collet when deep heavy cuts are made or when the end mill gets dull. As a precaution, to warn you that this is happening, you can make a mark with a felt tip pen on the revolving end mill shank where it meets the collet face. Observing this mark during the cut will give you an early indication if the end mill is changing its position in the collet.

One often performed operation with end mills is the cutting of keyways in shafts. It is very important that such a keyway be centrally located in the shaft. A very accurate method of doing this is by positioning the cutter with the help of the machine dials. After clamping the shaft in a vise, or possibly to the table, the quill is lowered so that the cutter is along side the shaft but not touching it. Then, with the spindle motor off and the spindle rotated by hand, the table is moved until a paper feeler strip is pulled from your hand (Figure 5). At this point, the cutter is approximately .002 in. from the shaft. Zero the cross slide micrometer dial compensating for the .002-in. Raise the quill so that the cutter clears the workpiece. Move the cross slide a distance equal to half the shaft diameter plus half the cutter diameter. This will locate the cutter centrally over the shaft. Lock the cross slide table

Section B Vertical Milling Machines

Figure 1. Using an end mill to mill steps (Lane Community College).

Figure 2. Using an end mill to square stock (Lane Community College).

Figure 3. Using an end mill to machine a pocket (Lane Community College).

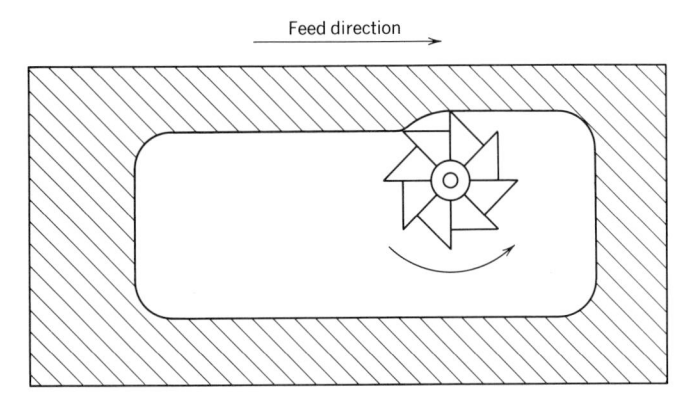

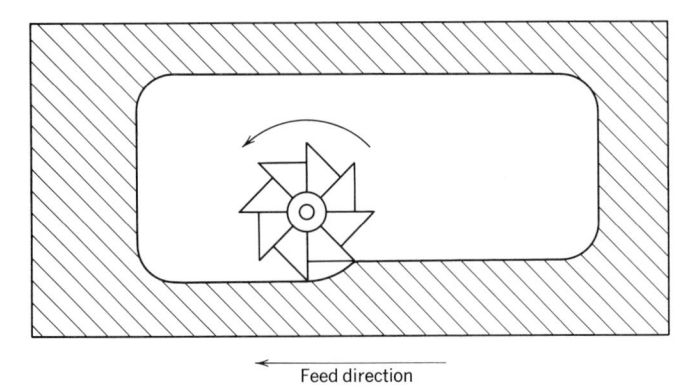

Figure 4. Feed direction is against cutter rotation.

movement into place. Raise the quill to its top position and lock it there. Move the table longitudinally to position the cutter where the keyway is to begin. Start the spindle motor and raise the knee until the cutter makes a circular mark equal to the cutter diameter (Figure 6). Zero the vertical travel micrometer dial. Move up a distance equal to half the cutter diameter plus an additional .005-in. and lock the knee. Cut the keyway the required length.

Unit 3 Vertical Milling Machine Operations   47

Figure 5. Setting an end mill to the side of a shaft with the aid of a paper feeler (Lane Community College).

Figure 6. Cutter centered on work and lowered to make a circular mark (Lane Community College).

Figure 7. Milling a slot and then the dovetail (Lane Community College).

Figure 8. First a slot is milled and then the T-slot cutter makes the T-slot (Lane Community College).

Figure 9. Setting up a workpiece for an angular cut with a protractor (Lane Community College).

To machine a T-slot or a dovetail into a workpiece, two operations are performed. First a slot is cut with a regular end mill and then a T-slot cutter or a single angle milling cutter is used to finish the contour (Figures 7 and 8). Angular cuts on workpieces can be made by tilting the workpiece in a vise with the aid of a protractor (Figure 9) and its built-in spirit level.

Figure 10. Machining an angle with an end mill (Lane Community College).

Figure 12. Cutting an angle by tilting the workhead and using the end teeth of an end mill (Lane Community College).

Figure 11. Using a shell mill to machine an angle (Lane Community College).

Figure 13. Cutting an angle by tilting the workhead and using the peripheral teeth of an end mill (Lane Community College).

Machining the angle can be performed with an end mill (Figure 10) or with a shell mill (Figure 11). Another possibility for machining angles is the tilting of the workhead (Figures 12 and 13).

Accurate holes can be drilled at any angle that the head can be swiveled to. These holes can be drilled by using the sensitive quill feed lever or the power feed mechanism (Figure 14) or, in the case of vertical holes, the knee can be raised.

Holes can be machine tapped by using the sensitive quill feed lever and the instant spindle reversal knob (Figure 15). When an offset boring head is mounted in the spindle, precisely located and accurately dimensioned holes can be bored (Figure 16). Circular slots can be milled when a rotary table is used (Figure 17). Precise indexing can be performed when a dividing head is mounted on the milling machine. Figure 18 shows the milling of a square on the end of a

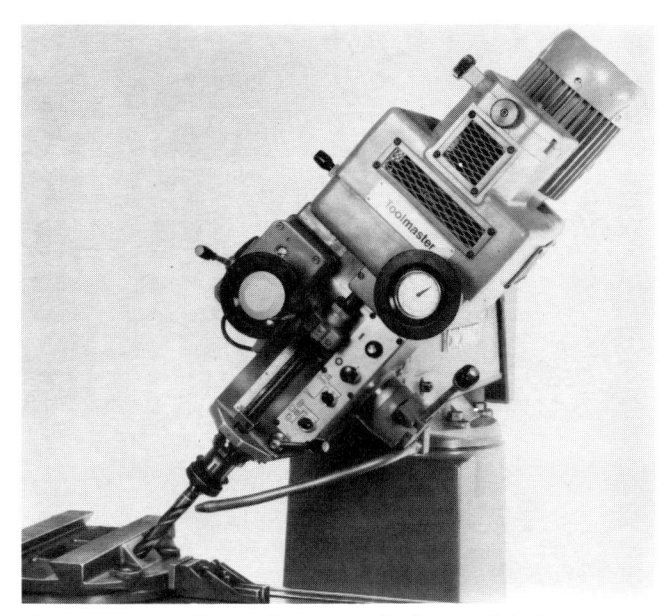

Figure 14. Drilling of accurately located holes (Courtesy of Cincinnati Milacron).

Figure 16. Boring with an offset boring head (Courtesy of Cincinnati Milacron).

Figure 15. Tapping in a vertical milling machine (Courtesy of Cincinnati Milacron).

Figure 17. *(right)* Using a rotary table to mill a circular slot (Courtesy of Cincinnati Milacron).

Figure 18. A dividing head in use (Courtesy of Cincinnati Milacron).

Figure 19. A shaping head used to cut a square corner hole (Lane Community College).

Figure 20. Using a right angle milling attachment (Courtesy of Cincinnati Milacron).

shaft. On many vertical milling machines a shaping attachment is mounted on the rear of the ram. This shaping attachment can be brought over the machine table by swiveling the ram 180 degrees. Shaping attachments are used to machine irregular shapes on or in workpieces such as the square corner hole shown in Figure 19. When a right angle milling attachment (Figure 20) is mounted on the spindle, it is possible to machine hard to get at cavities at often difficult angles on workpieces.

self-evaluation

SELF-TEST 1. How is an end mill centered over a shaft prior to cutting a keyway?

Unit 3 Vertical Milling Machine Operations   51

2. Why should the feed direction be against the cutter rotation when milling a cavity?

3. What can cause an end mill to work itself out of a collet while cutting?

4. Describe two methods of cutting angular surfaces in a vertical milling machine.

   a.

   b.

5. How are circular slots milled?

6. What milling machine attachment is used to mill a precise square or hexagon on a shaft?

7. How can a square hole in a workpiece be machined on the vertical milling machine?

8. When is a right angle milling attachment used?

9. Why should workpieces be laid out before machining starts?

10. Why are two operations necessary to mill a T-slot?

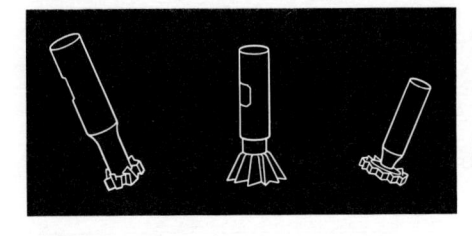

unit 4
cutting tools for the vertical milling machine

Vertical milling machines are very versatile tools. To utilize this versatility, a number of different cutting tools are available. To make an intelligent choice, a machinist needs to know the characteristics and limitations of different cutting tools.

objective After completing this unit, you will be able to identify and select from commonly used vertical milling machine cutting tools.

information

The most frequently used tool on a vertical milling machine is the end mill. Figure 1 is an illustration of the cutting end of a four flute end mill. End mills are made as right-hand cut or left-hand cut. Identification is made by viewing the cutter from the cutting end. A right-hand cutter rotates counterclockwise. The helix of the flutes can also be left or right hand; a right-hand helix turns to the right. Figure 1 shows a right-hand cut, right-hand helix end mill.

The end teeth of an end mill can be different, depending on the cutting that is to be performed (Figure 2). Two flute end mills are center cutting, which means they can make their own starting hole. This is called plunge cutting. Four flute end mills may have center cutting teeth or a gashed or center drilled end. End mills with center drilled or gashed ends cannot be used to plunge cut their own starting holes. These end mills only cut with the teeth on their periphery. End mills can be single end (Figure 3) or double end (Figure 4). Double end-type end mills are usually more economical because of the savings in tool material in their production.

End mills are manufactured with two, three, four or more flutes and with straight flutes (Figure 5), slow, regular, and fast helix angles. A slow helix is approximately 12 degrees, a regular helix is 30 degrees, and the fast helix is 40 degrees or more when measured from the cutter axis. Most general purpose cutting is done with a regular helix angle cutter (Figure 6).

Aluminum is efficiently machined with a fast helix end mill and highly polished cutting faces to minimize chip adherence (Figure 7). The end mills illustrated so far are made from high speed steel. High speed steel end mills are available in great variety of styles, shapes, and sizes as stock items. High speed steel end mills are relatively low in cost when compared with carbide tipped or solid carbide end mills. But to machine highly abrasive or hard and tough materials or in production milling, carbide tools should be considered.

Carbide tools may have carbide cutting tips brazed to a steel shank (Figure 8). This two flute carbide tipped end mill is designed to cut steel. It has a negative axial rake angle and a slow left-hand helix. Figure 9 is a carbide tipped four flute end mill designed to machine nonferrous materials such as brass, cast iron, and aluminum. This end mill has a positive radial rake with a right-

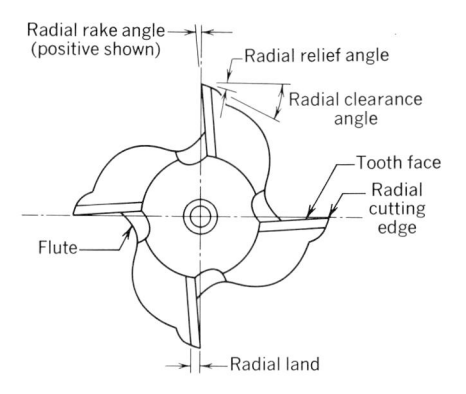

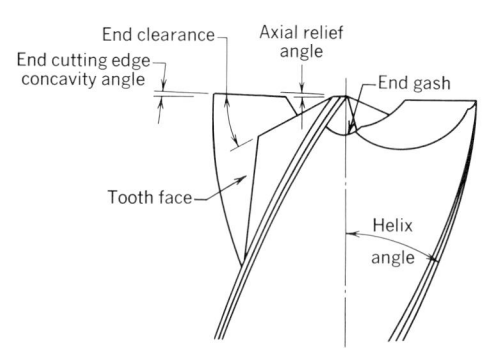

Figure 1. End mill nomenclature (Copyright © National Twist Drill & Tool Div., Lear Siegler, Inc.).

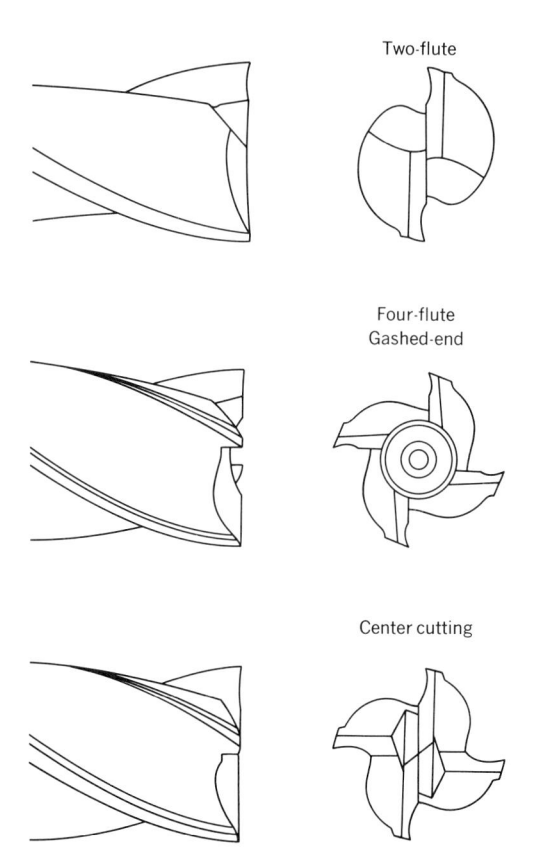

Figure 2. Types of end teeth on end mills (Copyright © National Twist Drill & Tool Div., Lear Siegler, Inc.).

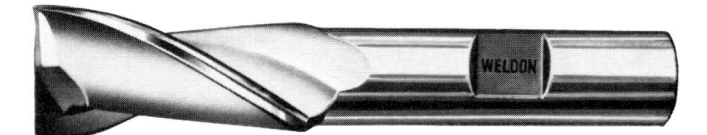

Figure 3. Single end helical teeth end mill (Photo courtesy of The Weldon Tool Company, Cleveland, Ohio).

Figure 4. Two flute, double end, helical teeth end mill (Photo courtesy of The Weldon Tool Company, Cleveland, Ohio).

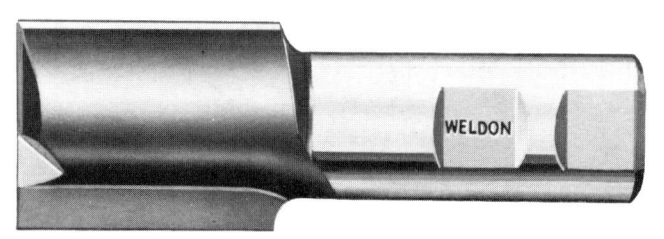

Figure 5. Straight tooth, single end end mill (Photo courtesy of The Weldon Tool Company, Cleveland, Ohio).

Figure 6. Four flute, double end end mill.

Figure 7. Forty-five degree helix angle aluminum end mill (Photo courtesy of The Weldon Tool Company, Cleveland, Ohio).

Figure 8. Two flute, carbide tipped end mill (Courtesy of Brown & Sharpe Mfg. Co.).

Figure 9. Four flute, carbide tipped end mill (Courtesy of Brown & Sharpe Mfg. Co.).

Figure 10. Inserted blade end mill.

Figure 11. Roughing mill (Copyright © Illinois Tool Works, Inc., 1976).

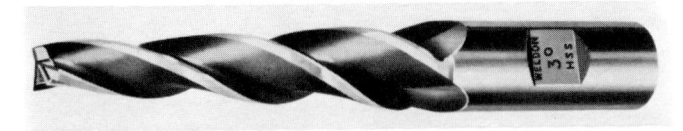

Figure 12. Three flute, tapered end mill (Photo courtesy of The Weldon Tool Company, Cleveland, Ohio).

Figure 13. Two flute, single end, ball end mill (Photo courtesy of The Weldon Tool Company, Cleveland, Ohio).

hand helix. Another kind of carbide tipped end mill uses throwaway inserts (Figure 10). Each of the carbide inserts has three cutting edges; when all three cutting edges are dull, the insert is replaced with a new one. No sharpening is required.

Different kinds of carbide grades are available to provide the correct tool material for different work materials. When large amounts of material need to be removed, a roughing end mill (Figure 11) should be used. These end mills are also called hogging end mills and have a wavy thread form cut on their periphery. These waves form many individual cutting edges. The tip of each wave contacts the work and produces one short compact chip. Each succeeding wave tip is offset from the next one, which results in a relatively smooth surface finish. During the cutting operation, a number of teeth are in contact with the work. This reduces the possibility of vibration or chatter.

Tapered end mills (Figure 12) are used in mold making, die work, and pattern making, where precise tapered surfaces need to be made. Tapered end mills have included tapers ranging from 1 degree to over 10 degrees.

Ball-end end mills (Figure 13) have two or more flutes and form an inside radius or fillet between surfaces. Ball-end end mills are used in tracer milling and in die sinking operations. Round bottom grooves can also be machined with them. Precise convex radii can be machined on a milling machine with corner rounding end mills (Figure 14). Dovetails are machined with single angle milling cutters (Figure 15). The two commonly available angles are 45 degrees and 60 degrees. T-slots in machine tables and workholding devices are machined with T-slot cutters (Figure 16). T-slot cutters are made in sizes to fit standard T-nuts.

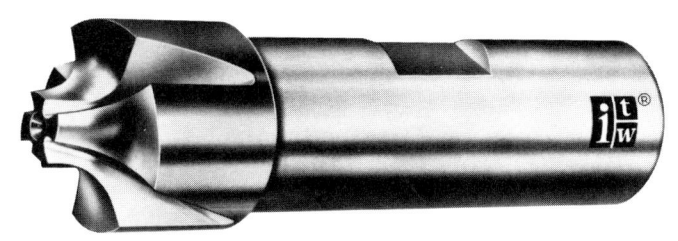

Figure 14. Corner rounding milling cutter (Copyright © Illinois Tool Works Inc., 1976).

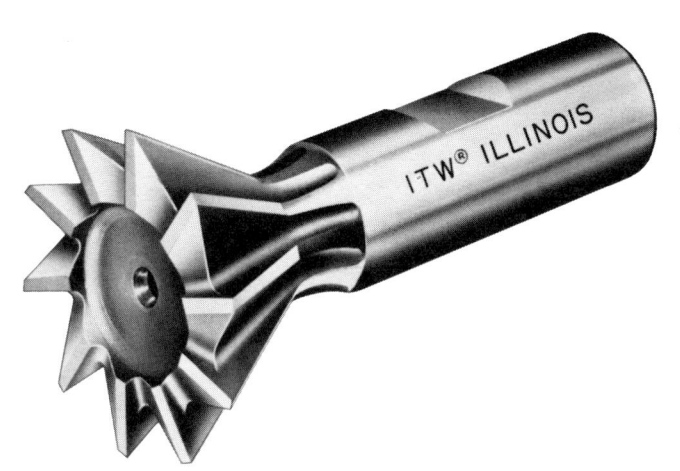

Figure 15. Single angle milling cutter (Copyright © Illinois Tool Works Inc., 1976).

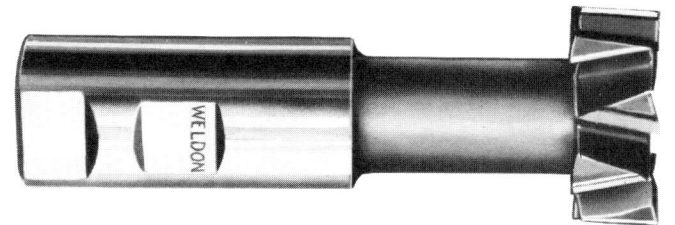

Figure 16. T-slot milling cutter (Photo courtesy of The Weldon Tool Company, Cleveland, Ohio).

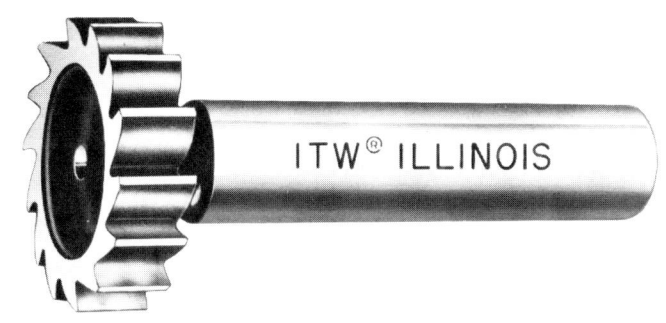

Figure 17. Woodruff keyslot milling cutter (Copyright © Illinois Tool Works Inc., 1976).

Figure 18. Shell end mill (Copyright © Illinois Tool Works Inc., 1976).

Woodruff key slots are cut into shafts to retain a woodruff key as a driving and connecting member between shafts and pulleys or gears. Woodruff key slot cutters (Figure 17) come in many different standardized sizes. When larger flat surfaces need to be machined, a shell end mill (Figure 18) can be used. Shell end mills are more economical to produce because less of the costly tool material is needed to make one than for a solid shank end mill of the same size. To obtain rapid metal removal, shell end mills are made as a roughing type mill (Figure 19) with a wavy thread forming many cutting edges. Shell mills are also made with carbide inserts (Figure 20). The ease with which new sharp cutting edges can be installed makes this a very practical, efficient cutting tool. The great number of different carbide grades available makes it possible to select a grade suitable for all work materials. An inexpensive face milling cutter is a fly cutter (Figure 21). A fly cutter can be made with a high speed steel or carbide tipped tool bit sharpened to have the correct clearance and rake angles.

All of these cutting tools have to be mounted and driven by the machine spindle. End mills or other straight shank tools can be held in collets. The most rigid type of collet is a solid collet (Figure 22). This collet has been precision

Figure 19. Shell-type roughing mill (Copyright © Illinois Tool Works Inc., 1976).

Figure 20. Shell mill with carbide inserts (Lane Community College).

Figure 21. Fly cutter (Lane Community College).

Figure 22. Solid collet (Lane Community College).

ground with a hole that is concentric and the exact size of the tool shank. Driving power to the tool is transmitted by one or two setscrews engaging in flats on the tool shank.

Another type of frequently used collet is the split collet (Figure 23). The shank of the tool is held by the friction created when the tapered part of the collet is pulled into the taper of the spindle nose. When a heavy side thrust is created through a deep cut, a large feedrate, or a dull tool, helical flute end mills have a tendency to be pulled out from the collet. With the solid collet, the setscrews prevent any end movement of the cutting tool.

Unit 4 Cutting Tools for the Vertical Milling Machine 57

Figure 23. Split collet (Lane Community College).

Figure 24. Quick-change adapter and tool holders (Lane Community College).

To speed up frequent tool changes, a quick-change tool holder (Figure 24) is used. Different tools can be mounted in their own holder, preset to a specific length, and interchanged with a partial turn of a clamping ring. Shell end mills are mounted and driven by shell mill arbors (Figure 25). All of the tools and tool holders will perform satisfactorily only if they are cleaned and inspected for nicks and burrs and corrections are made before mounting them in the spindle.

Figure 25. Shell mill arbor (Photo courtesy of The Weldon Tool Company, Cleveland, Ohio).

self-evaluation

SELF-TEST

1. How is a right-hand cut end mill identified?

2. What is characteristic of end mills that can be used for plunge cutting?

3. What is the main difference between a general purpose end mill and one designed to cut aluminum?

4. When are carbide tipped end mills chosen over high speed steel end mills?

Section B Vertical Milling Machines

5. To remove a considerable amount of material, what kind of end mill is used?

6. Where are tapered end mills used?

7. Why are tools with carbide inserts used?

8. How are straight shank tools held in the machine spindle?

9. How are shell end mills driven?

10. Why are quick-change tool holders used?

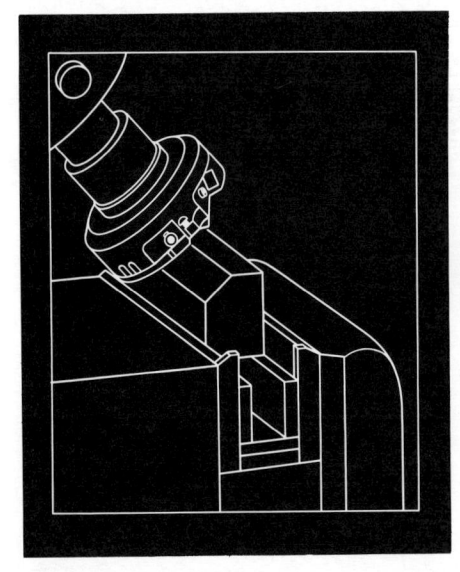

unit 5
setups on the vertical milling machine

Before any cutting takes place a milling machine has to be prepared for the operations to be performed. This preparation involves the alignment of workholding devices and the work head.

objective After completing this unit, you will be able to align vises and toolheads, locate the edges of workpieces, and find the centers of holes.

information

The two most commonly used workholding methods on the vertical milling machine are the fastening of workpieces to the machine table and the holding of workpieces in a machine vise. When a workpiece is fastened to the machine table, it must be aligned with the axis of the table.

Machine tables are accurately machined and the table travels parallel to its outside surfaces and also parallel to its T-slots. Workpieces can be aligned by placing them against stops that fit snugly into the T-slots (Figure 1), or by measuring the distance from the edge of the table to the workpiece in a few places (Figure 2). More accurate alignments can be made when a dial indicator is used to indicate the edge of a workpiece (Figure 3). When a vise is used to hold the workpiece, the solid jaw of the vise should be indicated to assure its alignment with the axis of the table travel. For precise machining operations, the toolhead needs to be aligned squarely to the top surface of the machine table. To align the toolhead, follow the recommended procedure.

1. Fasten a dial indicator in the machine spindle (Figure 4). The dial indicator should sweep a circle slightly smaller than the width of the table.
2. Lower the quill until the indicator contact point is deflected .015 to .020 in. Lock the quill in this position.
3. Tighten the knee clamping bolts. If this is neglected, the knee will sag on the front.
4. Now set the indicator bezel to read zero.
5. Loosen the head clamping bolts one at a time and retighten each one to create a slight drag. This slight drag makes fine adjustments easier.
6. Rotate the spindle by hand until the indicator is to the left of the spindle in the center of the table, and note the indicator reading.
7. Rotate the spindle 180 degrees so that the indicator is to the right of the spindle, and note the indicator reading at that place. Be careful that the indicator contact does not catch and hang up when crossing the T-slots.
8. Split the difference between the left-hand reading and the right-hand reading in two by turning the head tilting screw.
9. Check and compare the indicator reading at

Figure 2. Measuring the distance from the edge of the table to the workpiece (Lane Community College).

Figure 1. Work aligned by locating against stops in T-slots (Lane Community College).

Figure 3. Aligning a workpiece with the aid of a dial indicator (Lane Community College).

Figure 4. Aligning the toolhead square to the table with a dial indicator (Lane Community College).

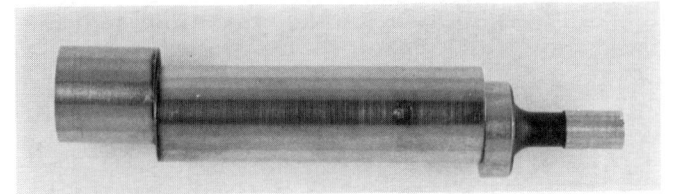

Figure 5. Offset edge finder (Lane Community College).

Figure 6. Work approaches the tip of the offset edge finder (Lane Community College).

the left side of the table. If both readings are the same, tighten the head clamping bolts. If the readings differ, repeat step 8.

10. After the head clamping bolts are tight, make another comparison on both sides. Often the tightening of the bolts changes the head location and additional adjustments need to be made.
11. Now the head needs to be aligned in relation to the width of the table. The procedure is the same as for the lengthwise alignment. A final check should be made to be sure all clamping bolts are tight.

When the workpiece edges are aligned parallel with the table travel and the toolhead is aligned square with the table top, it becomes necessary to align the spindle axis with the edges of the workpiece. A commonly used tool to locate edges on a milling machine is an offset edge finder (Figure 5). An offset edge finder consists of a shank and a tip that is held against the shank by an internal spring. The shank is usually ½ in. in diameter and the tip is either .200 or .500 in. in diameter. To use an edge finder, it is mounted in a collet. The spindle should revolve at 600 to 800 RPM. The tip should be eccentric to the shank. The workpiece is now moved slowly toward the tip of the edge finder until it just touches (Figure 6). Continue to advance the work very slowly, reducing the run out or eccentricity of the tip. Suddenly the tip will walk off sideways. At this point the spindle axis is exactly one half of the tip diameter away from the edge of the workpiece. If the tip diameter is .200 in., then the centerline of the spindle is .100 in. away from the workpiece edge. Set the micrometer dial of the just adjusted machine axis .100 in. from zero. Repeat the approach to the workpiece a few more times while observing the micrometer dial position until you feel secure in locating an edge with an edge finder. Repeat the edge finding process for the other machine axis. If your machine has any backlash, remember to locate all positions in the same direction as that used when locating with the edge finder.

If an edge finder is not available, an edge can be located with the aid of a dial indicator. The dial indicator is mounted in the spindle. Rotate the spindle by hand and set the indicator contact point as close to the spindle centerline as possible. Lower the spindle so that the indicator contact point touches the workpiece edge and registers a .010 to .020 in. deflection (Figure 7). A slight rotating movement of the spindle forward and backward is used to locate the lowest read-

Figure 7. Indicator used to locate the edge of a workpiece (Lane Community College).

Figure 8. Indicator against parallel to locate the edge of a workpiece (Lane Community College).

Figure 9. Dial indicator locating the center of a hole (Lane Community College).

ing on the dial indicator. Set the dial indicator to register zero. Raise the spindle so that the indicator contact point is ½ in. above the workpiece and turn the spindle 180 degrees from the way it was when the indicator was zeroed. Hold a precision parallel against the edge of the workpiece so it extends above the workpiece. Lower the spindle until the indicator contact point is against the parallel (Figure 8). Read the indicator value; use a mirror to read the indicator when it faces away from you. Now turn the table hand wheel and move the table to where the indicator pointer is halfway between the reading against the parallel and the zero on the indicator dial. Set the dial on zero and check the position of the spindle, as in Figure 7. Again, make a halfway correction until both readings are the same.

To pick up the center of an existing hole, the indicator is mounted in the spindle and swiveled so that the contact point touches the side of the hole (Figure 9). The spindle is rotated and table adjustments are made until the same reading is obtained in a complete circle. The spindle should be centered first in one table axis and then in the other.

self-evaluation

SELF-TEST

1. How can workpieces be aligned when they are clamped to the table?

2. How is a vise aligned on a machine table?

3. When is the toolhead alignment checked?

4. Why is it important that the knee clamping bolts are tight before aligning a toolhead?

5. Why does the toolhead alignment need to be checked again after all the clamping bolts are tightened?

6. How can the machine spindle be located exactly over the edge of a workpiece?

7. With a .200 in. edge finder tip, when do you know that the spindle axis is .100 in. away from the edge of the workpiece?

8. What is the recommended RPM to use with an offset edge finder?

9. When locating a number of positions on a workpiece, how can you eliminate the backlash in the machine screws?

10. How is the center of an existing hole located?

WORKSHEET
Machining Holes in Vise Body (Figure 10)

1. Align the workhead square to the machine table.
2. Align and fasten a machine vise on the table so its jaw is parallel to the long axis of the table.
3. Mount the vise body in the machine vise with the bottom surface against the solid jaw of the machine vise.

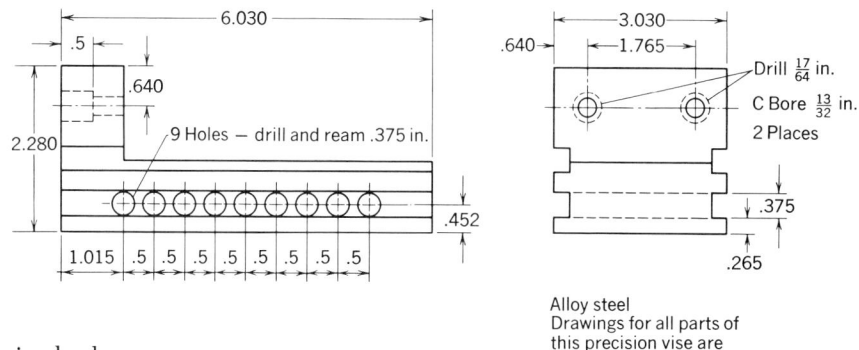

Figure 10. Machining the holes in the vise body.

Alloy steel
Drawings for all parts of this precision vise are in the appendix

4. Mount an edge finder in a spindle collet and align the spindle axis with the base surface of the vise body.

5. Move the table the required .452 in. distance and lock the table cross slide.

6. Now pick up the outside of the solid jaw of the vise body.

7. Move to the first hole location 1.015 in. from the outside edge.

8. Center drill this hole.

9. Use a ¼ in. diameter twist drill and drill this hole 1½ in. deep.

10. Repeat steps 8 and 9 for the remaining eight holes. Accurate positioning is done with the micrometer dials.

11. Remove the workpiece from the machine vise. Turn it over so that the just drilled holes are down and the bottom surface of the vise body is again against the solid jaw.

12. Use the edge finder to pick up the two sides, as for the first drilling operation.

13. Position the spindle over the first hole location, again with the first hole on the solid jaw side.

14. Center drill this hole.

15. Drill this hole with a ¼ in. diameter drill deep enough to meet the hole from below.

16. Switch to an $\frac{11}{32}$ in. diameter drill and drill completely through the vise body (Figure 11). The ¼ in. hole acts as a pilot hole to let the $\frac{11}{32}$ in. drill come out in the correct place on the bottom side.

17. Change from the $\frac{11}{32}$ in. drill to a ⅜ in. diameter machine reamer and ream this hole completely through (Figure 12).

18. Repeat steps 14 to 17 for the remaining eight holes.

19. Reposition the workpiece so it is upright in the machine vise with the solid jaw of the vise body up.

20. With an edge finder, pick up the edges of the workpiece.

Figure 11. Drilling $\frac{11}{32}$ in. diameter holes in the vise body (Lane Community College).

Figure 12. Reaming $\frac{3}{8}$ in. diameter holes in the vise body (Lane Community College).

Figure 13. Drill and counterbore holes in the solid jaw of the vise body (Lane Community College).

21. Position for the two hole locations and drill the $\frac{17}{64}$ in. diameter holes with their $\frac{13}{32}$ in. diameter counterbores (Figure 13).
22. Remove all burrs.

unit 6
using end mills

After the selection of a workholding device and cutting tool has been made, a decision on cutting speed, feed, depth of cut, and coolant needs to be made.

objectives After completing this unit, you will be able to:
1. Select cutting speeds and calculate RPM for end mills.
2. Select and calculate feedrates for end mills.
3. Use end mills to machine grooves and cavities.

information

One factor in efficient cutting with end mills requires the intelligent selection of the correct cutting speed. The cutting speed of a cutting tool is influenced by the tool material, work material, condition of the machine, rigidity of the setup, and use of coolant. Commonly used tool materials are high speed steel and cemented carbide. After the cutting speed has been selected for a job, the cutting operation should be observed carefully so that speed adjustments can be made before a job is ruined. Table 1 gives starting values for some commonly used materials. As a rule, lower speeds are used for hard materials, tough materials, abrasive materials, heavy cuts, minimum tool wear, and maximum tool life. Higher speeds are used to machine softer materials, for better surface finishes, with smaller diameter cutters, for light cuts, for frail workpieces, and delicate setups. When you calculate an RPM to use on a job, use the lower cutting speed value to start. The formula for this is:

$$\text{RPM} = \frac{\text{CS} \times 4}{\text{D}}$$

D is the diameter of the cutting tool in inches.

Cutting fluids should be used when high speed steel cutters are used. The cutting fluid dissipates the heat generated while cutting. It reduces the heat by acting as a lubricant between the tool and chip. Higher cutting speeds can be used with cutting fluids. A stream of coolant also washes the chips away. Water base coolants have very good cooling qualities and oil base coolants produce very good surface finishes. Most milling with carbide cutters is done dry unless a large enough flow of coolant at the cutting edge can be maintained to keep the cutting edge from being intermittently heated and cooled. Intermittent heating and cooling of a carbide tool usually results in thermal cracking and premature tool failure.

Some materials, such as cast iron, brass, and plastics, are commonly machined dry. A stream of compressed air can be used to cool tools and to keep the cutting area clear of chips, but precautions have to be taken to prevent flying chips from injuring anyone.

Another very important factor in efficient machining is the feed. The feed in milling is calculated by starting with a desired feed per

Table 1
Cutting Speeds and Starting Values for Some Commonly Used Materials

Work Material	Tool Material	
	High Speed Steel	Cemented Carbide
Aluminum	300-800	1000-2000
Brass	200-400	500-800
Bronze	65-130	200-400
Cast Iron	50-80	250-350
Low carbon steel	60-100	300-600
Medium carbon steel	50-80	225-400
High carbon steel	40-70	150-250
Medium alloy steel	40-70	150-350
Stainless steel	30-80	100-300

tooth. The feed per tooth determines the chip thickness. The chip thickness affects the tool life of a cutter. Very thin chips dull the cutting edges very rapidly. Commonly used feed per tooth values for end mills are given in Table 2. Usually the feed per tooth is the same for HSS and carbide end mills.

The values in Table 2 are only intended as starting points and may have to be adjusted up or down, depending on the machining conditions of the job at hand. The highest possible feed per tooth will usually give the longest tool life between sharpenings. Excessive feeds will cause tool breakage or the chipping of the cutting edges. When the feed per tooth for a cutter is selected, the feedrate can be calculated. The feedrate on a milling machine is expressed in inches per minute (IPM) and is the product of the feed per tooth (F) times the revolution per minute (RPM) times the number of teeth in the cutter (n). The formula for feedrate is:

$$\text{Feedrate} = f \times \text{RPM} \times n$$

As an example, to use the values given in Tables 1 and 2, calculate the RPM and feedrate for a $\frac{1}{2}$ in. diameter HSS two flute end mill cutting aluminum.

$$\text{RPM} = \frac{\text{CS} \times 4}{D} = \frac{300 \times 4}{1/2} = \frac{1200}{.5} = 2400$$

Feedrate = f × RPM × n = .005 × 2400 × 2 = 24 IPM

The third factor to be considered in using end mills is the depth of cut. The depth of cut is limited by the amount of material that needs to be removed from the workpiece, by the power available at the machine spindle, and by the rigidity of the workpiece, tool, and setup. As a rule, the depth of cut for an end mill should not exceed one half of the diameter of the tool. But if deeper cuts need to be made, the feedrate needs to be reduced to prevent tool breakage. The end

Table 2
Feeds for End Mills (Feed per Tooth in Inches)

Cutter Diameter	Aluminum	Brass	Bronze	Cast Iron	Low Carbon Steel	High Carbon Steel	Medium Alloy Steel	Stainless Steel
$\frac{1}{8}$	.002	.001	.0005	.0005	.0005	.0005	.0005	.0005
$\frac{1}{4}$	.002	.002	.001	.001	.001	.001	.0005	.001
$\frac{3}{8}$	.003	.003	.002	.002	.002	.002	.001	.002
$\frac{1}{2}$	.005	.003	.003	.0025	.003	.002	.001	.002
$\frac{3}{4}$	.006	.004	.003	.003	.004	.003	.002	.003
1	.007	.005	.004	.0035	.005	.003	.003	.004
$1\frac{1}{2}$	.008	.005	.005	.004	.006	.004	.003	.004
2	.009	.006	.005	.005	.007	.004	.003	.005

mill must be sharp and should run concentric in the end mill holder. The end mill should be mounted with no more tool overhang than necessary to do the job.

A problem that occasionally arises when using end mills to machine grooves or slots is a slot with nonperpendicular sides. Grooves with leaning sides are caused by worn spindles, excessive tool projection from the spindle, dull end mills, or excessive feedrates. The leaning slot is produced by an end mill that is deflected by high cutting forces (Figure 1). To reduce the tendency of the tool to cut a leaning slot, reduce the feedrate, use end mills with only a short projection from the spindle, and use end mills with straight or low helix angle flues.

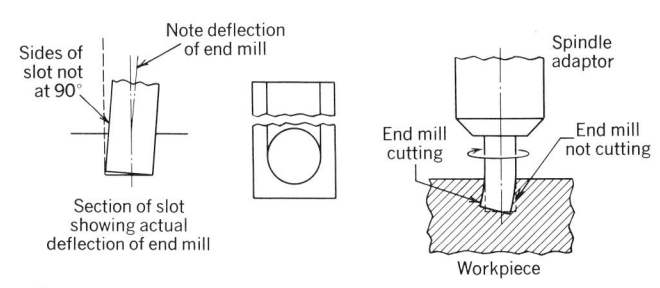

Figure 1. The causes of a leaning slot in end milling.

self-evaluation

SELF-TEST

1. When are the lower cutting speeds recommended?

2. When are the higher cutting speeds used?

3. Should you always use calculated RPM?

4. When are cutting fluids used?

5. When should machining be performed dry?

6. How is the tool life of an end mill affected by the chip thickness of a cut?

7. What is normally considered the maximum depth of cut for an end mill?

8. What are the limitations on the depth of cut?

9. Calculate the RPM for a $\frac{3}{4}$ in. diameter HSS end mill to machine brass.

10. Calculate the feed rate for a two flute ¼ in. diameter carbide end mill to machine medium alloy steel.

_____

WORKSHEET Milling Slot and Cavity in Vise Body (Figure 2).

1. Fasten workpiece in vise; use parallels so the workpiece extends above the vise for measuring while it is being machined. The workpiece should be layed out.

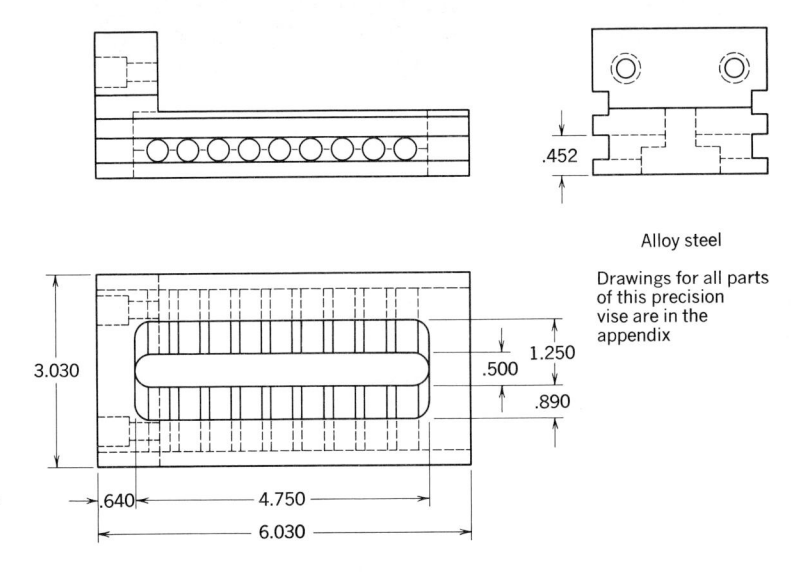

Figure 2. Milling a slot and cavity in the vise body.

Alloy steel

Drawings for all parts of this precision vise are in the appendix

2. Use a 7/16 in. diameter center cutting end mill to rough out the ½ in. wide slot. Fasten this end mill in the spindle.

3. Calculate the RPM.

4. Calculate the feedrate.

5. Align the end mill over the slot and lock all machine axes that will not move during the cutting operation.

6. Rough out the slot; use coolant while cutting.

7. Change to a ½ in. end mill.

8. Make a ⅛ in. deep trial cut, then measure the location of the slot in the workpiece. Make any necessary adjustments to put the slot into the center of the workpiece.

9. Finish machine the ½ in. slot.

10. Rough out the cavity (Figure 3), leaving 1/32 in. for the finish cut all around. Keep the cutting pressure against the tool.

11. Measure the depth of the cavity with a depth mike. Then make the depth adjustment necessary for the finish cut.

12. Make a cut with the cutter just touching the side of the cavity, measure the distance with a micrometer, then adjust for one final cut the correct distance in from the outside edge.

Figure 3. Milling a cavity in the vise body (Lane Community College).

13. Machine each side of the cavity, repeating the operations performed in step 12.
14. Deburr all edges.
15. This completes the machining of the vise body prior to heat treating and finish grinding.

unit 7
using the offset boring head

The offset boring head is used on a vertical milling machine to make accurately located holes with precisely controlled diameters.

objective After completing this unit, you will be able to use an offset boring head to bore a hole pattern to the specified dimensions and within the tolerances given.

information

Most holes in a machine shop are made with a drill; when a higher accuracy is required, these holes are reamed. But the reaming and drilling of holes is limited to standard sized holes for which these tools are available. Also, drilled and reamed holes may wander off from a desired location during the machining process. To produce standard or nonstandard holes at specific locations, an offset boring head can be used. An offset boring head can only be used to enlarge an existing hole. The hole should be drilled to approximately $\frac{1}{16}$ in. smaller in diameter than the finished size of the hole, but it is better to leave more material in the hole, especially if the hole is rough, than to come to the finished size and have a hole that is not cleaned up completely.

The hole can be drilled on a drill press or on the vertical milling machine. The workpiece is fastened to the machine table or other workholding device. If the hole is to be bored through the workpiece it should be supported on parallels. The parallels are spaced far enough apart so they will not be interfering with the penetrating boring tool (Figure 1). The next step is to position the workpiece so that the hole to be bored is aligned with the centerline of the machine spindle by moving the machine table. Before mounting the offset boring head in the spindle, clean the spindle hole and the tool shank.

The offset boring head has two main parts (Figure 2). One is the body, which is fastened in the spindle by its shank. The second part is the adjustable tool slide that holds the boring tool. The tool slide can be precisely moved with the micrometer tool adjustment screw. After tool slide adjustments are made, tightening the locking screw will prevent additional tool movements. A number of holes in the tool slide give different locations where boring tools can be clamped, depending on the diameter of the hole to be bored.

When using any boring head, it is very important to determine the amount of tool movement produced when the tool adjustment screw is rotated one graduation. Some boring heads are graduated to where one graduation movement will produce a .002 in. change in diameter. Others give only a .001 in. change in diameter for a single graduation movement. Boring tools are made in many different sizes and lengths (Figure 3). The best boring tool to use is the one with the largest diameter that will fit the hole to be bored and the shortest shank that will do the job.

When a boring tool is mounted in a boring head, it is very important that the cutting edge is on the centerline of the boring head and in line with the axis of the tool slide movement (Figure 4). Only in this position are the rake angles and clearance angles correct as ground on the tool. This is also the only position when the tools cutting edge moves the same distance as the tool slide when adjustments are made. Boring tools are made from high speed steel, carbide tipped or solid carbide. The kind of tool material and the workpiece material determine the cutting speed

Figure 1. Workpiece supported on parallels. Note the clearance for the penetrating boring tool (Lane Community College).

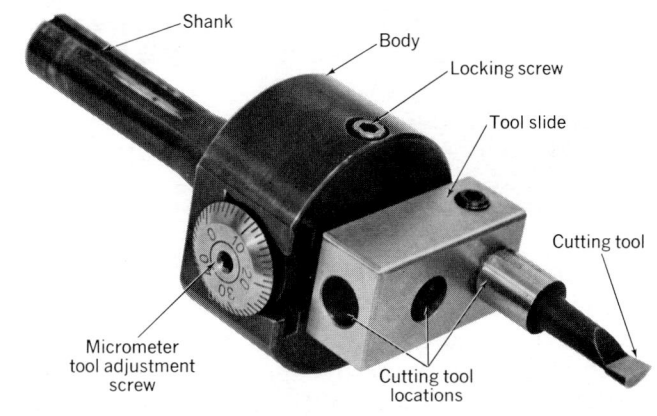

Figure 2. Offset boring head (Lane Community College).

Unit 7 Using the Offset Boring Head

Figure 3. Set of boring tools for the offset boring head (Lane Community College).

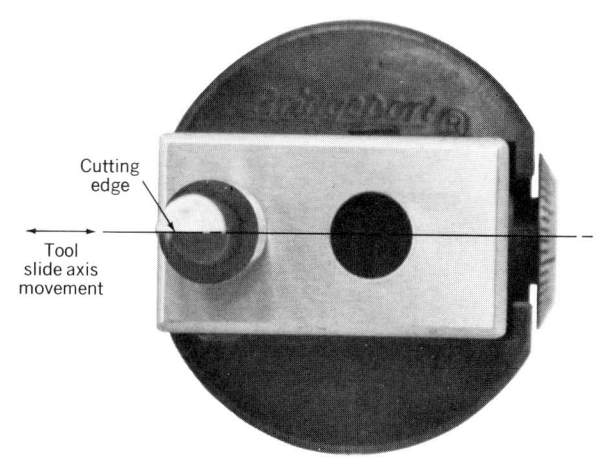

Figure 4. Boring tool cutting edge is on the centerline of the boring head (Lane Community College).

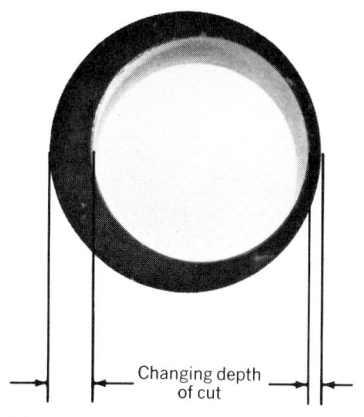

Figure 5. When the hole is eccentric to the spindle centerline, it will cause a variable depth of cut for the boring tool (Lane Community College).

that should be used. But the rigidity of the machine spindle and the setup often require a lower than calculated RPM because the imbalance of the offset boring head creates heavy machine vibrations.

The quill feed on many vertical milling machines is limited to .0015, .003, and .006 in. of feed per spindle revolution. Roughing cuts should be made at the higher figure and finishing cuts should be made with the two lower values. Roughing cuts are usually made with the tool feeding down into the hole. Finishing cuts are made with the tool feeding down and often the tool is fed back up thru the hole by changing the feed direction at the bottom of the hole. Because of the tool deflection, a light cut will be made without resetting the tool on that second cut. When cuts are made with the tool only feeding down but not out, the spindle rotation is stopped before the tool is withdrawn from the hole. If the spindle rotates while the quill is raised, a helical groove will be cut into the wall of the just completed hole, possibly spoiling it.

To obtain a predictable change in hole size for a given tool slide adjustment, certain conditions have to be met. The depth of cut of the boring tool needs to be the same around the circumference of the hole and not like the varying depth of cut illustrated in Figure 5. Roughing cuts should be taken until the hole is round and concentric with the spindle centerline. The depth of cut needs to be equal for successive cuts. As an example, assume that a hole has been rough bored to be concentric with the spindle axis. The tool is now resharpened and fastened in the tool slide. The tool slide is advanced until the tool just touches the wall of the hole. After raising the tool above the work, the tool is moved 20 graduations, or a distance that should increase the hole diameter by .020 in. The spindle is turned on, and with a feed of .003 in. per revolution, the cut is made thru the hole. The spindle is stopped and the tool is withdrawn from the hole. Measuring the hole shows the diameter to have increased by only .015 in. What has happened is that the tool was deflected by the cutting pres-

Figure 6. A radius is machined on a workpiece with the offset boring head (Lane Community College).

sure to produce a hole .005 in. smaller than expected. The tool is now advanced to again give an increase of .020 in. in the hole diameter. With the same feed as for the last cut, the hole is bored.

Measuring the hole again shows the hole to be .020 in. larger. Additional cuts made with the same depth of cut and the same feed will give additional .020 in. diameter increases. If the depth of cut is increased, more tool deflection will take place, resulting in a smaller than expected diameter increase. If the depth of cut is decreased, the tool will be deflected less, resulting in a larger than expected diameter.

When the same depth of cut is maintained but the feed per revolution is increased, higher cutting pressures will produce more tool deflection and a smaller than expected hole diameter. With an equal depth of cut and a smaller feed per revolution, less cutting pressure will produce a larger than expected hole diameter. Another factor that affects the hole diameter with a given depth of cut is tool wear. As a tool cuts, it becomes dull. A dull tool will produce higher cutting pressures with a resultant larger tool deflection.

An offset boring head can also be used to machine a precise radius on a workpiece (Figure 6). The workpiece is positioned the specified distance from the spindle axis. A scrap piece of metal is clamped to the table opposite the workpiece. As cuts are being made with the offset boring head, the tool cuts on both the workpiece and the scrap piece. The diameter of the cuts is measured between the pieces being machined.

Figure 7. A boring and facing head (Lane Community College).

With a boring and facing head (Figure 7), it is possible to machine flat surfaces with the same tool that was used to bore a hole to size. The tool can be moved sideways while the spindle is rotating.

self-evaluation

SELF-TEST 1. When is an offset boring head used?

Unit 7 Using the Offset Boring Head 73

2. Why is the workpiece normally placed on parallels?

3. Why is the locking screw tightened after tool slide adjustments have been made?

4. Why does the tool slide have a number of holes to hold boring tools?

5. Why is it important to determine the amount of tool movement for each graduation on the adjustment screw?

6. What would be the best boring tool to use on a job?

7. How important is the alignment of the tool's cutting edge with the axis of the tool slide?

8. What factors affect the size of the hole obtained for a given amount of tool adjustment?

9. Name three causes for changes in boring tool deflection.

   a.

   b.

   c.

10. What determines the cutting speed in boring?

WORKSHEET (Figure 8)
1. Cut off material; allow for machining ends square.
2. Machine ends square to the sides and to size.
3. Lay out the 1 in. hole.
4. Drill a $\frac{7}{8}$ to $\frac{15}{16}$ in. diameter hole.
5. Mount the workpiece in the vise on the vertical milling machine.
6. Use an edge finder to locate the hole position.
7. Set up a boring tool in the boring head and make a cleanup cut.

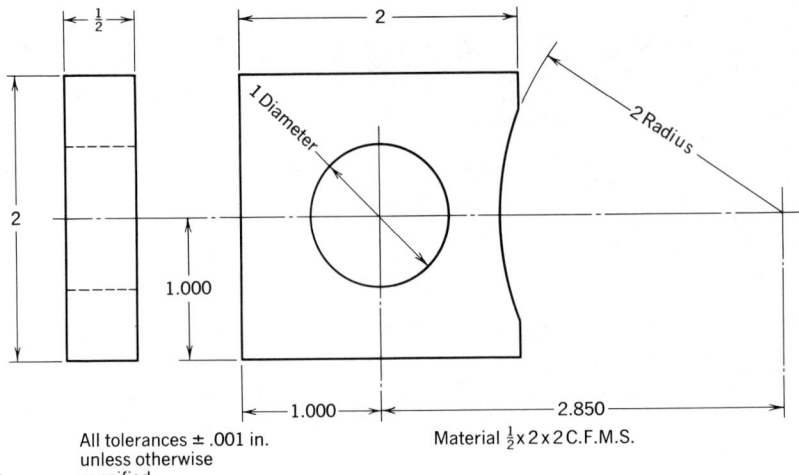

Figure 8. Offset boring head exercise.

8. Make adjustments with the boring head to get predictable hole size changes when cutting.//
9. Bore the hole to its finished size.
10. Move to the center position of the 2 in. radius, using the machine dials.
11. Mount a piece of scrap steel approximately $1\frac{7}{8}$ in. away from this center position. This is needed to measure the diameter of the 2 in. radius.
12. Reset the boring tool so it will just touch the 2 × 2 in. workpiece, then feed the tool another .030 in. out for the first cut.
13. Make this cut.
14. Measure the diameter between the two pieces of steel and then make light cuts, like the last one, until the 2 in. radius is achieved.
15. Deburr the workpiece and present it to your instructor for evaluation.

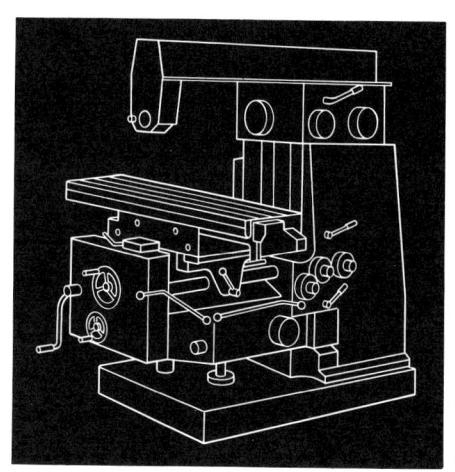

section c
horizontal milling machines

The first horizontal milling machine was probably that of Eli Whitney, invented around 1820 (Figure 1). Whitney used his machine in making parts for firearms. The milling cutter of that period was more like a rotary file than the modern coarse tooth cutter. It is interesting that this very early machine was equipped with a power feed to move the table beneath the cutter. The height of the spindle to the table surface was fixed. The table was fitted with gibs to hold the table down and to make adjustment for wear, similar to recent practices. Another milling machine utilizing a rack and pinion to move the table beneath the cutter was produced in the same area in Connecticut at about the same time by Robert Johnson. A little more than 10 years later, in Britain, James Nasmyth produced a substantial milling machine specialized for milling the flats on machine nuts using an indexing plate.

These first milling machines all had a fixed relationship between the spindle and the table. The next development appeared a few years later with a machine that permitted the cutter spindle to be raised or lowered and clamped in place. About the same time, formed cutters were devised to machine contours into metal surfaces.

By 1850, the horizontal milling machine had become more rugged and precise and had provision for moving the machine table in cross feed. By this time the milling cutter was evolving from a "rotary file" to a coarse-toothed cutter capable of making substantial chips.

The next major developments in the horizontal milling machine came in 1861 with Joseph Brown's universal milling machine (Figure 2). In a single machine design he included a "knee" to move the table assembly up and down in relationship to the cutter, a spiral indexing head connected to the table feed screw, and a table that could be swiveled so that spiral milling, such as the flutes of twist drills, could be machined easily.

The next 40 years showed refinement of horizontal milling machines toward convenience in control positions and improved cutter support (Figure 3). This was also a period of great cutter development with nearly all of the current forms completed by 1900.

Figure 1. One of the earliest milling machines produced by Eli Whitney (Courtesy of DoAll Company).

76   Machine Tools and Machining Practices

Figure 2. First universal milling machine by Brown & Sharpe in 1861 (Courtesy of the MIT Press, Massachusetts Institute of Technology).

Figure 3. By the 1880s machine controls were arranged for convenience and arbor support was provided. This is Cincinnati's first milling machine (Courtesy of Cincinnati Milacron).

Figure 4. By 1908 the horizontal milling machine was a powerful production tool (Courtesy of Cincinnati Milacron).

Section C Horizontal Milling Machines

Scientific experimentation with cutter designs for metal removal efficiency began in the first decade of the twentieth century. With the advent of individual electric motors, massive high power milling machines began to appear with heavy support for the outboard end of the cutter arbor.

The period of the milling machine as a powerful production machine had begun (Figure 4). Machines had been developed with geared constant speed drives and independently controlled feeding rates. Machine accessories like the universal dividing head also reached a high level of development at the turn of the century. The development of the horizontal milling machine, as we know it today, was essentially complete by the time of World War I.

Figure 5. A large planer-type milling machine with two tables. A setup can be made on one table while machining is taking place on the other (Courtesy of The Ingersoll Milling Machine Company).

Since 1910, with the advent of the mass-produced automobile, highly specialized forms of milling machines have been developed that are characterized by great massiveness, high power, and multiple cutting heads, with sophisticated tooling and automatic table cycles. Hydraulic control on horizontal and vertical milling machines appeared about the same time.

The first milling machines could be termed bed-type machines because the part was moved past the cutter with only one table motion. Later machines were designed so that the spindle assembly could be raised or lowered relative to the table. A bed-type milling machine is one in which the position of the milling spindle, but not the height of the table can be changed. There are many configurations of bed-type milling machines. One type has the general appearance of a knee and column. One type of very large milling machine is known as a planer mill (Figure 5) or adjustable rail milling machine. The example shown is equipped with two separate tables, so that one table can be set up while machining is taking place on the other table. These machines often have 250 hp to their spindles. Other bed-type milling machines that employ two or more cutting heads are called duplex or triplex milling machines (Figure 6). These are commonly used in high production setups.

A particularly common form of bed milling machine is called the manufacturing milling machine (Figure 7). On this machine the spindle assembly is positioned and secured at the correct height and parts are passed, typically in a fixture, under the cutter. These machines are usually equipped with means for automatic cycling of the table; and they often have twin fixtures so that one part can be added while the other part is being machined. This is called reciprocal milling. This design of machine can also be found with hy-

Figure 6. Triplex bed-type milling machine (Courtesy of Cincinnati Milacron).

Figure 7. (Plain manufacturing-type milling machine (Courtesy of Cincinnati Milacron).

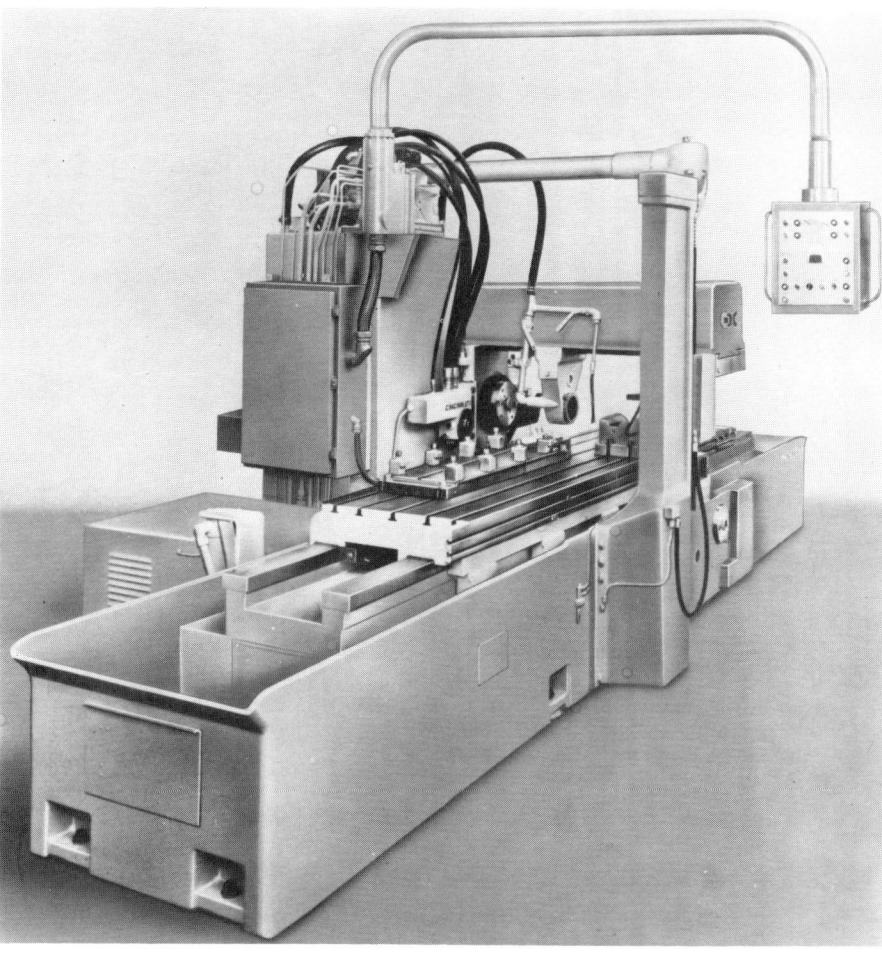

Figure 8. Plain tracer controlled milling machine (Courtesy of Cincinnati Milacron).

draulic tracer controls that move the spindle carrier and cutter vertically in response to a stylus following a cam. These are termed tracer-type manufacturing milling machines (Figure 8).

Another bed-type milling machine also has a table traverse motion, with a spindle assembly that can be moved vertically (Figure 9). This type of machine is found with either horizontal or vertical head configuration and, by general appearance, is often mistaken for a knee and column-type milling machine.

Knee and column milling machines are derived from the heritage of Joseph Brown's universal milling machines in the 1860s. Universal means that the machine table can swivel on its horizontal axis (Figure 10) so that the work can be presented to the cutter at an angle in conjunction with a suitable dividing head permitting helical milling. The plain knee and column milling machine (Figure 11) omits the table swiveling feature in the interest of greater machine rigidity. The more jointed connections, the greater the possibility of uncontrolled movement or "chatter." There is one type of knee and column milling machine used in manufacturing that is capable of vertical table positioning and longitudinal table travel, but it does not have transverse or cross feeding capability. On this type of machine (Figure 12), the spindle bearings are carried in a quill so

Figure 9. Bed-type horizontal milling machine with transverse table motion. This type mill is also found in a vertical spindle design (Courtesy of Cincinnati Milacron).

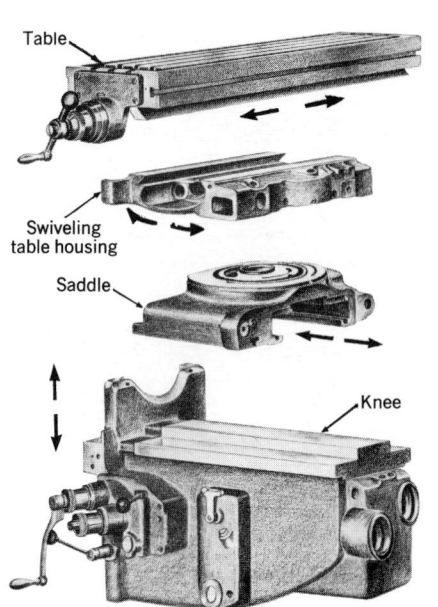

Figure 10. The main features of the knee, saddle, and table assembly on the universal knee and column milling machine (Courtesy of Cincinnati Milacron).

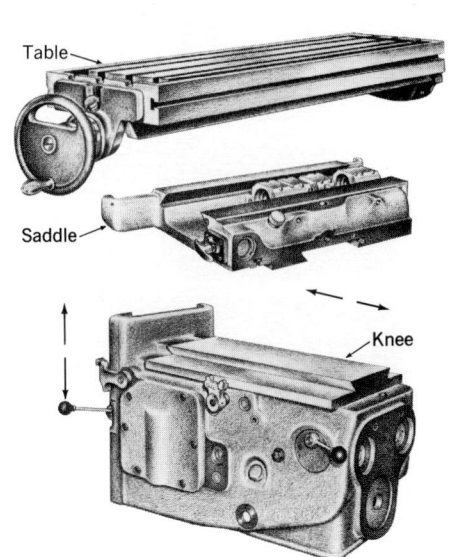

Figure 11. The feature of the knee, saddle, and column milling machine. The swiveling table housing is omitted in this design (Courtesy of Cincinnati Milacron).

Figure 12. Small, plain, automatic knee and column milling machine (Courtesy of Cincinnati Milacron).

that the cutter can be positioned traversely over the part and locked into position. Eliminating the cross feeding saddle adds to the machine rigidity.

On the bed-type manufacturing milling machines, the table is typically moved by a hydraulic-mechanical means that includes backlash control. This control permits the cut to be made easily in either direction without the danger of having the milling cutter suddenly grab the work and take up the backlash, as can happen with most ordinary nut and screw table feeds. This sudden taking up of backlash results in many broken cutter teeth if it is not controlled by the method of milling or by special devices. One of these devices is called a backlash eliminator (Figure 13) which, when engaged, automatically takes up the backlash by applying a preload between two nuts following the leadscrew. Another method, employed mainly with numerically controlled machine tools, is a ring ball nut, commonly called ball screw, arrangement that, by design, is essentially free of backlash.

Since much of the concern in the development of the milling machine and its cutters has been toward the highest possible machine rigidity by various means such as minimum number of moving components, overarm supports, and devices like backlash eliminators, another technique should be mentioned. Machine castings often vary in their ability to absorb vibration, even with the most careful design of internal webbing. Another means to attack the problem has been the "tuning out" of vibration by special vibration dampening devices. Figure 14 shows the milling machine overarm equipped with a device to reduce the resonance of vibration

Section C Horizontal Milling Machines

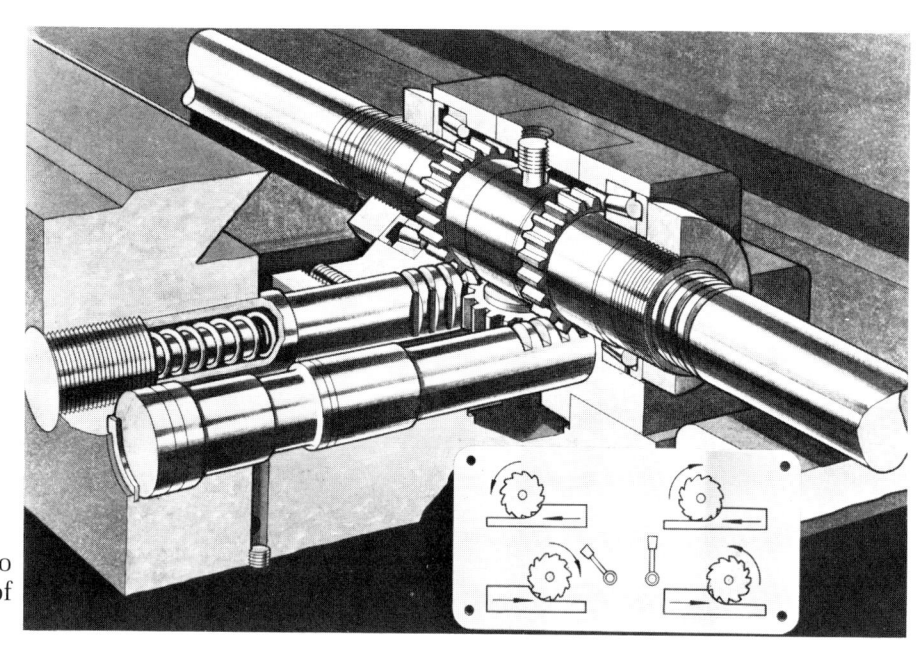

Figure 13. Backlash eliminator to permit "climb milling" (Courtesy of Cincinnati Milacron).

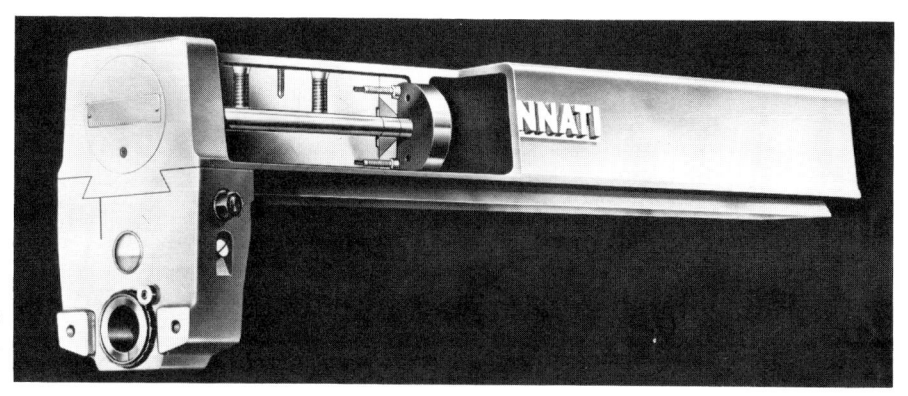

Figure 14. The overarm can reduce cutting vibration (Courtesy of Cincinnati Milacron).

passing through the casting. This capability permits increased cutting loads before chatter sets in.

A number of attachments are available to increase the capabilities of milling machines, particularly for toolroom applications where a few parts are made or for limited production where the expense of a special machine would not be warranted. The vertical milling attachment (Figure 15) is used on horizontal milling machines to obtain the capability of both vertical and angular machining. The universal milling attachments (Figure 16) permit an additional motion so that spiral milling may be done on a plain table milling machine in addition to vertical and angular cuts.

Another attachment is the independent overhead spindle with an angular swivel head (Figure 17), which is powered separately from the horizontal machine spindle. This device replaces the standard overarm and can be used in conjunction with the horizontal spindle as needed to machine angular relationships without moving the workpiece. When not needed, it can be swiveled out of the way, and the regular overarm brackets can be attached.

Figure 15. A vertical milling attachment with quill feeding capability (Courtesy of Cincinnati Milacron).

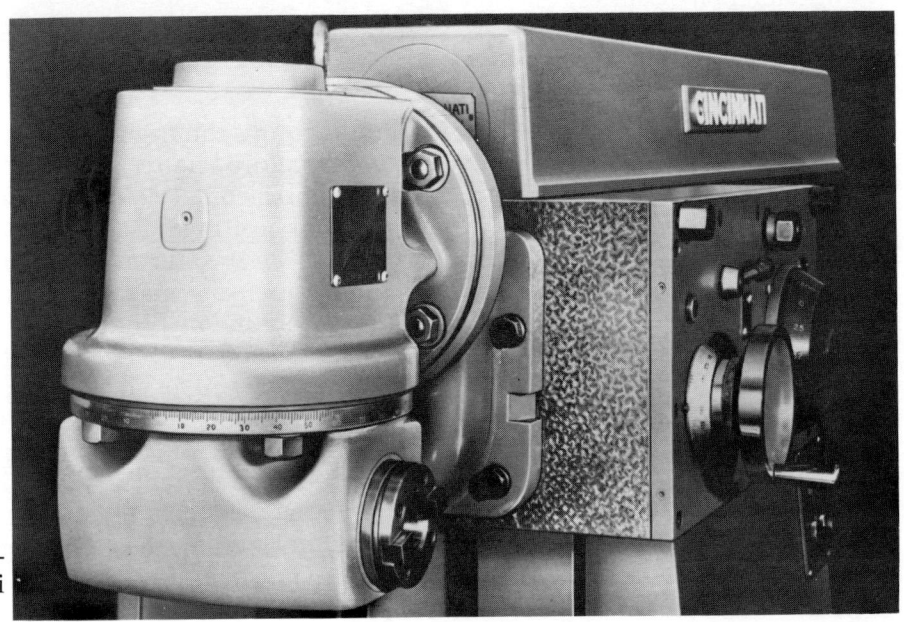

Figure 16. Universal milling attachment (Courtesy of Cincinnati Milacron).

Section C Horizontal Milling Machines

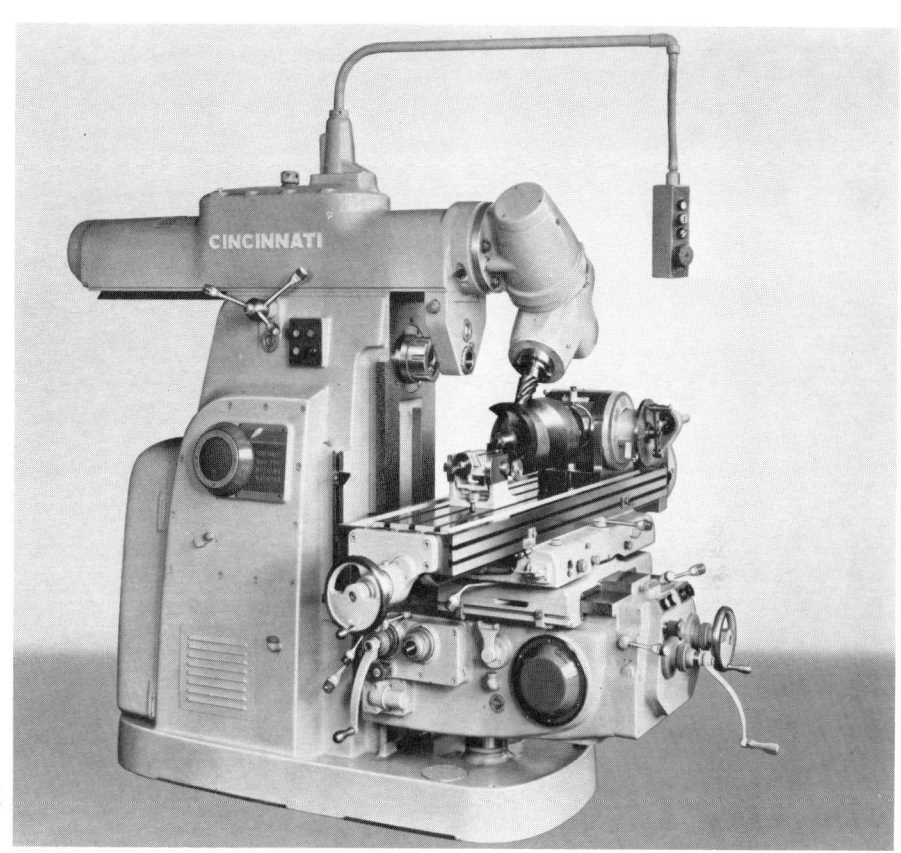

Figure 17. Independent overhead spindle with angular head (Courtesy of Cincinnati Milacron).

Figure 18. A slotting attachment for the horizontal milling machine (Courtesy of Cincinnati Milacron).

A slotting attachment (Figure 18) is also available for horizontal milling machines to utilize a single point tool for operations like internal keyway cutting, where a vertical slotter is not available. It may be set at an angle as well as being set vertically.

Devices for tool holding, such as arbors and adapters for horizontal milling, will be studied in this section. Table-mounted attachments such as rotary tables and indexing heads and workholding devices such as clamps and vises will also be studied.

Milling machines can be equipped with accessory measuring equipment, called direct readouts (DRO), to reduce the chance of operator error when machining expensive complex parts (Figure 19). These measuring systems can be switched to present information in either inch or metric form, which is a great time-saver and eliminates the possibility of making errors in conversion between the two systems.

The horizontal milling machine, in combination with its wide array of accessories, is an extremely versatile machine tool with nearly 160 years of development in its various forms. It should be pointed out that since the advent of numerical control in 1953, many of the functions of the horizontal and vertical milling machines are being displaced, especially on the production of highly complex parts with large numbers of interrelated or repeated dimensions. When you are milling a part with more than 100 related hole posi-

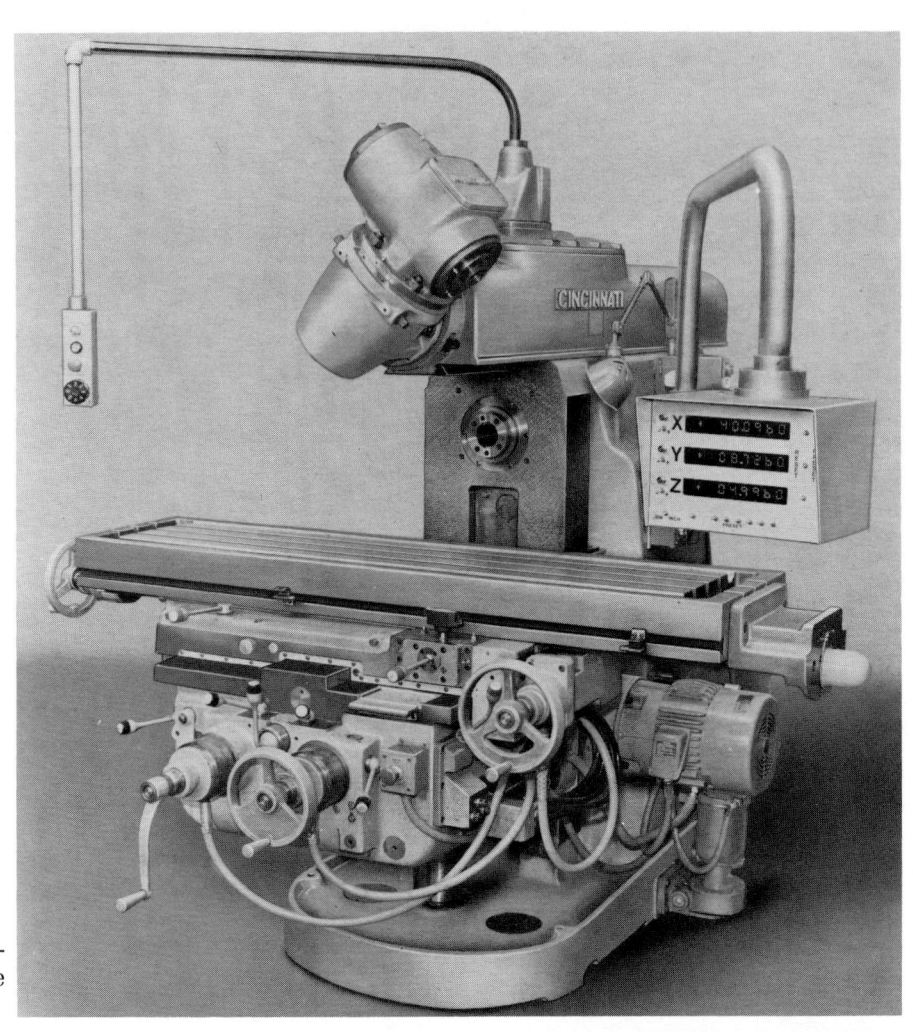

Figure 19. Direct readout (DRO) fitted to a horizontal milling machine (Courtesy of Cincinnati Milacron).

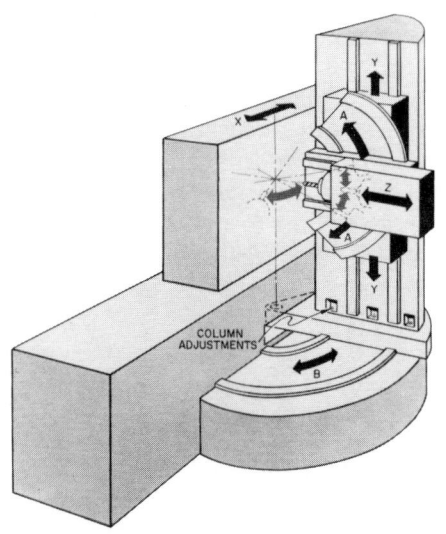

Figure 20. Diagram of five axes of machine motion for complex milling under numerical control (Courtesy of Cincinnati Milacron).

tions and depths, it is very difficult to avoid making at least one mistake. Consequently, much of the work that was formerly done by milling machines and by the accessories shown is now numerically programmed, even to make a single complex part.

It is important for you to learn how to use vertical and horizontal milling machines competently because the cutting and locating principles apply also to the most complex numerically controlled machine tool. Few companies are willing to risk damage to an expensive and complex numerically controlled machine by an operator without a background in conventional milling practice. An operator must be able to determine readily when there is something going wrong with the cutting operation and make appropriate corrections by replacing tools or manually overriding the machine control. Numerically controlled milling centers are sometimes equipped with five programmable machine axes or motions that can permit the spindle to be presented to the work at any angle between the horizontal and vertical besides moving up and down (Figure 20) under the guidance of a programmed tape or by direct computer control.

In this section you will observe specific safety precautions that

Section C Horizontal Milling Machines

relate to horizontal milling, identify components and functions of horizontal mills, and perform routine maintenance. Various mounting systems used to drive milling cutters will be studied, and you will be able to match cutters to their respective applications. You will calculate RPM and feedrates for milling cutters and set the values into the machine controls. You will learn about a variety of workholding methods and alignment procedures and how to mill a square workpiece. In addition, you will use side milling cutters in various combinations and you will use face milling cutters to machine flat surfaces.

Horizontal and vertical milling machines are as basic as the lathe, particularly where one-of-a-kind or small quantities of workpieces are involved. Both of these machine types will be in use for a long time to come. It is important for you to learn to set up and use these machines quickly, accurately, and safely.

review questions

1. The first milling machines were an example to what type of design?

2. What could the universal milling machine do that was not done easily before?

3. When would a plain horizontal milling machine be preferred over a universal milling machine?

4. What purpose is served by a backlash eliminator?

5. Why is direct readout an especially important accessory at the present time?

unit 1
horizontal milling machine safety

Safe operation of any machine tool does not come accidentally, but must be learned. Safety on a machine involves both the person operating the machine and the equipment being used.

objective After completing this unit, you will be able to identify safe and dangerous practices on and around milling machines.

information

A safe worker is properly dressed for the job he is doing. This includes rolled up sleeves or, preferably, short sleeves. Loose fitting clothing may catch in rotating machinery and pull the operator into a dangerous situation. Rings, bracelets, and watches can be very dangerous because they can catch in a cutter and tear off a finger or pull your hand or arm into the cutter with obvious consequences. Long or loose hair can be caught in a cutter or even be wrapped around a rotating smooth shaft, resulting in a quick, painful scalping. Persons with long hair should wear a cap or hairnet in the machine shop.

A milling machine should not be operated while wearing gloves because of the danger of getting caught in the machine. When gloves are needed to handle sharp-edged materials, the machine should be stopped. Eye protection should be worn at all times in machine shops. Eye injuries can be caused by flying chips, tool breakage, or cutting fluid sprays. Keep your fingers away from the moving parts of a milling machine such as the cutter, gears, spindle, or arbor. Do not reach over a rotating spindle or cutter.

Safe operation of a machine tool requires that you think before you do something. Before starting up a machine, know the location and operation of its controls. Operate all controls on the machine yourself; do not have another person start or stop the machine for you. Chances are good that he will turn a control at the wrong time. While operating a milling machine, observe the cutting action at all times so that you can stop the machine immediately when you see or hear something unfamiliar. Always stay within reach of the controls while the machine is running. An unexpected emergency may require quick action on your part. Never leave a running machine unattended.

Before operating the rapid traverse control on a milling machine, loosen the locking devices on the machine axis to be moved. Check that the hand wheels or hand cranks are disengaged, or they will spin and injure anyone near them when the rapid traverse is engaged. The rapid traverse control will move any machine axis that has its feed lever engaged singularly or simultaneously. Do not try to position a workpiece too close to the cutter with this control, but approach the final 2 in. by using the hand wheels or hand cranks.

A person concentrating on a machining operation should not be approached quietly from behind, since it may annoy and alarm him and he may ruin a workpiece or injure himself. Do not lean on a running machine; moving parts can hurt you. Signs posted on a machine indicating a dangerous condition or a repair in progress should only be removed by the person making the repair or by a supervisor.

Measurements should only be taken on a milling machine after the cutter has stopped rotating and after the chips have been cleared away. Milling machine chips are dangerously sharp and often hot and contaminated with cutting fluids. They should not be handled with bare hands. Chips should be removed with a brush. Compressed air should not be used to clean off chips from a machine because it will make small missiles out of chips that can injure a person, even one who's quite a distance away. A blast of air will also force small chips into the ways and sliding surfaces of the milling machine where they will cause scoring and rapid premature wear. Cleaning chips and cutting fluids from the machine or workpiece should only be done after the cutter has stopped turning. Before and during the operation of a milling machine, keep the area around the machine clean of chips, oil spills, cutting fluids, and other obstructions to prevent the operator from slipping or stumbling. Wash your hands and arms thoroughly after being splashed with cutting fluid to prevent getting dermatitis, a skin disease.

Many milling machine attachments and workpieces are heavy; use a hoist to lift them on or off the table. Do not walk under a hoisted load; the hoist may release and drop the load on you. If a hoist is not available, ask for assistance. Any lifting you do should be done with your back straight, using the leg muscles to lift or lower loads. Injuries can be caused by improper setups or the use of wrong tools. Use the correct size wrench when loosening or tightening nuts or bolts, preferably a box wrench or a socket wrench. An oversized wrench will round off the corners on bolts and nuts and prevent sufficient tightening or loosening; a slipping wrench can cause smashed fingers or other injuries to the hands or arms. Milling machine cutters have very sharp cutting edges; handling cutters carefully and with a cloth will avoid cuts on the hands.

All machine guards should be checked to see that they are in good condition and in place to increase milling machine safety. Workpieces should be centered in a vise with only enough extending out to permit machining. Clean the working area after a job is completed. A clean machine is a safer working place than one buried under chips.

self-evaluation

SELF-TEST

1. Why is it dangerous to wear loose clothing, loose, long hair, and rings and bracelets around machinery?

2. When can gloves be worn?

3. When and why should eye protection be worn?

4. Name two hazards that could injure you near a revolving cutter.

    a.
    b.

5. Name three precautions that an operator of a running milling machine should observe.

    a.

Unit 1 Horizontal Milling Machine Safety

    b. _____

    c. _____

6. Under what conditions are measurements made on a milling machine?

   _____

7. What danger exists when compressed air is used on a milling machine?

   _____

8. What is the safest way for you to lift a milling machine attachment if you are alone and a hoist is not available?

   _____

9. The wrenches used on a milling machine should fit the nuts and bolts; why?

   _____

10. Name two general precautions that should be observed around a milling machine.

    a. _____

    b. _____

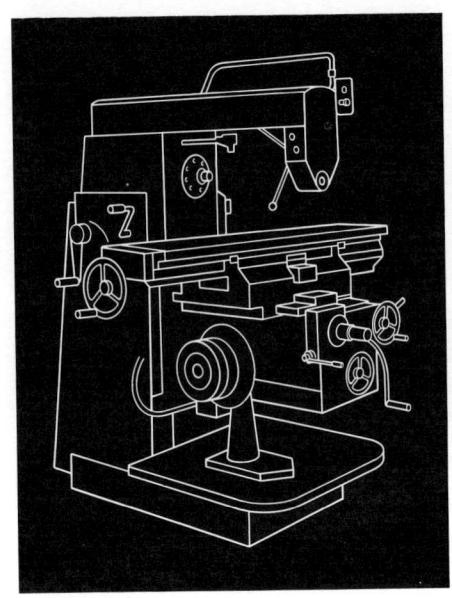

unit 2
plain and universal horizontal milling machines

Machine tool components and parts are identified by names. Anyone operating and maintaining machine tools should be familiar with these names and with the location of these parts.

objectives After completing this unit, you will be able to:
1. Identify the important parts of a horizontal milling machine.
2. Perform routine maintenance.

information

The major components of a horizontal milling machine are the column, knee, saddle, table, spindle, and overarm. The table of the plain horizontal milling machine does not swivel as does the universal milling machine. Figure 1 is an illustration of a horizontal milling machine with the major parts identified. The column is the main part of the milling machine. The face of the column is machined to provide an accurate guide for the vertical travel of the knee. The column also contains the main drive motor and the spindle. The spindle holds and drives the various cutting tools, chucks, and arbors. The spindle is hollow, the front end has a tapered hole with a standard milling machine taper. The front end of the spindle is called the spindle nose.

The overarm is mounted on top of the column and supports the arbor through an arbor support. The overarm slides in and out and can be clamped securely in any position. The knee can be moved vertically on the face of the column. The knee supports the saddle and the saddle provides the sliding surface for the table. The saddle can move toward and away from the column to give crosswise movement of the machine table. The table provides the surface on which the workpieces are fastened. T-slots are machined along the length of the top surface of the table to align and hold fixtures and workpieces. Hand wheels or hand cranks are used to manually position the table. Micrometer collars make possible positioning movements as small as .001 in.

Power feed levers control automatic feeds in three axes, the feedrate being adjusted by the feed change crank (Figure 2). On many milling machines, the power feed only operates when the spindle is turning. Two safety stops at each axis

Unit 2 Plain and Universal Horizontal Milling Machines

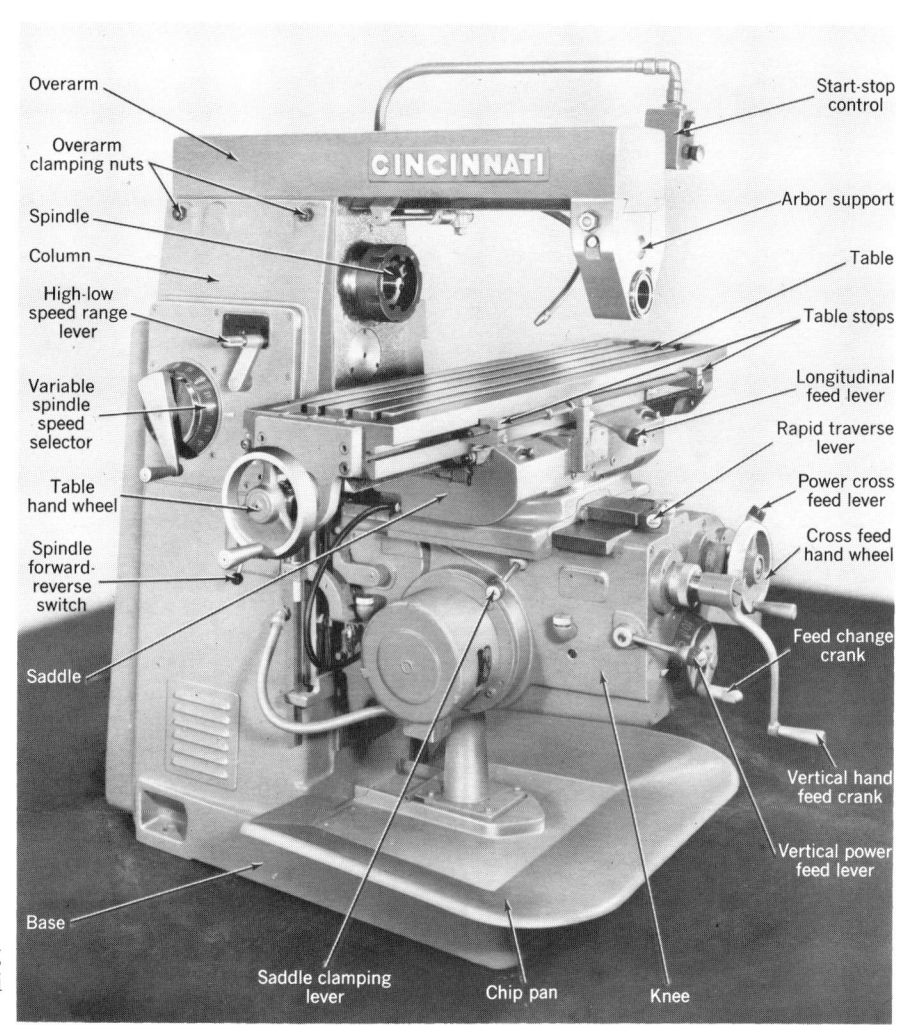

Figure 1. Horizontal milling machine (Courtesy of Cincinnati Milacron).

travel limit prevent accidental damage to the feed mechanisms by providing automatic kickout of the power feed. Two adjustable trip dogs for each axis allow the operator to preset specific power feed kickout travel distances. Rapid positioning of the table is accomplished with the rapid traverse lever (Figure 3). The direction of the rapid advance is dependent on the position of the respective feed lever. Locking devices on the table, saddle, and knee are used to prevent unwanted movements in any or all of these axes. The locking devices should be released only in the axis in which power feed is used. The spindle can rotate clockwise or counterclockwise, depending on the position of the spindle forward-reverse switch. Spindle speeds are changed to a high or low range with the speed range lever (Figure 4). The variable spindle speed selector makes any spindle speed possible between the minimum and maximum RPM available in each

Figure 2. Feed change crank (Courtesy of Cincinnati Milacron).

Figure 3. Rapid traverse lever (Courtesy of Cincinnati Milacron).

Figure 4. Speed change levers (Courtesy of Cincinnati Milacron).

speed range. The speed range lever has a neutral position between the high and low range. When the lever is in this neutral position, the spindle can easily be rotated by hand during machine setup.

The universal milling machine (Figure 5) closely resembles a plain horizontal milling machine. The main difference between these machines is that the universal machine has an additional housing that swivels on the saddle and supports the table. This allows the table to be swiveled 45 degrees in either direction in a horizontal plane. The universal milling machine is especially designed to machine helical slots or grooves as in twist drills and milling cutters. Other than these special applications, a universal mill and a plain milling machine can perform the same operations.

The size of a horizontal milling machine is usually given as the range of movement possible and the power rating of the main drive motor of the machine. An example would be a milling machine with a 28 in. longitudinal travel, 10 in. cross travel, and 16 in. vertical travel with a 5 hp main drive motor. As the physical capacity of a machine increases, more power is also available at the spindle through a larger motor.

Before any machine tool is operated, it should be lubricated. A good starting point is to

Figure 5. Universal milling machine (Courtesy of Cincinnati Milacron).

wipe clean all sliding surfaces and to apply a coat of a good way lubricant to them. Way lubricant is a specially formulated oil for sliding surfaces. Dirt, chips, and dust will act like a lapping compound between sliding members and cause excessive machine wear. Most machine tools have a lubrication chart that outlines the correct lubricants and lubrication procedures. When no lubrication chart is available, check all oil sight gages for the correct oil level and refill, if necessary. Too much oil causes leakage. Lubrication should be performed progressively, starting at the top of the machine and working down. Machine points that are hand oiled should only receive a small amount of oil at any one time, but this should be repeated at regular intervals, at least daily. Motor or pulley bearings should not get too much grease, since this may destroy the seals.

Before operating a milling machine you should be familiar with all control levers. Do not use force to engage or disengage controls or levers. Check that the locking levers are loosened on all moving slides. Check that all operating levers are in the neutral position before the machine is turned on. On a variable speed drive milling machine, change spindle speeds only while the spindle motor is running. On geared models, the spindle has to be stopped. Stop the spindle motor when shifting from one range to another. All power feed levers should be in their neutral position before feed changes are made. Spindle rotation should be reversed only after the machine has come to a complete standstill.

self-evaluation

SELF-TEST

1. Go to a plain or universal horizontal milling machine and locate the following parts.

   overarm
   column
   saddle clamping lever
   speed change lever
   powerfeed levers for
     longitudinal feed,
     crossfeed and
     vertical feed
   rapid traverse lever
   switch for spindle
     ON-OFF
   arbor support
   switch for coolant pump

   table
   knee
   feed change lever
   spindle nose
   saddle
   knee clamping lever
   spindle forward-
     reverse switch
   trip dogs for all
     three axes

2. Lubricate a plain horizontal milling machine.

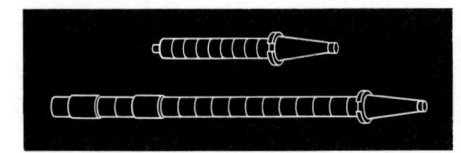

unit 3
types of spindles, arbors, and adaptors

Milling cutters, in order to be used, have to be mounted on a spindle, arbor, or in an adaptor. This unit describes different mounting methods used for milling cutters.

objective After completing this unit, you will be able to identify different mounting systems used to drive milling cutters.

information

The spindle of the milling machine holds and drives milling cutters. Cutters can be mounted directly on the spindle nose, as with face milling cutters (Figure 1), or by means of arbors and adaptors. These arbors and adaptors have tapered shanks that fit into the tapered hole or socket in the spindle nose. The tapers used for mounting milling cutters are divided into two general classes: (1) self-holding tapers, and (2) self-releasing or steep tapers.

Self-holding tapers have a small included angle of 5 degrees or less. When the shank is firmly seated in a socket, it will stay in place because of the high frictional forces between the contacting surfaces. Self-holding taper assemblies often are very difficult to take apart; that is why spindle noses are now made with a self-releasing taper.

Self-releasing tapers have a large included angle generally over 15 degrees. This steep taper permits easy and quick removal of arbors from the spindle nose. Most manufacturers have adapted the standard national milling machine taper. This taper is 3½ inches per foot (IPF) or about 16½ degrees. National milling machine tapers are available in four standard sizes, numbered 30, 40, 50, and 60, the most common being number 50. Self-releasing taper type shanks must be locked in the spindle socket with a draw-in bolt. Positive drive is obtained through two keys

Figure 1. Mounting a face mill on the spindle nose of a milling machine (Courtesy of Cincinnati Milacron).

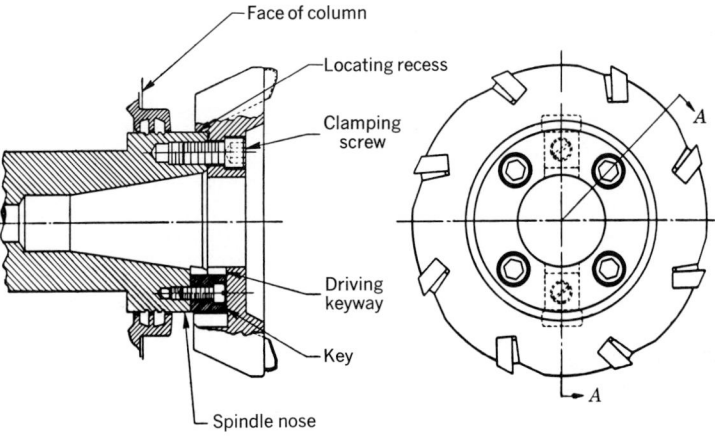

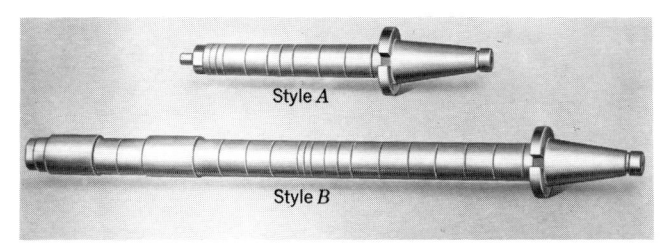

Figure 2. Arbors, styles *A* and *B* (Courtesy of Cincinnati Milacron).

in the spindle nose that engage in keyways in the flange of arbors and adaptors.

Two common arbor styles are shown in Figure 2. Style *A* arbor has a cylindrical pilot on the end opposite the shank. The pilot is used to support the free end of the arbor. Style *A* arbors are used mostly on small milling machines. But they are also used on larger machines when a style *B* arbor support cannot be used because of a small diameter cutter or interference between the arbor support and the workpiece.

Style *B* arbors are supported by one or more bearing collars and arbor supports. Style *B* arbors are used to obtain rigid setups in heavy duty milling operations.

Style *C* arbors are also known as shell end mill arbors or as stub arbors (Figure 3). Shell end milling cutters are face milling cutters up to 6 in. in diameter. Because of their relatively small diameter, these cutters cannot be counterbored so that they can be mounted directly on the spindle nose, as are face mills, but they are mounted on shell end mill arbors. Figure 4 shows how arbors are mounted in the milling machine. The draw-in bolt is screwed into the arbor as far as it will go, then the arbor is pulled into the spindle nose by tightening the draw-in bar lock nut. Note that the cutters are mounted close to the spindle and the first bearing support is close to the cutter. Spacing collars and a shim are used to get an exact width in the straddle milling setup. Keys provide positive drive to the cutters and bearing collars. The bushing fit in the arbor supports can be adjusted to the bearing collars and pilot size. Bushing adjustments have to be made very carefully, because too loose a fit causes inaccuracy or chatter. Too tight a fit causes excessive friction and heat, which damage the bushing and bearing collar. When the spindle turns at high RPM, more clearance is needed than at low RPM.

The bearing collars have a larger diameter than the spacing collars for easy positioning of the arbor supports. All collars are manufactured to very close tolerances with their ends or faces being parallel and also square to the hole. It is very important that the collars and other parts fitting on the arbor are handled carefully to avoid damaging the collar faces. Any nicks, chips, or dirt between the collar faces will misalign the cutter or deflect the arbor and cause cutter run-out. The arbor nut should be tightened or loosened only with the arbor support in place. Without the arbor support, the arbor can easily be sprung and permanently bent.

Adaptors are used on milling machines to mount cutters that cannot be mounted on arbors. Adaptors can be used to hold and drive taper shank tools (Figure 5).

Collets used with these adaptors increase the range of tools that can be used in a milling machine having a given size spindle socket. The spring chuck adaptor (Figure 6), with different size removable spring collets, is used with straight shank tools such as drills and end mills. With a quick-change adaptor (Figure 7) mounted on the spindle nose, a number of milling machine operations such as drilling, end milling, and boring can be performed without changing the setup of the part being machined. The different tools are mounted on quick-change adaptors that are ready for use (Figure 8).

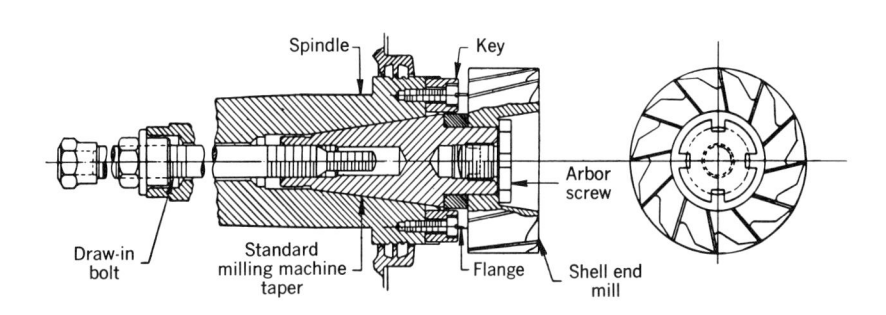

Figure 3. Style *C* arbor; shell end mill arbor (Courtesy of Cincinnati Milacron).

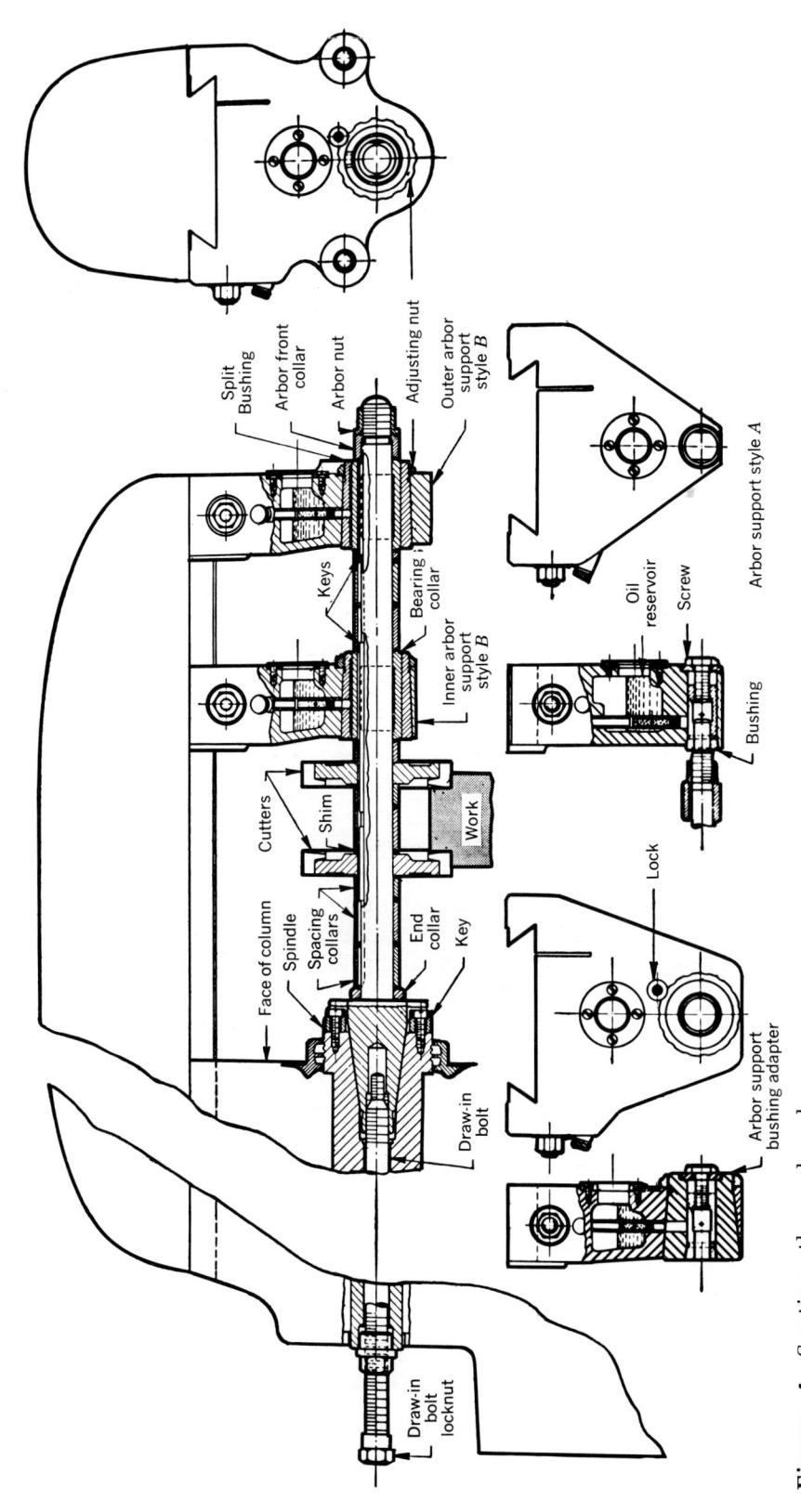

Figure 4. Section through arbor showing location of arbor collars, keys, bearing collars, and various arbor supports (Courtesy of Cincinnati Milacron).

Unit 3 Types of Spindles, Arbors, and Adaptors

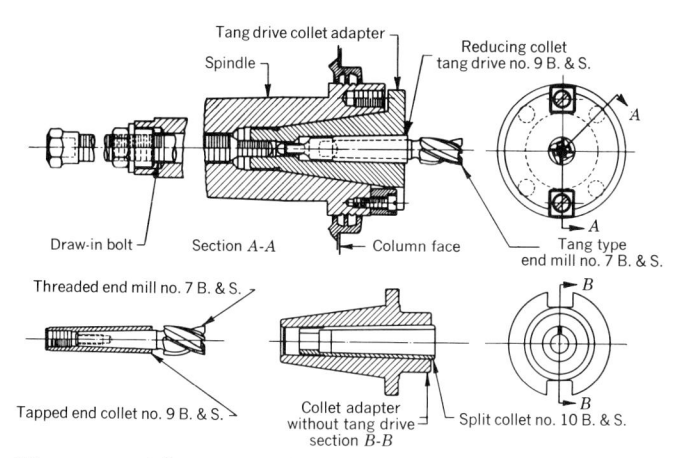

Figure 5. Adaptors and collets for self-releasing and self-holding tapers (Courtesy of Cincinnati Milacron).

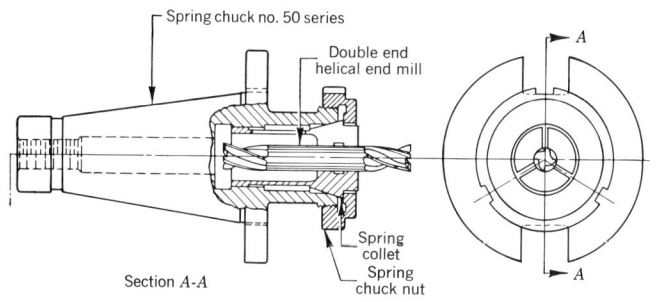

Figure 6. Spring chuck adaptor (Courtesy of Cincinnati Milacron).

Figure 7. Quick-change adaptor mounted on spindle nose (Courtesy of Cincinnati Milacron).

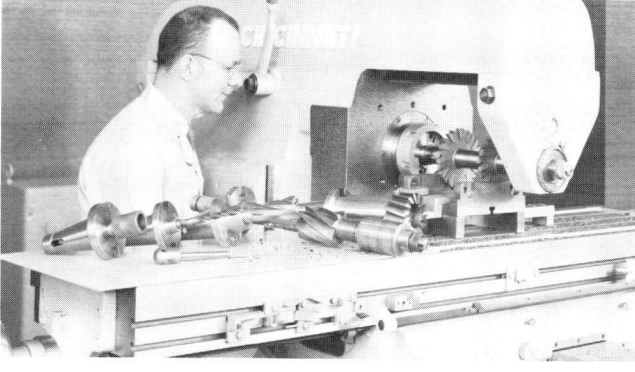

Figure 8. A number of tools mounted on quick-change tool holders ready to use (Courtesy of Cincinnati Milacron).

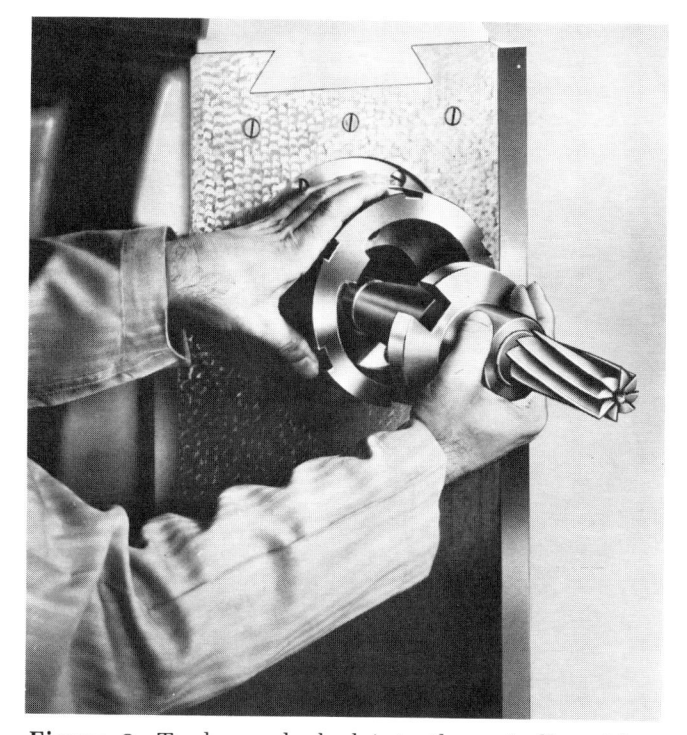

Figure 9. Tools are locked into the spindle with a partial turn of the clamp ring (Courtesy of Cincinnati Milacron).

Figure 9 shows the easy method of changing tools in a machine with a quick-change adaptor. To remove arbors and adaptors that are held with a draw-in bar, use the following procedure.

1. Loosen the locknut on the draw-in bolt one turn.
2. Tap the end of the draw-in bolt with a lead hammer. This releases the arbor shank from the spindle socket.

3. Arbors are heavy; you may need someone to hold the arbor while you unscrew the draw-in bolt from the rear of the machine.

Arbors should be stored in an upright position. Long arbors laying on their sides, if not properly supported, may bend. Rules for mounting arbors, adaptors, and cutters are:

Before inserting a tapered shank into the spindle socket, clean all mating parts and check for nicks and burrs. Nicks and burrs should be removed with a honing stone.

The cutter, spacing collars, and bearing collars should be a smooth sliding fit on the arbor. Nicks should be stoned off.

Use an arbor length that does not give much arbor overhang beyond the outer arbor support. Arbor overhang may cause vibration and chatter.

Mount cutters as close to the column as the work permits.

Cutters are sharp; handle carefully with shop towels.

When changing cutters on an arbor, place the cutter and spacers on a smooth and clean area on the worktable to avoid damage to their accurate bearing or contact surfaces.

Tighten the arbor nut with a wrench that fits accurately after the arbor support is in place. Do not use a hammer to tighten the arbor nut. Overtightening will spring or bend the arbor.

self-evaluation

SELF-TEST 1. What kind of cutters are mounted directly on the spindle nose?

2. Milling machine spindle sockets have two classes of taper. What are they?

3. What is the amount of taper on a national milling machine taper?

4. Why is it important to carefully clean the socket, shank, and arbor spacers prior to mounting them on a milling machine?

5. When is a style A arbor used?

6. Where should the arbor supports be in relation to the cutter?

7. What is a style C arbor?

8. Why should the arbor support be in place before the arbor nut is tightened or loosened?

9. Why are milling machine adaptors used?

10. If an arbor extends some distance beyond the outer arbor support, what can happen?

unit 4
arbor-driven milling cutters

Milling cutters are the cutting tools of the milling machines. They are made in many different shapes and sizes. These various cutters are mostly designed for a specific application. A milling machine operator should be capable of matching a cutting tool to the required application.

objective After completing this unit, you will be able to identify 10 milling cutters and list their names and common applications.

information

Most milling cutters are designed to perform specific kinds of operations. To make an intelligent decision as to which cutter to use, one should be able to identify milling cutters by sight and to know their capabilities and limitations. Most milling cutters are made from high speed steel; large cutters have inserted blades or teeth. More and more cutters are made with cemented carbide cutting edges. Milling cutters can be divided into profile sharpened cutters and form relieved cutters. Profile sharpened cutters are resharpened by grinding a narrow land (Figure 1) back of the cutting edges. Form relieved cutters are resharpened by grinding the face of the tooth parallel to the axis of the cutter. Cutters are classified also as being arbor driven or of the shank type. From the many different milling cutters available, this unit deals only with the more commonly used arbor-driven cutters.

Milling cutters are manufactured for either right-hand or left-hand rotation and with either right-hand or left-hand helix. The hand of a milling cutter is determined by looking at the front end of a spindle mounted cutter; a right-hand cutter requires a counterclockwise rotation (Figure 2), and a left-hand cutter rotates clockwise. The hand of the helix is determined by looking at the teeth or flutes from the cutter end. Flute to

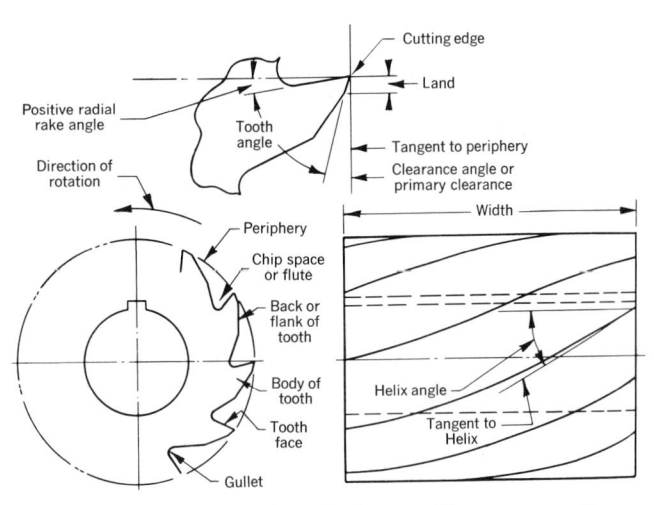

Figure 1. Nomenclature of plain milling cutter (Courtesy of Cincinnati Milacron).

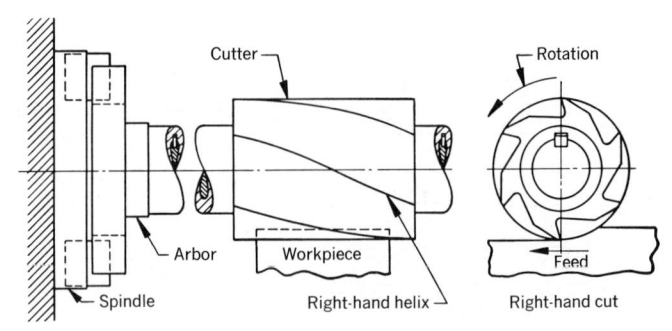

Figure 2. Plain milling cutter with right-hand helix and right-hand cut (Courtesy of Cincinnati Milacron).

the right make a right-hand helix; to the left, they are a left-hand helix.

PLAIN MILLING CUTTERS

Plain milling cutters are designed for milling plain surfaces where the width of the work is narrower than the cutter (Figure 3). Plain milling cutters less than $\frac{3}{4}$ in. wide have straight teeth. On straight tooth cutters, the cutting edge will cut along its entire length at the same time. Cutting pressure increases until the chip is completed. At this time the sudden change in tooth load causes a shock that is transmitted through the drive and often leaves chatter marks or an unsatisfactory surface finish. Light duty milling cutters have a large number of teeth, which limits their use to light or finishing cut because of insufficient chip space for heavy cutting. Heavy duty plain mills (Figure 4) have fewer coarse teeth, which makes for strong teeth with ample chip clearance. The helix angle of heavy duty mills is about 45 degrees. The helical form enables each tooth to take a cut gradually, which reduces shock and lowers the tendency to chatter. Plain milling cutters are also called slab mills. Plain milling cutters with a helix angle over 45 degrees are known as helical mills (Figure 5). These milling cutters produce a smooth finish when used for light cuts or on intermittent surfaces. Plain milling cutters do not have side cutting teeth and should not be used to mill shoulders or steps on workpieces.

Figure 3. Light duty plain milling cutters (Copyright © Illinois Tool Works Inc., 1976.)

Figure 4. Heavy duty plain milling cutter (Copyright © Illinois Tool Works Inc., 1976).

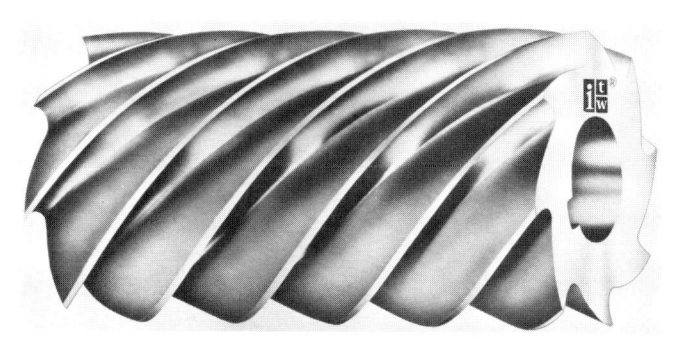

Figure 5. Helical plain milling cutter (Copyright © Illinois Tool Works Inc., 1976).

Figure 6. Side milling cutter (Lane Community College).

Figure 7. Stagger tooth milling cutter (Copyright © Illinois Tool Works Inc., 1976).

SIDE MILLING CUTTERS

Side milling cutters are used to machine steps or grooves. These cutters are made from $\frac{1}{4}$ to 1 in. in width. Figure 6 shows a straight tooth side milling cutter. To cut deep slots or grooves, a staggered tooth side milling cutter (Figure 7) is preferred because the alternate right-hand and left-hand helical teeth reduce chatter and give more chip space for higher speeds and feeds than are possible with straight tooth side milling cutters. To cut slots over 1 in. wide, two or more side milling cutters may be mounted on the arbor simultaneously. Shims between the hubs of the side mills can be used to get any precise width cutter combination or to bring the cutter again to the original width after sharpening.

Half side milling cutters are designed for heavy duty milling where only one side of the cutter is used (Figure 8). For straddle milling, a right-hand and a left-hand cutter combination is used.

Plain metal slitting saws are designed for slotting and cutoff operations (Figure 9). Their sides are slightly relieved or "dished" to prevent binding in a slot. Their use is limited to a relatively shallow depth of cut. These saws are made in widths from $\frac{1}{32}$ to $\frac{5}{16}$ in.

To cut deep slots or when many teeth are in contact with the work, a side tooth metal slitting saw will perform better than a plain metal slit-

Figure 8. *(top)* Half side milling cutter (Copyright © Illinois Tool Works Inc., 1976).

Figure 9. *(top right)* Plain metal slitting saw (Copyright © Illinois Tool Works Inc., 1976).

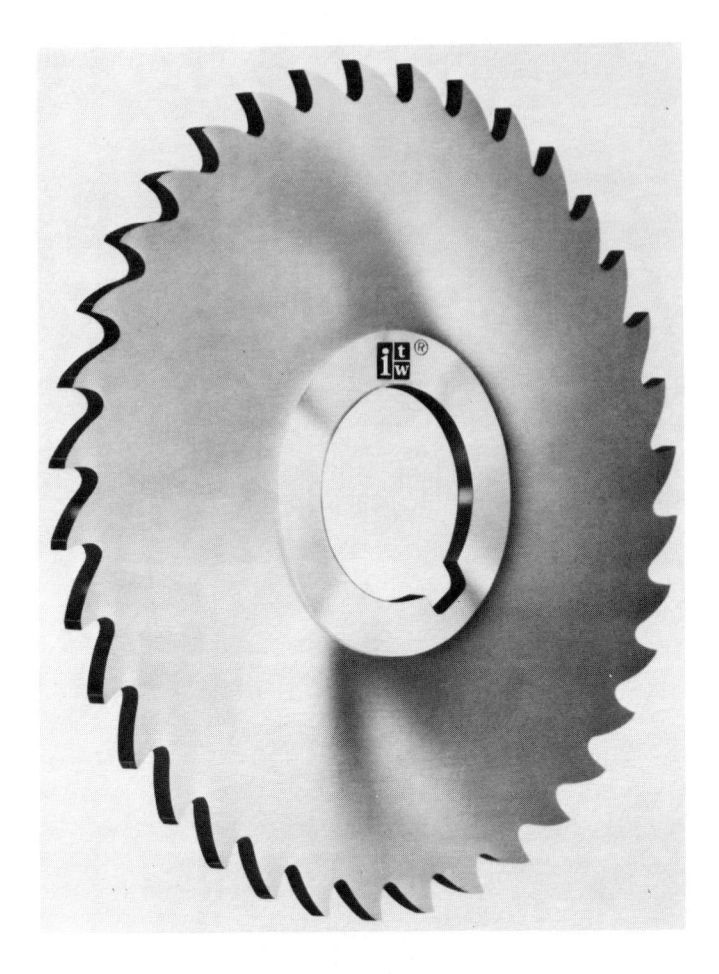

Figure 10. *(bottom right)* Side tooth metal slitting saw (Copyright © Illinois Tool Works Inc., 1976).

ting saw (Figure 10). These saws are made from $\frac{1}{16}$ to $\frac{3}{16}$ in. wide.

Extra deep cuts can be made with a staggered tooth metal slitting saw (Figure 11). Staggered tooth saws have greater chip carrying capacity than other saw types. All metal slitting saws have a slight clearance ground on the sides toward the hole to prevent binding in the slot and the scoring of the walls of the slot. Stagger tooth saws are made from $\frac{3}{16}$ to $\frac{5}{16}$ in. wide.

Angular milling cutters are used for angular milling such as cutting of dovetails, V-notches, and serrations. Single angle cutters (Figure 12) form an included angle of 45 or 60 degrees, with one side of the angle at 90 degrees to the axis of the cutter.

Double angle milling cutters (Figure 13)

Figure 11. Staggered tooth metal slitting saw (Copyright © Illinois Tool Works Inc., 1976).

Figure 12. Single angle milling cutter (Copyright © Illinois Tool Works Inc., 1976).

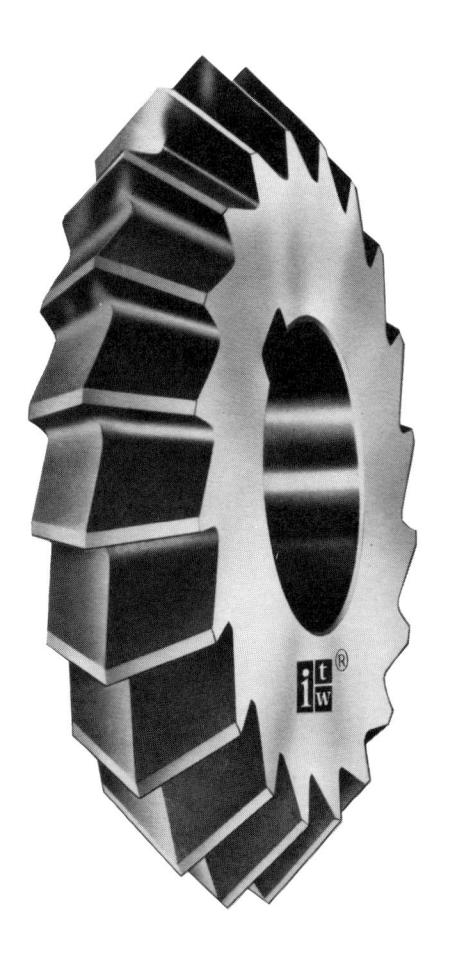

Figure 13. Double angle milling cutter (Copyright © Illinois Tool Works Inc., 1976).

usually have an included angle of 45, 60, or 90 degrees. Angles other than those mentioned are special milling cutters.

Convex milling cutters (Figure 14) produce concave bottom grooves or they can be used to make a radius in an inside corner. Concave milling cutters (Figure 15) make convex surfaces. Corner rounding milling cutters (Figure 16) make rounded corners. The cutters illustrated in Figures 14 to 17 are form relieved cutters.

Involute gear cutters (Figure 17) are commonly available in a set of eight cutters for a

Figure 14. Convex milling cutter (Copyright © Illinois Tool Works Inc., 1976.)

Figure 15. Concave milling cutter (Copyright © Illinois Tool Works Inc., 1976).

Figure 16. *(left)* Corner rounding milling cutter (Copyright © Illinois Tool Works Inc., 1976).

Figure 17. Involute gear cutter (Lane Community College).

given pitch, depending on the number of teeth for which the cutter is to be used. The ranges for the individual cutters are as follows.

Number of cutter	Range of teeth
1	135 to rack
2	55 to 134
3	35 to 54
4	26 to 53
5	21 to 25
6	17 to 20
7	14 to 16
8	12 and 13

These eight cutters are designed so that their forms are correct for the lowest number of teeth in each range. If an accurate tooth form near the upper end of a range is required, a special cutter is needed.

self-evaluation

SELF-TEST

1. What are the two basic kinds of milling cutters with reference to their tooth shape?

2. What is the difference between a light duty and a heavy duty plain milling cutter?

3. Why are plain milling cutters not used to mill steps or grooves?

4. What kind of cutter is used to mill grooves?

5. How does the cutting action of a straight tooth side milling cutter differ from a stagger tooth side milling cutter?

6. Give an example of an application of half side milling cutters.

7. When are metal slitting saws used?

8. Give two examples of form relieved milling cutters.

9. When are angular milling cutters used?

10. When facing the spindle, in which direction should the right-hand cutter be rotated in order to cut?

unit 5
setting speeds and feeds on the horizontal milling machine

A milling machine can be operated efficiently or inefficiently. To be an efficient machine operator, you must understand the relationships between different work materials and the cutting tools used to machine them. To use these cutting tools correctly and economically, you must know about cutting speeds and feeds.

objectives After completing this unit, you will be able to:
1. Select cutting speeds for different materials and calculate the RPM for different milling cutters.
2. Select and calculate feedrates for different materials and milling cutters.
3. Set speeds and feeds on a horizontal milling machine.

information

To get maximum use from a milling cutter, it is important that it is operated at the correct cutting speed. Cutting speed is expressed in surface feed per minute and varies for different work materials and cutting tool materials. The cutting speed of a milling cutter is the distance which the cutting edge of a cutter tooth travels in 1 min. Table 1 is a table of cutting speeds for a number of commonly used materials.

The cutting speeds given in Table 1 are only intended to be starting points. These speeds represent experience in instructional settings where cutter and machine conditions are often less than ideal. Different hardnesses within each materials group account for the wide range of cutting

Table 1
Cutting Speeds for Milling

Material	Cutting Speed SFM	
	High Speed Steel Cutter	Carbide Cutter
Free machining steel	100-150	400-600
Low carbon steel	60-90	300-550
Medium carbon steel	50-80	225-400
High carbon steel	40-70	150-250
Medium alloy steel	40-70	150-350
Stainless steel	30-80	100-300
Gray cast iron	50-80	250-350
Bronze	65-130	200-400
Aluminum	300-800	1000-2000

speeds. Generally speeds are lower for hard materials, abrasive materials, and deep cuts. Speeds are higher for soft materials, better finishes, light cuts, frail workpieces, and light setups. It is very important that the cutting speed is not too high for the material being machined or the cutting edges will dull rapidly. Too slow a cutting speed will not damage a cutter, but it will be inefficient. It is good practice to use the lower cutting speed value to start and then, if the setup allows, increase it to the higher speed. To use these cutting speed values on a milling machine, they have to be expressed in RPM. The formula used to convert cutting speed into RPM is:

$$\text{RPM} = \frac{\text{CS} \times 4}{\text{D}}$$

CS = cutting speed found in Table 1

4 = constant

D = diameter of cutter in inches

As an example of how to use the formula, calculate the RPM for a 3 in. diameter high speed steel cutter to be used on cast iron.

$$\text{RPM} = \frac{\text{CS} \times 4}{\text{D}} = \text{RPM} = \frac{50 \times 4}{3} = \frac{200}{3} = 67$$

In this example the low end of the cutting speed range was used to get a starting point. Other factors such as the use of coolant and machine rigidity also influence the selection of a cutting speed. Tool materials, other than high speed steels such as cast alloys or cemented carbides, can be used at higher cutting speeds because they retain a sharp cutting edge at elevated temperatures. As a rule, when high speed steel cutting speeds are 100 percent, then cast alloys can be 150 percent and cemented carbides can be 200 to 600 percent of that figure.

These are the general variations that affect the selection of speed for milling cutters. As you become more experienced and begin to deal with a wide range of materials, it will be very useful to refer to references such as the *Machining Data Handbook* and the general section in *Machinery's Handbook* on feeds and speeds. Determining accurately the speeds and feeds for milling is complicated by the fact that the cutting edge does not remain in the work continuously and that the chip being made varies in thickness during the cutting period. The type of milling being done (slab, face, or end milling) also makes a difference, as does the way that the heat is transferred from the cutting edge.

The second important item in efficient milling machine operation is the feed. It is expressed as a feedrate and given in inches per minute (IPM). Most milling machines have two different drive motors, one to power the spindle and one to power the feed mechanism. These two motors make independent changes of the spindle speed and the feedrate possible. The feedrate is the product of the feed per tooth times the number of teeth on the cutter times the RPM of the spindle. You have already determined how to calculate the RPM of a cutter and, by counting the number of teeth in a cutter and knowing the feed per tooth, you can determine the feedrate. Table 2 is a chart of commonly used feeds per tooth (FPT).

As you can see in Table 2, there is only a slight difference in the feed per tooth allowance between high speed steel cutters and carbide cutters. Calculate the feedrate for a 3 in. diameter six tooth helical mill cutting free machining steel, first for a high speed steel tool and then for a carbide tool. The formula for feedrate is

$$\text{FPT} \times \text{N} \times \text{RPM}$$

EXAMPLE.

FPT = feed per tooth

N = number of teeth on cutter

RPM = revolutions per minute of cutter

For a starting point, calculate the RPM (refer to Table 1).

$$\text{RPM} = \frac{\text{CS} \times 4}{\text{D}} = \frac{100 \times 4}{3} = \frac{400}{3} = 134 \text{ RPM}$$

Table 2 gives an FPT of .002 in.

The cutter has six teeth.

The feedrate is .002 × 6 × 134 = 1.608 IPM for the high speed steel cutter.

EXAMPLE.

For a carbide cutter the RPM is

$$\text{RPM} = \frac{\text{CS} \times 4}{\text{D}} = \frac{400 \times 4}{3} = \frac{1600}{3} = 534 \text{ RPM}$$

The FPT is .003 in.

Table 2
Feed in inches per tooth (instructional setting)

Type of Cutter	Aluminum HSS	Aluminum Carbide	Bronze HSS	Bronze Carbide	Cast Iron HSS	Cast Iron Carbide	Free Machining Steel HSS	Free Machining Steel Carbide	Alloy Steel HSS	Alloy Steel Carbide
Face mills	.007 to .022	.007 to .020	.005 to .014	.004 to .012	.004 to .016	.006 to .020	.003 to .012	.004 to .016	.002 to .008	.003 to .014
Helical mills	.006 to .018	.006 to .016	.003 to .011	.003 to .010	.004 to .013	.004 to .016	.002 to .010	.003 to .013	.002 to .007	.003 to .001
Side cutting mills	.004 to .013	.004 to .012	.003 to .008	.003 to .007	.002 to .009	.003 to .012	.002 to .007	.003 to .009	.001 to .005	.002 to .008
End mills	.003 to .011	.003 to .010	.003 to .007	.002 to .006	.002 to .008	.003 to .010	.001 to .006	.002 to .008	.001 to .004	.002 to .007
Form relieved cutters	.002 to .007	.002 to .006	.001 to .004	.001 to .004	.001 to .005	.002 to .006	.001 to .004	.002 to .005	.001 to .003	.001 to .004
Circular saws	.002 to .005	.002 to .005	.001 to .003	.001 to .003	.001 to .004	.002 to .006	.001 to .003	.001 to .004	.005 to .002	.001 to .004

The cutter has six teeth.

The feedrate is .003 × 6 × 534 = 9.612 IPM for the carbide cutter.

To calculate the starting feedrate, use the low figure from the feed per tooth chart and, if conditions permit, increase the feedrate from there. The most economical cutting takes place when the most cubic inches of metal per minute are removed and a long tool life is obtained. The tool life is longest when a low speed and high feed rate is used. Try to avoid feedrates of less than .001 in. per tooth, because this will cause rapid dulling of the cutter. Exceptions to this limit are small diameter end mills when used on harder materials. The depth and width of cut also affect the feedrate. Wide and deep cuts require a smaller feedrate than do shallow, narrow cuts. Roughing cuts are made to remove material rapidly. The depth of cut may be ⅛ in. or more, depending on the rigidity of the machine, the setup, and the horsepower available. Finishing cuts are made to produce precise dimensions and acceptable surface finishes. The depth of cut on a finishing cut should be between .015 and .030 in. A depth of cut of .005 in. or less will cause the cutter to rub instead of cut and also results in excessive cutting edge wear.

Cutting fluids should be used when machining most metals with high speed steel cutters. A cutting fluid cools the tool and the workpiece. It lubricates, which reduces friction between the tool face and chip, and reduces friction. Cutting fluids prevent rust and corrosion and, if applied in sufficient quantity, will flush away chips. Cutting fluids will, through these characteristics, increase production through higher speeds and produce better surface finishes. Most milling with carbide cutters is done dry unless a large constant flow of cutting fluid can be directed at the cutting edge. An interrupted coolant flow on a carbide tool causes thermal cracking and results in subsequent chipping of the tool.

self-evaluation

SELF-TEST

1. What is cutting speed?

2. Why should cuts be started at the low end of the cutting speed range?

3. If the cutting speed is 100 FPM with an HSS tool, what would it be with a carbide tool for the same material?

4. What is the effect of too low a cutting speed?

5. How is the feedrate expressed on a milling machine?

6. How is the feedrate calculated?

7. Why is a feed per tooth given instead of a feed per revolution?

8. What is the effect of too low a feedrate?

9. What RPM is used with a 3 in. diameter HSS cutter on low carbon steel?

10. What is the feedrate for a 4 in. diameter, five tooth carbide face mill machining alloy steel?

unit 6
workholding and locating devices on the milling machine

A very important factor in milling is the method used to hold a workpiece while it is being machined. Considerable ingenuity on the machine operator's part is required to select a workholding method suitable to the job.

objective After completing this unit, you will be able to select workholding devices for common milling machine jobs.

information

Large and irregularly shaped workpieces often are fastened directly to the machine table top. T-slots, which run lengthwise along the top of the table, are accurately machined and parallel to the sides of the table. These T-slots are used to retain the clamping bolts. Workpieces can also be aligned when snug fitting parallels are set into the T-slot and the workpiece is pushed against these parallels while the work is being clamped. Figure 1 shows the workpiece clamped to the table with T-slot bolts and clamps. The bolts are placed close to the workpiece and the block supporting the outer end of the clamp is the same height as the shoulder being clamped. Figure 2 illustrates a good clamping arrangement. When the bolt is closer to the work than to the clamp support block, maximum leverage is obtained. The support block should never be lower than the work being clamped.

When workpieces with finished or soft surfaces are clamped, care must be taken to protect those surfaces from damage by clamping. A shim

Figure 1. Work clamped to the table with T-slot bolts and clamps (Courtesy of Cincinnati Milacron).

should be placed between the work surface and the clamp (Figure 3). Before placing rough castings or weldments on a machine table, protect the table surface with a shim (Figure 4). This

Unit 6 Workholding and Locating Devices on the Milling Machine 111

Figure 2. Clamping bolt close to the work gives effective clamping (Lane Community College).

Figure 3. Highly finished surfaces should be protected from clamping damage (Lane Community College).

Figure 4. Protect the machine table surface from rough workpieces (Lane Community College).

Figure 5. Workpiece supported under the clamp (Lane Community College).

shim can be paper, sheet metal, or even plywood, depending on the accuracy of the machining to be performed.

A workpiece should have a support directly underneath where a clamp exerts pressure (Figure 5). Clamping an unsupported workpiece may cause it to bend or spring, and it will bend back after clamping pressure is released. If the workpiece material is brittle, clamping pressure may break it.

Workpieces tend to move on the table from the cutting pressure against them. This movement can be prevented by clamping a stop block on the table and placing the workpiece against it

Figure 6. Stop block prevents work slippage (Lane Community College).

Figure 7. Examples of screw jacks (Courtesy of Cincinnati Milacron).

Figure 8. Work set up and clamped on table (Courtesy of Cincinnati Milacron).

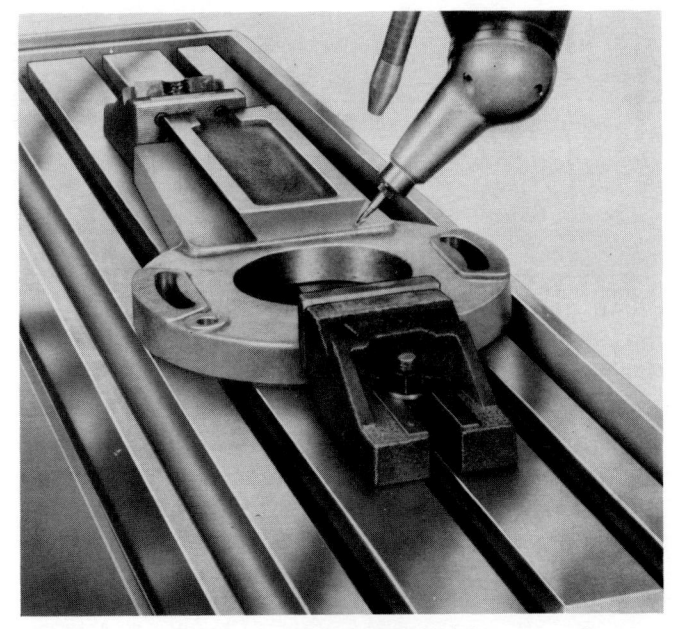

Figure 9. Quick action jaws holding workpiece (Courtesy of Cincinnati Milacron).

(Figure 6). Different kinds of supports are used in clamping work. Figure 7 shows an assortment of screw jacks that can be raised or lowered to any height and then are locked in that position. Often solid blocks or combination of blocks are used to give the correct height of supports (Figure 8).

Another method of holding work on the table is with quick action jaws (Figure 9). These individual jaws can be located anywhere on the machine table. They are tightened by turning screws that move the jaws outward and also give a downward pull on the workpiece.

Probably the most common method of workholding on a milling machine is a vise. Vises are simple to operate and can quickly be adjusted to the size of the workpiece. A vise should be used to hold work with parallel sides if it is within the size limits of the vise, because it is the fastest and

Figure 10. Plain vise (Courtesy of Cincinnati Milacron).

Figure 11. Swivel vise (Courtesy of Cincinnati Milacron).

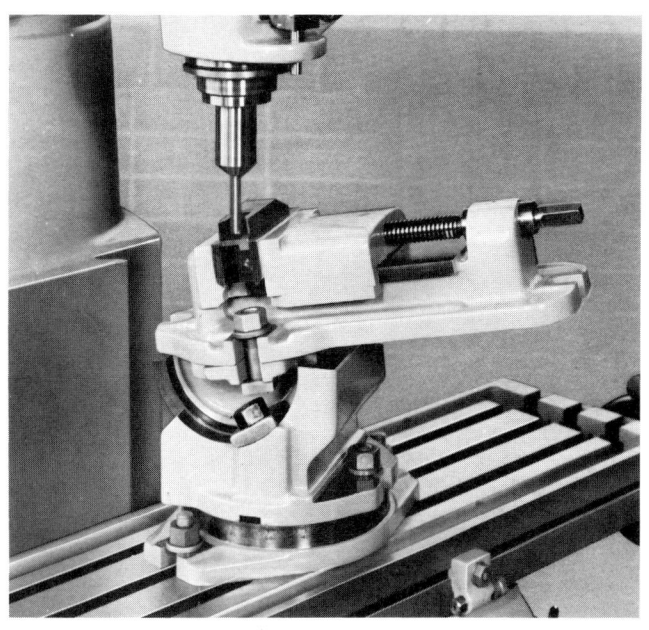

Figure 12. Universal vise (Courtesy of Cincinnati Milacron).

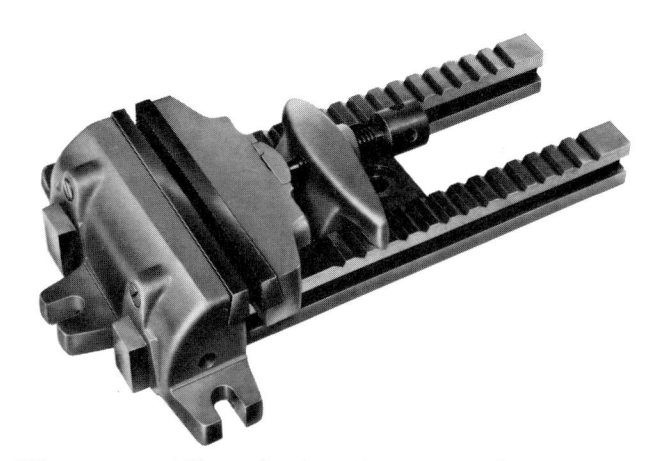

Figure 13. All-steel vise (Courtesy of Cincinnati Milacron).

most economical workholding method. The plain vise (Figure 10) is bolted to the machine table. Alignment with the table is provided by two slots at right angles to each other on the underside of the vise. These slots are fitted with removable keys that align the vise with the table T-slots either lengthwise or crosswise. A plain vise can be converted to a swivel vise (Figure 11) by mounting it on a swivel plate. The swivel plate is graduated in degrees. This allows the upper section to be swiveled to any angle in the horizontal plane. When swivel bases are added to a plain vise, the versatility increases, but the rigidity is lessened.

For work involving compound angles, a universal vise (Figure 12) is used. This vise can be swiveled 90 degrees in the vertical plane and 360 degrees in the horizontal plane.

The strongest setup is the one where the workpiece is clamped close to the table surface. Castings, forgings, or other rough workpieces can be securely fastened in an all-steel vise (Figure 13). The movable jaw can be set in any

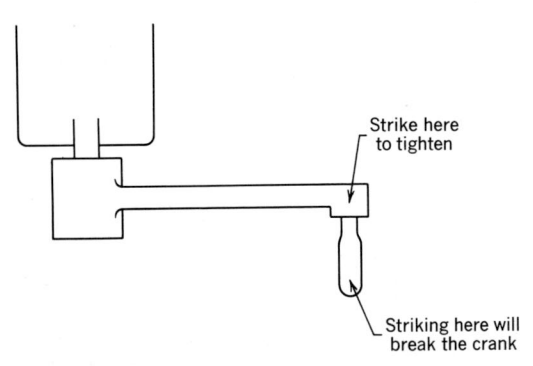

Figure 14. Tightening a vise (Courtesy of Cincinnati Milacron).

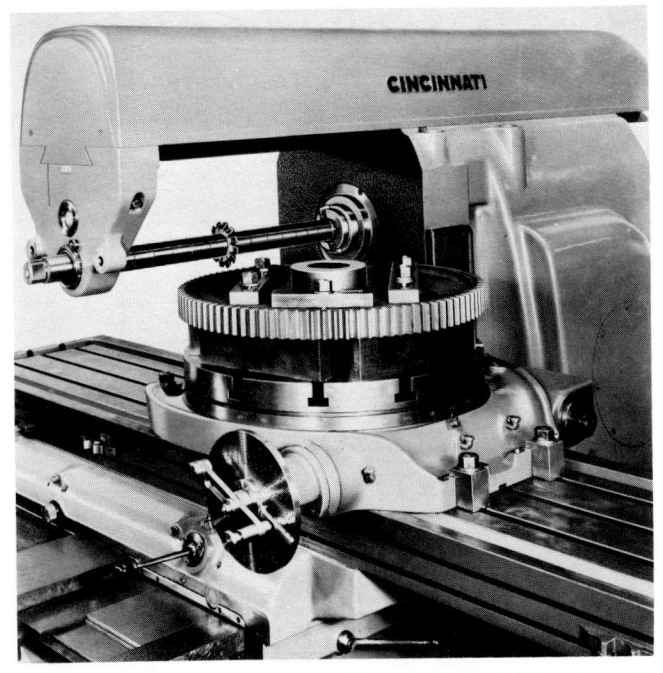

Figure 15. Rotary table (Courtesy of Cincinnati Milacron).

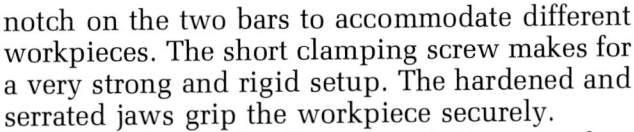

Figure 16. Dividing head and foot stock (Courtesy of Cincinnati Milacron).

Figure 17. Dividing head used to drill equally spaced holes (Courtesy of Cincinnati Milacron).

notch on the two bars to accommodate different workpieces. The short clamping screw makes for a very strong and rigid setup. The hardened and serrated jaws grip the workpiece securely.

Air or hydraulically operated vises are often used in production work, but in general toolroom work, vises are opened and closed by cranks or levers. To hold workpieces securely without slipping under high cutting forces, a vise must be tightened by striking the crank with a lead hammer (Figure 14).

A rotary table or circular milling attachment (Figure 15) is used to provide rotary movement to a workpiece. The rotary table can be used for angular indexing, milling circular grooves, or to cut radii. The rotary table is shown in Figure 15 in a gear cutting operation.

The dividing head (Figure 16) is used to divide the circumference of a workpiece into any number of equally spaced divisions. Work is held between centers, in collets, or in a chuck. The supporting member opposite the dividing head is the foot stock. The dividing head can be swiveled from below a horizontal line to beyond the vertical. The dividing head can also be used to drill equally spaced holes in workpieces held

Figure 18. Round shaft being held in vee-blocks (Lane Community College).

Figure 19. Milling fixture used for many identical parts (Courtesy of Cincinnati Milacron).

in a chuck (Figure 17). Round workpieces can be securely fastened in a set of vee blocks (Figure 18). To prevent the shaft from bending under cutting pressure, a screw jack such as those shown in Figure 7 can be used to support the shaft halfway between the vee blocks. If a number of identical workpieces are to be machined, a milling fixture (Figure 19) may be the most efficient way of holding them. A fixture is used when the savings resulting from its use are greater than the cost of making the fixture.

self-evaluation

SELF-TEST

1. In relationship to the workpiece, where should the clamping bolt be located?

2. What precautions should be taken when clamping finished surfaces?

3. When are screw jacks used?

4. What is the reason for clamping a stop block to the table?

5. What are quick action jaws?

6. What is the difference between a swivel vise and a universal vise?

7. When is an all-steel vise used?

8. When is a rotary table used?

9. When is a dividing head used?

10. When is a fixture used?

unit 7
plain milling on the horizontal milling machine

Plain milling is the operation of milling a flat surface in a plane parallel to the cutter axis. It involves the selection of a workholding device, milling cutter, speed, feed, and depth of cut.

objectives After completing this unit, you will be able to:
1. Align workholding devices.
2. Mill flat surfaces to size.
3. Mill surfaces square to each other.

information

Preparing a machine tool prior to machining is called setting up the machine. Before a setup can be made, the machine should be cleaned, especially all sliding surfaces such as the ways and the machine table. After wiping the table clean, use your hand to feel for nicks or burrs. If you find any, use a honing stone to remove them. Workpieces must be fastened securely for the machining operation. They can be held in a vise, clamped to an angle plate, or clamped directly to the table. Odd shaped workpieces may be held in a fixture designed for that purpose. On a universal milling machine it is good practice to check the alignment of the table before mounting a vise or fixture on it.

TABLE ALIGNMENT ON A UNIVERSAL MILLING MACHINE

1. Clean the face of the column and the machine table.
2. Fasten a dial indicator to the table with a magnetic base or other mounting device (Figure 1).
3. Preload the indicator to approximately ½ revolution of its dial and set the bezel to zero.
4. Move the table longitudinally with the hand wheel to indicate the column.
5. If the indicator hand moves, loosen the locking bolts on the swivel table and adjust the table one half the distance of the indicated difference.

Figure 1. Aligning the universal milling machine table (Lane Community College).

Figure 2. Fixture alignment keys (Lane Community College).

6. Tighten the locking bolts and reindicate the column; make another adjustment if needed.

Never indicate the table with the indicator mounted on the column, as this would always show alignment.

A good machine vise is an accurate and dependable workholding device. When milling only the top of a workpiece, it is not necessary that the vise be square to the column or parallel to the table travel. When the job requires that the outside surface is parallel to a step or groove in the workpiece, however, the vise has to be precisely aligned and positioned on the table.

The base of the vise should be located with keys (Figure 2) that fit snugly into the T-slots on the milling machine table. This normally positions the solid jaw of the vise parallel with or square to the face of the column. Before mounting a vise or other fixture on a machine table, inspect the base carefully for small chips and nicks and remove any that you find. When the base is clean, fasten the vise to the table. Whenever possible, position the vise so that the cutting pressure will be against the solid jaw (Figure 3). Often references are made to the "solid jaw" of a vise. The solid jaw will not move or change when the vise is tightened, although the movable jaw will align itself to some degree with the work and should never be indicated. A

Figure 3. Cutting pressure against solid jaw (Lane Community College).

number of different methods of aligning a vise on a table are shown below.

ALIGNING A VISE PARALLEL WITH THE TABLE TRAVEL

1. Fasten a dial indicator with a magnetic base to the arbor (Figure 4) and preload indicator contact point to one half revolution of the dial. Set bezel to zero.
2. Move the table so that the indicator slides along the solid jaw. Record any indicator movement.
3. Loosen the holddown bolts and lightly retighten. Lightly tap the vise with a lead or soft faced hammer to move the vise one half the distance of the indicator movement. Be sure the solid jaw moves away from the indicator contact point; tapping the jaw against the indicator may damage the indicator movement.
4. Tighten the holddown bolts securely and reindicate. Often the tightening of bolts or nuts will again move the vise.

ALIGNING A VISE AT A RIGHT ANGLE TO THE TABLE TRAVEL

1. Fasten a dial indicator with a magnetic base to the arbor (Figure 5) and preload the indicator.
2. Move the table with the cross feed hand wheel and indicate the solid jaw.
3. Loosen the vise holddown bolts and make any necessary correction.

Figure 4. Aligning vise parallel to table (Lane Community College).

Figure 5. Aligning vise square to table travel (Lane Community College).

4. Indicate the solid jaw again to check the alignment. *Always* take another indicator reading after securely tightening the clamping bolts. Often the final tightening will move a vise, fixture, or workpiece.

If no indicator is available to align a vise on a table, a combination square may be used, as shown in Figure 6. The beam of the square is slid along the machined surface of the column until contact is made with the solid jaw of the vise. Two strips of paper used as feeler gages help in locating the contact point. A soft headed hammer or lead hammer is used to tap the vise into position.

ALIGNING A VISE AT AN ANGLE OTHER THAN 90 DEGREES TO THE TABLE TRAVEL

Occasionally a vise has to be mounted on the table at an angle other than square to the table travel. This can be done with a protractor (Figure 7). Paper strips are used as feeler gages, the angular setting being correct when both strips contact the protractor blade and the vise jaw at the same time. This is not a precise method of setting an angle because of the limitations in setting an angle accurately with a protractor, maintaining the level of the protractor blade, and accurately sampling the "drag" on the paper strips.

Before machining a workpiece to size on a milling machine, several important decisions need to be made. One consideration is how to hold the workpiece while it is being machined. Large workpieces can be clamped directly to the table (Figure 8). A bar bolted to the table behind the workpiece is a safety stop that prevents the workpiece from moving when the cutting pressure is against it. Many workpieces can be held securely in a machine vise (Figure 9). If the workpiece is high enough, seat it on the bottom of the vise. If it is not, use parallels to raise it. Remember that friction between the vise jaws and the workpiece holds the workpiece. The more contact area there is the better.

SELECTING THE CUTTER

For flat surfaces use a plain milling cutter that is wider than the surface to be machined. The diameter of the milling cutter should be as small as practical. Figure 10 illustrates the difference

Figure 7. Using a protractor to align a vise on table (Lane Community College).

Figure 6. Using a square to align vise on table (Lane Community College).

Figure 8. Workpiece clamped to table (Lane Community College).

Figure 9. Workpiece held in vise (Lane Community College).

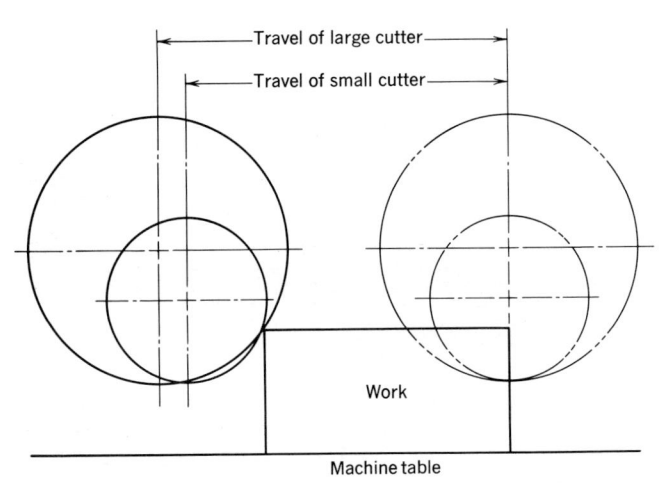

Figure 10. Different travel distances between different diameters of cutters.

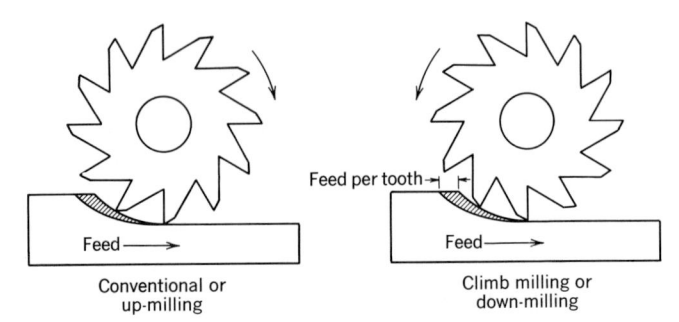

Figure 11. Conventional and climb milling.

in distance that a small diameter and a larger diameter cutter travel when machining the same length workpiece. A smaller diameter cutter is more efficient because it uses less machining time. Whatever the diameter is of the cutter used, it is important that sufficient clearance remains between the arbor support and the vise (Figure 9). Each cut taken brings the two parts closer together and, if much material is to be removed, it becomes necessary to reset the workpiece on higher parallels to avoid a collision between vise and arbor support.

Figure 11 shows one cutter operated in a conventional milling mode and the other cutter in a climb milling operation. In conventional milling modes (sometimes called up-milling), the workpiece is forced against the cutter with the teeth of the cutter trying to lift the workpiece up, especially at the beginning of a cut. In climbing (or down-milling), the cutter tends to hold the workpiece down.

Climb milling should only be performed on machines equipped with an antibacklash device. Backlash, which is play between the table drive screw and nut assembly, would let the cutter pull the workpiece under it, break the cutter, and ruin the workpiece. Remember that every cutter can be operated in an up-milling or down-milling fashion. The only difference is from which side of the workpiece the cut is started. The cutter should be located on the arbor as close to the spindlenose as the location of the workpiece permits. The cross travel of the saddle is limited. The arbor support should also be as close to the cutter as possible for a rigid setup.

Prior to assembly of the spacing collars and cutter on the arbor, all pieces should be cleaned. Keys should always be used to drive the cutter. Do not depend on the friction between the spacers and cutter. The drive keys should extend into the spacing collars on both sides of the cutter. Tighten and loosen the arbor locknut only when the arbor support is in place, and do not use a hammer on the wrench. Use only a sharp cutter to minimize cutting pressures and to get a good surface finish. Resharpen a cutter when it becomes slightly dull. A slightly dull cutter can be resharpened easily and quickly.

SETTING UP THE MACHINE

After the cutter for the job has been selected, the speed and feed can be calculated and set on the machine. The depth of cut depends largely on the amount of material that is to be removed.

Good milling practice is to take a roughing cut and then a finish cut. Better surface finishes and higher dimensional accuracy are achieved when roughing and finishing cuts are made. The depth of the roughing cut often is limited by the horsepower of the machine or the rigidity of the setup. A good starting point for roughing is .100 to .200 in. deep. The finishing cut should be .015 to .030 in. deep. Depth of cut less than .015 in. deep should be avoided, because a milling cutter, especially in conventional or up-milling, has a strong rubbing action before the cutter actually starts cutting. This rubbing action causes a cutter to dull rapidly. Assuming that a cut .100 in. deep has to be taken, the following steps outline the procedure to be used.

1. Loosen the knee locking clamp and the cross slide lock.
2. Turn on the spindle and check its rotation.
3. Position the table so that the workpiece is under the cutter.
4. Raise the knee slowly by turning the vertical hand feed crank until the cutter just touches the workpiece. If the cutter cuts a groove, you have gone too far and should try again on a different place on the workpiece.
5. Set the micrometer dial on the knee feedscrew on zero.
6. Lower the knee by approximately one half revolution of the hand feed crank. If the knee is not lowered, the cutter will leave tool marks on the workpiece in the following operation.
7. Move the table longitudinally until the cutter is clear of the workpiece. Move to the side of the workpiece, which will result in making a conventional milling cut.
8. Raise the knee past the zero mark to the 100 mark on the micrometer dial.
9. Tighten the knee lock and the cross slide lock. *Always* prior to starting a machining operation, tighten all locking clamps except the one that would restrict table movement while cutting. This aids in making a rigid chatterfree setup.
10. The machine is now ready for the cut. Turn on the coolant (Figure 12). Move the table slowly into the revolving cutter until the full depth of cut is obtained before engaging the power feed.
11. When the cut is completed, disengage the power feed, stop the spindle rotation, and turn off the coolant before returning the table to its starting position. If the revolving cutter is returned over the newly machined surface, it will leave cutter marks and mar the finish. To prevent this, lower the table before returning the cutter to the starting point.
12. After brushing off the chips and wiping the workpiece clean, the workpiece should be measured while it is still fastened in the machine. If the workpiece is parallel at this time, additional cuts can be made if more material needs to be removed.

When machining rough castings or forgings, some thought has to be given to the setup of the workpiece, especially if the stock to be removed is limited. Figure 13 is an example of a bar that should be square but, instead, it is a parallelogram. If all the material were to be removed from sides 1 and 4 to make these sides square to sides 2 and 3, the block would be undersize between

Figure 12. Coolant cools the cutter and washes away chips (Lane Community College).

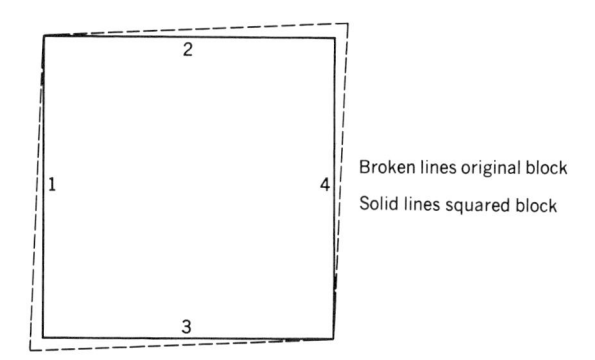

Broken lines original block
Solid lines squared block

Figure 13. Machining a square from a parallelogram.

sides 1 and 4. If the block were to be shimmed so that some material would be removed from all four sides (the dotted lines in Figure 13), the resulting square would be of a larger size.

self-evaluation

SELF-TEST

1. Why is the solid jaw used to align a vise on a milling machine table?

2. Which is more accurate, using a precision square or a dial indicator to square a vise on a machine table?

3. What is the purpose of the keys used on the base of machine vises?

4. Should you mount the indicator on the column to align the table on a universal milling machine?

5. Should the solid vise jaw be in a specific position in relation to the direction of the cut?

6. Is a large or small diameter cutter more efficient?

7. What is the difference between conventional milling and climb milling?

8. How deep should a finish cut be?

9. Why are all table movements locked except the one being used during machining?

10. Why is the cutter rotation stopped while the table is returned over the newly cut surface to its starting position?

WORKSHEET
Machining a Rectangular Block (Figure 14)

1. Saw off material that is between $\frac{1}{8}$ and $\frac{1}{4}$ in. larger than the finished size of the workpiece.

2. Fasten the workpiece centrally in the clean vise and tap the vise screw handle with a soft headed hammer. Tightening the vise usually raises

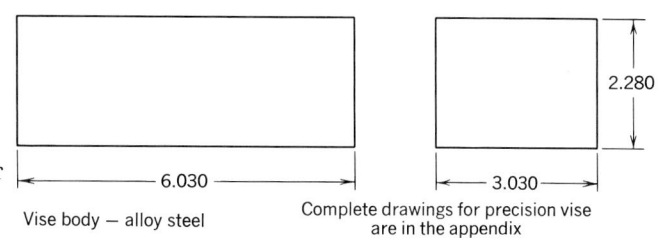

Figure 14. Machining a rectangular block.

Vise body — alloy steel: 6.030 × 2.280, 3.030
Complete drawings for precision vise are in the appendix

the workpiece slightly, so tap the workpiece with a soft headed (preferably lead) hammer to reseat it on the bottom of the vise.

3. Select a sharp plain milling cutter that is wider than the workpiece and put it on the arbor, taking care to have the cutter as near the spindle as possible and the arbor support as close to the cutter as feasible. Make sure the arbor key extends into the arbor spacers on each side of the cutter. Tighten the arbor nut after mounting the arbor support on the overarm.

4. Select and set the cutter speed. What RPM will you use?

$$\text{RPM} = \frac{\text{CS} \times 4}{\text{D}} = \underline{\qquad}$$

5. Check that the spindle rotation is correct.

6. Calculate the feedrate to use and set it on the machine. Feed rate equals feed per tooth × number of teeth × RPM. What is the feedrate?

7. Set the depth of cut. Arrange your cutting plan to remove about equal amounts of material from opposite sides of the workpiece. Tighten all movement locking clamps except the one that would restrict table movement while cutting.

8. Turn on the coolant.

9. Start the cut by turning the table hand wheel; then engage the power feed. Observe the cutting operation closely so that you can turn off the power feed at the first sign of trouble. Do not remove chips or reach into the cutting area while the cutter is revolving.

10. When the cut is finished, disengage the power feed, stop the spindle, and return the table to its starting position.

11. Remove the chips with a brush and remove all burrs with a file; then remove the workpiece from the vise. Side 1 is now finished.

12. Clean the vise and workpiece and reposition the workpiece as shown in Figure 15, on parallels, so that about $\frac{1}{2}$ in. extends out from the vise. Inserting a small rod ($\frac{1}{4}$ in. diameter) between the moving jaw and the workpiece assures positive contact between side 1 and the solid jaw of the vise. The workpiece may not touch on both parallels, even when it is tapped down with a soft headed hammer, because the sides are not square to each other.

13. Set the depth of cut for side 2. Take only one half of the material that is to be removed with this cut. This side will have to be remachined if later measuring shows it not to be square to side 1.

14. Take this cut repeating procedures from steps 9 to 11.

Figure 15. Setup to machine side 2 (Lane Community College).

Figure 16. Setup to machine side 3 (Lane Community College).

15. Fasten the deburred workpiece in a clean vise (Figure 16). Side 1 is again against the solid jaw. Side 2 is toward the bottom of the vise. Side 3 is now to be machined.
16. The depth of cut for side 3 should be clean up cut as was side 2.
17. Repeat steps 9 to 11.
18. With a micrometer, measure dimensions "A" and "B" (Figure 17). If dimension "A" is equal to dimension "B," or within less than .004 in., the solid jaw is square to the base and you can skip step 19 and go to step 20. If the difference is more than .004 in., go to step 19.
19. If dimension "A" is larger than "B", use a thin shim and place it between the work and solid jaw at point "C" (Figure 18). With the shim in place and the workpiece securely fastened, take a cut .020 in. deep on

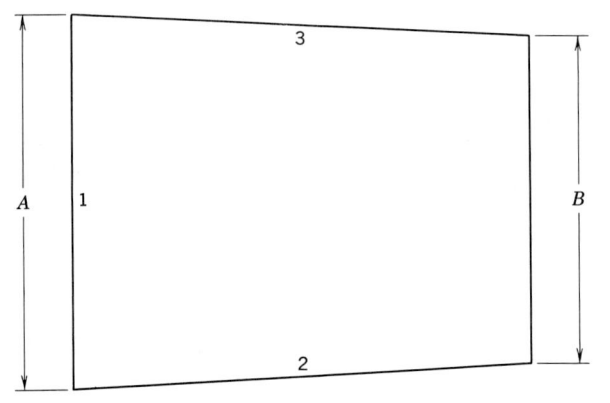

Figure 17. Measuring for squareness.

Figure 18. Location of shim to square up work (Lane Community College).

side 2 and also on side 3. Measure dimensions "A" and "B" again to see if the shim corrected the prior difference. If necessary, repeat this operation with different size shims until the workpiece is parallel between sides 2 and 3.

20. Sides 1, 2, and 3 are square to each other, and only side 4 is yet to be machined (Figure 19). Set the workpiece as deep in the vise as practical. Use parallels to raise it above the vise if that is necessary. Because sides 2 and 3 are parallel, no rod is needed between the work and moving jaw. Set the depth of cut and repeat steps 9 to 11.

21. The workpiece is now square and parallel on four sides. Measure the dimensions of these sides and make additional cuts if more material has to be removed. In squaring a workpiece, do not remove all the excess material with the first cut, because additional corrective cuts often have to be made.

22. The ends of the workpiece are now machined square to the sides. Figure 20 shows how a relatively short workpiece can be set up in a vise to machine an end square to the sides. An accurate square is used to check the squareness of the workpiece in relation to the base of the vise. The vise clamps the workpiece only very lightly so that the workpiece can be aligned with the square by tapping it. Double-check the squareness by applying the square to the opposite side of the work, also. The workpiece should always be in the vise. If that cannot be done, use a

Figure 19. Setup to machine side 4 (Lane Community College).

Figure 20. Setup of a workpiece to machine an end square (Lane Community College).

Figure 21. Work clamped off center needs a spacer (Lane Community College).

Figure 22. Use of angle plate to mill ends of workpieces (Lane Community College).

spacer of equal thickness to get the vise jaws to close parallel to each other (Figure 21). The ends of a workpiece should be machined with the solid jaw parallel to the arbor axis and with the cutting pressure against the solid jaw. If the vise jaws are parallel to the table travel, there is the danger that the cutting pressure will push the workpiece out of the vise. The ends of workpieces can be safely machined when the work is clamped against an angle plate (Figure 22). The squareness of a workpiece in relation to the table surface can also be measured with a dial indicator. Fasten the dial indicator to the arbor with contact point touching the vertical work surface to be measured, then raise or lower the knee and adjust the workpiece until the indicator hand shows the workpiece to be vertical. Always test the workpiece again for squareness after the final tightening of the vise or other holding device, because the workpiece alignment is often disturbed. When both ends are to be squared, remove only enough material from the first side to clean up that end. This will leave adequate stock to machine the second side square and to the desired overall length.

This completes this milling exercise. Clean the machine and return all tools and attachments to their places.

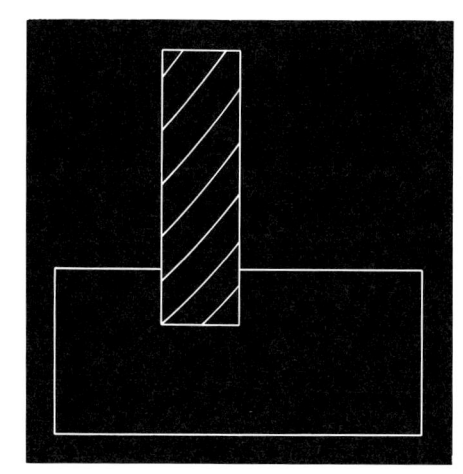

unit 8
using side milling cutters on the horizontal milling machine

Some milling cutters are used to mill only flat surfaces; others can be used to machine grooves and steps. Side milling cutters fall into this second category. A milling machine operator often makes the selection of the cutter, feed, speed, and workholding device to perform side milling operations.

objectives After completing this unit, you will be able to:
1. Set up side milling cutters and cut steps and grooves.
2. Use side milling cutters for straddle milling.
3. Use side milling cutters in gang milling.

information

Milling cutters with side cutting teeth are used when grooves and steps have to be machined on a workpiece. The size and kind of cutter to be used depends largely on the operation to be performed. Full side mills with cutting teeth on both sides are used when slots or grooves are cut (Figure 1).

Use only sharp cutters; they will leave a better surface finish and use less power. Cutters should be resharpened when they are only slightly dull. At this point, resharpening takes very little time. Cutters that are used until the cutting edges are worn down need an expensive reconditioning operation.

Cutters usually make a slot that is slightly wider than the nominal width of the cutter. Cutters that wobble because of dirt or chips between the arbor spacers will cut slots that are considerably oversize. A slot will become wider when more than one cut is made through it. If a slot needs to be .375 in. wide, a $\frac{3}{8}$ in. cutter probably cannot be used because it may cut a slot .3755 to .376 in. wide. A trial cut in a piece of scrap metal will tell the exact slot width. It may be necessary to use a $\frac{5}{16}$ in. wide cutter and make two or more cuts.

The width of a slot from a given cutter is also affected by the amount of feed used. A very slow feed will tend to let the cutter cut more clearance for itself. A fast feed crowds the cutter in the slot. A cutter tends to cut a wider slot in soft material than it will in harder material. The width of slots are measured with vernier calipers or with adjustable parallels. If it is a keyway, the key itself can be used as a plug gage to test the slot width.

Half side mills with cutting teeth on one side and on the periphery can be used when a step is milled where the cutter is in contact with two

Figure 1. Full side milling cutter machining a groove (Lane Community College).

Figure 2. Half side milling cutter machining a step (Lane Community College).

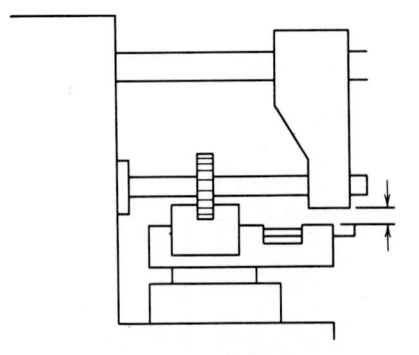

Figure 3. Check for clearance.

Figure 4. Work laid out for milling (Lane Community College).

sides only (Figure 2). The diameter of the cutter to be used on a job depends on the depth of the slot or step. As a rule, the smallest diameter cutter that will do the job should be used, as long as sufficient clearance remains between arbor and work and between arbor support and vise (Figure 3).

A good machinist will mark his workpiece with layout lines before he fastens it in a vise. The layout should be an exact outline of the part to be machined. The reason for making the layout prior to machining is that reference surfaces are often removed by machining. After the layout has been made, make diagonal lines on the portion to be machined away. This helps in identifying on which side of the layout line the cut is to be made (Figure 4). The cutter can be accurately positioned on the workpiece with the hand feed cranks and the micrometer dials. When the finished outside surface of a workpiece must not be marred or scratched by a revolving cutter, a paper strip is held between the workpiece and the cutter (Figure 5). The power is turned off and the spindle is rotated by hand. Make sure the paper strip is long enough so your hands are not near the cutter. Carefully move the table toward the revolving cutter. When the cutter pulls the paper strip from your fingers, the cutter is about .002 in. from the workpiece. At this time set the cross feed dial on zero. Lower the knee until the cutter clears the top of the workpiece. Then, by using the cross feed hand wheel, position the work where the cut is to be made. The same method will work in positioning a cutter above a workpiece and establishing the zero position for the depth of cut without actually touching the workpiece with the cutter (Figure 6).

A quicker, but not as accurate, method is illustrated in Figure 7. A steel rule is used to position a side mill a given distance from the outside edge of a workpiece. The end of the rule is held firmly against the side cutting edge of a tooth. The distance is indicated by the edge of the workpiece. The micrometer dials of the cross feed and knee controls should be zeroed when the cutter contacts the side and top of the workpiece. When these zero positions are established, additional cuts can easily be made by positioning from these points with the micrometer dials.

After the first cut is made, the distance of the side of the step or groove to the outside of the workpiece should be measured. A measurement made on both ends shows if the cut is parallel to the sides of the workpiece. If the step is not parallel, the vise needs to be aligned or, on a universal milling machine, the table may need aligning.

Figure 6. Using a paper strip to set the depth of cut (Lane Community College).

Figure 5. Setting up a cutter by using a paper strip (Lane Community College).

Figure 7. Positioning a cutter using a steel rule (Lane Community College).

When the table direction is reversed, compensation for backlash needs to be made. Backlash can be observed when the feed screw is turned, but the table does not start moving until all play between the drivenuts and feedscrew is removed. When possible, measurements should be made while the workpiece is still fastened in the milling machine because additional cuts can then be made without additional setup work. No measurements should be made while the cutter is revolving. Carefully remove all burrs from steps and grooves with a file prior to measuring (Figure 8).

Before making a final cut, if you are not sure of your dimensions, advance the revolving cutter until the cutter just nicks the corner of the workpiece (Figure 9). Stop the spindle and make a measurement at point "X" with a micrometer. If the location is correct, finish the cut. If the dimension is wrong, adjust the table accordingly. A small nick left on a corner is often covered up when the workpiece is chamfered.

Side milling cutters are combined to perform straddle milling operations (Figure 10). The width of the spacers between the cutters controls the width of the workpiece. It is important that the diameters of the cutters are the same if steps of equal depth are to be produced. In gang milling, a number of milling cutters are combined to cut special shapes and contours (Figure 11). The depth of the steps is determined by the difference in diameter of the various cutters. The RPM of the spindle is calculated for the largest diameter cutter in the gang.

Figure 9. Taking a trial cut (Lane Community College).

Figure 8. Measuring depth of a step (Lane Community College).

Figure 10. Straddle milling (Lane Community College).

Figure 11. Gang milling (Courtesy of Cincinnati Milacron).

Figure 12. Left-hand and right-hand helical flutes on wide cuts (Courtesy of Cincinnati Milacron).

When wide cuts are made, interlocking tooth cutters with right-hand and left-hand helical flutes are used to offset the heavy side thrust (Figure 12).

Interlocking side milling cutters are used when grooves of a precise width are machined in one operation (Figure 13). Shims inserted between the individual cutters make precise adjustments possible. The overlapping teeth leave a smooth bottom groove. Cutters, which have become thinner through sharpening, can also be readjusted to their full width by adding shims.

Before any machining is started on a workpiece, you should have a plan of the sequence of operations that you will perform. This plan should include the answers to questions such as:

How is the workpiece held while it is being machined?

Do some machining operations come before others?

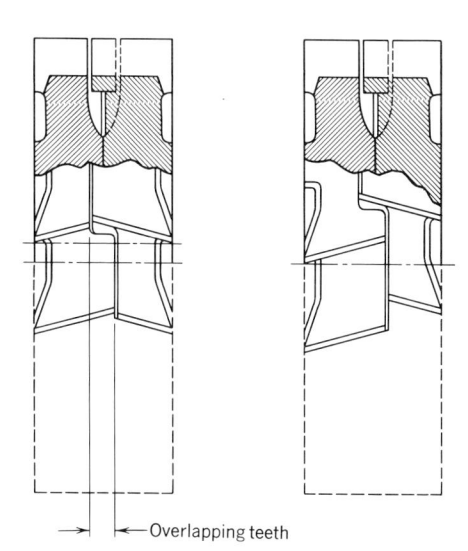

Figure 13. Interlocking side milling cutters.

Is the setup strong enough to withstand the cutting forces?

self-evaluation

SELF-TEST 1. When are full side milling cutters used?

Section C Horizontal Milling Machines

2. When are half side milling cutters used?
   _____

3. What diameter side milling cutter is most efficient?
   _____

4. Is a groove the same width as the cutter that produces it?
   _____

5. Why should a layout be made on workpieces?
   _____

6. How can a side milling cutter be positioned for a cut without marring the workpiece surface?
   _____

7. Why should measurements be made before removing a workpiece from the workholding device?
   _____

8. How is the width of a workpiece controlled in a straddle milling operation?
   _____

9. What determines the depth of the steps in gang milling?
   _____

10. When are interlocking side mills used?
    _____

WORKSHEET
Machining a Step and Grooves into a Block (Figure 14)
PROCEDURE

1. Use a steel block that has been squared and machined to length and make a layout of the grooves and the step to be machined.

2. Mount a vise on the milling machine table and align the solid jaw so it is square to the column face.

3. To machine the large step on the project, mount the workpiece in the vise. Support the workpiece on parallels so that about $1\frac{5}{16}$ in. extend above the vise top. The step is to be $1\frac{1}{4}$ in. deep; the extra $\frac{1}{16}$ in. is to clear the vise jaws.

4. Select a side milling cutter with a diameter large enough to make a cut $1\frac{1}{4}$ in. deep.

5. Mount the cutter on the arbor so that the cutting pressure will be against the solid jaw. Check for sufficient clearance between arbor support and vise. Remember the cut is $1\frac{1}{4}$ in. deep. Check the position of the cutter on the arbor by moving the table cross feed to see if the full width of the cut can be made.

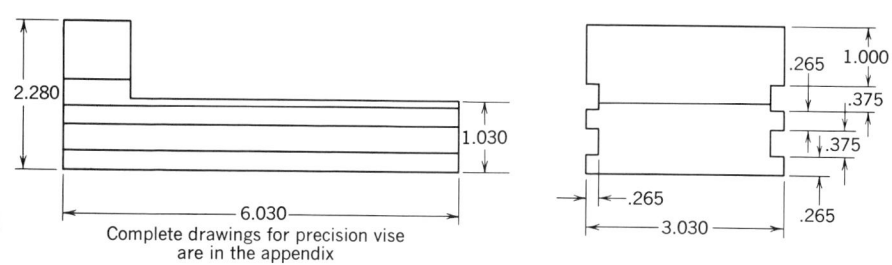

Figure 14. Machining a step and grooves into a block.

Complete drawings for precision vise are in the appendix

6. Select and set the cutter speed. What RPM will you use?
$$\text{RPM} = \frac{\text{CS} \times 4}{\text{D}} = \underline{\hspace{3cm}}.$$

7. Check that the spindle rotation is correct.

8. Calculate the feedrate to use and set it on the machine. Feedrate equals feed per tooth × number of teeth × RPM. What is the feedrate?

9. Set the depth of cut. A roughing cut and a finishing cut will be made. The roughing cut should be $\frac{11}{32}$ in. away from the finished size.

10. Move the table to set the width of the cut to $\frac{1}{2}$ in. The width of the cut should be less than the width of the cutter.

11. Tighten the knee and cross slide locking clamps.

12. Turn on the coolant.

13. Start the cut manually, then engage the power feed (Figure 15). Observe the cutting operation carefully and be prepared to disengage the power feed at the first sign of trouble.

14. When the cut is completed, return the table to its starting position.

15. Move the table into position to take another $\frac{1}{2}$ in. wide cut.

16. Repeat taking cuts until you are within $\frac{1}{4}$ in. of your layout lines. Then stop the spindle and measure the width of the remaining section.

Figure 15. Milling a step (Lane Community College).

Calculate the amount of oversize. Adjust the machine table by the amount of oversize using the cross feed micrometer collar.

17. Make this cut. Set the cross feed micrometer collar to zero. This is also the reference point for the finishing cut. This completes the roughing cut.

18. Return the machine table to the starting point for the first cut. Set the depth of cut for the finishing cut.

19. Make the finishing cuts in the same manner as the roughing cuts, except that you *stop the cutter* before returning it over the finished surface. Returning a revolving cutter over a surface that was just machined will leave cutter marks.

20. After completing the final cut, stop the machine. Remove the chips and deburr the workpiece. Measure the workpiece while it is still clamped in the vise. If another cut is needed, it should be made next. This completes the milling of the step. Remove the workpiece from the vise.

21. Turn the vise and align the solid jaw parallel with the table.

22. Clamp the workpiece with the long outside surface against the solid jaw of the vise (Figure 16).

23. Select a sharp side milling cutter to machine the grooves. Exchange the cutter on the arbor with the new cutter.

24. Position the cutter on the side of the workpiece by using a paper strip. Zero the cross feed micrometer collar.

25. Position the cutter over the groove location using the micrometer collar. The cutter should now be aligned with the layout lines.

26. Set the depth of cut. Only one cut is made to get the groove depth.

27. Calculate and set the speed for this cut.

28. Calculate and set the feedrate.

29. Turn on the coolant and start the cut manually, then engage the power feed.

Figure 16. Setup for milling grooves (Lane Community College).

30. At the completion of this cut, turn off the feed, coolant, and spindle, and return the table to its starting position.

31. Move the table the required distance to line up the cutter for the second groove. Use the micrometer adjustments of the machine to get accurate locations.

32. Cut the second groove.

33. Stop the machine, then return the table to its starting position.

34. Clean off the chips, deburr the just machined grooves, and remove the workpiece from the vise.

35. Clean the vise and workpiece. Reposition the workpiece in the vise with the grooved side down and with the same side against the solid jaw as for the last operation. Clamp the workpiece securely.

36. Positioning of the workpiece under the cutter should be done with the aid of the micrometer adjustments. The grooves on the top side should be exact duplicates of the grooves on the bottom side. Remember to remove the backlash that exists when you change the rotation of any of the feedscrews.

37. Cut both grooves.

38. Clean and deburr the workpiece and remove it from the vise. Clean the machine.

39. This completes the horizontal milling machine work on this workpiece.

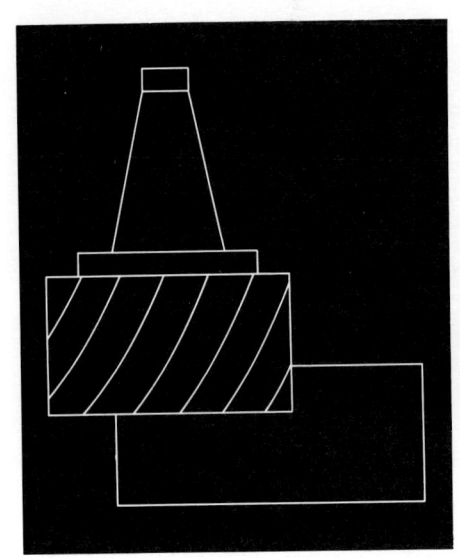

unit 9
using face milling cutters on the horizontal milling machine

Face milling cutters are used to machine flat surfaces at a right angle to the cutter axis. They are very efficient tools when large quantities of material are to be removed.

objectives After completing this unit, you will be able to:
1. Identify face milling cutters.
2. Use face milling cutters to machine flat surfaces.

information

Face milling cutters are the most commonly used milling cutters with inserted teeth. Face milling cutters up to 6 in. in diameter are called shell end mills. Face milling cutters are made with inserted teeth because it would be very difficult to make such large cutters in one piece, and because of the high cost of the cutting tool material; the cutter body can be used almost indefinitely and only the cutting inserts need to be exchanged. Figure 1 names the major parts of a face milling cutter and explains commonly used terms. Face milling cutters are usually mounted directly on the spindle nose and held in position by four capscrews. The back of the cutter is counterbored to fit closely over the outer diameter of the spindle nose. The driving keys of the spindle nose fit the keyways in the back of the face mill (Figure 2). Another method of mounting face mills is the flat back mount (Figure 3). It uses a centering plug located in the machine spindle taper to center the cutter. When a face mill is mounted on the machine spindle, it is very im-

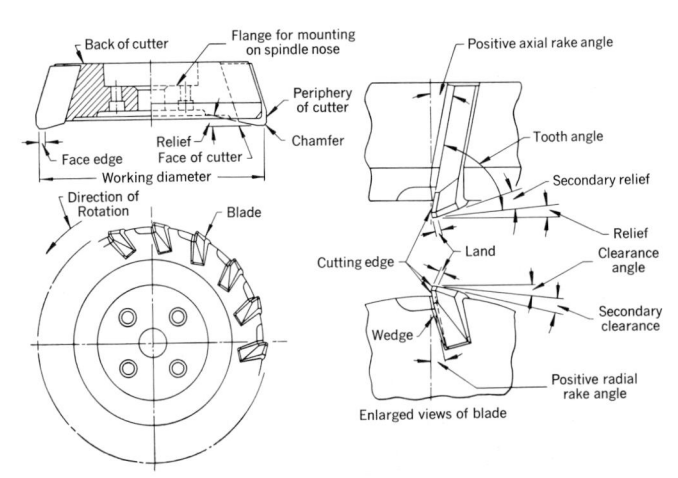

Figure 1. Identification of face milling cutter (Courtesy of Cincinnati Milacron).

portant that all mating surfaces between the cutter and spindle are clean and free of any nicks or burrs. Even the smallest particle between the mating surfaces will make the required align-

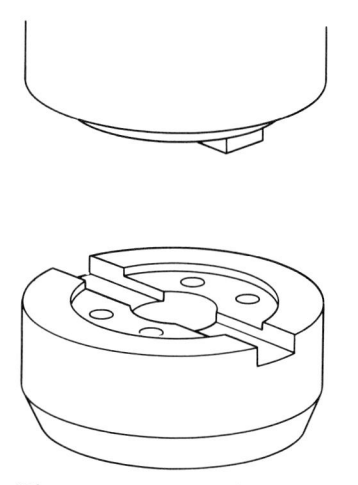

Figure 2. National standard drive (Photo courtesy of Kennametal Inc., Latrobe, Pa.).

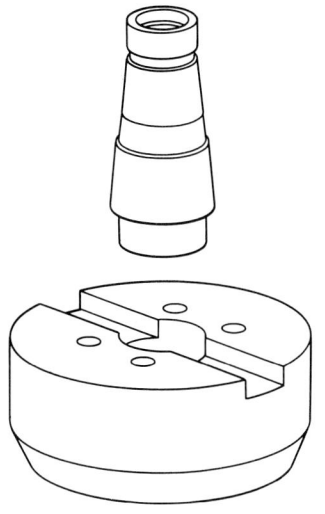

Figure 3. Flat back mount (Photo courtesy of Kennametal Inc., Latrobe, Pa.).

ment of the cutter impossible. After the cutter has been mounted on the machine, a check with a dial indicator should show the cutter to run true or to be out of alignment less than .001 in. Face mills are made as light duty cutters with a large number of teeth for finishing cuts or as heavy duty cutters with fewer teeth and heavier bodies for roughing operations.

The body of face mills is usually made from heat-treated alloy steel. The cutting inserts are high speed steel in older face mills, but new face mills are made with carbide inserts. Major differences in face mills appear in the rake angles of the cutting surfaces. High speed milling cutters are normally used with positive rake angles (Figure 4). Positive rake angles produce a good surface finish, increase cutter life, and use less power in cutting, both with HSS and carbide inserts. Positive rake angles are very effective in the machining of tough and work hardening materials. Cutting pressures, which may deflect thin-walled workpieces, are smaller with positive rake cutters than with negative rake cutters under the same conditions. Zero or negative rake cutters are used only with carbide inserts. These inserts are very strong and will give good service under heavy impact or interrupted cutting conditions (Figure 5). Negative rake inserts create high cutting pressures, which tend to force the workpiece away from the cutter. Negative rake inserts should not be used on work hardening materials or ductile materials such as aluminum or copper.

The lead angle, which is the angle of the cutting edge measured from the periphery of the

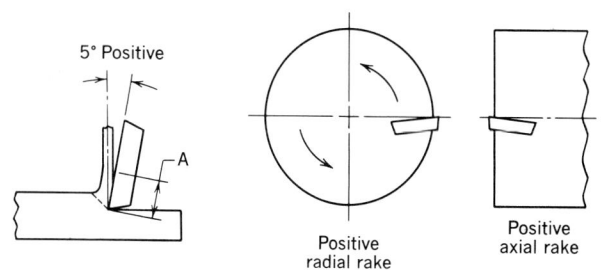

Figure 4. Positive rake angles (Photo courtesy of Kennametal Inc., Latrobe, Pa.).

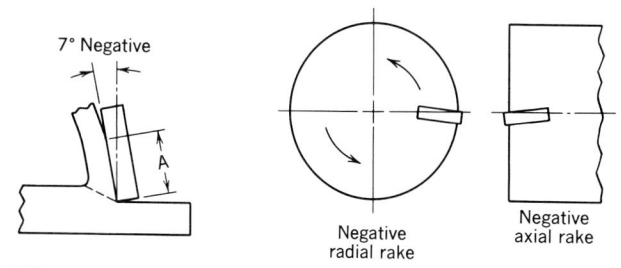

Figure 5. Negative rake angles (Photo courtesy of Kennametal Inc., Latrobe, Pa.).

cutter, varies from 0 to 45 degrees, depending on the application. Small lead angles of 1 to 3 degrees can be used to machine close to square shoulders. A small lead angle cutter, when used with a square insert, will have sufficient clearance on the face of the cutter to prevent its rubbing. Figure 6 shows the effect of a small lead

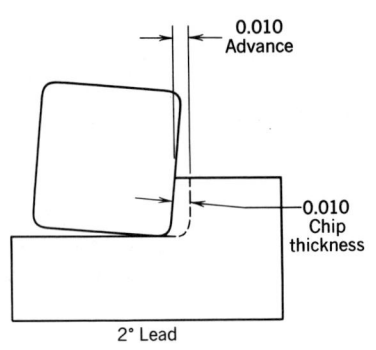

Figure 6. Small lead angle (Photo courtesy of Kennametal Inc., Latrobe, Pa.).

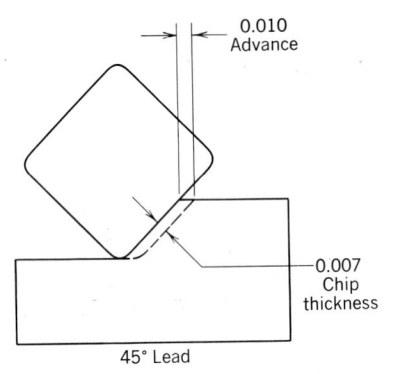

Figure 7. Large lead angle (Photo courtesy of Kennametal Inc., Latrobe, Pa.).

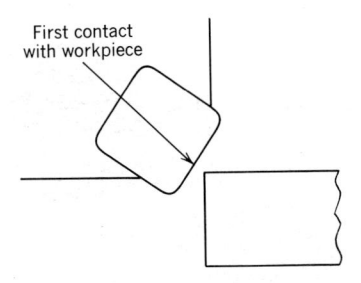

Figure 8. Lead angle effect.

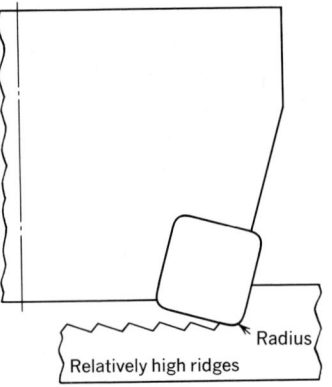

Figure 9. Feed lines on surface (Photo courtesy of Kennametal Inc., Latrobe, Pa.).

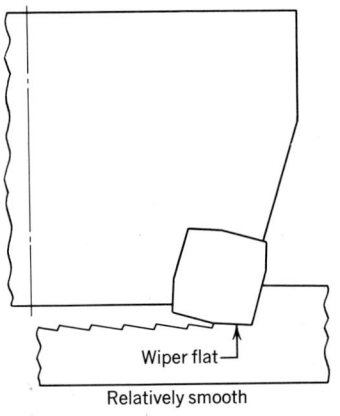

Figure 10. Effect of wiper flat (Photo courtesy of Kennametal Inc., Latrobe, Pa.).

angle on chip thickness. With a .010 in. feed per insert the chip is also .010" thick and as long as the depth of cut. Figure 7 uses the same feed and depth of cut, but the chip is quite different. The chip now is longer than the depth of cut and its thickness is only .007 in. A thinner chip gives an increased cutting edge life, but it does limit the effective depth of cut. The maximum practical lead angle is 30 degrees. Figure 8 shows another beneficial effect of a large lead angle; the cutter contacts the workpiece at a point away from the tip of the cutting edge. The cutter does not cut a full size chip on initial impact, but eases into the cut. The same thing happens on completion of the cut as the cutter eases out of the work.

The surface finish produced by a cutter depends largely on nose radius of the insert and the feed used. Figure 9 is an exaggerated view of the ridges left by the cutter. The finish can be improved by the use of an insert with a wiper flat (Figure 10). The flat should be wider than the feed being used.

The cutting action of face mills is also affected by its diameter in relation to the width of the workpiece. When it is practical, a cutter should be chosen that is larger than the width of cut. A good ratio is obtained when the cut is two thirds as wide as the cutter diameter. Cuts as wide as the cutter diameter should not be taken, because of the high friction and rubbing of the cutting edge before the material starts to shear. This rubbing action results in rapid cutting edge wear. Tool life can be extended when the cutting tool enters the workpiece at a negative angle (Figure 11). The work makes its initial tool contact away from the cutting tip at a point where the insert is stronger. When possible, arrange the width of cut to obtain the cutting action in Figure 11. Figure 12 illustrates a workpiece where the cutter enters at a positive angle. Contact takes place at the weakest point of the insert, and the cutting edge may chip. Figure 12 also shows a climb milling operation. In climb milling, the cutter enters the work material and produces the thick part of the chip first, and then thins out toward the finished surface. Climb milling should be performed when possible because it reduces the tool wear caused in conventional milling by the rubbing action of the tool against the work before the actual cutting starts.

Cutting fluid is important when a face mill with high speed steel inserts is used. A good cutting fluid absorbs heat generated in cutting and provides a lubricant between the chips and the cutting tool. In face milling, cutting fluid should be applied where the cutting is being done, and it should flood the cutting area. Cooling a carbide face mill is very difficult because of the intermittent nature of the cutting process and the fanning action of the revolving cutter. If the cooling of the carbide insert is intermittent, thermal cracking, which causes rapid tool failure, takes place. Unless a sufficient coolant flow at the cutting edge is maintained, it is better to machine dry. A mist-type coolant application is sometimes used to provide lubrication between the chip and cutting tool. This reduces friction and generates less heat.

To machine a flat surface with a face mill, it is necessary that the face of the cutter and the travel of the table are exactly parallel to each other. If there is any looseness in the table movements or cutter run-out, it will appear either as back drag or as a concave surface. Back drag appears as light cuts taken behind the main cut by the trailing cutting edges of the cutter. A concave surface is generated when the cutter is tilted in relation to the table so that the trailing cutting edges are not in contact with the work surface. A slight tilt, where the trailing cutting edges are .001 to .002 in. higher than the leading cutting edges, gives satisfactory surfaces in most applications.

Effective face milling operations are often the result of many small factors. Rigidity of the setup is of main concern. The workpiece must be

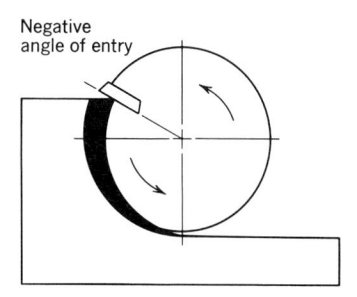

Figure 11. Work contacts tool away from cutting tool tip (Photo courtesy of Kennametal Inc., Latrobe, Pa.).

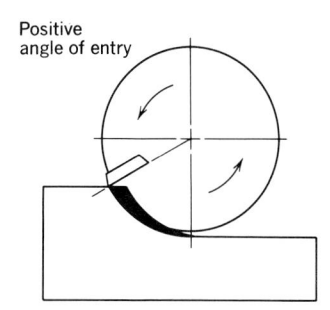

Figure 12. Work contacts tool on the cutting tool tip (Photo courtesy of Kennametal Inc., Latrobe, Pa.).

Figure 13. Handle tools with care, they are sharp (Courtesy of Cincinnati Milacron).

supported where the cutting takes place. Rigid stops prevent the workpiece from being moved by the cutting pressures. The table should be as close to the spindle as possible. The gibs on the machine slides need to be adjusted to prevent looseness. Locking devices should be used on machine slides. Efficient machining requires sharp cutting edges. Sharp cutting edges are dangerous to the operator — handle cutters with a rag (Figure 13). Dull cutting edges create excessive heat and leave a poor surface finish. Worn tools produce higher cutting pressures, which may deflect the workpiece. They also use additional power. When dull tools are used on work hardening materials, both the tool and workpiece may be ruined. The speed of the cutter affects the surface finish of the workpiece. Too low a speed leaves a poor surface because of the build-up on the cutting edge. Too high a speed results in excessive tool wear. If the feed is too low, the rubbing action wears down the cutting edge. If the feed is too high, the cutting edges will chip or break.

self-evaluation

SELF-TEST

1. How are face mills mounted on a milling machine?

2. What is a shell end mill?

3. What is the difference between light duty and heavy duty face mills?

4. Give some reasons for using positive rake angles on face mills.

5. Give some reasons for using negative rake angles on face mills.

6. What is the effect of a large lead angle over a small one?

7. Why should the width of cut in face milling be narrower than the face mill diameter?

8. Should cutting fluids be used in face milling?

9. What causes a concave work surface in face milling?

10. Give four factors that help in efficient face milling.

WORKSHEET
Machining a Flat Surface With a Face Mill

1. Clean the milling machine table and the underside of the workpiece.

2. Set the workpiece on the milling machine table with an overhang ¼ in. greater than the amount of material to be machined off.

3. Use clamps to fasten the workpiece to the table.

4. With a dial indicator mounted on the column, check the location of the surface to be machined. Make adjustments if necessary.

5. Clamp a stop block on the table to prevent any work movement.

6. Clean the spindle nose and the cutter; back and mount the cutter on the spindle nose.

7. Calculate and set the speed. What RPM will you use?
$$RPM = \frac{CS \times 4}{D} = \underline{\hspace{2in}}.$$

8. Check for correct spindle rotation.

9. Calculate the feedrate to use and set it on the machine. Feed rate equals feed per tooth × number of teeth × RPM. What is the feedrate?

10. Raise the knee until the centerline of the cutter is even with the centerline of the workpiece, then lock the knee. If the workpiece is wider than the diameter of the cutter, let one fourth of the cutter diameter extend above the top of the workpiece.

11. Start the spindle.

12. Carefully feed the saddle with the workpiece toward the cutter until a strip of paper held against the surface to be cut is torn by the cutter. Set the crossfeed micrometer dial on zero and lock it. Use a long paper strip, so that your fingers are not near the cutter.

13. Move the table longitudinally until the cutter clears the workpiece. The cutter should now be on the side of the workpiece that allows the cutting pressure to be downward on the workpiece.

14. Set the depth of cut and lock the saddle.

15. Turn on the coolant and start the power feed.

16. At the end of the cut, stop the feed, coolant, and spindle rotation before returning the table to its starting position.

section d
dividing the circle

The precise division of the circle has been a major concern for many centuries, from the first instruments for navigation and later for clocks and instruments for surveying. A dividing engine for cutting clock gears was made in 1670. It used a plate with holes for the basis of its divisions. By the middle of the eighteenth century, it was possible to compare distances along an arc within .001 in. and to estimate to about one third that increment, or about .0003 in. The basis was set for the evolution of dividing engines into practical devices to be used for manufacturing parts requiring accurate division.

Indexing devices are used on vertical and horizontal milling machines and on other types of machine tools such as jig boring machines, planers, slotters, and shapers. These devices are also used for many types of inspection.

The indexing devices used range from quite simple to highly complex and are selected to match the requirements of the job. The use of collet fixtures and hand circular milling tables are more popular on vertical mills, while the use of dividing heads and power-fed rotary tables are more common on horizontal milling machines.

One of the simplest indexing devices is a collet index fixture (Figure 1), which can be set up either horizontally or vertically on the milling machine table. These are quick and easy to use for milling polygons and other holding and dividing operations.

Another device that is more complex is the circular milling table. The table is marked in degrees and uses either a vernier to read out minutes or a hand wheel graduated in minutes of arc (Figure 2). Large circular milling tables (Figure 3) are frequently equipped to be driven in rotation by being coupled to the machine table drive mechanism and thus can be used for dividing by hand setting or for circular milling by hand or by power feed (Figure 4).

The circular milling attachment may be equipped with an indexing attachment (Figure 5) when indexing requirements are more exacting. The indexing attachment is especially useful for the accurate division of parts that have relatively large diameters.

Devices that have been developed directly from the line of the "dividing engines" are known as rotary tables. These are based on either a precise worm to worm wheel relationship (Figure 6) accurate to plus or minus 2 sec of arc or, in its most refined form, on an

Figure 1. Collet index fixture (Courtesy of Hardinge Brothers, Inc.).

Figure 2. Circular milling table (Courtesy of Cincinnati Milacron)

Figure 3. A circular milling table with power feed is termed a circular milling attachment (Courtesy of Cincinnati Milacron).

optical circle read with a combination of microscope and vernier to a final reading of 1 sec of arc (Figure 7). Rotary tables are used on master machine tools such as jig boring (Figure 8) and jig grinding machines and for use in inspection (Figure 9). For calibration and inspection of these master tools, a precision index (Figure 10) is used and it is accurate to $\frac{1}{10}$ sec of arc. The second of arc is approxi-

Figure 4. Circular milling attachment being used to mill a circular T-slot (Courtesy of Cincinnati Milacron).

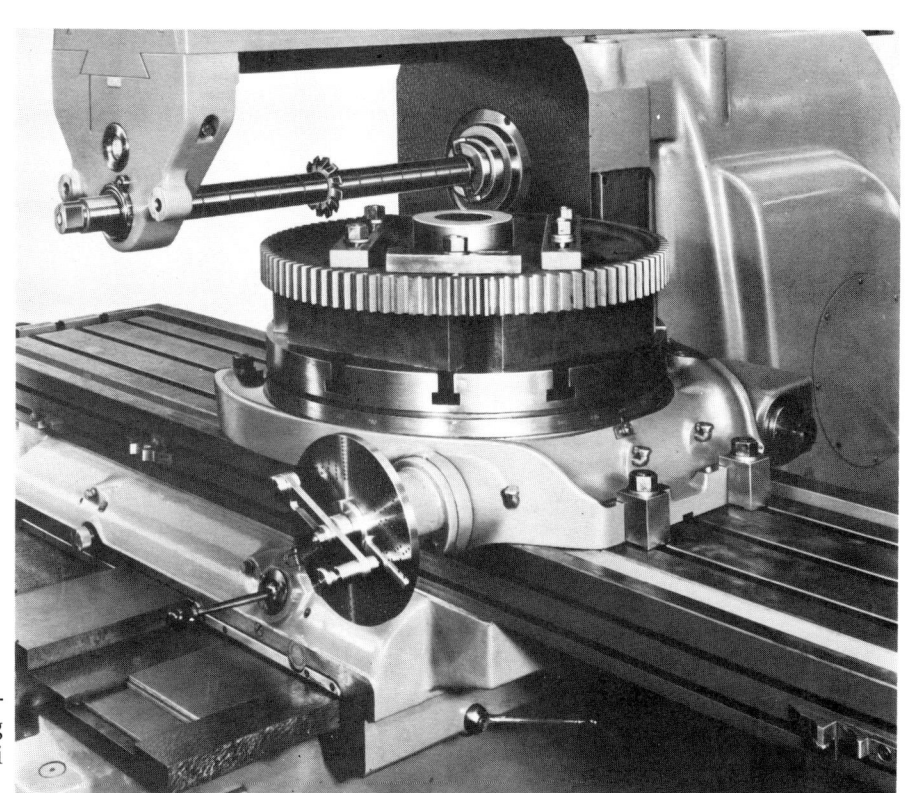

Figure 5. Circular milling attachment equipped with an indexing attachment (Courtesy of Cincinnati Milacron).

Figure 6. Ultraprecise rotary table (Courtesy of Moore Special Tool Co., Inc., Bridgeport, Conn.).

Figure 7. Optical dividing table (Courtesy of American SIP Corporation).

Figure 8. Rotary table with tailstock and adjustable center being used for inspection (Courtesy of Moore Special Tool Co., Inc., Bridgeport, Conn.).

Figure 9. Using the rotary table on a jig boring machine. (Courtesy of Moore Special Tool Co., Inc., Bridgeport, Conn.).

Figure 10. Precision index—a master tool for master tools (Courtesy of Moore Special Tool Co., Inc., Bridgeport, Conn.).

mately the angle subtended by the diameter of a common pencil at a range of a mile.

Many types of horizontal and universal dividing heads have been developed, since they are applied to the manufacture of parts on milling machines and other manufacturing machines. Horizontal spindle indexing devices such as the direct indexing head has the spindle directly connected with the indexing crank (Figure 11). The division equals the number of holes in hole circles, available on index plates to fit the head (50 holes maximum).

A gear cutting attachment that has the addition of 40:1 worm and worm wheel (Figure 12) gives capability of numbers up to 60, any even number, and those numbers divisible by 5 from 60 to 120.

The spiral milling head is the gear cutting attachment with the addition of a driving connection (Figure 13) to couple through gears to the leadscrew. It permits the milling of spiral forms, such as the flutes on helical reamers and slab milling cutters.

The universal spiral dividing head is an indexing device that performs the same function as the spiral milling head with provision for coupling to the leadscrew of the milling machine. In addition, the spindle assembly is carried in a swivelling block that permits the spindle to be tilted from minus 5 degrees to past the vertical by 5 to 10 degrees. A typical application of the swivel feature is shown in Figure 14. Dividing heads are also useful for cutting graduations in combination with devices like the slotting attachment (Figure 15).

Spiral indexing heads are coupled to the machine leadscrew by the use of a standard universal dividing head driving mechanism

Figure 11. Direct indexing head (Courtesy of Cincinnati Milacron).

Figure 12. Gear cutting attachment (Courtesy of Cincinnati Milacron).

Figure 13. Spiral milling head (Courtesy of Cincinnati Milacron.)

Section D Dividing the Circle

Figure 14. *(top)* Swiveling block on universal spiral dividing heads permits angular indexing (Courtesy of Cincinnati Milacron).

Figure 15. *(right)* Dividing heads are also useful for making graduations on conical surface (Courtesy of Cincinnati Milacron).

(Figure 16), which has change gears to control the direction of rotation, and the overall ratio to the dividing head. Figure 17 shows a combination of a driving mechanism and a universal spiral dividing head set up on a universal horizontal milling machine for fluting a large drill with a formed milling cutter.

The short and long lead attachment is used on universal general purpose milling machines equipped with a feature to disengage the leadscrew clutch so the table may be driven by the main shaft through gearing in the attachment gearbox (Figure 18). This permits the selection of over 13,000 leads ranging from .010 to 1000 in., or by an additional gear change from .025 to 3000 in. An example of the type of milling that can be done with this device in combination with other accessories is shown in Figure 19.

The wide range dividing head makes the regular 40:1 ratio dividing head, supplied with standard or high number index plates, adequate for the great majority of indexing work. However, there are combinations where a wide range dividing head (Figure 20), having an additional 100:1 ratio compounded with the standard 40:1 ratio, becomes necessary. This gives a minimum increment of 6 sec of arc per division. Dividing heads are also useful for inspection purposes in the same way that the rotary table is used (Figure 21).

As with the development of many specialized machine tools for production purposes, the indexing has been synchronized with the rotation of the machine spindle to create a hobbing attachment (Figure 22) for the horizontal milling machine. This type of device is rare, but it shows the basis for one type of production gear-making machine that will be shown in the next section.

Devices to divide the circle, both for the production of other instruments and for the production of gears for quiet, efficient transmission of power, are really at the core of an industrialized society. It is important for you to be able to select these devices

Figure 16. Standard universal dividing head driving mechanism (Courtesy of Cincinnati Milacron).

Figure 17. Spiral milling (Courtesy of Cincinnati Milacron).

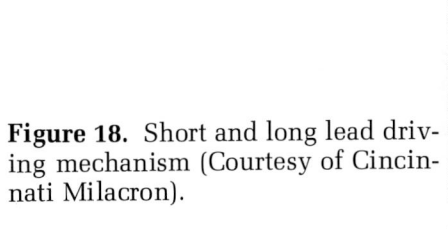

Figure 18. Short and long lead driving mechanism (Courtesy of Cincinnati Milacron).

Figure 19. A combination of attachments for the milling of a leadscrew (Courtesy of Cincinnati Milacron).

Figure 20. Dividing head with wide range divider (Courtesy of Cincinnati Milacron).

Figure 21. Inspecting a cam using a dividing head (Courtesy of Cincinnati Milacron).

Figure 22. A special hobbing attachment for the horizontal milling machine (Courtesy of Cincinnati Milacron).

according to your requirements and make competent use of them. Indexing is a task that you can expect to use often as a machinist.

In the following units of instruction, you will be able to identify the main components of common indexing devices, calculate indexing movements for simple indexing, and calculate the indexing movements necessary for angular indexing.

review questions
1. Why is the ability to divide the circle important to navigation?
2. What purpose is served by the use of an indexing attachment as part of the circular milling attachment?
3. Why are rotary tables usually considered master tools?
4. Differentiate among four types of horizontal spindle indexing devices.
5. What important feature is necessary on a milling machine to employ a short and long lead attachment?

unit 1
indexing devices

Indexing devices are precision milling machine attachments. They will accurately rotate a workpiece a full or partial turn. Indexing devices are used when cutting gears, splines, keyways, or holes that must be specific angular distances apart.

objectives After completing this unit, you will be able to identify the main components of indexing devices.

information

The dividing head, which is also called an index head, is used to give rotary motion to workpieces in milling operations. Its important part is the housing, which contains the spindle (Figure 1). The spindle has a worm wheel attached to it, and this worm wheel meshes with a worm. The worm is turned thru a set of gears with the index crank or by a gear train connected to the milling machine leadscrew. The spindle axis can be swiveled to allow the holding of workpieces in a horizontal or vertical position. Once the spindle axis is adjusted, the clamping straps are tightened to lock the spindle in this position. On most dividing heads the worm can be disengaged from the worm wheel. The worm is disengaged by turning an eccentric collar. When the worm is disengaged, the spindle can easily be turned by hand while setting up or while direct indexing. In direct indexing, the plunger is used to engage the direct indexing pin into a hole of the direct indexing hole circle on the spindle nose (Figure 2). The spindle nose has a tapered hole to hold taper shank tools and chucks. On some dividing heads the spindle nose is threaded to hold screw on chucks. After the spindle has been rotated, the spindle lock is tightened to prevent any spindle movement while cutting. When the worm is engaged, spindle rotation is obtained by turning the index crank. The most commonly used index ratio between the index crank and the spindle is 40:1. This means that the index crank needs to be turned 40 revolutions to get 1 revolution of the spindle. Another index ratio found on dividing heads is 5:1. The index crank has a plunger that moves an index pin in and out of a hole on the index plate. The index plate has a number of hole circles with equally spaced holes in each circle. The index plate is used to obtain precise partial revolutions of the index crank.

Often the index plate has a different set of hole circles on the reverse side. For some dividing heads, high number index plates are available to obtain a great number of different hole circles. When only a partial revolution of the index crank is made, the sector arms are set apart a distance equal to that partial turn to avoid counting the spaces for each indexing turn. The index plate stop engages in serrations on the circumference of the index plate. When it is loosened, the index plate can be rotated in small increments.

A wide range index head (Figure 3) that permits indexing of 2 to 400,000 divisions is available. The wide range divider has a large

Section D Dividing the Circle

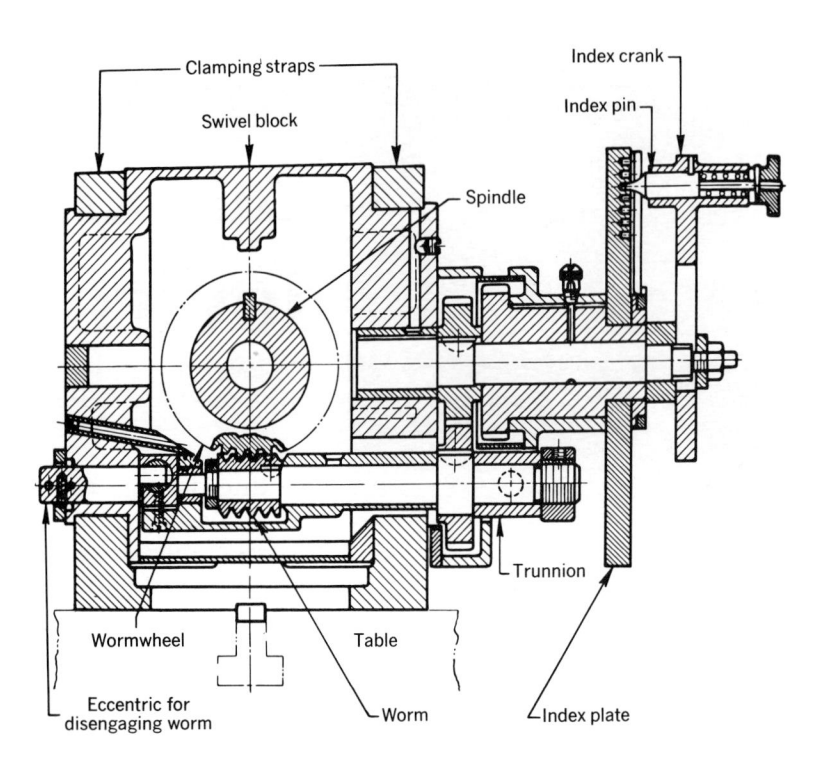

Figure 1. Section through dividing head showing worm and worm shaft (Courtesy of Cincinnati Milacron).

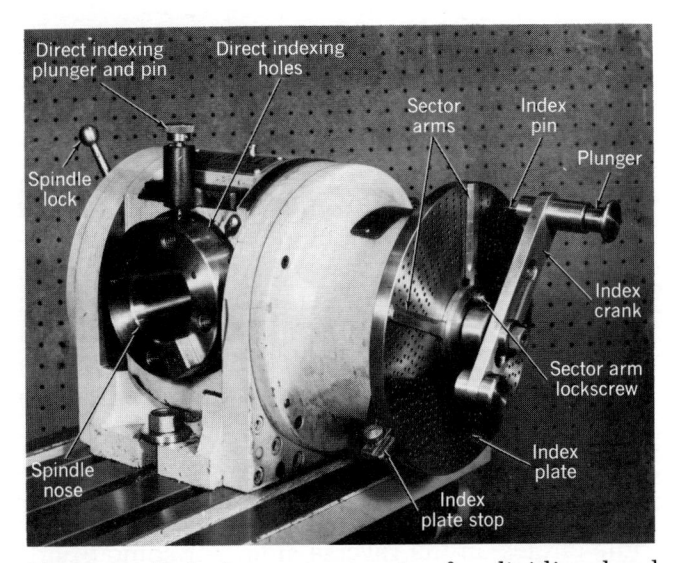

Figure 2. Indexing components of a dividing head (Lane Community College).

Figure 3. Wide range divider permits indexing of 2 to 400,000 divisions (Courtesy of Cincinnati Milacron).

index plate, index crank, and sector arm, as does a regular dividing head, with a 40:1 ratio. In front of the large index plate a small index plate, index crank, and sector arms are mounted. The small index crank is geared to the worm of the large index crank with an additional 100:1 ratio to get a total ratio of 4000:1 (40:1 times 100:1). It takes 4000 revolutions of the small index crank to rotate the spindle one full revolution. The small index plate has 2 hole circles of 100 and 54 holes. When the index crank is moved 1 space in the 54 hole circle on the small plate, the work is rotated through 6 sec of arc.

The footstock (Figure 4) supports one end of the workpiece. The footstock center can be adjusted toward and away from the dividing head to permit the removal of workpieces and to make up for different length workpieces. The footstock

Figure 4. Footstock (Lane Community College).

Figure 5. Adjustable center rest (Lane Community College).

center can also be adjusted vertically to allow the leveling of workpieces. Tapered workpieces are supported by swiveling the axis of the footstock center horizontally. Long or slender workpieces are supported with an adjustable center rest (Figure 5).

Another commonly used indexing device is the rotary table (Figure 6). Index ratios of rotary tables vary from 120:1, 80:1, or 40:1. Usually the table is graduated in degrees, while the hand wheel has 1 min graduations. With an index plate, very accurate spacings can be made. To determine the index ratio of an indexing device, turn the index crank 10 complete revolutions and measure the resulting spindle revolution. If the spindle has rotated $\frac{1}{4}$ turn, the index ratio is 40:1.

The index crank should be carefully adjusted so that the index pin slips easily into the holes in whatever hole circle is used. Once an indexing operation is started, the index crank should only be turned in one direction (usually

Figure 6. Rotary table (Lane Community College).

clockwise). Turning the index crank clockwise and then counterclockwise will allow the backlash between the worm and worm wheel to affect the accuracy of the indexing.

self-evaluation

SELF-TEST 1. When are indexing devices used?

2. What makes indexing devices so accurate?

Section D Dividing the Circle

3. When is the worm disengaged from the worm wheel?

4. When is the hole circle on the spindle nose used?

5. What is a commonly used index ratio on dividing heads?

6. Why does the index plate have a number of different hole circles?

7. What is the purpose of the sector arms?

8. What does the spindle lock do?

9. How can divisions be made that are not possible with a standard index plate?

10. Why should the index crank be rotated in one direction only while indexing?

unit 2
direct and simple indexing

Most indexing operations performed in machine shops fall into the direct and simple indexing categories. The accuracy of the indexing operation determines which dividing method to use. Examples of indexing operations are the cutting gear teeth, splines, keyways, or the machining of square or hexagon shapes.

objectives After completing this unit, you will be able to do the calculations for direct and simple indexing.

information

Direct indexing is the easiest method of dividing a workpiece into a number of equal divisions. The number of divisions obtainable by direct indexing is limited by the number of holes in the direct indexing hole circle in the spindle nose. Hole circles available have 24, 30, or 36 holes. To perform direct indexing rapidly, the worm should be disengaged from the worm wheel to allow the spindle to be turned by hand. When a 24 hole circle is used, equal divisions of 2, 3, 4, 6, 8, 12, and 24 spaces can be made. As an example, let us assume a hexagon needs to be machined with direct indexing. The 24 holes divided by 6 equal 4 holes. This means that for each cut, the spindle has to be turned a distance of 4 more holes from the preceding cut. When making this 4 hole advance, do not count the hole the index pin is in. As soon as the index pin is seated in the hole, tighten the spindle lock to prevent any spindle movement while the machining takes place. To avoid using the wrong hole when rotating the spindle, take a felt pen or other marker to identify which holes to use prior to the machining.

Simple indexing, also known as plain indexing, involves the turning of the index crank to rotate the spindle. On most dividing heads, 40 turns of the index crank results in one revolution of the workpiece. To obtain a specific number of spaces on the circumference of a workpiece, 40 is divided by that number to get the number of whole or partial turns of the index crank for each division. To make 20 equal divisions on a workpiece, divide 40 by 20, which gives $\frac{40}{20} = 2$. The 2 represents 2 complete turns of the index crank. To cut 80 teeth on a gear, write $\frac{40}{80} = \frac{1}{2}$, or $\frac{1}{2}$ revolution of the index crank for each tooth.

When a partial turn of the index crank is needed, an index plate with a number of different hole circles is used. Index plates with the following holes are available: 24, 25, 28, 30, 34, 37, 38, 39, 41, 42, 43, 46, 47, 49, 51, 53, 54, 57, 58, 59, 62, and 66. To get a $\frac{1}{2}$ revolution of the index crank, any hole circle divisible by 2 can be used. If the 30 hole circle is used, the index pin is to be advanced 15 holes each time.

To make 27 divisions, proceed as follows. Divide 40 by 27, which is $\frac{40}{27} = 1\frac{13}{27}$. The 1 represents one complete turn of the index crank. The denominator 27 could be the hole circle to use and the numerator 13 the number of holes to be advanced. A check of the available hole circles shows that a 27 is not there, but a 54 is, and that is a multiple of 27. To keep the fraction of $\frac{13}{27}$ intact and to increase the denominator to 54, multiply both the numerator and denominator by 2. It now looks like this: $\frac{13 \times 2}{27 \times 2} = \frac{26}{54}$ and means 26 holes in the 54 hole circle. For each division then make 1 turn plus 26 holes in the 54 hole circle.

When 52 divisions are required, divide 40 by 52, $\frac{40}{52}$, and reduce it to the lowest fraction possible: $\frac{40}{52} \frac{4}{4} = \frac{10}{13}$ There is not a 13 hole circle, but a 39 is available. Now raise the fraction to a denominator of 39 by multiplying by 3, $\frac{10 \times 3}{13 \times 3} = \frac{30}{39}$, or 30 holes in a 39 hole circle. The highest indexing accuracy is achieved when the hole circle with the greatest number of holes is used that will accommodate the denominator of the fraction.

To obtain 51 divisions, write $\frac{40}{51}$. There is a 51 hole circle available, which makes no calculation necessary. It takes 40 holes in the 51 hole circle for each division.

It would be very awkward to count the number of holes for each indexing operation. This is why sector arms are found on the index plate. The sector arms can be adjusted to form different angles by loosening the lockscrew. One side of each sector arm is beveled. The number of

Figure 1. Dividing head sector arms set for indexing 11 spaces (Lane Community College).

holes of the partial turn needed are located within these beveled sides. The sector arms in Figure 1 are set for an 11 hole movement in the hole circle. The hole that the pin is in is not counted.

To obtain spacings other than those available with standard index plates, high number index plates can be used. Another choice would be the use of a wide range divider. When change gears are available for older dividing heads, compound or differential indexing can be employed to make divisions not obtainable with standard index plates. Consult *Machinery's Handbook* for the procedures used for compound or differential indexing.

self-evaluation

SELF-TEST

1. What is the difference between direct and simple indexing?

2. How can you avoid using a wrong hole in the index plate while direct indexing?

3. If the direct indexing hole circle has 24 holes, what are the different divisions you can make with it?

Unit 3 Angular Indexing 159

4. How are the sector arms used on a dividing head?

5. If both a 40 and a 60 hole circle can be used, which is the better one?

6. Calculate how to index for 6 divisions. Use the hole circles from the text.

7. Calculate how to index for 15 divisions. Use the hole circles from the text.

8. Calculate how to index for 25 divisions. Use the hole circles from the text.

9. Calculate how to index for 47 divisions. Use the hole circles from the text.

10. Calculate how to index for 64 divisions. Use the hole circles from the text.

unit 3 angular indexing

Work may be indexed to produce a given number of spaces on the circumference, or the spacing can be indicated as an angular distance measured in degrees.

objective After completing this unit, you will be able to perform the calculations for angular indexing.

information

Indexing by degrees can be done on a dividing head by the direct and simple indexing method. One complete revolution of the dividing head spindle is 360 degrees. If the direct indexing hole circle has 24 holes, the angular spacing between each hole is $\frac{360}{24}$ degree = 15 degrees. Any division requiring 15 degree intervals can be made. To drill 2 holes at a 75 degree angle to each other, divide 75 by 15 ($\frac{75}{15}$) = 5. The 5 represents 5 holes on the 24 hole circle. Remember to not count the hole that the index pin is in. If angles other than 15 degrees are to be indexed, the simple indexing method is used. To obtain one 360 degree revolution of the dividing head spindle, it takes 40 turns of the index crank. One turn of the index crank produces a $\frac{360}{40}$ degrees = 9 degree movement of the dividing head spindle. Any hole circle on the index plate that is divisible by 9 can be used to index by degrees.

With a 27 hole circle it takes 3 holes for 1 degree, or 1 hole is 20 min.

With a 36 hole circle it takes 4 holes for 1 degree, or 1 hole is 15 min.

With a 45 hole circle it takes 5 holes for 1 degree, or 1 hole is 12 min.

With a 54 hole circle it takes 6 holes for 1 degree, or 1 hole is 10 min.

When indexing in degrees, the formula for the number of turns of the index crank is $\frac{\text{degrees required}}{9}$. As an example, two cuts must be made 37 degrees apart.

$\frac{37}{9} = 4\frac{1}{9}$, which is 4 complete turns and $\frac{1}{9}$ turn. The $\frac{1}{9}$ is expanded to fit a 54 hole circle $\frac{1}{9} \times \frac{6}{6} = \frac{6}{54}$. The total movement required is 4 turns and 6 holes in the 54 hole circle.

Indexing can also be done in minutes, with the formula $\frac{\text{minutes required}}{540}$. The denominator of 540 is obtained by multiplying 9 degrees by 60 min, the number of minutes in one revolution of the index crank. When using this formula, it becomes necessary to convert the degrees and partial degrees into minutes. As an example, two holes have to be drilled at 8 degrees 50 min from each other. The 8 degrees 50 min equals 530 min. Putting this value into the equation, we get a fraction of $\frac{530}{540}$ To use a 54 hole circle, this fraction has to be reduced.

$$\frac{530}{540} \div \frac{10}{10} = \frac{53}{54}$$

The result is an index crank movement of 53 holes in a 54 hole circle.

When the required minutes are not evenly divisible by 10, the required spacing may only be approximate with a slight error. As an example, calculate the index crank movement for a 1 degree 35 min spacing. Hole circles available are: 38, 39, 41, 42, 43, 46, 47, 49, 51, 53, 54, 57, 58, 59, 62, and 66. Convert the mixed number into minutes: 60 min + 35 min = 95 min. This value is entered into the equation $\frac{95}{540} = \frac{1}{5.685}$. There is no hole circle with 5.685 holes available, so use the trial and error method in expanding this fraction.

$$\frac{1}{5.685} \times \frac{7}{7} = \frac{7}{39.7949}$$

$$\frac{1}{5.685} \times \frac{8}{8} = \frac{8}{45.4799}$$

$$\frac{1}{5.685} \times \frac{9}{9} = \frac{9}{51.1649}$$

$$\frac{1}{5.685} \times \frac{10}{10} = \frac{10}{56.85}$$

It appears that the 51.1649 is the closest number to the available hole circle of 51. Each hole in the 51 hole circle is $\frac{540}{51}$, or 10.588 min spaced from the adjacent one. Moving 9 holes in the 51 hole circle gives an angle of 95.292 min or 95 min 17 sec. This is an error of only 17 sec.

High number index plates or a wide range divider will give a greater choice of accurate angular spacings.

self-evaluation

SELF-TEST

1. If there are 24 holes in the direct indexing plate, how many degrees are between holes?

2. How many holes movement is necessary to index 45 degrees using the direct indexing method?

3. How many degrees in the movement produced by one complete turn of the index crank?

4. Which hole circles on the index plate can be used to index by whole degrees?

5. How many turns of the index crank are necessary to index 17 degrees?

6. What fraction of one degree is represented by 1 space on the 18 hole circle?

7. What fraction of one degree is represented by 1 space on the 36 hole circle?

8. What fraction of one degree is represented by 1 space on the 54 hole circle?

9. How many minutes movement is produced by one turn of the index crank?

10. How many turns of the index crank are necessary to index 54 degrees 30 min, using the 54 hole circle?

section e
gears

Making gears for light duty purposes such as clock mechanisms goes back for several hundred years, but the history of machines that can produce gears capable of transmitting substantial power only dates back to about 1800. Since that time thousands of patents have been taken out on devices for cutting gears and gear cutting tools.

The ability to produce gears economically did not appear until about 1850 and was the basis for the development of the high speed printing press, the sewing machine, and many other useful products that began to appear in quantity after the Civil War. These early gear cutting machines (Figure 1) typically used a formed cutter made to mill out the space between the teeth to leave a tooth of correct shape. Large machines of this period often used a template to guide a tool in a slide to form the shape of the tooth.

As the demands increased for faster production of gears in the 1860s, the development of highly specialized production gear making machinery began to emerge. It was observed that a worm made as a cutter, having straight sides in the same tooth form as a rack, would generate an involute gear if the two were turned together so that the gear would move one space while the worm made one revolution (Figures 2 and 3). This is the basis for a specialized variety of milling machine known as the hobbing machine. By 1900, this was a well-developed and highly productive method of producing accurate gears. Hobbing machines come in both horizontal spindle (Figure 4) and vertical types. The larger machines usually have a vertical spindle for holding the gear blanks. Some of these machines are used for gears 16 ft in diameter and larger (Figure 5).

About 1900, another important production gear making machine appeared. It used a cutter shaped like a mating gear that was reciprocated across the gear to be cut and rotated with the blank as the cutting progressed. This machine is called a gear shaper (Figure 6). This machine can be used to generate external or internal spur or helical gears and has the advantage of being able to machine the gear form next to a shoulder, as is often the case in gear clusters in transmissions. This type machine can also be used to generate unusual forms of gears (Figure 7).

As the automobile became popular, it was also important to be able to produce quiet, accurate gears for the rear ends of cars. Machines like this hypoid gear generator (Figure 8) were developed to meet that need.

As more rapid means were needed to produce gears of accurate form for high production applications, methods such as broaching

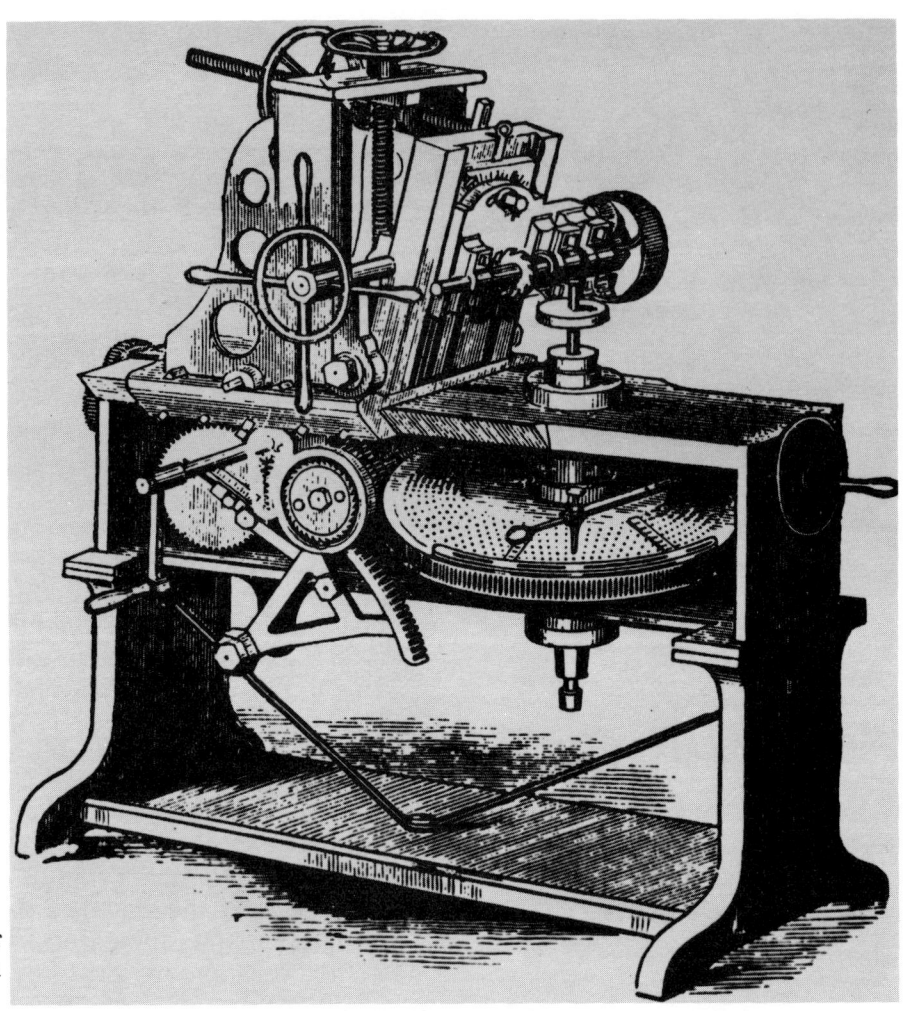

Figure 1. Joseph Brown's 1855 gear cutting machine (Courtesy of Brown & Sharpe Mfg. Co.).

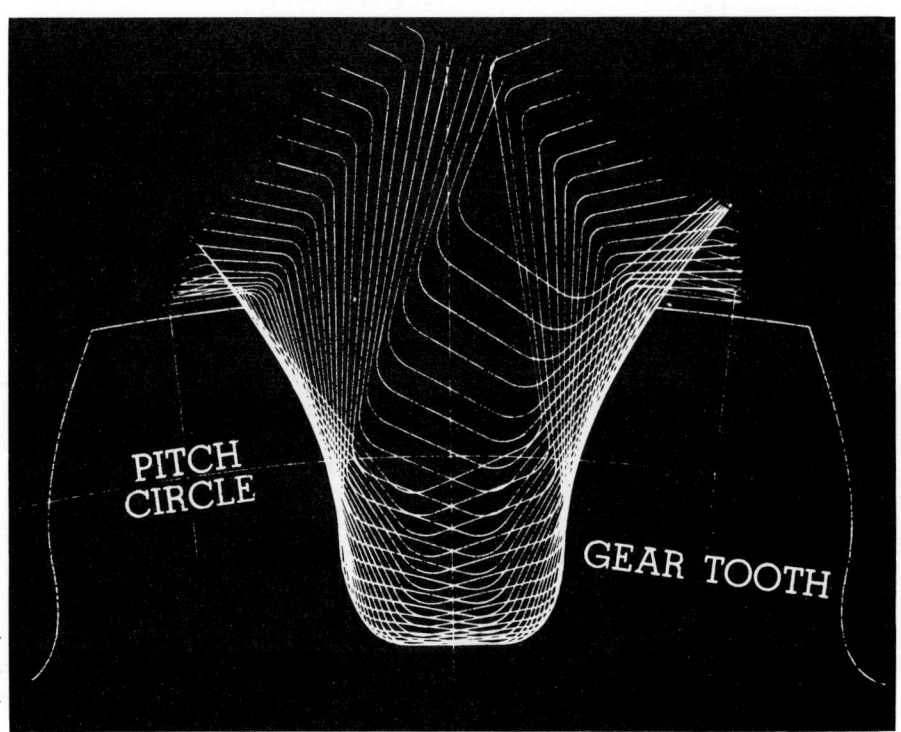

Figure 2. Schematic view of the generating action of a hob (Reprinted with permission of Barber-Colman Company).

Section E Gears   165

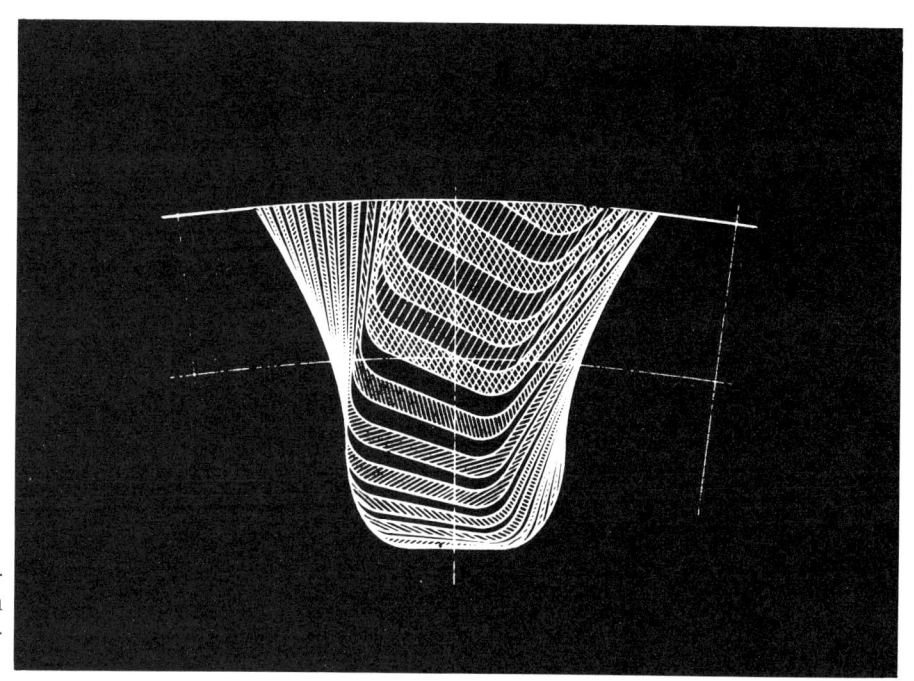

Figure 3. Schematic view of the performance of an individual hob tooth (Reprinted with permission of Barber-Colman Company).

Figure 4. Small horizontal spindle hobbing machine (DeAnza College).

were developed. This method is especially adapted to producing individual or nonclustered gears (Figure 9) with straight teeth or with helix angles under 21 degrees.

For commercial use, gear teeth are usually finished after they are cut. The type of finishing method used depends greatly on the

Figure 5. A huge vertical spindle gear hobbing machine. Notice the operator with the pendant control at the upper left (Courtesy of Ex-Cell-O Corporation).

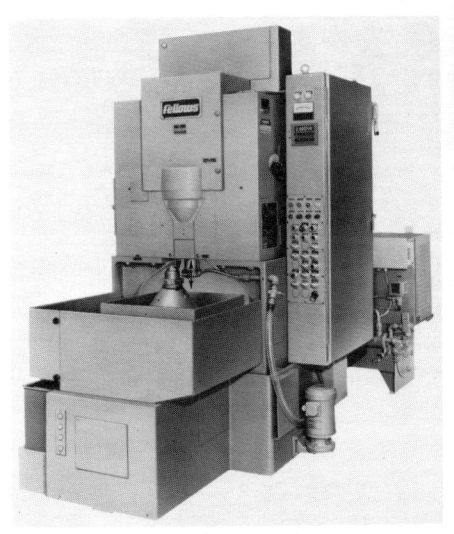

Figure 6. A gear shaper. The cutter moves up and down with the same twist as the helix angle of the gear (Courtesy of Fellows Corportion).

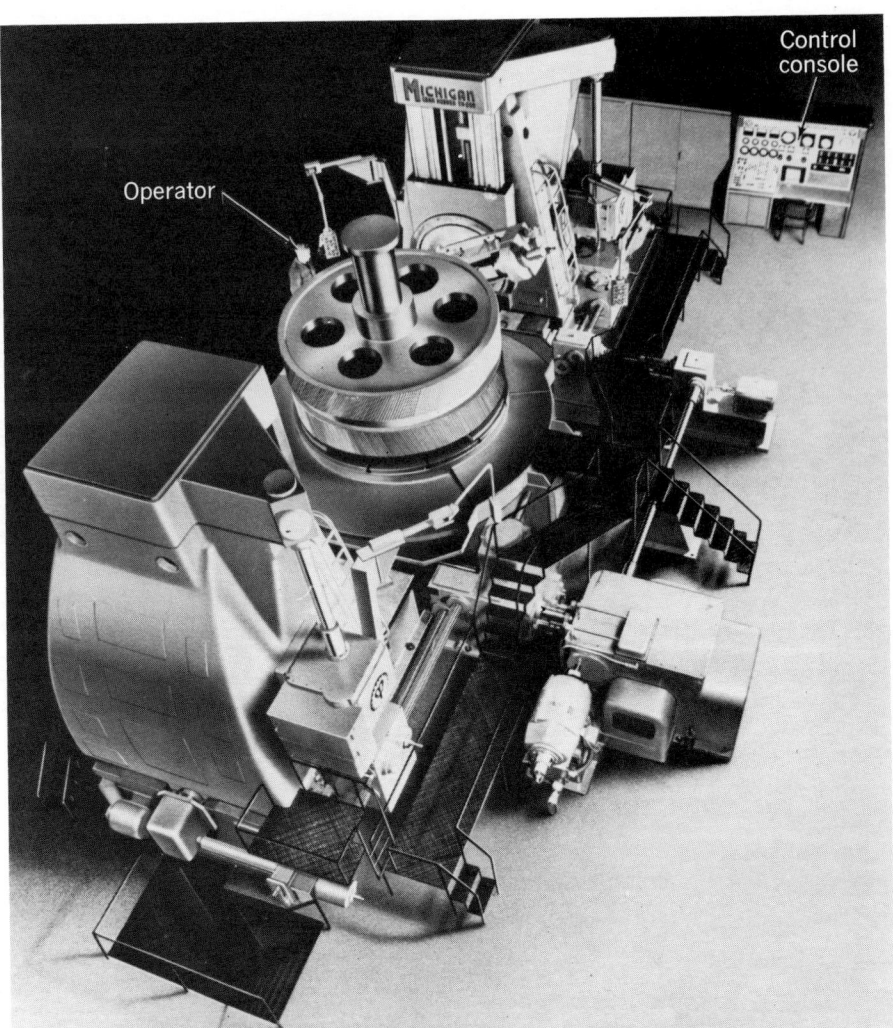

Figure 7. The gear shaper can produce unusual gear shapes (Courtesy of Fellows Corporation).

Figure 8. Hypoid gear generator (Courtesy of Gleason Works).

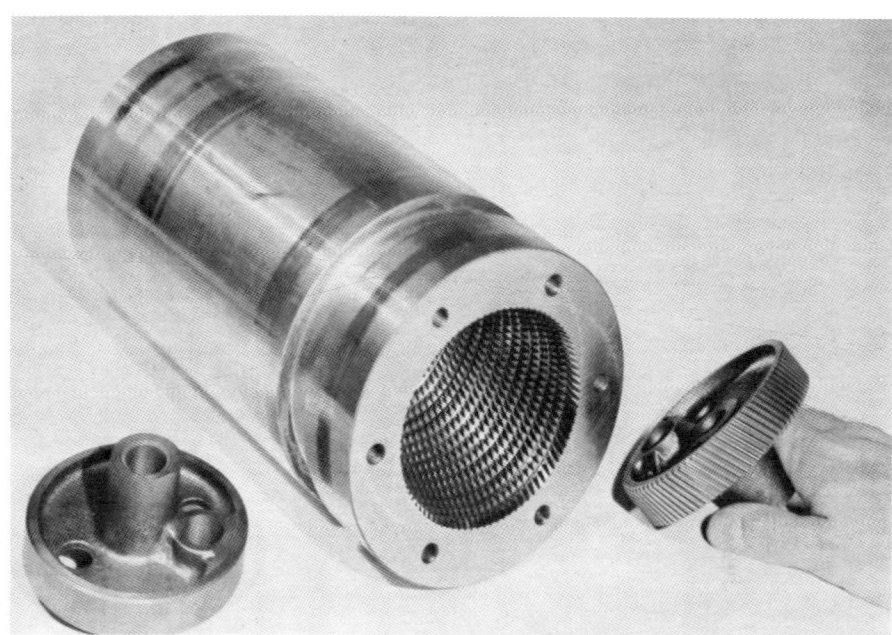

Figure 9. Pot broaching tool that produces a helical gear (Courtesy of National Broach and Machine Division, Lear Siegler, Inc.)

Figure 10. Gear shaving (Courtesy of National Broach and Machine Division, Lear Siegler, Inc.).

service condition of the gear, its accuracy requirements, noise limitation, and hardness requirements.

Gear shaving can be employed on gears that are not harder than Rockwell C30, where surface finishes in the order of 32 microinches are adequate for the application. In some cases, shaving can produce finishes of 16 microinches.

One method uses a shaving cutter rolled in mesh with the work gear at an angle (Figure 10). This is called diagonal shaving.

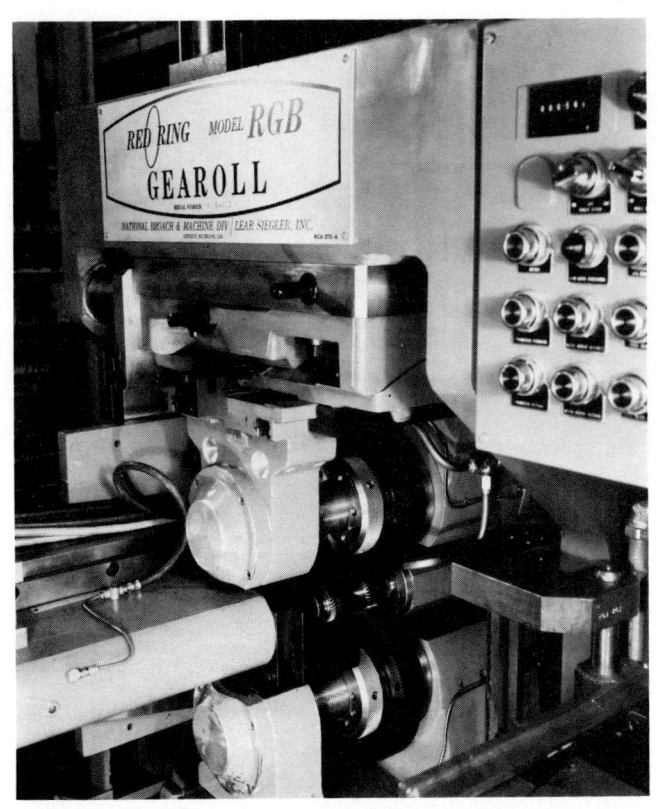

Figure 11. Roll finishing of unhardened gears (Courtesy of National Broach and Machine Division, Lear Siegler, Inc.).

Figure 12. Finishing a gear by form grinding (Courtesy of National Broach and Machine Division, Lear Siegler, Inc.).

Another method for use with gears that are not hardened is roll finishing (Figure 11). This is a cold forming process that can produce finishes as fine as 10 microinches on relatively small gears, up to about 4 in. pitch diameter.

For hardened gears, it is necessary to use more expensive processes such as gear grinding. Gear grinders either use a method in which the grinding wheel is formed to the shape of the gear space (Figure 12) or a method in which the final shape is generated by relative motion between the grinding wheel and the already cut and hardened gear (Figure 13). The surface finishes produced by grinding generally range from 20 to 30 microinches.

For still finer finishes or hardened gears in the 10 to 15 microinch range, gear honing is employed (Figure 14). The honing tool is typically an abrasive impregnated mating gear rotated with the gear in a crossed axis relationship. Honing is also readily done on internal gears.

The inspection of gears is a highly refined technology employing very complex optical mechanical and electronic devices (Figure 15).

The gears that are produced in the general machine shop utilizing formed gear cutters are typically either for emergency use or for slow speed applications where unhardened gears are satisfactory

Figure 13. Finishing a gear by generating the ground surface (Courtesy of Fellows Corporation).

Figure 14. Finishing a hardened gear by honing. The honing tool has the helical teeth (Courtesy of National Broach and Macine Division, Lear Siegler, Inc.).

Figure 15. Inspecting gear profiles (Courtesy of Fellows Corporation).

and where accuracy and low noise requirements are not an important factor.

It is useful for machinists to know how to produce and measure spur gears for this type of service, especially in situations where mechanical maintenance is done. Producing spur gears by milling is often satisfactory, but making acceptable helical gears by spiral milling is unlikely and should probably only be done as a machine shop exercise for its instructional value.

The following units of instruction will define the common types of gears, gear tooth parts, and calculations, how to set up for milling a spur gear, and how to measure the dimensions of the resulting gear.

review questions

1. What type of cutters were used in early gear production machines?
2. A hobbing machine uses a cutter that is a spiral form of a rack, yet it produces a curved tooth form. What is the cutting process called when the resulting shape is not in the form of the cutting tool?
3. What type of gear making machine can produce unusual gear forms?
4. Can gear broaching be used for other than straight tooth gears?
5. Name four types of gear finishing processes. Which are used with unhardened gears and which are used with hardened gears?

unit 1
introduction to gears

Gears are used to transmit power and motion from one rotating shaft to another. Many different kinds of gears are manufactured, and this unit will explain the uses of some of the more common ones.

objective After completing this unit, you will be able to identify different gear designs and some of the materials used for gears.

information

Gears provide positive, no-slip power transmission and are used to increase or decrease the turning effort or speed in machine assemblies. When two gears are running together, the one with the larger number of teeth is called the gear and the one with the smaller number of teeth is called the pinion. Gears are generally used when shaft center distances are short, to provide a constant speed ratio between shafts, or to transmit high torques.

SPUR GEARS

Gears fall into several categories; the first is gears that connect parallel shafts. The best known of these gears is the spur gear. Spur gears have a cylindrical form with straight teeth cut into the periphery (Figure 1). When teeth are cut on a straight bar, a gear rack is made. A gear rack converts rotary gear motion into a linear movement. When the teeth on a cylindrical gear are at an angle to the gear axis, it is a helical gear (Figure 2). On a helical gear, several teeth are in mesh simultaneously with the mating gear, which provides a smoother operation than with spur gears. Because of the angle of the teeth, both radial and thrust loads are imposed on the gear support bearings. To offset this thrust effect, double helical gears are used. These gears are herringbone gears (Figure 3), which consist of a right-hand and a left-hand helix. The hand of a helical gear is determined by facing the gear; if the teeth rotate to the right it is a right-hand helix.

Common reduction ratios for gears are 1:1 to 5:1 for spur gears, 1:1 to 10:1 for helical gears, and 1:1 to 20:1 for herringbone gears. These same three gear types are also manufactured as internal gears (Figure 4). An internal gear has a greater tooth strength than that of an equivalent exter-

Figure 1. Spur gear and gear rack (Lane Community College).

Figure 2. Helical gears (Lane Community College).

Figure 3. Herringbone gear (Lane Community College).

Figure 4. Helical internal gears (Lane Community College).

nal gear. An internal gear rotates in the same direction as its mating pinion. Internal gears permit close spacing of parallel shafts. Internal gears mesh with external pinions.

BEVEL GEARS

The second category concerns gears that connect shafts at any angle, providing the shaft's axis would intersect if extended. Gears used to transmit power between intersecting shafts are often bevel gears (Figure 5). Bevel gears are conical gears and may have straight or spiral teeth. Spiral bevel gears are smoother running and usually will transmit more power than straight bevel gears because more than one pair of teeth is in contact at all times.

Mating bevel gears with an equal number of teeth producing a 1:1 ratio are called miter gears (Figure 6). When the angle of the two shafts is other than 90 degrees, angular bevel gears are used. Bevel gears are used to provide ratios from 1:1 to 8:1. Face gears have teeth cut on the end face of a gear. The teeth on a face gear can be straight or helical. Ratios for face gears range from 3:1 to 8:1.

Figure 5. Bevel gears (Lane Community College).

Figure 6. Miter gears (Lane Community College).

Figure 7. Worm and worm gear (Lane Community College).

HELICAL GEARS

The third category is for nonintersecting, nonparallel shafts. Helical gears of the same hand, cut with a 45 degree helix angle, will mesh when two shafts are at 90 degree or right angles from each other. By changing the helix angle of the two gears, the angle of the shafts in relation to each other can be changed. This type of gear arrangement is called a crossed helical gear drive and is used for ratios of 1:1 to 100:1.

WORM GEARS

Worm and worm gears (Figure 7) are used for power transmission and speed reduction on shafts that are at 90 degrees to each other. Worm gear drives operate smoothly and quietly and give reduction ratios of 3:1 to over 100:1. In a worm gear set, the worm acts as the pinion, driving the worm gear. Two basic kinds of worm shapes are made: the single-enveloping worm gear set with a cylindrical worm, and the double-enveloping worm gear set where the worm is hourglass shaped. The worm of the double enveloping worm gear set is wrapped around the worm gear and gives a much larger load carrying capacity to the worm gear drive. Worms are made with single thread or lead, double lead thread, triple lead thread, or other multiple lead threads. The number of leads in a worm is determined by counting the number of thread starts at the end of the worm. A single thread worm is similar to a single tooth gear when the ratio of a worm gear set is determined. A worm gear is made to run with the worm of a given number of leads. A double thread worm gear must be run with a double lead worm and cannot be interchanged with a single lead worm.

Another commonly used gear form is hypoid gears (Figure 8). Hypoid gears are similar in form to spiral bevel gears except that the pinion axis is offset from the gear axis. Hypoid gear ratios range from 1:1 to 10:1.

GEAR MATERIALS

Gear materials fall into three groups: ferrous, nonferrous, and nonmetallic materials. In the ferrous materials, steel and cast iron are used often. Steel gears, when hardened, carry the greatest load relative to their size. Steel gears can be hardened and tempered to exacting specifications. The composition of the steel can be changed. When the carbon content is increased, the wear resistance also increases. Lowering the carbon content gives better machinability. Cast iron is low in cost and is easily cast into any desired shape. Cast iron machines easily and has good wear resistance. Cast iron gears run relatively quietly and have about three quarters of the load carrying capacity of an equal size steel gear. One drawback in cast iron is its low impact strength, which prohibits its use where severe shockloads occur. Other ferrous gear materials are ductile iron, malleable iron, and sintered metals.

Nonferrous gear materials are used where corrosion resistance, light weight, and low cost production are desired. Gear bronzes are very tough and wear resistant. Gear bronzes make very good castings and have a high machinability. Lightweight gears are often made from aluminum alloys. When these alloys are anodized, a hard surface layer increases their wear resistance.

Low cost gears can be produced by die casting. Most die cast gears are completely finished when ejected from the mold with the exception of the removal of the flash on one side. Die casting materials are zinc base alloys, aluminum base alloys, magnesium base alloys, and copper base alloys.

Nonmetallic gears are used primarily because of their quiet operation at high speed. Some of these materials are layers of canvas im-

Figure 8. Hypoid gears (Lane Community College).

pregnated with phenolic resins; they are then heated and compressed to form materials such as formica or micarta. Other materials include thermoplastics, such as nylon. These nonmetallic materials exhibit excellent wear resistance. Some of these materials need very little lubrication. When plastic gears are used in gear trains, excessive temperature changes must be avoided to control damaging of dimensional changes. Good mating materials with nonmetallic gears are hardened steel and cast iron.

In many instances, gear sets are made up with different gear materials. Many worm drives use a bronze worm gear with a hardened steel worm. Cast iron gears work well with steel gears. To equalize the wear in gear sets, the pinion is made harder than the gear. Even wear in gear sets can be obtained when a gear ratio is used that allows for a "hunting tooth." For example, a gear set ratio of approximately 4:1 is needed. This is possible by using an 80 tooth gear in mesh with a 20 tooth gear. In this gear arrangement, the same tooth of the pinion will mesh with the same tooth of the gear in every revolution. If an 81 tooth gear were used, the teeth of the pinion will not equally divide into it, but each tooth of one gear will mesh with all of the mating teeth one after the other, distributing wear evenly over all teeth.

self-evaluation

SELF-TEST

1. Name two types of gears used to connect parallel shafts.

    a. _____

    b. _____

2. What are some advantages and some disadvantages of helical gears?

    a. _____

    b. _____

3. What is the direction of rotation of a pinion in relation to an internal gear when they are meshed together?

4. Two helical gears of the same hand and a 45 degree helix angle are in mesh. What relationship exists between the axis of the two shafts?

5. What is the gear reduction in a worm gear set when the worm has a double lead thread and the worm gear has 100 teeth?

6. Can a worm gear set ratio be changed by substituting a single start worm with a triple start worm?

7. What kind of gear material gives the greatest load carrying capacity for a given size?

8. What kind of material can be used for gears to run quietly at high speed?

9. How can the wear on the gear teeth be equalized when a large and a small gear are running together?

10. What kind of gear materials give corrosion resistance to gear sets?

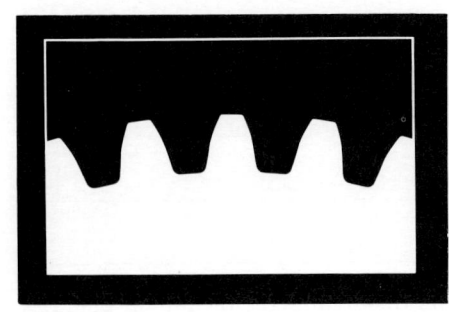

unit 2
spur gear terms and calculations

Most spur gears are mass produced on production gear making machines. Occasionally a spur gear needs to be made on a milling machine. Before a gear can be cut, its dimensions are calculated. This unit will help you to learn the names of gear tooth parts and how to calculate their sizes.

objective After completing this unit, you will be able to identify gear tooth parts and to calculate their dimensions.

information

Spur gear terms are illustrated in Figure 1. The definitions of these terms are as follows.

Addendum. The radial distance from the pitch circle to the outside diameter.

Dedendum. The radial distance between the pitch circle and the root diameter.

Circular thickness. The distance of the arc along the pitch circle from one side of a gear tooth to the other.

Circular pitch. The length of the arc of the pitch circle from one point on a tooth to the same point on the adjacent tooth.

Pitch diameter. The diameter of the pitch circle.

Outside diameter. The major diameter of the gear.

Root diameter. The diameter of the root circle measured from the bottom of the tooth spaces.

Chordal addendum. The distance from the top of the tooth to the chord connecting the circular thickness arc.

Chordal thickness. The thickness of a tooth on a straight line or chord on the pitch circle.

Whole depth. The total depth of a tooth space equal to the sum of the addendum and dedendum.

Working depth. The depth of engagement of two mating gears.

Clearance. The amount by which the tooth space is cut deeper than the working depth.

Backlash. The amount by which the width of a tooth space exceeds the thickness of the engaging tooth on the pitch circles.

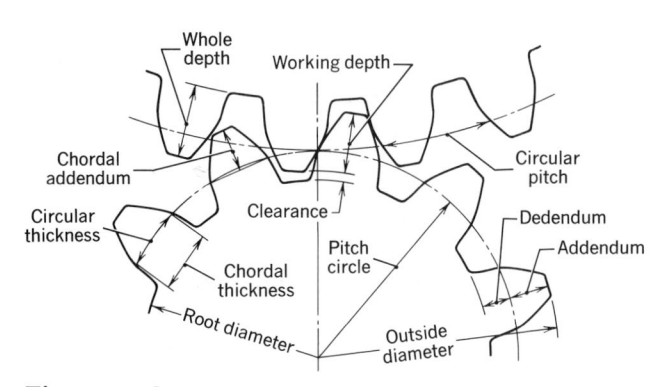

Figure 1. Spur gear terms.

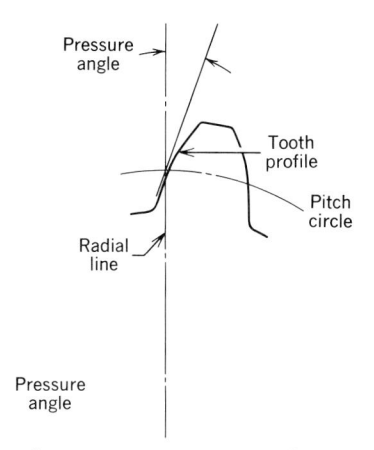

Figure 2. Pressure angle.

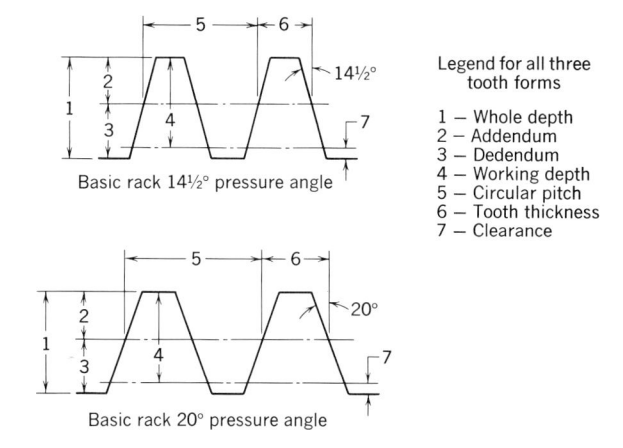

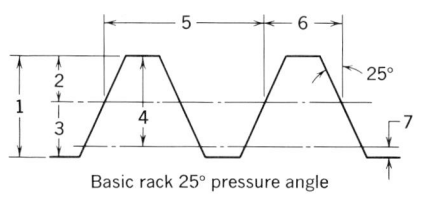

Figure 3. Comparison of tooth shape on gear rack with different pressure angles.

Diametral pitch. The number of gear teeth to each inch of pitch diameter.
Pressure angle. The angle between a tooth profile and a radial line at the pitch circle (Figure 2).
Center distance. The distance between the centers of the pitch circles.

Three spur gear tooth forms are generally used with pressure angles of 14½, 20, and 25 degrees. The 14½ degree tooth form is being replaced and made obsolete by the 20 and 25 degree forms. Figure 3 illustrates these three pressure angles as applied to a gear rack with all teeth being the same depth. The larger pressure angle makes teeth with a much larger base, which also makes these teeth much stronger. The larger pressure angles also allows the production of gears with fewer teeth. Any two gears in mesh with each other must be of the same pressure angle.

Information in this unit will include constants for 14½ degree pressure angle gear calculations because many existing gears and gear cutters are of this tooth form. When gear tooth measurements are to be made with gear tooth calipers, the chordal tooth thickness and also the chordal addendum must be calculated. Most of the gear tooth dimensions that you will calculate in this unit can be found in tables in the *Machinery's Handbook*, by Oberg and Jones.

The following symbols are used to represent gear tooth terms in spur gear calculations.

P = diametral pitch
D = pitch diameter
D_o = outside diameter
N = number of teeth in gear
t = tooth thickness — circular
a = addendum
b = dedendum
c = clearance
C = center distance
h_k = working depth
h_t = whole depth
t_c = tooth thickness — chordal
a_c = addendum — chordal

Table 1 gives the formulas to calculate the gear dimensions in the following examples.

EXAMPLE 1. Determine the dimensions for a 30 tooth gear, 14½ degree pressure angle, and a 2.500 in. pitch diameter.

1. Number of teeth $N = 30$.
2. Pitch diameter $D = 2.500$ in.
3. Diametral pitch $P = \dfrac{N}{D} = \dfrac{30}{2.500} = 12$

Section E Gears

Table 1
Spur Gear Formulas

To Find	Spur Gear Formulas 14½ degree Pressure Angle	20 and 25 degree Pressure Angles
Addendum, a	$a = \dfrac{1.0}{P}$	$a = \dfrac{1.0}{P}$
Dedendum, b	$b = \dfrac{1.157}{P}$	$b = \dfrac{1.250}{P}$
Pitch diameter, D	$D = \dfrac{N}{P}$	$D = \dfrac{N}{P}$
Outside diameter, D_o	$D_o = \dfrac{N + 2}{P}$	$D_o = \dfrac{N + 2}{P}$
Number of teeth, N	$N = D \times P$	$N = D \times P$
Tooth thickness, t	$t = \dfrac{1.5708}{P}$	$t = \dfrac{1.5708}{P}$
Whole depth, h_t	$h_t = \dfrac{2.157}{P}$	$h_t = \dfrac{2.250}{P}$
Clearance, c	$c = \dfrac{.157}{P}$	$c = \dfrac{.250}{P}$
Center distance, C	$C = \dfrac{N_1 + N_2}{2 \times P}$	$C = \dfrac{N_1 + N_2}{2 \times P}$
Working depth, h_k	$h_k = \dfrac{2}{P}$	$h_k = \dfrac{2}{P}$
Chordal tooth thickness, t_c	$t_c = D \sin\left(\dfrac{90 \text{ degrees}}{N}\right)$	$t_c = D \sin\left(\dfrac{90 \text{ degrees}}{N}\right)$
Chordal addendum, a_c	$a_c = a + \dfrac{t^2}{4D}$	$a_c = a + \dfrac{t^2}{4D}$
Diametral pitch, P	$P = \dfrac{N}{D}$	$P = \dfrac{N}{D}$
Center distance, C	$C = \dfrac{D_1 + D_2}{2}$	$C = \dfrac{D_1 + D_2}{2}$

4. Addendum $a = \dfrac{1}{P} = \dfrac{1}{12} = .083$ in.

5. Dedendum $b = \dfrac{1.157}{P} = \dfrac{1.157}{12} = .096$ in.

6. Tooth thickness
$t = \dfrac{1.5708}{P} = \dfrac{1.5708}{12} = .131$ in.

7. Clearance $c = \dfrac{.157}{P} = \dfrac{.157}{12} = .013$ in.

8. Whole depth $h_t = \dfrac{2.157}{P} = \dfrac{2.157}{12} = .179$ in.

9. Working depth $h_k = \dfrac{2}{P} = \dfrac{2}{12} = .166$ in.

10. Chordal tooth thickness

$t_c = D \sin\left(\dfrac{90 \text{ degrees}}{N}\right) =$

$2.500 \times \sin\left(\dfrac{90 \text{ degrees}}{30}\right) =$

$2.500 \times \sin 3 \text{ degrees} =$

$2.5 \times .0523 = .1307$ in.

11. Chordal addendum $a_c = a + \dfrac{t^2}{4D} =$

$.083 + \dfrac{.131^2}{4 \times 2.5} = .083 + \dfrac{.017}{10} =$

$.083 + .0017 = .0847$ in.

12. Outside diameter $D_o = \dfrac{N+2}{P} = \dfrac{30+2}{12} =$

$\dfrac{32}{12} = 2.6666$ in.

EXAMPLE 2. Determine the gear dimensions for a 45 tooth gear, 8 diametral pitch, 20 degree pressure angle.

1. Number of teeth $N = 45$
2. Diametral pitch $P = 8$
3. Pitch diameter $D = \dfrac{N}{P} = \dfrac{45}{8} = 5.625$ in.
4. Addendum $a = \dfrac{1}{P} = \dfrac{1}{8} = .125$ in.
5. Dedendum $b = \dfrac{1.250}{P} = \dfrac{1.250}{8} = .1562$ in.
6. Tooth thickness $t = \dfrac{1.5708}{P} = \dfrac{1.5708}{8}$

$= .1963$ in.

7. Clearance $c = \dfrac{.250}{P} = \dfrac{.250}{8} = .031$ in.
8. Whole depth $h_t = \dfrac{2.250}{P} = \dfrac{2.250}{8} = .281$ in.
9. Working depth $h_k = \dfrac{2}{P} = \dfrac{2}{8} = .250$ in.
10. Outside diameter $D_o = \dfrac{N+2}{P} = \dfrac{45+2}{8} =$

$\dfrac{47}{8} = 5.875$ in.

11. Chordal tooth thickness

$t_c = D \sin \dfrac{90 \text{ degrees}}{N} = 5.625 \times$

$\sin \dfrac{90 \text{ degrees}}{45} = 5.625 \times \sin 2 \text{ degrees} =$

$5.625 \times .0349 = .1963$ in.

12. Chordal addendum $a_c = a + \dfrac{t^2}{4D} =$

$.125 \text{ in.} + \dfrac{.1963^2}{4 \times 5.625}$

$= .125 \text{ in.} + .0017 \text{ in.} = .1267$ in.

The center distance between gears can be calculated when the number of teeth in the gears and the diametral pitch is known. Two gears in mesh make contact at their pitch diameters.

EXAMPLE 3. Determine the center distance between gears with 25 and 40 teeth and a diametral pitch of 14.

Center distance $C = \dfrac{N_1 + N_2}{2 \times P} = \dfrac{25 + 40}{2 \times 14} = \dfrac{65}{28} =$

2.3214 in.

Another method of finding the center distance if the pitch diameters are known is to add both pitch diameters and divide that sum by 2.

EXAMPLE 4. What is the center distance of two gears when their pitch diameters are 2.500 and 3.000 in., respectively?

Center distance $C = \dfrac{D_1 + D_2}{2} = \dfrac{2.500 + 3.000}{2} =$

$\dfrac{5.500}{2} = 2.750$ in.

self-evaluation

SELF-TEST 1. What are commonly found pressure angles for gear teeth?

_____

2. Why are larger pressure angles used on gear teeth?

_____

Section E Gears

3. What is the center distance between two gears with 20 and 30 teeth and a diametral pitch of 10?

4. What is the center distance between two gears with pitch diameters of 3.500 and 2.500 in.?

5. What is the difference between the whole depth of a tooth and the working depth of a tooth?

6. What relationship does the addendum and the dedendum have with the pitch diameter on a tooth?

7. What is the outside diameter and the tooth thickness on a 50 tooth gear with a diametral pitch of 5?

8. What is the diametral pitch of a gear with 36 teeth and a pitch diameter of 3.000 in.?

9. What is the outside diameter, whole depth, pitch diameter, and dedendum for a 40 tooth, 8 diametral pitch, 20 degree pressure angle gear?

10. What is the outside diameter, clearance, whole depth, tooth thickness, and pitch diameter for a 48 tooth, 6 diametral pitch, $14\frac{1}{2}$ degree pressure angle gear?

unit 3
cutting a spur gear

When a spur gear is cut on a milling machine, specific operations must be performed. This unit deals with the order of these operations and how they are accomplished.

objective After completing this unit, you will be able to set up for and machine a spur gear on a milling machine.

information

Gears cut on a milling machine are limited in their possible uses because of the quality obtainable and the availability of the correct cutter. But in an emergency or for experimental purposes, satisfactory gears can be cut on a milling machine. Gears produced on gear generating machines are usually produced more economically and with higher accuracy. Before a gear is cut, the gear cutter must be selected. Involute gear cutters are available in sets numbered from 1 to 8 with pressure angles of either $14\frac{1}{2}$ or 20 degrees. These eight cutters are made with eight different tooth forms for each diametral pitch, depending on the number of teeth for which the cutter is to be used. Each cutter is designed to cut a number of teeth. See Table 1.

Each cutter is made with the correct tooth shape for the lowest number of teeth in its range. This means that a Number 5 cutter has the exact tooth shape for a 21 tooth gear, and gears with 22 to 25 teeth have only an approximate tooth shape. Gears with 22 to 25 teeth will work when used with another gear of the same pitch, but they will not run as smoothly as a 21 tooth gear. More accurate tooth forms within each cutter's range can be obtained by half number gear tooth cutters, such as a Number $5\frac{1}{2}$ cutter, which cuts 19 and 20 teeth. Special cutters are made for a specific number of teeth in a gear or for low number of teeth in a pinion, from 6 to 11 teeth. Gear tooth cutters are marked on their side with the information as to their number, the diametral pitch, the number of teeth the cutter is designed for, the pressure angle, the whole depth the tooth is to be cut and, usually, the manufacturer's trademark.

Let us assume that a gear is to be made with 48 teeth and a 12 diametral pitch with a $14\frac{1}{2}$ degree pressure angle. The first step is to calculate the gear blank dimensions. Determine the dimensions for the:

Outside diameter $= D_o = \dfrac{N + 2}{P} = \dfrac{50}{12} =$ 4.167 in.

Width of gear $= .750$ in.

Whole depth of cut $= h_t = \dfrac{2.157}{P} = \dfrac{2.157}{12} =$ .180 in.

Number of cutter to use $= 48$ teeth $=$ Number 3

Table 1

Number of Cutter	Number of Teeth Cut	Number of Cutter	Number of Teeth Cut
1	135 to rack	5	21 to 25
2	55 to 134	6	17 to 20
3	35 to 54	7	14 to 16
4	26 to 34	8	12 and 13

Then make the gear blank from the required material. You have to make it possible to hold the gear blank while cutting teeth. This can be done on a mandrel or with a hub on the gear. If more than one gear of the same size is needed, teeth can be cut on a piece of bar stock, and the individual gears are later parted off in a lathe or cutoff machine.

Prepare the milling machine for gear cutting by mounting the dividing head on the milling machine table. Make sure that the dividing head spindle axis is parallel to the table surface. Mount the footstock on the machine table. Check its axis for being parallel to the machine table.

Lubricate the hole in the gear blank before pushing the gear blank on the mandrel in the arbor press.

Fasten a driving dog at the large end of the mandrel and mount it between the dividing head and footstock centers. Clamp the dog tail in the driving slot at the dividing head spindle. With a dial indicator, test the height of the mandrel at the foot stock and at the dividing head (Figure 1). Remember to allow for the taper on the mandrel. Make footstock center height adjustments if necessary.

Move the saddle close to the column and mount the gear cutter on the machine arbor near the centerline of the gear blank. Keeping the work and cutter as close to the column as possible gives a more rigid setup. Mount the cutter so that the cutting pressure will be against the dividing head.

Align the cutter so it is centrally located over the gear blank (Figure 2). Calculate the RPM to use and adjust the machine to it. With the cutter

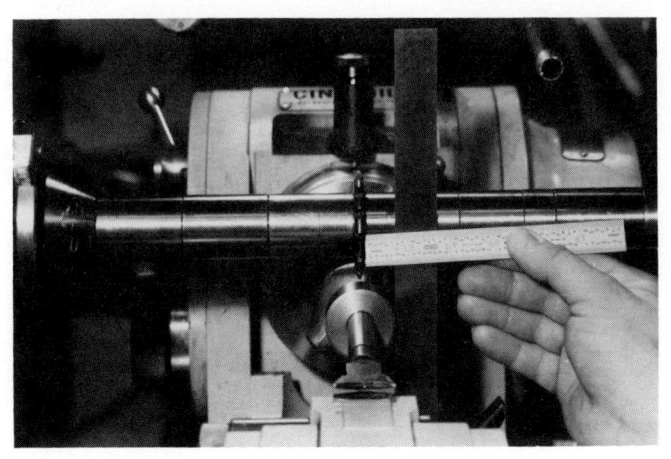

Figure 2. Aligning the cutter centrally over the gear blank (Lane Community College).

rotating, raise the knee until the cutter just touches the gear blank, then set the micrometer dial to zero. Move the table longitudinally until the cutter clears the workpiece and turn off the machine.

Calculate the dividing head movement for the number of teeth to be cut. The formula is $\frac{40}{N} = \frac{40}{48} = \frac{5}{6}$. Any hole circle divisible by 6 can be used. Using the 54 hole circle we get $\frac{5}{6} \times \frac{9}{9} = \frac{45}{54}$, or 45 holes in the 54 hole circle. Adjust the index crank to the correct hole circle. Set the sector arms for the wanted number of holes, remembering not to count the hole the index pin is in. Rotate the index crank clockwise one revolution and seat the index pin in the hole circle number. Lock the dividing head spindle. This will be the location for the first tooth space. If you develop the habit of always making the first cut with the index pin in the numbered hole of the hole circle, it will be easy to double-check your location by coming back to the original location.

Set the depth of cut, allow for a finishing cut of approximately $\frac{1}{32}$ in. Start the spindle and hand feed the table to the cutter and let the cutter just mark the edge of the gear blank. Back the cutter away from the gear blank. Index to the next position and repeat marking the gear blank around the complete circumference (Figure 3). Count the number of spaces. If you made an indexing error, it can be corrected without having ruined a gear. When the number of spaces is correct, make all the roughing cuts. Adjust the table feed trip dogs to stop the table feed at the

Figure 1. Checking the axis of the mandrel to be parallel with the table surface (Lane Community College).

Unit 3 Cutting a Spur Gear 183

Figure 3. Marking all the tooth spaces around the circumference of the gear blank (Lane Community College).

Figure 4. Finish cut the gear teeth (Lane Community College).

completion of the cut on one side and when the cutter clears the gear blank at the starting side. Adjust the feed rate to obtain an acceptable finish. Using a center rest under the gear blank during the cutting operation helps to control chatter.

Raise the knee for the finish cut. Cut two spaces and measure the tooth between these spaces for its thickness with a gear tooth vernier caliper.

Make the remaining finishing cuts (Figure 4). Remove all burrs with a file or wire brush.

Remove the mandrel from the gear and clean the machine.

This completes the cutting of a spur gear in the milling machine.

self-evaluation

SELF-TEST

1. What are two factors against cutting spur gears in the milling machine?

   a. _____

   b. _____

2. How many gear cutters are in a standard set?

3. Which cutter is used to cut a gear with 38 teeth?

4. Can a $14\frac{1}{2}$ degree pressure angle gear be in mesh with a 20 degree pressure angle gear?

5. A Number 6 gear cutter has the correct tooth shape for a gear with how many teeth?

6. What information is marked on the side of gear cutters?

7. Why should the work and cutter be mounted close to the column?

8. How many holes should be located within the sector arms?

9. Why is the gear blank marked with the cutter prior to cutting the teeth?

10. When is a center rest used?

WORKSHEET

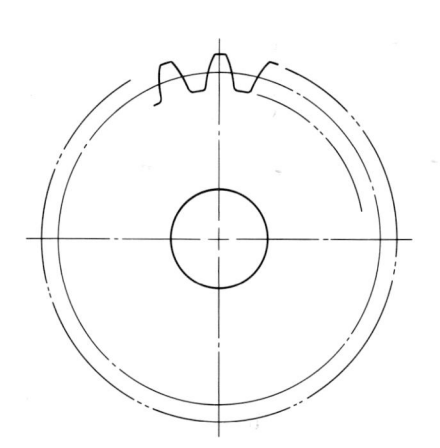

Spur gear — Material MS
  32 Teeth
  12 Diametral pitch
  14½° pressure angle
  ¾ in. wide
Calculate the:
  1. Outside diameter _____
  2. Pitch diameter _____
  3. Whole depth _____
  4. Chordal addendum _____
  5. Chordal thickness _____

Figure 5. Spur gear.

1. Calculate the gear dimensions required for the spur gear in Figure 5.
2. Machine the gear blank in a lathe.
3. Press the gear blank on a mandrel.
4. Set up the dividing head on the milling machine.
5. Mount the gear cutter on the arbor and set the RPM and feedrate.
6. Make sure the gear blank axis is parallel to the table surface.
7. Center the cutter over the gear blank axis.
8. Zero the knee crank dial when the revolving cutter touches the gear blank surface.
9. Set the depth of cut for the roughing cut.
10. Calculate the index movement necessary and set the sector arms to the required number of holes.
11. Mark all gear spaces lightly on the gear blank by indexing through all 32 spaces, then double-check their number.
12. Make the first roughing cut, then adjust and set the table feed trip dogs at the beginning and end of the cut.
13. Make all roughing cuts.
14. Set depth for finish cut and cut two spaces.
15. Measure the tooth thickness.
16. Complete the finish cut on all spaces.
17. Remove and deburr the gear.
18. Clean the milling machine and accessories.

unit 4
gear inspection and measurement

When gears are made, their important dimensions must be measured to assure a useful quality product at their completion. This unit explains commonly used gear measurement methods.

objective After completing this unit, you will be able to measure gears by two methods: the gear tooth vernier caliper and the micrometer with two pins.

information

A machinist making spur gears in a machine shop should be able to measure gears by two common methods. These methods are with a gear tooth vernier caliper and with two pins and a micrometer. These gear measurements either are made on the pitch diameter or they involve the pitch diameter. When a gear tooth vernier caliper is used in gear measurement, the outside diameter of the gear is important because the height adjustment of the vernier caliper is set to the chordal addendum (Figure 1). Any changes in the outside diameter would affect the chordal addendum for a given pitch diameter. A gear tooth vernier caliper measures the tooth thickness on the pitch diameter, but the measurement is the chordal tooth thickness, which is less than the circular tooth thickness. The chordal addendum is larger than the addendum on a gear tooth. Dimensions of the addendum, chordal addendum, circular tooth thickness, and chordal tooth thickness can be found in the *Machinery's Handbook* or they may be calculated. As an example, let us determine the dimensions required to measure a 6 diametral pitch gear with 12 teeth.

Outside diameter
$$D_o = \frac{N + 2}{P} = \frac{12 + 2}{6} = \frac{14}{6} = 2.3333 \text{ in.}$$

Addendum $a = \frac{1}{P} = \frac{1}{6} = .1666 \text{ in.}$

Pitch diameter $D = \frac{N}{P} = \frac{12}{6} = 2.000 \text{ in.}$

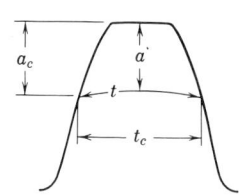

a = Addendum
a_c = Chordal addendum
t = Circular tooth thickness
t_c = Chordal tooth thickness

Figure 1. Illustration of the difference between the standard and chordal addendum and tooth thickness.

Circular tooth thickness
$$t = \frac{1.5708}{P} = \frac{1.5708}{6} = .2618 \text{ in.}$$
Chordal addendum
$$a_c = a + \frac{t^2}{4D} = .1666 + \frac{.2618^2}{4 \times 2}$$
$$= .1666 + \frac{.0685}{8} = .1666 + .0085 = .1751 \text{ in.}$$
Chordal tooth thickness
$$t_c = D \sin \frac{90 \text{ degrees}}{N} = 2 \times \sin \frac{90 \text{ degrees}}{12}$$
$$= 2 \times \sin 7.5 = 2 \times .1305 = .261 \text{ in.}$$

With these dimensions known, the gear can be measured with a gear tooth vernier caliper. The vertical scale is set to the chordal addendum of .175 in. and, with the vernier caliper resting on the top of the tooth (Figure 2), the tooth thickness should measure .261 in. If the outside diameter of the gear is .002 in. less than the calculated dimension, the vertical scale reading of the vernier caliper should be reduced to .174 in. This would assure that the tooth thickness measurement is made on the pitch diameter.

When any gear tooth measurements are made, especially while the gear is still on the milling machine, any burrs on the gear teeth must be carefully removed, or inaccurate measurements will result.

Another very accurate method of measuring gears is with two pins or wires and a micrometer (Figure 3). The diameter of the pins is calculated by dividing 1.728 by the diametral pitch. The constant 1.728 is the most commonly used for external spur gears. The dimension "M" (Figure 4) is found in the *Machinery's Handbook* calculated for different gear pressure angles and different teeth numbers conforming to the Van Keuren standard. In most of these tables, the dimensions are given for a diametral pitch of 1. For other diametral pitches divide the value in the table by the diametral pitch used. Internal spur gears can also be measured with this method.

As an example, determine the dimension "M" for a 45 tooth gear 10 diametral pitch, 14½ degree pressure angle P.

The pin diameter $\frac{1.728}{10} = .1728$ in.

The dimension "M" for 1 diametral pitch is 47.4437 in.

For a 10 diametral pitch divide $\frac{47.4437}{10} = 4.74437$ in. 4.74437 in. is the dimension for a gear without any backlash allowance. Gears need some backlash and, if .005 in. backlash is required, the dimension "M" must be recalculated. Again, we consult the *Machinery's Handbook* and find that for a 45 tooth gear, each .001 in. tooth thickness reduction at the pitch line re-

Figure 2. Measuring a spur gear with a gear tooth vernier caliper (Lane Community College).

Figure 3. Measuring a spur gear with two pins and a micrometer (Lane Community College).

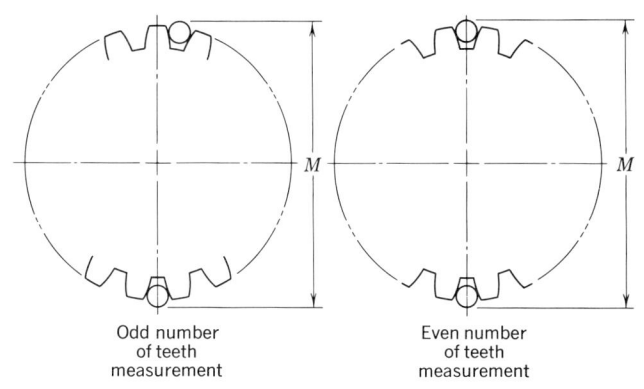

Figure 4. Measuring of gear dimensions over two pins.

quires dimension "M" to be reduced by .0031 in.

To obtain .005 in. backlash, multiply .0031 in. by 5, which equals .0155 in. Then reduce the measurement over wires, which is 4.74437 in. by .0155 in., to get a dimension of 4.72887 in.

Another gear inspection method available to a machinist is an optical comparator. The optical comparator employs a light, a lens, a mirror, and a viewing screen to enlarge the shadow outline of a gear being inspected (Figure 5). An enlarged transparent drawing of the part equal to the magnification power of the lens is matched to the shadow on the screen for a direct comparison of

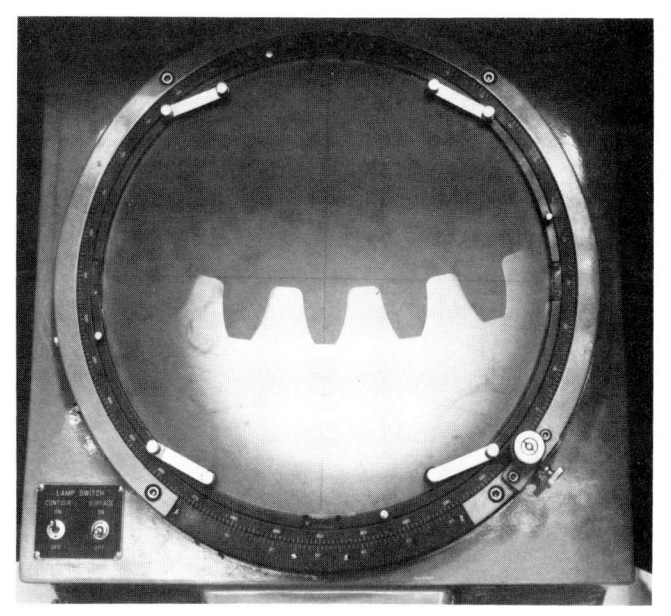

Figure 5. Using an optical comparator to check gear tooth dimensions (Lane Community College).

the enlarged shadow. Special testing and gear measuring tools are made to check the involute profile of a gear tooth, the gear tooth spacing, or the accuracy of the helix on a helical gear, or to check the meshing of a gear with a master gear to determine runout.

self-evaluation

SELF-TEST

1. What are two common gear measurements performed by a machinist?

   a. _____

   b. _____

2. Which tooth thickness is measured by a gear tooth vernier caliper?

   _____

3. The vertical scale of a gear tooth vernier caliper is set to what dimension?

   _____

4. Which is larger, the chordal addendum or the standard addendum?

   _____

   _____

5. Which is larger, the chordal tooth thickness or the circular tooth thickness?

6. Name two ways of determining the chordal thickness of a gear tooth?

7. When a gear is measured with a micrometer and two pins, can any size pin be used?

8. How is the dimension determined when measuring over pins?

9. When a dimension is given for 1 diametral pitch but the gear measured is 12 diametral pitch, how is the measurement determined?

10. How is an optical comparator used in gear measurement?

section f
shapers and planers

The shaper and planer are presented together because, in their basic forms, they are machines designed to generate flat surfaces by linear travel of the tool past the workpiece in the case of the shaper, or linear travel of the workpiece past a cutting tool in the case of the planer. Both machines are quite versatile and can be used for a remarkable variety of functions.

The planer was designed out of the necessity to be able to generate reliable flat surfaces. This planer (Figure 1) was introduced by a Welshman, Richard Roberts, in 1817. The cutting power was supplied by hand and a strong machinist could surface about 1 square foot of a cast iron workpiece in an hour's time. This machine was the most important basic machine tool to the subsequent manufacturer of other machine tools and is still very important for precise generation of the basic geometry of highly accurate machine tools such as jig borers.

Figure 1. Roberts' planer of 1817. This machine made the production of flat surfaces possible and is the basic machine tool for the manufacture of other machine tools (Courtesy of DoAll Company).

THE HORIZONTAL SHAPER The horizontal shaper (Figure 2) is an extremely versatile machine tool, particularly in those situations where small numbers of parts are to be made and a variety of angles are to be machined into the part (Figure 3).

One of the more common uses of shapers is the production of dovetail assemblies (Figures 4a and 4b) for applications such as toolslides used on machine tools like lathes, shapers, and planers.

Another interesting application of the shaper is in the production of internal features in a workpiece (Figure 5). The making of internal keyways on limited production parts, particularly in sizes and shapes where the use of an ordinary keyway broach and shop press is not feasible, makes the shaper a difficult machine to do without. This internal capability is called slotting and can be combined with the use of the index head (Figure 6) to produce a variety of internal shapes. In some instances, specialized tooling is made up (Figure 7) to make unusual internal and external parts (Figure 8). Sophisticated hydraulic tracing systems have also been applied to shapers to produce internal contours on a production basis (Figure 9a, 9b, and 9c).

Contour work on shapers is by no means limited to internal work. External contouring is done on the shaper by a wide variety of means. The most basic form of contouring takes place when the

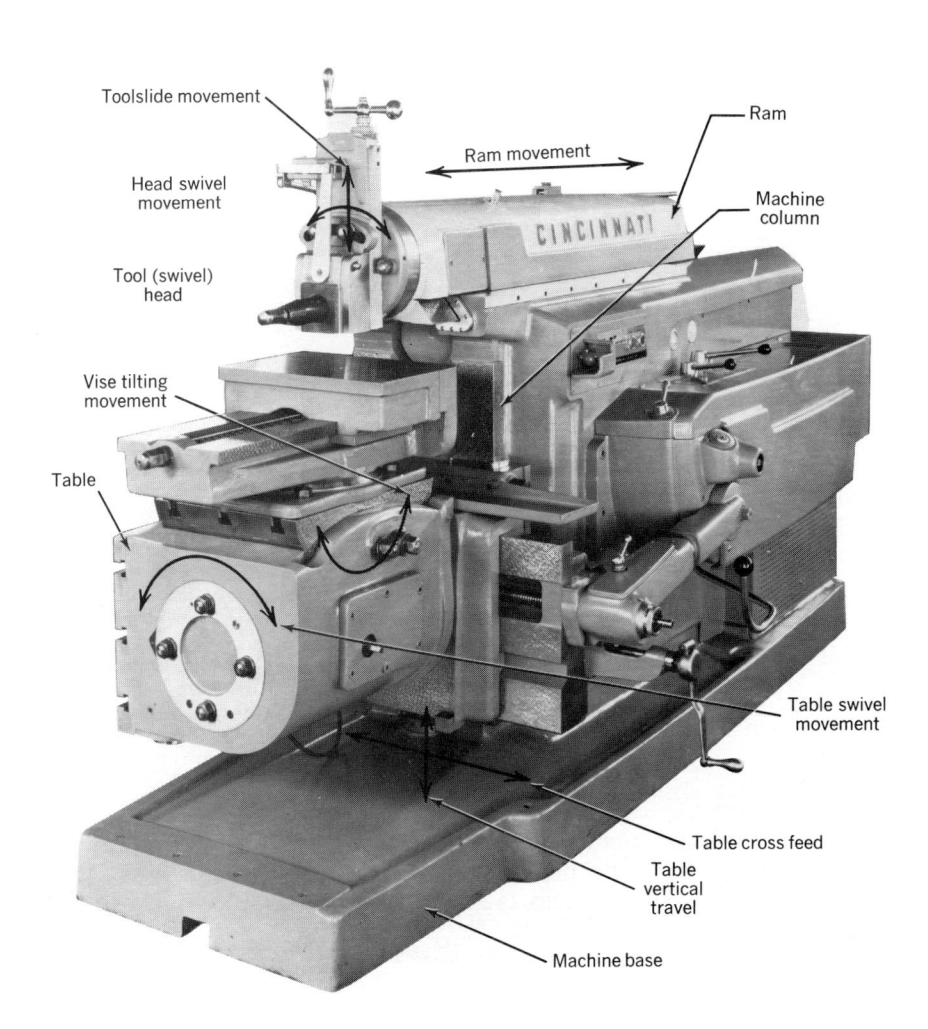

Figure 2. The horizontal shaper is a versatile machine tool capable of many motions. The model shown is a universal shaper (Courtesy of Cincinnati Incorporated). We are grateful to Cincinnati Incorporated for access to materials in their archives. Their last production of shapers was in 1969. They furnish replacement parts for currently installed Cincinnati shapers. Their present production is mechanical and hydraulic press brakes and shears, shear conveyors, press brake tooling, coil processing equipment, and powder metal compacting presses.

Figure 3. The workpiece in the universal shaper can be rotated about three axes to obtain compound angles on the workpiece (Courtesy of Cincinnati Incorporated).

Figure 4a. The toolhead of the shaper can be readily swivelled to produce dovetails for machine assemblies. A toolslide is being machined.

Figure 4b. An internal dovetail is being machined (Courtesy of Cincinnati Incorporated).

Figure 5. Making of internal keyways is an important ability of the shaper. This Kennedy keyway has to be held to close size tolerance to work effectively (Courtesy of Cincinnati Incorporated).

Figure 6. The dividing or indexing head in combination with the shaper can be used for making complex internal shapes (Courtesy of Cincinnati Incorporated).

Figure 7. A specially tooled shaper for complex internal parts (Courtesy of Rockford Machine Tool Company).

operator manually controls the downfeeding of the tool and the cross feeding of the work to follow a layout line marked on the part (Figure 10). Another unique method of contouring on the shaper is the use of a strong wide template with a roller follower that moves the shaper table up and down while the cross-feeding is taking place (Figure 11). This type of system must be strongly constructed because the weight of the table and workpiece and much of the force of the cut must be taken through the template and roller. Hydraulic

Section F Shapers and Planers

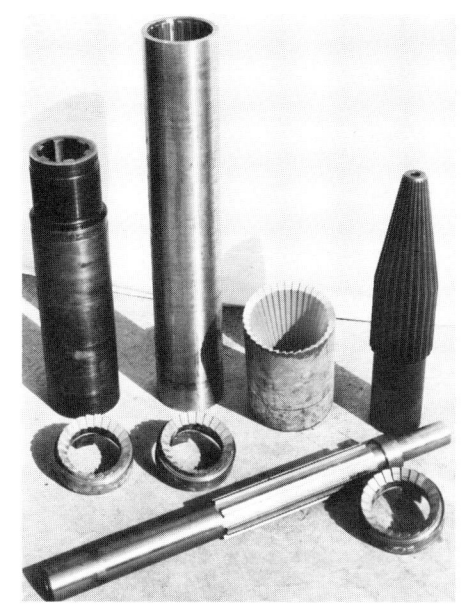

Figure 8. A selection of complex parts on which the shaper was used with special tooling (Courtesy of Rockford Machine Tool Company).

Figure 9a. A horizontal shaper with indexing system and hydraulic tracing controls for a production application. (Courtesy of Rockford Machine Tool Company).

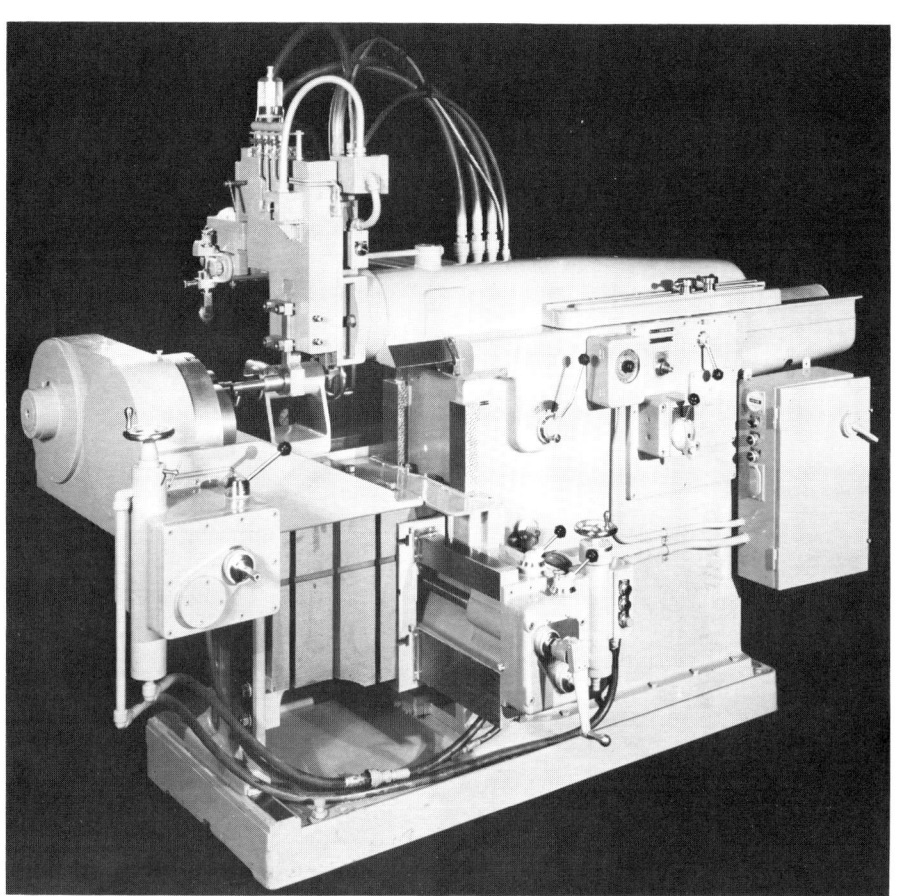

Figure 9b. Tracer stylus with master in place between centers. The master rotates at the same rate as the internal part shown in Figure 9c (Courtesy of Rockford Machine Tool Company).

Figure 9c. The internal part being machined in the hydraulic tracing set up (Courtesy of Rockford Machine Tool Company).

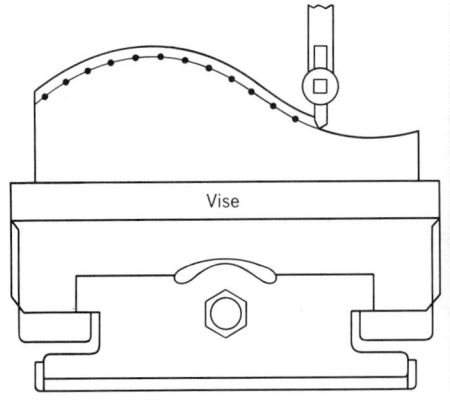

Figure 10. Contouring by hand in the shaper to a layout line on the workpiece (Courtesy of Cincinnati Incorporated).

Figure 11. Contouring can be done with this device, which moves the table up and down during cross feeding (Courtesy of Cincinnati Incorporated).

systems are also used in external contouring with the shaper (Figures 12 and 13). Although the shaper is most often considered to be a toolroom machine, it can be adapted to production functions by the addition of specialized components (Figures 14a, 14b, and 15).

Figure 12. (Top) Two-dimensional contour cutting in the shaper by using a hydraulic tracing device with a template. Note the massive chip. Cuts like this can be made on heavy duty shapers using large, well-supported cutting tools (Courtesy of Cincinnati Incorporated).

Figure 13. (Top right) Horizontal shaper with hydraulic tracing attachment for three-dimensional duplicating. The stylus on the left contacts the master on the table. The toolhead hydraulically follows the stylus movements during the ram travel to produce a part that duplicates the master (Courtesy of Rockford Machine Tool Company).

Figure 14a. Special tooling for holding and working on the breech end of a gun barrel (Courtesy of Rockford Machine Tool Company).

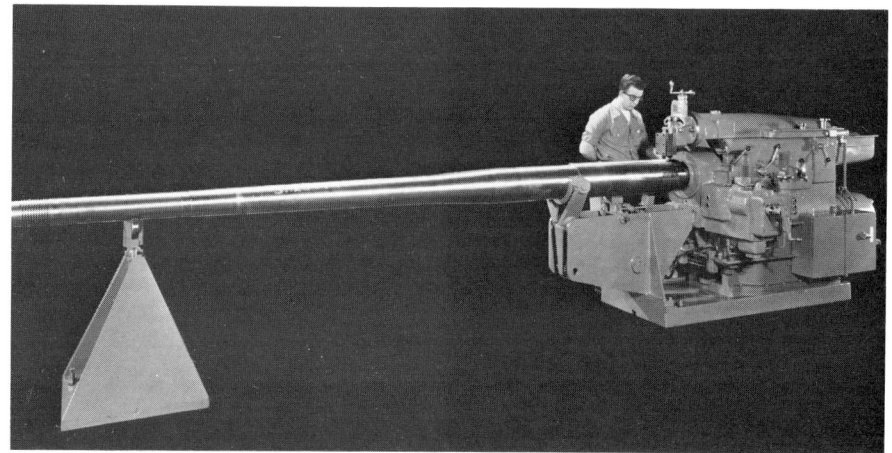

Figure 14b. Details of the indexing fixture and toolhead (Courtesy of Rockford Machine Tool Company).

195

Figure 15. Tooling applied to the shaper for making a bevel cut on both ends of precut steel plates on a production basis (Courtesy of Rockford Machine Tool Company).

THE VERTICAL SHAPER (SLOTTER) The vertical shaper or slotter (Figure 16) is used often in job shops and toolrooms. This machine functions much like its horizontal counterpart except that it is supplied with a rotary table as standard equipment. This rotary table can be moved both traversely and longitudinally. The ram can also be tilted 10 degrees from the vertical. This makes the machine especially versatile for making complex internal shapes. It is also much easier for the operator than an arrangement with a dividing head on a horizontal shaper, as shown in Figure 6, because the work is quite visible. These machines can also be equipped with tracing equipment for the production of contours (Figure 17).

PLANERS Planers have evolved a great deal since the manually operated model of Roberts in 1817. Figure 18 shows a century of progress in planer design. This machine is termed a double housing planer because the upright columns are solidly joined together at the top, which promotes rigidity. A more recent double housing planer (Figure 19) has the same functional features, but much of the mechanism is concealed. The size of a double housing planer is specified by the largest workpiece that can be machined, for example 72 in. × 72 in. × 34 ft, which means an object no wider or higher than 72 in. nor longer than 34 feet could be handled by a planer of that size specification.

The openside planer (Figure 20) has a single housing or column that permits machining of parts of a width that would not pass between the housing of a double housing planer. They are, however, specified in the same way, such as 48 in × 48 in. × 20 ft, for

Figure 16. A slotter being used for internal slotting in a casting (Courtesy of Rockford Machine Tool Company).

Figure 17. A slotter with a hydraulic tracer installed (Courtesy of Rockford Machine Tool Company).

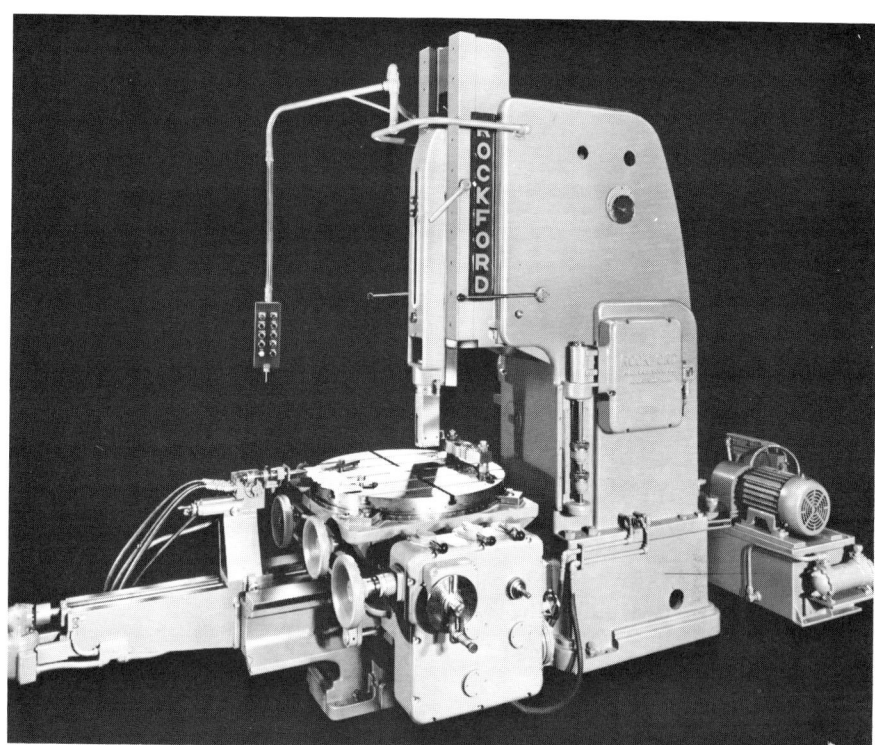

Figure 18. A century after Roberts, the planer had nearly evolved to its present form (Courtesy of Giddings & Lewis).

example. Openside planers may also be found with accessory support columns that can be installed or removed as necessary to accommodate wide parts (Figure 21). Planers also have a good deal of variation in length to meet the needs of different users (Figure 22).

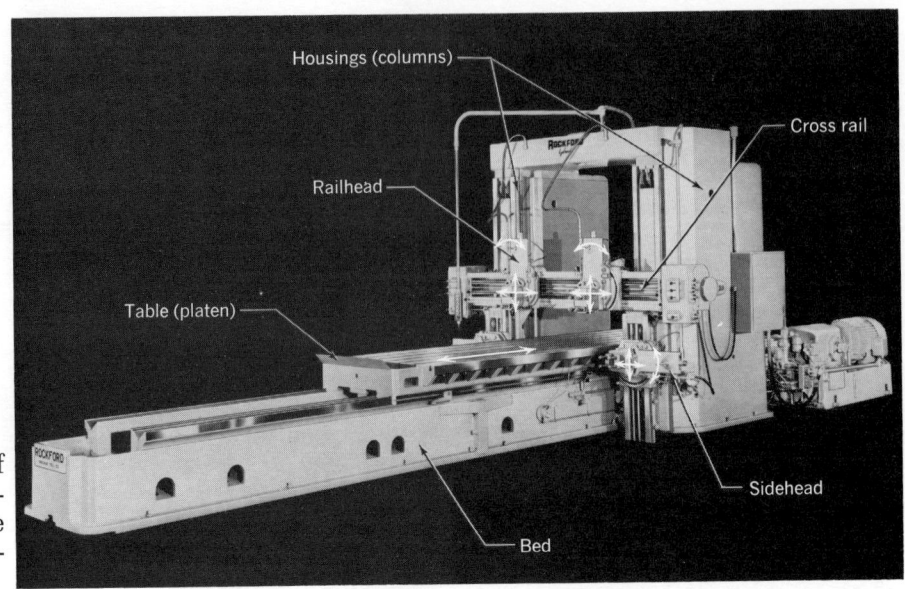

Figure 19. The present evolution of the planer with a hydraulically driven table or platen. This is a double housing planer (Courtesy of Rockford Machine Tool Company).

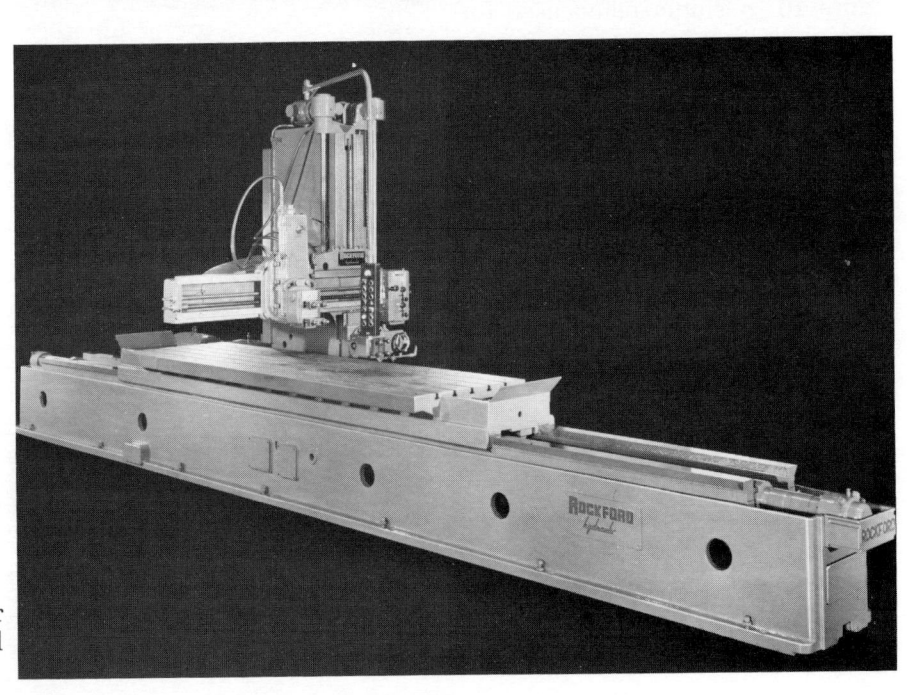

Figure 20. An openside planer (Courtesy of Rockford Machine Tool Company).

The smaller openside-type planers (under 42 in × 42 in.) are called shaper-planers (Figure 23) by one manufacturer. These, as other planers, can be equipped with either single or dual railheads or with or without a sidehead (Figure 24).

The advantage of multiple heads is important for productivity. These heads may be used in combination to make simultaneous cuts on the workpiece (Figure 25). In addition to multiple heads for single point tools, accessory milling heads may be obtained for planers to perform milling operations, as shown in Figure 26. From this type of adaptation of the planer a type of milling machine has been derived called the planer-mill, which has evolved into what is now frequently called an adjustable rail mill (Figure 27). Other adaptations of cutting heads on the basic design of the planer have resulted in the planer-grinder (Figure 28).

Figure 21. An openside planer with an accessory support column (Courtesy of Rockford Machine Tool Company).

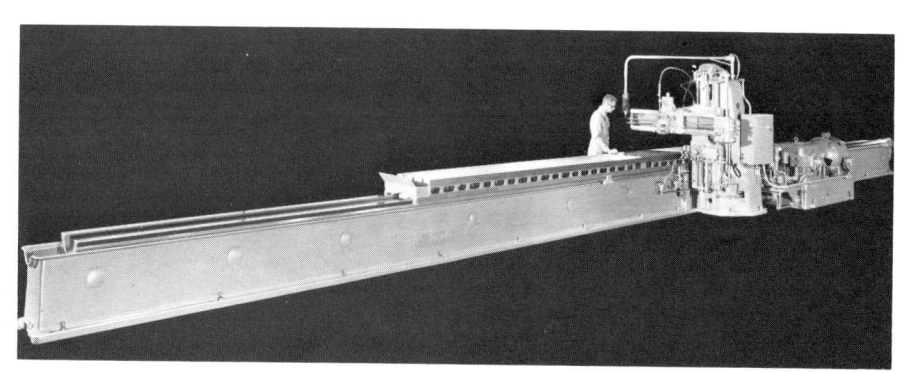

Figure 22. Long bed openside planer (Courtesy of Rockford Machine Tool Company).

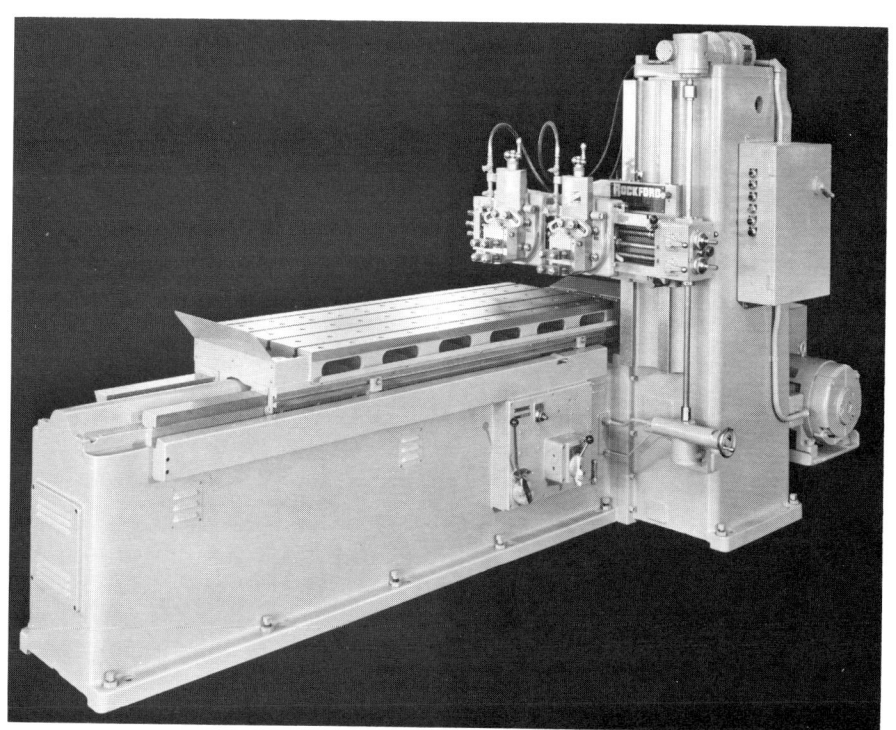

Figure 23. Smaller openside planers are often called shaper-planers (Courtesy of Rockford Machine Tool Company).

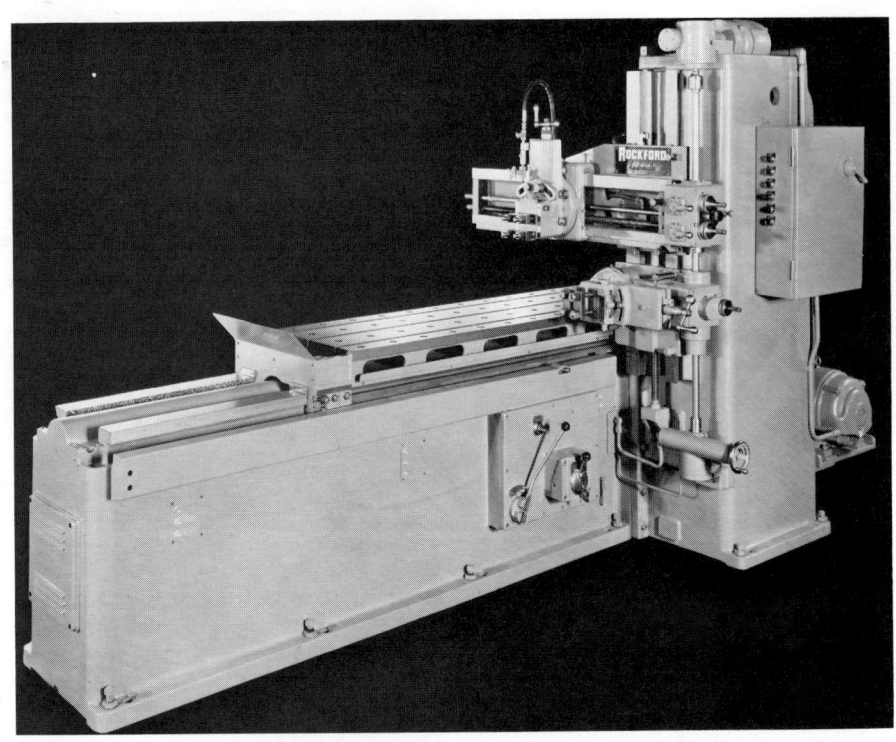

Figure 24. Shaper-planer with sidehead (Courtesy of Rockford Machine Tool Company).

Figure 25. The ability to make simultaneous cuts is a great advantage on the planer (Courtesy of Rockford Machine Tool Company).

One of the most important basic uses of the planer is the machining of castings which are machine tool components (Figure 29). Single point generating of a surface with sharp tools is considered by some manufacturers of machine tools to result in less residual stress in the casting than in milling the surface, which could result in a more accurate machine tool over its life of service.

Figure 26. An accessory milling head on the planer (Courtesy of Rockford Machine Tool Company).

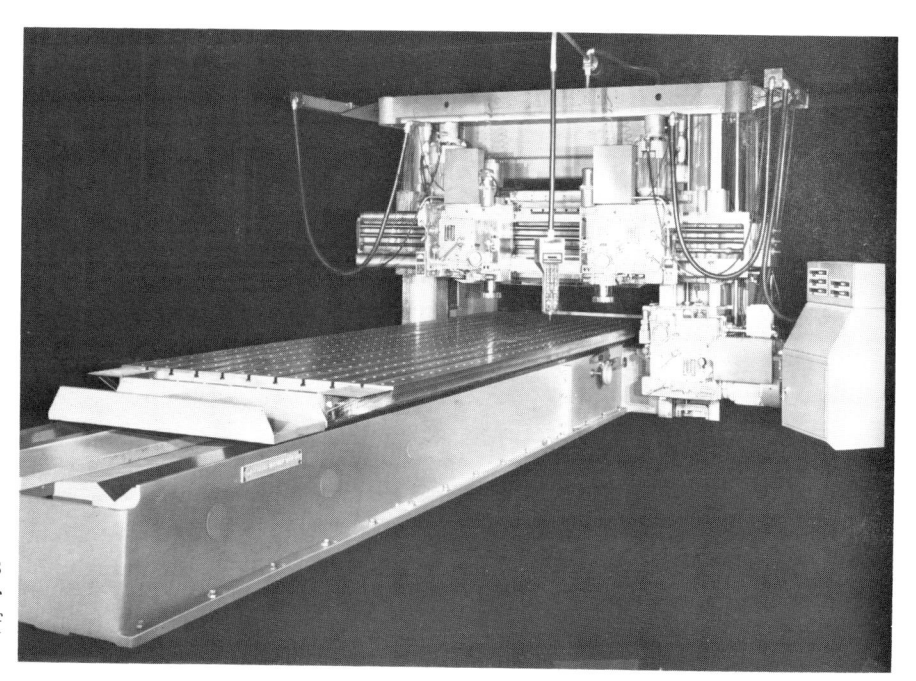

Figure 27. The planer design has been evolved into the planer mill or adjustable rail mill (Courtesy of Rockford Machine Tool Company).

The massiveness of planers often permits the use of form tools having a great deal of surface area. Figure 30 shows a smooth form cut being made over a large area of tool contact without observable chatter. The work is being held in a planer vise.

As is true with the shaper, there are many adaptations that can be done with planers to extend their capabilities. One particularly

Figure 28. A planer-grinder (Courtesy of Rockford Machine Tool Company).

Figure 29. Machining a component casting of a machine tool (Courtesy of Rockford Machine Tool Company).

difficult machining task is the machining of an accurate sector. Figures 31a and 31b show how an adaptation of an openside planer can be made to machine this type of workpiece.

Internal slotting is another operation that can be done on the planer (Figure 32). It is often very difficult to machine internal features such as keyways deep inside of castings.

Figure 30. Using a large form tool on a planer. Work is held in a planer vise (Courtesy of Rockford Machine Tool Company).

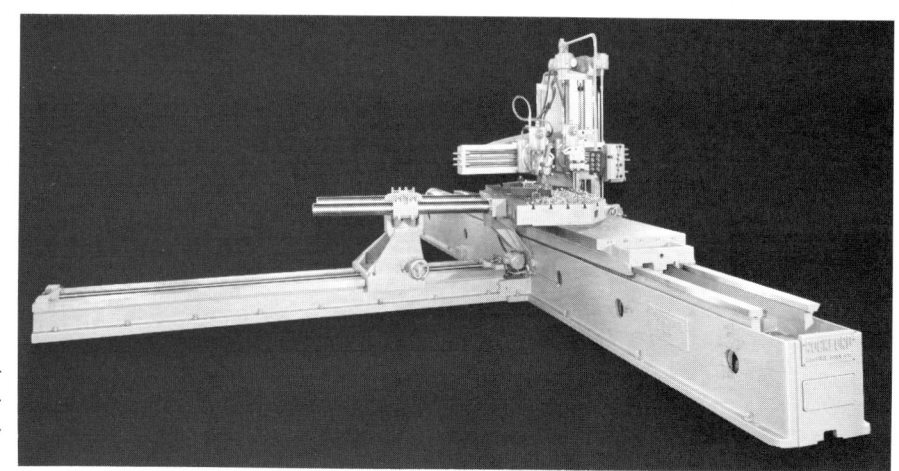

Figure 31a. Planer adapted for accurate machining of a sector (Courtesy of Rockford Machine Tool Company).

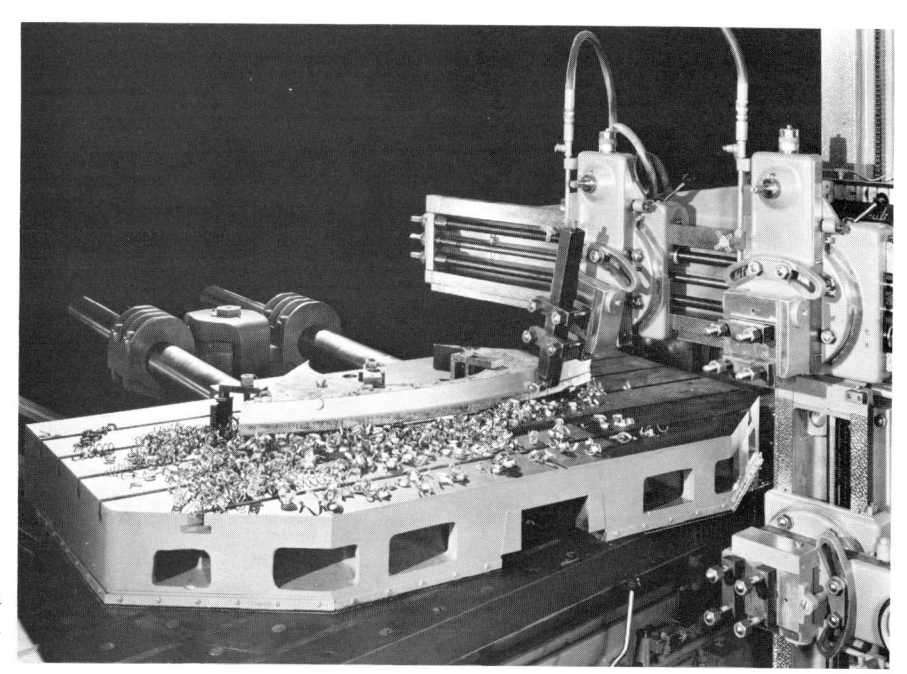

Figure 31b. Detail of machining of a sector (Courtesy of Rockford Machine Tool Company).

203

Figure 32. Planer equipped with special tool holder for internal slotting (Courtesy of Rockford Machine Tool Company).

Figure 33a. Planer being used to trace a vee section (Courtesy of Rockford Machine Tool Company).

As with the shaper, the planer is adaptable to a wide range of hydraulic tracing devices. The most basic of these devices is the template-type duplicator, which uses a template that represents the cross section of the work to be machined (Figures 33a and 33b). This type of tracer can also be combined with synchronized rotary

Section F Shapers and Planers

Figure 33b. The template is mounted immediately behind the toolhead (Courtesy of Rockford Machine Tool Company).

Figure 34. Tracing coupled with synchronized rotation being used to machine a supercharger rotor (Courtesy of Rockford Machine Tool Company).

movement to produce complex regular shapes like this rotor to be used in a positive displacement engine supercharger (Figure 34). Three-dimensional duplicating is also a function that can be done on the planer (Figure 35). In this case, the master pattern is moved

206   Machine Tools and Machining Practices

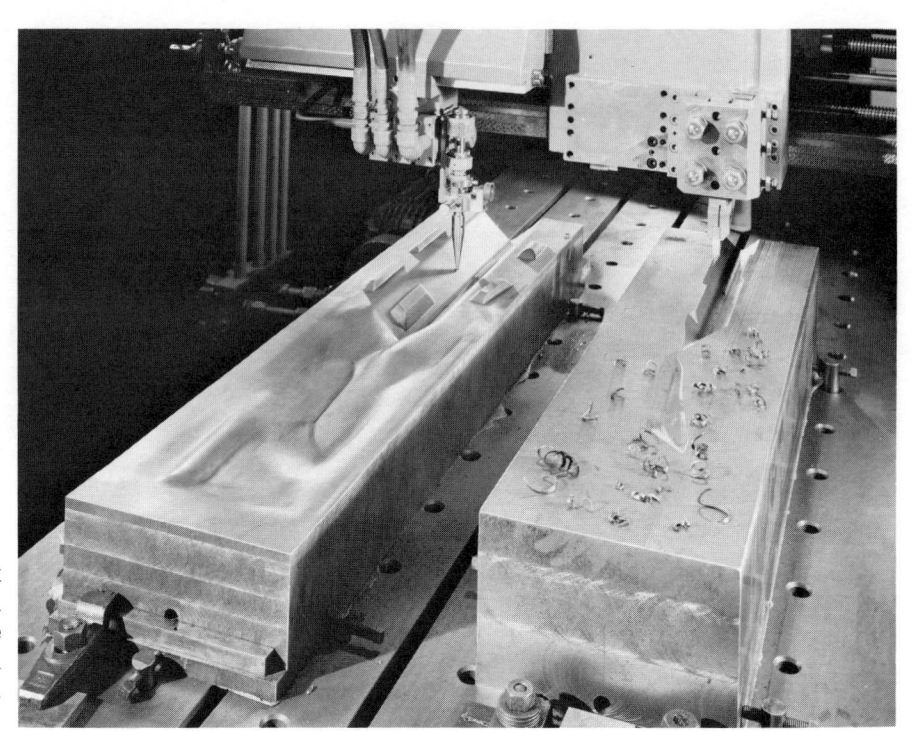

Figure 35. The stylus on the left controls the depth of the cutting tool as a machine table travels to produce three-dimensional duplication (Courtesy of Rockford Machine Tool Company).

Figure 36. Gang reproduction of contours on propeller blades using a three-axis hydraulic duplicator (Courtesy of Rockford Machine Tool Company).

past a stylus that hydraulically controls the depth of the cutting tool. This method of duplication can be extended to a number of workpieces simultaneously (Figure 36).

In summary, the shaper, while not as popular a machine tool as the milling machine or lathe, is still very difficult to do without in shops where prototype and repair work is being done. It is a very versatile machine utilizing very simple and easily modified single edge cutting tools. For small lot machining of parts with complex angles and internal features, it is often the machine tool of choice and, therefore, it is important to know how to use it.

The planer is a machine that is basic to the production of machine tools. Much of the work that was formerly machined on planers with single point tools is now being done on adjustable rail milling machines, or planer-mills, in the interest of faster metal removal rates.

Both of these machines are interesting and challenging to operate, especially in the setup phase. It has been said that 80 percent of planer work is in the setup. In the following units of instruction, we will deal with issues of safety, setup, and operation in these interesting machine tools.

review questions

1. Why was the development of the planer of critical importance to industrial development?
2. Why is the shaper a difficult machine to do without in job shops and toolrooms?
3. The ram on a slotter or vertical shaper can be angled about 10 degrees from the vertical. Why is this an important feature?
4. What method of machining might you be able to use to machine a supercharger rotor other than that shown in Figure 34?
5. The planer, while not a popular machine, is considered a critically important machine tool. Why is this?

unit 1
features and tooling on the horizontal shaper

Horizontal shapers are important, not because of their great popularity, but because they are so versatile. Before operating the horizontal shaper, you should be familiar with the operating controls and components, the mechanisms of shaper operation, the workholding mechanisms, and the cutting tools commonly used on shapers.

objectives After completing this unit, you will be able to:
1. Name components and operating controls on shapers.
2. Describe two differing types of shaper tables.
3. Explain the functions of two basically different shaper drive systems.
4. Describe mechanisms that relate to tool positioning functions.
5. Describe at least three differing types of shaper workholding procedures.
6. Explain differences between shaper tools prepared for steels and cast iron.

information

OPERATING CONTROLS AND COMPONENTS OF THE HORIZONTAL SHAPER

Figure 1 provides detailed nomenclature of the components and control features of a horizontal shaper. You will refer to this figure several times in this unit.

Base and Column
The base of a shaper (Figure 2) provides a platform for the main casting, frequently referred to as the column of the shaper that contains the drive mechanisms. It also serves as a reservoir for lubricants or hydraulic oil to operate shaper mechanisms. On the front of this casting are ways machined accurately square with other ways that provide guidance for the ram of the shaper. This relationship is critical to the accuracy of the machine.

Cross Rail
The cross rail assembly rides on a saddle that moves vertically on the face of the column driven by the vertical leadscrew. The cross rail is maintained accurately square to the column and contains drive mechanisms for moving the table assembly in a horizontal direction. On the cross rail is mounted an apron that is moved horizontally along the rail by a leadscrew.

Apron and Table
The apron is the attachment point for the table. The top and side of the table are provided with

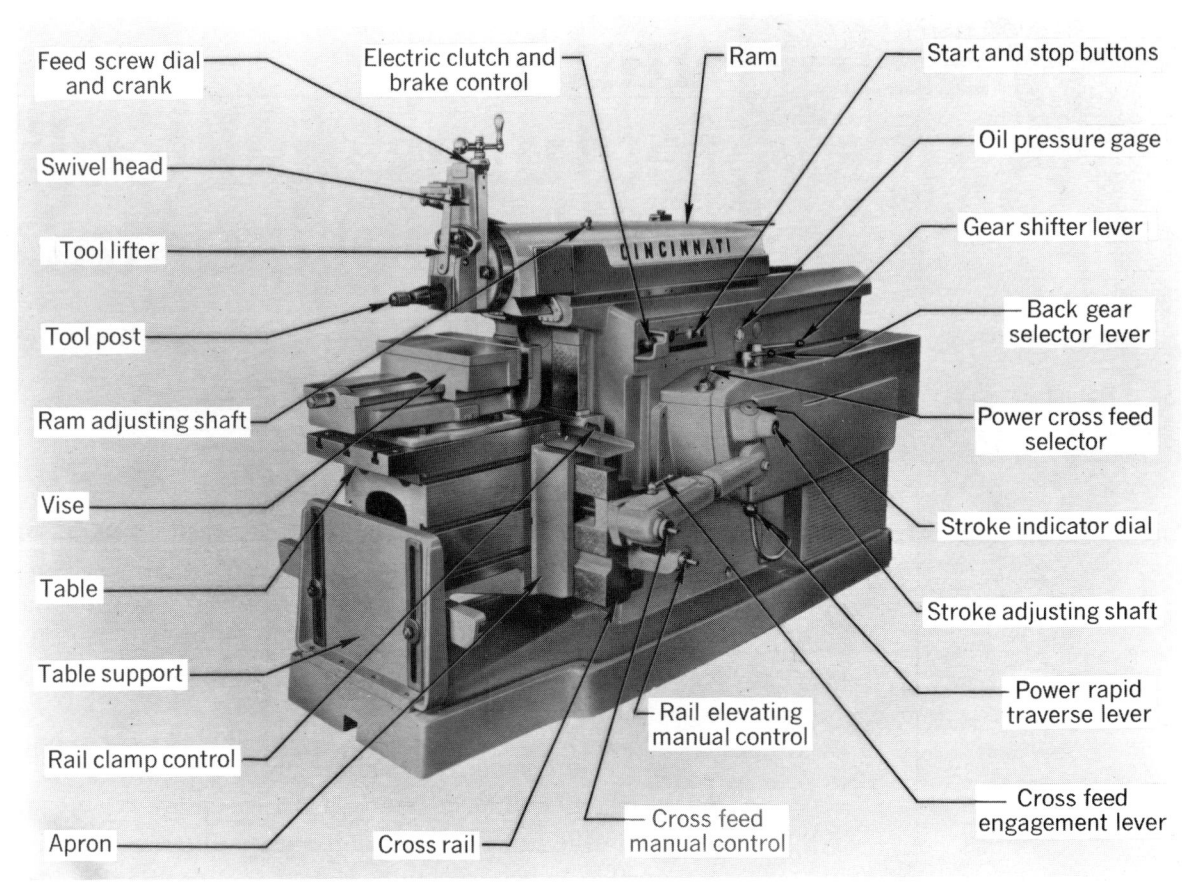

Figure 1. Detailed nomenclature of a plain horizontal shaper. Shapers are quite similar in location of basic controls despite drive differences (Courtesy of Cincinnati Incorporated).

T-slots for the attachment of vises, fixtures, and other tooling needed for shaper work. Since there are so many attachment points and such great mass and leverage relative to the column, a table support is typically provided as an "outrigger" to support the overhang of the table and to transmit machining forces back to the machine base.

Shaper tables are supplied in two designs. The shaper shown in Figure 1 has a plain table that provides for motion of the workpiece only in vertical and cross feed (other than the swiveling action of the vise). The universal table shaper has two additional motions (Figure 3) that allow for machining of complex angles, as shown in Figure 3 of the section introduction. The table is mounted to the apron with a trunion, which permits full rotation of the table, and one face of the table has a swivel plate with limited motion. When this feature is not needed for a particular job, it is often rotated to the side in the interest of easier setup and greater rigidity.

Ram
The ram (Figures 1 and 2) moves in ways on the top of the shaper. It is designed to be as rigid as possible, consistent with reasonable lightness to allow rapid reversals of direction without having the machine move along the floor. This will be discussed later as part of safety considerations.

Toolhead
On the face of the ram is mounted the toolhead or swivel head. This head is typically capable of being swiveled on the face of the ram through a full 180 degrees or more. The base of the toolhead is graduated. The toolhead consists of several components (Figure 4). It has a leadscrew that drives an accurately made toolslide. Fitted to the face of the toolslide is an apron that can be turned and secured at an angle to the slide. This apron is part of an assembly that consists of the clapper box, a hinge pin, a clapper block, and a

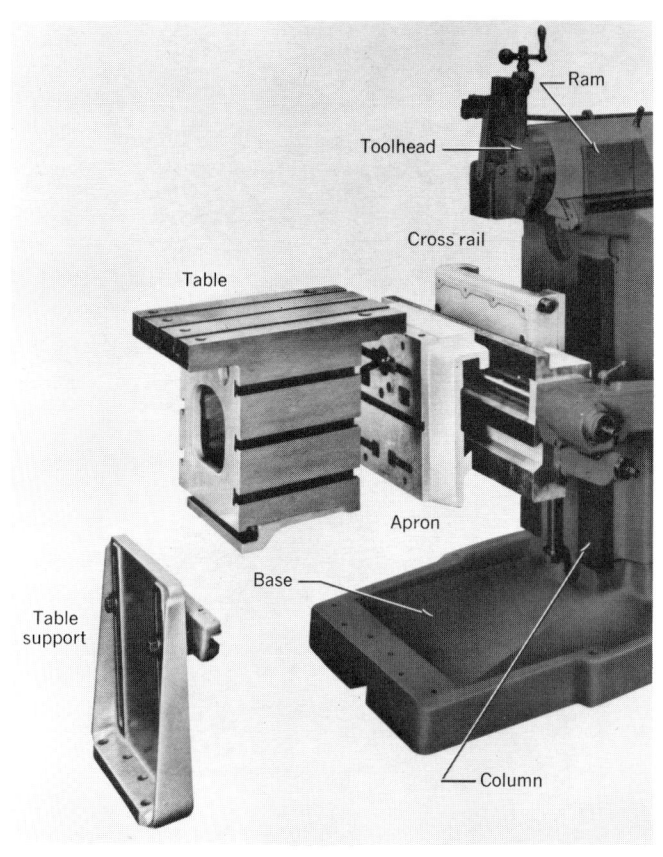

Figure 2. Components of the plain horizontal shaper table assembly (Courtesy of Cincinnati Incorporated).

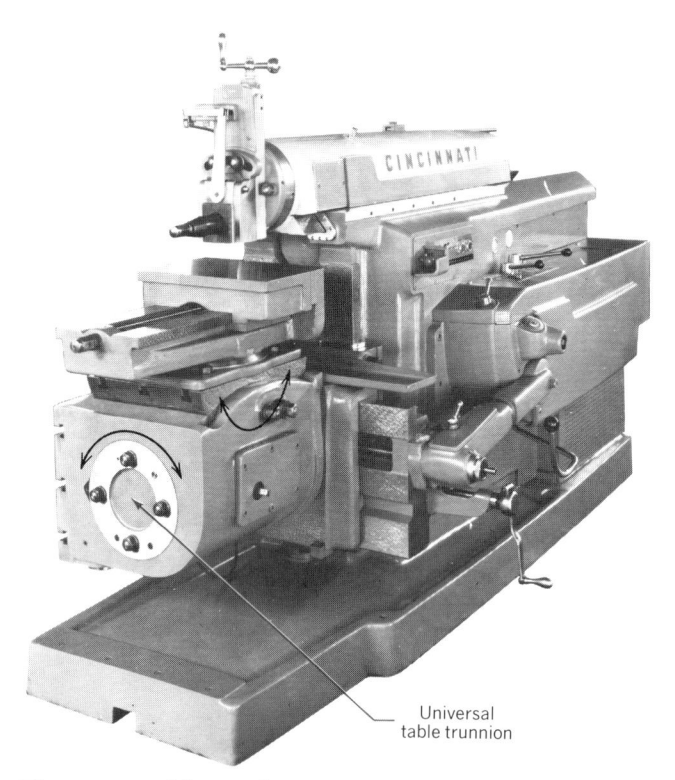

Figure 3. Additional workpiece motions are available on the universal shaper (Courtesy of Cincinnati Incorporated).

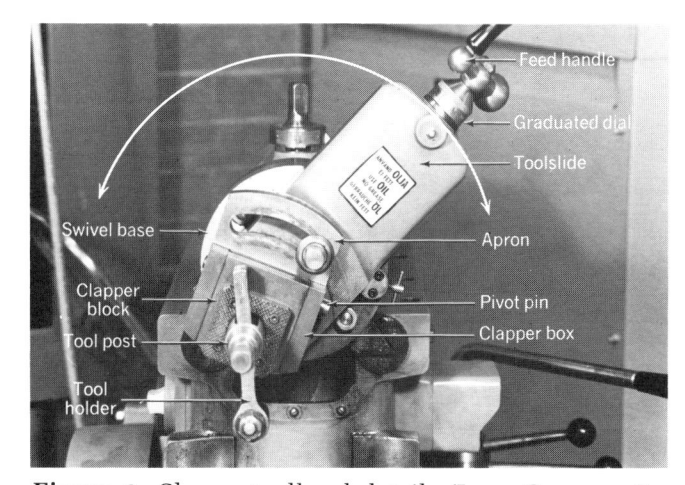

Figure 4. Shaper toolhead details (Lane Community College).

tool post for holding the cutting tool or tool holder. The function of the clapper box is to allow the cutting tool to tilt up on the return (feeding) stroke without damage to the tool. On some shapers, a tool lifter (Figure 1) is provided, programmed to lift the tool actively on the return stroke of the ram. The tool lifter is of great importance when using carbide cutting tools or for making deep, narrow slots.

TYPES OF SHAPER DRIVES

The Crank Shaper
The plain and universal shapers shown in the previous figures have been examples of crank or mechanically driven shapers. In this type of shaper the motor drives through a gearbox and clutch to a crank gear in the center of the machine (Figure 5). The crank gear (Figure 6) has a movable pivot on a crank block that, in turn, fits accurately in a sliding block on the rocker arm. The crank block is moved toward and away from the center of the crank gear by the stroke adjustment shaft, thereby increasing or decreasing the movement of the rocker arm and its attached ram. The ram in turn can be changed in position relative to rocker arm by means of the ram adjustment shaft. Examination of the top diagram in

Figure 5. The power train of the mechanical or crank-type shaper (Courtesy of Cincinnati Incorporated).

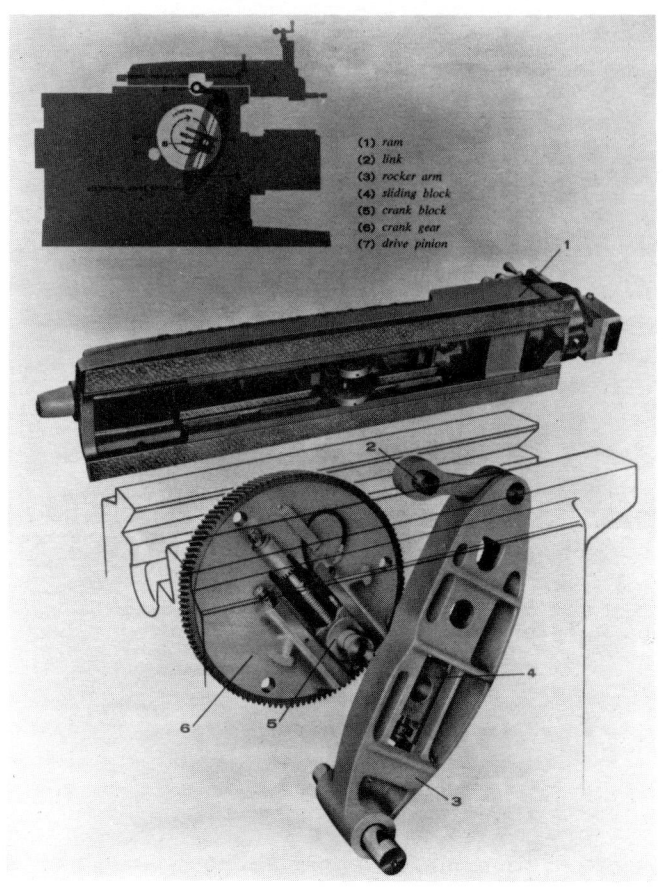

Figure 6. Details of the crank mechanism of a mechanical shaper (Courtesy of Cincinnati Incorporated).

(1) ram
(2) link
(3) rocker arm
(4) sliding block
(5) crank block
(6) crank gear
(7) drive pinion

Figure 6 shows the working relationship between the reciprocating drive components. As you can see, the return of the ram is made during the part of the cycle where the sliding block is closest to the stationary pivot point. This relationship accelerates the return stroke in relationship to the forward or cutting stroke, thus saving noncutting time. The ratio of the time of the cutting stroke to the return stroke will be about 3:2. We will go into this further in the next unit.

The Hydraulic Shaper

Hydraulically driven shapers (Figure 7) do not differ greatly in appearance from the crank-type mechanical shaper. Hydraulically operated shapers have nearly constant cutting tool velocity, and many have an arrangement for high speed ram return independent of the cutting stroke speed. Figure 8 is a schematic diagram of the hydraulic circuit, which includes the high speed return feature. Examine this diagram closely and trace the flow of oil through each circuit.

The pump supplies hydraulic oil at a constant volume to the system, which is regulated by the speed control valve. Oil not bypassed by this valve is admitted to the start-stop valve, which also distributes oil to each side of the drive piston under the command of a pilot valve, which is tripped by the ram reverse dogs. The pilot valve also activates the high return valve if it has been selected by the operator to speed up the return of the ram. In addition, the pilot valve actuates the tool lifter (if fitted) and the power feed during the return stroke. Lubrication for the ram is filtered separately and passed through this valve as well.

SHAPER FEEDS

Table Feeds

Any shaper, whether simple or sophisticated, can be hand fed vertically or horizontally for table travel. Most shapers, even the simplest table top models, have a power feed for the horizontal travel of the table. On crank shapers the drive for the feed is in the form of either an adjustable eccentric (Figure 9) or a cam arrangement, which is attached to the hub of the crank gear. The reciprocating motion that results from this arrangement is passed to a ratchet mechanism connected to the cross feed

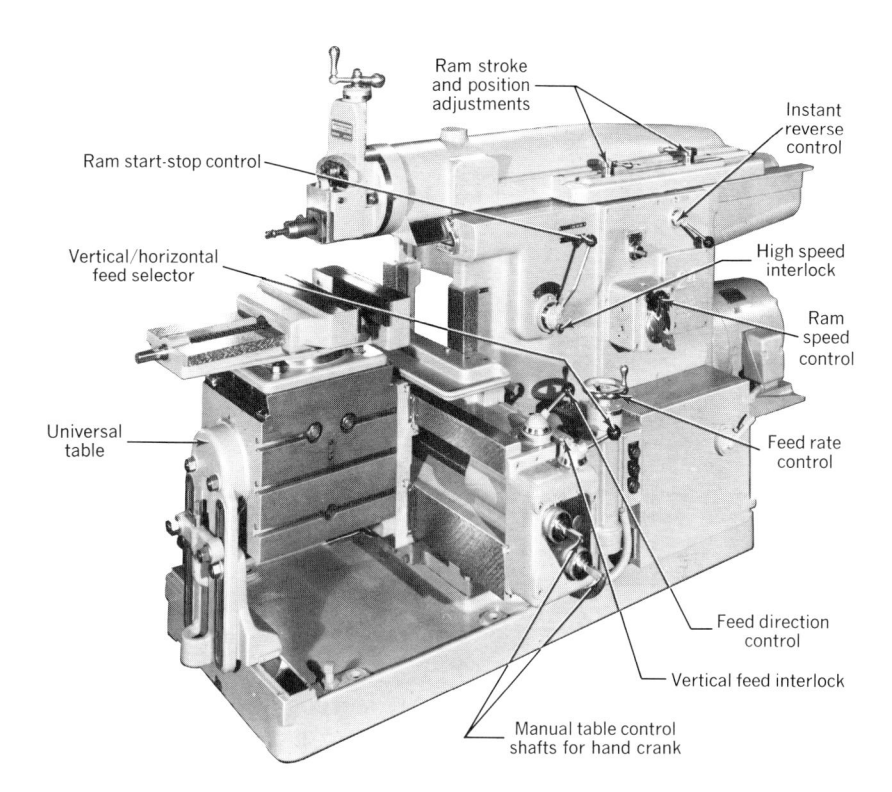

Figure 7. Twenty-four in. hydraulic shaper. Shaper size is designated by the maximum length of available ram stroke. The example shown is equipped with a universal table (Courtesy of Rockford Machine Tool Company).

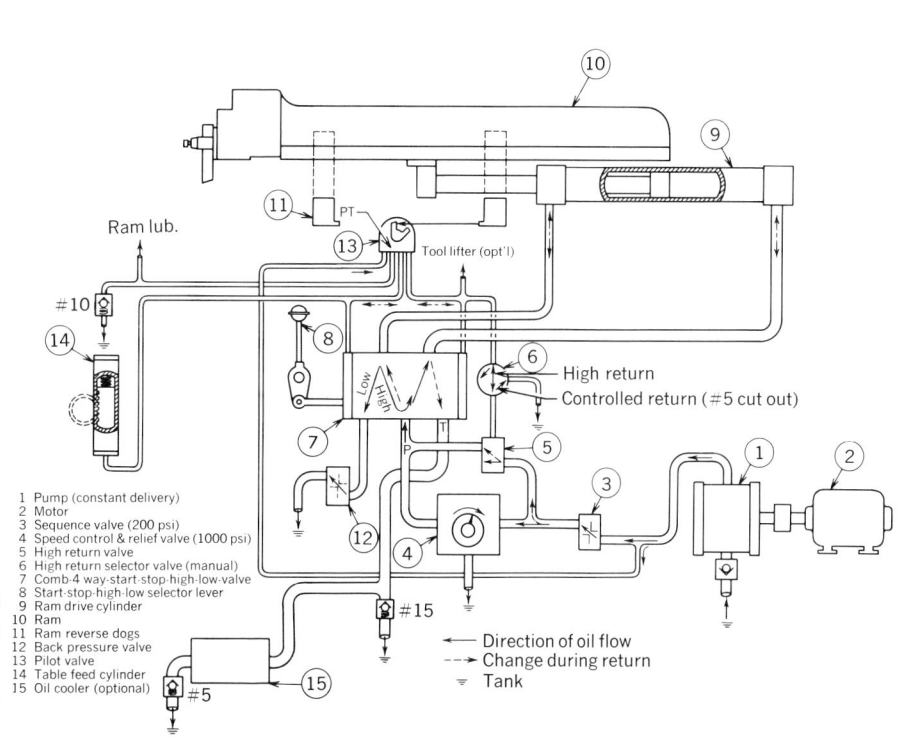

Figure 8. Schematic diagram of the hydraulic circuit in a hydraulic shaper with high speed return (Courtesy of Rockford Machine Tool Company).

leadscrew. On many shapers, vertical feed can be selected as an alternative to cross feed. On hydraulically operated shapers the table feeds are accomplished by a hydraulic table feed cylinder (Figure 8). Most larger shapers are also equipped with a rapid traverse, which is a device to move the table either vertically or horizontally at a relatively high rate of speed. This is very useful

Figure 9. Adjustable eccentric on crank gear shaft to drive feed rachet (Lane Community College).

for getting back to start a new cut or for use in setup for moving rapidly to a position for "indicating in" the vise or other machine components.

Head Feeds
Feeding of the toolhead is a manual function on most horizontal shapers. Some shapers are also equipped with a power feed to the toolhead, often cam operated or hydraulically driven (Figure 10).

SHAPER SIZE

Shapers are size classified by the working length of the maximum cutting stroke that the machine can make. They are also classified by the working capability such as light, standard, or heavy duty. In the smaller sizes like 7 in., the machine would weigh only a few hundred pounds and would be capable of being mounted on a bench top or a sheet metal base.

A 28 in. heavy duty shaper would weigh on the order of 7500 lb and should be placed on a heavily reinforced floor. In sizes up to about 16 in., the shaper can be expected to accept a cubic workpiece of about its rated size, as 16 in. × 16 in. × 16 in. In the larger sizes, the maximum table to ram distance usually does not exceed that of a 16 in. shaper.

Heavy duty shapers are capable of making massive chips and are equipped with motors on the order of 10 hp.

Figure 10. Shaper head swivelled at a steep angle to make a keyway cut in a long shaft. This machine is also equipped with a power feed on the toolhead (Courtesy of Rockford Machine Tool Company).

WORKHOLDING DEVICES ON THE SHAPER

Vises
By far, the most common method of holding workpieces in the shaper is with a swivel base, single screw vise that is standard equipment on most shapers (Figure 11). This type of vise can be used in a variety of ways to hold workpieces either square to the travel of the ram or parallel to ram travel. Sometimes the combined arrangement of the vise and the toolhead can result in an arrangement to do what would not ordinarily be thought of as a shaper job (Figure 10). Auxiliary workholding can also be used in conjunction with a shaper vise to hold workpieces that have features of unequal length that would be difficult to hold by other methods (Figure 12). Another accessory device that can be used with a shaper vise is a holddown wedge (Figure 13), which can be used to force the workpiece against the solid or reference jaw of the vise to help insure that the

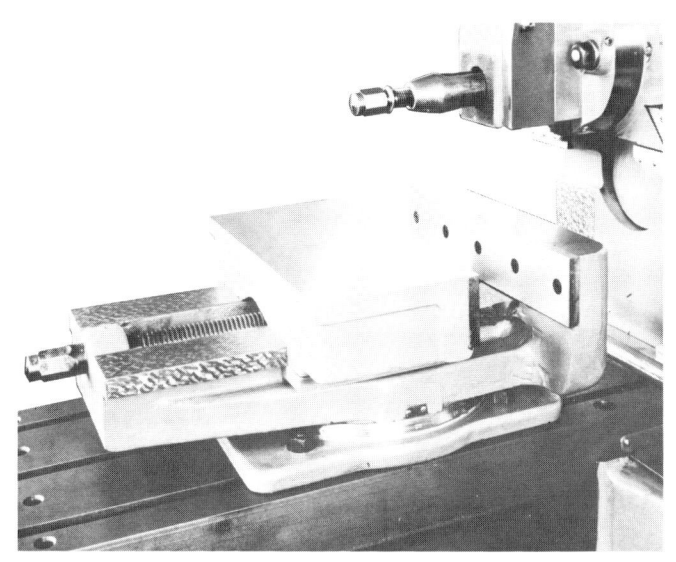

Figure 11. Swivel base single screw shaper vise (Courtesy of Cincinnati Incorporated).

Figure 12. Auxiliary workholding method to extend the usefulness of the vise (Courtesy of Rockford Machine Tool Company).

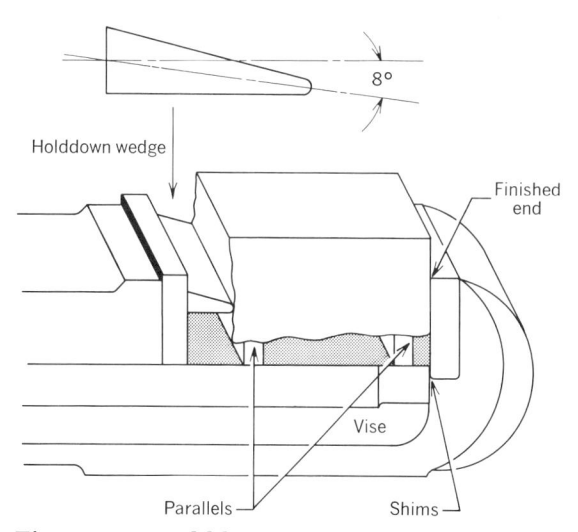

Figure 13. Holddown wedge for squaring to a finished surface (Courtesy of Cincinnati Incorporated).

surface being cut is perpendicular to the finished surface.

An important thing to remember, especially when using the standard vise with its carefully finished jaws, is that rough, ferrous castings usually have a hard, scaly outside surface before they are machined, and this surface can badly mar the vise jaws unless they are protected. Emery cloth (Figure 14) is one way that this is done. Other compressible materials also can be used for this protection.

When shaping is to be done on a side parallel to a finished surface, a cylinder of soft metal such as 1020 steel, brass, or aluminum can be used to provide line contact on the edge of the work. This prevents the movable jaw from providing a lifting movement to the work as the screw is tightened. The movable jaw on most vises has some free play that permits the jaw to tilt slightly as it is tightened. As the work is tightened with this arrangement (Figure 15) it is tapped down onto the parallels with a babbitt hammer. Then, after tightening, the parallels are checked to see that they are firmly in contact with the bottom of the workpiece. Sometimes paper strips are used on top of the parallels to check this contact; if the strip can be pulled out, you do the job over.

Since much of shaper work done in vises has the force directed toward the movable jaw, and some of this work is quite critical, especially in operations like slotting (Figure 16), some vises are made with additional clamping for the movable jaw to reduce the tendency to tilt.

When the force of the cut is parallel to the

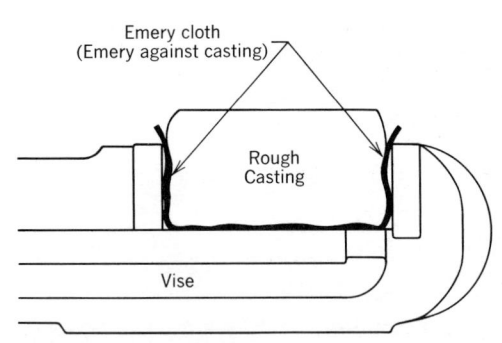

Figure 14. Protecting vise jaws from rough castings (Courtesy of Cincinnati Incorporated).

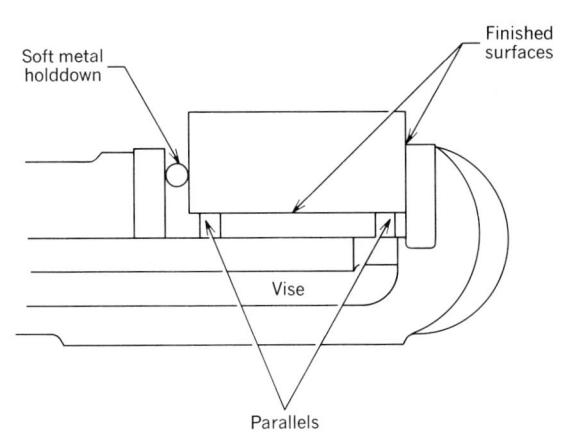

Figure 15. Method for shaping parallel to a finished surface (Courtesy of Cincinnati Incorporated).

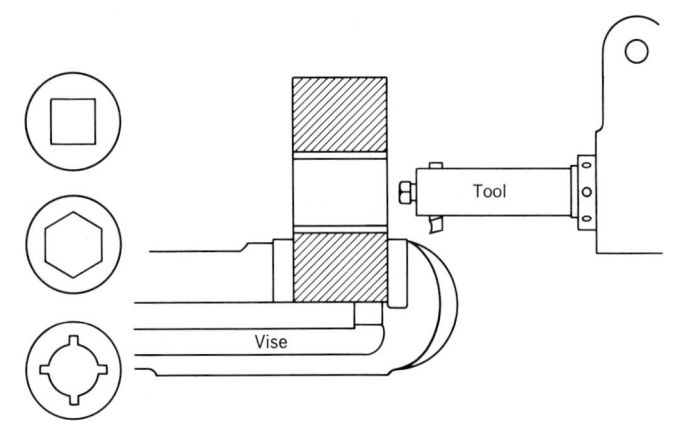

Figure 16. Slotting makes special accuracy demands on vise holding (Courtesy of Cincinnati Incorporated).

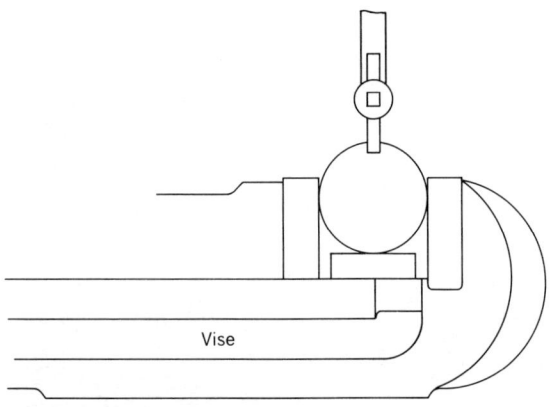

Figure 17. Keyway cutting demands adequate workholding because of limited contact (Courtesy of Cincinnati Incorporated).

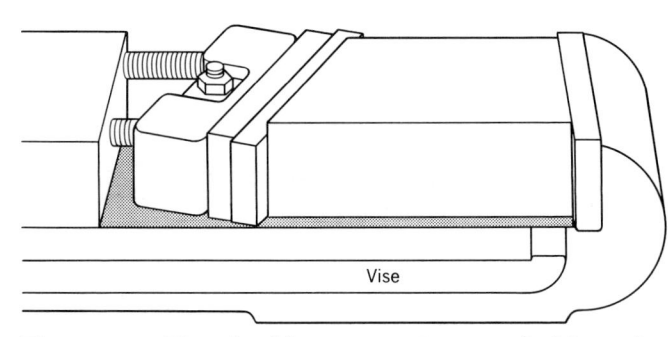

Figure 18. The double screw vise can hold workpieces that are not parallel (Courtesy of Cincinnati Incorporated).

jaw as is common in making keyways on long parts (Figures 17 and 10), it is important to be sure that the vise has an adequate grip on the part. Single screw shaper vises are supplied with a handle of correct length for adequate tightening by an average sized man without using a hammer on the handle.

The double screw vise (Figure 18) is often used with both shapers and planers to hold tapered or out-of-parallel work. This type of vise can be had with a swivel base but, for heavy work, they can be supplied with a plain base and multiple screws (Figure 19). The jaws can be supplied either plain or serrated, but the serrated jaws, as shown in Figure 19, are more common.

Direct Attachment to the Table

Some shaper work is best accomplished by direct attachment of the workpiece to the table with straps and T-bolts. When this is done, it is important to make sure that the clamping force in the setup is directed more toward the work (Figure 20a) than the riser block (Figure 20b) that is used to level the clamping strap. Shapers and planers

Figure 19. Multiple screw vise with serrated jaws has exceptional holding ability (Courtesy of Cincinnati Incorporated).

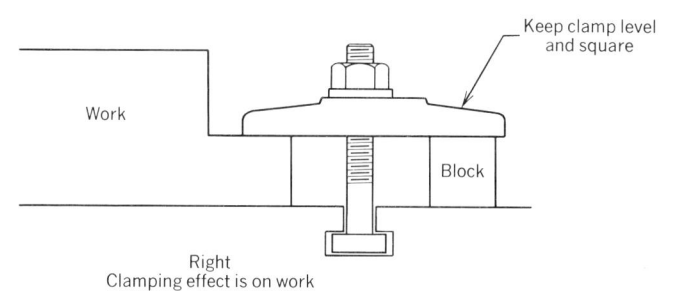

Figure 20a. Right method for direct clamping to the table (Courtesy of Cincinnati Incorporated).

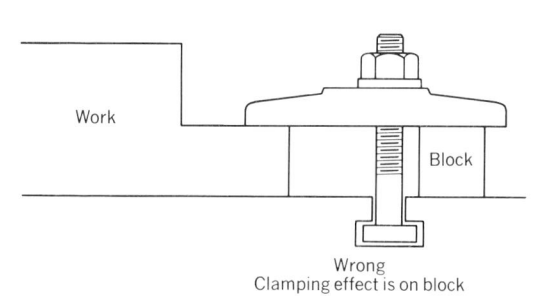

Figure 20b. Wrong method for direct clamping to the table (Courtesy of Cincinnati Incorporated).

are relatively less forgiving of poor setup than milling machines because of the high forces that are transmitted through the work upon tool impact.

It is best to strive for some excess in fastening the work than to take a chance on having the work come loose on impact. Figure 21 is an example of a secure direct clamping setup, employing adjustable table stops.

Clamping is not limited to work on the top of the table. It is useful to keep in mind that workpieces also can be attached to the side of the table (Figure 22).

On some workpieces the T-bolt can be used directly with nuts to hold down the workpiece. With recessed workholding machining can take place right up to an obstruction, as shown in the surfacing of this machine vise (Figure 23).

Thin work can be directly attached to the table with poppets and toe dogs (Figures 24 and 25) combined with the use of a stop. The poppet must be aligned straight with the toe dog, as viewed from the top, to work properly. A poppet designed to hold in a T-slot is sometimes called a bunter.

Using Fixtures with the Shaper

A fixture is a workholding device that is usually directly fastened to a machine table (Figure 26) or spindle. It typically serves as an intermediate device to locate and fasten parts more rapidly than would be the case for individual setup. Another example of a fixture would be the indexing center (Figure 27) used in combination with the tilting base on the universal table shaper to make a forming die.

Other fixtures would include devices like angles for attaching angular workpieces for shaping, as shown in Figure 28.

SHAPER CUTTING TOOLS

The tools used on the shaper are very similar to the single point tools used in the lathe. The nomenclature for the cutting geometry is the same as that used for lathe tools. Most of the tools used in the shaper are of high speed steel, with preference toward grades that have increased amounts of vanadium for better shock resistance. Cemented carbide tools are also used on occasion, but the impact characteristic of the shaper and a cutting speed, which at a maximum is less than 150 surface feet per minute, makes the carbide less the tool of choice except on hard and abrasive materials. Carbides are brittle and do not readily stand being drawn back along the cut, even with a properly set and working clapper box, so for carbide use, the tool lifter described

Figure 21. Use of stops and straps to make secure direct table clamping (Courtesy of Cincinnati Incorporated).

Figure 22. The side of the table may also be used for direct attachment (Courtesy of Rockford Machine Tool Company).

Figure 23. With recessed holding, machining can be done right up to an obstruction (Courtesy of Rockford Machine Tool Company).

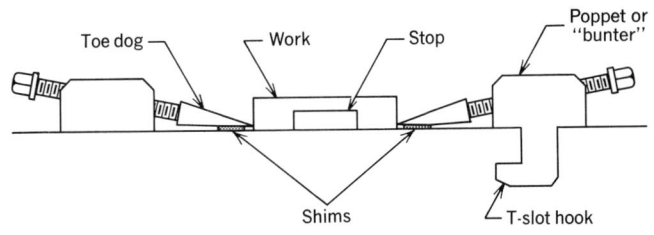

Figure 24. Using a bunter with toe dogs for a thin workpiece (Courtesy of Cincinnati Incorporated).

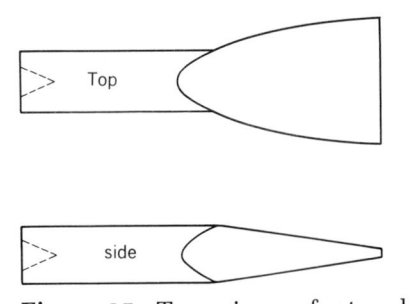

Figure 25. Two views of a toe dog.

earlier is important. In past years, forged tools were used, especially with the larger shapers. At the present time, their use is rare, replaced by tool bits in holders or by large tool bits ground for the application and directly clamped into the tool post.

218

Figure 26. Using a fixture on a plain shaper (Courtesy of Cincinnati Incorporated).

Figure 27. Indexing center setup as a fixture on the tilt table of a universal shaper (Courtesy of Cincinnati Incorporated).

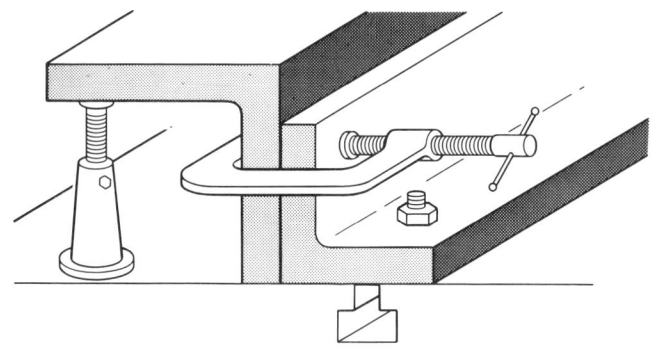

Figure 28. Angle fixture with a planer jack for shaping angles (Courtesy of Cincinnati Incorporated).

Preparing Tools for Shaper Use
Tools for use in the shaper (and the planer) are only generally similar to lathe tools. They are characterized by clearances back of the cutting edges on the order of 4 degrees. The reason for these minimal clearances is to provide maximum support for the cutting edges. In addition, on the planer or shaper, feeding does not take place during the cut; hence no additional clearance is needed to allow for the feedrate, as is necessary on the lathe. Figure 29 illustrates typical cutting tools recommended for use with low carbon steels. On work in steels the roughing tool should be used to bring the work to within .010 to .015 in. (.25 to .3 mm) of final dimension. Where considerable stock is removed and where the setup is substantial, the depth of cut and feed rate should be the maximum possible consistent with the power and rigidity of the machine, the strength of the tool, and the strength of the workpiece. Where it is necessary to reduce cutting forces, the feedrate should be lessened before a reduction in depth of cut is made. Finishing the last .010 to .015 in. should be done in one cut, whenever possible, to save time.

On cast iron, the fundamental difference between shaper and planer tools prepared for steels is on the face of the tool. The side and back rake angles of the tools are less, but the clearances remain at about 4 degrees (Figure 30). Since dovetail work is mainly done in castings, recommendations for dovetail tools (Figure 31) are given for cast iron applications. Dovetail tools for steel should have additional rake. On cast iron, it is important to have the roughing cut completely penetrate the scale on the first cut. If the tool rides up on the scale it will dull very quickly. As with steel, you should strive for maximum depth of cut consistent with the strength of workpiece, workholding, and tool. For cast iron, the roughing cut should be carried to .005 to .010 in. (.12 to .25 mm) of finished size. Finishing cuts on flat surfaces in cast iron and free cutting steels are generally done with wide tools and a cross feed rate about three quarters of tool width. This is

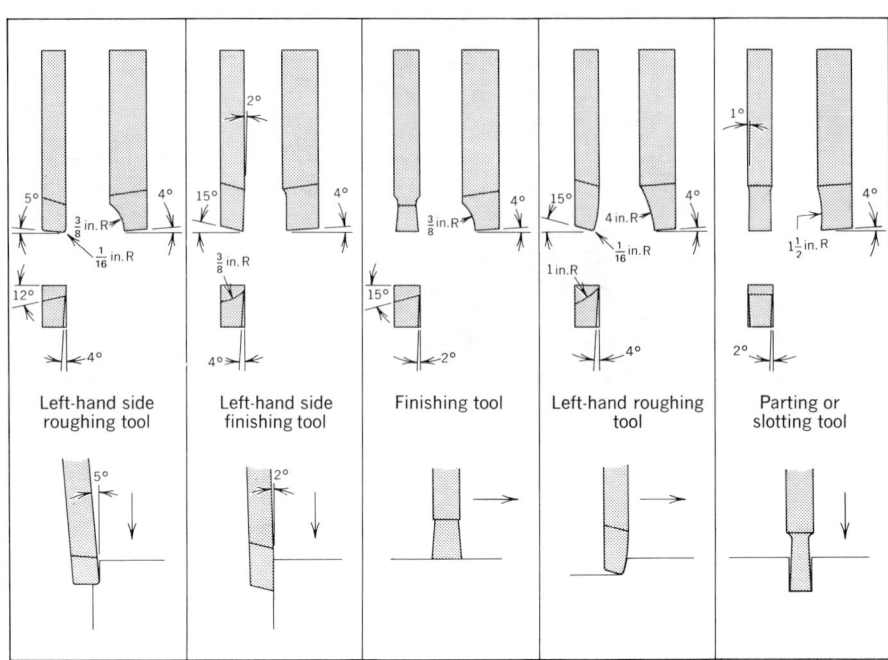

Figure 29. Shaper and planer tool geometry for cutting low carbon steels (Courtesy of Cincinnati Incorporated).

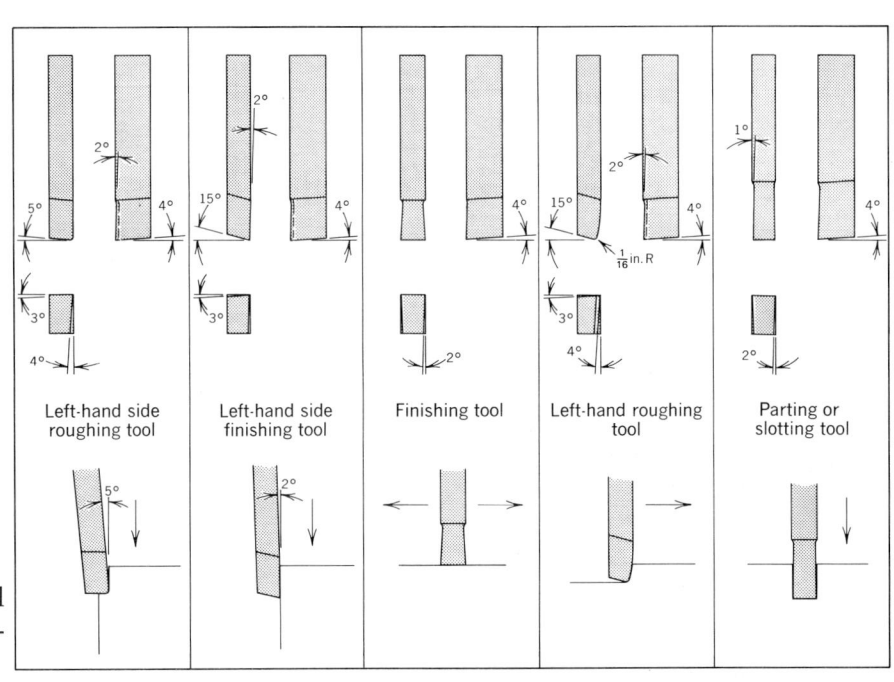

Figure 30. Shaper and planer tool geometry for cutting cast iron (Courtesy of Cincinnati Incorporated).

one instance where finishing is usually quite fast and the time for more than one finishing cut is minimal where very accurate work is called for.

If chatter is experienced during the roughing cut in either steel or cast iron, it may be due to poor adjustment of machine gibs or to excessive overhang of the tool, toolslide, or workpiece. With regard to the cutting tool proper, this may be reduced by limiting the radius on the point of the tool to $\frac{1}{32}$ or less. It is also important to have the tool mounted as close into the tool post as possible and the toolside overhang minimal (Figures 32a and 32b). Tools used on shapers and planers should be kept sharp and the cutting edges should be honed after grinding to give the longest possible tool life.

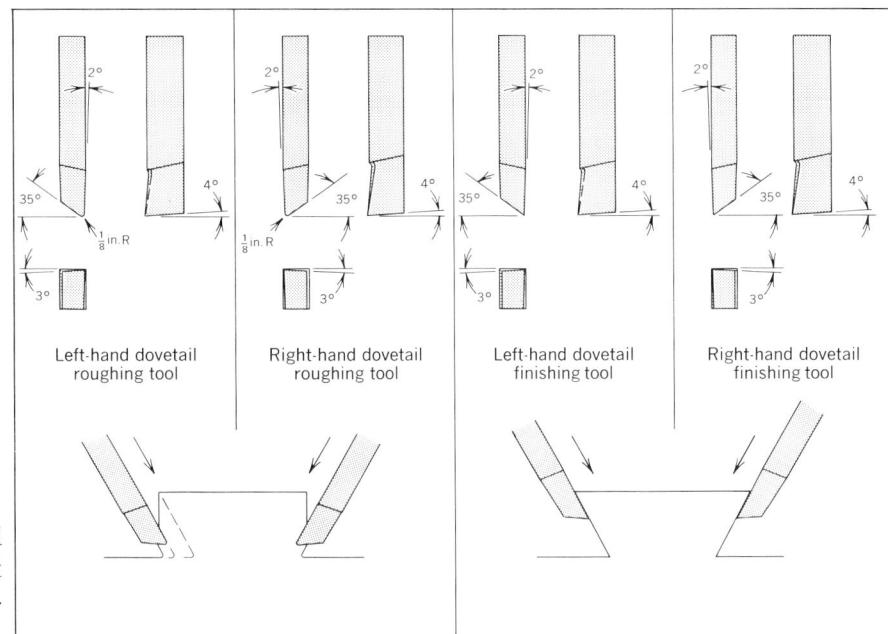

Figure 31. Shaper and planer tool geometry for dovetail work in cast iron (Courtesty of Cincinnati Incorporated).

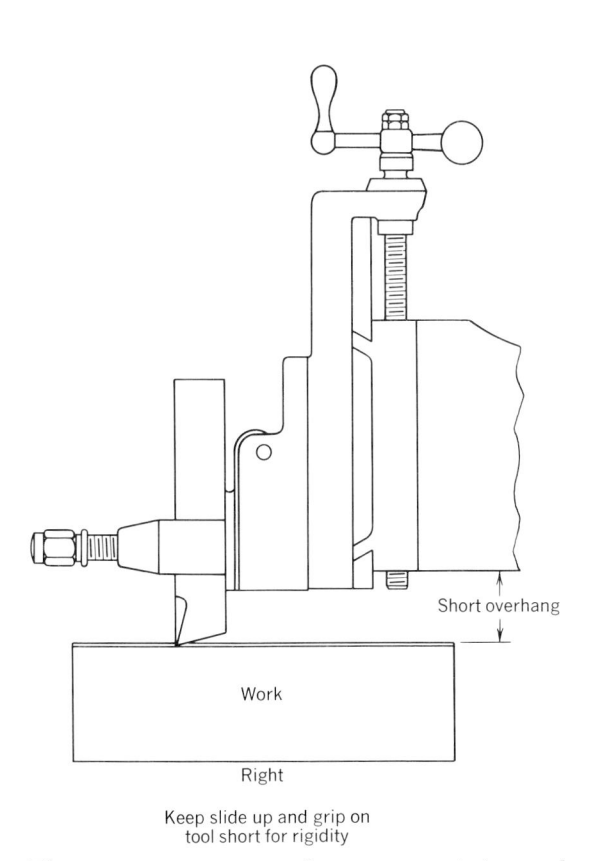

Figure 32a. Keep overhang to a minimum by using the highest possible table position for the job (Courtesy of Cincinnati Incorporated).

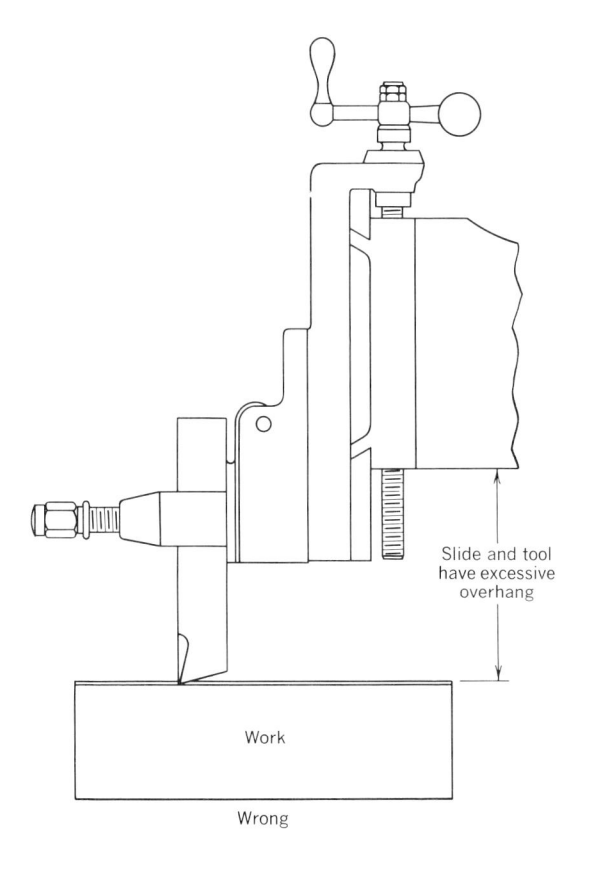

Figure 32b. Excessive overhang of slide and tool may cause chatter (Courtesy of Cincinnati Incorporated).

Figure 33. Small planer-type tool holder (Lane Community College).

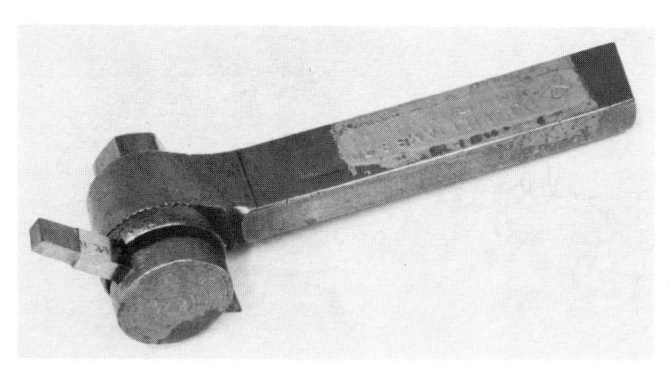

Figure 34. Tool holder can be set at various angles (Lane Community College).

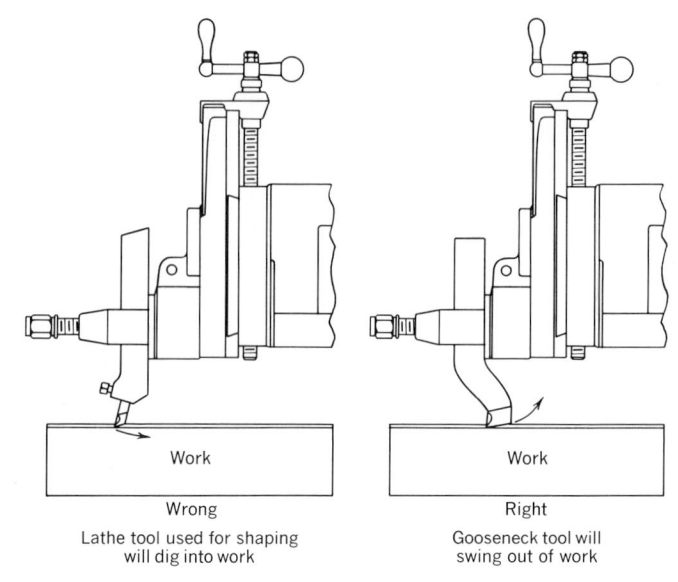

Wrong — Lathe tool used for shaping will dig into work

Right — Gooseneck tool will swing out of work

Figure 35. Using a gooseneck tool can avoid chatter (Courtesy of Cincinnati Incorporated).

Figure 36. Extension tool holder for internal shaping (Lane Community College).

Tool Holding

A common form of tool holder used with shapers and planers is shown in Figure 33. This type of tool holder permits a tool bit to be held straight as shown or to be angled (Figure 34) relative to the shank of the tool holder. For relatively heavy cuts, the tool is typically placed in the holder with the tool bit side toward the ram of the shaper. For relatively light cutting, and where maximum visibility is also important, the tool is often mounted in front, on the side away from the ram. The reason for mounting the tool on the ram side of the holder is to reduce or prevent chatter and "digging in" by giving the tool and holder room to deflect under load. This type of deflection decreases the load instead of increasing it. This is why lathe tool holders with built-in rake angles are poor for use in shapers and why gooseneck tools often relieve problems with chatter (Figure 35), particularly with worn, out-of-adjustment machines.

The extension shaper tool holder (Figure 36) is another useful tool for internal shapes. It allows the tool to be extended into existing holes for making internal keyways and other forms.

self-evaluation

SELF-TEST

1. Name the shaper features indicated in Figure 37.

   A _____   F _____

   B _____   G _____

   C _____   H _____

   D _____   I _____

   E _____   J _____

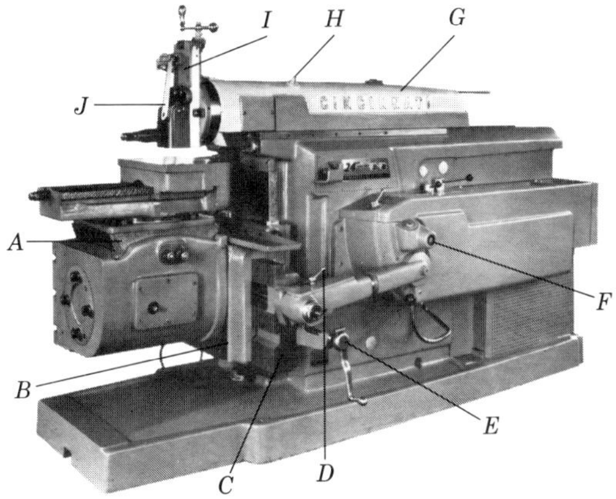

Figure 37. (Courtesy of Cincinnati Incorporated).

2. What is the function of the clapper box on the toolhead?

3. Explain the basic differences between crank and hydraulic shaper drives.

4. Describe the table feeding functions on a crank type shaper.

5. How is the size of the shaper specified? Would a 28 in. shaper hold a workpiece of the same proportion as a 12 in. shaper?

6. What are the two basic types of vises used on shapers?

7. Explain a holding method for securing thin parts to a shaper table.

8. What advantage is gained by using a fixture over direct attachment to the machine table?

9. What is the fundamental difference between shaper and planer tools prepared for cast iron and those prepared for steel cutting?

10. Why would you choose to place a cutting tool on the ram side of a planer tool holder instead of on the forward side?

unit 2
cutting factors on shapers and planers

The characteristics of cutting tools for shapers and planers are similar, so a discussion of cutting factors related to both types of machines is made in this unit.

objectives After completing this unit, you will be able to:
1. **Select a grade of high speed steel or cemented carbide suitable for use on shapers or planers.**
2. **Name a source of machining data useful for interrelating machine tools, cutting tools, and workpiece characteristics.**
3. **State generally the machining characteristics sought for maximum productivity on shapers and planers.**
4. **Determine the number of strokes per minute needed to match cutting speed on crank shapers.**
5. **Establish feedrates that are reasonable for both roughing and finishing on shapers and planers.**

information

CUTTING CHARACTERISTICS OF SHAPERS AND PLANERS

Shapers and planers have the common characteristic of noncontinuous cutting, with shock to the tool at the start of each cutting stroke. For this reason, the cutting tools must have more adequate support beneath the cutting edges than is the case for most turning tools. As previously discussed, carbide tools are not frequently applied to shapers because of the relatively low attainable cutting velocities (below 150 SFPM maximum). With modern planers, speeds up to more than 300 SFPM can be obtained, which makes the use of carbide tools attractive for productivity. Carbide tools used in planing and shaping typically have negative back and side rake in the interests of maximum support for the cutting edges. The selection of carbide cutting tools for planer and shaper use is typically grade C-2 for the straight tungsten carbides and C-6 for the steel cutting grades. These are relatively less brittle grades and will stand the impacts of interrupted cutting better than the more brittle but very hard C-4 and C-8 grades.

The selection of high speed steel tools used in shaping and planing is also in the direction of the less brittle tools. High speed steel grades M-2 and M-3 are the most common recommendations. These grades have a relatively high vanadium content, which improves impact resistance. The high cobalt grades are to be avoided because of brittleness.

INTERRELATIONSHIPS, SPEED, FEED, DEPTH OF CUT, AND MATERIAL CONDITIONS

The generalities that are made for roughing and finishing for lathes do not readily apply to shaper and planer operations. Speeds for roughing are generally much lower than those used in turning, and they are tied substantially to the depth of cut as well. Finishing done on lathes is generally at a higher speed than roughing, but on shapers and planers the reverse is true. Speeds are generally cut about one third from the rate recommended for shallow roughing cuts, mainly because finishing tools for either cast iron or the softer steels are usually wide and fed nearly full width each pass. This full tool contact can lead to chatter at higher speeds.

Another major factor that affects cutting speed is the metallurgical condition of the workpiece, as usually reflected in hardness measurements.

The United States Army and Air Force made extensive studies on machinability of a large variety of workpiece materials using various cutting tool combinations on most types of machine tools. These findings are collected in a large book of more than 1000 pages, *Machining Data Handbook*, published by Metcut Research Associates in 1972.

For planing application (which includes shapers) very few generalities can be made. For most steels, whether low or high carbon, in the Brinell hardness range of 175 to 225, the recommended cutting speed is about 60 SFPM for depth of cut of .100 in. The feedrate recommendations, however, vary with carbon content. The high carbon steel feedrate is reduced about 40 percent from the rate recommended for low and medium carbon steels. For gray cast iron in annealed condition, as compared with steels, the cutting speed recommendation is substantially higher for the same depth of cut and with a substantially larger feedrate. But, at the other extreme, some conditions of cast iron call for cutting speeds as low as 10 SFPM, pearlitic or acicular plus free carbides, for example.

As depth of cut is increased to .500 in., the cutting speed recommendation in most materials is reduced to about 60 percent of that recommended for a .100 in. depth.

There are many material factors, such as workpiece hardness and inclusions in the workpiece, that are especially important in planing, where many castings are machined. Since there are also many cutting factors, such as workholding, workpiece strength, and tool contact area, compounded by great variation in machine rigidity and power, it is helpful to have a source like the *Machining Data Handbook* for guidance. Most information in the handbook is derived from actual cutting practice and serves as a good starting point when setting up planers, shapers, and other machine tools.

A generality that can be made is that the combination of depth of cut and feedrate should be as much as can be tolerated by the weakest link of the cutting tool, workholding, part strength, or machine tool rigidity and power, consistent with the required surface characteristics required on the part such as flatness and finish. Where reductions must be made, it is better to reduce the feedrate than to reduce the depth of cut. With the increased cutting depth you have the work spread out over more cutting edge, and thinner, more easily controlled chips are produced for a given volume of metal being removed in a given time.

DETERMINING STROKE REQUIREMENTS ON A CRANK SHAPER

The duration of the cutting stroke of the crank shaper is about 220 degrees of crank rotation (Figure 1). This leaves 140 degrees for the return stroke or a ratio of 3 cutting to 2 return. The sum of this ratio is 5, so three fifths of the time is spent cutting and two fifths of the time is spent in the return strokes. Since cutting speed (CS) is given in feet per minute, the length of the stroke (L) times the number of strokes per minute (N) must also be expressed in the same unit. This is done by dividing $\frac{L \times N}{12}$ = feet per minute of tool movement.

Since the cutting stroke occupies three fifths of the time, it is necessary to allow for this fact in the equation.

$$CS = \frac{L \times N}{\frac{3}{5} \text{ of } 12} = \frac{L \times N}{7.2 \text{ (or 7)}}$$

Now, since you know the cutting speed you intend to use and the stroke length that you will be using for each operation, you can convert:

$$N = \frac{CS \times 7}{L}$$

As an example, consider the problem of machining a 10 in. long piece of medium carbon hot rolled steel with a rated cutting speed of 60 SFPM:

CS = 60

L = 11 (10 in. plus 1 in. allowance for overtravel)

therefore:

$$N = \frac{60 \times 7}{11} = \frac{420}{11} = 38.18$$

You would then set a value as close to 38 strokes per minute as available on the machine.

If your shop makes considerable use of the shaper, the formula $N = \frac{CS \times 7}{L}$ should be committed to memory.

Another somewhat easier approximation can be used while you are setting up. Since you know that about 40 degrees more of crank pin travel is required for the cutting stroke than the return stroke, you have a little more than 10 percent more for the time of the cutting stroke than the average that you would have if the machine had a theoretical uniform speed in both directions. If you multiply the cutting speed by 12, you will have the inches per minute of cutting speed. If you multiply the stroke length by 2, you will have the distance the tool must move in both forward and reverse in a minute to cover the distance, assuming uniform velocity.

$$N = \frac{CS \times 12}{L \times 2} \quad \text{or} \quad \frac{CS \times 6}{L}$$

Now increase the number of strokes indicated by 10 percent because of the additional crank pin travel above the average time to raise the cutting velocity to average speed.

$$N = \frac{CS \times 6}{L} + 10 \text{ percent}$$

For the example just given:

$$\frac{60 \times 6}{11} = \frac{360}{11} = 32.7 + 10 \text{ percent } (3.3) = 36$$

which reasonably approximates the more exact procedure. If this approach is easier for you, it is close enough to be useful.

For hydraulically driven shapers, the situation is very simple. The cutting speed can be entered directly into the machine controls.

ESTABLISHING FEEDRATES

For roughing, in either planing or shaping, the feed rate should be about one tenth the depth of cut, where a deep roughing cut ($\frac{1}{2}$ in., for example) is made. Where a relatively shallow roughing cut is made, like .100 in., feedrates from one third to one half the depth of cut are more usual.

For finishing, in either planing or shaping, the feedrate to obtain a finished surface is dependent mainly on the shape of the tool. For most steels and cast iron, the finishing tool should be essentially square on the nose and be fed up to three quarters of the tool width on each

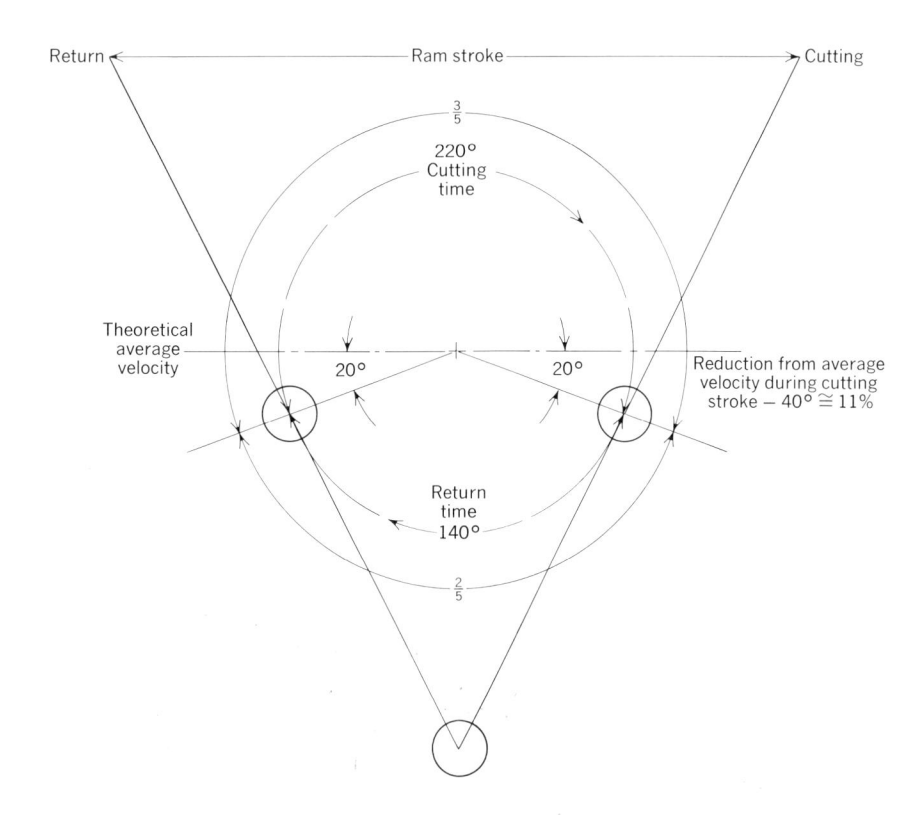

Figure 1. Timing characteristics of the crank shaper.

stroke. To summarize the relationship of feed to finishing speed, the finishing speed should be somewhat faster than that recommended for deep roughing cuts and slightly slower than the recommended speed for shallow roughing cuts.

self-evaluation

SELF-TEST

1. Why are carbide planer and shaper cutting tools usually specified with negative back and side rake?

2. What grades of high speed steel and cemented carbide tools are generally specified for use on planers and shapers? Why are these grades specified?

3. Finishing speeds on the planer and shaper are often lower than some roughing speeds. Why is this the case?

4. Name a source of machining information useful to determining speeds and feeds on planers and shapers and other machine tools.

5. What general statement can be made about machining factors including

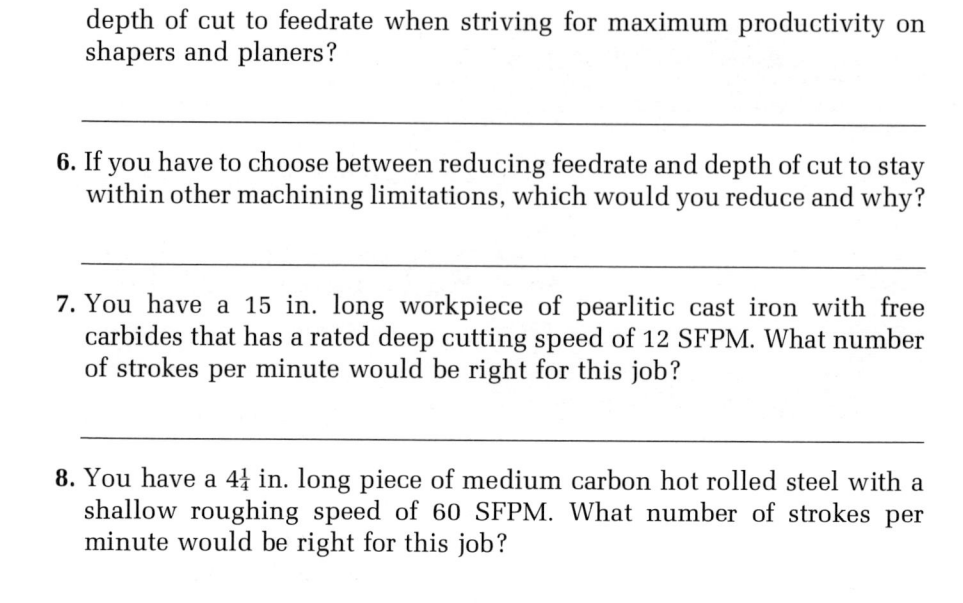

depth of cut to feedrate when striving for maximum productivity on shapers and planers?

6. If you have to choose between reducing feedrate and depth of cut to stay within other machining limitations, which would you reduce and why?

7. You have a 15 in. long workpiece of pearlitic cast iron with free carbides that has a rated deep cutting speed of 12 SFPM. What number of strokes per minute would be right for this job?

8. You have a 4¼ in. long piece of medium carbon hot rolled steel with a shallow roughing speed of 60 SFPM. What number of strokes per minute would be right for this job?

9. For roughing cut, on planers and shapers, what would be an expected feedrate to match a .450 in. depth of cut for deep roughing and a .090 in. cut for a shallow roughing cut, assuming that the overall machining system is adequate for the jobs?

10. If you are making a .005 in. deep finishing cut on a planer or large shaper in cast iron using a square nose finishing tool ½ in. wide, what feedrate would you attempt to set into the machine for making this cut?

unit 3
shaper safety and using the shaper

Before using the shaper you must have a background of information about shaper tools, cutting speeds and feeds, and how to calculate the required strokes per minute. You must also be aware of both general safety considerations and those specific to the shaper before applying power to the machine tool. This unit covers safety requirements and operating procedures for several types of cutting operations on the shaper.

objectives After completing this unit, you will be able to:
1. List at least six safety checks that need to be made relative to the shaper.
2. Demonstrate that you have complied with the lubrication requirements specific to the shaper in your shop.
3. Set the stroke length and stroke rate, and indicate in the vise on a plain shaper (or universal shaper, as necessary).
4. Calculate and enter stroke rates for a specific cutting job.
5. Prepare worksheets for shaper operations.
6. Surface a workpiece in the shaper.
7. Square a workpiece in the shaper.
8. Make angular cuts in the shaper.

information

SHAPER SAFETY

The shaper has one very major hazard in its use that has to be considered at all times. The travel of the ram back and forth is very unyielding to objects that get in the way. You must insure that no tools are left in the way and that you keep your hands and the rest of your body away from the working area. Do not stand, or permit anyone else to stand, in front or in back of the ram. Be sure that the machine is completely stopped, including the motor, before making measurements or any adjustments that require you to place any part of yourself within danger of being struck by the ram. Allow no one but yourself to operate the machine controls.

Be sure that the stroke rate and stroke length are in balance. A fast stroke rate and a long stroke can cause the machine to "walk" or even to "self-destruct" if the clutch is engaged under these circumstances. "Drift" the clutch to bring the machine up to speed to be safe.

Large, powerful shapers can eject large, hot chips with great force. It is important to set up a screen of substantial fireproof material in front of the ram beyond the end of the cutting stroke to stop chips from being a hazard to yourself and to others in the shop.

It is important to review your workholding methods before making a cut. Any workholding method that is the least bit questionable should be checked by your instructor and approved before cutting begins. If you are using a vise, it is also important to check the fastening of the vise to the worktable. Some vises also have a special lock pin that fits into the machine table for added security.

It is also important to remove all handles from the stroke adjusting shaft, the ram positioning clamp, and the cross feed shaft before operating the shaper. The vise handle should be positioned downward or removed entirely.

Other hazards of the shaper are similar to those of most other machine tools. Safety glasses must be worn in the machine shop at all times to comply with federal and state safety regulations.

If there are any doubts about safety or compliance with safety regulations in your shop, the instructor must be consulted before starting the cut. In many instruction labs, the rules specify that the instructor be present the first time a student operates this specific machine tool.

When you have finished using the shaper, the machine must be cleaned and, in the interest of safety for the next operator, the following steps should be taken.

1. Remove the cutting tool.
2. Remove the tool holder.
3. Set the toolhead to vertical and retract the toolslide (this is very important!).
4. Set the stroke length to minimum.
5. Set the stroke rate to the minimum setting.
6. Check that the clutch is disengaged.

LUBRICATION

Before setting up the horizontal shaper, the lubrication requirements should be satisfied. On some larger shapers a lubrication pump is directly coupled to the main drive motor or, if the shaper is hydraulically driven, part of the hydraulic oil will be diverted into a special lubrication channel that meters oil to the ways of the ram. These larger machines are usually equipped with a sight glass to check on oil flow. On shapers without these automatic features, it is necessary to service lubrication points *before* the shaper is operated. Shaper lubrication is usually done with an *oil* (not a grease) especially compounded for sliding lubrication and fed through automotive-type fittings with a pressure lubricating gun. Check the specific requirements of the machine you will be using and follow the steps recommended for lubrication.

SETTING UP TO MAKE A CUT IN THE HORIZONTAL SHAPER

"Indicating In" the Vise

Prepare Worksheet 1 (based on the following information) to match the characteristics of the shaper that you have available. Before the shaper is operated, you must be completely familiar with the controls of your specific machine. With Figure 1 as a guide, locate the same controls or control functions on your machine.

In the following sequences, you will see basic setups being made on the machine in Figure 1. After studying the sequence shown, use the worksheets to match the operation with the characteristics of the machine you have available. You may come out with a different number of steps. Have your worksheet complete for each individual sequence before you attempt to operate the machine.

Initially, the machine is checked to see that the toolhead is vertical, that the toolslide is retracted flush with its own base, and that cutting tools are removed. Then the plate that indicates the number of strokes per minute (Figure 2) is examined and the transmission controls are set for the fewest number of strokes per minute (Figure 3). A check is made to see that there are no tools on the machine that could interfere with ram travel, and that no wrenches remain on the stroke adjusting shaft or on the ram positioning shaft or lock. The clutch lever is also checked to see that it was not left in engagement by the last operator. The areas both in front of the ram and behind the ram should be checked for safe operation of the ram. The next step is the setting of the stroke length. The procedure varies greatly with different designs of shapers. Most large crank shapers are adjusted in the rearmost position of the ram. The shaper shown is adjusted at the center position, and some bench shapers are adjusted with the ram forward. Hydraulic shapers are arranged to set both stroke length and position by means of adjustment dogs that trip a pilot valve. The motor is then started and the clutch (or operating valve) is "drifted" to move the ram gently to the required position for adjusting stroke length (Figure 4). Since this operation is

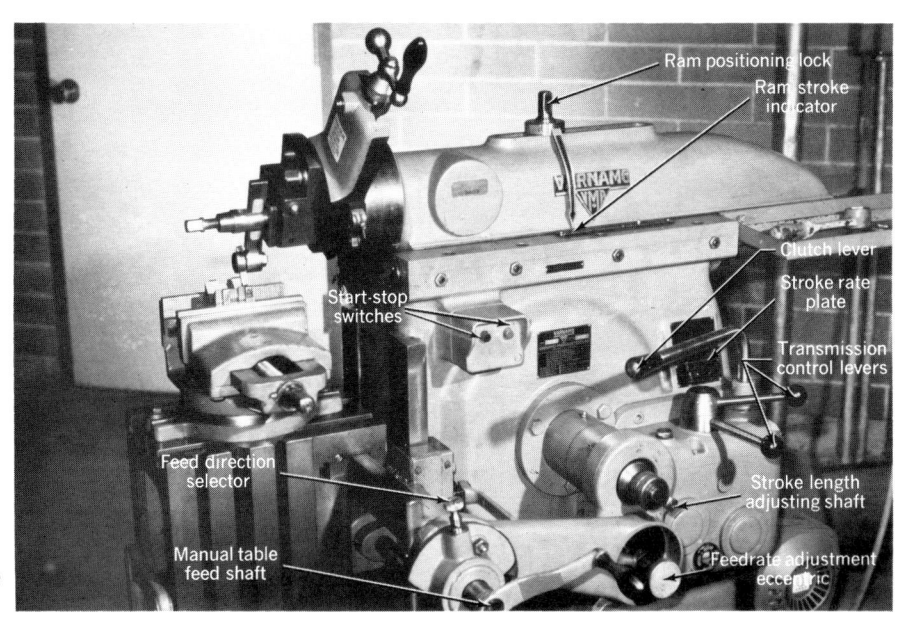

Figure 1. Shaper controls (Lane Community College).

Figure 2. Stroke rate plate (Lane Community College).

Figure 3. Transmission controls being set (Lane Community College).

Figure 4. Adjusting the stroke length. A knurled ring unlocks the adjustment shaft (Lane Community College).

"indicating in," the stroke length set should be about an inch less than the length of the jaw of the vise you will be using. If your machine has a universal table, the table must be "indicated in" on the cross feed direction before the vise is indicated.

Next, the ram is unlocked and positioned (Figure 5) so that an indicator attached to the ram or held in the tool post is over the correct part of

Figure 5. Positioning the ram. On light machines this is done by releasing the ram and sliding it into position (Lane Community College).

Figure 6. "Indicating in" the shaper vise (Lane Community College).

Figure 7. Setting the clapper box angle (Lane Community College).

the vise jaw. On the machine shown, the indicator would be placed central to the vise jaw. Then the table of the shaper is brought up and locked and the cross feed adjusted until the solid jaw contacts the indicator tip, with at least a quarter turn of indicator travel, below the top of the jaw. Now, with the motor started, the indicator is tracked slowly across the face of the fixed jaw by drifting the clutch to check alignment, and adjustment is made as necessary (Figure 6) until minimum indicator travel is obtained. If the machine is a universal table type, and you are working on the tilt table side, first check for parallel on the tilt table by indicating along precision parallels resting on the bottom of the vise jaw.

Surfacing a Workpiece

Prepare Worksheet 2 to list the steps necessary for making a surface cut based on the following information.

PREPARING THE WORKPIECE TO TOOL RELATIONSHIP. A square sectioned rectangular workpiece of hot rolled steel is being used for an example. This is mounted in the vise on parallels so that the top surface extends somewhat above the vise jaws. In mounting a rough workpiece, it may be necessary to use shims to obtain a solid hold on the part. The table is unlocked and lowered as necessary to clamp a tool in the tool post clear of the workpiece. As necessary, the stroke length control is adjusted and the ram is repositioned so that at least $\frac{1}{2}$ in. of pretravel and $\frac{1}{4}$ in. of travel are obtained beyond the workpiece. The pretravel gives the clapper box time to close fully before the tool strikes the workpiece on the cutting stroke. Next, the apron carrying the clapper box is adjusted (Figure 7) to allow the tool to swing clear of the work on the return stroke. The cutting tool is then reset to vertical (Figure 8) or, in a heavy cutting situation, just a little off the vertical to allow the tool to swing away from the work if the feeding load is excessive (Figure 9). The table is then brought nearly to tool level and clamped firmly (Figure 10). Then the table support is raised and secured (Figure 11).

SETTING THE SPEED AND FEED. The number of strokes per minute must be determined to obtain the rated cutting speed of the material for the type of cut (shallow roughing).

$$\frac{CS \times 7}{L}$$

Figure 8. Positioning the tool holder after setting the clapper box (Lane Community College).

Figure 10. Raising and clamping the table (Lane Community College).

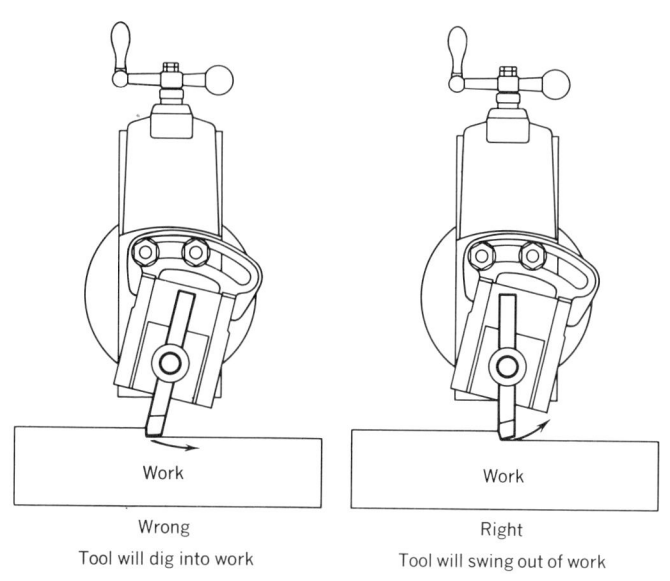

Work — Wrong
Tool will dig into work

Work — Right
Tool will swing out of work

Figure 9. On heavy cuts, leave room for the tool to escape (Courtesy of Cincinnati Incorporated).

Figure 11. Securing the table support (Lane Community College).

Remember to add an inch to your workpiece length to allow for overtravel.

The calculated number of strokes are set into the machine by shifting the transmission lever to the nearest indicated setting (Figures 2 and 3). If the shaper you are using is hydraulic, the cutting speed can be set directly.

As discussed in the previous unit, the rate of feed for roughing should be as much as possible consistent with the machining factors: tool strength, workpiece strength, workholding, machining conditions, and available power. This balance must also result in usable part surface in terms of finish and flatness for the operations that follow.

The shaper illustrated uses an adjustable eccentric (Figure 12) that increases the feedrate in increments of .005 in. as the radius from center is increased. The ratchet mechanism driven by the eccentric has a control to set the direction of feed or, when the knob is turned 90 degrees to a feeding direction, the ratchet mechanism is disconnected. On the part being machined, the rough finished surface is to serve as a reference for other cuts. For the conditions shown here, a suitable roughing feedrate would be about .015 to .020 in. per stroke.

ESTABLISHING THE DEPTH OF CUT AND MACHINING THE SURFACE. The toolslide is then advanced (Figure 13) to establish the depth of cut. Remember that the table support has already been

Figure 12. Setting the feedrate with an adjustable eccentric (Lane Community College).

Figure 14. Making the first cut (Lane Community College).

Figure 13. Setting and locking the depth of cut (Lane Community College).

positioned so the depth of cut for this surfacing is done by means of toolslide movement. After the toolslide is positioned, it must be secured with the slide locking screw.

In positioning the table in cross feed, before making the cut, it is desirable to leave room for the feeding of several strokes before the tool reaches the work. This gives time to stop the machine if something does not seem right. When you engage the clutch for the first time, ease in the clutch gently to make sure you have not made a mistake in setting the stroke rate. The first surfacing cut is then made (Figure 14). This surface will be designated as side 1.

Squaring the Workpiece

Prepare Worksheet 3 to list the steps necessary for squaring a workpiece based on the following. With a completed surface from the previous cut, this surface is used as a reference for squaring the workpiece. Before this is done, the workpiece is removed and burrs are carefully removed. The vise and parallels are cleaned and the parallels are reinstalled. Then the workpiece is replaced in the vise with the machined side toward the fixed jaw of the vise. A piece of unhardened round bar stock is obtained and placed between the workpiece and the movable jaw of the vise (Figure 15), and the vise is tightened. The round bar establishes line contact with the workpiece and forces the finished surface into full contact with the fixed vise jaw.

When the second cut is made it establishes a surface square to the first cut if everything was done correctly. This assumption should be checked with a square. After the part has been deburred and the vise and parallels have been recleaned, the just completed surface (2) is placed against the parallels with the original surface (1) again toward the fixed vise jaw (Figure 16). This is done by turning the part end for end in the process of turning it over. Again, the round

Figure 15. Round bar forces reference surfaces together (Lane Community College).

Figure 16. Turn the part over end for end to prepare for side 3 (Lane Community College).

bar is used between the movable jaw and the workpiece. As the vise is tightened, the workpiece is tapped down onto the parallels with a babbitt hammer or other soft-faced hammer with a weighted head. The parallels must be snug beneath the workpiece with the jaw fully tightened.

Now the exposed surface of side 3 is machined. The relationship between the newly machined surface (3) and the previous surface (2) should be checked with a micrometer on the machine, if possible, to insure that the two surfaces are parallel. This may also be checked off the machine and the part reset before making a final cut to the required dimension. Since the relationship of side 1 to side 2 was established as square, surface 3 should also be square to surface 1, but there is still the possibility of other errors, such as taper along the part axis.

The fourth side is machined after checking, deburring, and cleaning the vise by placing the completed side 3 toward the solid vise jaw. Again, it is important that the parallels be snug as the vise jaw is tightened. Check the fit of the parallels before machining the fourth side to the required dimensions.

Squaring the End of the Workpiece
Prepare Worksheet 4 to list the steps necessary for squaring the end of a workpiece, based on the following information. For this operation, the vise is set crosswise to the ram travel direction (Figure 17). If the work is to be done from the end of the vise, the fixed jaw of the vise needs to be "indicated in" by moving the vise with the table cross feed under the indicator. It is also desirable

Figure 17. Setting the vise crosswise to the ram travel (Lane Community College).

to indicate across parallels resting on the bottom of the vise opening, and then across the top edge of the vise jaw to see that this surface is also parallel.

For relatively short workpieces, the part can be set vertically in the vise using the combination square, as shown in Figure 18. Since the reference edge is the top of the fixed vise jaw in this case, this surface must be checked for parallel to the table travel.

The stroke length needs to be reset and the number of strokes per minute recalculated and the ram repositioned before making the cut (Figure 19). If the part is too long to be squared in this way, it can be held out the end of the vise and cut with a side tool (Unit 1, Figure 29). If squaring is done in this fashion, vise squareness is critical.

Section F Shapers and Planers

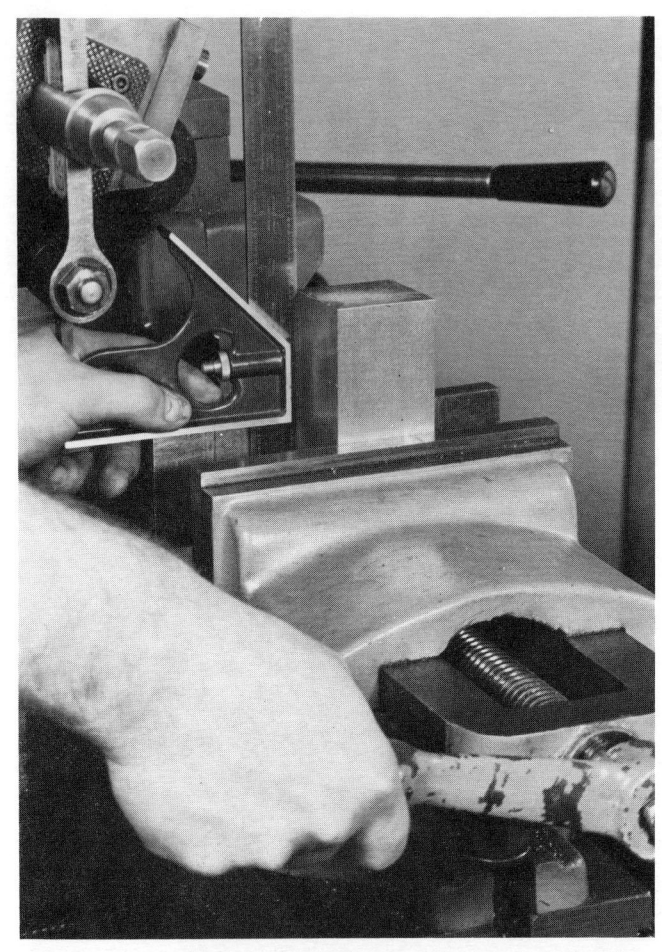

Figure 18. Setting a workpiece with the combination square (Lane Community College).

Figure 19. Squaring the end of the part (Lane Community College).

Figure 20. Setting the tool for roughing out the excess material (Lane Community College).

Making Angular Cuts

Prepare Worksheet 5 to list the steps necessary for machining a V-shaped feature in a workpiece, based on the following information. Angular cuts are easy to make in the shaper. The example shown is the sequence for making a 90 degree V in the previously squared workpiece. The machine is set up and indicated as in preparing the part for surfacing. Initially a roughing tool is used to remove excess material. For this operation the toolhead is set vertically, and the apron is set to permit the tool to lift away from the cut on the return stroke. The tool and tool holder are placed with the tool mounted on the forward side (Figure 20) because for this operation visibility is quite important.

For roughing out the form, cuts are made by combinations of toolhead and table cross feed to hand contour the part just short of a layout line (Figure 21).

After the excess material has been roughed out, the toolhead is repositioned to 45 degrees (Figure 22) and the clapper box is reset to permit the tool to lift from the cut on the return stroke (Figure 23). Then the finishing cut is made along the 45 degree axis (Figure 24), using the toolhead feedscrew. Since it is important to have the V central to the workpiece, the cross feed must not be moved once the finishing cut is established. When the bottom of the V is reached and the ram is stopped, the slide is retracted and the part removed and *reversed end for end*; the finishing cut is taken on the opposite face, using the same exact machine position. In this way, the part will be accurately symmetrical.

Figure 21. Roughing out the V by hand contouring (Lane Community College).

Figure 22. Setting the toolhead to 45 degrees (Lane Community College).

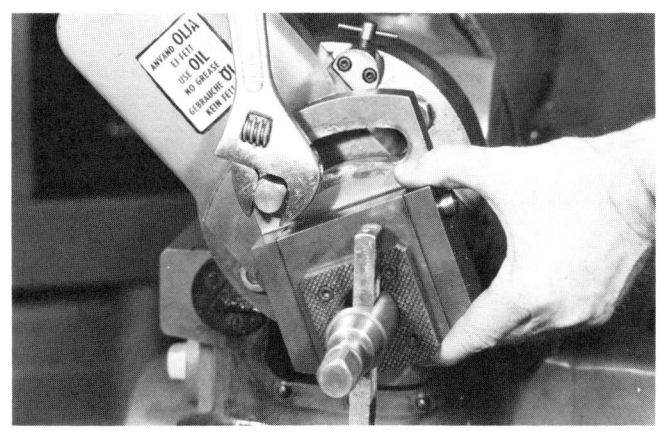

Figure 23. Resetting the clapper box apron (Lane Community College).

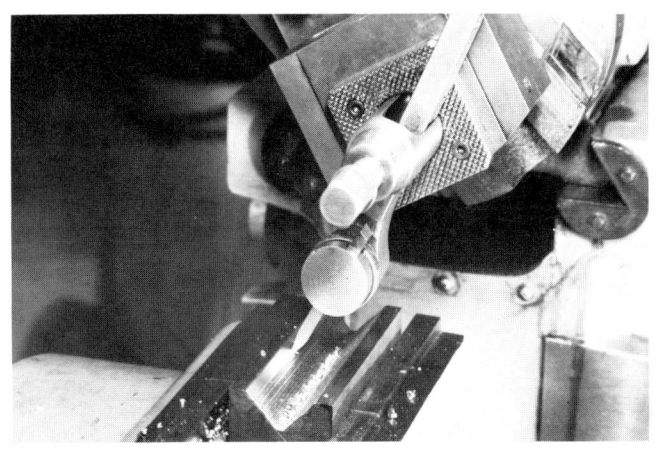

Figure 24. Finishing one side of the V. Turn the part end for end to finish the opposite side (Lane Community College).

Figure 25. Vee block workpiece cut in two being finished on the ends (Lane Community College).

After the straight and angular cuts have been made on the vee block workpiece, this workpiece can be sawed in two and the ends finished as shown in Figure 25.

Securing the Machine
The last step is to secure the machine. It should be cleaned, the tooling put away, and the following steps observed for the safety of the next operator.

1. the cutting tool and holder should be removed.
2. The toolhead should be set to vertical and the toolslide retracted flush with its base.

3. The stroke length should be set to minimum.
4. The stroke rate should be set to minimum.
5. The clutch must be left disengaged.

self-evaluation

WORKSHEET 1 Examples are shown in the appendix to this volume. List the steps necessary to "indicating in" the shaper and machine vise to prepare to make a horizontal cut on the machine that you have available.

1. _____
2. _____
3. _____
4. _____
5. _____
6. _____
7. _____
8. _____

WORKSHEET 2 Examples are shown in the appendix to this volume. List the steps necessary to surfacing a workpiece after the machine has been indicated in.

1. _____
2. _____
3. _____
4. _____
5. _____
6. _____
7. _____
8. _____
9. _____
10. _____
11. _____

Unit 3 Shaper Safety and Using the Shaper

12. _____

13. _____

14. _____

WORKSHEET 3 Examples are shown in the appendix to this volume. List the steps necessary for squaring a workpiece in a shaper after one flat surface has been obtained.

1. _____
2. _____
3. _____
4. _____
5. _____
6. _____
7. _____
8. _____
9. _____
10. _____
11. _____
12. _____

WORKSHEET 4 Examples are shown in the appendix to this volume. List the steps necessary to squaring the end of a workpiece after the sides have been squared.

1. _____
2. _____
3. _____
4. _____
5. _____
6. _____

Section F Shapers and Planers

7. _____

8. _____

WORKSHEET 5 Examples are shown in the appendix to this volume. List the steps necessary to machining a V shaped part in the shaper.

1. _____
2. _____
3. _____
4. _____
5. _____
6. _____
7. _____
8. _____
9. _____
10. _____
11. _____
12. _____
13. _____
14. _____
15. _____
16. _____

SELF-TEST

1. What is the single most critical hazard in working with the horizontal shaper?

    _____

2. Name at least six safety items relative to the shaper.

    _____

3. How is lubrication generally accomplished on large mechanical and hydraulic shapers?

    _____

Unit 3 Shaper Safety and Using the Shaper

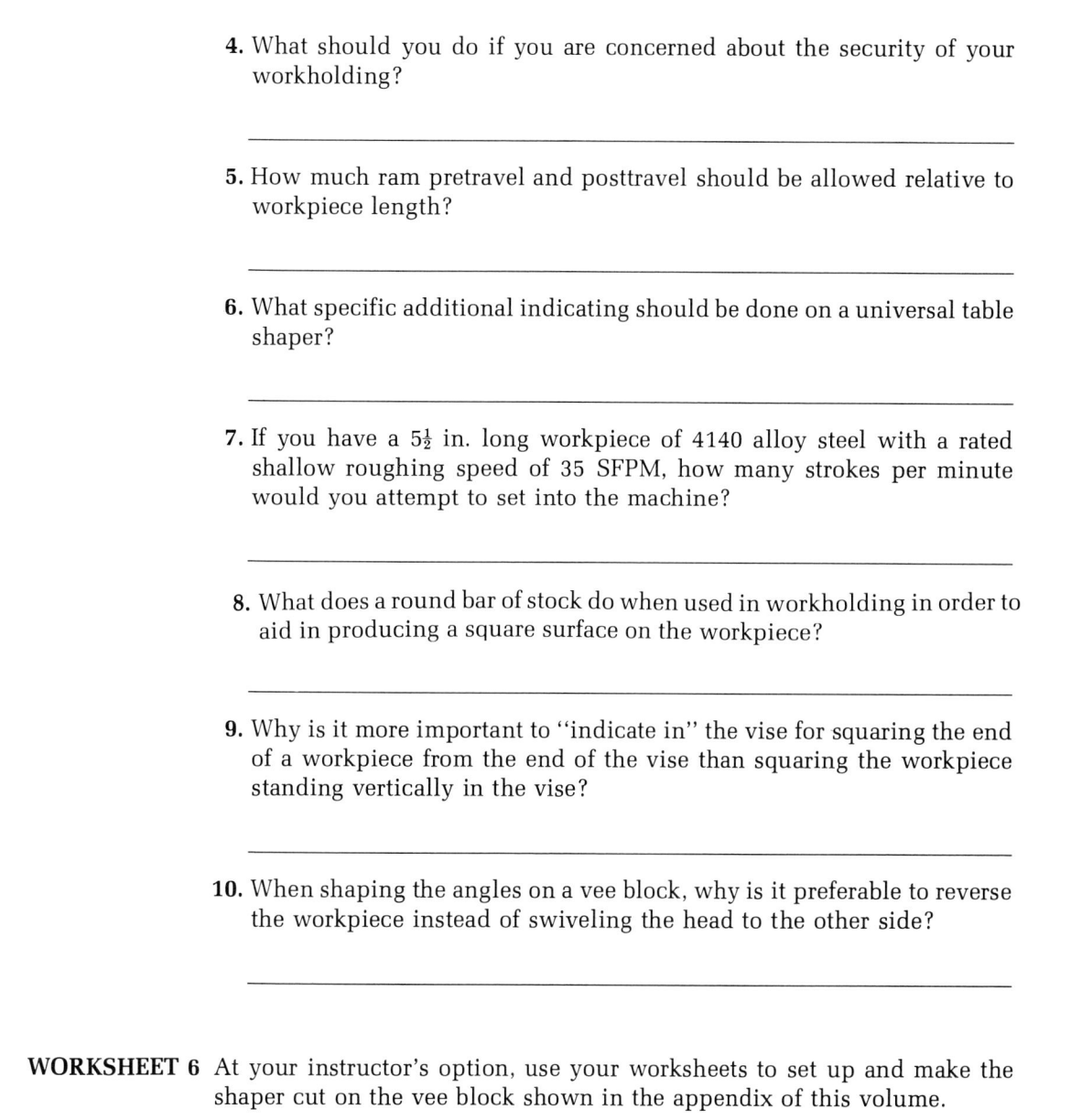

4. What should you do if you are concerned about the security of your workholding?

   _____

5. How much ram pretravel and posttravel should be allowed relative to workpiece length?

   _____

6. What specific additional indicating should be done on a universal table shaper?

   _____

7. If you have a 5½ in. long workpiece of 4140 alloy steel with a rated shallow roughing speed of 35 SFPM, how many strokes per minute would you attempt to set into the machine?

   _____

8. What does a round bar of stock do when used in workholding in order to aid in producing a square surface on the workpiece?

   _____

9. Why is it more important to "indicate in" the vise for squaring the end of a workpiece from the end of the vise than squaring the workpiece standing vertically in the vise?

   _____

10. When shaping the angles on a vee block, why is it preferable to reverse the workpiece instead of swiveling the head to the other side?

    _____

WORKSHEET 6 At your instructor's option, use your worksheets to set up and make the shaper cut on the vee block shown in the appendix of this volume.

unit 4
features and tooling of the planer

This unit will make you more aware of the specifics of planer components, controls, drives, and workholding, preliminary to setting up and operating the planer.

objectives After completing this unit, you will be able to:
1. Identify the basic operating controls of a small hydraulic planer.
2. Name the component parts of a planer.
3. Describe the functions and motions of the major planer components.
4. State differences between two types of planer drives.
5. Name at least four kinds of straps used with planer setups.
6. Explain the purposes of various kinds of "planer furniture."

information

You have already seen the basic motions of planers and a variety of work done by these machines in the introductory part of this section. Also, you have seen the basic design differences between double housing and openside planers.

At this point, we need to describe controls and components, as well as the common type of planer drives, workholding, and support devices.

Figure 1 provides detailed nomenclature of the control features of a small openside planer. Refer back to this figure as needed during the following discussion.

MAJOR COMPONENTS

The Bed
The bed of the planer is a massive casting, heavily ribbed inside that serves as a platform for the table. The top surface of the bed has ways that provide guidance and support for the planer table (platen) that carries the workpiece. Contained within this bed are driving components, either mechanical or hydraulic, for moving the table.

The Table
The table of the planer moves along the precision ways of the bed on a film of pressurized oil. The table on high speed planers is fitted with laminated plastic bearing surfaces that permit high sliding velocity without the danger of galling between the table and the bed. The table (Figure 2) is equipped with T-slots for the attachment of clamping devices and anchor holes for the fitting of stops and poppets. Pockets are provided at the table ends to keep chips contained when cleaning out the T-slots. On some very large planers, two tables are fitted, so that one can be used for machining while the other is being set up.

Housing or Column
This portion of the machine is solidly attached to the bed and is fitted with accurate machine vertical ways for the attachment of the rail and

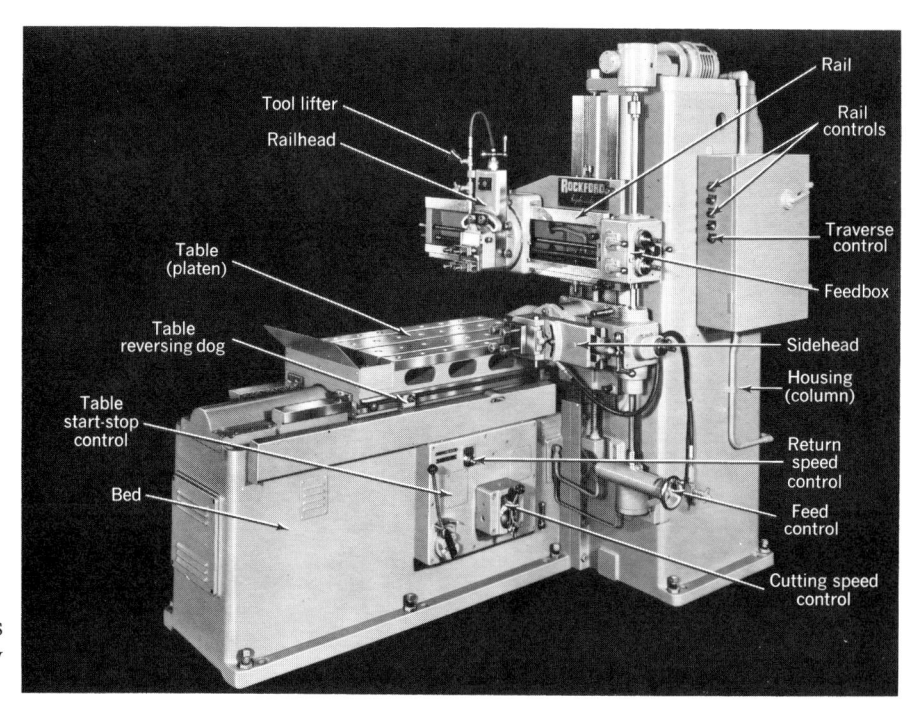

Figure 1. Components and controls of a small hydraulic planer (Courtesy of Rockford Machine Tool Co.).

sidehead. This portion contains the feed drive mechanism.

Rail
The rail, which is adjustable up and down, slides on the face of the column by means of an apron, which can be firmly clamped to the column. The rail is designed to be accurately parallel to the table surface. The rail serves as the mounting and guidance function for the toolhead which, when mounted on the rail, is called a railhead.

Heads
The railhead and the sidehead have similar design features. They are each mounted on a saddle that rides accurately along the rail for the cross feeding function or up and down the face of the column in the case of the sidehead. On the saddle is a swivel base that permits the head to be angled (Figure 3), like the toolhead on the shaper. The heads are usually equipped to power feed the toolslides as needed. On the face of the toolslides is the clapper box, which can be angled relative to the toolslides, again like the shaper. The railhead, or railheads on larger machines, operate either independently or together. The sidehead is independent of the railhead. Most planers are equipped with tool lifters to pivot up the clapper box on the return stroke. Tool lifters are essential where carbide cutting tools are used on the planer.

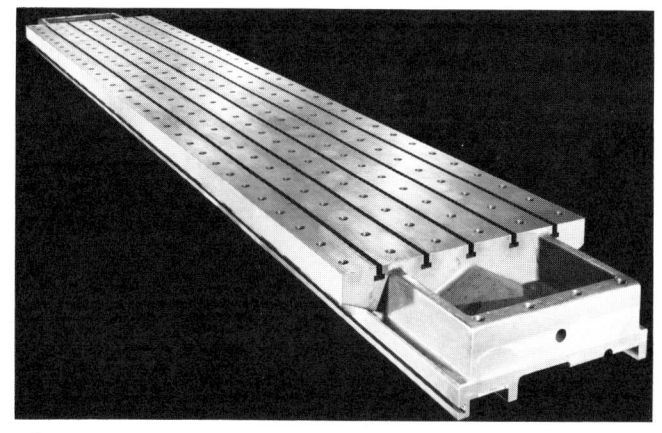

Figure 2. Planer table showing T-slots, holes, and chip pocket (Courtesy of Rockford Machine Tool Co.).

PLANER DRIVE SYSTEMS

There are two basic methods of transmitting drive to the planer table: rack and pinion (with variations like herringbone and spiral bevel mechanisms), and hydraulic. The mechanical types of machines are often driven by direct current motors, which permit quick reversal and rapid return.

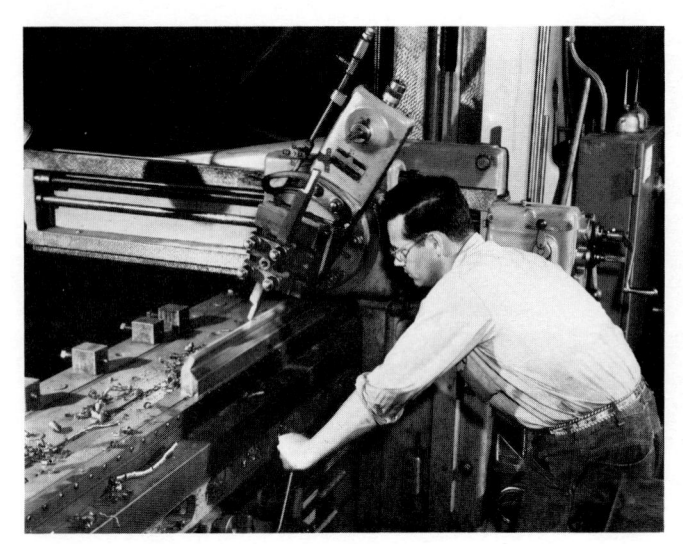

Figure 3. Railhead setup for angular cutting (Courtesy of Rockford Machine Tool Co.).

Small hydraluic planers, like the one shown in Figure 1, have very similar hydraulic circuits (Figure 4) to the hydraulic shaper shown in a previous unit. For some larger models (used in the following unit), an additional speed range is provided (Figure 5). With this type of design, very slow and very high speed table movement (up to 300 FPM) is possible.

On planers, whether they are hydraulic or mechanical, the return stroke is usually much faster than the cutting stroke.

WORKHOLDING DEVICES ON THE PLANER

Vises

The planer can be fitted with the same types of vises that are found on shapers, plain swivel base, and single or multiple screw types. The double screw type is the most usual. The principles of holding material in vises is also the same as used with the shaper.

Table Holding Methods

Most planer work is done on relatively long pieces held either directly to the planer table or in a fixture attached to the planer table (Figure 6). Virtually every planer setup requires the use of a stop (Figure 7) to prevent endwise movement of the workpiece. Stops are fixed or adjustable or, when an adjustable stop is turned sideways so the screw is not used, it becomes a stop pin. A similar device to the adjustable stop, but with the threaded part at an angle, is called a poppet. This is typically used together with an additional piece with either a pointed end called a planer pin or a spadelike tip called a toe dog. Planer pins will be seen in use in the next unit. This combination is especially useful for holding

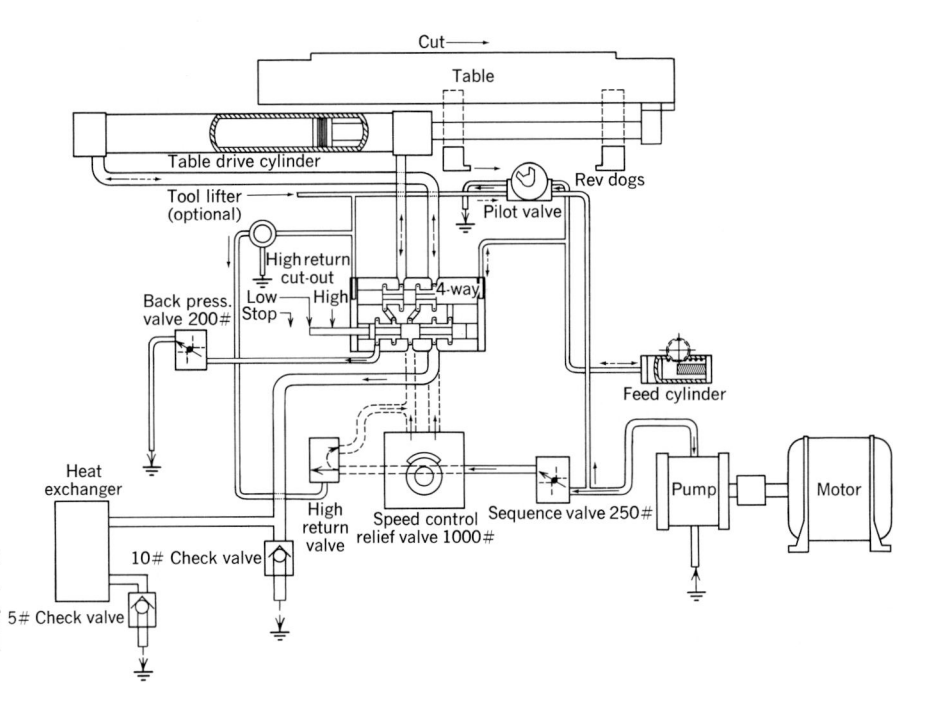

Figure 4. The hydraulic circuit of a small hydraulic planer is very similar to that of an hydraulic shaper (Courtesy of Rockford Machine Tool Co.).

Figure 5. Three speed range hydraulic planer drive system. (1) Reversing hydraulic table drive pump. (2) Constant speed main drive motor. (3) Hydraulic oil reservoir. (4) Control valve actuated by table dogs. (5) Double acting hydraulic cylinder. (6) Single acting hydraulic cylinder. (7) Table or platen. (8) Double acting piston. (9) Piston rod. (10) Anchor bolts. (11) Feed cylinder (can be set to feed at either end of the stroke). (12) Feed shaft (drives the toolheads). (13) Feed adjustment handwheel. (14) Lubricating pump for ways. (Courtesy of Rockford Machine Tool Co.).

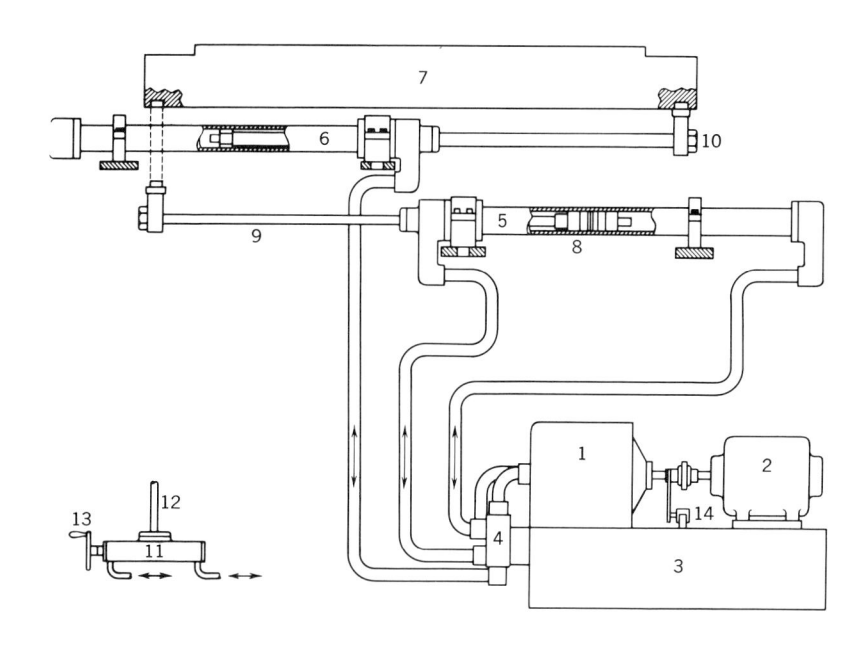

Figure 6. Planer fixture for a narrow workpiece (Courtesy of Rockford Machine Tool Co.).

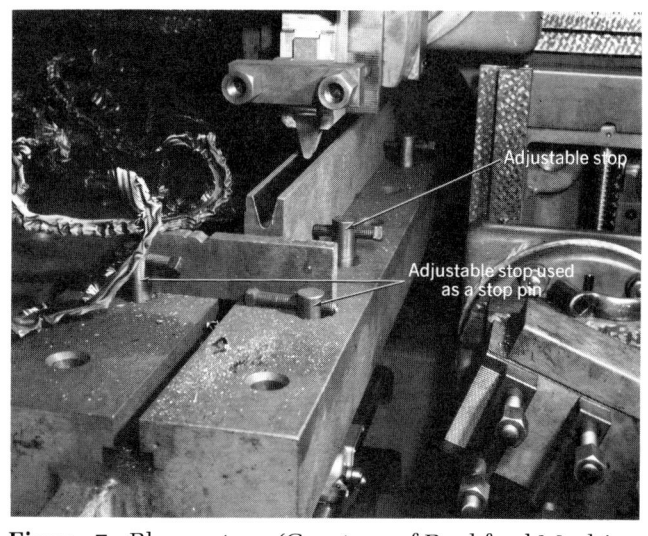

Figure 7. Planer stops (Courtesy of Rockford Machine Tool Co.).

down thin parts for planing and shaping (Figure 8).

Other holddown methods are common to table-type machine tools that include a variety of strap clamps of differing designs (Figure 9). Other "planer furniture," as these workholding devices are termed, include planer jacks (Figure 10), which may be used in conjunction with T-bolts, step blocks, and risers to support overhanging workpieces. Some setups require unusual adaptations to secure the work (Figure 11).

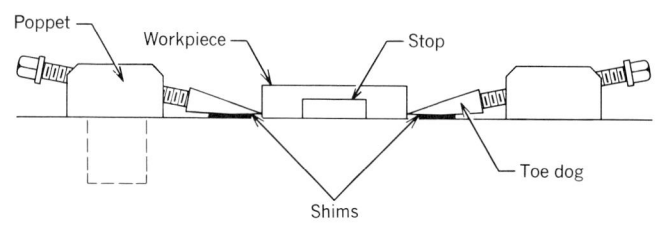

Figure 8. Holding thin work for planing (Courtesy of Cincinnati Incorporated).

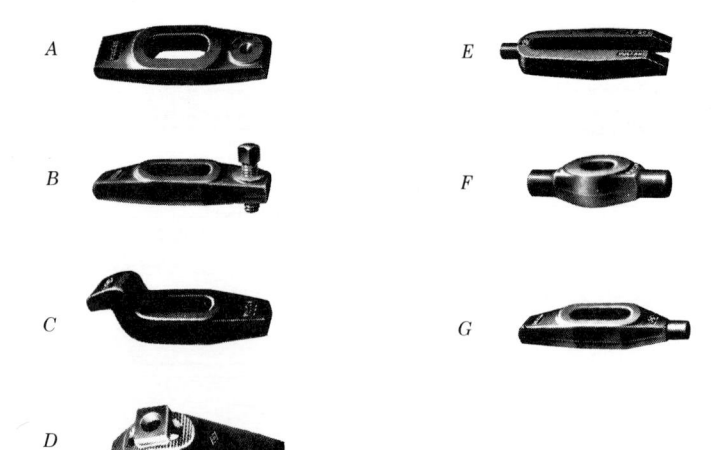

Figure 9. Strap clamps of various designs: (A) Plain slot pattern; (B) Adjustable step pattern; (C) Gooseneck pattern; (D) Universal adjustable pattern; (E) "U" pattern; (F) Double end finger pattern; (G) Single end finger pattern (Courtesy of J. H. Williams Division of TRW Inc.).

Figure 10. A variety of "planer furniture" for the support of overhanging workpieces (Ames Research Center — NASA).

PLANER CUTTING TOOLS

The geometry of planer cutting tools is the same as that covered in the units entitled "Features and Tooling on the Horizontal Shaper" and "Cutting Factors on Shapers and Planers." Typically, planer tools are much more massive because of the large size and power of the typical planer, as compared to the largest horizontal shaper. Planers of 100 HP are not uncommon. These planers can generate cutting forces as much as 18 tons.

Figure 11. Planer workholding often requires adaptability (Courtesy of Rockford Machine Tool Co.).

self-evaluation

SELF-TEST

1. What is the function of the rail on a planer?

2. Is the planer equipped with a clapper box? Does it swivel?

Unit 4 Features and Tooling of the Planer   247

3. Are planer heads often equipped with power feed on the toolslide?

_____

4. Describe the two most common types of planer drive systems.

    a. _____

    b. _____

5. Name at least four types of holddown straps commonly used in planer setups.

    a. _____

    b. _____

    c. _____

    d. _____

6. What is the purpose for the large number of holes in the top surface of the planer table?

_____

7. When the planer is equipped with two railheads, can these be operated independently from one another?

_____

8. You are considering making a planing setup to machine the full width of a 4 foot long piece of steel that is 6 in. wide and $\frac{3}{8}$ in. thick. How would you make a setup to machine this part?

_____

9. You are setting up to machine a part that has a great deal of overhang from its mounting base. What "planer furniture" would you use for support?

_____

10. Why are planer tools usually much more massive than shaper tools?

_____

11. Identify the components and controls of the planer in Figure 12.

    A _____

    B _____

    C _____

    D _____

    E _____

    F _____

G _____
H _____
I _____
J _____

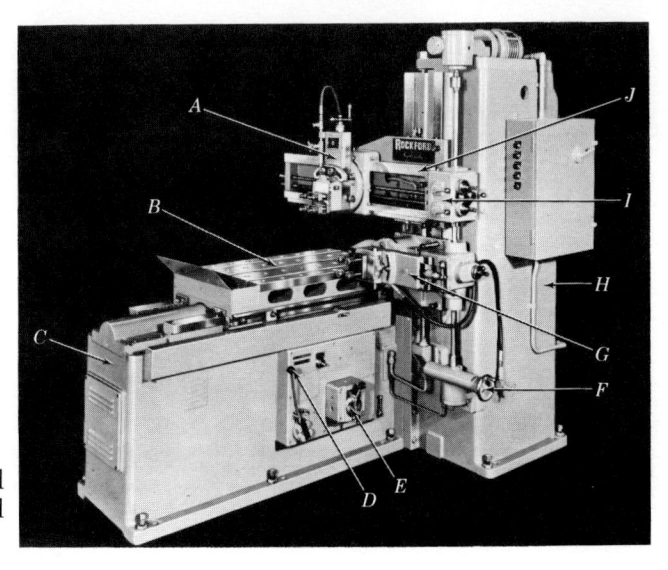

Figure 12. Self-test components and controls (Courtesy of Rockford Machine Tool Co.).

unit 5
planer safety and using the planer

The planer is an interesting machine tool to use because it demands substantial judgment from the machinist, particularly in devising part handling and workholding methods.

Unit 5 Planer Safety and Using the Planer

objectives After completing this unit, you will be able to:
1. Describe the two major safety hazards of work on planers.
2. Describe a method of handling a heavy workpiece.
3. Describe the use of standard items of "planer furniture."
4. List specific procedural steps in preparing a planer for use.
5. List sequences of action necessary to using the planer for out-of-balance workpieces.
6. Define the process for making accurate height measurements of a planer workpiece.

information

SAFETY ON THE PLANER

The planer demands attention to safety practices for two compelling reasons. First, the planer is usually a very powerful machine tool with a heavy, fast moving table and with numerous opportunities for the careless operator to get caught between the workpiece and machine components. The operator should make no adjustments other than setting table speeds and cutting feeds while the machine table is moving. The second major hazard around planers is common to all large machine tools; that is, they accept large and heavy workpieces. It is important to be aware of the hazards of rigging and placing large parts. You should always consider the necessity of keeping your body out from under suspended parts and leaving room to withdraw your hands. Keep your body positioned so that your hands can be withdrawn instantly from danger when handling large parts.

Other safety concerns with planers should be considered. If you are using a rack and pinion drive planer, safety dogs should be attached to the end of the table to prevent the possibility of a "runaway" table. The floor around the planer should be kept free of oil to avoid slipping and free of any obstruction that could trip the operator.

All fastening devices that are not part of the setup should be removed from the table.

PROTECTING THE TABLE

The table on any machine tool should be treated like fine furniture. Burrs and irregularities from the workpiece can deface the surface and cause later inaccurate work. A good way to prevent this is to use a special hard fibrous plastic material to cover the area where the workpiece will be placed (Figure 1).

HANDLING AND PLACING THE WORKPIECE

Workpieces used on planers are usually of a size and weight that requires a crane for handling. Before the part is picked up, the operator attaches clamps and a sling prior to bringing the crane to the job (Figure 2). It is important to estimate the center of gravity of the workpiece and attach the cabling so that the eyes in the cable are free to swivel on the clamps. Otherwise the leverage applied during the lifting could "wrench" the clamps from the workpiece.

Before the workpiece is placed on the table, it is given a final check (Figure 3) for irregularities and burrs. These are filed off and the part and table surface are wiped free of grit (Figure 4). Castings and forgings that are set on the platen should have the "rock" removed by shimming under the high side, thus enabling the workpiece to be solid on the platen.

FASTENING THE WORKPIECE

In this case, two identical weldments are to be machined top and bottom and brought to a specified height. After the second weldment is prepared and placed, the excess table protecting material is cut away to make room for the work-holding devices. The first step in attachment in most planing jobs is the placing of a stop (Figure 5). Since planing involves tool impact, the end movement of the workpiece is an ever present risk. The workpiece must be in contact with the stop before the remainder of the clamping is done. In this case, this is assured by the use of a

Section F Shapers and Planers

Figure 1. Protecting the planer table (Ames Research Center — NASA).

Figure 2. Rigging the workpiece (Ames Research Center — NASA).

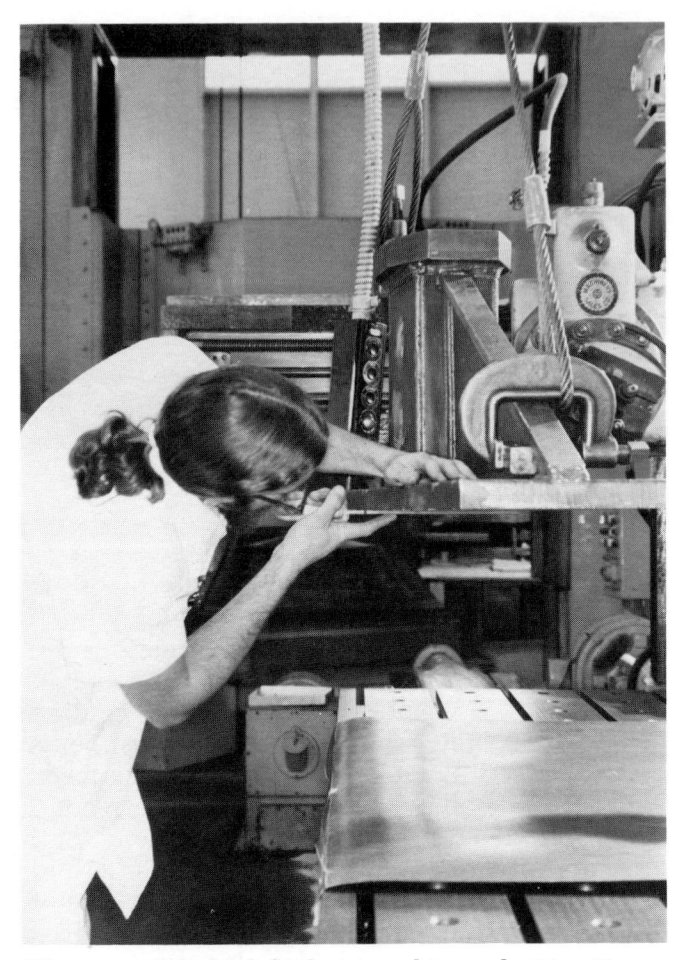

Figure 3. Checking for burrs and irregularities (Ames Research Center — NASA).

Figure 4. Removing grit from workpiece and table surface (Ames Research Center — NASA).

Figure 5. Placing stops (Ames Research Center — NASA).

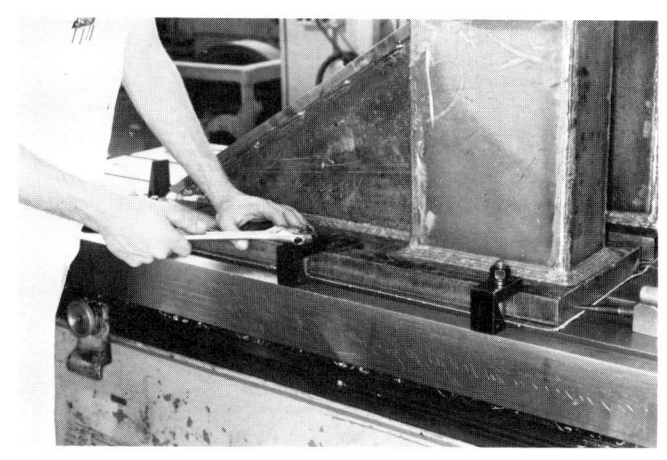

Figure 7. Straps and step blocks being applied (Ames Research Center — NASA).

Figure 6. Poppet and planer pin being used to force workpiece to the stop (Ames Research Center — NASA).

Figure 8. U straps being used for common centerline holding (Ames Research Center — NASA).

poppet and a planer pin applied to the other end of the workpiece to force it against the stop (Figure 6). When this has been done, the part is fastened down with T-bolts, straps, and step blocks on the outside (Figure 7), and the combination of T-bolts and straps on the inside (Figure 8); then the whole setup is rechecked to insure that the fasteners are secure and of approximately even tension (Figure 9).

PREPARING THE TOOLHEAD AND SETTING THE RAIL HEIGHT

For the surfacing work that is to be done, the toolhead is set vertically (Figure 10). The apron does not have to be angled away from the cut as it

Figure 9. Checking for even tension on the whole setup (Ames Research Center — NASA).

Figure 10. Setting the toolhead to vertical (Ames Research Center — NASA).

Figure 11. Placing a carbide-type tool holder (Ames Research Center — NASA).

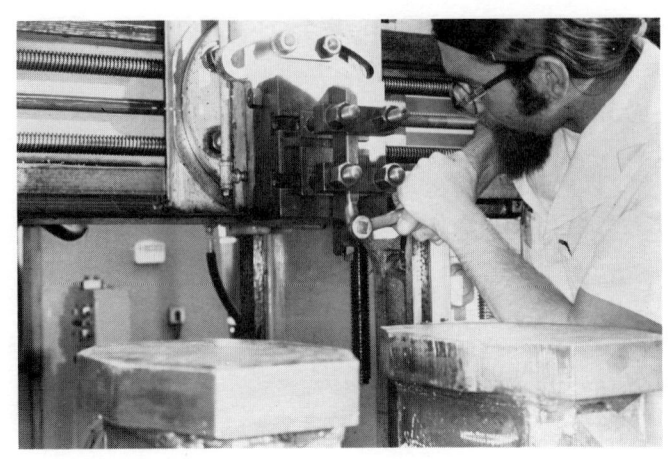

Figure 12. Installing the roughing tool (Ames Research Center — NASA).

Figure 13. Positioning the rail (Ames Research Center — NASA).

Figure 14. Tighten the back rail clamps first (Ames Research Center — NASA).

would be on most shapers, because this machine is equipped with a hydraulic tool lifter that will tilt the clapper box on the return stroke. Next, the tool holder is placed (Figure 11). The cuts to be taken on this job are relatively small; consequently, a carbide-type (0 degree angle) lathe tool holder can be used with an inserted tool bit. It is important that the tool bit be installed firmly with as little extension from the tool holder as possible (Figure 12).

The rail is lowered and then raised just slightly (Figure 13) to take up backlash on the adjustment screw so that the extension of the toolhead to reach the workpiece is minimal. On the openside planer the clamps at the back of the column should be clamped first (Figure 14). The

Figure 15. Final securing of the rail (Ames Research Center — NASA).

Figure 16. Setting the pretravel stop (Ames Research Center — NASA).

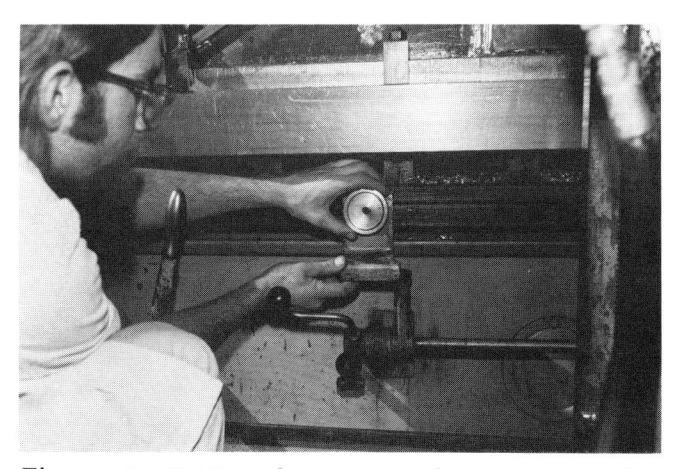

Figure 17. Setting the overtravel stop (Ames Research Center — NASA).

Figure 18. Setting the table into motion (Ames Research Center — NASA).

Figure 19. Setting the cutting speed (Ames Research Center — NASA).

full array of locking bolts that secure the cross rail apron to the front of the column should be tightened evenly (Figure 15).

SETTING THE TABLE STOPS AND CUTTING SPEED

With the cutting and return speeds set at a minimum, the machine should be started and the table stops set (Figures 16 and 17). The stop on the entrance side of the cut should be set with adequate clearance to allow for feeding of the tool before the next cut begins.

Once the stops have been set, the table is set into motion (Figure 18) and the cutting and return speeds are set (Figures 19 and 20). This is

Figure 20. Setting the table return speed (Ames Research Center — NASA).

done by adjusting valves on two sides of a hydraulic pump that controls hydraulic cylinders with pistons attached to the table or platen. One side controls the cutting rate, and the other side independently controls the rate of return.

One thing should be noted that relates to the setup of most of the larger machine tools. Setups often cross over into other shifts, so the person who sets up may not be the person who runs the job. Since there are usually other jobs waiting and idle time is very expensive, the work usually continues until it is complete.

SETTING THE FEEDRATE AND POSITIONING THE TOOL

Next, with the table in motion, the feedrate is set (Figure 21). The feed on this planer is hydraulic and the valve is adjusted until the movement of each cycle on the dial above is what the operator is seeking, in this case .045 in.

Then the planer is stopped; the head is manually fed across and the toolslide fed down to set the tool height for the roughing cut (Figure 22). The toolslide is then locked in place (Figure 23), the toolhead is moved manually to clear the work on the operator's side (Figure 24), the table travel is started, and the feed begins (Figure 25).

MAKING THE CUTS

The roughing cut is started (Figure 26). Cutting oil is added from a safe distance by the use of a squeeze bottle to direct the stream where it is needed. Do not try to use a brush for this operation!

Figure 21. Adjusting the feedrate (Ames Research Center — NASA).

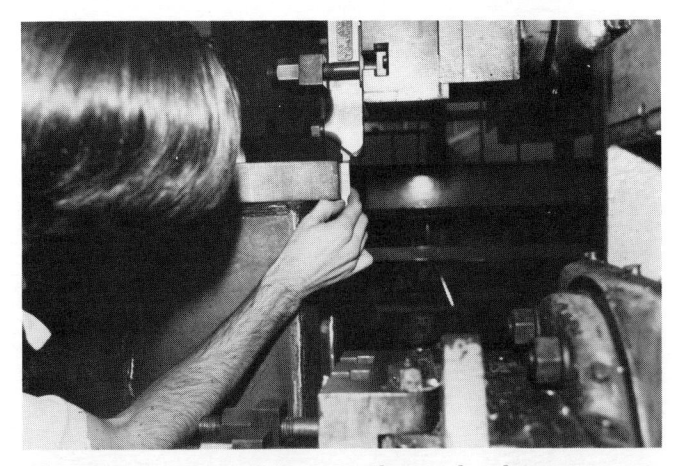

Figure 22. Setting the roughing depth (Ames Research Center — NASA).

After the roughing cut is complete for both parts on the table, the toolslide is raised and the toolhead is returned to have a finishing tool installed (Figure 27). This tool has a broad, slightly

Figure 23. Locking the toolslide (Ames Research Center — NASA).

Figure 24. Setting the starting position (Ames Research Center — NASA).

Figure 25. Starting the feed (Ames Research Center — NASA).

Figure 26. Making the roughing cut — apply oil from a safe distance (Ames Research Center — NASA).

Figure 27. Installing the finishing tool (Ames Research Center — NASA).

radiused surface (Figure 28). The finishing tool is positioned in the same way as the roughing tool, and the toolslide is firmly clamped (Figure 23). Before the finishing operation is started, the entire surface of the work to be finished is coated with cutting oil. The depth of cut with the finishing tool is about .005 in. and the feedrate set in is about two-thirds of the tool width. This tool form, depth of cut and feedrate results in a smooth, wide chip (Figure 29). The table speed was the same for finishing as for roughing, since the roughing depth of cut was quite shallow.

MAKING THE SETUP ON THE OTHER SIDE

After the roughing and finishing cuts have been completed on the first side, the part is deburred on the edges (Figure 30), the workpiece is unfas-

Figure 28. The finishing tool has a broad contact surface (Ames Research Center — NASA).

Figure 29. The finishing chip of a broad nose tool (Ames Research Center — NASA).

Figure 30. Deburring the part edges (Ames Research Center — NASA).

Figure 31. Removing the workpiece after the first side is finished (Ames Research Center — NASA).

Figure 32. Rerigging and inverting the part (Ames Research Center — NASA).

tened, and the crane is brought back to remove the workpiece (Figure 31) to a wooden pallet next to the machine. Here it is rigged again and turned over (Figure 32). This operation must be done carefully, both for the safety of the operator and to prevent damaging the newly machined surface. Then the part is returned to the table (Figure 33), but no protective material is used because the side facing the table is machined flat and smooth. Since this part would not stand on the present side, the crane is kept in place while part of the fastening is taking place. A bar is placed across the stops in front of the part and poppets and planer pins are applied to the rear (Figure 34). Once the part is located in the front-to-back direction, a back clamp is applied (Figure

Figure 33. Repositioning the part. Note that protective material is omitted (Ames Research Center — NASA).

Figure 34. Using stops, poppets, and pins to position the part. Crane is still attached (Ames Research Center — NASA).

Figure 35. Placing the back clamps — a critical step (Ames Research Center — NASA).

Figure 36. Removing the crane (Ames Research Center — NASA).

Figure 37. Seating the workpiece (Ames Research Center — NASA).

35). It is important that no additional clamping take place until after the crane is removed (Figure 36); otherwise, the part might not seat correctly on the table.

Now the additional clamping is done on the base of the workpieces (Figure 37).

Since the part has so little base to support the forces of machining on the present top surface, it is necessary to use auxiliary support. This is done with a combination of riser blocks, planer jacks, step blocks, and clamps (Figure 38). When this type of auxiliary support is required, it is important to determine if the jacks are causing distortion of the workpiece. This can be done by using an indicator to check the parts for parallel (Figures 39 and 40).

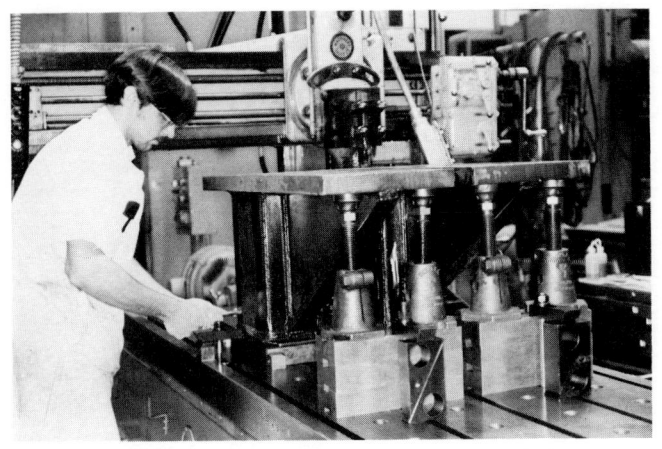

Figure 38. Auxiliary support installed and the setup checked for uniform workholding tension (Ames Research Center — NASA).

Figure 39. Indicating for parallel to the table (Ames Research Center — NASA).

Figure 40. Indicating for parallel to the table (Ames Research Center — NASA).

Figure 41. Setting the height gage to zero to check part dimension (Ames Research Center — NASA).

Once any adjustments have been made, the height gage and indicator can be zeroed on the table top (Figure 41) and an accurate assessment of the existing height of the part and the amount to be removed can be determined.

The workpiece is now ready to machine with the same sequence of operations used on the first side.

self-evaluation

SELF-TEST 1. Describe two or more hazards that apply to planer operations.

a. _____

b. _____

Unit 5 Planer Safety and Using the Planer   259

    c. _____

2. What can be done to protect the planer table from damage by rough workpieces?

_____

3. Why must the eye of a cable sling be able to swivel freely?

_____

4. What is done to force the workpiece against a fixed stop on the planer?

_____

5. Why is the cross rail usually lowered and then raised slightly before it is clamped?

_____

6. What important planer function must be allowed for in setting the table stop on the starting side of the cut?

_____

7. For a broad nose finishing tool, how much feed in proportion to the tool width is recommended?

_____

8. Describe the special precautions needed when setting up a part that requires auxiliary support.

_____

9. Describe the procedure for leveling and measuring the height of a part that requires auxiliary support.

_____

10. Why must planer jacks occasionally be adjusted after the workpiece has already been completely clamped?

_____

WORKSHEET   List the steps that you would follow in setting up a planer to machine the flat surfaces of an I-beam.

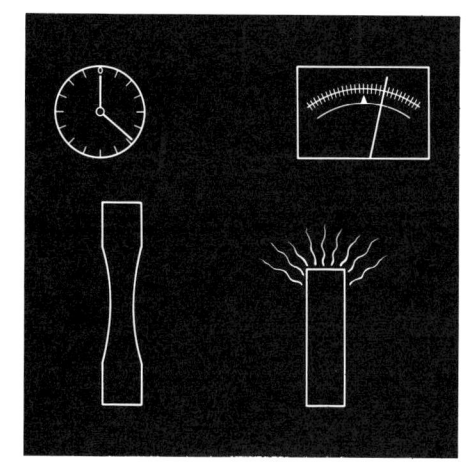

section g
physical properties and the heat treating of metals

When the only available steel for heat treating was a medium or high carbon steel, only a limited understanding of the properties of metals was needed by the heat treater. The metal simply had to be heated to a cherry red and quenched in water or brine and then tempered, usually by the color method, to give it the proper toughness for its intended purpose. Present requirements for tool steels, however, demand a variety of tool steels, including deep hardening steels, for making products such as die blocks (Figure 1) that must be hardened in thick sections. Water hardening steels will only harden thin sections or the surface of the metal about $\frac{1}{8}$ in. deep. A great variety of tool steels are used for purposes such as shock (Figure 2) or heat resistance.

Heat treating is no longer a simple process; it consists of some very complex processes and procedures that must be followed to avoid having difficulties and failures such as quench cracking (Figure 3). A good heat treater should know and understand the behavior of metals and the means by which the mechanical properties of metals can be measured. These mechanical properties and the way they are measured are first in the order of importance in this section, and tensile tests or notch toughness tests are used to determine some of these properties (Figures 4a and 4b). You will also look into the basic structure of metals from the atomic to the crystalline structures to see how they are arranged to make up the particular substances that we know of as metals (Figure 5).

Figure 1. Large parts such as this die block must be heat treated to exact specifications (Lane Community College).

You will learn:

1. How different metallic elements alloy themselves together and how they arrange their atoms into certain crystals.

262   Machine Tools and Machining Practices

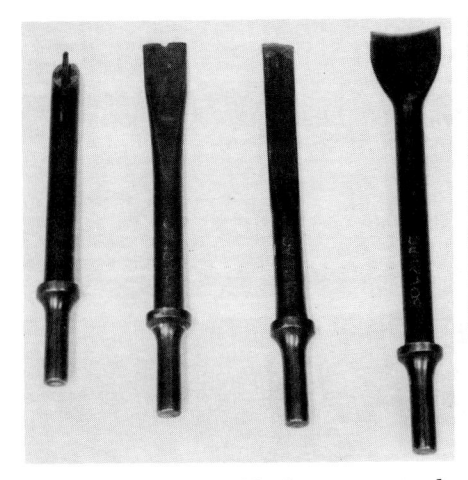

Figure 2. (Top) Air hammer tools that must resist shock require special steels (Lane Community College).

Figure 3. (Top right) Quench cracking occurred in this part because the proper heat treating procedures were not followed (Lane Community College).

Figure 4a. Tensile tests are made on this machine (Photo courtesy of W. C. Dillon & Company, Inc.).

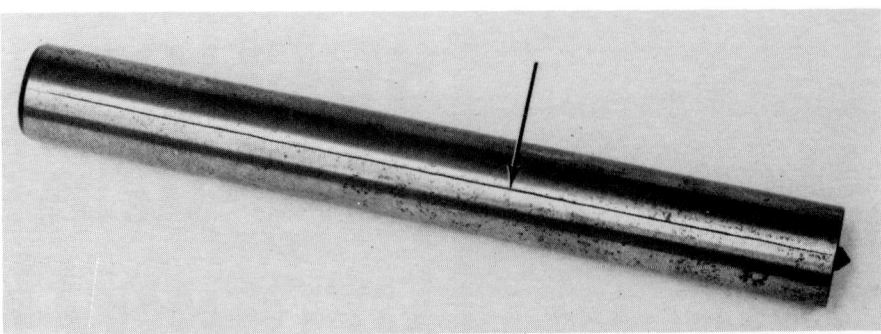

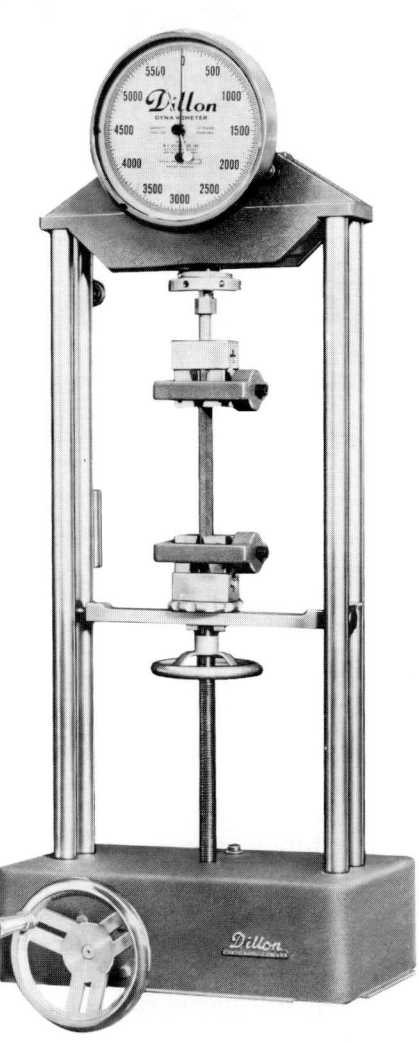

Figure 4b. Notch toughness is measured on this Charpy-Izod testing machine (The Tinius Olsen Testing Machine Co., Inc.).

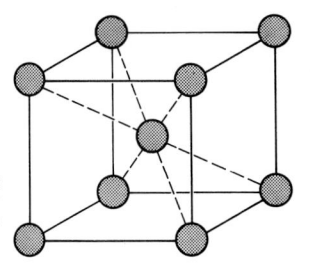

Figure 5. A body-centered crystal of ferrite, which is the low temperature phase of iron.

2. How different materials such as carbon and iron can combine to form an entirely new substance that is much harder than the original soft iron and carbon.
3. How to graph various percentages of these elements and how to determine their structure.
4. The method used to determine and measure hardenability of steels or depth of hardening.

When you have studied all these concepts, you will be shown how to do the actual job of heat treating of a specific project with a specific type of tool steel. You will be able to predict the outcome of its hardness and depth of hardening by the use of tables and graphs and by understanding the type of steel that you are using.

review questions

1. How did heat treaters determine the correct temperature for hardening before temperature gages (thermocouples) were used?

2. What medium was used for quenching carbon steel?

3. Would you consider carbon steel deep hardening? Why?

4. What can happen if the proper heat treating process is not carried out?

5. What should a heat treater know besides how to harden and temper various steels?

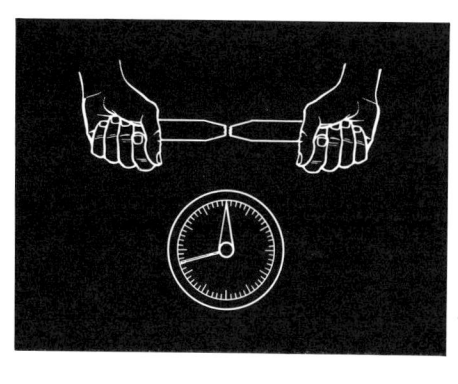

unit 1
mechanical and physical properties of metals

The mechanical properties of a material determine its usefulness for a particular job. An understanding of the nature of the mechanical properties and of how they are measured will help in the selection of materials in the shop. This unit introduces the concepts of creep, scaling, brittle transition, and corrosion. An understanding of these characteristics will help you to be more aware of problems in some severe environments.

Metals conduct both heat and electricity. They also expand when heated and contract when cooled. You will compare the conductivity and expansion of some metals.

objectives

After completing this unit, you will be able to:

1. Correctly define and describe the mechanical and physical properties of metals.
2. Describe the various testing machines and their uses, including the formulas and calculations needed.
3. Prepare specimens for the tensile tester and make tests and evaluations.
4. Prepare specimens for the Izod-Charpy tester and make tests and evaluations.
5. Conduct an experiment to demonstrate differences in thermal conductivity between two metals.
6. Perform an experiment that demonstrates scaling characteristics of a mild steel and a stainless steel.

information

Perhaps you will remember the list and definitions of the mechanical properties of metals (brittleness, ductility, elasticity, hardness, malleability, plasticity, toughness, and strength) in Volume I, Section D, Unit 3, "Steel Finishing Processes." Other properties are fatigue strength, oxidation corrosion of metals at high temperature, creep, thermal conductivity, and expansion of metals. Each mechanical property may be tested and evaluated to determine the usefulness of the metal or heat treatment for a particular application.

HARDNESS

The property of hardness as tested by Brinell and Rockwell instruments is seen as the resistance to penetration. Microhardness testers that also measure resistance to penetration are used in metallurgical laboratories. An example is the Tukon instrument (Figure 1) that uses the knoop indenter and scale. Some of the advantages are:

1. **Built-in microscopic inspection and measurement of the indentation.**
2. **Greater sensitivity of the tester.** Smaller, thinner sections may be tested (Figure 2).
3. **A *Polaroid** camera may be attached directly onto the microscope** (Figure 3) for making micrographs of the test area.

Elastic hardness is measured by an instrument called a Shore Scleroscope (Figure 4a and 4b), which measures the height of rebound of a small diamond-tipped hammer after it falls by its own weight from a measured height. Hardness as related to resistance to cutting and abrasion is

*Polaroid, a trademark of Polaroid Corporation.

Figure 2. Indentations in teeth of hacksaw blade. Note that indentations get smaller at cold worked edge. In mounting the two sets of saw teeth were interlaced together to prevent edge from rounding when the sample was polished (Courtesy of Wilson Instrument Division of Acco).

measured in the shop with the file test. This test is used mainly to accept or reject the part for a machining operation.

STRENGTH

The strength of a metal is its ability to resist changing its shape or size when external forces are applied. There are three basic types of stresses: tensile, compressive, and shear (Figure 5). When we consider strength, the type of stress the material will be subjected to must be known. Steel has equal compressive and tensile strength, but cast iron has low tensile strength and high compressive strength. Shear strength is less than tensile strength in virtually all metals. See Table 1.

The strength of materials is expressed in terms of pounds per square inch (PSI). This is called unit stress (Figure 6). The unit stress equals the load divided by the total area.

$$\text{Unit stress} = \frac{\text{Load}}{\text{Area}}$$

When stress is applied to a metal, it changes shape. For example, a metal in compressive stress will shorten and metal in tension is stretched longer. This change in shape is called strain and is expressed as inches of deformation

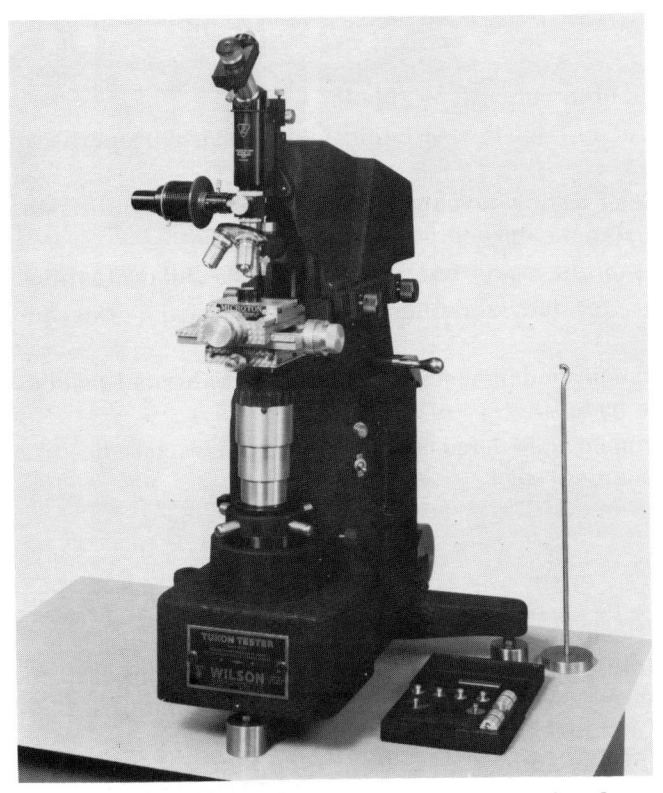

Figure 1. Tukon hardness tester (Courtesy of Wilson Instrument Division of Acco).

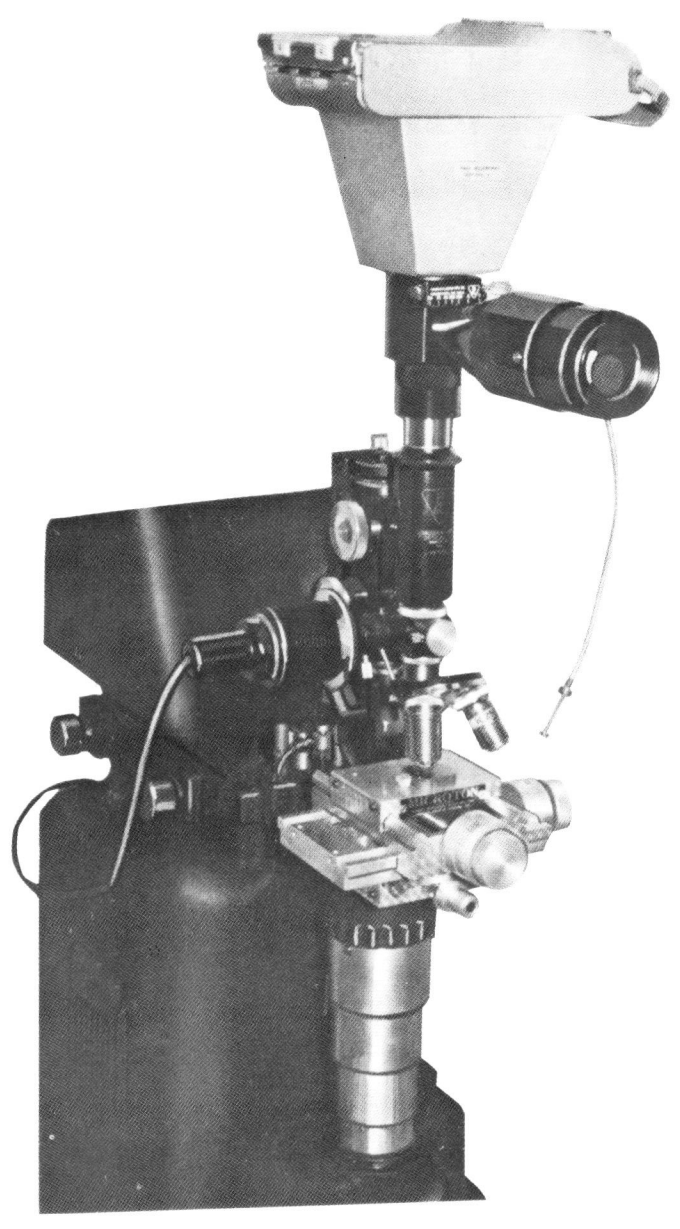

Figure 3. Camera mounted on the microscope in order to take photomicrographs (Courtesy of Wilson Instrument Division of Acco).

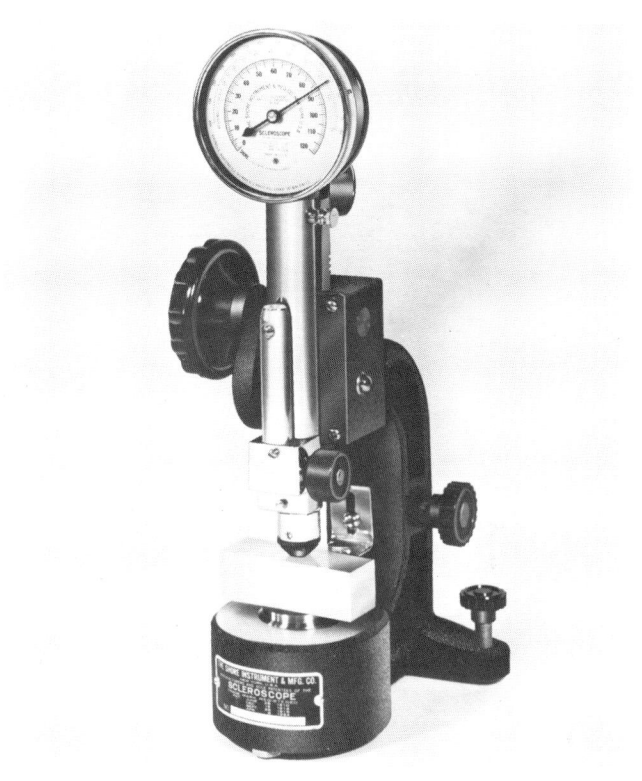

Figure 4a. Model D scleroscope. Small parts may be tested with this model. Scleroscopes are very simple to operate and do not mar finished surfaces (Shore U.S.A. Trademark #757760, Scleroscope U.S.A. Trademark #723850).

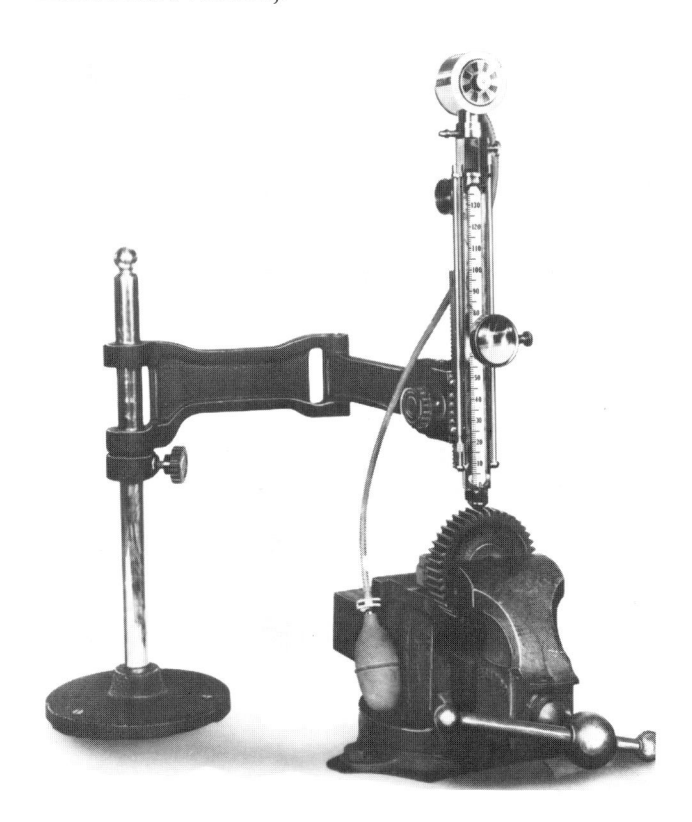

per inch of material length. As stress increases, strain increases by direct proportion within the elastic range.

Metals are pulled on a machine called a tensile tester (Figure 7). A specimen of known

Figure 4b. The clamping stand is used with the Model C-2 scleroscope to make a hardness test on a gear (Shore U.S.A. Trademark #757760, Scleroscope U.S.A. Trademark #723850).

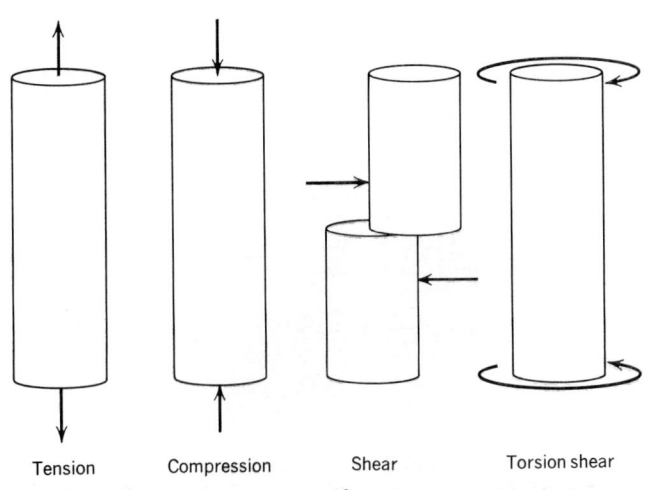

Figure 5. The three types of stresses.

Tension — Compression — Shear — Torsion shear

dimension is placed in the machine and loaded until it breaks (Figure 8). Instruments are sometimes used that make a continuous record of the load and the amount of strain (Figure 9). This information is put on a graph called a stress-strain diagram (Figure 10).

ELASTICITY

The ability of a metal to strain under load and then return to its original size and shape when unloaded is called elasticity. The elastic limit (proportional limit) is the greatest load a material can withstand and still spring back into its original shape when the load is removed. The elastic limit is easy to identify on any stress-strain diagram. It is the end of the straight line portion of the stress-strain curve (Figure 11).

Table 1
Material Strength

Material	Modulus of Elasticity (PSI)	Allowable Working Unit Stress — Tension	Compression	Shear	Extreme Fiber in Bending	Elastic Limit (PSI) Tension	Compression
Cast iron	15,000,000	3,000	15,000	3,000		6,000	20,000
Wrought iron	25,000,000	12,000	12,000	9,000	12,000	25,000	25,000
Steel, structural	29,000,000	20,000	20,000	13,000	20,000	36,000	36,000
Tungsten carbide	50,000,000					80,000	120,000

Ultimate Strength (PSI) Tension	Compression	Shear
20,000	80,000	20,000
50,000	50,000	40,000
65,000	65,000	50,000
100,000	400,000	70,000

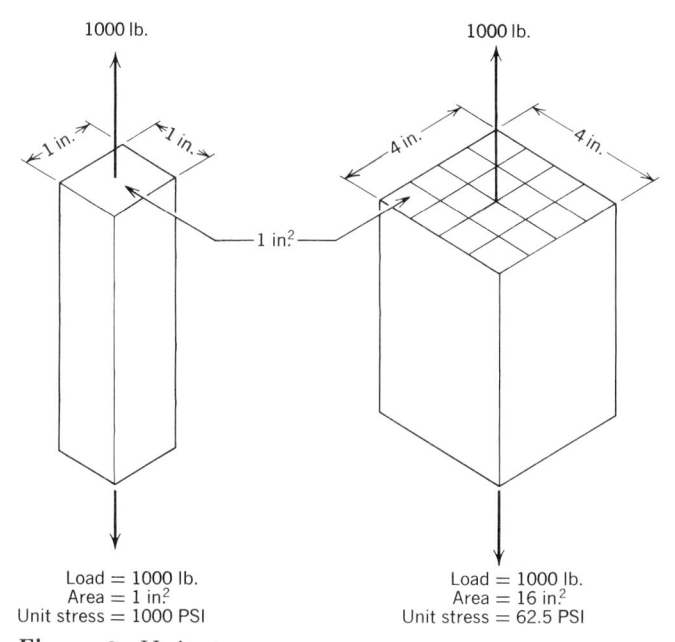

Figure 6. Unit stress.

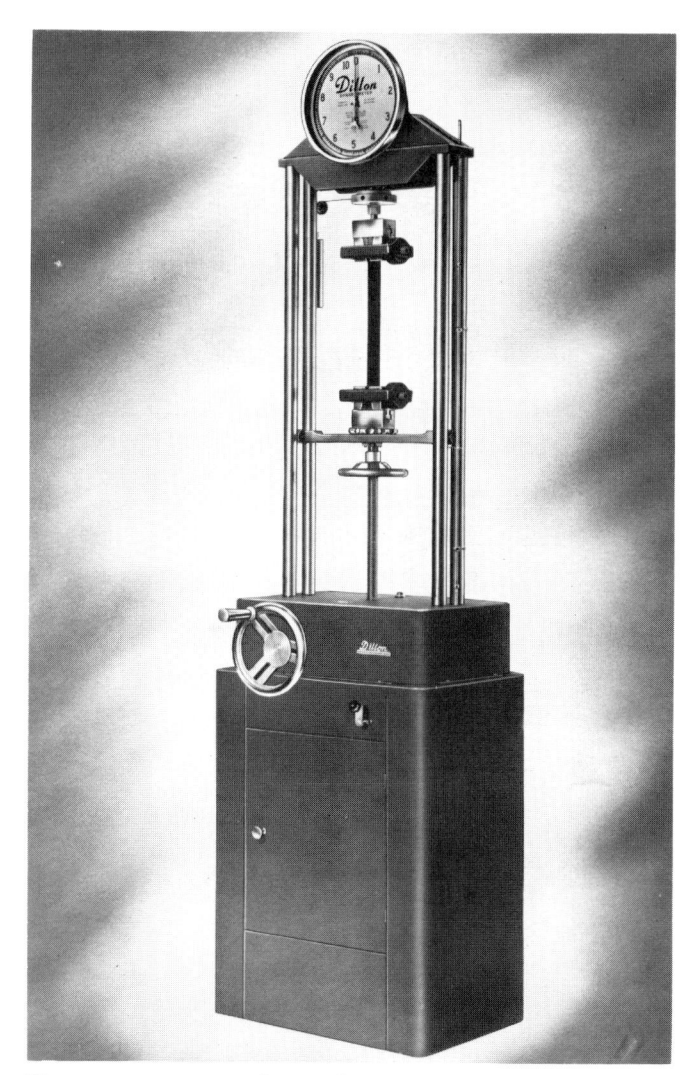

Figure 7. Universal tensile tester (Photo courtesy of W. C. Dillon & Company, Inc.).

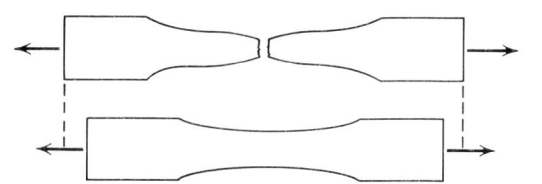

Figure 8. A tensile specimen of a ductile material before pull and after pull.

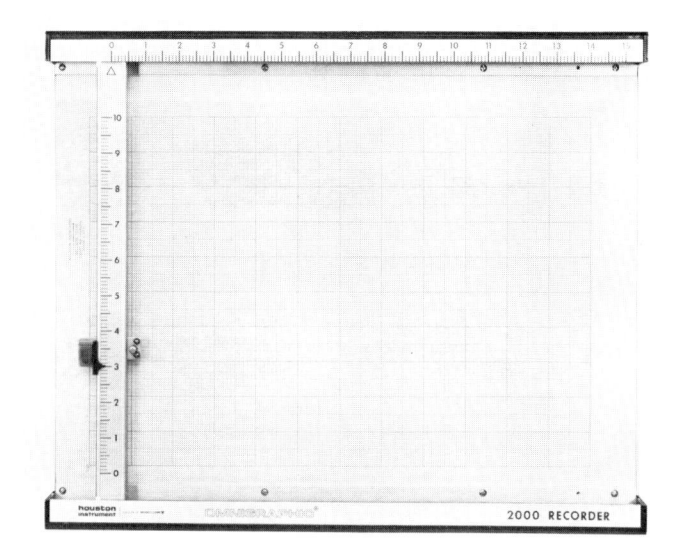

Figure 9. Model 2000 XY recorder. Stress-strain curves are recorded on machines such as this (Omnigraphic® is a registered trademark of Houston Instrument. Courtesy of Houston Instrument).

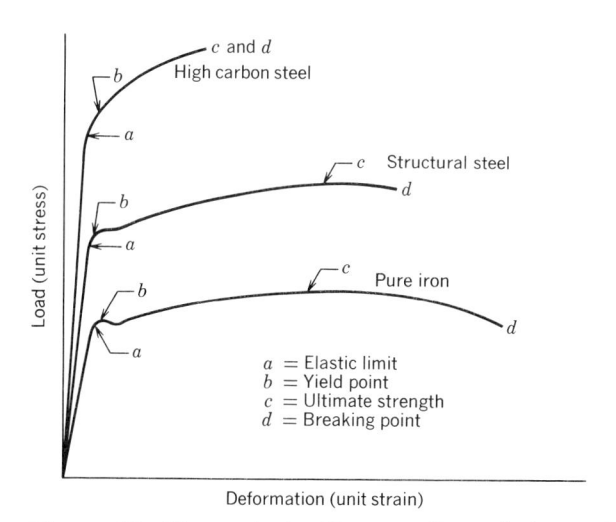

Figure 10. Stress-strain diagram. Several stress-strain curves are shown on this diagram.

The yield point (yield strength) is a point slightly higher than the elastic limit and, for most cases, they can be considered the same. The allowable (safe) load for a metal in service should be well below the elastic limit or yield strength.

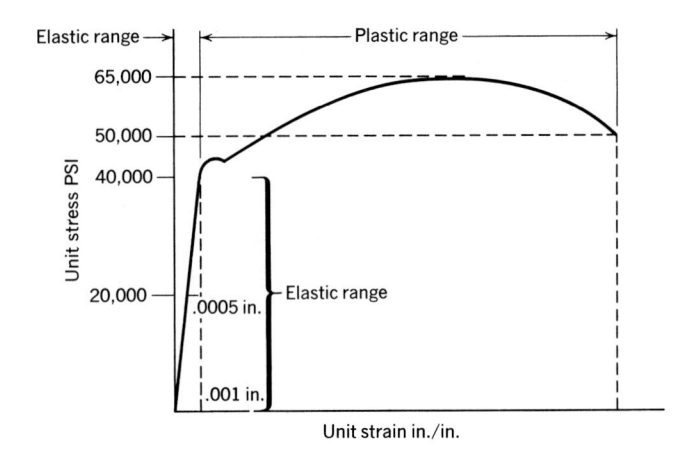

Mechanical properties charts in metals handbooks contain data such as yield point, ultimate strength, and hardness. Table 2 is one such chart for SAE 1015 steel. Table 3 is a similar chart for SAE 1095 steel. Note the contrast (under mass effect) between hardness, yield point, and ultimate strength between these two steels. The percentages of carbon and other elements are given, as well as grain size and critical temperatures. The particular steel in Table 2 is used for carburizing and case hardening, so related in-

Figure 11. Stress-strain diagram for a ductile steel.

Table 2
Mechanical Properties for SAE 1015 Steel. (*Source.* Bethlehem Steel Corporation, *Modern Steels and Their Properties*, Seventh Edition, Handbook 2757, 1972.)

SINGLE HEAT RESULTS

	C	Mn	P	S	Si	
Grade	.13/.18	.30/.60	.040 Max	.050 Max	—	Grain Size
Ladle	.15	.53	.018	.031	.17	6-8

Critical Points, F: Ac_1 1390 Ac_3 1560 Ar_3 1510 Ar_1 1390

SINGLE QUENCH AND TEMPER

Carburized at 1675 F for 8 hours; pot-cooled; reheated to 1425 F; water-quenched; tempered at 350 F.
   1-in. Round Treated   Case Depth .048 in.   Case Hardness HRC 62

MASS EFFECT

Size Round in.	Tensile Strength psi	Yield Point psi	Elongation % 2 in.	Reduction of Area, %	Hardness HB
Annealed (Heated to 1600 F, furnace-cooled 30 F per hour to 1340 F, cooled in air.)						
1	56,000	41,250	37.0	69.7	111
Normalized (Heated to 1700 F, cooled in air.)						
½	63,250	48,000	38.6	71.0	126
1	61,500	47,000	37.0	69.6	121
2	60,000	44,500	37.5	69.2	116
4	59,250	41,800	36.5	67.8	116
Mock-Carburized at 1675 F for 8 hours; reheated to 1425 F; quenched in water; tempered at 350 F.						
½	106,250	60,000	15.0	32.9	217
1	75,500	44,000	30.0	69.0	156
2	70,750	41,375	32.0	70.4	131
4	67,250	39,000	30.5	69.5	121

As-quenched Hardness (water)

Size Round	Surface	½ Radius	Center
½	HRC 36.5	HRC 23	HRC 22
1	HRB 99	HRB 91	HRB 90
2	HRB 98	HRB 84	HRB 82
4	HRB 97	HRB 80	HRB 78

Table 3
Mechanical Properties for SAE 1095 Water Quenched Steel. (Source. Bethlehem Steel Corporation, *Modern Steels and Their Properties,* Seventh Edition, Handbook 2757, 1972.)

SINGLE HEAT RESULTS

	C	Mn	P	S	Si	
Grade	.90/1.03	.30/.50	.040 Max	.050 Max	—	Grain Size
Ladle	.96	.40	.012	.029	.20	50% 5-7 50% 1-4

Critical Points, F: Ac_1 1350 Ac_3 1365 Ar_3 1320 Ar_1 1265

MASS EFFECT

Size Round in.	Tensile Strength psi	Yield Point psi	Elongation % 2 in.	Reduction of Area, %	Hardness HB
Water-quenched from 1450 F, tempered at 900 F.					
½	191,500	135,500	12.3	31.7	375
1	182,000	121,000	13.0	37.3	363
2	179,750	113,000	12.7	33.8	352
4	167,250	94,500	12.5	31.4	331
Water-quenched from 1450 F, tempered at 1000 F.					
½	172,000	111,000	12.4	44.1	321
1	165,000	102,500	16.0	41.4	311
2	154,750	98,500	15.7	39.1	302
4	150,000	81,000	15.7	35.3	285
Water-quenched from 1450 F, tempered at 1100 F.					
½	144,000	99,000	17.2	44.9	293
1	143,000	96,500	16.7	43.7	293
2	140,000	90,000	17.5	43.6	285
4	131,250	78,000	18.7	41.1	262

As-quenched Hardness (water)

Size Round	Surface	½ Radius	Center
½	HRC 65	HRC 55	HRC 48
1	HRC 64	HRC 46	HRC 44
2	HRC 63	HRC 43	HRC 40
4	HRC 63	HRC 38	HRC 30

formation is supplied. The steel in Table 3 is a tool steel.

PLASTICITY

A perfectly plastic substance such as modeling clay will not return to its original dimensions when the load is removed, regardless of how small the load. Metals undergo plastic flow when stressed beyond their elastic limits. Therefore, the area of the stress-strain curve beyond the elastic limit in Figure 11 is called the plastic range. It is this property that makes metals so useful. When enough force is applied by rolling, pressing, or hammer blows, metals can be formed, when hot or cold, into useful shapes. Many metals tend to work harden when they are cold formed, which increases their usefulness in most cases. They must be annealed for further cold work when a certain limit has been reached.

BRITTLENESS

A material that will not deform plastically under load is said to be brittle. Excessive cold working causes brittleness. Cast iron does not deform plastically under a breaking load and is therefore brittle.

A very sharp "notch" that concentrates the load in a small area can also reduce plasticity (Figure 12). Notches are common causes of premature failure in parts. Weld undercut, sharp shoulders on machine shafts, and sharp angles

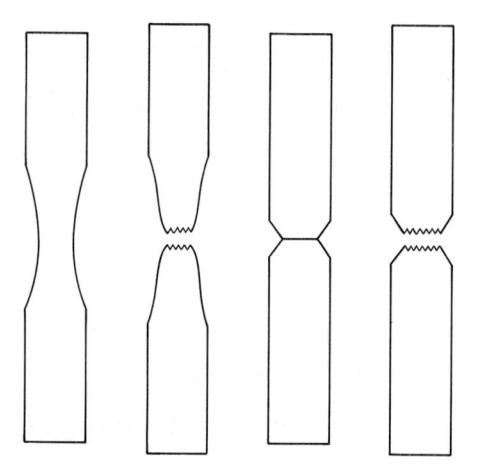

Figure 12. Notching and its effect on plasticity in an otherwise ductile metal can behave in a brittle manner when a stress raiser is present.

on forgings and castings are examples of unwanted notches (stress raisers).

STIFFNESS (MODULUS OF ELASTICITY)

Stiffness is expressed by the modulus of elasticity, also called Young's Modulus. Within the elastic range, if the stress is divided by the corresponding strain at any given point, the result will be the modulus of elasticity for that material.

$$\text{Modulus of elasticity in PSI} = \frac{\text{Stress}}{\text{Strain}}$$

The modulus of elasticity for some common materials is given in Table 1.

As an example of stiffness, two rods of equal dimensions are suspended horizontally on one end with equal weights hanging on the other end. Of course, both rods will deflect the same amount if they are made of the same steel. Even if one rod were made of mild steel and the other of hardened tool steel, both would still deflect the same amount within the elastic range. The reason is that all steels have about the same modulus of elasticity. If one rod were made of tungsten carbide, the results would be quite different; the carbide rod would deflect much less than the steel, since its modulus of elasticity is considerably higher than that of steel.

DUCTILITY

The property that allows a metal to deform permanently when loaded in tension is called ductility. Any metal that can be drawn into a wire is ductile. Steel, aluminum, gold, silver, and nickel are some ductile metals.

The tensile test is used to measure ductility. Tensile specimens are measured for area and length between gage marks before and after they are pulled. The percent of elongation (increase in length) and the percent of reduction in area (decrease of area at the narrowest point) are the values for ductility. A high percent elongation (about 70 percent) and a reduction in area indicate a high ductility. The method for calculating these values is shown in Figure 13. A metal showing less than 20 percent elongation would have low ductility.

MALLEABILITY

The ability of a metal to deform permanently when loaded in compression is called malleability. Metals that can be hammered or rolled into sheets are malleable. Most ductile metals are also malleable, but some very malleable metals such as lead are not ductile. Some very malleable metals are lead, tin, gold, silver, iron, and copper.

NOTCH TOUGHNESS

Notch toughness is the ability for a metal to resist rupture from impact loading when there is a

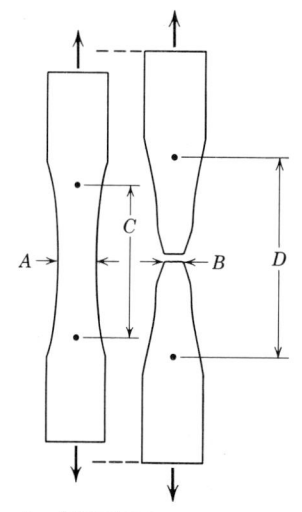

A = Original area
B = Area after tensile pull
C = Original gage length
D = Gage length after pull

Elongation (%) = $\frac{D-C}{C}$ x 100

Reduction in area (%) = $\frac{A-B}{A}$ x 100

Figure 13. Elongation and reduction in area.

Figure 14. Izod-Charpy testing machine (The Tinius Olsen Testing Machine Co., Inc.).

Figure 15a. The vertical mounting position for the Izod test specimen is shown as it is clamped in the vise. The inset shows the underside of the striking bit (The Tinius Olsen Testing Machine Co., Inc.).

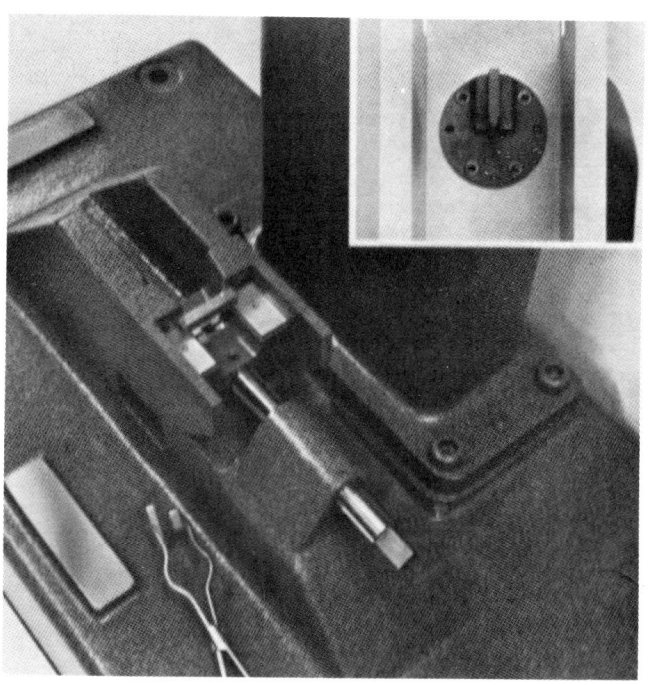

Figure 15b. The horizontal mounting position for the Charpy test specimen is shown as it is clamped in the vise. The inset shows the underside of the striking bit (The Tinius Olsen Testing Machine Co., Inc.).

notch or stress raiser present. The devices used to measure toughness are the Izod-Charpy testing machines (Figure 14). The method of loading distinguishes between the Izod and Charpy methods (Figures 15a and 15b), but the results are practically the same. The base (Figure 16) has two leveling pads for adjusting the machine. The hammer straddles the anvil support and the striking bit is in the hammer. Standard test specimens (Figure 17) are used for either the Izod or Charpy tests. The testing machine consists of a vise where the test specimen is clamped. A weight on a swinging arm is allowed to drop (Figure 18). Note that the specimens are of standard geometry. The pendulum drops, strikes the specimen, and continues to swing forward. But it will not swing up as high as the starting position. The difference between the pendulum's beginning height and ending height indicates how much energy was absorbed in breaking the specimen. This energy is measured in foot-pounds. Tough metals absorb more foot-pounds of energy than brittle metals and, therefore, the pendulum will not swing as far for tough metals.

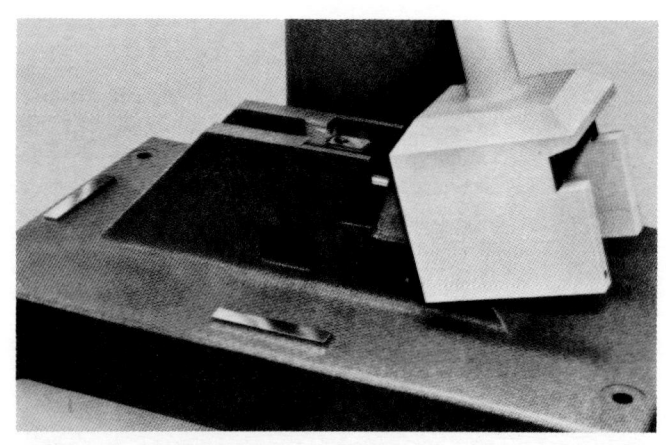

Figure 16. Base showing leveling pads. Hammer is dropping and about to strike the specimen (The Tinius Olsen Testing Machine Co., Inc.).

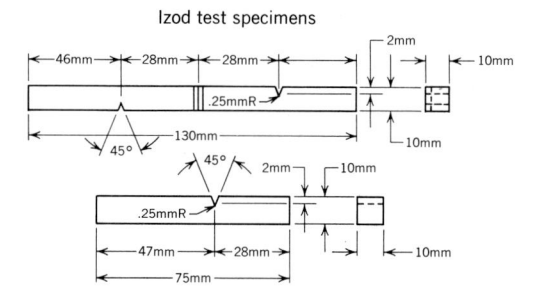

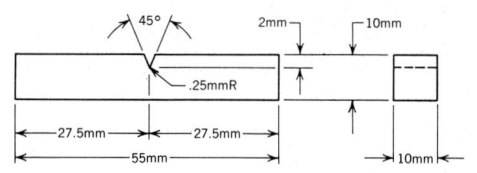

Figure 17. Test specimen specifications for Charpy and Izod tests (The Tinius Olsen Testing Machine Co., Inc.).

FATIGUE

When metal parts are subjected to repeated loading and unloading, they may fail at stresses far below their yield strength with no sign of plastic deformation. This is called a fatigue failure. When designing machine parts that are subject to vibration or cyclic loads, fatigue strength may be more important than ultimate tensile or yield strength.

Fatigue-testing machines put specimens

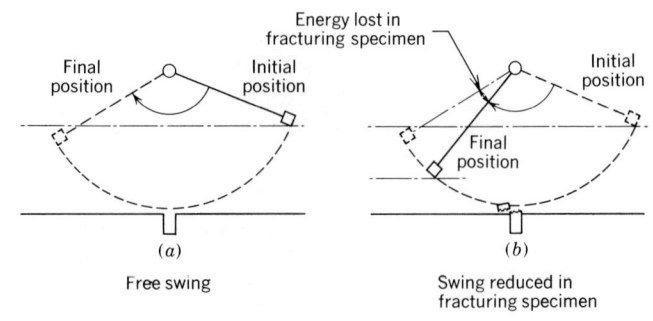

Figure 18. The method by which (Izod) impact values are determined. (a) Free swing. (b) Swing reduced in fracturing specimen.

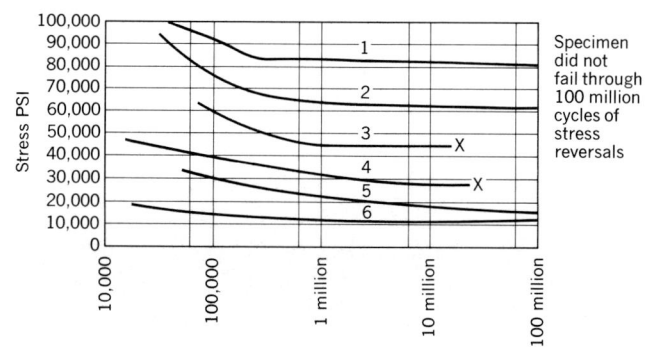

1. High carbon steel, oil quench, temper to 860° (460° C)
2. Heat treated alloy steel, oil quench, temper to 1200° F (649° C)
3. 50% carbon steel, heat treated
4. Structural steel
5. 2024 aluminum alloy
6. Gray cast iron

Figure 19. Relation between fatigue limit and tensile strength. The fatigue limit of steel is approximately 45 to 50 percent of its tensile strength up to about 200,000 lb. Repeated stresses in excess of this fatigue limit causes ultimate failure.

through many cycles of loading at a given stress. The results of repeated tests at different stresses can be plotted on a graph called a stress cycle diagram (Figure 19). The fatigue limit is the maximum load in pounds per square inch that can be applied an infinite number of times without causing failure. But 10 million loading cycles is usually considered enough to establish fatigue limits.

Fatigue life can be enhanced by smooth design. Avoiding undercut in welds, sharp corners, and shoulders, and deep tool marks in machined

Table 4
Creep Strengths for Several Alloys

Alloy	70° F Tensile Strength (PSI)	800° F — Stress for 1 percent Elongation per 10,000 h	1200° F — Stress for 1 percent Elongation per 100,000 h	1500° F — Stress to Failure
.20 percent carbon steel	62,000	35,100	200	1500
.50 percent molybdenum .08 percent to 20 percent carbon steel	64,000	39,000	500	2600
1.00 percent chromium .60 percent molybdenum .20 percent C steel	75,000	40,000	1500	3500
304 stainless steel 19 percent chromium 9 percent Nickel	85,000	28,000	7000	15,000

CREEP STRENGTH (PSI)

parts will help eliminate stress raisers (notches) and thereby increase fatigue life.

CREEP STRENGTH

Creep is a continuing slow plastic flow at a stress below the yield strength of a metal. Creep is usually associated with high temperatures though sometimes occurring at normal (room) temperature. As the temperature increases, creep becomes more of a problem. Creep strength for a metal is given in terms of an allowable amount of plastic flow (creep) per 1000-h period. Table 4 gives creep strength for some alloys. Note that strengths are given for 1 percent creep per 10,000 h at 800° F (427° C) and for 1 percent creep per 100,000 h at 1200° F (649° C). Stress to failure is also given.

SCALING

When metals are subjected to high temperatures, they often form heavy coatings of oxide. This coating is called scale. If the metal has a low resistance to scaling and if it is allowed enough time, it will eventually be converted entirely to scale. Resistance to scaling is usually achieved in steels by the addition of chrome or nickel or both. These elements tend to form an oxide skin with a high melting temperature. This skin protects the metal and retards further scaling. Some stainless steels resist scaling at high temperatures, for example.

CORROSION RESISTANCE

The ability of a metal to resist attack by chemical action is called corrosion resistance. Some metals corrode easily. Iron, for instance, needs only water to corrode (rust). Other metals, such as gold, show a tremendous resistance to almost any chemical environment. Some more common metals that are noted for corrosion resistance are nickel, chrome, cobalt, and manganese. Some alloys, such as stainless steels, have an excellent ability to resist attack by the environment. This is due to the formation of a thin film of protective oxide that forms on the surface.

METALS AT LOW TEMPERATURES

As the temperature decreases, the strength, hardness, and modulus of elasticity increases for almost all metals. The effect of temperature drop on ductility separates metals into two groups: those that become brittle at low temperatures and those that remain ductile. Figure 20 shows examples of these two groups.

Metals of the group that remain ductile show a slow, steady decrease in ductility with temperature drop. Metals in the group that become brittle at low temperatures show a temperature range

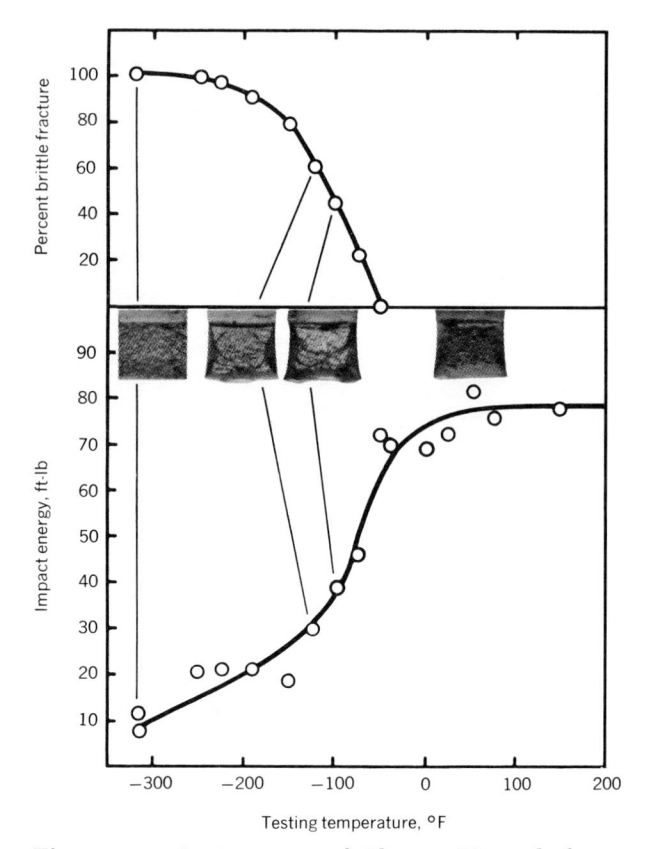

Figure 20. Appearance of Charpy V-notch fractures obtained at a series of testing temperatures with specimens of tempered martensite of hardness around 30 Rockwell C (Courtesy Republic Steel Corporation).

where ductility and, most important, toughness drop rapidly. This range is called the *transition zone*. The Izod-Charpy impact test is the most common method used to determine the transition zone of metals. When the notch-bar specimens show half brittle failures and half ductile failures, the transition temperature has been reached. When designing parts for low temperature service, the operating temperature should be well above the transition temperature. Nickel is the most effective alloying element for lowering the transition temperature of steels. The following are some examples of operating temperatures for a few alloys and metals.

1. For operating temperatures as low as −50° F (−46° C).
   a. Killed low-carbon steel.
   b. Three percent nickel low-carbon steel.
2. For operating temperatures as low as −150° F (−101° C).
   a. Six percent nickel low-carbon steel.
   b. Stainless steels with 8 percent nickel or more.
3. For operating temperatures below −150° F (−101° C).
   a. Stainless steels with at least 8 percent nickel.
   b. Nine percent nickel steel.
   c. FCC metals such as aluminum or monel.

EXPANSION AND CONDUCTIVITY OF METALS (PHYSICAL PROPERTIES OF METALS)

Metals conduct heat better than nonmetals. Silver conducts heat the best of all metals. This ability to conduct heat and the ability to conduct electricity are related. If silver is the best heat conductor, it is also the best electrical conductor. Figure 21 compares the thermal conductivity of some common metals and alloys. Note that in all cases the pure metals are better conductors than their alloys. Pure copper would be a better choice for electrical wiring than a copper alloy. A pure aluminum auto radiator would conduct heat away from the water inside better than an aluminum alloy radiator.

THERMAL EXPANSION

In almost all cases solids become larger when heated and smaller when cooled. Each substance expands and contracts at a different rate. This rate is expressed in inches per inch per degree F, and is called the coefficient of thermal expansion. Figure 22 compares the coefficient of thermal expansion of some common metals and alloys.

SOME CASES WHERE A KNOWLEDGE OF THERMAL EXPANSION HELPS

Knowing the thermal expansion coefficient of steel allows the engineer to calculate the sizes of the expansion joints in bridges and other steel structures. Heat treaters must be aware of differing expansion and contraction rates when heating and quenching steels. The internal expansion rate is often lower than the external rate when a piece of steel is being heated rapidly. The stresses caused by uneven heating can cause cracking in metals with low ductility.

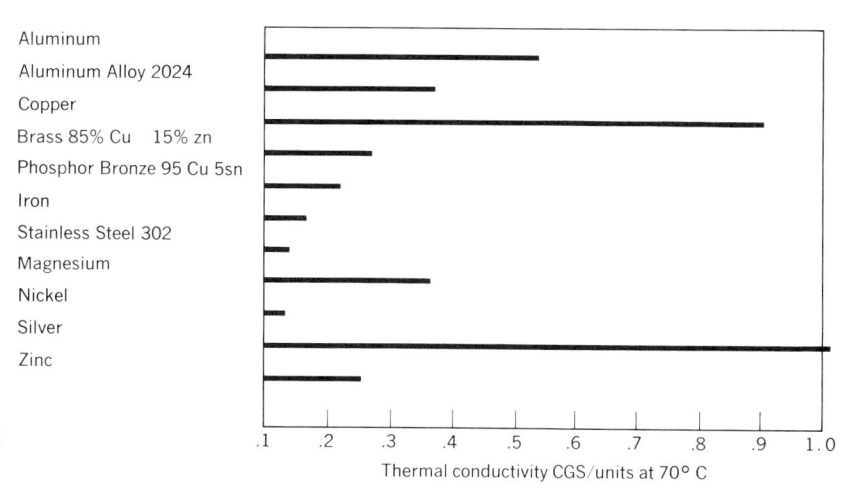

Figure 21. Comparison of thermal conductivity.

Thermal conductivity CGS/units at 70° C

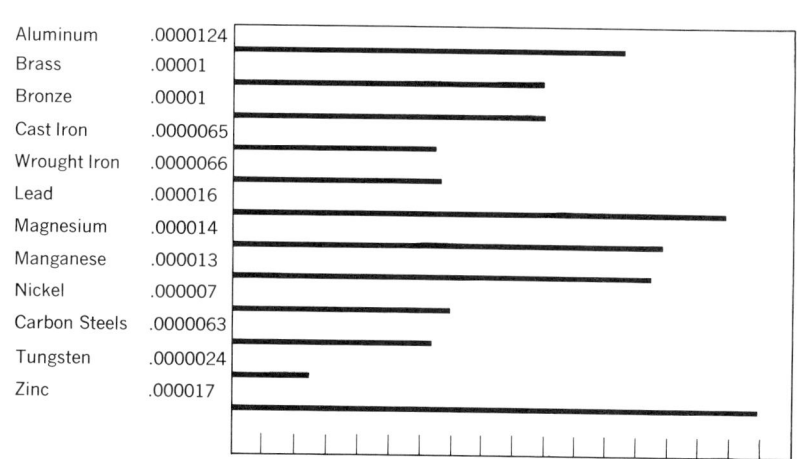

Figure 22. Coefficient of thermal expansion per degree Fahrenheit per unit length.

If a mechanic must remove a bronze bushing from a housing, or if heat is inadvertently applied to a bushing area, such as by welding in the vicinity of the bushing, it may be loosened as a result of applying heat (Figure 23). The thermal expansion coefficient of bronze is almost twice that of steel. If the bushing and housing are heated, the bronze will try to expand at almost twice the rate of steel for the same amount of heat. The steel restricts the bronze from expansion. The bronze then is stressed above its elastic limit and into the plastic range, where the bushing is deformed to a slightly smaller diameter. When both steel and bronze are cool, the bushing is now smaller than the steel bore, and so it is easily removed from the hole.

A machinist has turned a bearing fit with a plus or minus .0001 in. tolerance on a 4 in. steel shaft. The shaft is still hot from turning when the operator measures it. After a coffee break, he returns and checks his work to find that it is .0025 in. under size. What happened? The shaft cooled down 100° F (37.8° C) to room temperature. The following formula is used to calculate the amount of contraction after cooling to room temperature.

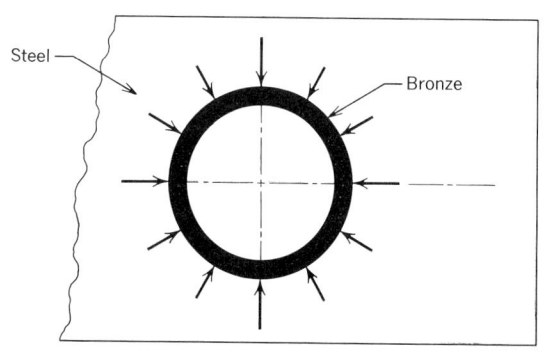

Figure 23. An application of thermal expansion of metals.

Coefficient of expansion × diameter × rise in temperature F = expansion

or

.0000063 × 4 in. diameter × 100° F = .0025 in.

If a 1 in. diameter steel shaft would expand .0000063 in. for 1° F rise in temperature, it would expand .00063 in. for a 100° F rise. If the shaft were 4 in. diameter and had a 100° F rise, the expansion would be .0025 in. If machined at that temperature, it would contract the same amount when cooled. All lathe operators are familiar with the lengthwise expansion of turned workpieces that cause the dead center to tighten and heat up.

Temperature differences on workpieces, especially thin parts, can cause them to "crawl" on the milling machine table. The heat generated by milling with carbide cutters, when no coolant is used, is most likely to cause this problem.

self-evaluation

SELF-TEST

1. Does creep occur within the elastic or plastic range of steels?

2. Explain the relative rate of failure in creep. Are creep failures sudden or do they take years to fail?

3. What happens below the transition temperature of a metal?

4. What happens to the properties of hardness, strength, and modulus of elasticity with a decrease in temperature?

5. Name an alloying element that can be added to steel to lower its transition temperature.

6. Describe the three categories of hardness and how they can be measured.

   a. _____

   b. _____

   c. _____

7. Name the three basic stresses.

   a. _____

   b. _____

   c. _____

Unit 1 Mechanical and Physical Properties of Metals

8. A 2 in. square bar in tension has a load of 40,000 lb. What is the unit stress?

9. Explain the property of ductility.

10. Explain the property of malleability.

11. In what ways can fatigue strength be improved?

12. What is the correlation between metals for electrical and thermal conductivity?

13. In what state does a metal conduct best, alloyed or unalloyed?

14. How is the rate of thermal expansion for a particular material expressed?

15. Why should a machinist be very aware of thermal expansion in the metal he is using?

WORKSHEET 1 Given a tensile testing machine and a prepared specimen, you will:

1. Learn to use the tensile testing machine
2. Calculate elongation, reduction in area, and unit stress of a pulled specimen

Procedure

1. Prepare a specimen of mild steel as shown in Figure 24. Punch the gage marks exactly 2 in. apart.
2. With a micrometer measure the width at the narrowest point and the thickness.
3. Record this information.
4. Set up the tensile tester with flat gripping jaws and a 0-10,000 lb dial if it is the interchangeable type.
5. Place the specimen in the tensile testing machine and pull it to rupture.
6. Record the yield point.

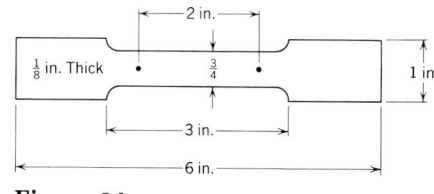

Figure 24

7. Record the breaking load.

8. Remove the specimen and fit the broken pieces together. Measure the width and thickness of the specimen at the narrowest point; measure the length between gage marks and record the information.

9. Calculate the elongation and reduction in area using the formulas given in Figure 13.

10. Calculate the unit stress using the following formula.

$$\text{Unit Stress} = \frac{\text{Load}}{\text{The original area}}$$

Note The original area is equal to the width times the thickness at the narrowest point before pulling.

Conclusion Do you think this was a ductile metal? Why?

Data Tensile specimen
    Length between gage marks =
    Thickness =
    Width =
    Area =
Tensile specimen after pull
    Length between gage marks =
    Breaking load =
    Yield point =
    Thickness =
    Width =
    Area =
Results
    Ultimate unit stress =
    Reduction in area, percent =
    Elongation, percent

WORKSHEET 2 Given an Izod-Charpy testing machine and prepared specimens, you will:

1. Determine the relative notch toughness of a carbon steel in its annealed and hardened state.

Procedure 1. Prepare two Izod or Charpy specimens of annealed or as-rolled carbon steel, SAE 1080 to 1095 per specifications in Figure 16 or Figure 17.

2. Harden one specimen in a water quench from 1500° F (815.6° C) and temper to 400° F (204.4° C).

3. Test both specimens for hardness. Record your results.

4. Test both specimens on the Izod-Charpy machine. Record your results.

Unit 1 Mechanical and Physical Properties of Metals   281

Conclusion Which metal shows greater toughness? Which is most brittle?

Data   Specimen 1 (soft)   Specimen 2 (hard)
Hardness
Ft-lb

WORKSHEET 3 Given a convenient heat source such as a Bunsen burner or propane torch, a strip of copper, and an equally shaped strip of stainless steel, you will:

1. Demonstrate the difference in thermal conductivity between copper and stainless steel.

Procedure 1. Set up the burner so the flame is at the ends of the strips arranged as in Figure 25.

2. Mark the opposite ends with a 200° F (93.3° C) temperature crayon. In lieu of the crayon, a wooden match will work as shown in Figure 25.

3. Note which crayon mark melts first or which match lights first.

Conclusion Which metal has the highest thermal conductivity?

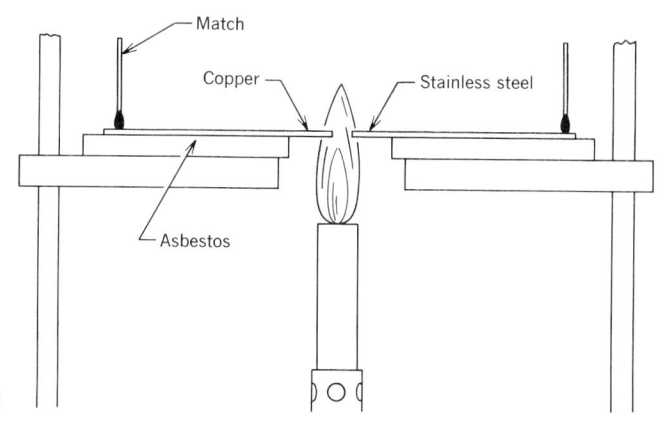

Figure 25

Note You may demonstrate the difference between thermal conductivity of other metals such as steel and stainless steel if you use closely controlled conditions.

WORKSHEET 4 Given a furnace, one piece of mild steel about $\frac{1}{16} \times 1 \times 4$ in., and one piece of stainless steel about $\frac{1}{16} \times 1 \times 4$ in., you will:

1. Determine the relative scaling characteristics of the two metals.

Procedure 1. Place the two specimens over a Bunsen burner or in an electric furnace and allow them to remain at a yellow heat for 1 h.

2. Check and record observations at 10 min intervals.

Conclusion What happens to the mild steel?

What happens to the stainless steel?

Which one would you suggest for high temperature service?

unit 2
the crystal structure of metals

What are the forces that hold metals together? Why do metals behave as they do? These and other questions relating to the atomic and crystalline structure of metals will be discussed in this unit.

objectives After completing this unit, you will be able to:
1. Explain various conditions and phases of crystalline structure of metals.
2. Explain various aspects of solid solutions.
3. Conduct a Metcalf experiment and determine approximate grain size in steel samples.

information

The great utility of metals is due to their elastic behavior to a certain level of stress followed by a plastic behavior at higher levels of stress. Along with ceramic materials, which are brittle, or polymers such as wood or leather, metals play a unique and useful role in the economy of man.

Matter is composed of atoms that are too small to be seen even with the aid of microscopes. Atoms of different materials vary only in the number and arrangement but not the type of their parts. Matter composed of a single kind of atom is called an *element*.

ATOM

An atom resembles a miniature solar system, with its chief parts as shown in Figure 1. The *nucleus* of the atom consists of *protons* and *neutrons*. The protons have a positive electrical charge. Neutrons weigh essentially the same as protons but are neutral in charge. Revolving at high speed around the nucleus are much smaller particles, called *electrons*. Electrons are negatively charged, which means that they are strongly attracted to the positively charged nucleus.

Each atom has preferred electron paths, called *shells*. The number, arrangement, and spin of the electrons in these shells in combination with the positively charged nucleus determine the kind of atom and its characteristics.

The electrons on the outer shell called the *valence* electrons are the most important in determining chemical and physical properties.

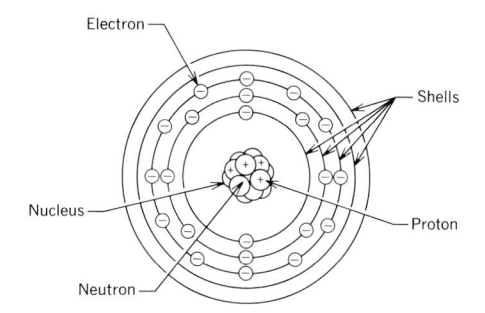

Figure 1. The atom.

Metal atoms are easily stripped of their valence electrons and thus form positive ions.

BONDING

With this information we can now go on to find out how metals are held together. There are four possible types of bonding arrangements that hold the atoms together. They are ionic, covalent, metallic, and Van de Waals forces.

Ionic bonding is the attraction of negative and positive ions. Sodium chloride (NaCl) is an example of ionic bonding (Figure 2), where a metal (sodium) loses its single valence electron to a nonmetal (chlorine) to complete its valence shell. The sodium atom is now positively charged and the chlorine atom is negatively charged. The resultant structure of salt (Figure 3) is rather weak, since the electrostatic attractions are very selective and directional.

Covalent bonding, or shared valence electrons, is a very strong atomic bond, depending on the number of shared electrons. Covalent bonding is found primarily between nonmetallic elements such as carbon (in diamonds, for instance). As in ionic bonding, the structure is rigid and directional.

The oxygen atom has six electrons in the second shell. This shell desires to have eight electrons, so two electrons are shared by each atom, making a satisfactory arrangement (Figure 4). When one or more atoms are combined through bonding, the arrangement is called a *molecule.*

The metallic bond is where the electrons in the valence shell separate from their atoms and exist in a cloud surrounding all the positively charged atoms. These positively charged atoms are arranged in a very orderly pattern. The atoms are held together because of their mutual attrac-

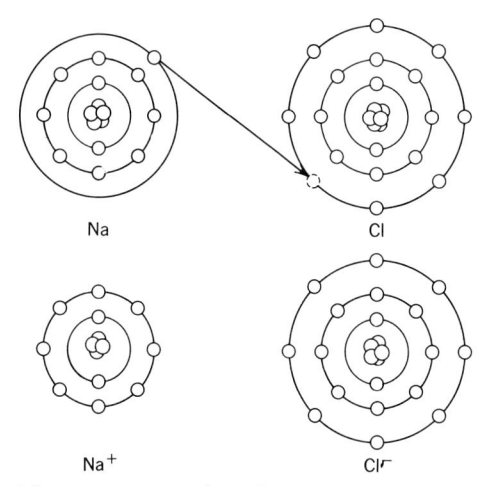

Figure 2. Ionic bonding.

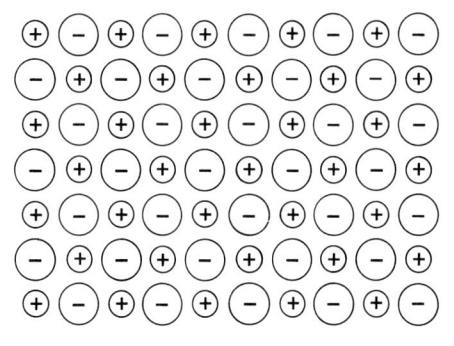

Figure 3. Lattice structure of salt (sodium chloride).

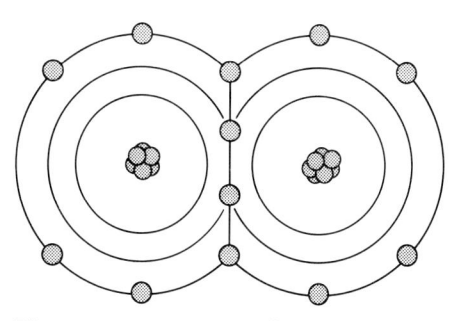

Figure 4. Oxygen molecule has a covalent bond.

tion for the negative electron cloud (Figure 5). The free movement of electrons accounts for the high electrical and heat conductivity of metals and for their elasticity and plasticity.

Van der Waals bonding is found in neutral atoms such as the inert gases. There is only a very weak attractive force, and it is important only at very low temperatures.

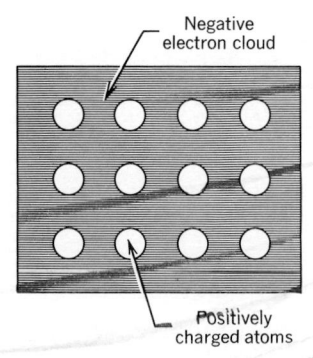

Figure 5. Metallic bond.

METALS AND NONMETALS

Approximately three quarters of all the elements are considered to be metals. Some of these are metalloids or transitional elements such as silicon or germanium. Some of the properties that an element must have to be considered a metal are:

1. Crystalline structure — grain structure.
2. High thermal and electrical conductivity.
3. Can be deformed plastically.
4. Metallic luster or reflectivity.

Table 1 lists some common metals and their chemical symbols and crystalline structure. Nonmetals are characterized by their brittleness, noncrystalline structure or lack of luster.

CRYSTALLINE LATTICE STRUCTURES

Metals solidify into six main lattice structures.

1. Body-centered cubic (BCC).
2. Face-centered cubic (FCC).
3. Close-packed hexagonal (CPH).
4. Cubic.
5. Body-centered tetragonal.
6. Rhombohedral.

The crystalline lattice structures begin to grow first by the formation of seed crystals or nuclei as the metal solidifies. The number of nuclei or grain starts formed determines the fineness or coarseness of the metal grain structure. Slow cooling promotes large grains and fast cooling promotes smaller grains. The grain grows outward from this dendrite crystal until it meets another dendrite crystal, which is also growing. Where these grains meet, grain boundaries are set up.

Table 1
Properties of Some Common Metals

Symbol	Element	Crystal Structure
Al	Aluminum	FCC
Sb	Antimony	Rhombohedral
Be	Beryllium	CPH
Bi	Bismuth	Rhombohedral
Cd	Cadmium	CPH
C	Carbon (graphite)	Hexagonal
Cr	Chromium	BCC (above 26° C)
Co	Cobalt	CPH
Cu	Copper	FCC
Au	Gold	FCC
Fe	Iron (alpha)	BCC
Pb	Lead	FCC
Mg	Magnesium	CPH
Mn	Manganese	Cubic
Mo	Molybdenum	BCC
Ni	Nickel	FCC
Nb	Niobium (columbium)	BCC
Pt	Platinum	FCC
Si	Silicon	Cubic, diamond
Ag	Silver	FCC
Ta	Tantalum	BCC
Sn	Tin	Tetragonal
Ti	Titanium	CPH
W	Tungsten	BCC
V	Vanadium	BCC
Zn	Zinc	CPH
Zr	Zirconium	CPH

As solidification is taking place, the arrangement of the crystalline lattice structure takes on a characteristic pattern. Each unit cell builds on another to form crystalline needle patterns that resemble small pine trees. These structures are called dendrites (Figure 6).

Figure 7 represents the growth of the dendrite from the nucleus to the final grain when metal is solidifying from the melt. The nucleus can be a small impurity particle or a unit cell of the metal.

GRAIN BOUNDARY

As the crystal structures grow in the various grains in different directions, it can be seen that at the grain boundaries, the atoms are jammed together in a misfit pattern (Figure 8). This strained condition also makes the grain boundaries stronger than the adjacent grain lattice

Figure 6. A three-dimensional dendrite crystal on the surface of solidified tin.

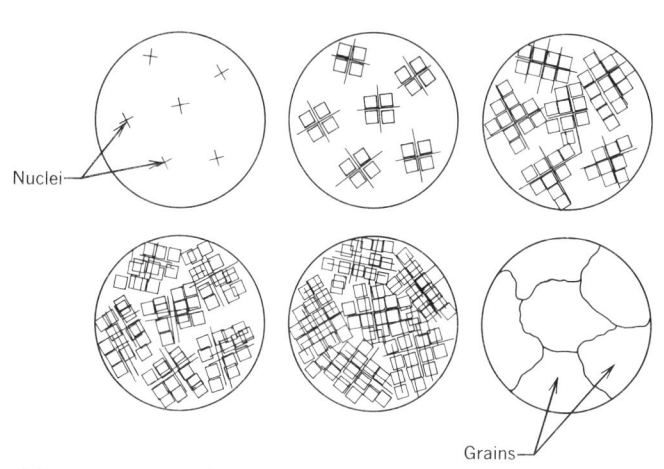

Figure 7. The formation of grains during solidification.

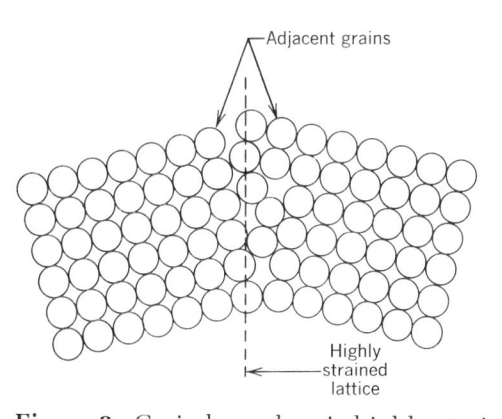

Figure 8. Grain boundary is highly strained.

structure at low temperatures (under a red heat), but weaker at high temperatures (yellow or white heat). Grain boundaries are only about one or two atoms wide, but their strained condition causes them to etch differently, so they may be observed with the aid of a microscope.

BODY-CENTERED CUBIC LATTICE (BCC)

This cubic lattice is made up of atoms at each corner of the cube and one in the very center. Steel under 1333° F (723° C) has this arrangement and it is called alpha iron or ferrite. Other metals such as chromium, columbium, barium, vanadium, molybdenum, and tungsten crystallize into this lattice structure. These cubes are identified within the lattice structure, as seen in Figure 9. Body-centered cubic metals (Figure 10) show a lower ductility but a higher yield strength than face-centered cubic metals.

FACE-CENTERED CUBIC LATTICE (FCC)

Atoms of calcium, aluminum, copper, lead, nickel, gold, platinum, and some other metals arrange themselves with an atom in each corner of the cube and one atom in the center of each cube face. When steel becomes nonmagnetic in the critical range, it rearranges its atoms to this struc-

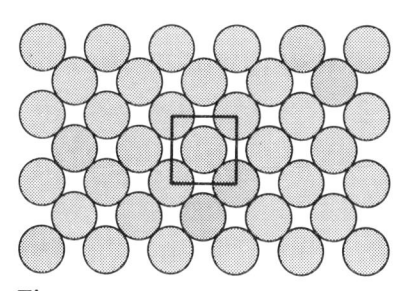

Figure 9. Lattice structure showing body-centered cubic formation.

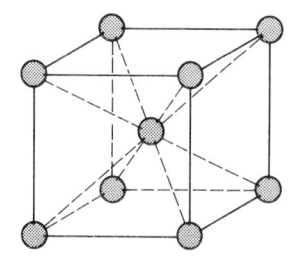

Figure 10. Idealized body-centered cubic structure.

ture and is called gamma iron or austenite (Figure 11).

CLOSE-PACKED HEXAGONAL LATTICE (CPH)

This structure (Figure 12) is found in many of the least common metals. Beryllium, zinc, cobalt, titanium, magnesium, and cadmium are examples of metals that crystallize into this structure. Because of the spacing in the lattice structure, rows of atoms do not easily slide over one another in CPH. For this reason these metals have lower plasticity and ductility than cubic structures.

The metal manganese has a simple cubic structure. Manganese is used as an alloying element in steel. Antimony is used as an alloying element with zinc and tin. Antimony has a rhombohedral crystal structure.

When carbon steel is quenched from the austenitizing temperature, the FCC structure attempts to change to the BCC structure. Since there is a solid solution of carbon and iron at the austenitizing temperature, the lattice contains the smaller carbon atoms in the interstices, and complete conversion to BCC is not possible. This is due to interference of the carbon atoms, since there is not enough room in BCC to hold all of them. The result is an elongated body-centered cubic crystal, body-centered tetragonal. This distortion of the lattice is what causes the hardness of martensite, which is a body-centered tetragonal structure.

CRYSTALLINE CHANGES DURING HEATING

When metals are heated slowly all the way to the melting point, certain changes take place. Most nonferrous metals, such as aluminum, copper, and nickel, go through no change in the crystalline lattice structure before becoming a liquid. However, this is not the case with ferrous metals (metals containing iron).

Iron is a special type of metal that does go through a crystalline change as it is heated to the liquid stage. Iron in a cool condition is BCC (Figure 13) but, when it is heated to about 1700° F (927° C) in its pure form it changes to FCC (Figure 14). A material that changes its crystalline lattice structure under certain conditions is called *allotropic*.

GRAIN SIZE CONSIDERATION

The size of the grain has a great effect on the mechanical properties of the metal. The effects of gain growth caused by heat treatment are easily predictable. Temperatures, alloying elements, and soaking time all affect grain growth.

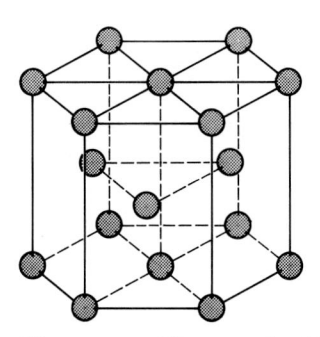

Figure 12. Close-packed hexagonal structure.

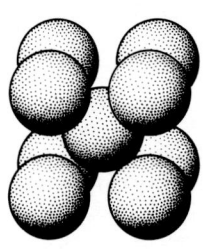

Figure 13. Body-centered cubic (BCC).

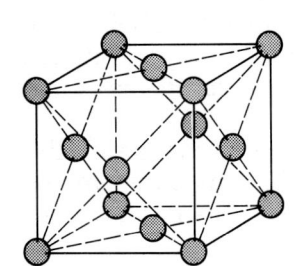

Figure 11. Idealized face-centered cubic structure.

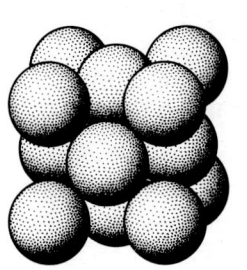

Figure 14. Face-centered cubic (FCC).

Generally, a small grain is preferable over a large grain. The small grain has more tensile strength, greater hardness, will distort less during quenching, and will be less susceptible to cracking. Fine grain is best for tools and dies. However, a large grain has increased hardenability, which is often preferable for carburizing and also for steels subjected to extensive cold working.

All metals will experience grain growth at high temperatures. However, there are some metals that can reach relatively high temperature [about 1800° F (982° C)] with very little grain growth but, as the temperature rises, experience a rapid growth rate. These metals are referred to as fine grained.

GRAIN SIZE CLASSIFICATION

There are several methods of determining grain size as seen under a microscope. The method explained here is one that is widely used. The size of grain is determined by a count of the grains per square inch under 100× magnification. Figure 15 is a chart that represents the actual size of the grains as they appear under 100× magnification. Specified grain size is generally the austenitic grain size. A properly quenched steel should show a fine grain.

In describing a piece of steel it is often necessary to specify the grain size. The actual method is done on a comparison between the specimen and grain size classification chart. The chart includes eight different grain sizes. A steel is considered fine grained if it is predominately 5 to 8 inclusive, and coarse grained if it is predominately 1 to 5 inclusive. If 70 percent of the grain size falls into the given range, it is considered acceptable. Two size classifications may be necessary if there is a wide variation in a section of metal. When austenitic grain size is specified, as it is in most mechanical property tables, the accepted method of determining it is the McQuaid-Ehn test. This test consists of carburizing a specimen at 1700° F (927° C) and then slow cooling it to develop a carbide network at the grain boundaries. The specimen is then polished, etched, and compared to the grain size standards at 100× magnification.

SOLUTIONS, LIQUID AND SOLID

When two or more metals are heated to or above their melting points and combined, they usually become a solution that is considered an alloy. The metal composing the greatest percentage is the solvent and the metal composing the smaller percentage is the solute. Some molten metals will not dissolve at all in other molten metals; they separate or form a mixture. We are accustomed to thinking of solutions in terms of liquids such as salt or sugar solutions. There are also limits of solubility; water will dissolve only so much salt or sugar. Oil *will dissolve* in water but to a very limited extent. As you will see, there is a similarity between liquid metal solutions and other liquid solutions.

Solutions may also be found in solid metals, but the changes are made within the lattice and grain structure in solids. The atoms are not quite so free to move about as they are in liquids. Atoms move only to a limited extent and are much slower in solids. The rate of movement depends on the temperature.

TYPES OF SOLUTION

The dissolving of one material into another can take place in two ways: substitutional and interstitial, as illustrated in Figure 16. A substitutional solid solution is a solution of two or more elements with atoms that are of relatively the same size. This requirement is necessary in that the alloying atoms must replace the regular atoms in the lattice structure and not just fit in the spaces between the regular atoms as it does in interstitial solutions.

When one metal is *completely soluble* in another, such as copper and nickel, both metals must have the same lattice structure, atoms of the same relative size, and a chemical desire to combine. This type of solution is called *continuous solid solution*. Figure 17 is a diagram showing how nickel and copper combine into a continuous solid solution. When the above factors vary, metals take on varying degrees of solubility. Copper-silver is an example of this. Their atom size differs somewhat more than copper-nickel, and their chemical desire to combine is much less.

It can be seen from Figure 18 that solubility is *continuous* at each end, but insoluble in the middle. This type of solution is called *terminal solid solution*.

To summarize, as a general rule, the more alike the two metals are chemically and physi-

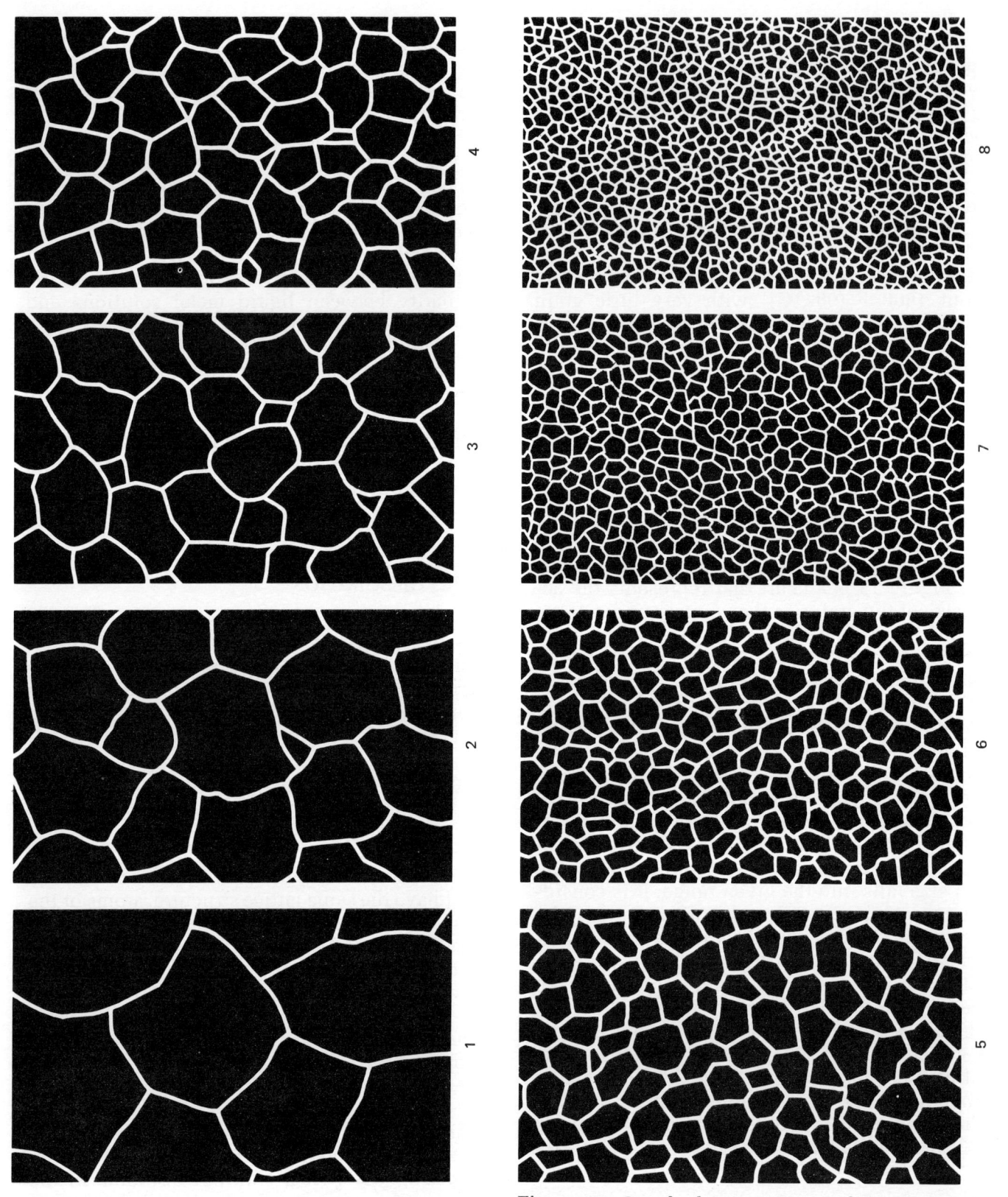

288　Section G Physical Properties and the Heat Treating of Metals

Figure 15. Standard grain size numbers. The grain size per square inch at 100X magnification (Photograph courtesy of Bethlehem Steel Corporation).

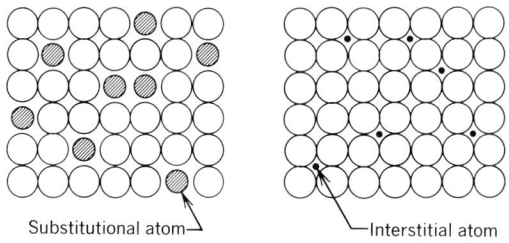

Figure 16. Substitutional solid solution and interstitial solid solution.

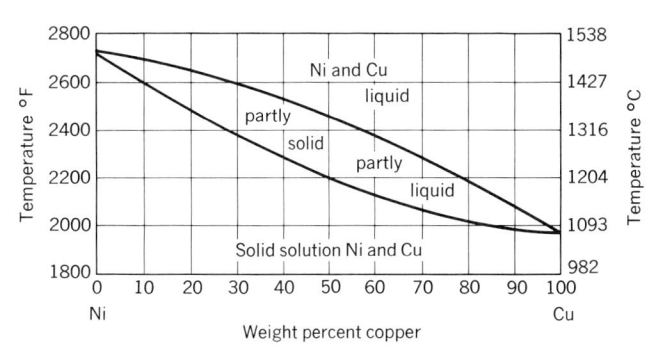

Figure 17. Copper-nickel phase diagram.

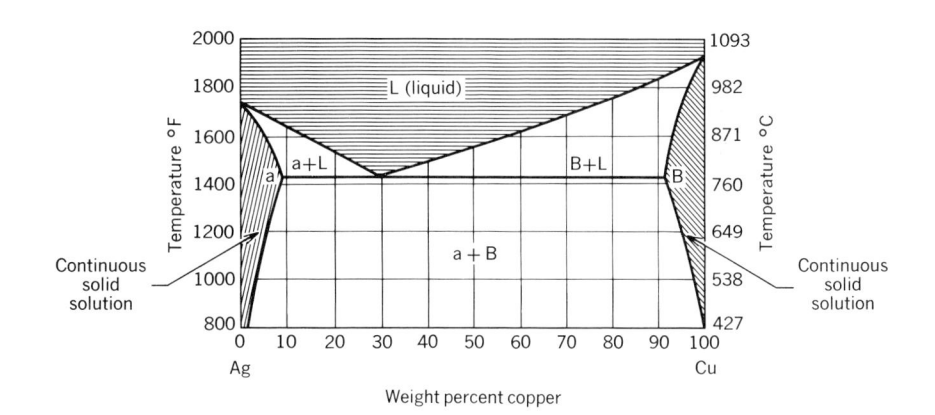

Figure 18. Silver-copper phase diagram.

cally, the more they tend to form continuous solid solutions.

1. The size of the atom of the alloying metals cannot differ by more than 15 percent.
2. The chemical characteristics should be similar.
3. The metals must crystallize in the same pattern, such as BCC, FCC, or CPH.

Insoluble solid solutions that become insoluble below the transformation temperature often separate into a lamellar (platelike) structure that is a mixture.

Interstitial solid solutions are made up of alloying elements or atoms that differ greatly in size, as Figure 16 demonstrates. The alloying atoms must be small enough to fit within the lattice structure of the base material. It has been determined that the alloying atom should be about one half the size of the base atom. Common elements that are able to form interstitially with iron are carbon, nitrogen, oxygen, hydrogen, and boron. These elements are also important in their ability to combine chemically with the base metal to form compounds, such as combining iron (Fe) and carbon (C) to form iron carbide (Fe_3C). Iron carbide and other compounds such as iron nitrides and chromium nitrides have the potential for greatly increasing hardness, strength, heat resistance, and abrasion resistance of metals. The development of the carbides, nitrides, and borides metals has aided the aircraft and space industry in their building programs.

A third condition is the combining of substitutional and interstitial atoms with the same base metal. Figure 19 illustrates how this would be done. Many alloys are like this. It makes possible the strength, hardness, and heat treat ad-

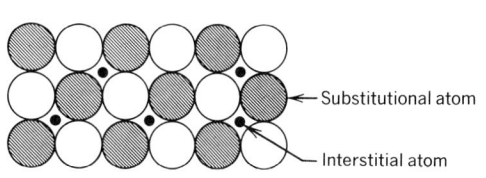

Figure 19. Substitutional and interstitial solid solution.

vantages of both processes within the same material.

In both the substitutional and interstitial solid solutions, the strengthening of the material is accomplished through the distortion of the lattice structure caused by the alloyed atoms. The lattice distortion creates a strain around the slip planes and grains of the material, resulting in the increase of strength and hardness.

self-evaluation

SELF-TEST

1. Briefly describe the structure of an atom and the importance of the valence electrons.

2. How does the metallic bond work, and what effect does it have on metals?

3. Name five crystalline lattice structures found in metals.

4. The growth from the nucleus of the grain that resembles small pine trees is called a _____.

5. Are the grain boundaries a continuation of the regular lattice structure from one grain to another? Explain.

6. When austenitized carbon steel is quenched, why is the BCC crystal elongated into a body-centered tetragonal crystal structure?

7. What is meant by the term "allotropic" when applied to iron or steel?

8. Explain the relative advantages of fine grained steel as compared to coarse grained steel.

9. Will all metals dissolve in other metals when they are molten? Explain.

10. List the types of solid solutions and other related conditions of solids.

WORKSHEET Given a piece of SAE 1095 steel ⅜ in. diameter by 4 in., and a set of number stamps, you will:

1. Show the effect of heat on fine and coarse grain materials.

2. Determine approximate grain size of steel samples used.

The Metcalf Experiment When high carbon steels (about 1 percent carbon) are overheated to 1800° F (1024° C) and quenched in water, the grain structure becomes very coarse and weak. If the specimen is quenched from 1300° F (704.4° C) to 1500° F (815.6° C), however, the grain structure is very fine and much stronger. In their normal state the grains are fairly coarse, but the steels are soft and tough. This experiment will allow you to observe visually varying grain sizes.

By using the metal selected you will be able to compare the grain growth of a fine grain to a coarse grain structured metal.

Procedure 1. Take the material and make notches or vee grooves in it every ½ in., about .050 in. deep. See Figure 20. File or machine a flat ⅛ wide on one side and stamp a number on each section.

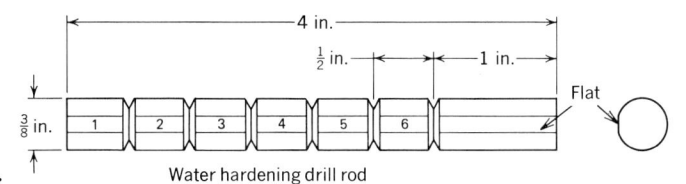

Figure 20. Metcalf experiment.

2. Heat one end (#1) with a torch to 1800° F (1024° C), a bright orange color, and make each section progressively cooler so #6 is gray or black and #4 is a dull red.

3. Quench in water until cool.

4. Place the sample in a vise with one section extending into the jaws.

5. Use a pipe or tube and break off each section. *Wear safety glasses.*

6. Compare the grain size of your specimens to the standard grain size chart in the unit. Use a 100× power magnifier or microscope when viewing the specimen.

7. If a hardness tester is available, test the hardness of each section and list on the chart. Take the results of your experiment to your instructor for evaluation.

SAE 1095 steel

Grain size	#1	#2	#3	#4	#5	#6
Approximate temperature	___	___	___	___	___	___
Rc	___	___	___	___	___	___

Conclusions

1. Which piece of metal can be considered fine grained, coarse grained? Explain why.

2. How does grain size affect tensile strength?

3. What conclusion can you draw from the experiment relating grain size to hardness? To toughness?

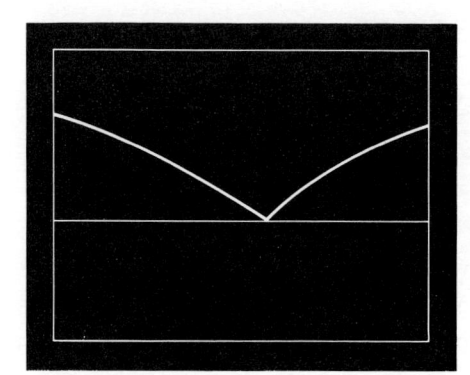

unit 3
phase diagrams for steels

Phase diagrams are a useful means to explain and understand the behavior of metals. Many very complex phase diagrams for various alloys are used by metallurgists, but only several of the simple types will be used in this unit. The iron-carbon diagram is basic to an understanding of heat treating iron and steel.

objectives After completing this unit, you will be able to:
1. Demonstrate your understanding of phase diagrams by completion and recognition.
2. Encapsulate, polish, and etch specimens for microscopic study.
3. Be able to establish relative carbon content by microscopic evaluation.

information

Matter may exist in three states or phases: solid, liquid, or gaseous. Some substances are capable of changing within the solid state to other phases or crystalline structures. The ability to change into different phases is called allotropy. Iron is an allotropic element and changes from face-centered cubic to body-centered cubic during cooling.

When two metals are alloyed together, the temperature of phase changes will be different for every combination of the two metals. The temperatures and compositions of phase changes can be graphed so that all possible combinations of two pure metals are represented. Type I alloys are completely soluble in both the liquid and solid states, and their solid phase has a substitutional lattice. See Figure 1. Type II alloys are soluble in the liquid state but insoluble in the solid state (Figure 2). Alloys that are not solid solutions are usually mixtures of various forms of both constituents.

One pure metal (A) is represented on the left and the other pure metal (B) is represented on the right. Along the bottom of the graph the percent of concentration of one metal increases while the other decreases. The vertical lines represent temperature. The freezing (solidification) temperatures for pure metals A and B are points 1 and 2. For alloys of the two metals, solidification begins at line 3 (liquidus line) and is completed at line 4 (solidus line), as shown in Figure 1. Point 5 is the lowest melting alloy on the graph. The metallurgical name for this is the eutectic point, which means low melting. The freezing of a 70-30 composition alloy begins at point 6 and ends at the solidus line. Notice that at the eutectic point there is no such mushy area. The eutectic composition begins and finishes freezing at the same temperature. The solidus line (4) is also called the eutectic line, since the eutectic composition is made at this temperature.

Eutectic compositions or grains often show a lamellar or alternating platelike microstructure (Figure 3). This structure is found in carbon steel. Most metals show at least a little ability to dissolve other metals at room temperature, but eutectic solids are basically mixtures.

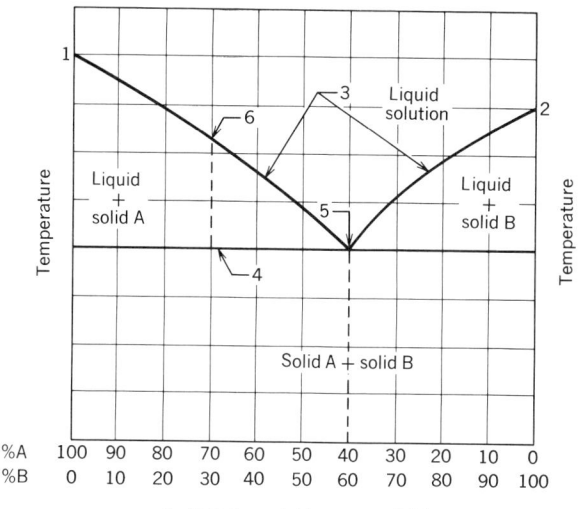

1. Freezing point for pure metal A.
2. Freezing point for pure metal B.
3. Liquidus line.
4. Solidus line.
5. Eutectic.
6. Seventy to thirty percent alloy.

Figure 2. Type II alloy diagram.

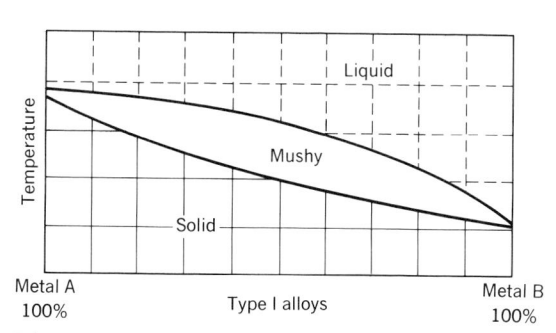

Figure 1. Type I silver-gold constitutional diagram.

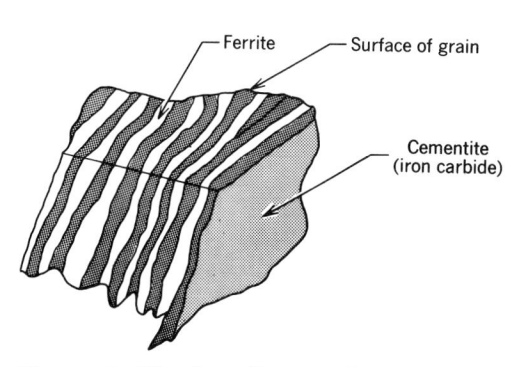

Figure 3. The lamellar pearlite microstructure.

PHASE CHANGES OF IRON

Iron, being an allotropic element, can exist in more than one crystal structure, depending on temperature. When a substance goes through a phase change after cooling, it releases heat. When a phase change is reached after heating, it absorbs heat. This characteristic is the basis for the construction of cooling curve graphs. If a continuous record is kept of the temperature of cooling iron, we can construct a graph that will resemble Figure 4.

Each flat segment of the cooling curve represents a phase change. These flat portions are caused by the release of heat from phase changes. At 2800° F (1538°C), iron changes to a solid body-centered cubic structure. All of the remaining changes involve a solid of one lattice structure transforming into a solid of another lattice structure. At 2554° F (1401°C) BCC delta iron changes to FCC austenite. Austenite transforms to BCC ferrite at 1666° F (908° C). The next change is not a phase change at all, but it does give off heat. This is the change from nonmagnetic ferrite to magnetic ferrite.

THE IRON CARBON DIAGRAM

In the preceding phase diagrams of alloy systems, none of the metals went through any solid state phase changes. There are some new lines that must be added to account for these solid phase transformations (Figure 5). Many of the lines, the liquidus, solidus, and the eutectic point, are similar. This diagram differs from previous ones in that the diagram ends on the right at 6.67 percent carbon (C) instead of 100 percent carbon. The rest of the diagram from 6.67 to 100 percent carbon would give no useful information about steels and cast irons. But a more important reason for ending the diagram at this point is that 6.67 is the percent of carbon by weight in the compound iron carbide.

THE DELTA IRON REGION

The area at the left of the diagram between 2800° F (1538° C) and 2554° F (1410° C) in Figure 5 describes the solidification and transformation of delta iron. This area has no commercial value in heat treating and therefore is only of passing interest.

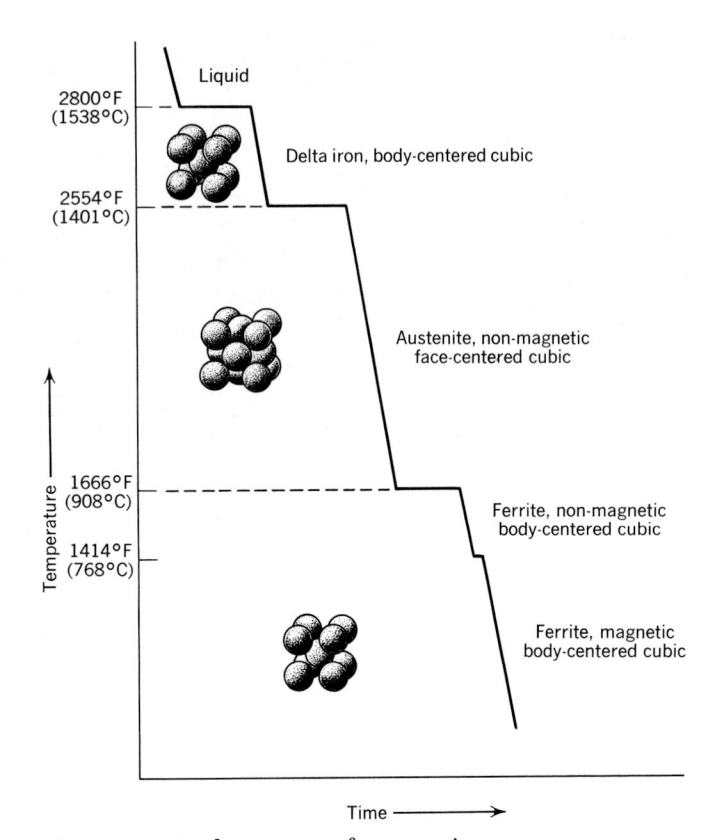

Figure 4. Cooling curve for pure iron.

THE STEEL PORTION

Study the portion of the diagram outlined by dashed lines in Figure 5 and you will see how it resembles phase diagrams that you have already studied. A little different terminology is used for the lines, however. Since the metal is now a solid, the terms liquidus and solidus do not apply. Eutectic means low melting point, so this term is not correct, either. The word ending -oid means similar but not the same. The eutectoid point appears like a eutectic on the diagram, but it is the lowest temperature transformation point of solid phases, while eutectic is the lowest freezing point of a liquid phase. Instead of a liquidus line, we have the line that shows the transformation from austenite to ferrite, called the A_3 line, and the line showing the amount of carbon that is soluble in austenite, called the A_{cm} line. Instead of a solidus line we have the line that shows where austenite completes its transformation to ferrite and where pearlite is formed. This line is called the A_1 to the left of the eutectoid point and the $A_{3,1}$ to the right of the eutectoid point.

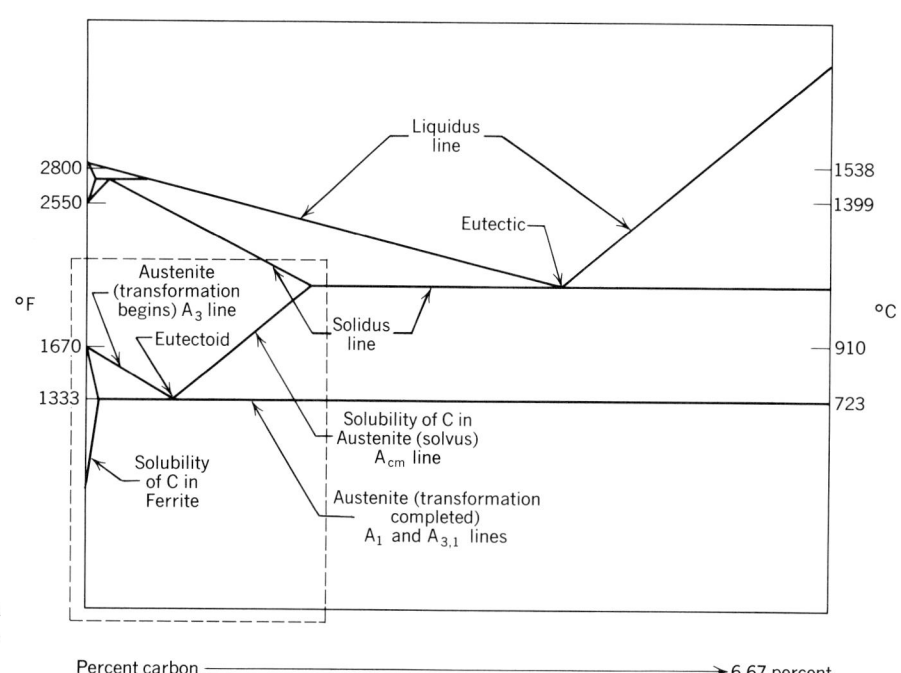

Figure 5. Simplified iron-carbon diagram in equilibrium or very slow cooling.

MICROSTRUCTURES

When a specimen of steel is ground, polished, and etched with Nital (a solution of 5 percent nitric acid dissolved in methanol), it may be viewed through a metallurgical microscope. The acid etches the iron carbide (cementite) dark and the ferrite is etched light in the pearlite. Each steel has an individual type microstructure and many variations are possible, depending on heat treatment.

Cementite or iron carbide (Fe$_3$C) is the hardest structure on the iron carbon diagram. It is 6.67 percent carbon by weight. The iron-carbon diagram (Figure 6) shows the composition of iron carbide or cementite on the right. Massive cementite in grain boundaries and cast iron etch white.

Austenite is a face-centered cubic iron. It has the ability to dissolve carbon interstitially to a maximum of 2 percent at 2065° F (1129° C). See Figure 6. The iron carbon diagram shows the solubility range of carbon in austenite. Austenite is not stable at room temperature except in some alloy steels, but decomposes into other microstructures.

Ferrite is a body-centered cubic iron. It will dissolve only .008 percent carbon at room temperature and a maximum of .025 percent carbon at 1330° F (721° C). Locate the narrow solid solution area at the left of the diagram (Figure 6). Ferrite forms an interstitial solid solution with carbon to a limited extent in this area. Ferrite is the softest structure that appears on the diagram. It appears light gray or white through the microscope (Figure 7).

Pearlite is the eutectoid mixture .8 percent carbon (Figures 8 and 9). It appears like a fingerprint and is actually thin alternating layers of ferrite and cementite. The cementite is dark and the ferrite is light. Pearlite forms at 1330° F (721° C) after cooling.

THE SLOW COOLING OF 1020 CARBON STEEL

For steel less than eutectoid, such as 1020 or .20 percent carbon steel during cooling, see Figure 10, line 1. Notice that at point *a*, the entire microstructure is austenite. After crossing the A$_3$ line (point *b*), ferrite begins to form along the austenite grain boundaries. Further cooling causes more ferrite to form. Because the solubility of carbon in ferrite is so low, excess carbon must be dissolved by the remaining austenite. At point *c*, most of the austenite has become ferrite. The remaining austenite has a carbon content of .8 percent carbon. As the steel crosses the A$_1$ line,

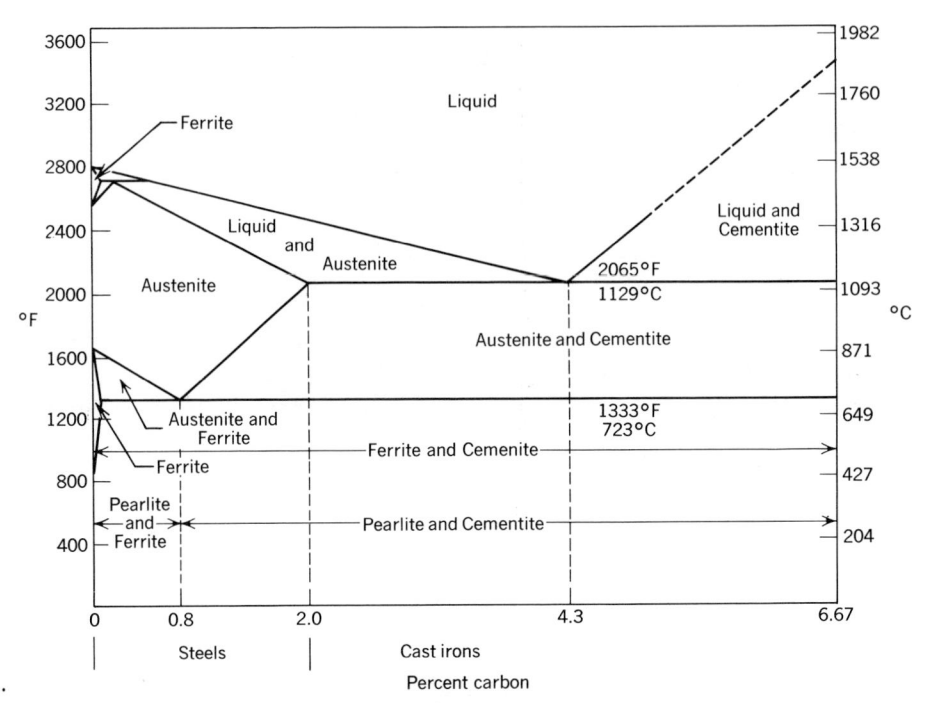

Figure 6. The iron-carbon diagram.

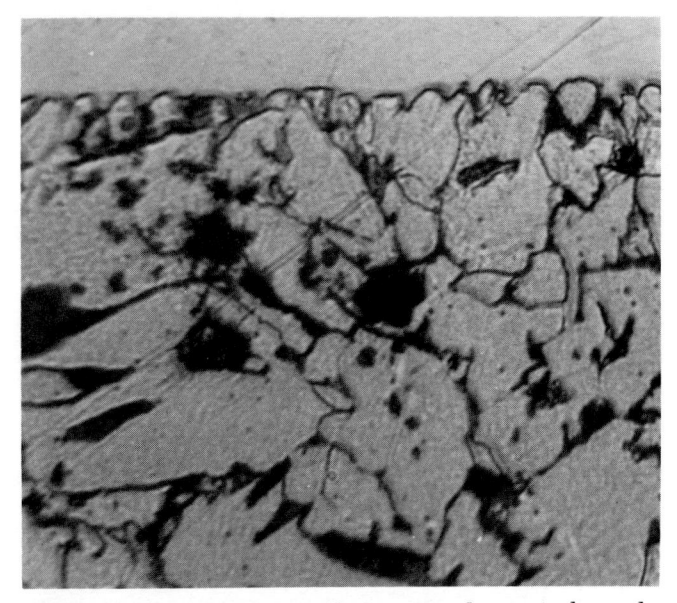

Figure 7. Low carbon steel showing the grain boundaries of mostly ferrite grains with isolated grains of fine pearlite (500×).

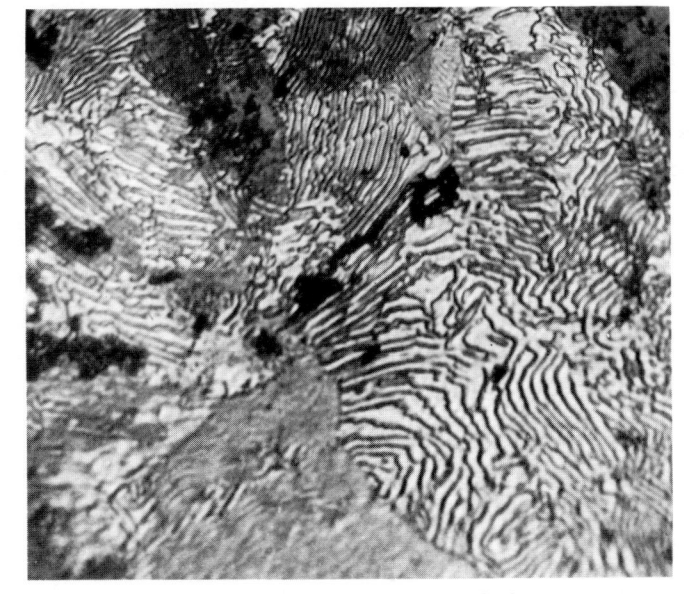

Figure 8. SAE 1090 steel slowly cooled (100 percent pearlite) (500×).

the remaining carbon-rich austenite transforms to the eutectoid or pearlite at point d; .8 percent carbon produces 100 percent pearlite, the iron-carbon eutectoid.

SLOW COOLING OF A 1095 CARBON STEEL

At point a (Figure 11) the 1095 or about 1 percent carbon steel is entirely austenite and interstitially dissolved carbon. When the steel cools to point b, it has crossed the A_{cm} line. The A_{cm} line indicates the limit of carbon solubility in austenite. Since the austenite can no longer hold the entire amount of carbon in solution, cementite begins to appear along the grain boundaries. At point c the solubility of the austenite has dropped even further, so that the additional cementite has formed a fairly continuous network between

Unit 3 Phase Diagrams for Steels   297

the austenite grains. The carbon content of the remaining austenite is now the eutectoid composition of .8 percent carbon. At point d the steel has crossed the $A_{3,1}$ line and the remaining austenite transforms to pearlite.

HEATING 1020 STEEL

There are some very important differences in the transformation of steels during cooling and heating. If in Figure 10, the 1020 steel is raised from room temperature to 1800° F (982° C), the following changes take place. At room temperature, the microstructure is ferrite and pearlite; no change takes place until the A_1 line is reached. Now the pearlite grains transform to austenite, but the remaining ferrite grains are not affected until the A_3 line is reached. At the A_3 line the remaining ferrite

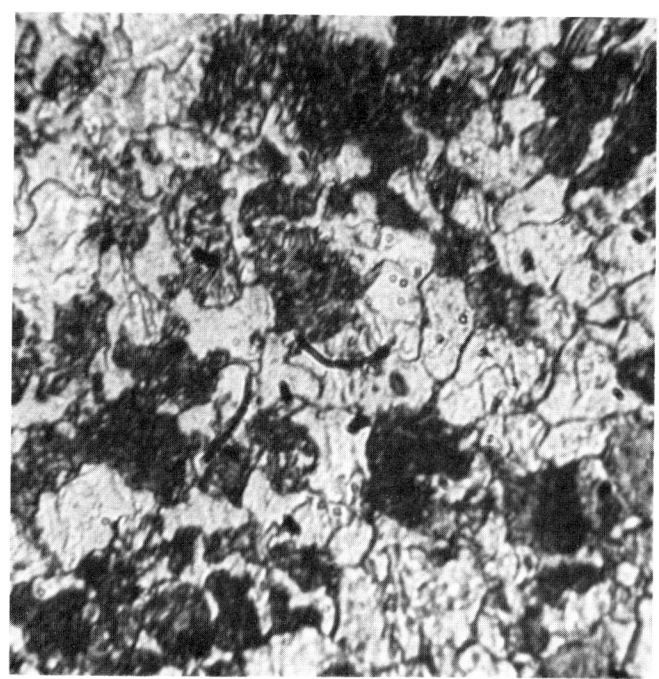

Figure 9. SAE 1030 steel (pearlite and ferrite) (500×).

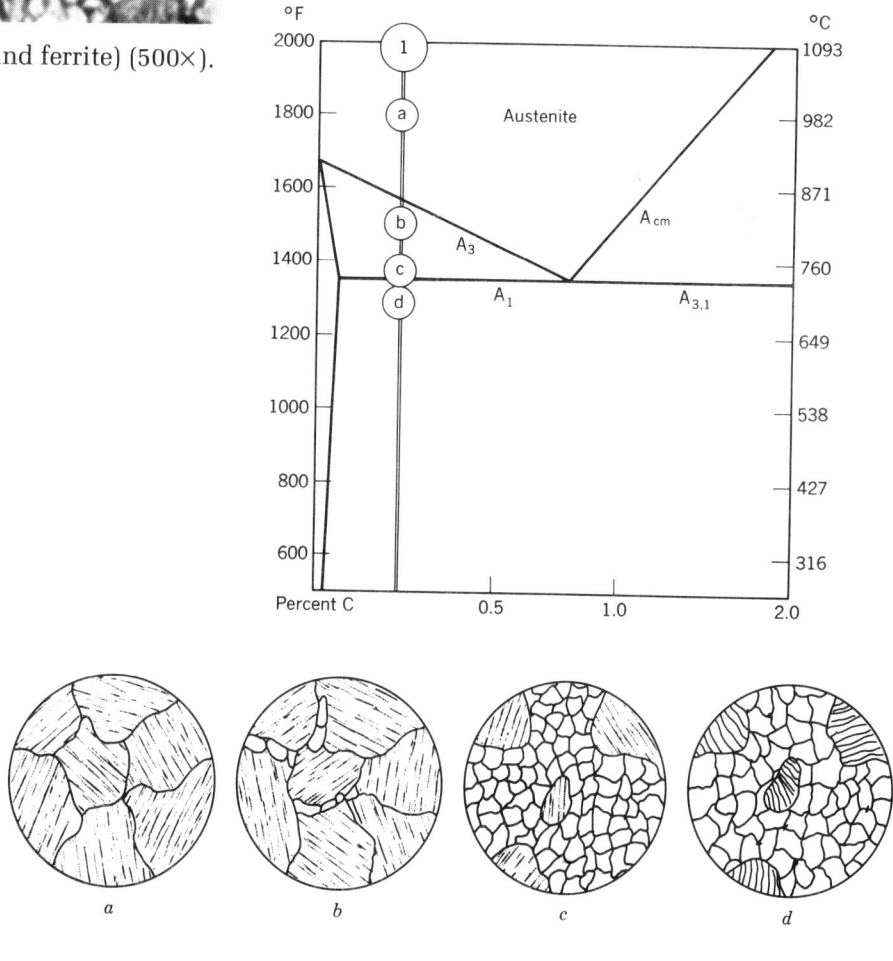

Figure 10. Microstructures at various temperatures of very slow cooling SAE 1020 steel.

Figure 11. Microstructures at various temperatures when cooling SAE 1095 steel.

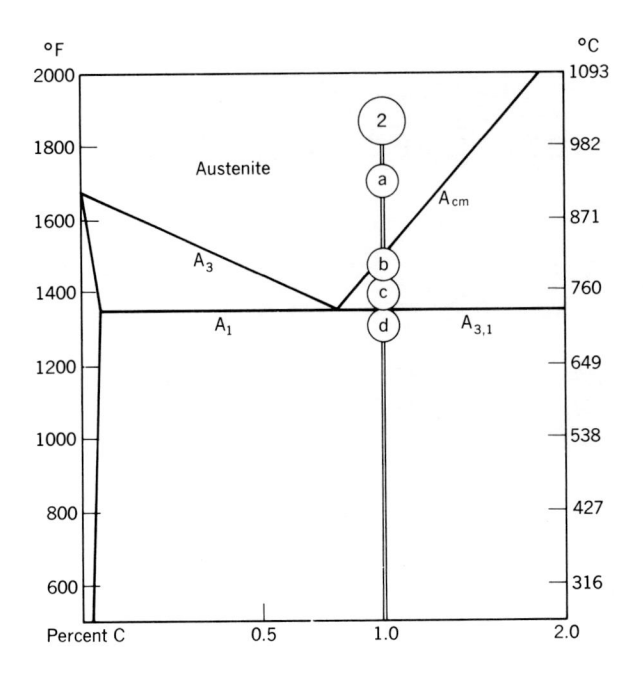

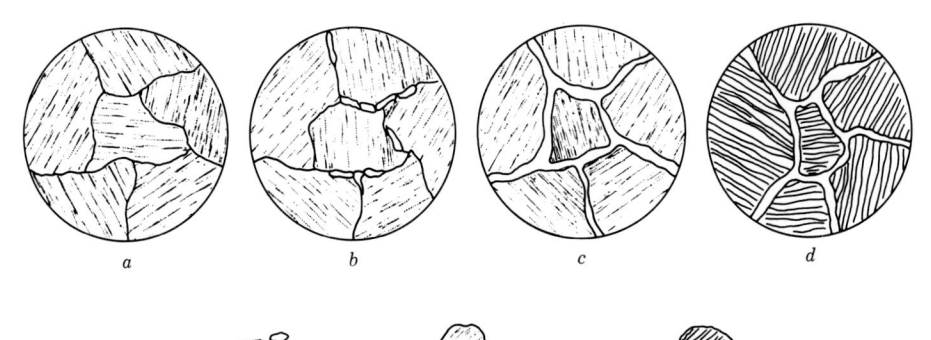

grains recrystallize to austenite. The structure becomes fine grained at this temperature because grain growth is interrupted by the diffusion of carbon. The austenite that was formed at a lower temperature from pearlite is richer in carbon than the new austenite. Consequently, carbon must migrate to the new austenite until the carbon content is uniform throughout. As the temperature increases beyond the A_3 line, grain size also increases dramatically.

Grain size has a marked effect on the strength, toughness, and plasticity of steels at room temperature. Large grain steels have lower strength and toughness and should be avoided when these properties are important. If the steel is to be cold formed, the increased plasticity of large grain size may be desirable, but not always; in some cold forming applications an "orange peel" effect develops from large grains.

When steels are overheated into the area of large grain growth, the grains remain large until transformation takes place, as in Figure 12. Cooling steel that has been overheated will not restore small grain size for heat treating. A reheating from normal temperature is necessary, and often a normalizing treatment is needed.

ALLOYING ELEMENTS AND THE IRON-CARBON DIAGRAM

Alloying elements move the transformation lines of the iron-carbon diagram. A common alloying element is chromium. Considering the austenite area of the iron-carbon diagram, it can be seen

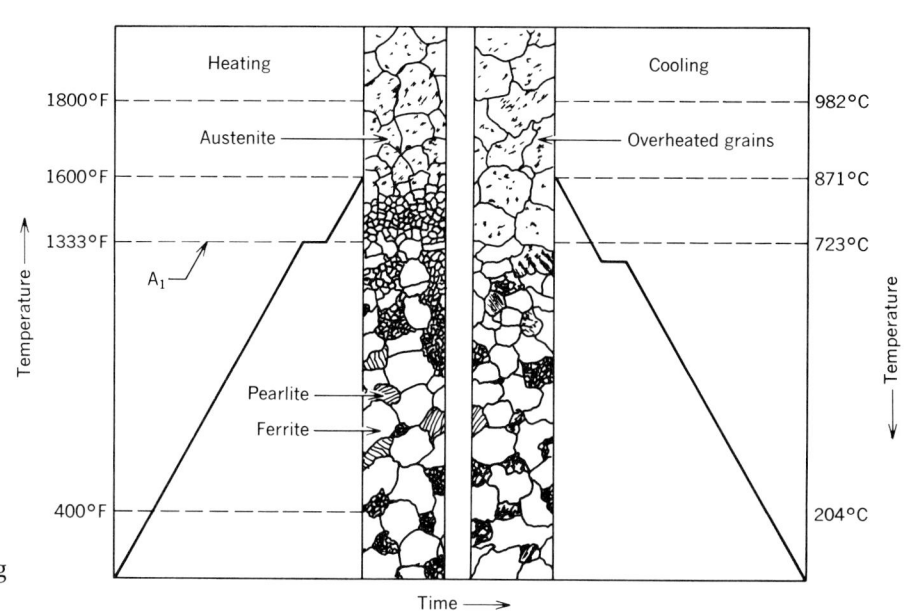

Figure 12. The effect of overheating of carbon steel and grain growth.

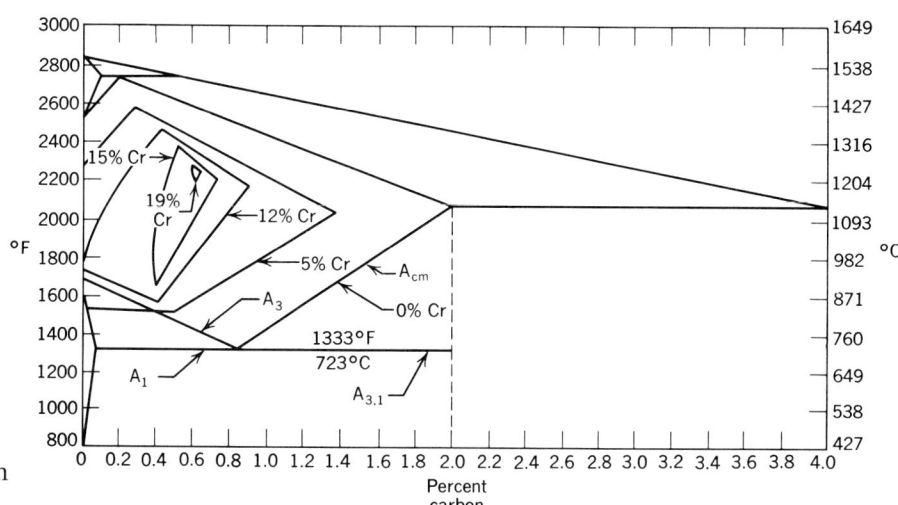

Figure 13. The effect of chromium on the austenite range of steel.

that the effect of increasing chromium is to decrease the austenite range (Figure 13). This will increase the ferrite range. Many other alloying elements are ferrite promoters also, such as molybdenum, silicon, and titanium.

Nickel and manganese tend to enlarge the austenite range and lower the transformation temperature (austenite to ferrite). A large percentage of these metals will cause steels to remain austenitic at room temperature. Examples are 18-8 stainless steel and 14 percent manganese steel.

Any alloy addition will move the eutectoid point to the left or, in other words, lower the carbon content of the eutectoid composition.

self-evaluation

SELF-TEST

1. Locate the following on the phase diagram (Figure 14).
   A. Eutectic point.
   B. Liquidus line.
   C. Solidus line.
   D. Area of liquid and solid.
   E. Area of 100 percent solubility.

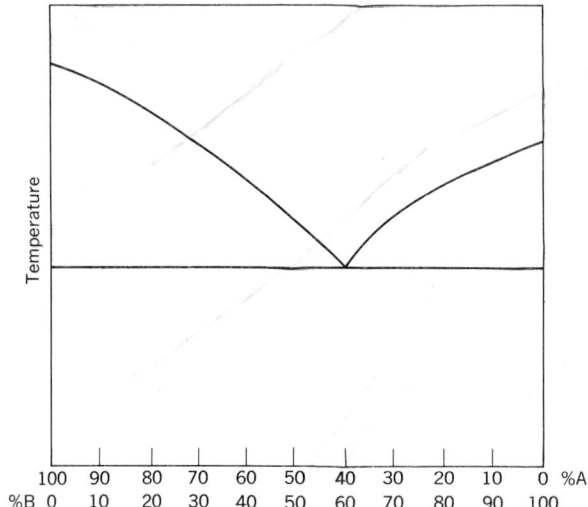

Figure 14

2. When a substance changes phases after cooling, there is a release of heat. How does this appear on a cooling curve graph?

3. What does eutectoid mean? How does this differ from eutectic?

4. What does the A_3 line indicate? The A_1? The A_{cm}?

5. What is the hardest structure found in the iron-carbon alloy system? How would it appear in pearlite under the microscope?

6. Give a definition of austenite.

7. What is the softest structure found in the iron-carbon alloy system? How would it appear under the microscope? How much carbon can this structure dissolve at normal temperatures and at 1330° F (721° C)?

8. What is the name for the eutectoid composition of carbon steel at room temperature? How would it appear under the microscope? What is it made of?

9. What phase do steels with less than .8 percent carbon begin to form at the A_3 line after slow cooling? At what temperature will this transformation be complete?

10. When steels are heated, at which line does pearlite form austenite and at which line does ferrite form austenite?

11. Grain size increases as temperature goes above the A_3 line. Do these grains decrease in size when steel is cooled toward the A_3 line?

12. What effect does the addition of alloying elements have on the eutectoid composition?

13. Name one ferrite promoter.

14. Name one alloying element that might make steel austenitic at room temperature.

15. How can having an understanding of the iron-carbon diagram help a person who does heat treating of iron and steel?

WORKSHEET Given a metallurgical microscope, nital etchant (5 percent nitric acid in alcohol), wash bottle of alcohol, electric polishing wheel and polishing compound, and mounted specimens of unidentified carbon, you will:

1. Identify the carbon content of several unknown specimens by microscopic examinations.

2. Learn to polish and etch specimens for microscopic viewing.

3. Learn to use a metallurgical microscope.

Procedure 1. Encapsulate the specimen in a mounting press and grind or sand any rough surfaces lightly (do not overheat) to make them level.

2. Sand the surface of the hand-held specimen on three or four progressively finer grit papers to 600 grit placed on a flat surface.

3. When the lines are all one direction, change to a finer grit and rotate the specimen to show the new lines. Be sure to clean the specimen thoroughly between each grit change and carefully avoid contaminating the felt on the polishing wheel.

4. Take a mounted specimen to the polishing wheel. Turn on the wheel and add a little slurry (polishing compound and water) to the felt surface. Gently press the specimen to the wheel surface and move it in the opposite direction of wheel rotation. Move the specimen around the

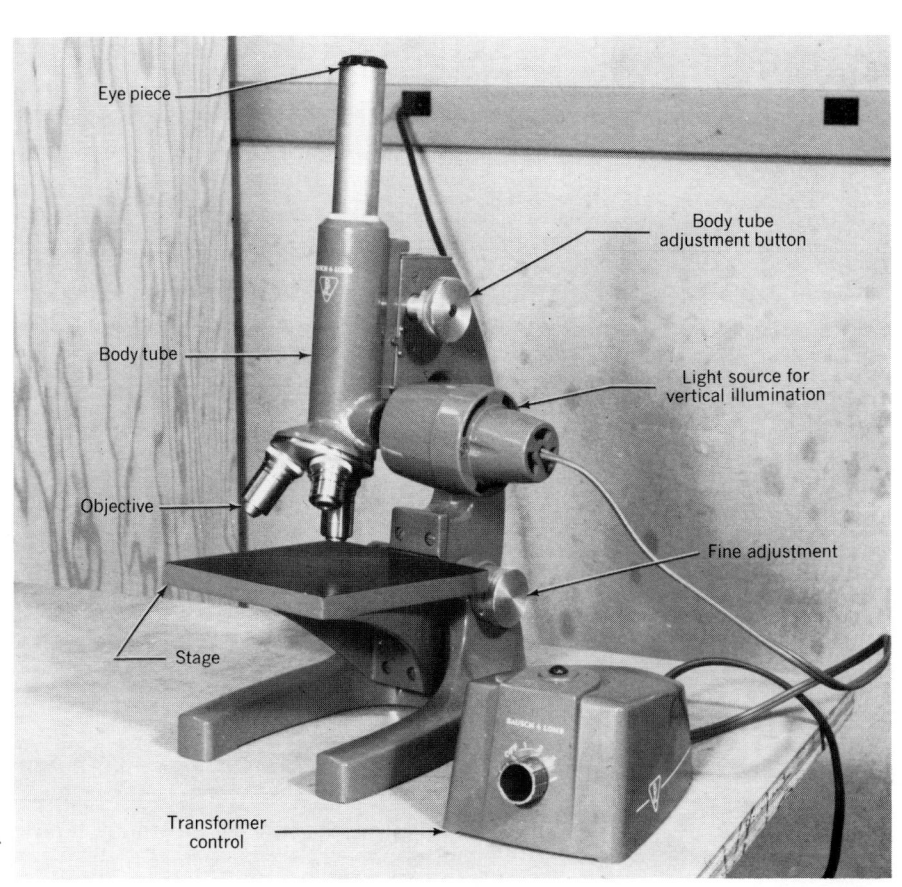

Figure 15. Metallurgical microscope.

wheel several times, then check the surface. When it becomes mirror bright, it is ready for etching.

5. Apply the nital on the surface of the specimen and count about 2 or 3 sec. Rinse with water and immediately follow with an alcohol rinse from the wash bottle. This will remove the water and prevent rusting. Allow the alcohol to dry.

6. Place the specimen under the microscope objective on the stage (Figure 15). Carefully lower the objective to the surface of the specimen. It should not quite touch. Use the fine adjustment button to raise the microscope until the specimen comes into focus.

7. If the microstructure is not visible, apply the nital again for a few seconds, rinse with water and then with alcohol wash.

8. Observe each specimen. Estimate the percentage of pearlite and ferrite. Now compare your observations with the following
    100 percent pearlite = .8 percent C
    75 percent pearlite = .6 percent C
    50 percent pearlite = .4 percent C
    25 percent pearlite = .2 percent C

Conclusion List your results.

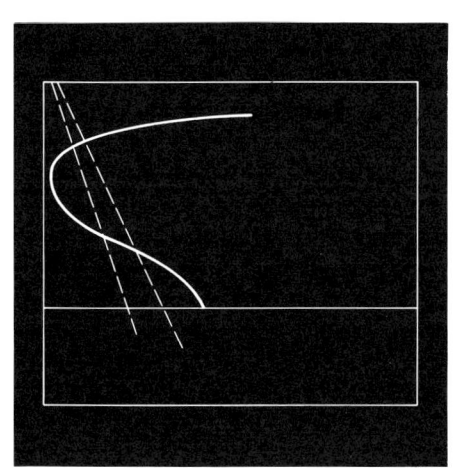

unit 4
I-T diagrams and cooling curves

When steel is held in the austenite phase at a constant temperature below its minimum stable temperature, it will transform into various transformation products. A graph is used to show these results. It is called an I-T diagram. Hardening procedures, rates of cooling, and various transformation products and how they are obtained are all explained by the study of this diagram. A working knowledge of transformation products is very important to the heat treater.

objectives After completing this unit, you will be able to:
1. **Determine the hardenability of steels and their quenching rates by using information gained from the I-T diagrams.**
2. **Recognize certain microstructures of transformation products pro-**

3. Estimate the hardness of a quenched steel by using the I-T diagram and microscope.

information

HARDENING PROCESS

The process of hardening steel is carried out by performing two operations. The first step is to heat the steel to the austenite range above the transformation temperature or, as it is called by heat treaters, the critical temperature. The second step is that of rapid cooling or quenching to near room temperature.

Austenitizing produces the solid solution of carbon and face-centered cubic iron austenite. The temperature used for austenitization is usually about 50° F (10° C) above the A_3 or $A_{3,1}$ lines (Figure 1). Alloys with .8 percent carbon or less will become 100 percent austenite at this temperature, while steel with more than .8 percent carbon will be austenite with some free cementite.

The second step, quenching, undercools the austenite to form a new structure at about 200° F (93.3° C). This structure is called martensite. The martensite is an extremely hard acicular or needlelike structure that, for most purposes, is too brittle to be of any use and must be tempered as an additional operation to toughen it.

TRANSFORMATION PRODUCTS

Austenite, when cooled below the transformation temperature, decomposes into various transformation products such as pearlite, bainite, or martensite. Austenite containing .8 percent carbon that is cooled quickly and held at 1300° F (704° C) does not begin to decompose or transform until after about 15 min has elapsed and does not completely decompose until after 5 h at that temperature (Figure 2). A very coarse pearlite structure has developed at this temperature and the material is very soft. If the austenite is quickly cooled to and held at a lower temperature of 1200° F (649° C), decomposition begins in about 5 sec and is completed after about 30 sec. The resultant pearlite is finer grained and harder than that formed at 1300° F (704° C). At a temperature of about 1000° F (538° C), the austenite decomposes extremely rapidly. It takes only about 1 sec before transformation begins and 5 sec to complete it. The resultant pearlite is extremely fine and its hardness is relatively high. This region of the S-curve where decomposition of austenite occurs is called the nose of the curve on an *isothermal transformation (I-T) diagram*. Isothermal means the same or constant temperature.

ISOTHERMAL TRANSFORMATION DIAGRAMS

An isothermal diagram is plotted by the use of small steel specimens of a specific kind of steel being tested that are heated to the austenitization temperature (Figure 3). They are next quenched in a liquid salt bath that is held at a constant temperature, such as 1200° F (649° C). At regular time intervals a specimen is removed and rapid quenched. Microscopic examination will then show martensite if transformation has not yet started, but martensite and pearlite if transformation has started, and only pearlite if transformation is complete. This procedure is repeated at other temperatures until the entire graph is plotted for that particular steel (Figure 4).

The vertical line on the left represents temperature and the horizontal line on the bottom

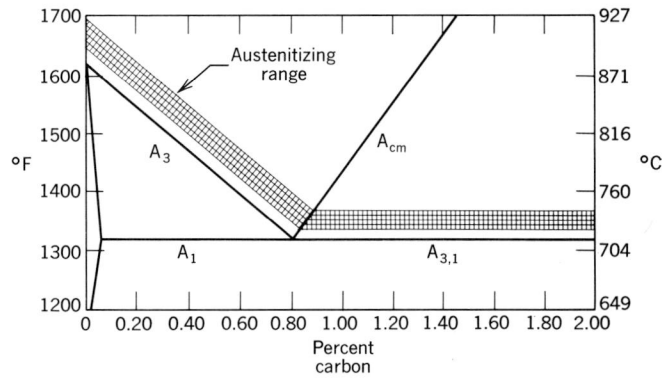

Figure 1. Diagram showing hardening temperature (austenitizing range).

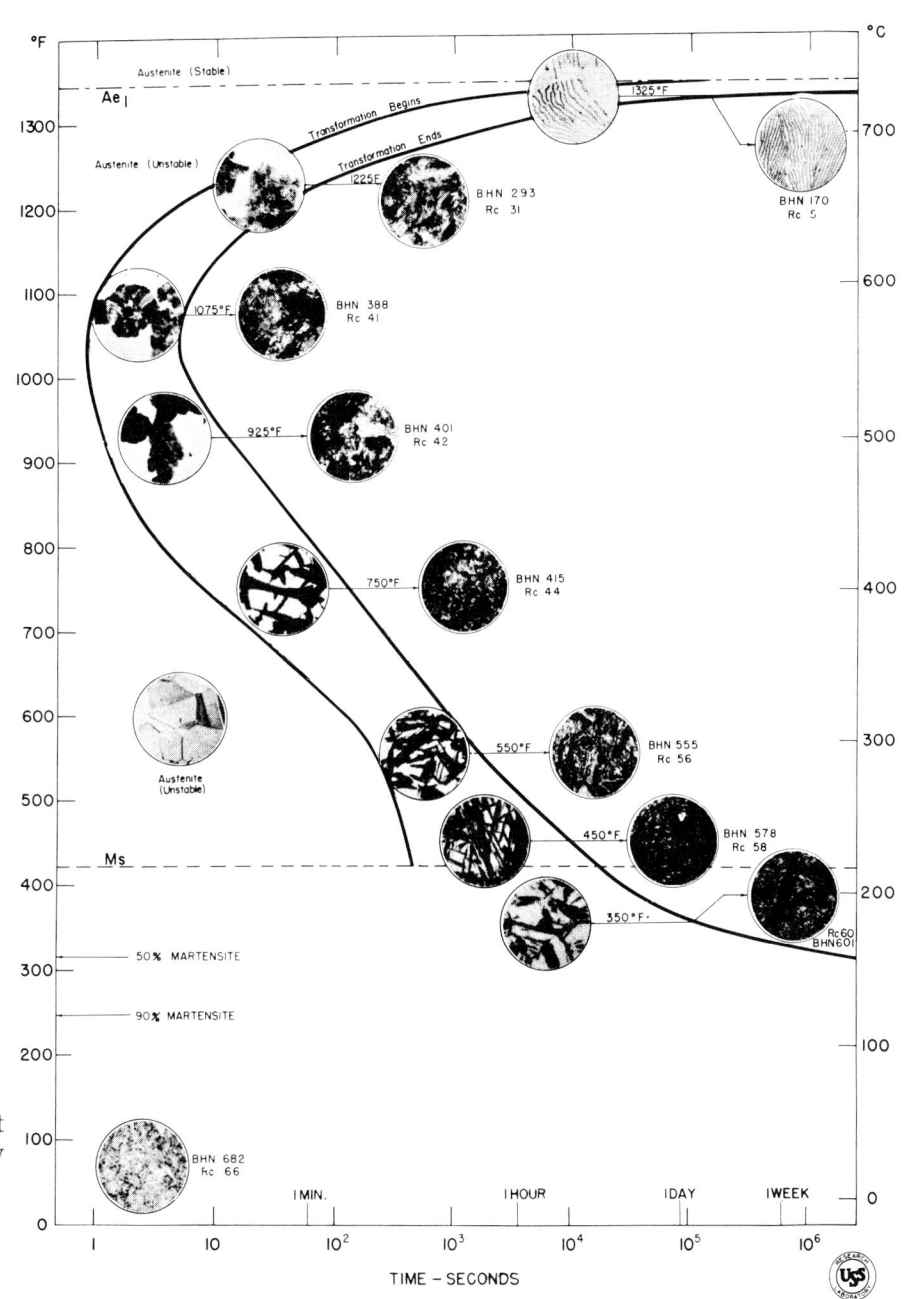

Figure 2. I-T diagram of .89 percent carbon steel (Copyright © 1951 by United States Steel Corporation).

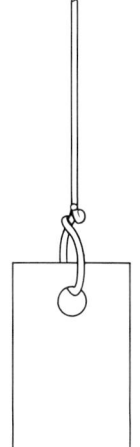

Figure 3. Isothermal test specimen.

represents time. It is plotted on a log scale that corresponds to one minute, one hour, one day, and one week.

COOLING CURVES

If a cooling curve (Figure 5) is superimposed on the I-T diagram, it can be seen that it must pass to the left (1) of the nose of the diagram to avoid an unwanted transformation from occurring. If the cooling rate is too slow, however, the cooling

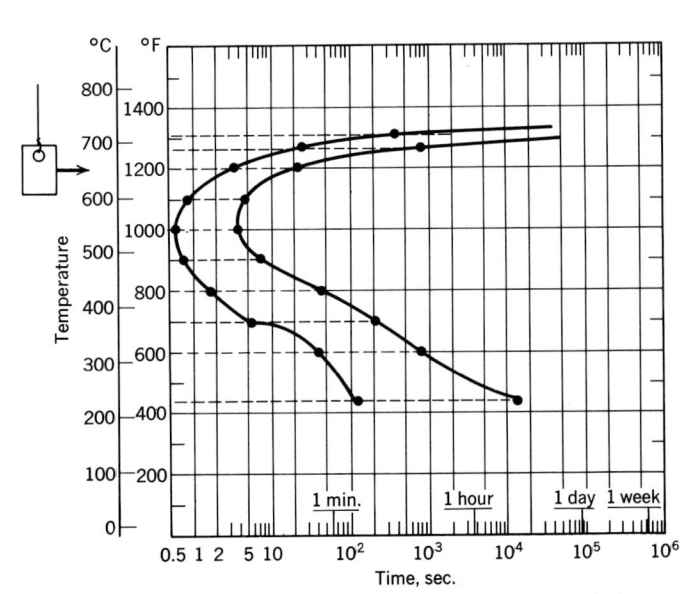

Figure 4. The method of plotting an isothermal diagram (Copyright © 1951 by United States Steel Corporation).

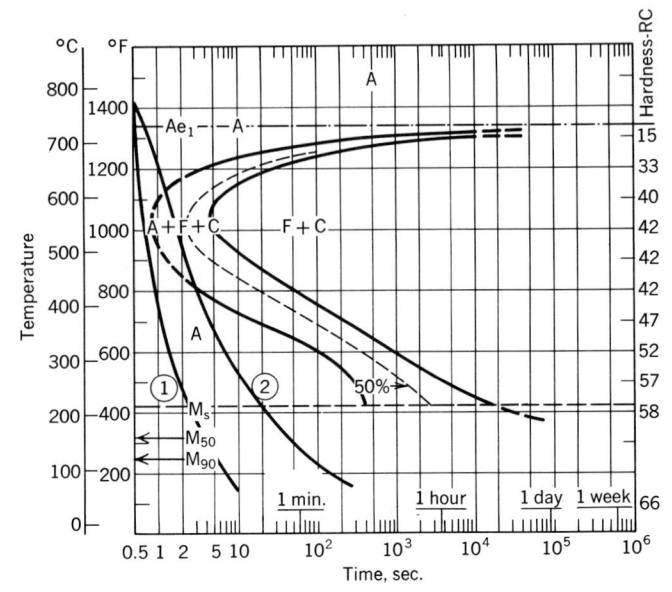

Figure 5. Two cooling curves shown on a SAE 1095 I-T diagram (Copyright © 1951 by United States Steel Corporation).

curve will cut into the nose of the diagram, showing that a partial or complete transformation has taken place at that point (2) and that fine pearlite plus martensite developed in the material instead of the desired martensite. Therefore, the rate of cooling for a hardening quench must be such that the nose of the diagram is well away from the cooling curve.

If the austenite is cooled to temperatures below the nose of the curve and held at these temperatures for a sufficient length of time, a transformation would produce a product that would be called bainite. If the austenite were cooled quickly to a temperature below the Ms line, the product would be called martensite.

The temperature at which martensite begins to form is called the Ms temperature. This transformation to martensite is complete at the Mf temperature. These temperatures vary considerably in steels and are a function of carbon content. The Ms and Mf temperatures are lower for high carbon steels than for low carbon steels.

THE CRITICAL COOLING RATE

Steels can have different cooling rates. An increase in carbon content moves the S-curve to the right or increases the time before transformation takes place. Grain size also has an effect on hardenability. Larger grain carbon steels transform more slowly. This also moves the S-curve to

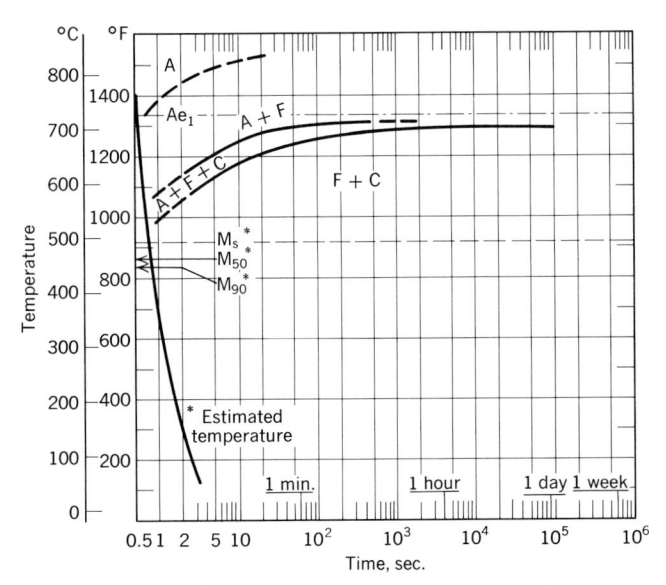

Figure 6. Cooling curve for SAE 1008 carbon steel (Copyright © 1951 by United States Steel Corporation).

the right. The addition of alloy to the steel also moves the S-curve to the right.

A low carbon steel cannot be hardened for practical purposes because the nose of the diagram is at or falls short of the zero time line and it would be impossible to avoid cutting into it with the quenching or cooling curve (Figure 6). However, with carbon steels above .30 percent, it is possible to quench rapidly enough to avoid

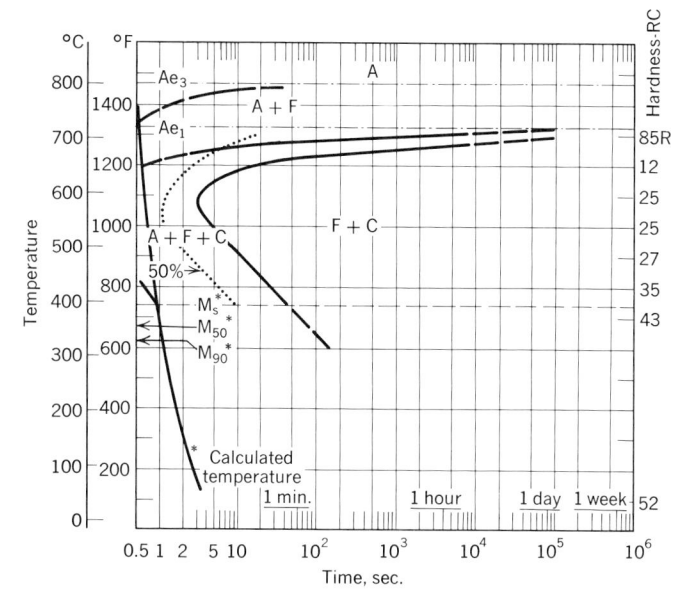

Figure 7. Cooling curve for 1034 modified Mn steel (Copyright © 1951 by United States Steel Corporation).

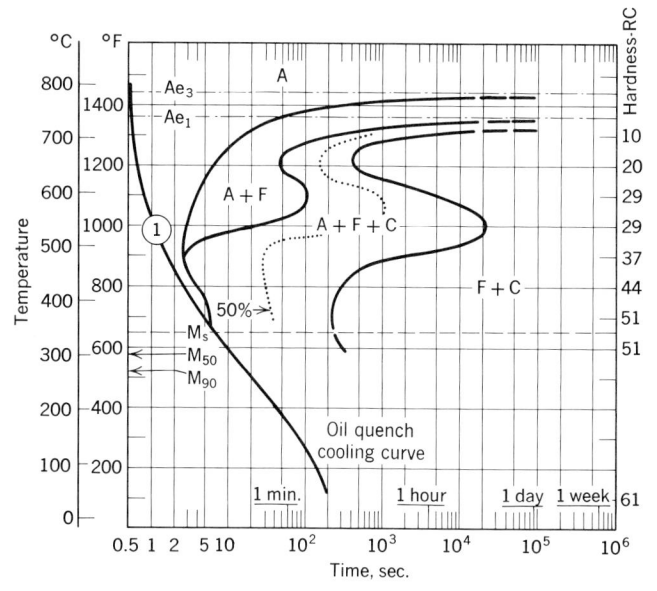

Figure 8. Cooling curve for SAE 4140 steel (Copyright © 1951 by United States Steel Corporation).

transformation to any great extent (Figure 7). Carbon steel of .83 percent must be quenched in water to make the quench rapid enough so that it would take place within the 1 or 2 sec needed to avoid cutting into the nose on the diagram.

Oil hardening steels with alloying elements such as chromium and molybdenum cause the nose of the diagram to move toward the right, thus increasing the time in which hardening can take place. Also, the shape of the nose is often changed on the diagram. These changes often allow a great deal of time for the quench to take place. In these aspects, oil hardening and air hardening (deep hardening) steels can be shown on the I-T diagram (Figure 8).

The surface area of the part and the thickness have a considerable effect on the cooling rates. A very thin part, such as a razor blade, with a large surface area would have a cooling rate that would be many times greater than a cube of steel 2 or 3 in. square. Therefore, a normally water hardened steel, when extremely small or thin, should be quenched in oil to achieve the proper cooling rate, while the steel block being of the same material would still not achieve critical cooling even in salt water.

Water quenched steels normally will harden only to a depth of approximately $\frac{1}{8}$ in., while the core is left quite soft. These are termed shallow hardening steels. A 1 or 2 in. thick, air cooled steel may harden completely to the core. This is

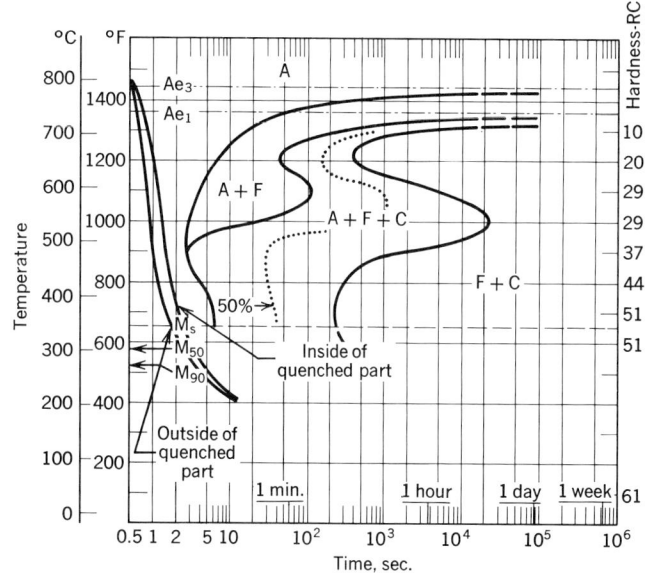

Figure 9. Internal and external cooling curves on the same part on this I-T diagram show how two different cooling rates can cause the high stresses to build up that sometimes result in quench cracks (Copyright © 1951 by United States Steel Corporation).

deep hardening steel. As the critical cooling time increases, the depth of hardening increases.

When quenching rates are drastic, such as in brine or water quench, the stresses set up in the part due to the different cooling rates from the interior and exterior of the part can cause warping and cracking (Figure 9). Water quench steels are particularly prone to this problem. The slower cooling rates of oil and air allow more uniform cooling and, because of this, these steels suffer less cracking and warping. For this reason, when large or heavy sections must be heat treated, the choice of oil or air cooled steels should be made (Figure 10).

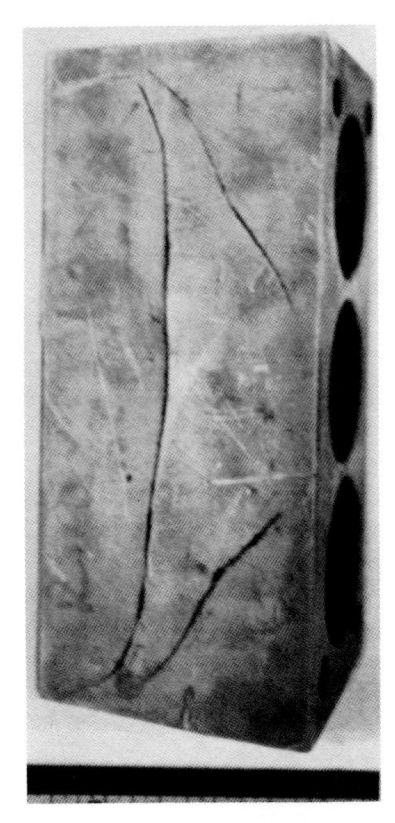

Figure 10. The design of this forming die, made of Type W1 tool steel, presents an almost impossible problem to the heat treater. Because of the blind holes and the thin section between them, the die cracked during heat treatment. Unless the die can be totally redesigned, the use of an air hardening steel is imperative (Photograph courtesy of Bethlehem Steel Corporation).

self-evaluation

SELF-TEST

1. What is the austenitizing temperature for a carbon steel?

2. How is the hard structure, martensite, produced? What is the major consideration when martensite is produced?

3. What are the Ms and Mf temperatures?

4. What is the critical cooling rate?

5. If the cooling curve of a .8 percent carbon steel cut partly through the nose of the I-T diagram, what would the resultant microstructure of the metal be?

6. How does the carbon content affect the position of the S-curve?

Unit 4 I-T Diagrams and Cooling Curves

7. What are the two steps needed in hardening steel to produce useful articles?

8. A thick section of W1 steel developed a quench crack when it was hardened. What do you think caused this? What steps can be taken to correct the problem?

9. How can you determine from studying an I-T diagram of very low carbon steel why little or no martensite can be produced when the steel is quenched from an austenitizing temperature?

10. How can you tell by studying an I-T diagram whether an alloy steel has deep hardening capabilities?

WORKSHEET Given metallurgical equipment such as a specimen mounting press, polishing equipment, abrasive cut-off saw, microscope, and two pieces of $\frac{3}{8}$ inch diameter SAE 1095 water hardening steel 3 inches long, you will:

1. Harden the two specimens by quenching them from the austenitizing temperature. One specimen in water and one in oil.

2. Determine their respective grain structures by microscopic observation.

3. Draw the approximate cooling curves for each quench.

Procedure

1. Heat both specimens to the austenitizing temperature and quench one in oil and one in water. Do not temper them. Make an identifying mark on each one immediately.

2. Use the abrasive cut-off saw with coolant to cut a small section out of each specimen.

3. Encapsulate in plastic and identify the specimen with a permanent mark.

4. Polish and etch $\frac{1}{2}$ to 2 sec with nital.

5. Set up the metallurgical microscope and observe the specimen that was quenched in water. What is its microstructure?

6. Observe the specimen quenched in oil. What is the microstructure?

7. Draw a cooling curve in the isothermal diagram (Figure 11) for each specimen based on the microstructures that you see and compare with those in Figure 2.

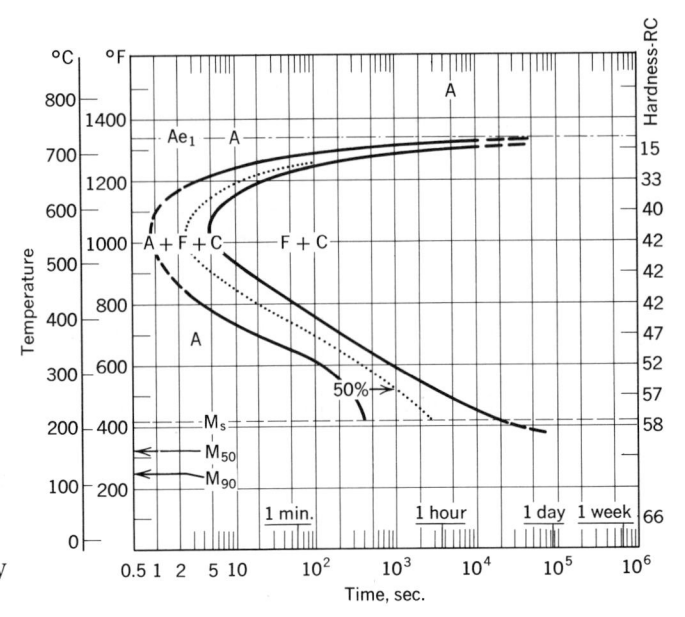

Figure 11. (Copyright © 1951 by United States Steel Corporation).

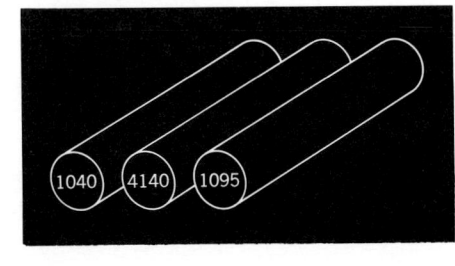

unit 5
hardenability of steels and tempered martensite

If all hardened parts were less than ½ in. thick, then plain water hardening carbon steels would be sufficient for most purposes. The fact is that a large proportion of hardened steel is made into devices having large sections that require deep hardening steels and techniques. The great difference in hardenability of different tool steels can be measured. This unit deals with these procedures.

objectives After completing this unit, you will be able to:
1. Explain the methods of determining and evaluating the depth of hardening (hardenability) of various tool steels.

2. Demonstrate and measure the hardenability of a shallow hardening steel.
3. Demonstrate the use of a mechanical properties chart for predicting the hardness of a hardened and tempered specimen.

information

THE JOMINY END-QUENCH HARDENABILITY TEST

This test is used to determine the depth of hardening of various types of tool steels. In conducting this test a 1 in. round specimen approximately 4 in. long is heated uniformly to the proper quenching temperature. The specimen is placed in a bracket in such a way that a jet of water at room temperature impinges on and is confined to the bottom face of the specimen without wetting the sides (Figure 1). It is allowed to remain in the water jet until the entire specimen has cooled. Later, longitudinal flat surfaces are ground on the side to remove decarburization, and Rockwell C scale readings are taken at $\frac{1}{16}$ in. intervals from the quenched end. Since the quenching effect is concentrated on the end surface and diminishes with the distance from the end, the measurement of hardness from the end corresponds to the measurement of hardenability at that depth of any piece of metal of that particular type. The data secured by this means are plotted on a graph.

From a study of the curves, it becomes apparent that initial surface hardness is a function largely of the carbon content and that depth hardness depends on the amount of carbon present, the alloy content, and the grain size. Manganese, chromium, and molybdenum are the chief elements which promote depth hardness, while nickel and silicon help to a lesser degree.

Figure 1. A Jominy end-quench hardenability test being performed (Courtesy of Pacific Machinery & Tool Steel Co.).

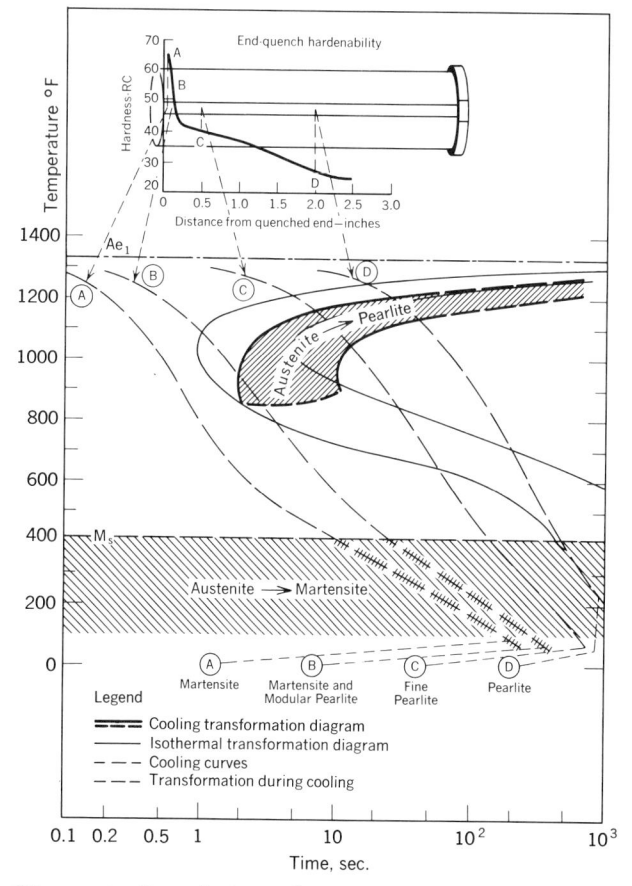

Figure 2. Correlation of continuous cooling and isothermal transformation diagrams with end-quench hardenability test data for eutectoid carbon steel (Copyright © 1951 by United States Steel Corporation).

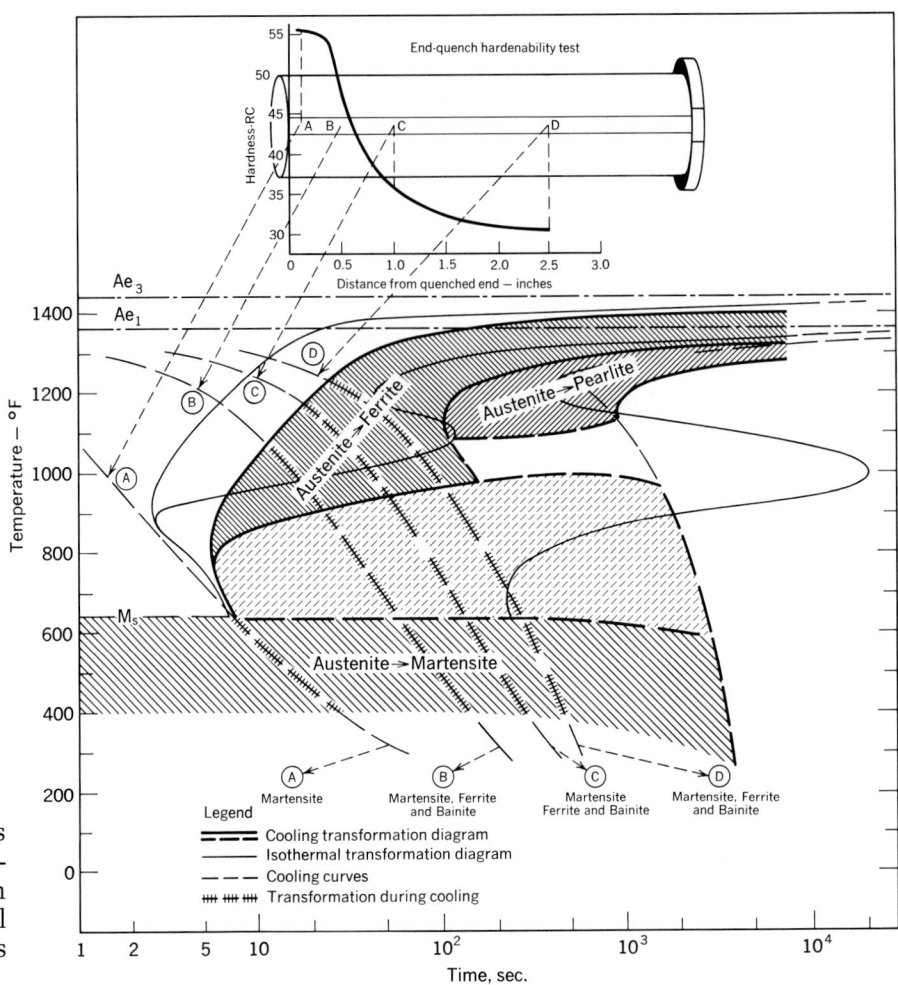

Figure 3. Correlation for continuous cooling and isothermal transformation diagrams with end-quench hardenability test data for 4140 steel (Copyright © 1951 by United States Steel Corporation).

Figures 2 and 3 compare depth of hardening between eutectoid (.83 percent) carbon steel and SAE 4140. Note that the rates of cooling become slower as the distance from the quenched end increases. These differing rates of cooling produce cooling curves in the nose area of the S-curve. It can readily be seen that transformation products other than martensite will be formed at certain distances back from the end (or near the center of an equally thick section). The effect of various types of cooling media on the hardenability or depth of hardening is shown in Figure 4.

The temperature at which martensite begins to form is called the Ms temperature. This temperature can be lowered considerably by increasing the carbon content. When .83 percent carbon steel is quenched to below the Ms temperature, or approximately 400° F (204° C), martensite begins to form. Just above 300° F (149° C), about 50 percent transformation to martensite (Figure 5, line 1) has transpired; just above the Mf temperature about 90 percent transformation has transpired; and at the Mf temperature or approximately 200° F (93° C), about 100 percent transformation has taken place. At this point eutectoid or any higher carbon steel is about as hard as it will get, which is about Rockwell C67.

TRANSFORMATION PRODUCTS

When cooling curves are such that they cut across the S-curve on the I-T diagram in various places, certain microstructures are formed. A soft, coarse pearlite develops when a very slow cooling takes place at about 1300° F (704° C). This would be the case when a part is furnace annealed (Figure 5, line 2). When a part is air cooled after heating in the furnace to 100° F (38° C) above its upper critical temperature, the

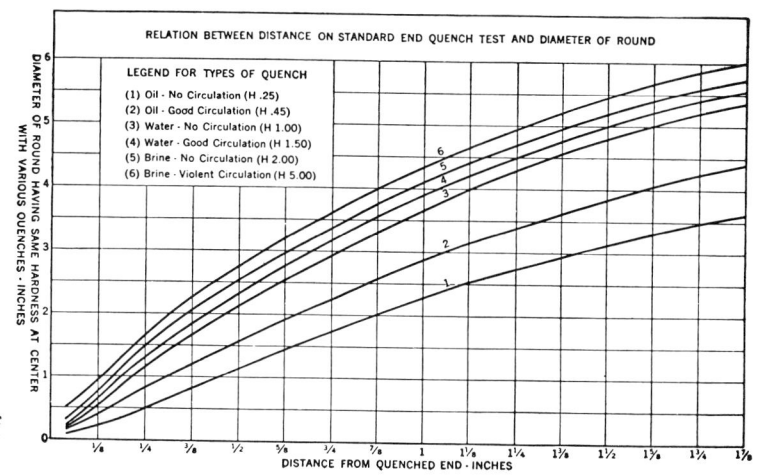

Figure 4. Cooling curves for various quenching media (Courtesy of Pacific Machinery & Tool Steel Co.).

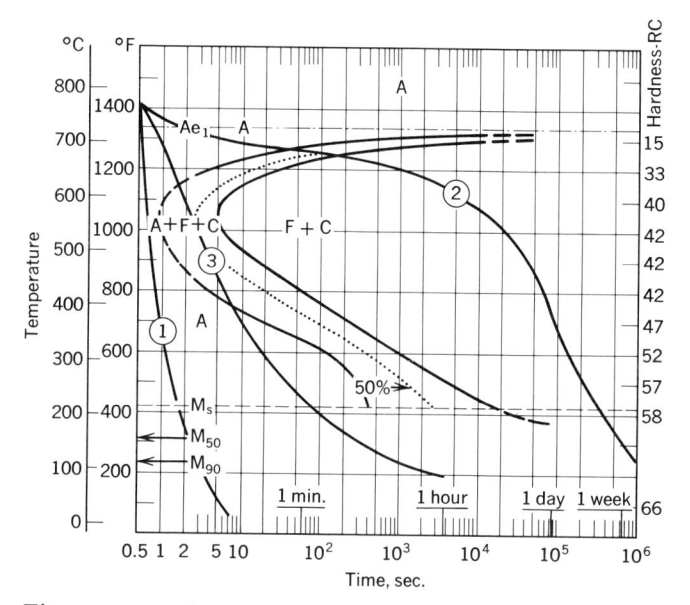

Figure 5. Cooling curves on an I-T diagram for eutectoid steel. Line 1 shows the quench for undercooling to produce martensite; line 2 shows an annealing curve; line 3 is a normalizing curve (Copyright © 1951 by United States Steel Corporation).

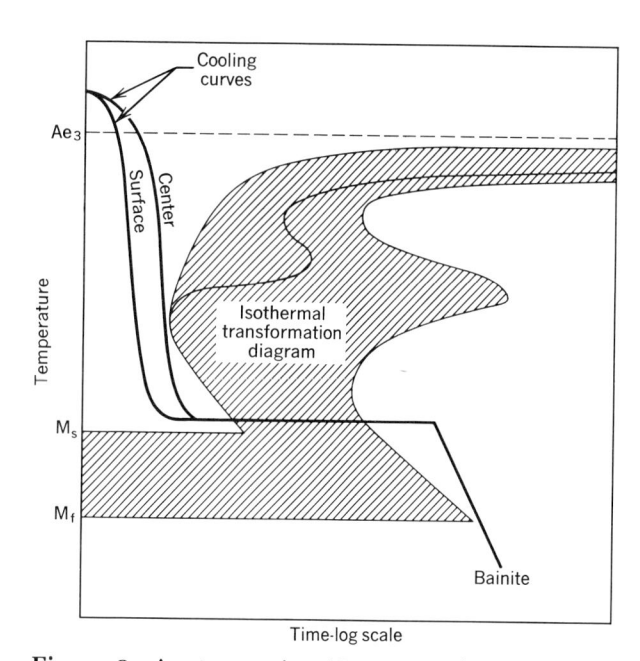

Figure 6. Austempering (Courtesy of Bethlehem Steel Corporation).

process is known as normalizing. The cooling curve for normalizing would be approximately through a medium pearlite or upper bainite section (Figure 5, line 3) of the S-curve in eutectoid steels, leaving a stronger structure than full anneal will produce.

Another method of hardening and tempering is a form of isothermal quenching called austempering (Figure 6), in which a part is austenitized and quenched into a lead or salt bath held at a temperature of approximately 600° F (316° C) to produce a desired microstructure of lower bainite. It is held at this temperature for several hours until a complete transformation has taken place. This type of hardening eliminates the second process that is normally used, which is tempering. Austempering produces a superior product that is much tougher than that developed in the conventional hardening and tempering process. There is one drawback, however; it is confined mostly to small or thin sections. Large, heavy sections cannot be austempered.

Another form of isothermal quenching is

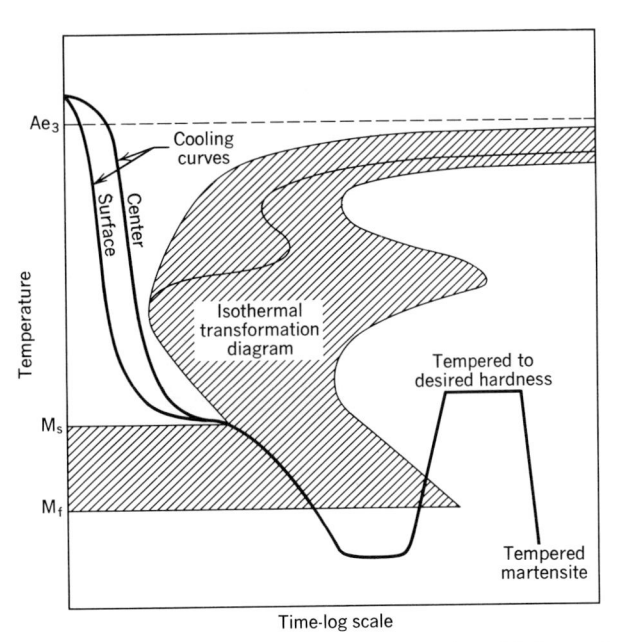

Figure 7. Martempering (Courtesy of Bethlehem Steel Corporation).

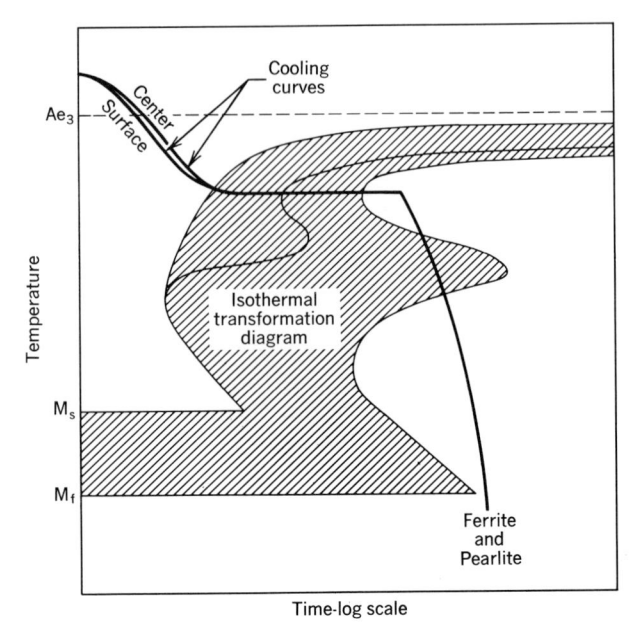

Figure 8. Isothermal annealing (Courtesy of Bethlehem Steel Corporation).

that of martempering (Figure 7) in which the austenitized part is brought to approximately 400° F (204° C) and held for a few minutes in order to bring the interior and exterior temperatures to an equalized temperature to avoid stresses. Then the quench is continued on to under 200° F (93° C) and the conventional tempering is carried out. Isothermal annealing is done by quenching to the desired annealing temperature from the austenitizing temperature and holding at the anneal temperature for a length of time (Figure 8).

TEMPERING PROCEDURES

Tempering, sometimes called drawing or temper drawing, is the process of reheating hardened martensitic steels to some temperature below the lower critical or Ac_1 line. Proper tempering of a hardened steel requires a certain length of time. With any selected tempering temperature the hardness drops rapidly at first and gradually decreases slowly over a period of time. For instance, within about a half an hour of tempering at 600° F (316° C), a part is reduced in hardness about 5 or 6 points on the Rockwell scale. Some heat treaters prefer to temper by use of color and, in this case, they must stop the tempering process when the proper color has been reached by cooling in water, thus causing the tempering time to be very short. This is not the best practice. Carbon steels and most of the low alloy steels should be tempered as soon as they are cool enough to be held in the hands. Carbon steels should not be tempered before they cool to this temperature range because in some steels the M_f temperature is quite low and untransformed austenite might be present. Part or all of this residual austenite will transform to martensite on cooling from the tempering temperature so that the final structure will consist of both tempered and untempered martensite. The brittle untempered martensite, together with the built-in internal stresses, can easily cause failure of the heat treated part.

In most cases toughness increases as the hardness decreases due to the temperature increase of tempering but, if the Izod-Charpy notch bar test is used as a measure of toughness, it has been found that a part tempered between 400° F (204° C) and 800° F (427° C) has a reduced notch toughness (Figure 9). It is true that ductility increases in this tempering range and, if parts and designs include stress raisers, this tempering range should be avoided. This is sometimes called the blue brittle tempering range.

Some alloy steels show a loss in notch toughness when tempered between 1000° F (538° C) and 1250° F (677° C) followed by slow

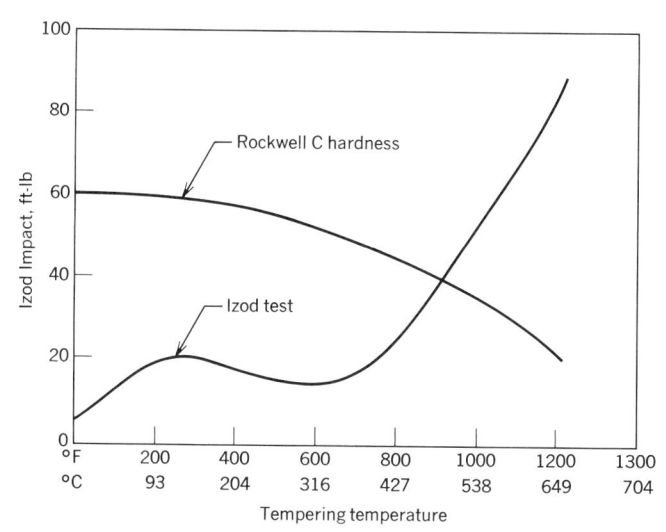

Figure 9. Graph showing the blue brittle tempering range.

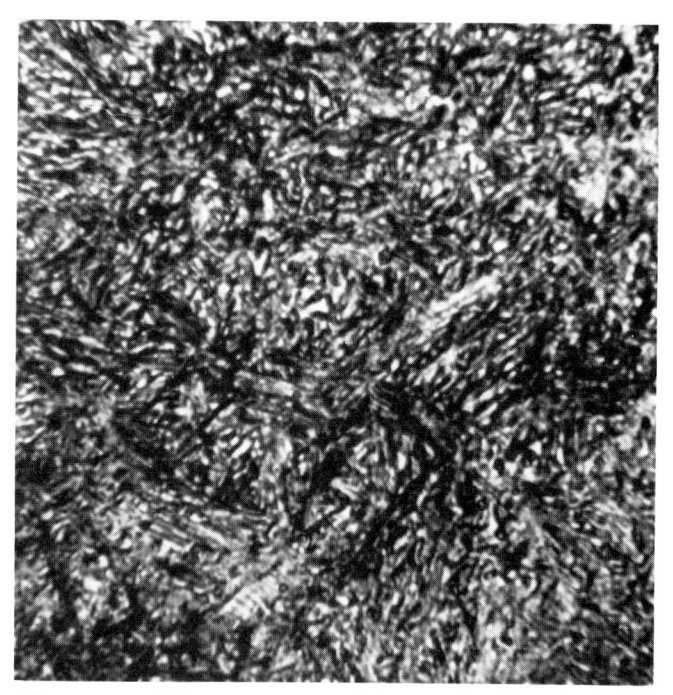

Figure 10. Microstructure of black martensite (500×).

cooling. This is called temper brittleness and can be avoided by quenching from the tempering temperature. Steels high in manganese, phosphorus, and chromium suffer from temper brittleness, while the addition of molybdenum will retard it.

TEMPERED MARTENSITE

Martensite is a supersaturated solution of carbon and iron. As heat energy is added, carbon gets the mobility to escape the iron lattice structure and form iron carbide. Tempering from 100° F (38° C) to 400° F (204° C) leaves low carbon, martensite, and submicroscopic iron carbide particles. This martensite etches dark but still retains the needlelike formation. It is sometimes called black martensite (Figure 10).

Higher tempering temperatures cause transformation of the low carbon martensite, 450° F (232° C) to 700° F (371° C), to lower bainite plus ferrite and an increase in carbide particle size. This etches to a black mass once known as trootsite. At a temperature between 700° F (371° C) and 1200° F (649° C), the carbide particles become visible and the ferrite becomes a more continuous network. This change is responsible for the marked increase in its toughness and ductility. Martensite tempered in this range begins to etch light (Figure 11). This structure was once known as sorbite. Many metallurgists use only the term "tempered martensite" for all these

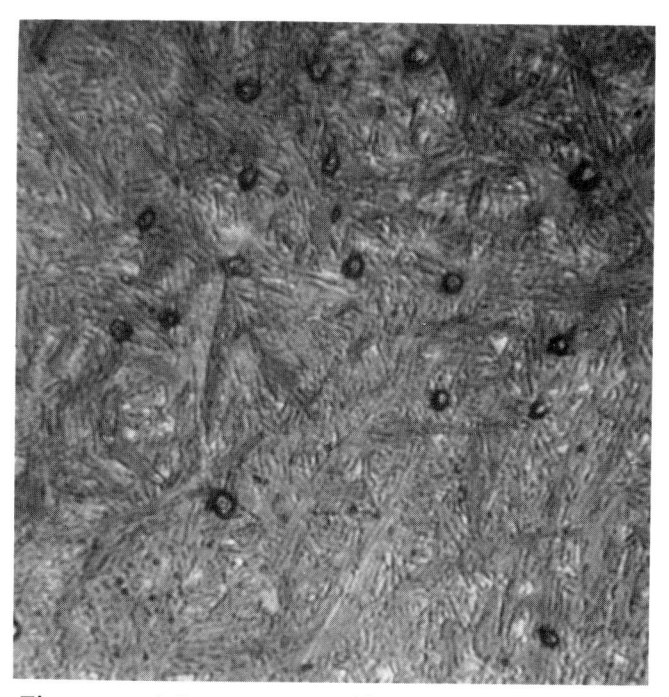

Figure 11. Microstructure of light martensite (500×).

conditions. Martensite and carbide particles (also pearlite), when held in a 1200°F (649°C) to 1300° F (704° C) temperature range, tend to form spheroidite (Figure 12).

Many alloy steels retain a great deal of austenite after quenching and consequently are not fully hardened. The air cool and oil quench steels have this characteristic. After tempering the retained austenite transforms to martensite, causing an increase in hardness. This tendency is so pronounced in the high alloy steels that tempering, sometimes double tempering, is necessary to develop full hardness.

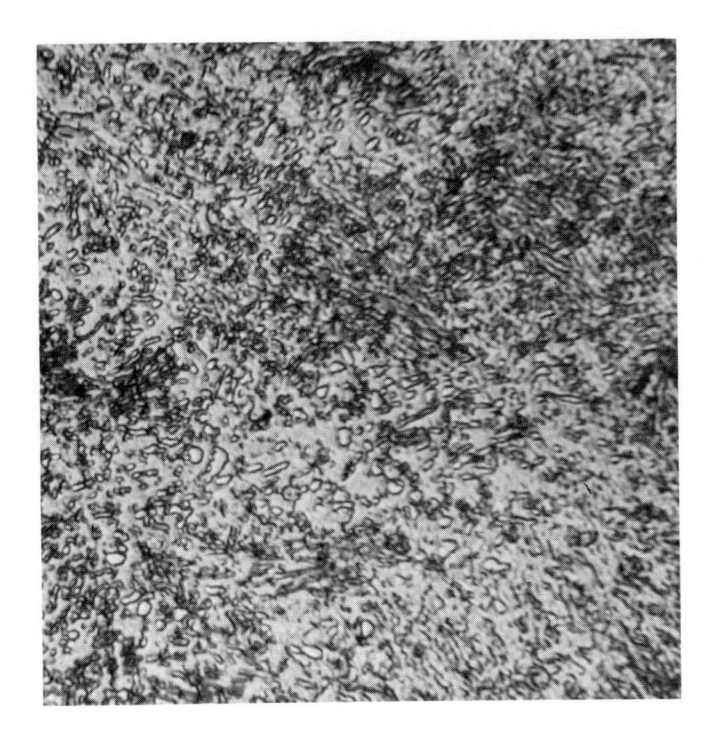

Mechanical properties charts are available in metals handbooks. These charts are useful to the heat treater for determining temperatures for heat treating. Tempering temperatures versus hardness, for instance, are readily determined for any carbon steel by using the mechanical properties chart. (See Figure 14 in Worksheet 2.)

Figure 12. Microstructure of spheroidite (500×).

self-evaluation

SELF-TEST

1. What test is used to determine hardenability?

2. Briefly explain how the test in question 1 is carried out.

3. What relationship does the test referred to in questions 1 and 2 have to the S-curve in the I-T diagram?

4. Refer to the graph in Figure 4. What effect does circulation of the quench seem to have on hardenability?

5. Approximately what is the maximum hardness of an austenitized steel of 1.50 percent carbon when quenched to the Mf temperature?

6. What type of microstructure develops in eutectoid steel when it is furnace annealed?

7. What is austempering? Name one advantage.

Unit 5 Hardenability of Steels and Tempered Martensite 317

8. When is the best time to temper? Explain.

9. Explain the difference between the blue brittle tempering range and temper brittleness in some steels.

10. How can you predict the final hardness of a hardened carbon steel that you are preparing to temper?

WORKSHEET 1 Given one ½ in. diameter 1040 (.40 percent carbon) piece of steel 3 in. long, Rockwell hardness tester, furnace, tongs, and a quenching tub of water, you will:

1. Show the difference in hardness and cooling rate between the surface and center of a ½ in. diameter steel specimen.

2. Demonstrate hardenability depth on a shallow hardening steel.

Procedure 1. Place the specimen in the furnace and set for 1550° F (843° C). Allow the specimen to reach the same color as the furnace bricks. Heat the end of the tongs so they will not quench the specimen when you remove it from the furnace.

2. Remove the specimen and quench it in still water.

3. Cut the specimen in half using the metallurgical cut-off saw. Be careful not to allow overheating of the specimen. Cut a short length, about $\frac{3}{8}$ in. long.

4. Next mark $\frac{1}{16}$ in. increments across the diameter of the freshly cut surfaces.

5. Rockwell test at the $\frac{1}{16}$ in. increments on center (Figure 13).

6. Record your results.

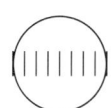

Figure 13. $\frac{1}{16}$ in. spaces

Note If your lab has a Jominy end-quench hardenability tester, make up a standard sample and make tests as explained in this unit.

Conclusion Why is there a difference in hardness between the surface and center?

WORKSHEET 2 Given three specimens of ½ in. round × 1 in. SAE 1095 steel, Rockwell hardness tester, heat treat furnace, water quenching tub, and tongs, you will:

1. Learn the effect of tempering on SAE 1095 carbon steels.
2. Learn to use the mechanical properties chart to estimate the physical properties resulting from tempering.

Procedure
1. Test the as-rolled specimens for hardness and compare with the chart (Figure 14).

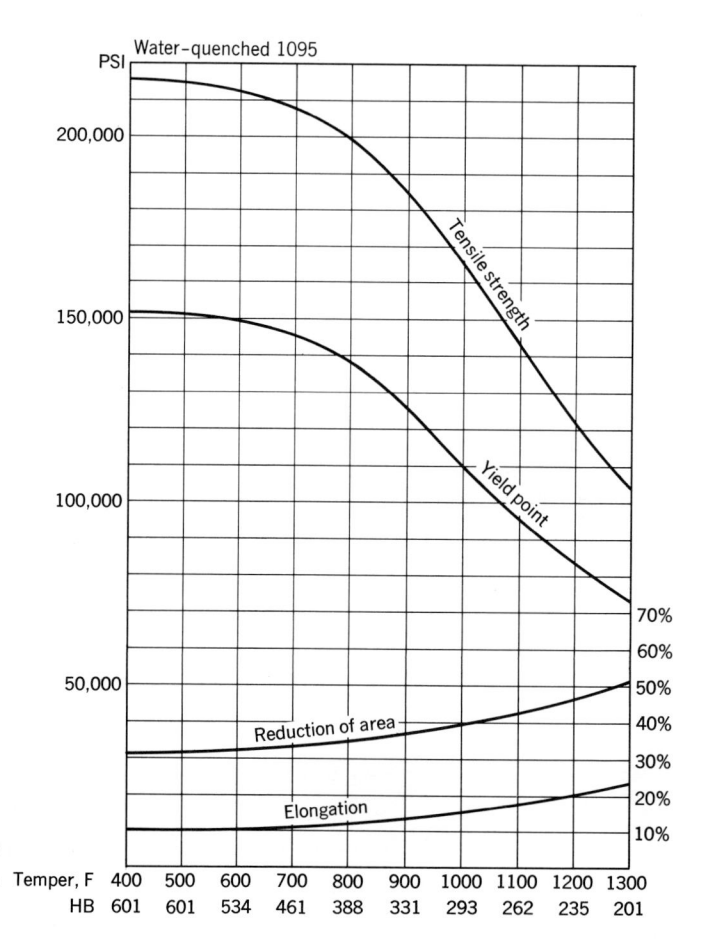

Figure 14. Mechanical Properties chart (Courtesy of Bethlehem Steel Corporation).

Temper, F	400	500	600	700	800	900	1000	1100	1200	1300
HB	601	601	534	461	388	331	293	262	235	201

2. Next place the specimens in a furnace heated to 1550° F (843° C). Allow the specimens to reach the same color as the furnace bricks. The steel should be soaked at the austenitizing temperature for a few minutes before quenching.
3. Heat the end of the tongs that will come in contact with the specimens. This will avoid cooling of the specimens before they can be quenched.
4. Remove the austenitized specimens and quench in water. Agitate the specimens for better quenching.
5. Check the hardness of the as-quenched steel. The mechanical properties chart gives the as-quenched hardness of 1095 steel to be 601 Brinell. Use conversion tables if your hardness tests are Rockwell.
6. Place a hardened specimen in a furnace at 400° F (204° C) for 20 min.

7. Remove, cool, and test for hardness. Repeat this procedure with the remaining specimens, one at 800° F (427° C) and one at 1000° F (538° C).

8. Compare your results with the mechanical properties chart provided.

Conclusion Did the results you acquired correspond with the chart? If not, how can you account for the difference?

unit 6
heat treating steels

Heat treating tool steels is a process made up of some very critical furnace operations. The proper steps must be taken for a particular grade of steel or a failure will almost surely be the result. Study these procedures and follow them so that you will be able to harden and temper your project successfully.

objectives After completing this unit, you will be able to:
1. Describe the proper heat treating procedures for most tool steels.
2. Correctly harden a SAE 4140 vee block or equivalent.
3. Correctly draw temper the SAE 4140 block to a prescribed hardness.

information

FURNACES

Electric, gas, or oil fueled furnaces are used for heat treating steels (Figure 1). They use various types of controls for temperature adjustment. These controls make use of the principle of the thermocouple (Figure 2). Temperatures generally range to 2500° F (1371° C). High temperature salt baths are also used for heating metals for hardening or annealing. One of the disadvantages of most electric furances is that they allow the atmosphere to enter the furnace, and the oxygen causes oxides to form on the heated metal. This causes scale and decarburization of the surface of the metal. A decarburized surface will not harden. One way to control this loss of surface carbon is to keep a slightly carbonizing atmosphere or an inert gas in the furnace.

One of the most important factors when heating steel is the rate at which heat is applied. When steel is first heated, it expands. If cold steel is placed in a hot furnace, the surface expands more rapidly than the still cool core. The surface will then have a tendency to pull away from the center, thus inducing internal stress. This can

Figure 1. Electric heat treating furnace. Part is being placed in furnace by heat treater wearing correct attire and using tongs (Lane Community College).

Figure 2. Thermocouple (Lane Community College).

Figure 3. Input controls on furnace (Lane Community College).

cause cracking and distortion in the part. Most furnaces can be adjusted for the proper rate of heat input (Figure 3) when bringing the part up to the soaking temperature (Figure 4). "Soaking" means holding the part for a given length of time at a specified temperature. Another factor is the time of soaking required for a certain size piece of steel. An old rule of thumb allows the steel to soak in the furnace for 1 hour for each inch of thickness, but there are considerable variations to this rule, since some steels require much more soaking time than others. The correct soaking period for any specific tool steel may be found in tool steel reference books.

QUENCHING MEDIA

In general six media are used to quench metals. They are listed here in their order of severity or speed of quenching.

1. Water and salt; that is, sodium chloride or sodium hydroxide. It is also called brine.
2. Tap water.
3. Fused or liquid salts.
4. Soluble oil and water.
5. Oil.
6. Air.

Liquid quenching media goes through three stages. The vapor-blanket stage occurs first because the metal is so hot it vaporizes the media. This envelops the metal with vapor, which insulates it from the cold liquid bath. This causes the cooling rate to be relatively slow during this stage. The *vapor transport cooling* stage begins

Figure 4. Temperature control (Lane Community College).

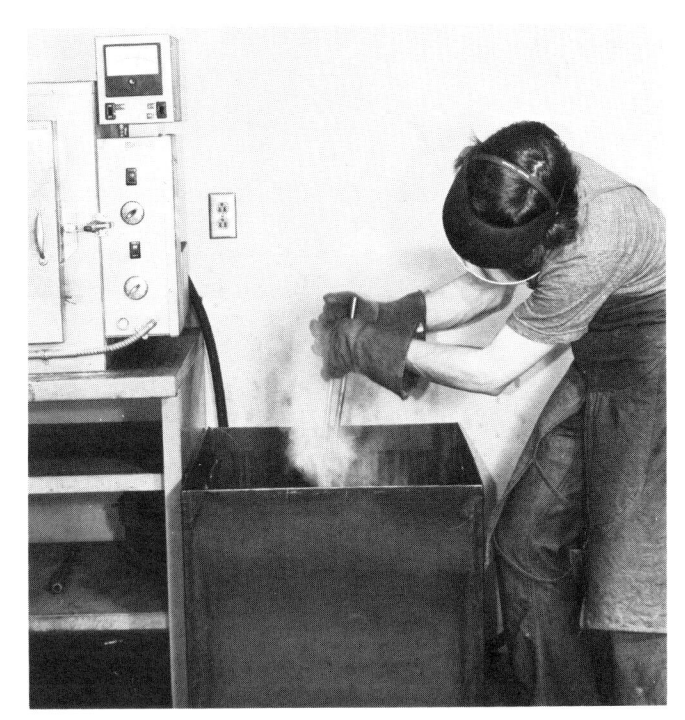

Figure 5. Heat treater is agitating part during quench (Lane Community College).

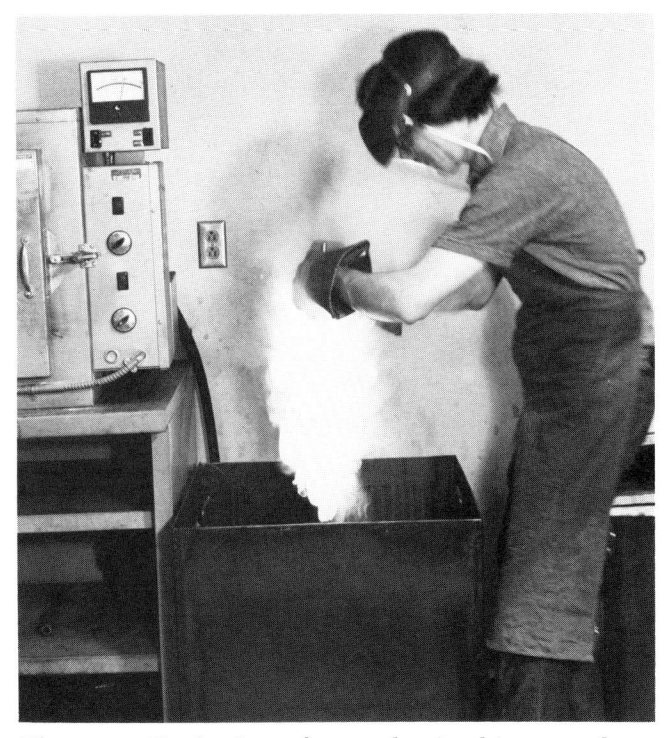

Figure 6. Beginning of quench. At this stage heat treater could be burned by hot oil if he is not adequately protected with gloves and face shield. (Lane Community College).

when the vapor blanket collapses, allowing the liquid medium to contact the surface of the metal. The cooling rate is much higher during this stage. The *liquid cooling* stage begins when the metal surface reaches the boiling point of the quenching medium. There is no more boiling at this stage, so heat must be removed by conduction and convection. This is the slowest stage of cooling.

It is important in liquid quenching baths that either the quenching medium or the steel being quenched should be agitated (Figure 5). The vapor that forms around the part being quenched acts as an insulator and slows down the cooling rate. This can result in incomplete or spotty hardening of the part. Agitating the part breaks up the vapor barrier. An up and down motion works best for long, slender parts held vertically in the quench. A figure eight motion is sometimes used for heavier parts.

Gloves and face protection must be used in this operation for safety (Figure 6). Hot oil could splash up and burn the heat treater's face if he is not wearing a face shield.

Molten salt or lead is often used for isothermal quenching. This is the method of quenching used for austempering. Austempered parts (Figure 7) are superior in strength and quality than those produced by the two-stage process of quenching and tempering. The final austempered part is essentially a fine, lower bainite microstructure (Figure 8). As a rule, only parts that are thin in cross section are austempered.

Another form of isothermal quenching is called martempering; the part is quenched in a lead or salt bath at about 400° F (204° C) until the outer and inner parts of the material are brought to the same uniform temperature. The part is next quenched below 200° F (93° C) to transform all of the austenite to martensite. Tempering is then carried out in the conventional manner.

Steels are often classified by the type of quenching medium that is used to meet the requirements of the critical cooling rate. For example, water quenched steels, which are the plain carbon steels, must have a rapid quench. Oil quench steels are alloy steels, and they must be hardened in oil. The air cooled steels are alloy steels that will harden when allowed to cool from the austenitizing temperature in still air. Air is the slowest quenching medium; however, its cooling rate may be increased by movement (by use of fans, for example).

Step or multiple quenching is sometimes used when the part consists of both thick and thin sections. A severe quench will harden the thin section before the thick section has had a chance to cool. The resulting uneven contraction often results in cracking. With this method the part is quenched for a few seconds in a rapid quenching medium, such as water followed by a slower quench in oil. The surface is first hardened uniformly in the water quench, and time is provided by the slower quench to relieve stresses.

TEMPERING

Furnace tempering is one of the best methods of controlling the final condition of the martensite to produce a tempered martensite of the correct hardness and toughness that the part requires. Tempering should be accomplished immediately after hardening. The part while still warm should be put into the furnace immediately. If it is left at room temperature for even a few minutes, it may develop a quench crack (Figure 9).

A soaking time should also be used when tempering, since it is in hardening procedures and the length of time is related to the type of tool steel used. A cold furnace should be brought up to the correct temperature for tempering. The residual heat in the bricks of a previously heated furnace may overheat the part, even though the furnace has been cooled down.

Some heat treaters, however, prefer to use the color system to determine the required temperature. With this system the tempering process

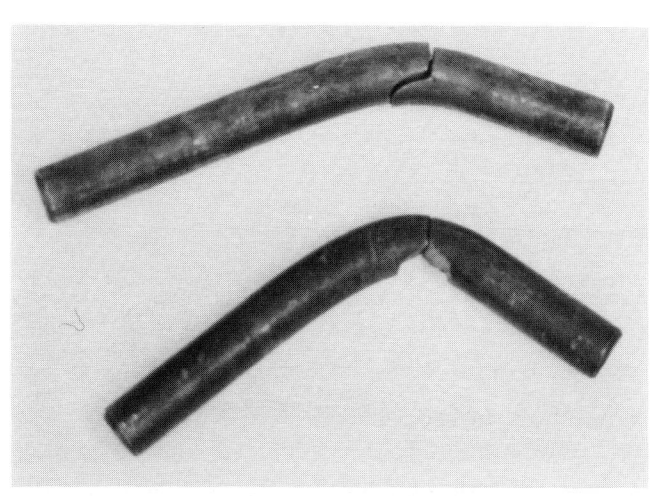

Figure 7. Austempered parts compared to the same kind of part hardened and tempered by the conventional method (Lane Community College).

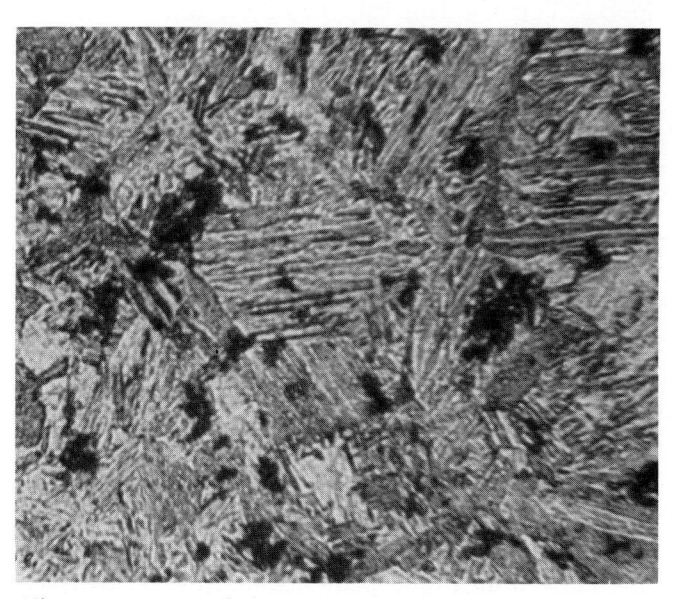

Figure 8. Lower bainite microstructure.

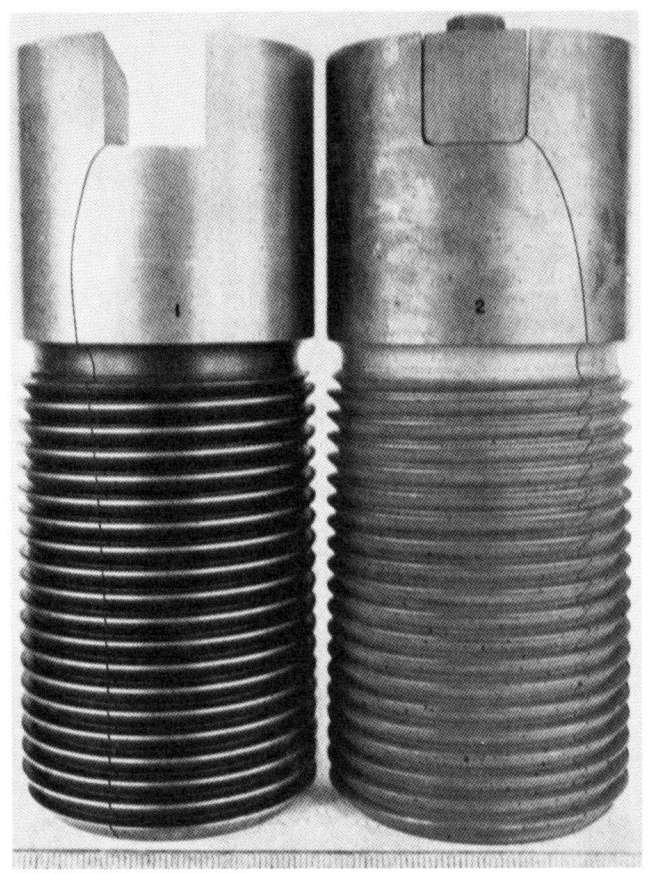

Figure 9. These two breech plugs were made of Type L6 tool steel. Plug #1 cracked in the quench through a sharp corner and was therefore not tempered. Plug #2 was redesigned to incorporate a radius in the corners of the slot, and a soft steel plug was inserted in the slot to protect it from the quenching oil. Plug #2 was oil quenched and checked for hardness (Rockwell C 62) and, after tempering at 900° F (482° C), it was found to be cracked. The fact that the as-quenched hardness was measured proves that there was a delay between the quench and the temper that was responsible for the cracking. The proper practice would be to temper immediately at a low temper, check hardness, and retemper to desired hardness (Photograph courtesy of Bethlehem Steel Corporation).

must cease when the part has come to the correct temperature, and the part must be dropped in water to stop further heating of the critical areas. There is no possible soaking time when this method is used.

Double tempering is used for some alloy steels such as high speed steels that have incomplete transformation of the austenite when they are tempered for the first time. The second time they are tempered, the austenite transforms completely into the martensite structure.

PROBLEMS IN HEAT TREATING

Overheating of steels should always be avoided, and you have seen that if the furnace is set too high with a particular type of steel, a coarse grain can develop. The result is often a poor quality tool, quench cracking, or failure of the tool in use. Extreme overheating causes burning of the steel and damage to the grain boundaries, which cannot be repaired by heat treatment (Figure 10); the part must be scrapped. The shape of the part itself can be a contributing factor to quench failure and quench cracking. If there is a hole, sharp shoulder, or small extension from a larger cross section, a crack can develop in these areas (Figure 11).

A part of the tool being held by tongs may be cooled to the extent that it may not harden. The tongs should therefore be heated prior to grasping the part for quenching (Figure 12). As mentioned before, decarburization is a problem in furnaces that do not have controlled atmosphere (Figure 13). This can be avoided in other ways, such as wrapping the part in stainless steel foil or covering it with cast iron chips.

A proper selection of tool steels is necessary to avoid failures in a particular application. If there is shock load on the tool being used, shock resisting tool steel must be selected. If there is to be heat applied in the use of the tool, a hot work type of tool steel is selected. See Table 1. If distortion must be kept to a minimum, an air hardening steel should be used.

Quench cracks have several characteristics that are easily recognized.

1. In general the fractures run from the surface

Figure 10. This tool has been overheated and the typical "chicken wire" surface markings are evident. The tool must be discarded (Lane Community College).

Figure 11. Drawing die made of Type W1 tool steel shows characteristic cracking when water quenching is done without packing the bolt holes (Photograph courtesy of Bethlehem Steel Corporation).

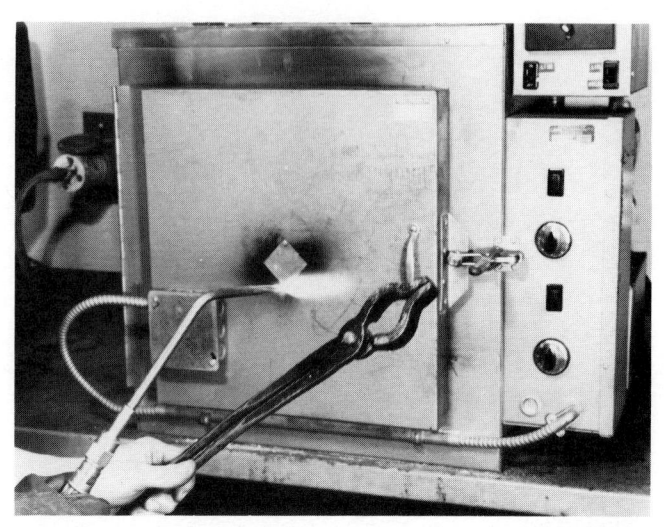

Figure 12. Heating the tongs prior to quenching a part (Lane Community College).

toward the center in a relatively straight line. The crack tends to spread open.

2. Since quench cracking occurs at relatively low temperatures, the crack will not show any decarburization.
3. The fracture surfaces will exhibit a fine crystalline structure when tempered after quenching. The fractured surfaces may be blackened by tempering scale.

Some of the most common causes for quench cracks are:

1. Overheating during the austenitizing cycle causing the normally fine grained steel to become coarse.
2. Improper selection of the quenching medium; for example, the use of water or brine instead of oil for an oil hardening steel.
3. The improper selection of steel.
4. Time delays between quenching and tempering.
5. Improper design. Sharp changes of section such as holes and keyways (Figure 14).
6. Improper angle of the work into the quenching bath with respect to the shape of the part, causing nonuniform cooling.

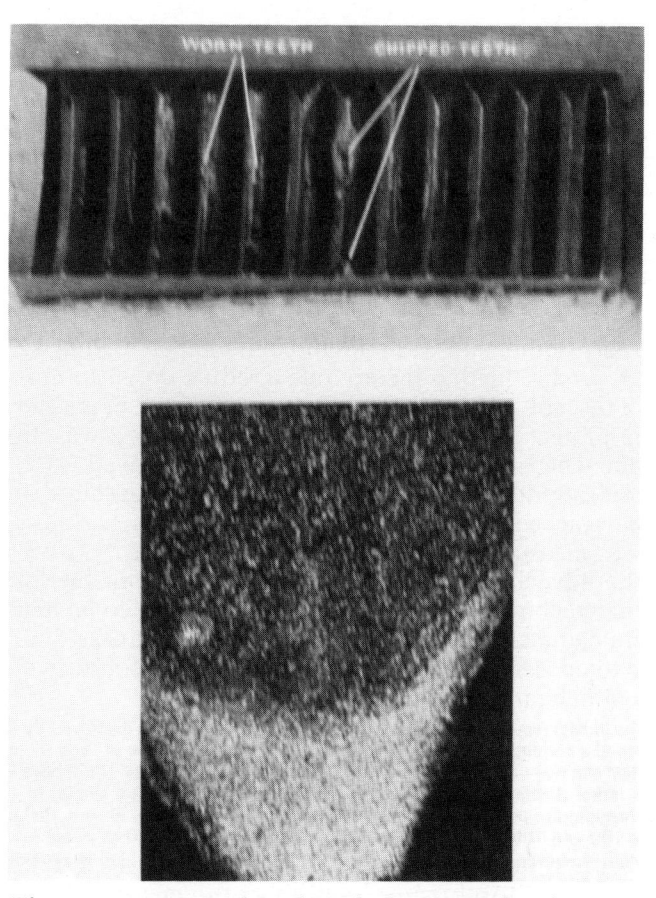

Figure 13. (Top) This thread chaser made of 18-4-1 high speed steel failed in service because of heavy decarburization on the teeth. (Below) Structure of one tooth at 150× magnification shows decarburized structure on the point of the tooth and normal structure below the decarburized zone (Photograph courtesy of Bethlehem Steel Corporation).

Table 1

Mass effect data for SAE 4140 steel. (*Source. Modern Steels and Their Properties*, Seventh Edition Handbook 2757, Bethlehem Steel Corporation, 1972.)

SINGLE HEAT RESULTS

	C	Mn	P	S	Si	Ni	Cr	Mo	
Grade	.38/.43	.75/1.00	—	—	.20/.35	—	.80/1.10	.15/.25	Grain Size
Ladle	.40	.83	.012	.009	.26	.11	.94	.21	7-8

MASS EFFECT

Size Round in.	Tensile Strength psi	Yield Point psi	Elongation % 2 in.	Reduction of Area, %	Hardness HB
Annealed (Heated to 1500 F, furnace-cooled 20 F per hour to 1230 F, cooled in air.)					
1	95,000	60,500	25.7	56.9	197
Normalized (Heated to 1600 F, cooled in air.)					
½	148,500	98,500	17.8	48.2	302
1	148,000	95,000	17.7	46.8	302
2	140,750	91,750	16.5	48.1	285
4	117,500	69,500	22.2	57.4	241
Oil-quenched from 1550 F, tempered at 1000 F.					
½	171,500	161,000	15.4	55.7	341
1	156,000	143,250	15.5	56.9	311
2	139,750	116,750	17.5	59.8	285
4	127,750	99,250	19.2	60.4	277
Oil-quenched from 1550 F, tempered at 1100 F.					
½	157,500	148,750	18.1	59.4	321
1	140,250	135,000	19.5	62.3	285
2	127,500	102,750	21.7	65.0	262
4	116,750	87,000	21.5	62.1	235
Oil-quenched from 1550 F, tempered at 1200 F.					
½	136,500	128,750	19.9	62.3	277
1	132,750	122,500	21.0	65.0	269
2	121,500	98,250	23.2	65.8	241
4	112,500	83,500	23.2	64.9	229

As-quenched Hardness (oil)

Size Round	Surface	½ Radius	Center
½	HRC 57	HRC 56	HRC 55
1	HRC 55	HRC 55	HRC 50
2	HRC 49	HRC 43	HRC 38
4	HRC 36	HRC 34.5	HRC 34

7. Failure to specify the correct size material to allow for cleaning up the outside decarburized surface of the bar before the final part is made.

It is sometimes desirable to stress relieve the part before hardening it. This is particularly appropriate for parts and tools that have been highly stressed by heavy machining or by prior heat treatment. If they are left unrelieved, the residual stresses from such operations may add to the thermal stress produced in the heating cycle and cause the part to crack even before it has reached the quenching temperature.

There is a definite relationship between grinding and heat treating. Development of surface temperatures ranging from 2000° F (1093° C) to 3000° F (1649° C) are generated during grinding. This can cause two undesirable effects on hardened tool steels: development of high internal stresses causing surface cracks to be formed, and changes in the hardness and metallurgical structure of the surface area.

Figure 14. (Top) Letter stamp made of Type S5 tool steel, which cracked in hardening through the stamped O. The other two form tools, made of Type T1 high speed steel, cracked in heat treatment through deeply stamped + marks. Stress raisers such as these deep stamp marks should be avoided. Although characters with straight lines are most likely to crack, even those with rounded lines are susceptible (Photograph courtesy of Bethlehem Steel Corporation).

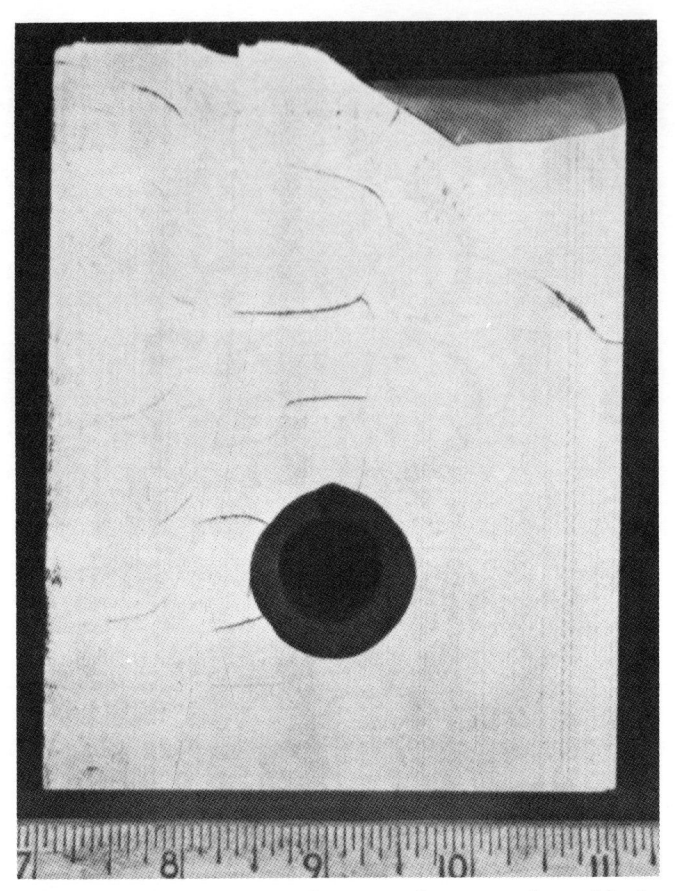

Figure 15. Severe grinding cracks in a shear blade made of Type A4 tool steel developed because the part was not tempered after quenching. Hardness was Rockwell C 64, and the cracks were exaggerated by magnetic particle test. Note the geometric scorch pattern on the surface and the fracture that developed from enlargement of the grinding cracks (Photograph courtesy of Bethlehem Steel Corporation).

One of the most common effects of grinding on hardened and tempered tool steels is that of reducing the hardness of the surface by gradual tempering where the hardness is lowest at the extreme surface but increases with distance below the surface. The depth of this tempering varies with the amount or depth of cut, the use of coolants, and the type of grinding wheel. If high temperatures are produced locally by the grinding wheel and the surface is immediately quenched by the coolant, a martensite having a Rockwell hardness of C65 to 70 can be formed. This gradiant hardness, being much greater than that beneath the surface of the tempered part, sometimes causes very high stresses that contribute greatly to grinding cracks. Sometimes grinding cracks are visible in oblique or angling light, but they can be easily detected when present by the use of magnetic particles of fluorescent particle testing.

When a part is hardened but not tempered before it is ground, it is extremely liable to stress cracking (Figure 15). Faulty grinding procedures can also cause grinding cracks. Improper grinding operations can cause tools that have been properly hardened to fail. Sufficient stock should be allowed for a part to be heat treated so that grinding will remove any decarburized surface on all sides to a depth of .010 to .015 in.

self-evaluation

SELF-TEST

1. Name three kinds of furnaces used for heat treating steels.

    a. _____

    b. _____

    c. _____

2. What can happen to a carbon steel when it is heated to high temperatures in the presence of air (oxygen)?

    _____

3. Why is it absolutely necessary to allow a soaking period for a length of time (that varies for different kinds of steels) before quenching the piece of steel?

    _____

4. Why should the part or the quenching medium be agitated when you are hardening steel?

    _____

5. Which method of tempering gives the heat treater the most control of the final product, by color or by furnace?

    _____

6. Describe two characteristics of quench cracking that would enable you to recognize them.

    a. _____

    b. _____

7. Name four or more causes of quench cracks.

    _____

8. In what ways can decarburization of a part be avoided when it is heated in a furnace?

    _____

9. Describe two types of surface failures of hardened steel when it is being ground.

    a. _____

b. _____

10. When distortion must be kept to a minimum, which type of tool steel should be used?

_____

WORKSHEET Given two heat treating furnaces, an oil quenching bath, and accessories, plus a previously machined set of SAE 4140 steel vee blocks, you will:

1. Correctly harden the vee blocks (Figure 16) or an equivalent project.

2. Correctly temper the vee blocks to RC 48 to 52 (HB 470 to 514).

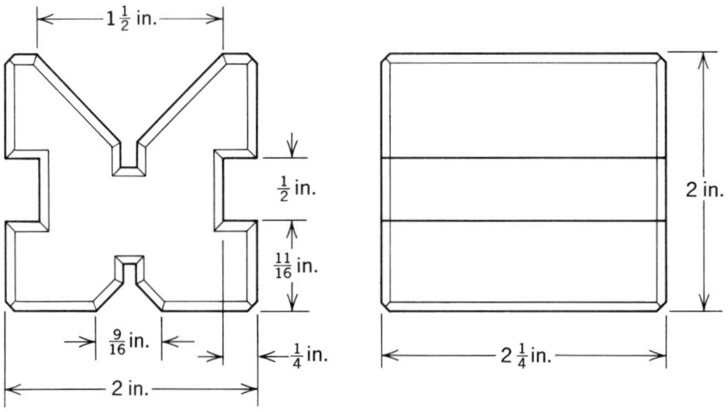

Figure 16. Vee blocks. Soaking times for hardening and tempering operations are based on the smallest cross section of the part. About ¾ to 1 in. may be considered as the least cross section of this vee block.

Procedure

1. Determine the correct hardening temperature from the mass effect data in Table 1.

2. Set the furnace thermocouple control to the correct temperature and turn it on.

3. With the tongs, place the vee blocks in the furnace on a fire brick.

4. Set the second furnace to the correct tempering temperature and turn it on. Consult the SAE 4140 mechanical properties chart (Figure 17) to determine this temperature to obtain a draw temper hardness of RC 48 to 52 (HB 470 to 514).

5. After the vee blocks are the same color as the furnace, allow a soaking time of 1 h per inch of least cross section (about ¾ to 1 in. or ¾ to 1 h).

6. Using face shield and gloves, heat the tongs on the gripping end. Remove one vee block from the furnace, close the door, and quickly plunge the block into the oil bath, agitating the block until it has cooled. It should still be warm to the touch. Repeat the process with the second vee block.

7. Place the two warm vee blocks immediately into the tempering furnace and hold them at that temperature for ½ h.

8. Remove the blocks and allow them to cool in air.

9. Check for hardness. Because of possible decarburization, the readings may be low. Check again after surface grinding.

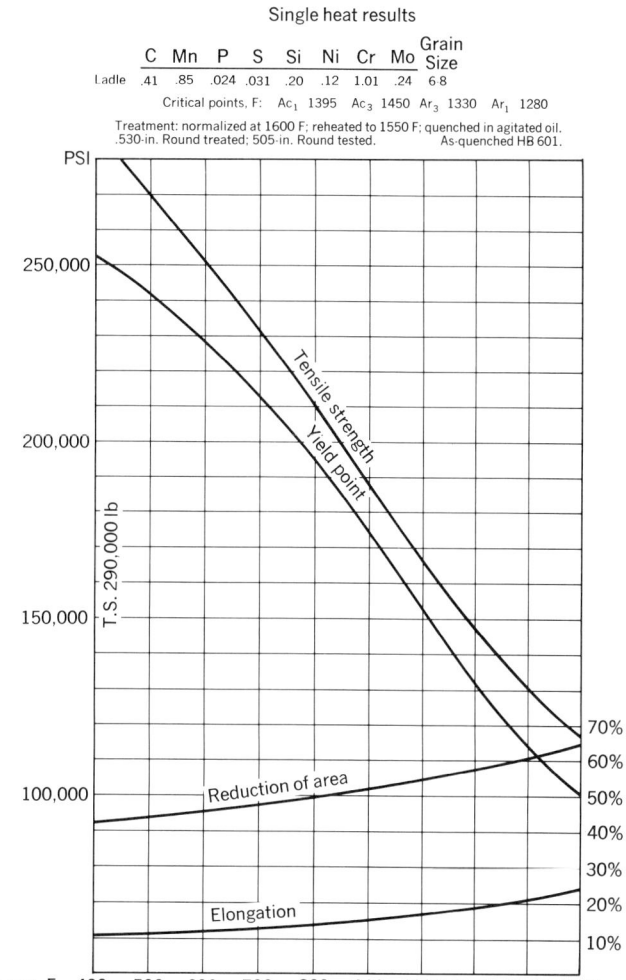

Figure 17. Mechanical properties chart for SAE 4140 steel (Courtesy of Bethlehem Steel Corporation).

Single heat results

	C	Mn	P	S	Si	Ni	Cr	Mo	Grain Size
Ladle	.41	.85	.024	.031	.20	.12	1.01	.24	6 8

Critical points, F: Ac_1 1395 Ac_3 1450 Ar_3 1330 Ar_1 1280

Treatment: normalized at 1600 F; reheated to 1550 F; quenched in agitated oil. .530-in. Round treated; .505-in. Round tested. As-quenched HB 601.

T.S. 290,000 lb

Temper, F	400	500	600	700	800	900	1000	1100	1200	1300
HB	578	534	495	461	429	388	341	311	277	235

Conclusion Is your vee block the same hardness that the mechanical properties chart indicated it would be at your selected tempering temperature? If not, what reason can you give for the discrepancy?

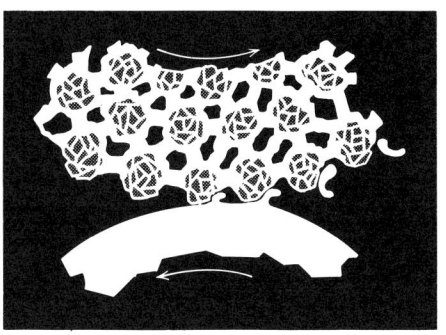

section h
grinding machines

It has been several thousand years since man first discovered that he could brighten up and sharpen his tools by rubbing them against certain stones or by plunging them into the sand several times.

About the middle of the nineteenth century, with the development of mechanical holding devices and of circular, rotating abrasive wheels, this became a precision process. Grinding received another push forward around 1900 with the discovery of silicon carbide and aluminum oxide, manufactured abrasives that freed grinding from the whims of natural abrasives.

But grinding still remained a polishing and finishing process, following some other primary metal cutting process, until only a few years ago. Then, in the development called *abrasive machining*, metal cutting with abrasives was rounded out to a full range of cutting, from the very fast metal removal to very fine surface finishing with extremely tight tolerances. It should be noted, of course, that rough snagging or grinding of castings and billets, to remove unwanted metal, had begun very soon after the discovery of artificial abrasives, but this was basically to get rid of excess metal. Abrasive machining, on the other hand, is usually considered to involve the removal of substantial amounts of stock to some kind of dimensional tolerance. Final finishing to exact dimensional tolerances may be done on the original grinder by changing the method of dressing the wheel or by changing wheels or, depending on the requirements for the part, by changing to another machine.

This concept of using abrasives (grinding wheels or coated abrasive belts — from raw piecepart to finished part) is still relatively new and has a long way to go to reach general shop acceptance. High stock removal by abrasive machining is still generally considered to be for the high production shops with big, powerful machines and high speed wheels. However, it is also true that almost any grinder, even the little 6 × 12 in. surface grinders found in almost every machine shop, is capable, with proper selection of grinding wheel and coolant, of removing more stock than it is usually called on to do. One of the French names for a grinding machine is *machine à rectifier*, which can be translated literally as a machine to correct something done wrong, or perhaps not done completely. Many shop owners and managers, whether or not they

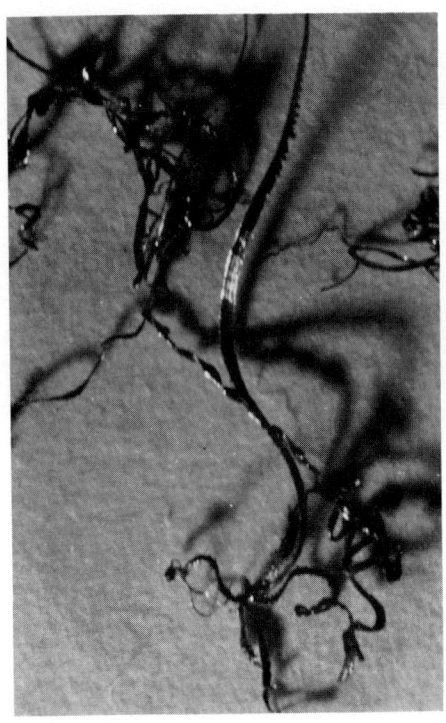

Figure 1. Chips from grinding, magnified about 350× (Courtesy of Bay State Abrasives, Dresser Industries, Inc.).

know the French, would agree with the idea. There are several aspects of the broad development of abrasive processes that need some elaboration.

First, these stages of development cannot be regarded as neatly separated time frames that can be separated and pigeonholed. They are anything but that. Although the farm grindstone, as a circular piece of natural sandstone with a hole in the middle and mounted on a wooden frame with a crank for hand operation, has all but disappeared from the American farm scene, it is a certainty that they are still found in many other parts of the world.

Second, it cannot be overemphasized that grinding and other abrasive processes are *cutting* processes. In spite of the name abrasive, which suggests rubbing and removal of stock by friction instead of by cutting, a grinding wheel or a coated abrasive is a cutting tool similar in function to a milling cutter or a planer tool. The significant difference is that instead of taking off large chips with one tooth or a few teeth, the grinding wheel or the belt takes off tiny chips with thousands or maybe millions of tiny cutting edges (Figure 1). Some idea of the number may be gleaned from the fact that grinding wheels operate at speeds of from 1 to 3 mi/min (6,000 to 18,000 surface feet per minute) and that a single abrasive grain usually measures less than $\frac{1}{8}$ in. in any direction. This means that every minute of contact with the grinding wheel, an abrasive cutting area from 6,000 to 18,000 ft long and as wide as the wheel, is passing over the area to be cut. Even with each bit of abrasive (called abrasive grain) taking only a very small chip, the possibilities for stock removal are considerable.

DEVELOPMENT OF GRINDING MACHINES AND GRINDING WHEELS

By 1860, a grinding machine capable of grinding cylinders and tapers had been developed (Figure 2). This design of machine made it possible to get both accurate dimensions and a superior finish on cylindrical parts. It represents a major step in the development of machine tools. By 1876, the concept of the cylindrical grinding machine was well developed (Figure 3) and had essentially all of the motions of the modern universal grinder. With this machine it was possible to grind steep tapers and cylindrical and tapered workpieces.

The grinding of flat surfaces also received the attention of the machine builders of the period. While the early cylindrical grinder was an offshoot of the lathe family, the early surface grinder was a development based on the milling machine (Figure 4).

The grinding wheels of this period were quite primitive, and many experiments were under way to develop more adequate wheels. Natural abrasives like corundum (an impure form of aluminum oxide) could not be depended on to provide consistent results from batch to batch of wheels, and the problems of holding the abrasive grains in an adequate bond were huge.

In the 1880s, Charles Norton suggested that a grinding wheel a foot in diameter and an inch in thickness could be used to an advantage. For his projection, he was ridiculed by his associates but, by 1900, his concept was embodied in a production cylindrical grinding machine (Figure 5). Norton's production grinder became a

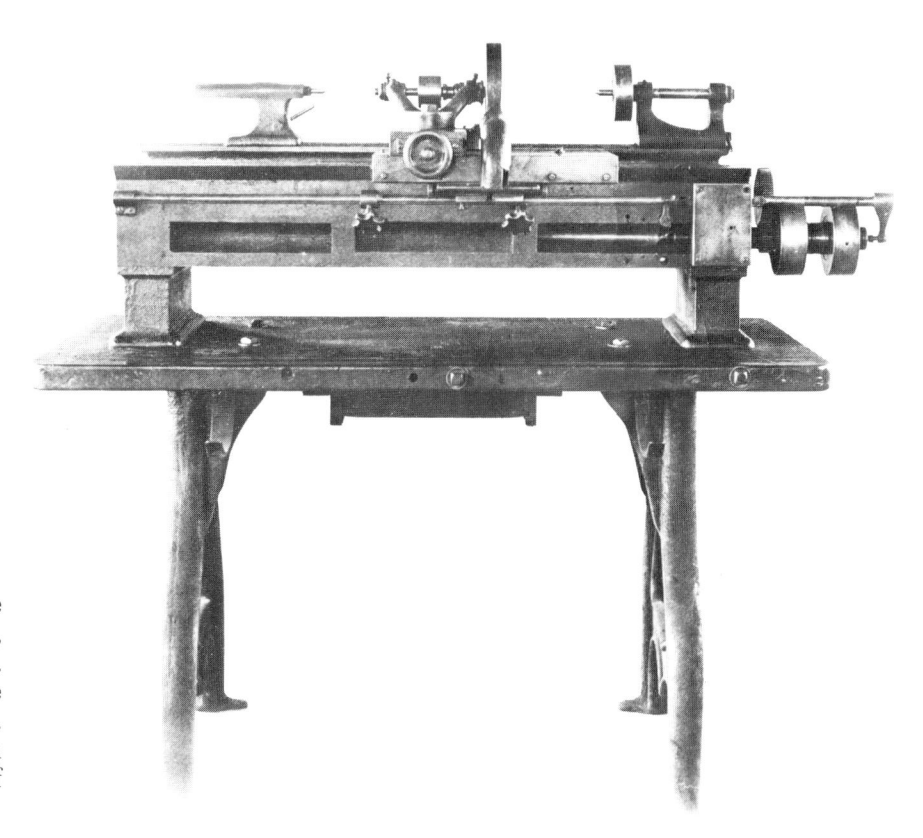

Figure 2. This is a back view of one of the earliest cylindrical grinders, made for the watchmaking industry. Note the mechanism that moves the grinding head along the workpiece. This is the same principle used with modern roll grinders (Courtesy of Norton Company).

reality because of the development of manufactured abrasives in the 1890s and the development of adequate bonds for holding the abrasive grains together.

In 1891, Dr. Edward Acheson heated some powdered coke and clay in a crude electric furnace and obtained crystals of silicon carbide (Figure 6), which would cut glass. At first he was able to produce only a few ounces per day, which sold for over $800 per pound, for the gemstone polishing market. Silicon carbide is now produced by the ton in large electric furnaces (Figure 7) and sells for a few cents per pound.

In 1897, Charles Jacobs was the principal in the discovery of aluminum oxide (Figure 8), which is made by fusing natural alumina or bauxite in an electric furnace and crushing the resulting glassy mass.

The idea of cementing or bonding together bits of abrasive to make a "solid" wheel developed early in the nineteenth century. Rubber appears to have been the first material tried, although it was soon outdistanced by vitrified or clay-type bonds that used a furnacing technique similar to that for making dishes and that is still the major bond, certainly in precision grinding. Silicate of soda, known as silicate bond, which required only low-temperature firing, and oxychloride bond, a mixture of oxide and chloride of magnesium, which makes a cold-setting cement, were also popular in the nineteenth century, but they have gone out of use now. In 1876, indeed, oxychloride was acclaimed as the strongest and best bond of all. Shellac came along about 1880 for certain fine finishing applica-

Figure 3. The Brown & Sharpe universal grinding machine of 1876 (Courtesy of Brown & Sharpe Manufacturing Company).

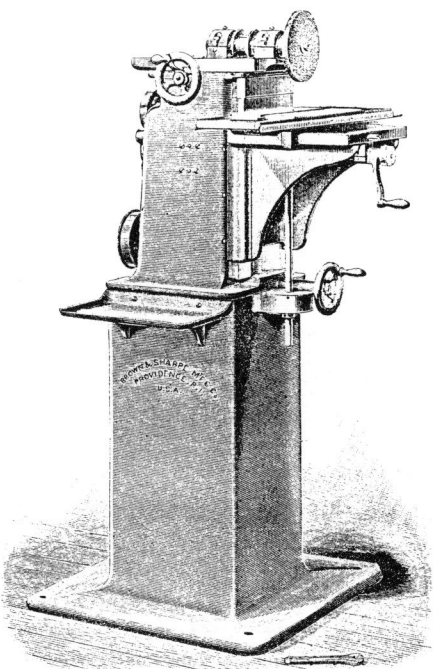

Figure 4. This surface grinder of 1887 was a close relative of the horizontal milling machine (Courtesy of Brown & Sharpe Manufacturing Company).

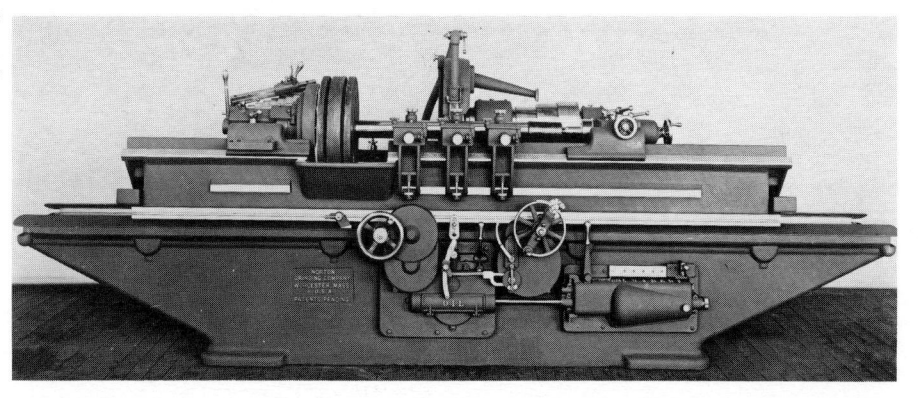

Figure 5. Production cylindrical grinding machine of 1900 (Courtesy of Norton Company).

tions, and it retains a limited popularity for finishing. Resin or resinoid bond, which has found wide use in foundry and other rough grinding applications, did not come along until about 1920.

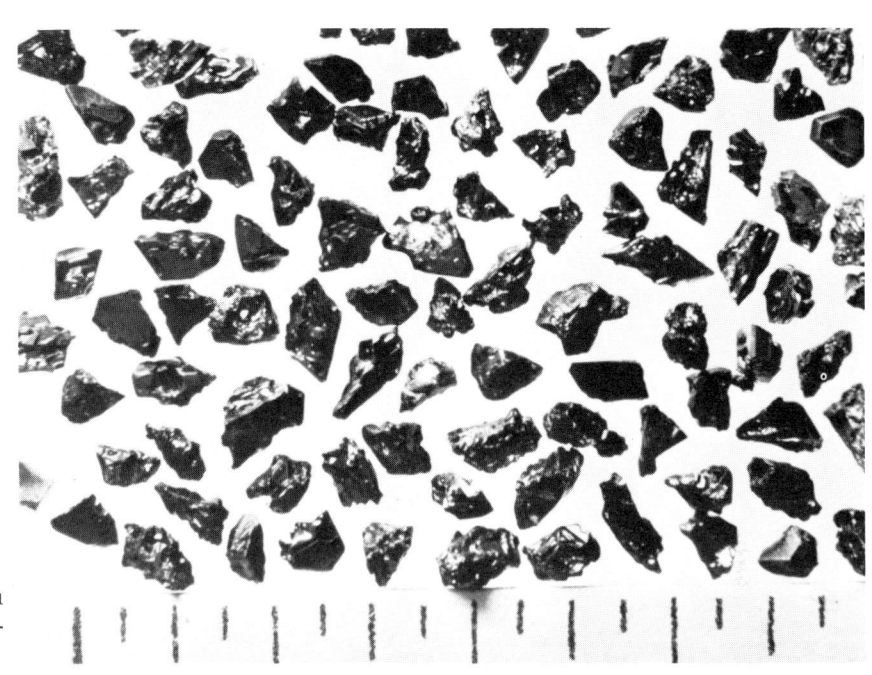

Figure 6. Large grains of silicon carbide (Courtesy of Exolon Company).

Figure 7. Silicon carbide production furnace where tons of coke, sand, sawdust, and salt are fused into the abrasive by electric heat in the range of 4000° F (Courtesy of Norton Company).

Today, it is likely that vitrified bonds and resinoid bonds share at least 90 percent of the market, with rubber and shellac taking most of the remainder. Nothing else is of much consequence. It is true, however, that there are many variations of each of these general groups, particularly in vitrified and resinoid bonds.

Some of the early grinding wheels were referred to as "solid" emery or corundum wheels. Possibly this was meant to indicate only that the wheel had been made without a core of steel or other such material; it certainly could not have suggested that the wheel was really solid and without pores. Every grinding wheel has some open

Figure 8. Large grains of aluminum oxide (Courtesy of Exolon Company).

Figure 9. Besly disc grinder of 1904. A coated abrasive disc is used as a grinding surface (Courtesy of Bendix Corporation).

spaces, particularly vitrified bonded wheels. In fact, vitrified wheels are porous enough that one way of feeding coolant is through a hollow wheel mount and then out through the wheel to the grinding zone. This does ensure that the coolant gets right into the grinding zone. With a flood coolant, even though it may *look* as though there is plenty of coolant covering the wheel-work contact area, the fan-like action of the wheel, revolving at 6000 surface feet and more per minute, may simply be blowing the coolant away, so that you are actually "grinding dry with coolant," as one authority puts it.

Once the development of consistent abrasives in commercial quantities and the problems in developing satisfactory bonding methods were solved, the development of highly productive grinding machines emerged quickly. The emergence of this class of production machine tool made the mass production of automobiles feasible. One of the early high production machines was the disc grinder (Figure 9). With this type of machine, thin parts, like piston rings, for example, could be surface ground for thickness quickly and inexpensively. Other developments of the period included magnetic workholding for surface grinding machines, with the invention of the magnetic chuck by O.S. Walker in 1896.

The need for production machines to satisfy the needs for high production of accurate highly finished parts for the emerging automotive industry led to rapid-fire developments in machine tools of all classes. One of these machines was the centerless grinding machine (Figure 10). This machine, invented about 1915 by L.R. Heim, was especially useful for the uniform sizing and finishing of automotive parts such as piston pins. The principle of the centerless grinder is quite interesting. The rubber bonded regulating wheel

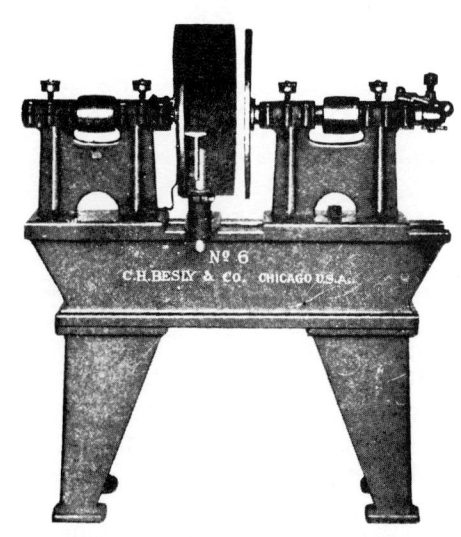

Figure 10. A modern centerless grinding machine used for grinding the OD of parts at high production rates (Courtesy of Cincinnati Milacron).

turns in the same direction as the grinding wheel, but at a much slower speed. The angle on top of the work rest blade causes the workpiece to be rotated (and effectively braked from spinning) by the regulating wheel so that there is a large differential speed between the work and the grinding wheel. It is also important that the center of the workpiece be above the common centers of the grinding and regulating wheels to obtain roundness on the part (Figure 11).

Another interesting grinding machine that emerged in the early 1900s was the vertical spindle rotary table grinder (Figure 12). With this machine, the grinding is done on the flat side of a cylinder wheel (Figure 13). In large models of this type of machine (Figure 14), the grinding is done by segments of abrasive held in a "chuck" in the form of a wheel attached to the spindle. Originally, this type of machine was used mainly for the grinding of glass optical parts, but it has developed to the point where it is one of the best ways to grind large volumes of metal on parts adapted for rotary table workholding. Rapid metal removal, on the order of $\frac{1}{2}$ in., is common on the larger machines.

SURFACE GRINDERS Since our historical development of grinding machines and grinding wheels closes on the vertical spindle rotary table grinder, it would be a good time to deal with other types of grinders, called surface grinders. The most basic form is the horizontal spindle, reciprocating table-type grinder (Figure 15), called Type I surface grinder. This is the type that most people think of when the words surface grinder are mentioned. There are certainly more of this type in existence that the other types (other than pedestal grinders) com-

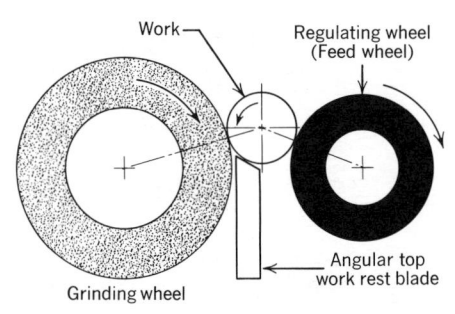

Figure 11. Principle of the centerless grinder. The grinding wheel travels at normal speeds and the regulating wheel travels at a slower speed to control the rate of spin of the workpiece (Courtesy of Bay State Abrasives, Dresser Industries, Inc.).

Figure 12. Blanchard vertical rotary surface grinder circa 1925 (Courtesy of Cone-Blanchard Machine Company, Windsor, Vermont).

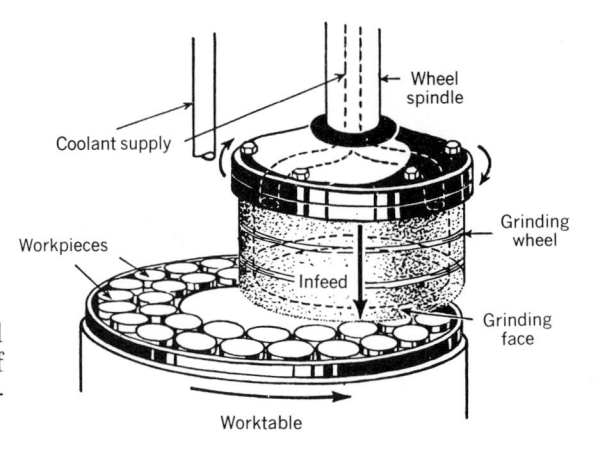

Figure 13. Principle of the vertical spindle rotary grinder (Courtesy of Bay State Abrasives, Dresser Industries, Inc.).

bined. The work is usually held by a magnetic chuck and is traversed under the rotating wheel with the table. The table in turn is mounted on a saddle, which provides cross motion of the table under the wheel. In some designs, the grinding head with the wheel, is moved across the work surface instead of carrying the table on a saddle (Figure 16). The size of these machines can vary greatly from as small as 4 by 8 in. surface grinding area to 6 by 16 ft and larger (Figure 17). The great majority of this type of grinder is 6 by 12 in.

Figure 14. A large vertical spindle rotary table grinding machine (Courtesy of Mattison Machine Works).

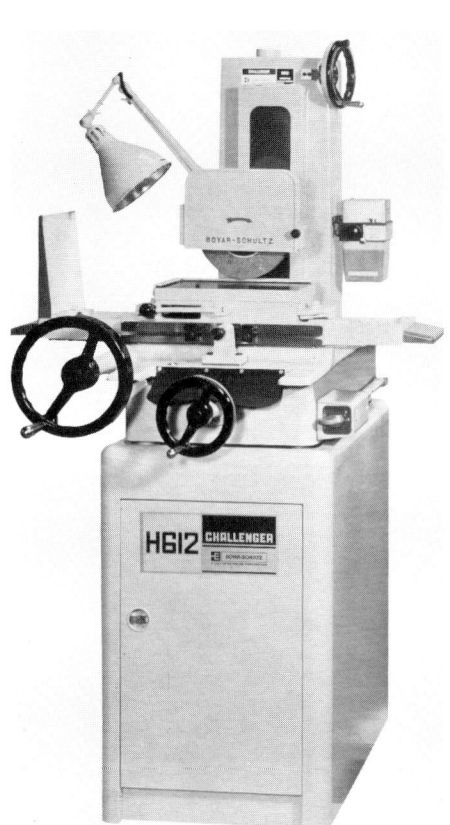

Figure 15. Type I surface grinder (Courtesy of Boyar-Schultz Corp.).

Figure 16. Principle of the Type I surface grinder with alternative method of cross feeding motion (Courtesy of Bay State Abrasives, Dresser Industries, Inc.).

Figure 17. A large Type I surface grinder (Courtesy of Mattison Machine Works).

Figure 18. Grinding contours with the horizontal spindle surface grinder. The workpiece is directly under the wheel and the diamond-plated form dressing block is just in front of it (Courtesy of Engis Corporation).

Figure 20. Principle of the Type II surface grinder. Sometimes circular parts are centered on this grinder; the resulting concentric scratch pattern is excellent for metal-to-metal seals of mating parts (Courtesy of Bay State Abrasives, Dresser Industries, Inc.).

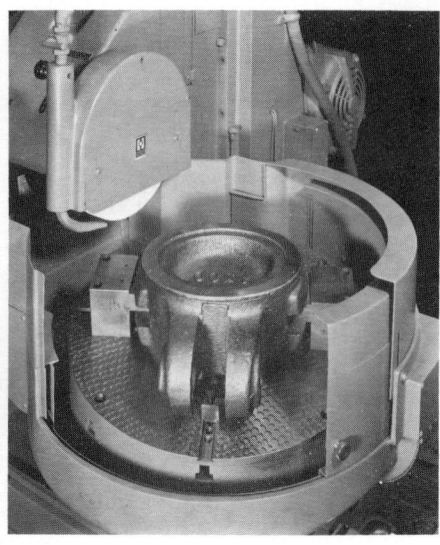

Figure 19. Horizontal spindle rotary table surface grinder (Type II). With the workpiece centered on the chuck, the wheel will produce a concentric scratch pattern. The table can also be tilted, making it possible to grind a part thinner in the middle than at the rim or vice versa (Courtesy of Heald Machine Division/Cincinnati Milacron Company).

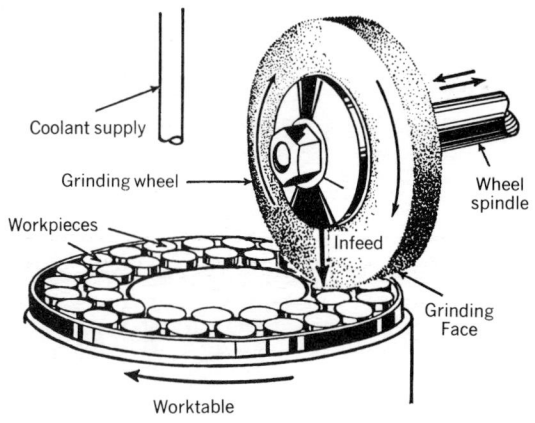

Most of the grinding on the Type I surface grinders is to produce a flat plane on the workpiece, usually to flatnesses less than .0002 in. overall. But this type of grinder can also be used to create contours in the work. The wheel can be dressed to the reverse form of that desired in the workpiece and the contour then ground into the part (Figure 18). This is covered in more detail in Unit 7 of this section.

The horizontal spindle rotary table grinder is another interesting form of the surface grinder (Figure 19). This is sometimes referred to as a Type II surface grinder. The motions of this type grinder are shown in Figure 20. Not shown in this figure is that the table's rotational axis may be tilted a few degrees for operations like the hollow grinding of circular saws. When sealing joints are ground in this fashion (Figure 19), the resulting circular scratch pattern makes for exceptionally good sealing.

A third type of surface grinding machine employs a vertical spindle and a reciprocating table (Figure 21). One form of this Type III design is the way grinder (Figure 22), which is well adapted for long and norrow work like the grinding of ways on other machine tools. These grinders are typically fitted with auxiliary spindles so

Section H Grinding Machines 341

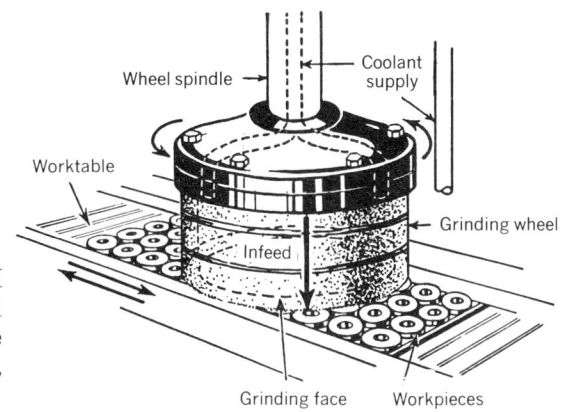

Figure 21. Principle of the Type III surface grinder, which has a vertical spindle and a reciprocating table (Courtesy of Bay State Abrasives, Dresser Industries, Inc.).

Figure 22. A combination way and surface grinder. This machine can remove a great deal of stock to good accuracy on machine ways and other long workpieces, usually eliminating hand scraping (Courtesy of Mattison Machine Works).

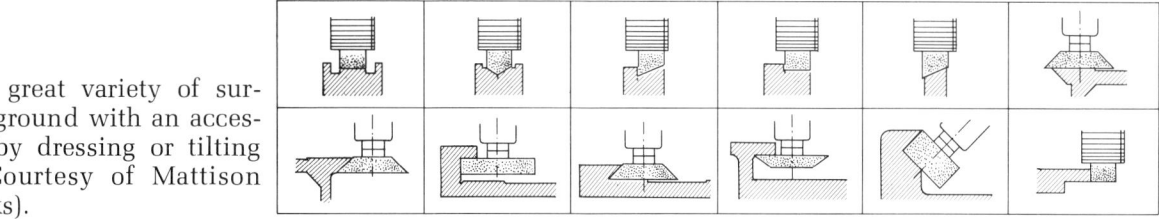

Figure 23. A great variety of surfaces may be ground with an accessory spindle by dressing or tilting the wheel (Courtesy of Mattison Machine Works).

that the complete configuration of the ways can be completed in a single machine setup (Figure 23).

A fourth and relatively uncommon type is the face grinder, which typically employs a segmented wheel mounted on a horizontal spindle, so that the end of the spindle is presented to the part. This machine is especially suited to the surfacing of wide vertical surfaces.

CYLINDRICAL GRINDERS The term cylindrical grinder covers a wide assortment of grinding machine tools, including those that grind workpieces mounted on centers; extremely heavy workpieces mounted on bearing journals;

centerless grinding; and internal grinding, either with the part held in a chuck or in a centerless fashion.

Center-Type Cylindrical Grinders The most basic form of cylindrical grinding is done with the workpiece mounted on centers (Figure 24). The operating principles of this type of machine are shown in Figure 25. For accuracy, the workpiece is rotated on dead centers on both ends, with the workpiece being driven by a plate that rotates concentric to the headstock center. The plain cylindrical grinder is capable of grinding tapered parts as well by the swiveling of the table around a vertical axis like the table of the universal milling machine. It is also capable of being used in a "plunge" mode, where the workpiece and the wheel are brought together **without** table traverse.

A number of design variations exist with the center-type cylindrical grinder, the most common being the **universal** cylindrical grinder (Figure 26). With this type of grinder, the wheel head assembly can be rotated to present the grinding wheel to the work at an angle. This makes the grinding of steep tapers easy to do. You will see more details of this type of machine in the units that follow. The grinding of slender workpieces on center-type grinders of either plain or universal design can be done with the application of accessory support (Figure 27) in the form of a steady rest.

Another interesting variation of the plain center-type cylindrical grinding machine employs a grinding wheel set at an angle to the centers (Figure 28). These machines are especially suited to grinding into shoulders, especially where the relationship between the diameter and the face is critical (Figure 29). This type of grinder is also capable of traversing the table in the same way as the plain and universal types. The angular wheel-type cylindrical grinding machine is typically a production grinder and, in recent years, many production type cylindrical grinders have been supplied with automatic controls (Figure 30) to improve productivity and to reduce operator errors.

Form grinding is also possible on center-type cylindrical grinders (Figure 31). In this type of grinding, the reverse of the form to be imparted into the workpiece is dressed into the grinding wheel, and then the part is ground by a direct plunge of the wheel into the

Figure 24. Working area of a plain cylindrical grinder (Photo courtesy of Diamond Abrasives Corporation).

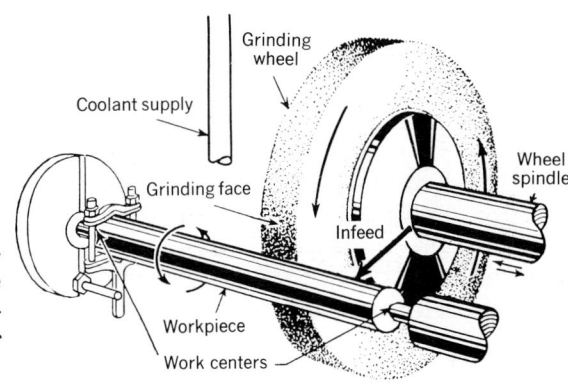

Figure 25. Principle of the cylindrical grinding machine showing the workpiece and wheel motions (Courtesy of Bay State Abrasives, Dresser Industries, Inc.).

Section H Grinding Machines 343

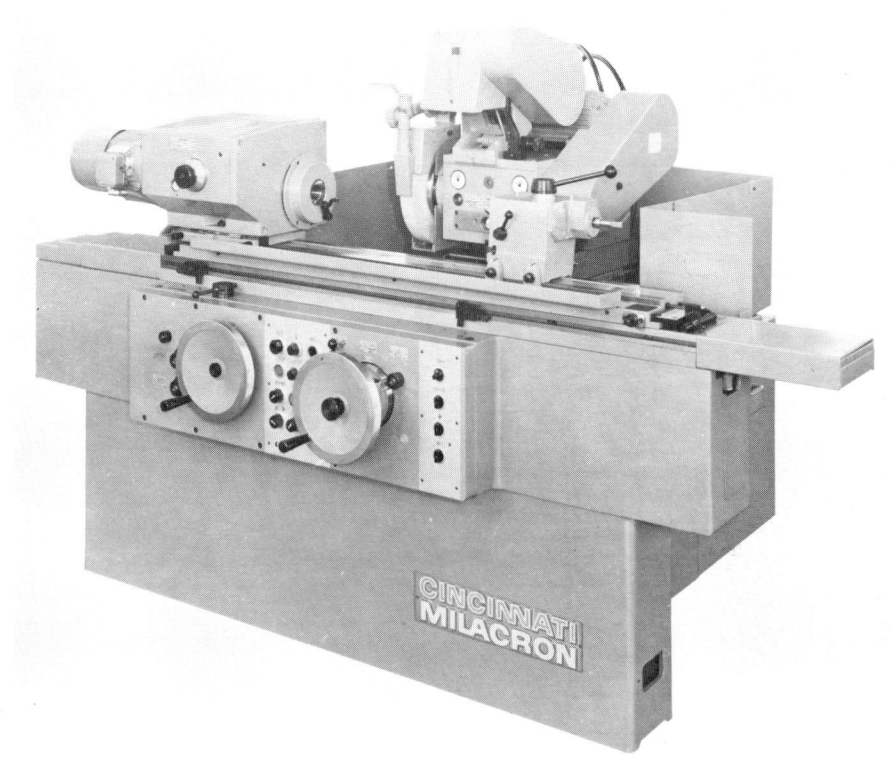

Figure 26. A universal cylindrical grinder (Courtesy of Cincinnati Milacron).

Figure 27. The use of a steady rest for grinding slender workpieces (Courtesy of Cincinnati Milacron).

344 Machine Tools and Machining Practices

Figure 28. Angular center-type grinder equipped with in-process gaging.

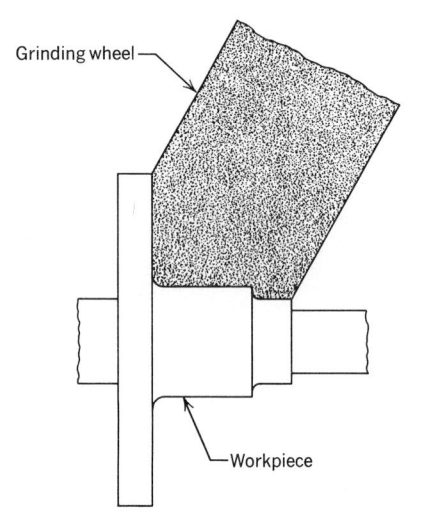

Figure 29. Typical application of the angular center-type grinding machine showing angled grinding wheel finishing both diameter and shoulder.

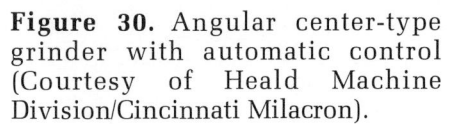

Figure 30. Angular center-type grinder with automatic control (Courtesy of Heald Machine Division/Cincinnati Milacron).

workpiece. This is a high production method widely used on complex parts like hydraulic valves.

The Roll Grinder The roll grinder is used to finish and resurface rolls used in both the hot and cold finishing of steels and other metals. These rolls are typically very heavy, so they are supported on journal bearings for grinding, just as they are when supported in the rolling mill where they are used. Also, because of the weight of the rolls, the roll

Figure 31. Cylindrical form grinding (Courtesy of Bendix Automation and Measurement Division).

Figure 32. Roll grinding machine used for grinding steel mill rolls or other very large workpieces (Courtesy of Landis Tool Co., Division of Litton Industries).

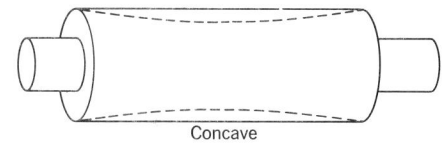

Figure 33. This steel mill roll has a convex surface (dotted lines) to compensate for the high pressures involved in cold rolling steel. The drawing is exaggerated for clarity.

Figure 34. This roll has a concave surface (dotted lines) for hot rolling steel. The drawing is exaggerated to show the slight variation.

grinding machine is designed (Figure 32) so that the roll rotates in a fixed position and the grinding head is moved along ways that are parallel to the roll. When the roll is to be used for rolling steel cold, where very high pressures are encountered, the machine is adjusted to grind a slightly convex cambered roll so the product comes out flat (Figure 33). In hot mill rolling the reverse is true and the roll is compensated to be slightly concave (Figure 34). For smaller rolling applications, some rolls are gound on centers in plain cylindrical grinders, or even sometimes on coated abrasive belt grinders that are able to grind the whole surface at one time.

The Centerless Grinder When someone speaks of the centerless grinding machine, it usually means a machine for work on the outside diameter of a cylindrical workpice, as shown in Figures 10, 11, and 35. These machines are usually used in high production applications, but they are by no means limited to simple cylindrical parts. Parts with differing diameters, like automotive valves, can be infed to a fixed stop (Fig-

Figure 35. Centerless grinding machine. The regulating wheel is on the left and parts for through-feed grinding can be loaded into the hopper (Courtesy of Landis Tool Co., Division of Litton Industries).

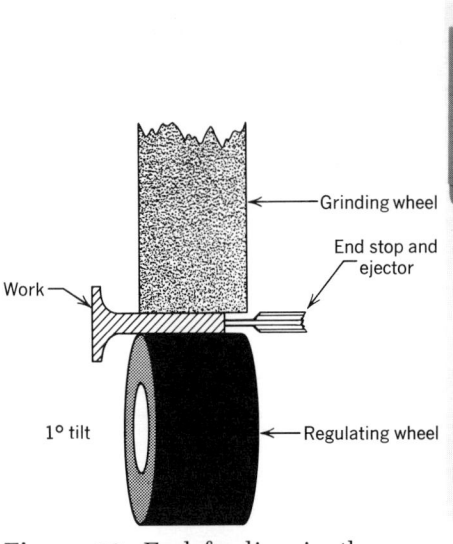

Figure 36. End feeding in the centerless grinder. The regulating wheel is set at a small tilt angle to keep the workpiece against the end stop (Courtesy of Bay State Abrasives, Dresser Industries, Inc.).

ure 36). It is also possible to do tapered parts by shaping both the grinding and regulating wheels to the reverse of the shape required (Figure 37), and it is even possible to centerless grind parts with center portions larger than the ends by loading the part down from the top with special feeding apparatus (Figure 38). Even headless threaded parts, like setscrews, can have the threads formed on centerless grinders.

Internal Cylindrical Grinding Internal cylindrical grinding (Figure 39) is usually done with a "mounted" grinding wheel having a shank that is held in the grinding spindle. The principles of this method of grinding are shown in Figure 40. The internal grinding wheel should be about three quarters of the finished hole diameter, mainly to have enough abrasive so that the interval between dressing is prolonged somewhat. With conventional wheels, dressing is frequent, sometimes even each pass in critical situations. In recent years a manufactured abrasive, cubic boron nitride (Figure 41) has been successfully applied in this type of application to reduce dressing requirements to the point where it is needed infrequently. This abrasive is close to the hardness of the diamond but, unlike the diamond, it works well in hardened steels. Mounted wheels of cubic boron nitride are also used in jig grinding machines to maintain uniform size between dressings. The use of this abrasive is not limited to internal grinding

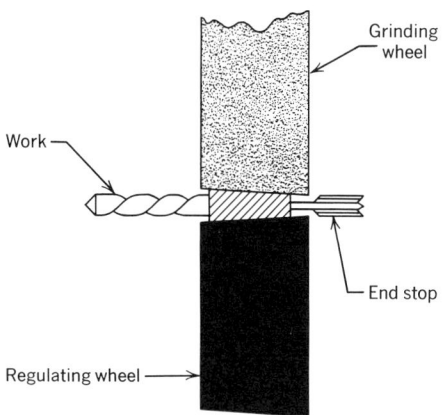

Figure 37. Centerless form grinding (Courtesy of Bay State Abrasives, Dresser Industries, Inc.).

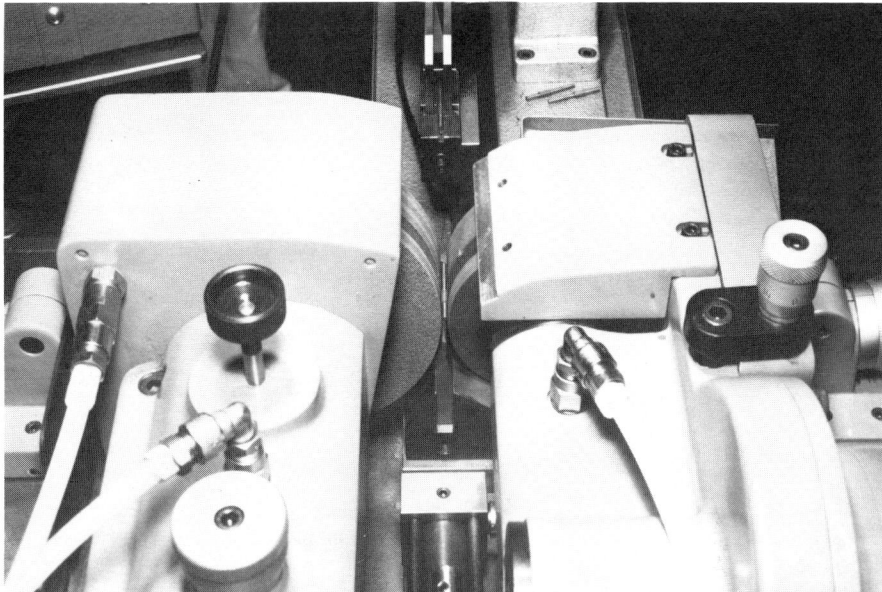

Figure 38. Centerless grinding both ends of a part.

Figure 39. Internal cylindrical grinding (Courtesy of Heald Machine Division/Cincinnati Milacron Company).

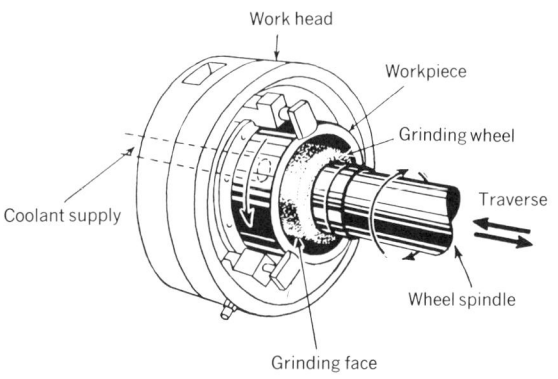

Figure 40. Principles of internal cylindrical grinding (Courtesy of Bay State Abrasives, Dresser Industries, Inc.).

applications; it can occasionally be found in surface grinding use as well.

Internal grinding is not limited to concentric workpieces. In many instances, special fixturing is applied to machines to hold irregularly shaped workpieces (Figure 42). Internal grinding attachments are also applied to universal grinding machines to accomplish this type of work.

The Cutter and Tool Grinder This machine (Figure 43) can be considered a type of cylindrical grinder, but it also can be used for certain classes of surface grinding. It is often called the **universal** cutter and tool grinder because it can be applied in so many ways to a variety of jobs. (A number of

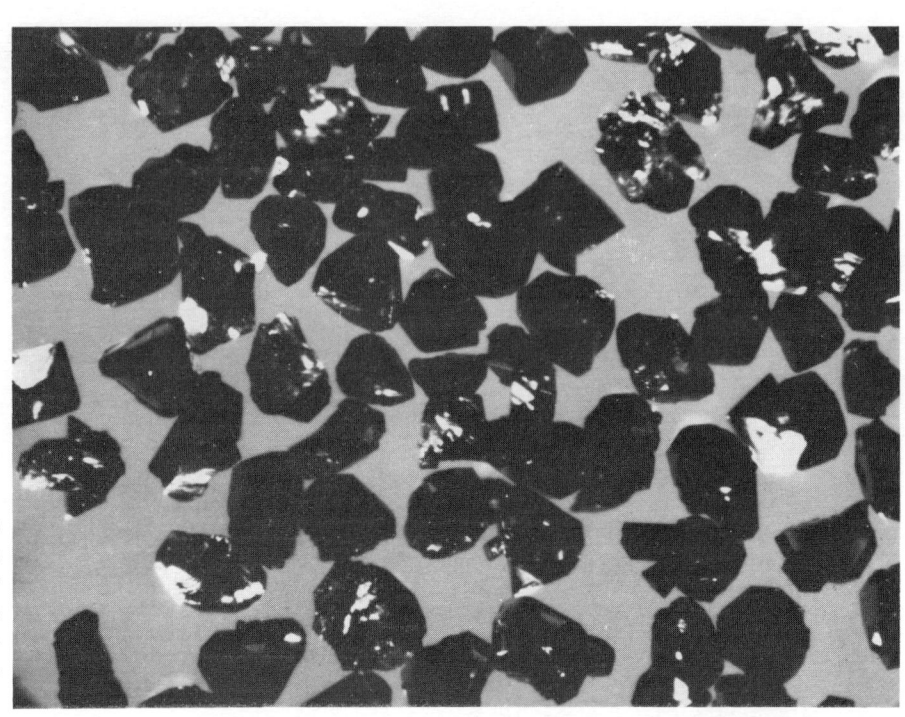

Figure 41. Cubic boron nitride (BOROZON™ CBN). This abrasive is almost as hard as diamond and very effective for grinding on tool steels (Courtesy of Specialty Materials Department, General Electric Company).

Figure 42. Internal grinding is not limited to cylindrical parts. This automotive connecting rod is an example of odd configurations that can be ground internally (Courtesy of Heald Machine Division/Cincinnati Milacron Company).

Figure 43. The cutter and tool grinder (Courtesy of Cincinnati Milacron).

ways of using this machine will be detailed in units in this section.) On this machine the spindle can be tilted and swiveled; straight, dish, or flaring cup wheels can be mounted. Wheels with diamonds (Figure 44) are commonly used for the sharpening of carbide tipped cutting tools (Figure 45). Even though the cost of a small diamond

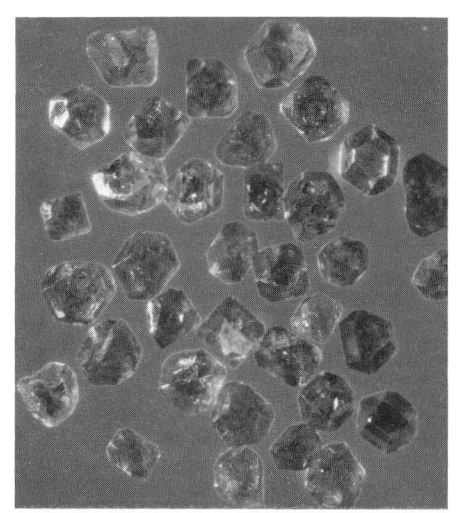

Figure 44. Manufactured diamond abrasive (Man-Made™ diamond), sometimes called synthetic diamond; it has the advantage over industrial grade natural diamond in that its mesh size can be controlled and varied to fit various materials (Courtesy of Specialty Materials Department, General Electric Company).

Figure 45. Sharpening of a large brazed carbide tipped face mill with a diamond cup wheel (Courtesy of Cincinnati Milacron).

Figure 46. Sharpening a plain milling cutter held between centers on the cutter and tool grinder (Courtesy of Cincinnati Milacron).

wheel may be several hundred dollars, the abrasive cost is ultimately lower than it would be using silicon carbide for the same application. Also, the size of the diamond wheel is maintained, which results in consistent size of the workpiece.

Most of the sharpening work on cutters is done with ordinary free cutting aluminum oxide grinding wheels on cutters mounted between centers (Figure 46).

Some unique types of cutter and tool grinders also exist (Figure 47). This grinder is often used for the sharpening of tapered, and tapered and ball ended milling cutters. This type of cutter is used in duplicating or die sinking milling machines.

MISCELLANEOUS GRINDING MACHINES

The disc grinder, which first appeared in 1904 (Figure 9), has evolved into a powerful, high production machine tool (Figure 48). This machine feeds parts between the faces of two grinding wheels. Various mehods are used for feeding parts to this type of machine. The type shown used a "ferris wheel" fixture that carries the parts between the wheels to be simultaneously ground to parallel and to a finished dimension. Other methods of feeding include the use of an oscillating arm to carry parts one at a time between the wheels or the use of guide bars to roll the parts between the wheels.

Gear grinders are another interesting type of grinding machine. There are a variety of principles employed with these machines, some of which were discussed earlier in this volume, with other gear producing machines. Basically these machines fall in two categories: form grinders (Figure 49), where the wheel is accurately dressed to the opposite shape of the tooth to be ground, and generating types (Figure 50), where the form results from the conjugate action of the wheel and workpiece.

Grinding is not always done with grinding wheels. Abrasive

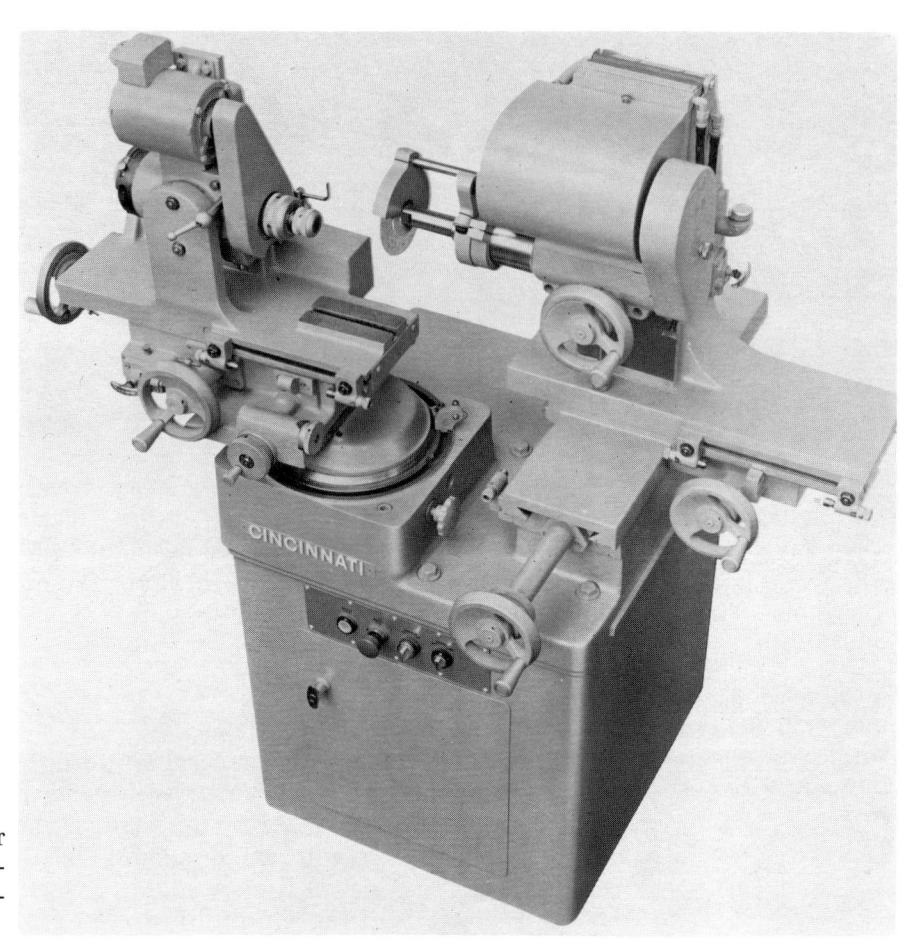

Figure 47. Cutter and tool grinder capable of generating helical surfaces (Courtesy of Cincinnati Milacron).

Figure 48. Double disc production grinding machine with feed wheel to carry the parts between the two opposed discs for grinding. The parts then drop out at the bottom (Courtesy of Bendix Corporation).

Figure 49. Form-type gear grinding machine (Courtesy of Ex-Cell-O Corporation).

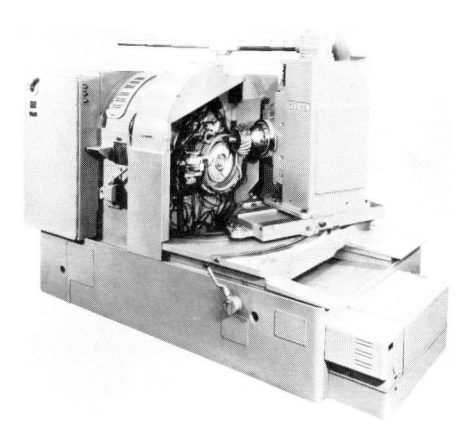

Figure 50. Hypoid gear grinding machine uses a cup wheel (Courtesy of Gleason Works).

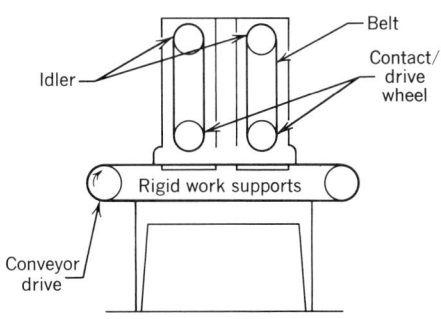

Figure 51. Diagram of a double belt coated abrasive grinder (Courtesy of 3M Company).

Other Abrasive Processes

belt machines (Figure 51) are coming into increasing use, especially where precise dimensions are not required.

Some abrasive processes use relatively slow moving abrasive materials to perform their work. One type uses free abrasive grains circulated on a hardened steel plate to generate flat surfaces or workpieces (Figure 52). This type is capable of comparatively large amounts of stock removal, because the abrasive tumbles along the hardened steel plate while it cuts the workpiece material. Another similarly appearing machine that works on a different principle is the lapping machine (Figure 53). In this machine, the cutting surface is prepared by imbedding abrasive grain into a relatively soft plate, which holds the abrasive, while relative motion is imparted to the workpiece. Exceptional flatness is obtained by this method, but the cutting rate is very slow. Temperature control is critical to accurate lapping.

Other relatively low velocity methods are used in the precise sizing of previously machined bores. The honing machine (Figure 54) can be used either for external or internal honing of workpieces within its size range, and it is often equipped with attachments (Figure 55) for mechanical movement of the workpiece along the honing mandrel.

Another relatively slow velocity finishing method is the "superfinisher" (Figure 56). This process uses a formed abrasive pad that is held in contact with the surface to be finished while it slowly turns in contact. In addition, a simultaneous slight side motion is also provided, which results in a highly finished accurate surface on the workpiece.

The vibratory deburring of parts is also a widely used abrasive

Figure 52. Free abrasive grinding machine has a hard, water cooled plate on which the abrasive is fed in a slurry. The abrasive grains do not become embedded as they do in lapping. Hence, the grains always roll around under the workpieces (Courtesy of Speedfam Corporation).

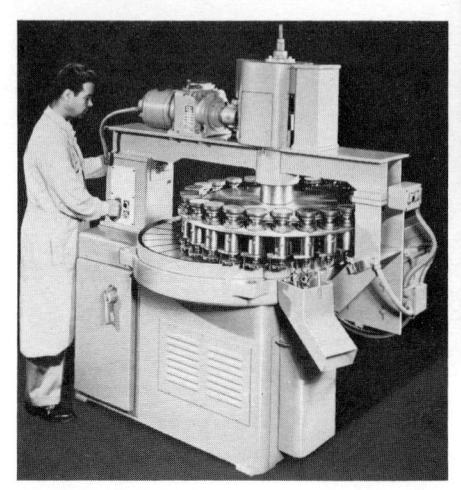

Figure 53. Lapping machine. The pressure plate on top holds the parts to be lapped by the abrasive embedded in the plate beneath (Courtesy of Lapmaster Division of Crane Packing Company).

Figure 54. Honing machine. Honing is a very popular method for finishing inside diameters of everything from bushings to cylinders of automobile engine blocks (Courtesy of Sunnen Products Company).

Section H Grinding Machines 353

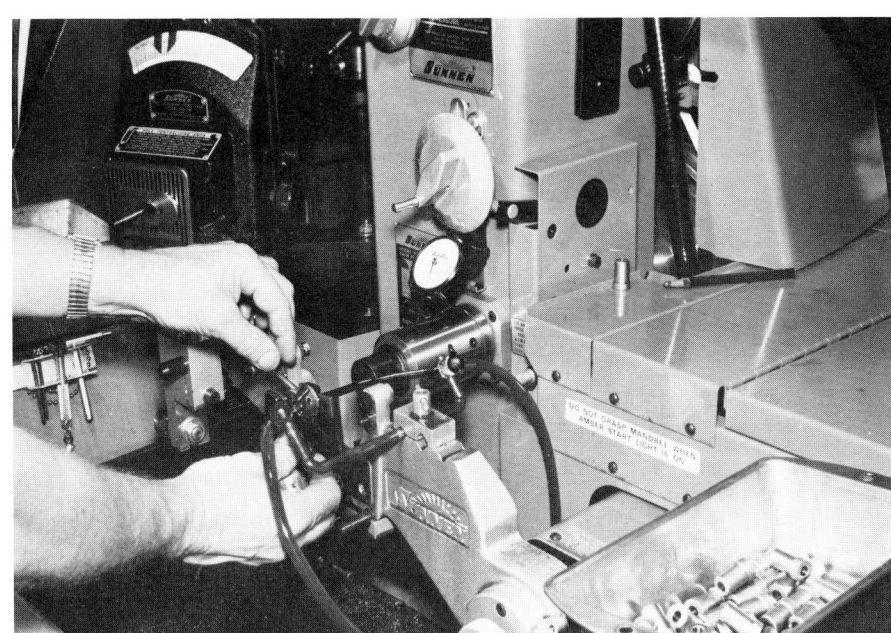

Figure 55. Using a honing machine with an attachment that reciprocates the workpiece.

Figure 56. Superfinishing an automotive crank (Courtesy of Taft-Peirce Mfg. Co.).

process (Figure 57). The tumbling abrasive can even deburr the interior of workpieces that are inaccessible to other methods.

Finally, there is a hybrid electrochemical and abrasive machine called the electro-chemical grinder (Figure 58). This is really a plating machine operated in reverse. The electro-chemical action (Figure 59) removes the material from the workpiece (anode) but, in the process, insulating oxides are formed. The abrasive serves mainly to remove the oxides so that the process of "deplating" can

Figure 57. This high production vibratory finisher shows the parts and abrasive coming from the vibrator to the front where the parts are unleaded; then the abrasive is returned to the vibrator by the conveyor (Courtesy of UltraMatic Equipment Co.).

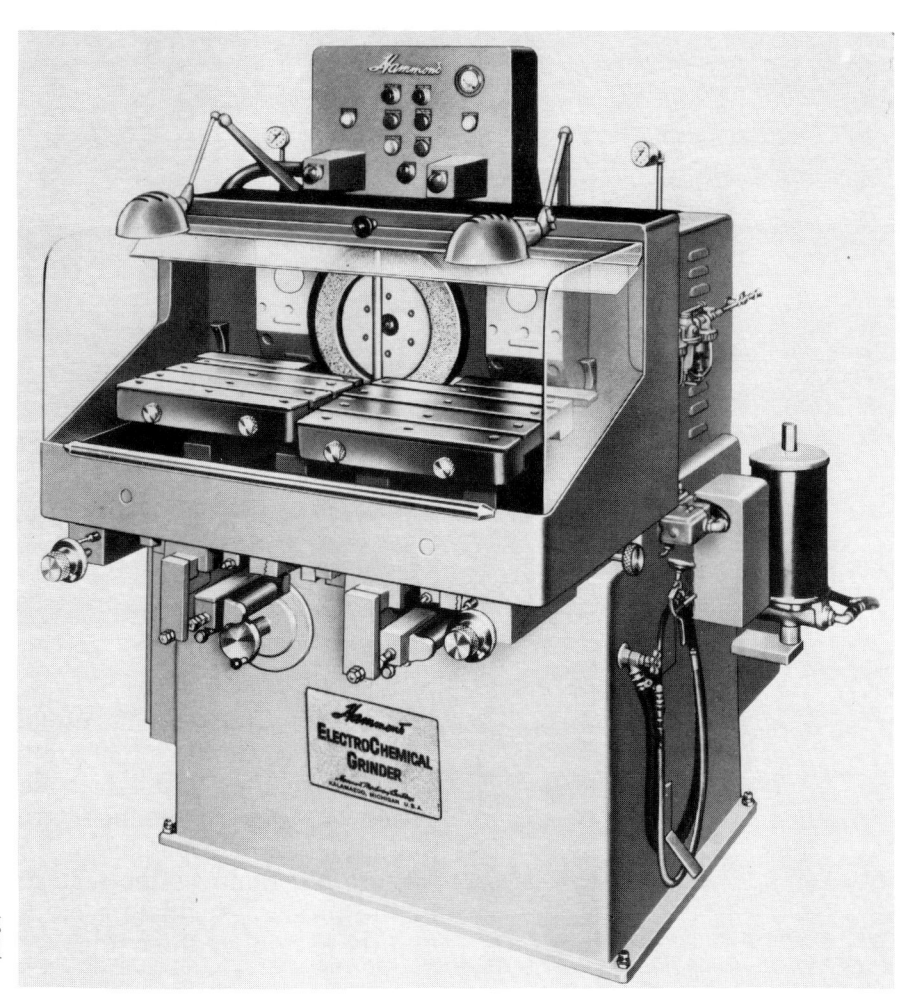

Figure 58. Electrochemical grinding machine (Courtesy of Hammond Machinery Builders Inc.).

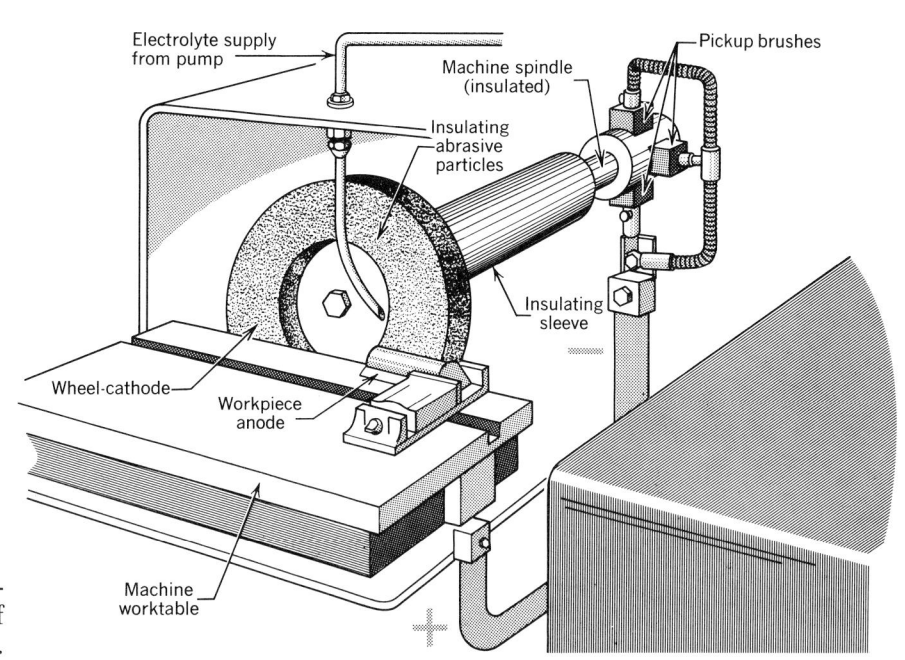

Figure 59. Principles of the electrochemical grinder (Courtesy of Hammond Machinery Builders Inc.).

continue. The abrasive wheel is usually metal bonded diamond, for conductivity, and the wheels last an extremely long time. This type of machine is used frequently in the sharpening of single point carbide lathe and planer tools.

review questions

1. The grinding machine was a comparative latecomer as a basic machine tool, well after lathes and milling machines. What advantages resulted from this timing? What disadvantages?
2. It has been said that the major advances in grinding have come from the improvement of abrasives instead of machines. Do you agree or disagree? Give some reasons for your point of view.
3. The abrasives industry, principally grinding, ranks very high among the critical defense industries in practically every country. Why?
4. The cutter and tool grinder is considered by many to be the most versatile grinding machine that exists. Why?
5. Is electrochemical grinding really an abrasive process?

unit 1
selection and identification of grinding wheels

A grinding wheel is at once both a many-toothed cutting tool and a tool holder. Selecting a grinding wheel is somewhat more complicated than selecting a milling cutter because there are more factors to be considered. These factors, including size, shape, and composition of the wheel, are expressed in a system of symbols (numbers and letters) that you must be able to interpret and apply.

objectives After completing this unit, you will be able to:
1. List the five principal abrasives with their general areas of best use.
2. List the four principal bonds with the types of applications where they are most used.
3. Identify by type number and name, from unmarked sketches or from actual wheels, the four most commonly used shapes of grinding wheels.
4. Interpret wheel shape and size markings together with the five basic symbols of a wheel specification into a description of the grinding wheel.
5. Given several standard, common grinding jobs, recommend the kind of abrasive, approximate grit size and grade, and bond.

information

Although six or seven factors must be considered in selecting a grinding wheel for a particular job, most of the decisions are almost automatic. For example, the size and shape of the wheel, are usually determined by the type and size of the grinder. The abrasive, of which there are five major kinds, is determined primarily by the material being ground. The abrasives are not equally efficient on all materials, and there are only a few areas where there would be much doubt about the choice. The size of the abrasive particles selected, of which the largest are perhaps $\frac{1}{8}$ to $\frac{1}{4}$ in. long and the smallest less than .001 in. long, depends on the amount of stock to be removed and the final finish desired.

Selecting the grade or hardness of the wheel, which is a measure of the force required to pull out the abrasive grains, is a difficult decision sometimes but, here again, it is typically a choice between one of two or, at most, three grades. Grain spacing or structure is most often standard, according to the grain or grit size and the grade of the wheel. Finally, there is the bond that holds the wheel together. And although there are four common bonds, the choice is usually made clear from the job.

All of these factors (wheel shape, wheel size, kind of abrasive, abrasive grain size, hardness, grain spacing, and bond) are expressed in symbols consisting of numbers and letters, most of which are easy to understand and interpret.

There are many different kinds of grinding wheels in practically any toolroom or machine shop or, for that matter, in any grinding production department or foundry. Keeping them straight and knowing which wheel to use for which job can be something of a problem. Some of the ways this could be done are as follows.

Color can be useful for identificaton, and it is used to some extent. It happens that one of the best abrasives for grinding tools is white when it is manufactured, and a more commonly used abrasive is gray or brown in color. Some wheels are green, pink, or black. Among a given number of grinding wheels, there are wheels of different diameters, thicknesses, or hole sizes. Different wheels are made of different sizes of abrasive grain or with different proportions of grain and the bond that holds the grain together in the wheel.

What has been developed over the years, mostly by the grinding wheel makers, is a code of numbers or letters that provide a paragraph of information about a given wheel in just a few letters and numbers. Within a given group of symbols, the order of listing is important.

Grinding wheels are designed for grinding either on the periphery (outside diameter), which is a curved surface, or on the flat side, but rarely on both. It is not a safe practice to grind on the side of a wheel designed for peripheral grinding. The shape of the wheel determines the type of grinding performed.

The shapes of grinding wheels are designated according to a system published in full in an American National Standard, *Specifications for Shapes and Sizes of Grinding Wheels*, whose number is ANSI B74.2-1974. The various shapes have been given numbers ranging from 1 to 28, but only five are important for you now. These are described below.

Type 1 (Figure 1) is a peripheral grinding wheel, a straight wheel with three dimensions: diameter, thickness, and hole, in that order. A typical wheel for cylindrical grinding is 20 in. (diameter) × 3 in. (thickness) × 5 in. (hole). Probably most wheels are of this type.

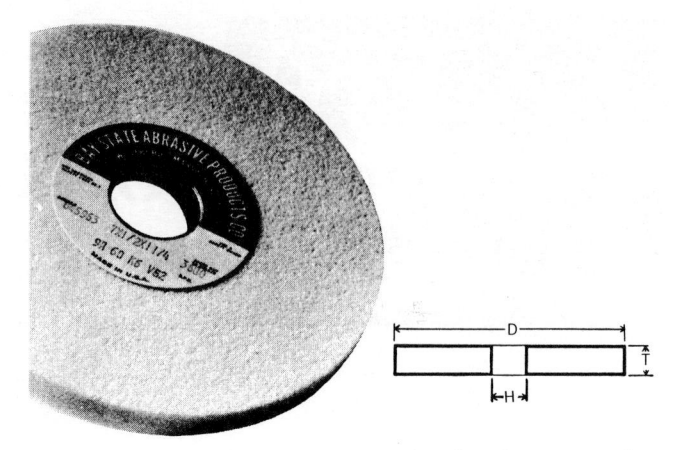

Figure 1. Straight or Type 1 wheel, whose grinding face is the periphery. Usually comes with the grinding face at right angles to the sides, in what is sometimes called an "A" face (Courtesy of Bay State Abrasives, Dresser Industries, Inc.).

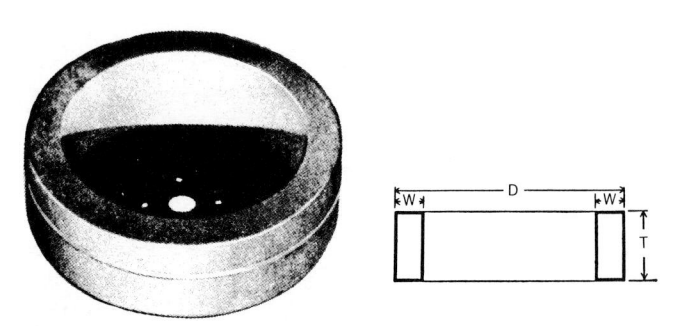

Figure 2. Cylinder or Type 2 wheel, whose grinding face is the rim or wall end of the wheel. Has three dimensions — diameter, thickness, and wall thickness (Courtesy of Bay State Abrasives, Dresser Industries, Inc.).

A *Type 2* or cylinder wheel (Figure 2) is a side grinding wheel, to be mounted for grinding on the side instead of on the periphery. This also has three dimensions, for example, 14 in. (diameter) × 5 in. (thickness) × 1½ in. (wall). This, of course, might also be called a 14 × 5 × 11 (14 in. D minus 2 times 1½ — the two wall thicknesses) but the wall thickness is more important than the hole size. Hence the change.

A *Type 6* or straight cup wheel (Figure 3) is a side grinding wheel with one side flat and the opposite side deeply recessed. It has four essential dimensions: the diameter, thickness, hole size (for mounting), and wall.

A *Type 11* or flaring cup wheel (Figure 4) is a side grinding wheel that resembles a Type 6,

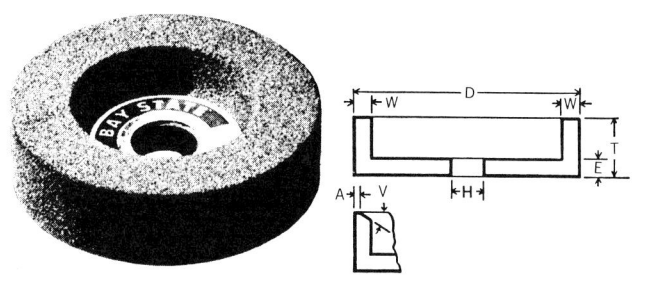

Figure 3. Straight cup or Type 6 wheel, whose grinding face is the flat rim or wall end of the cup (Courtesy of Bay State Abrasives, Dresser Industries, Inc.).

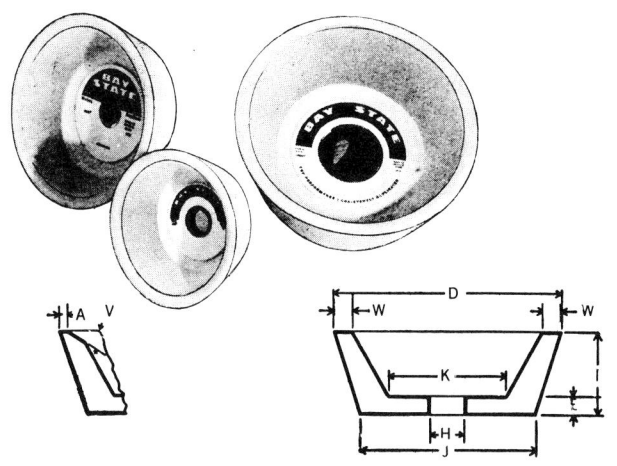

Figure 4. Flaring cup or Type 11 wheel, whose grinding face is also the flat rim or wall of the cup. Note that the wall of the cup is tapered (Courtesy of Bay State Abrasives, Dresser Industries, Inc.).

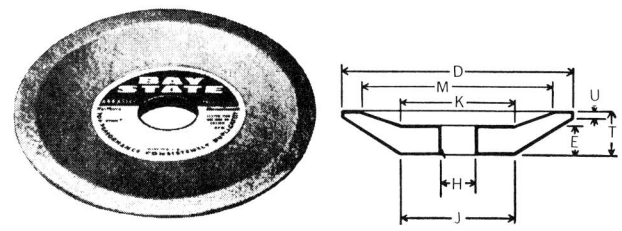

Figure 5a. Dish or Type 12 wheel, similar to Type 11, but a narrow, straight peripheral grinding face in addition to the wall grinding face. Only wheel of those shown that is considered safe for both peripheral and wall or rim grinding (Courtesy of Bay State Abrasives, Dresser Industries, Inc.).

Figure 5b. An assortment of mounted wheels most often used for deburring and other odd jobs (Courtesy of Bay State Abrasives, Dresser Industries, Inc.).

except that the walls flare out from the back to the diameter and are thinner at the grinding face than at the back. This introduces a couple of new dimensions, the diameter at the back, called the "J" dimension, and the recess diameter at the back, the "K" dimension. This is mentioned only to emphasize that the "D" dimension, the diameter, is always the largest diameter of any wheel.

The *Type 12* dish wheel (Figure 5a) is essentially a very shallow Type 11 wheel, mostly for side grinding. The big difference is the dish wheel has a secondary grinding face on the periphery, the "U" dimension, so that it is an exception to the rule of grinding *only* on the side or the periphery.

This factor of grinding on the side or the periphery of a wheel is important because it affects the grade of the wheel to be chosen. The larger the area, the softer the wheel should be. In peripheral grinding the contact is always between the arc of the wheel and either a flat (in surface grinding) or another arc (in cylindrical grinding). This makes for small areas of contact and somewhat harder wheels. On the other hand, if the flat side of the wheel is grinding the flat surface of a workpiece, then the contact area is larger and the wheel can be still softer.

It is important to understand that any grinding machine grinds either a flat surface like a planer or shaper, or a round or cylindrical surface, like a lathe or boring mill. The first group of grinders is collectively called surface grinders; the second group is called cylindrical grinders, whether the workpiece is held between centers or not, and whether the grinding is external or internal. Various forms can be cut into the grinding faces of peripheral grinding wheels, and these can then be ground into either flat or cylindrical surfaces.

Mounted wheels like the ones in Figure 5b can be used in a variety of ways around a shop. Often they are used in portable grinders for jobs like deburring or breaking the edges of workpieces where the tolerances are not too critical. They are also used in internal grinding.

STANDARD MARKING SYSTEM

The description of a grinding wheel's composition is contained in a group of symbols known as the *standard marking system*. That is, the basic symbols for the various elements are standard, but they are usually amplified by individual manufacturer's symbols, so that it does not follow that two wheels with the same basic markings, but made by two different suppliers, would act the same. However, it is a useful tool for anyone concerned with grinding wheels.

There are five basic symbols. The first is a letter indicating the kind of abrasive in the wheel, called the abrasive *type*. The second is a number to indicate the approximate size of the abrasive; this is commonly called *grit size*. In the third position, a letter symbol indicates the *grade* or relative hardness of the wheel. The fourth, *structure*, is a number describing the spacing between abrasive grains. The fifth is a letter indicating the *bond*, the material that holds the grains together as a wheel. Thus, a basic toolroom wheel specification (Figure 6) might be

A60-J8V

But, since most wheel makers have, for example, a number of different aluminum oxide or other types of abrasives and a number of different vitrified or other bonds, the symbol sometimes appears cluttered:

9A80-K7V22

This means that the wheel maker is using a particular kind of aluminum oxide (A) abrasive, which is indicated by the "9," and a particular vitrified (V) bond, which is indicated by the "22."

The wheel markings for diamond or cubic boron nitride wheels are a little different and are not standard enough for a simple explanation.

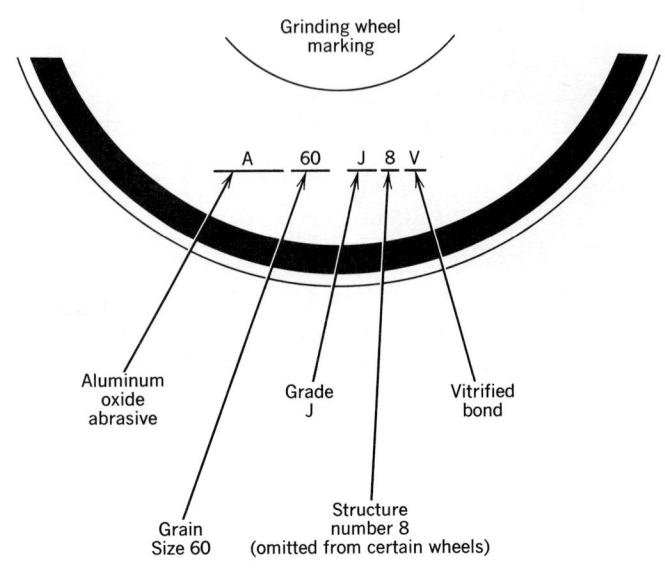

Figure 6. Sketch to illustrate the wheel specification that will be used as an illustration in following pages (McKee Editorial Assistance).

FIRST SYMBOL — TYPE OF ABRASIVE: (A60-J8V)

The symbol in the first position, as suggested above, denotes the type of abrasive. Basically, there are five:

A Aluminum oxide.
C Silicon carbide.
D Natural diamond.
MD or SD Manufactured diamond (sometimes called synthetic diamond).
B Cubic boron nitride.

The first two are "cents per pound" abrasives, cheap enough so that it is practical to make whole wheels of the abrasive. Both are made in electric furnaces that hold literally tons of material, and both are crushed and graded by size for grinding wheels and other uses.

Diamond, both natural and manufactured, and cubic boron nitride are expensive enough that most wheels are made of a layer of abrasive around a core of other material. However, they have made a place for themselves because they will grind materials that no other abrasive will touch and because they stay sharp and last so long that they are actually less expensive per piece of parts ground. Natural or mined diamond that is definitely diamond, but is of less than gem quality, is crushed and sized. Both manufactured diamond and cubic boron nitride are made by a combination of high heat (in the range of 3000° F) and tremendous pressure (1 million or more PSI). This heat, incidentally, is somewhat less than that for other manufactured abrasives, which require something in the range of 4000°F for fusing or crystallization.

Each of these abrasives generally has an area in which it excels, but there is no one abrasive that is first choice for all applications. Aluminum oxide is best for grinding most steels, but on the very hard tool steel alloys it is outclassed by cubic boron nitride. However, aluminum oxide (Al_2O_3) may have close to 75 percent of the market, because it is used in foundries for grinding castings and steel mills for billet grinding, and in other high volume applications. Silicon carbide, which does poorly on most steels, is excellent for grinding nonferrous metals and nonmetallic materials. Diamond abrasive excels on cemented carbides, although green silicon carbide is occasionally used. Green silicon carbide was the recommended abrasive for cemented carbides until the introduction of diamonds. Cubic boron nitride (CBN), finally, is superior for grinding high speed steels. CBN is a hard, sharp, cool cutting, long wearing abrasive.

There is a lot still unknown about why certain abrasives grind well on some materials and not on others, and this makes for some interesting speculation. All abrasives are harder than the materials ground. But their relative hardness is apparently only one of the factors in effectiveness. Aluminum oxide is three to perhaps five times harder than most steels, but it grinds them easily. Silicon carbide is harder than aluminum oxide, but not at all effective on steel; on the other hand, it grinds glass and other nonmetallics that are as hard or harder than many steels. Diamond, which is much harder than cubic boron nitride and many times harder than the hardest steel, does not grind steels well, either. The best theory is that there are chemical reactions between certain abrasives and certain materials that make some abrasives ineffective on some metals and other materials.

The basic symbol in the first position, A, C, D, or B, is often preceded and sometimes followed by a manufacturer's symbol that indicates which abrasive within the group is meant, for example, 9A, 38A, AA.

SECOND SYMBOL — GRIT SIZE (A60-J8V)

The symbol in the second position of the standard marking represents the size of the abrasive grain, usually called grit size. This is a number ranging from 4 to 8 on the coarse side to 500 or higher on the fine side. The number is derived from the approximate number of openings per linear inch in the final screen used to size the grain; the larger the number, the smaller the abrasive grain. Any standard grit size contains grain of smaller and larger sizes, whose amounts are strictly regulated by the federal government, because it would be very expensive to reduce the mix to just the size indicated.

While there is no real agreement as to what is coarse or fine, for general purposes anything from 46 to 100 might be considered medium, with everything 36 and lower considered as coarse and anything 120 and higher considered as fine. Selection of grit in any shop depends on the kind of work it is doing. Thus, where the job is to remove as much metal as fast as possible, 46 or 60 grit size would be considered very fine. On the other hand, in a shop specializing in fine finishes and close tolerances, 240 might be considered coarse.

A final point is in order about grit size. Most standard symbols end in a "0," particularly the three digit sizes 100 and finer. However, every abrasive grain manufacturer also makes some combinations that are not standard for special uses, and these usually end in a "1" or other low digit. Thus, a 240 grit is the finest that is sized by screening. Finer grits are sized by other means. On the other hand, 241 grit is a coarse 24 grit in a "1" combination.

Coarse grain is used for fast stock removal and for soft, ductile materials. Fine grain is used to obtain good finishes and for hard or brittle materials. Some materials are hard enough that fine grain removes as much stock as coarse, and neither removes very much. In a general machine shop or toolroom, most of the wheels used will be between 46 and 100 grit.

THIRD POSITION — GRADE OR "HARDNESS": (A60-J8V)

In the third position is a letter of the alphabet called grade. The later the letter, the "harder" the grade. Thus, a wheel graded F or G would be

grains until they have lost their sharpness. Ideally, a wheel should be self-sharpening. The bond should hold each grain only long enough for it (the grain) to become dull. In practice, this is difficult to achieve.

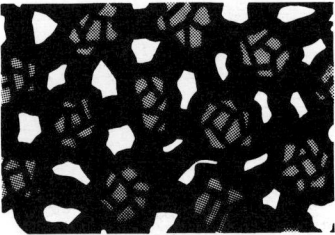

One thing, however, should be mentioned. Grade is a much less measurable thing than type or size of abrasive or, as you will see later, than bond. Grade depends on the formula for the mix used in the wheels, but it must be checked after the wheel is finished.

FOURTH POSITION — STRUCTURE: (A60-J8V)

Following the grade letter, in the fourth position of the symbol, is a number from 1 (dense) to 15 (open); this number describes the spacing of the abrasive grain in the wheel (Figure 8). The use of

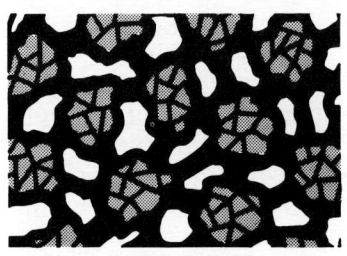

Figure 7. Three sketches illustrating (from top down) a soft, a medium, and a hard wheel. This is the "grade" of the wheel. The white areas are voids with nothing but air; the black lines are the bond, and the others are the abrasive grain. The harder the wheel, the greater the proportion of bond and, usually, the smaller the voids (Courtesy of Bay State Abrasives, Dresser Industries, Inc.).

Figure 8. Three similar sketches showing structure. From the top down, dense, medium, and open structure or grain spacing. The proportions of bond, grain, and voids in all three sketches are about the same (Courtesy of Bay State Abrasives, Dresser Industries, Inc.).

considered "soft," and one graded R to Z would be very "hard." Actually, these descriptions are put into quotation marks because the words are the best we have, but are still not quite accurate. What is being measured is the hold that the bond has on the abrasive grain (Figure 7), and the greater the proportion of bond to grain, the stronger the hold and the harder the wheel. Precision grinding wheels tend to be on the soft side, because it is necessary to have the grains pull out as they become dull; otherwise, the wheel glazes and its grinding face becomes shining, but the abrasive is dull. On high speed, high pressure applications like foundry snagging, the pressures to pull out the grains are much greater, and a harder wheel is needed to hold in the

structure is to provide chip clearance, so that the chips of ground material have some place to go and will be flung out of the wheel by centrifugal force or washed out by the coolant. If the chips remain in the wheel, then the wheel becomes loaded (Figure 9), stops cutting, and starts rubbing, and it has to be resharpened or dressed, which is the trade term.

Structure is also a result of grain size and proportion of bond, similar to grade. Quite often, large grain size wheels tend to have open structure, while smaller sized abrasive grain is often associated with dense structure. On the open side, say 11 to 12 and up, the openness is aided by the inclusion of something in the mix like ground up walnut shells, which will burn out as the wheel is fired and leave definite open spacing.

However, for many grit size and grade combinations, a "best" or standard structure has been worked out through experience and research, and so the structure number may be omitted.

FIFTH POSITION — BOND: (A60-J8V)

The fifth position of the wheel marking is a letter indicating the bond used in the wheel. This is always a letter, as follows: vitrified — V; resinoid — B (originally the bakelite bond); rubber — R (rubber was used well before resinoid); and shellac — E (originally the "elastic" bond, and also preceded by the now obsolete silicate bond). These are really general bond groups; each wheel maker uses extra symbols to indicate, for instance, which vitrified bond he has used in a particular wheel, and there is no standardization in these extra symbols.

The bond used has important influence on both the manufacturing process and on the final use of the wheel. Vitrified-bonded wheels are fired at temperatures between 2000 and 2500° F; for that reason, no steel inserts can be used. If such inserts are needed, they must be cemented

Figure 9. The wheel at the left is called "loaded" with small bits of metal imbedded in its grinding face. It is probably too dense in structure or perhaps has too fine an abrasive grain. Same wheel at the right has been dressed to remove all the loading (Photographs by courtesy of Desmond-Stephan Mfg. Co.).

in afterward. The others, also grouped together as organic bonds, are all baked at around 400° F, and inserts may be molded in without problems.

Vitrified, resinoid, and shellac wheels are all pressed in molds after mixing. Rubber bonded wheels, on the other hand, are mixed in a process similar to that of making dough for cookies; they are then reduced to thickness by passing the mass or grain-impregnated rubber between precisely spaced rolls. For that reason, it is possible to make much thinner wheels with rubber bond than can be made with any other bond. Thin grinding wheels with either resinoid or rubber bonds can be used in cutoff operations like very fine bandsaws, but the rubber wheels can be made thinner than the resinoid bonded wheels.

In general, vitrified wheels are used for precision grinding; resinoid wheels are used for rough grinding with high wheel speeds and heavy stock removal; and rubber or shellac wheels, are used for more specialized applications. The first two bonds monopolize over 90 percent of the market.

Bond also determines maximum safe wheel speed. Vitrified bonded wheels, with a few minor exceptions, are limited to about 6500 SFPM or a little over a mile a minute. The others run much faster, with some resinoid wheels getting up to 16,000 SFPM or more. Of course, all grinding wheels must be properly guarded to protect the operator in the unlikely case of wheel breakage. Grinding machines in general, however, are safe machines. Many will not operate unless the guards are properly in place and closed.

FACTORS IN GRINDING WHEEL SELECTION

It is generally considered that there are seven or eight factors to consider in the choice of a grinding wheel. Some people group together the elements of the amount of stock removal and of the finish required. Dividing them seems to be more logical.

These factors have also been divided into two groups — three concerned with the workpiece, which thus change frequently, and five concerned with the grinder, which are more constant.

Variable Factors
In the first group are things such as composition, hardness of the material being ground, amount of stock removal, and finish.

The composition of the material generally determines the abrasive to be used. For steel and most steel alloys, use aluminum oxide. For the very hard, high speed steels, use cubic boron nitride. For cemented carbides, use diamond. Whether this is natural or manufactured is too specialized a question to be discussed here, but the trend appears to be toward the use of manufactured diamond. For cast iron, nonferrous metals, and nonmetallics, use silicon carbide. On some steels, aluminum oxide may be used for roughing and silicon carbide for finishing. Between aluminum oxide and cubic boron nitride, because of the greatly increased cost, the latter is generally used only when it is superior by a wide margin, or if a large percentage of work done on the machine is in the very hard steel alloys like T15.

Material hardness is of concern in grit size and grade. Generally, for soft, ductile materials, the grit is coarser and the grade is harder; for hard materials, finer grit and softer grades are the rule. Of course, it is understood that for most machine shop grinding these "coarse" grits are mostly in the range of 36 to 60, and the finer grits are perhaps in the range of 80 to 120. Likewise, the "soft" wheels are probably something in the range of F, G, H, and perhaps I, and the "hard" wheels are in the range of J, K, or L, and maybe M or N. Too coarse a grit might leave scratches that would be difficult to remove later. And sometimes on very hard materials, coarse grit removes no more stock than fine grit, so you use fine grit. Too soft a wheel will wear too fast to be practical and economical. Too hard a wheel will glaze and not cut.

If **stock removal** is the only objective, then you can use a very coarse (30 and coarser) resinoid bonded wheel. However, in machine shop grinding, you are probably in the 36 to 60 grit sizes mentioned above, and definitely in vitrified bonds.

For **finishing** on production jobs, the wheel may be rubber, shellac, or resinoid bonded. But resinoid bonds are often softened by coolants and therefore are rarely used. And for finishing, fine grit sizes are usually preferred; however, as you will learn later when you study wheel dressing (sharpening or renewing the grinding face of a wheel), it is possible to dress a wheel so that a comparatively coarse grit like 54 or 64 will finish a surface as smoothly as 100 or 120 grit.

Fixed Factors

For any given machine the following five factors are likely to remain constant.

Horsepower of the machine, of course, is a fixed consideration. Grinding wheel manufacturers and grinding machine builders are constantly pressing for higher horsepower, because that gives the machine and the wheels the capacity to do more work, but that is a factor, usually, only in the original purchase of the machine. In wheel selection, this affects only grade. The general rule is the higher the horsepower, the harder the wheel that can be used.

The **severity of the grinding** also remains pretty constant on any given machine. This affects the choice of a particular kind of abrasive within a general group. Thus, you would probably use a regular or intermediate aluminum oxide on most jobs. But, in the toolroom, where pressures are low, you would probably want an easily fractured abrasive, white aluminum oxide. On the other end of the scale, for very severe operations like foundry snagging, you need the toughest abrasive you can get, probably an alloy of aluminum oxide and zirconium oxide. For most machine shop grinding, you will probably look first at white wheels.

The **area of grind contact** is also important but, again, it remains constant for a given machine. The rule is finer grit sizes and harder wheels for small areas of contact; and coarser grit sizes and softer wheels for larger areas of contact. All of this, of course, is within the grit size range of about 36 to 120, or 150 or 180, and within a grade range from E or F to L or N.

It is easy to understand that on a side grinding wheel, where a flat abrasive surface is grinding a flat surface, the contact area is large and the wheels are fairly coarse grained and soft (Figure 10). However, on peripheral grinding wheels, it is a different story. The smallest area is in ball grinding, where the contact area is a point, the point where the arc of the grinding wheel meets the sphere or ball. Thus ball grinding wheels are very hard and very fine grained, for instance, 400 grit, Z grade. In cylindrical grinding the contact area is the line across the thickness of the wheel, usually where the arc of the wheel meets the arc of the workpiece (Figure 11). Here grit sizes of 54, 60, or 80, with grades of K, L, or M, are common. A still larger area is in surface grinding with peripheral grinding where the line of contact is slightly wider because the wheel is cutting

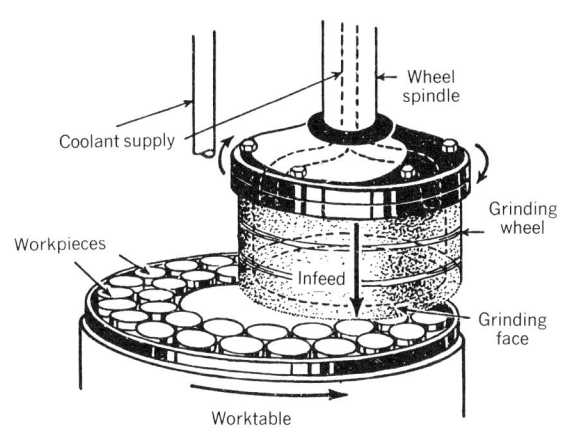

Figure 10. With the flat wall or rim of the wheel grinding a flat surface, as shown here, the wheel must be soft in grade and can have somewhat coarser abrasive grain. The area of contact between wheel and work is large (Courtesy of Bay State Abrasives, Dresser Industries, Inc.).

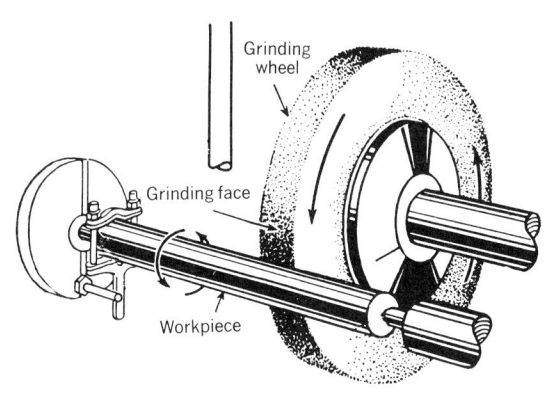

Figure 11. In center-type cylindrical grinding, as shown here, the arc of the grinding face meets the arc of the cylindrical workpiece, making the area of contact a line. This requires a "harder" wheel than in Figure 10 (Courtesy of Bay State Abrasives, Dresser Industries, Inc.)

into a flat surface (Figure 12). And a combination like 46 I or 46 J is not unusual. An internal grinding wheel where the OD of the wheel grinds the ID of the workpiece may have just a shade more area of grinding contact (Figure 13). And then when you get to side grinding wheels (cylinder Type 2), cup wheels, and segmental wheels, which are flats grinding flats, you get grit sizes and grades like 30 J or 46 J (see Figure 10). Of course, you have to realize that there may be other factors important enough to override contact area. For example, for grinding copper, you might use a grit size and grade like 14 J, in which

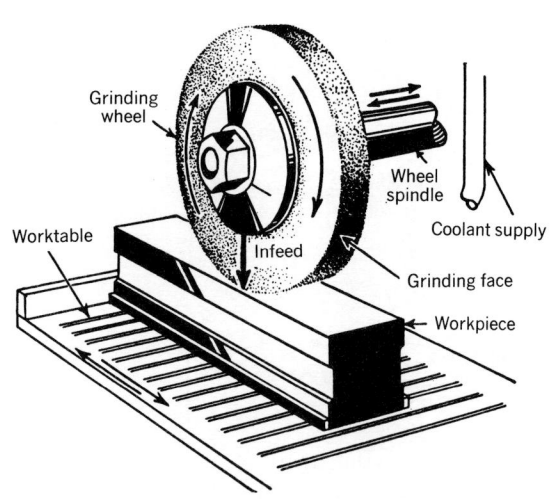

Figure 12. The contact area between the arc of the grinding face and the flat surface of the workpiece in surface grinding makes a somewhat wider line of contact than in cylindrical grinding (Courtesy of Bay State Abrasives, Dresser Industries, Inc.).

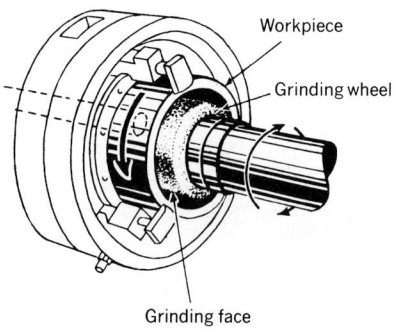

Figure 13. The contact area of the OD grinding face of the wheel and the ID surface of the workpiece creates a still larger area of contact and requires a somewhat "softer" grinding wheel than the two previous examples (Courtesy of Bay State Abrasives, Dresser Industries, Inc.).

Figure 14. The blotter on the wheel, besides serving as a buffer between the flange and the rough abrasive wheel, provides information as to the dimensions and the composition of the wheel, plus its safe speed in RPMs. This wheel is 7 in. in diameter × $\frac{1}{4}$ in. thick × $1\frac{1}{4}$ in. hole. It is a white aluminum oxide wheel, 100 grit, I grade, 8 structure, vitrified 52 bond. It can be run safely at up to 3600 RPMs. (Courtesy of Bay State Abrasives Division, Dresser Industries, Inc.)

the softness of the metal is probably the key factor.

Wheel speed is a factor that can be dealt with quickly. You must always stay within the safe speeds, which are shown on the blotter or label on every wheel of any size (Figure 14). Vitrified wheels generally have a maximum safe speed of 6500 SFPM or a little more; organic wheels (resinoid, rubber or shellac) go up to 16,000 SFPM or sometimes higher, but these speeds are generally set by the machine designer, and they are safe speeds for the recommended wheels.

Wet or dry grinding is a factor only in that using a coolant will usually permit the use of about one grade harder wheel than would be used for dry grinding, without as much concern about burning the workpieces. Burning is a discoloration of the workpiece surface caused by overheating. The most common cause is usually the use of a wheel that is too hard.

In any shop, however, unless you are really starting from scratch, there will be some information on what wheels have been used and how they have worked. If a factor seems to need changing, it will probably be grit size or grade. You must remember that the shop probably handles a range of work, and that it does not pay to switch wheels all the time. But change only one element, either grade or grit size, at a time.

self-evaluation

SELF-TEST

1. In the course of a week's grinding you might come up with some of each of the following to grind: bronze valve bodies, steel fittings, tungsten carbide tool inserts, and high speed steel tools. If you could pick the ideal abrasive for each metal, what would you use? List four abrasives. If you were limited to three, which one of the four could be eliminated most easily?

    a. _____

    b. _____

    c. _____

    d. _____

2. Straight (Type 1) and cylinder (Type 2) wheels both have three dimensions: diameter, thickness, and a third. What is the third dimension for each and why is it stated that way?

    _____

3. Five shapes of grinding wheels are described in this unit. Four are for side grinding and two are for peripheral grinding. List the wheels in the two groups either by name or shape number.

    a. _____

    b. _____

4. Tungsten used in the points of automobile engines is very expensive, which makes it necessary to use the thinnest abrasive cutting wheels possible, 6 × .008 × 1 in. What bond would be used and why?

5. Area of contact between wheel and workpiece is probably the most important factor in picking a wheel grade. Five different sets of grinding conditions are discussed, ranging from flat surfacing with a cylinder wheel to ball grinding. List the five in order by wheel grade, starting with the hardest.

    a. _____

    b. _____

    c. _____

    d. _____

    e. _____

6. Here are two wheel specifications, both for straight (Type 1) wheels: (a)

A14-Z3 B, and (b) C14-J6 V. Describe the composition of each wheel in a sentence or two, and suggest the material to be ground by each.

7. Here are two more specifications: (a) C36-K8V and (b) C24-H9V, one for peripheral grinding and one for side grinding. From these specifications, tell which is which.

8. Here is an actual wheel specification: 32A46-H8VBE. Describe the wheel's composition, stating at least the abrasive used, the size of the abrasive, the grade, structure and bond.

9. A wheel specification for cylindrical grinding of a hard steel fitting with a straight wheel is: A54-L5 V. If you were grinding a flat piece of the same steel with a straight wheel, what elements of the specification might change? Which way? For flat grinding of the same material with a segmental or a cup wheel, what further changes might be made?

10. Write one or two sentences about each of the following to show what elements of a wheel specification it affects.
    a. Material to be ground.

    b. Hardness of the material.

    c. Amount of stock to be removed.

    d. Kind of finish required.

unit 2
grinding wheel safety

An old saying among grinding wheel manufacturers is: A grinding wheel doesn't break, but it can be broken. Grinding machine builders go to great lengths to design guards to hold wheel fragments in case of breakage without cutting down on the ease of operation or the productivity of the grinder. They have succeeded to the extent that the modern grinder, as a machine tool, is a pretty safe piece of equipment. A modern grinder, treated with due care and respect, is a safe machine, but it does require a knowledge and the practice of safe methods of operation.

objectives After completing this unit, you will be able to:
1. List the steps in and, if possible, demonstrate checking a grinding wheel for soundness.
2. List and, if possible, demonstrate the preliminary steps in mounting a wheel and starting a machine safely.
3. List at least six things you should do or not do when grinding, aside from those covered above.
4. Demonstrate safe practices in handling and mounting grinding wheels.
5. Demonstrate safe personal practices around grinding machines.
6. Given a wheel diameter, calculate in revolutions per minute its maximum safe speed (6500 SFPM or less). Convert RPMs to SFPMs for given wheel diameters.

information

The "bible" of grinding wheel safety is an American National Standards Institute publication B7.1-1970, "Safety Code for the Use, Care, and Protection of Abrasive Wheels," which was the foundation for the development of OSHA (Occupational Safety and Health Act), federal regulations that now control safety and health in industrial plants. Basically, OSHA gives the employer responsibility for maintaining a clean, safe, healthful plant, and the employees the responsibility for complying with safety rules.

THE GRINDING MACHINE AND SAFETY

Here is the situation on practically all machine shop grinders. In the machine, spinning at 5000 FPM or more, is a vitrified abrasive wheel made of the same material as dishes. The wheel is susceptible to shocks or bumps. It can easily be cracked or broken. If that happens, even though the machine has been designed with a safety guard that will contain most of the pieces (Figures 1 and 2), there is a possibility of broken pieces from the wheel flying around the shop. That could be, at the least, unpleasant and even dangerous. Fortunately, there is not much chance that it will happen; but it could.

Other grinding wheels with organic bonds can be operated safely at speeds over 15,000 FPM, but these are mostly for rough grinding. The wheels are built for it, but the principles are the same. *Every grinding wheel, wherever used, has a safe maximum speed, and this should never be exceeded.*

Any shop grinding machine, properly handled, is a safe machine. It has been designed that way. It should be maintained to be safe. It should be respected for its possibility of causing injury, even though that possibility is low.

Much of the image of grinding wheels and their breakage comes from portable grinders, which are often not well maintained and are sometimes operated by unskilled and careless people.

DETERMINING WHEEL SPEED

Because of the importance of wheel speed in grinding wheel safety, it is important to know how it is calculated. This quality is expressed in terms of surface feet per minute (SFPM). This simply expresses the distance that a given spot (Figure 3) on a wheel travels in a minute. It is calculated by multiplying the diameter (in inches) by 3.1416, dividing the result by 12 to

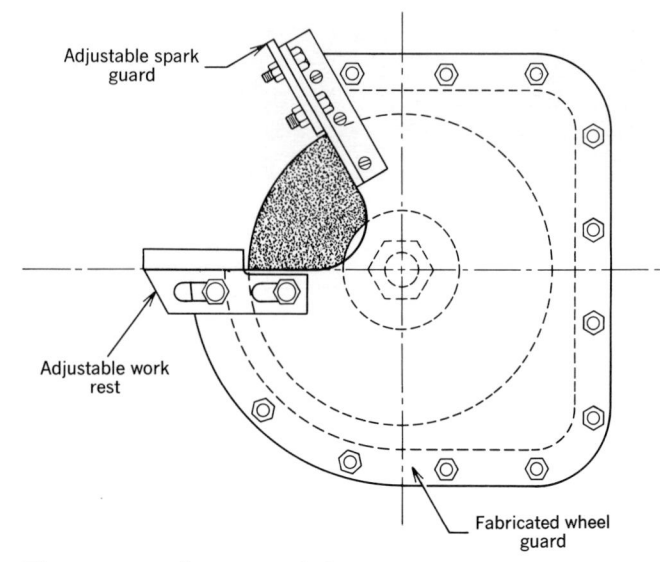

Figure 2. Safety guard for high speed wheel. The work is hand-held against the exposed peripheral grinding face of the wheel on top of the work rest. The squared corners tend to retain fragments in case of wheel breakage.

Figure 1. Safety guard on a surface grinder. Note that the guard is somewhat squared off and covers well over half the wheel (Courtesy of DoAll Company).

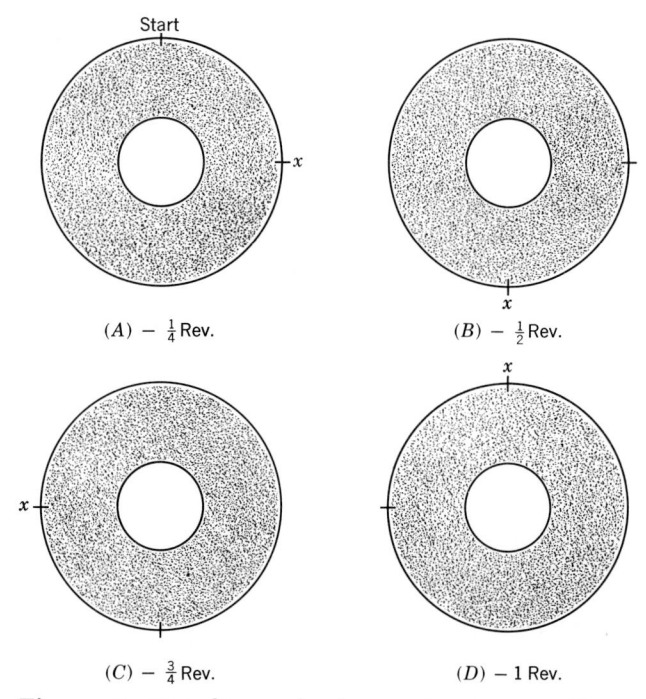

Figure 3. Grinding wheel speeds are usually expressed in surface feed per minute (SFPM). This figure is the distance that a given point (x) on the wheel travels at a given RPM in a minute. The sketch shows one complete revolution. (a) ¼ RPM. (b) ½ RPM. (c) ¾ RPM. (d) 1 RPM (McKee Editorial Assistance).

convert to feet, and multiplying that result by the number of revolutions per minute (RPM) of the wheel. Thus, a 10 in. diameter wheel traveling at 2400 RPM would be rated at approximately 6283 SFPM, under the safe speed of most vitrified wheels of 6500 SFPM.

To find the safe speed in RPMs of a 10 in. diameter wheel, the formula becomes

$$\frac{SFPM}{D \times 3.1416 \div 12}$$

or,

$$\frac{6500}{10 \text{ in.} \times 3.1416 \div 12} = 2483 \text{ RPM}$$

Most machine shop-type flat surface or cylindrical grinders are preset to operate at a safe speed for the largest grinding wheel that the machine is designed to hold. As long as the machine is not tampered with and no one tries to mount a larger wheel on the machine than it is designed for, there should be no problem.

It should be clear that with a given spindle speed (RPM) the speed in SFPM increases as the wheel diameter is increased (Figure 4), and it decreases with a decrease in wheel diameter. Maximum safe speed may be expressed in either way, but on the wheel blotter it is usually expressed in RPM.

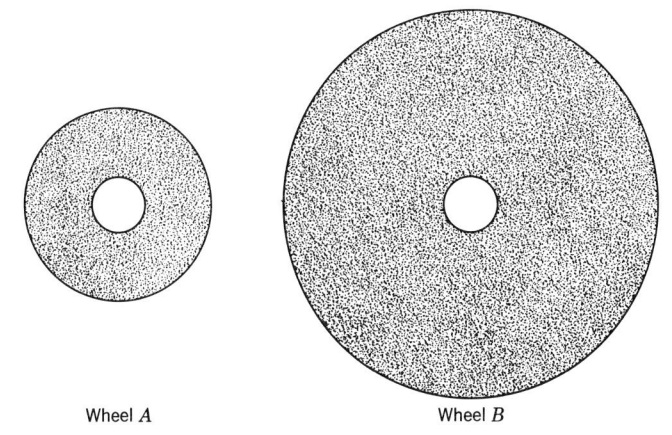

Figure 4. Wheel speed in SFPM, at a given RPM, increases as the wheel diameter increases. If the diameter of wheel B is twice that of wheel A, then the SFPM of wheel B is twice that of wheel A, at the same RPM (McKee Editorial Assistance).

Figure 5. Making a ring test on a small wheel (Lane Community College).

THE RING TEST

The primary method of determining whether or not a wheel is cracked is to give it the ring test. A crack may or may not be visible; of course, if there is a visible crack you need go no further and should discard the wheel immediately.

The test is simple. All that is required is to hold the wheel on your finger if it is small enough (Figure 5) or rest it on a clean, hard floor if it is too big to be held (Figure 6) and preferably strike it about 45 degrees either side of the vertical centerline with a wooden mallet or a similar tool. If it is sound, it should give forth a clear ringing sound. If it is cracked, it will sound dead. The sound of a vitrified wheel is clearer than the sound of any other, but there is always a different sound between a solid wheel and a cracked one.

There are a few permissible variations. Some operators prefer to hold the wheel on a stick or a metal pin instead of a finger. Some shops prefer to suspend large wheels by a sling (chain or cable covered by a rubber hose to protect the wheel) instead of resting them on the floor (Figure 7).

The important point is that the test be done when each lot of wheels is received from the supplier, just before a wheel is mounted on a grinder, and again each time before it is remounted. If a wheel sounds cracked, or even questionable, then discard it or set it aside to be checked by the supplier, and get another wheel which, of course, must also be ring tested.

WHEEL HANDLING

Table 1 provides a list of suggestions for safely and properly handling grinding wheels.

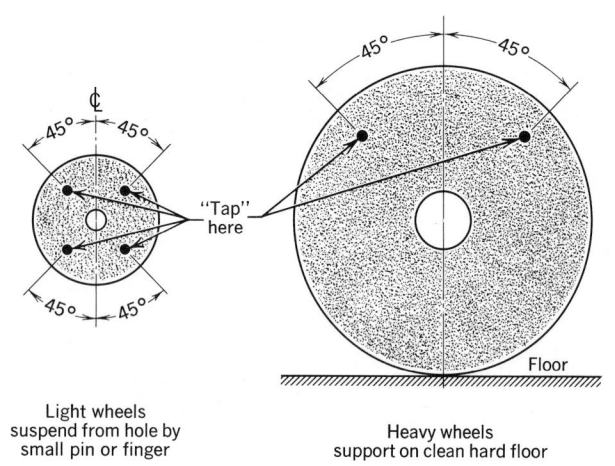

Figure 6. Sketch shows where to tap wheels for ring test, 45 degrees off centerline and 1 to 2 in. in from the periphery. After first tapping, rotate wheel about 45 degrees and repeat the test (Courtesy of Bay State Abrasives, Dresser Industries Inc.).

Figure 7. Wheel being lifted by a sling for ring testing (Courtesy of DoAll Company).

Table 1
Grinding Wheel Safety Rules

1. Handle and store grinding wheels carefully.
2. Wheels that have been dropped should never be used.
3. Inspect all wheels for cracks or chips before mounting. Ring test.
4. Do not alter a wheel to fit the grinding machine and do not force it onto the machine spindle.
5. Operating speed should never exceed the maximum allowable operating speed of the wheel.
6. Mounting flanges must have equal and correct diameter. The bearing surfaces must be clean for using mounting flanges.
7. If the wheel is supplied with a mounting blotter, use it.
8. The mounting nut must not be tightened excessively.
9. Adjust the work rest properly, no more than $\frac{1}{8}$ in. away from the grinding wheel.
10. Never grind on the side of a straight wheel.
11. Use a safety guard that covers at least half of the grinding wheel, and do not start the machine until the guard is in place.
12. Allow the grinding wheel to run at least one min. before using to grind (with the guard in place), and do not stand directly in line with the rotating grinding wheel.
13. Do not force the work into the wheel.
14. Always wear safety glasses.
15. Use the correct wheel for the material you are grinding.
16. Turn off the coolant and let the wheel run several min. to remove excess coolant. If this is not done, the wheel may become out-of-balance.

It should be clear that the wheels ought to be stacked carefully and separated from each other by corrugated cardboard or another buffer (Figure 8). Tools and other materials, particularly metals, should never be stacked on top of the wheels in transit or at the machine. In short, *handle with care.*

WHEEL STORAGE AT THE MACHINE

General wheel storage is a responsibility of shop management, but at the machine this is usually the operator's concern. Operator carelessness can ruin an otherwise good wheel storage plan. On the other hand, care by the operator can do much to help out a substandard shop plan.

If the shop provides you with a proper storage area, all you have to do is to use it and keep it clean. If the shop does not provide such an area, you may need some ingenuity. Here are some suggestions.

1. Keep on hand only the wheels you really need. Do not allow extra wheels to accumulate around your machine. They can become cracked or broken and may get in your way. If you do not need a wheel, take it back to the general storage.

2. Store wheels above floor level, either on a table or under it, on pegs in the wall (Figure 9), or in a cabinet. Be especially careful to protect wheels from each other and from metal tools or parts. Use corrugated cardboard, cloth, or even newspaper to keep wheels apart. *Keep wheels off the floor.*

MACHINE SAFETY REQUIREMENTS

Mounting flanges must be clean and flat, and equal in size, at least one third the wheel diameter (Figure 10). Between each flange and the wheel there should be a blotter, a circular piece of compressible paper to cushion the flange from the rough wheel and distribute the flange pressure. The side of the flange facing the wheel has a flat rim, but the rest is hollowed out so that there is no pressure at the hole, which is the weakest part of the wheel.

The wheel guard must be in place. Depending on the machine design, the safety guard covers half or more of the wheel (Figure 11). On older model grinders the guard tends to be circular in shape and fits closely around the wheel. However, the latest models have square shaped

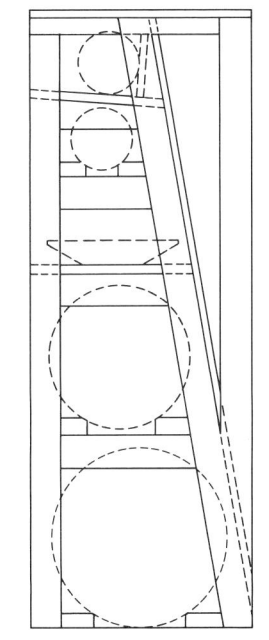

Figure 8. A recommended design for the safe storage of grinding wheels. Corrugated paper cushions should be placed between grinding wheels that are stored flat. The storage racks should be designed to handle the types and sizes of wheels used.

Figure 9. Storing extra wheels at the machine on pegs is often convenient and practical. The main requirement is to keep the wheels separate or protected, and off the floor (Courtesy of DoAll Company).

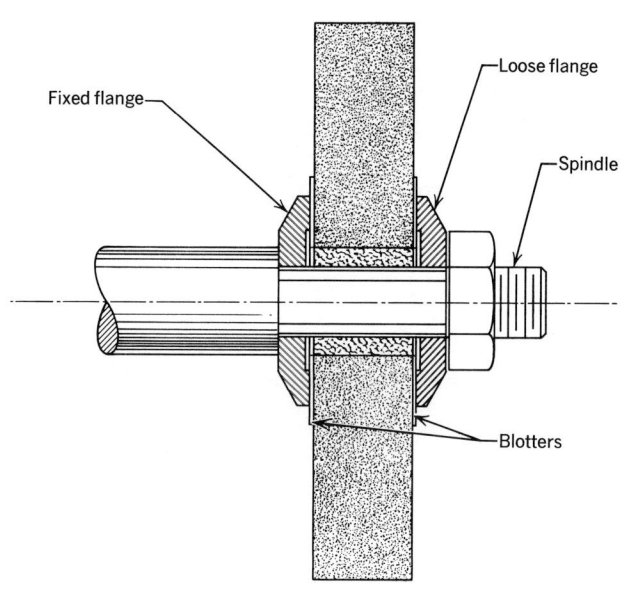

Figure 10. Typical set of flanges with flat rims and hollow centers with blotters separating the wheel and the flanges. Tightening the nut too much could spring the flanges and perhaps even crack the wheel.

Figure 11. Well-designed guard for a bench grinder. Note that the side of the guard, like the one on the surface grinder, is easily removed for access to the wheel (Lane Community College).

safety guards, sometimes called cavernous, on the theory that pieces of any wheel that breaks will be retained in the corners instead of being slung out against the machine or possibly the operator.

MOUNTING THE WHEEL SAFELY, STARTING THE GRINDER

Any time you mount a wheel on the grinder or start the machine, good safety practice requires a set routine like the following, which is covered in detail in Unit 8.

1. Ring test the wheel as shown in Figure 5, then check the safe wheel speed printed on the blotter with the spindle speed of the machine. The spindle speed must never exceed the safe wheel speed, which is established by the wheel manufacturer after considerable research.
2. The wheel should fit snugly on the spindle or mounting flange. Never try to force a wheel onto the spindle or enlarge the hole in the wheel. If it is too tight or too loose, get another wheel.
3. Be sure the blotters on the wheel are larger in diameter than the flanges, and be sure the flanges are flat, clean, and smooth. Smooth up any nicks or burrs on the flanges with a small abrasive stone. Do not overtighten the mounting nut. It just needs to be snug.
4. Always stand to one side when starting the grinder; that is, stand out of line with the wheel.
5. Before starting to grind, let the wheel run at operating speed for about a minute with the guard in place.

Steps 4 and 5 are essentially a double check on the ring test. If anything is going to happen to a wheel, it usually happens very quickly after it starts to spin.

OPERATOR RESPONSIBILITIES

In grinding, there are a number of operator duties that have long been a part of company safety policy and state safety regulations, and that, with the passage of the OSHA regulations, also became a matter of federal regulations. These are mostly designed for your protection, and not, as some operators have thought, to make the job more awkward or less productive.

1. Wear approved safety glasses or other face protection when grinding (Figure 12). This is the first and most important rule.
2. Do not wear rings, a wrist watch, gloves, long sleeves, or anything that might catch in a moving machine.

Figure 12. Approved safety glasses are required for all grinding. Today, as a matter of fact, they are usually required for anyone, even visitors, in the shop area (Courtesy of Sellstrom Manufacturing Co.).

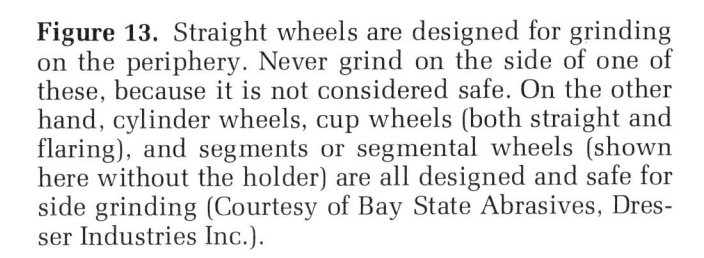

Figure 13. Straight wheels are designed for grinding on the periphery. Never grind on the side of one of these, because it is not considered safe. On the other hand, cylinder wheels, cup wheels (both straight and flaring), and segments or segmental wheels (shown here without the holder) are all designed and safe for side grinding (Courtesy of Bay State Abrasives, Dresser Industries Inc.).

3. Grind on the side of the wheel *only* if the wheel is designed for this purpose, such as Type 2 cylinder wheels, Types 6 and 11 cup wheels, and Type 12 dish wheels (Figure 13).
4. If you are grinding with coolant, turn off the coolant a minute or so before you stop the wheel. This prevents coolant from collecting in the bottom half of the wheel while it is stopped and throwing it out of balance.
5. Never jam work into the grinding wheel. This applies particularly in off-hand grinding on a bench grinder.

self-evaluation

SELF-TEST 1. A 12 in. vitrified wheel is available for mounting on a 2000 RPM grinder. Can this wheel be used safely on the grinder? Show the calculations to prove your answer.

2. List at least three important requirements for a ring test.

a. _____

b. _____

c. _____

3. There are at least three times when a wheel should be ring tested. What are they, and why is the test needed each time?

a. _____

b. _____

c. _____

4. Handling wheels carefully is most important. List at least four specific precautions to be taken in handling grinding wheels.

a. _____

b. _____

c. _____

d. _____

5. Wheel flanges are not always considered as important as they really are. Put down at least four points to be considered in their selection and care.

a. _____

b. _____

c. _____

d. _____

6. There are five or six essential steps to be followed in starting up a grinder. List them in order.

a. _____

b. _____

c. _____

d. _____

e. _____

f. _____

Unit 2 Grinding Wheel Safety

7. Providing a safe place to work is an employer's responsibility, but an operator also has safety responsibilities. List at least four things that an operator should or should not do to operate a grinder safely.

   a. _____

   b. _____

   c. _____

   d. _____

8. Explain the relationship of wheel diameter and spindle speed (RPM) to wheel speed in SFPM.

   _____

9. At the factory, grinding wheels are tested running free at 150 percent of safe operating speed. Why, then, isn't it all right to operate the wheels at the same higher speed?

   _____

10. Safe grinding practice requires that you **not** wear certain things (clothing and similar items) around a grinding machine. Name at least three.

    a. _____

    b. _____

    c. _____

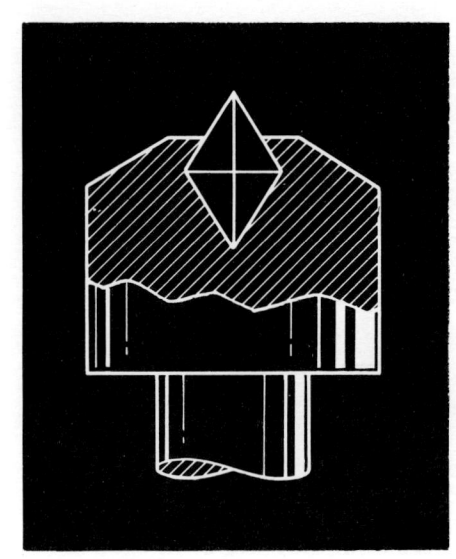

unit 3
care of abrasive wheels: trueing, dressing, and balancing

This unit focuses on the care of a grinding wheel *as a cutting tool* to keep it running true, that is, with every spot on the outside diameter (OD) of the wheel at the same distance from the center of the spindle, keeping it sharp, and keeping it in balance. All three conditions must be met if the wheel is to perform at its best as a cutting tool, producing dimensionally accurate parts with good to superior surface finish.

objectives After completing this unit, you will be able to:
1. Make a check list of steps to be followed in trueing a grinding wheel.
2. Make a similar list for dressing a wheel, including the variations required for a wheel to be used for roughing as against those for a wheel intended for finishing.
3. Define form dressing, listing at least three types of form dressers, and describe how each operates.
4. List steps to be followed in balancing a grinding wheel. Include frequency of balancing and wheels that must be balanced.

information

Ideally, a grinding wheel should be so well matched to the work that grains will pull out of the wheel as they become dull and bits of the work material never become lodged in the wheel. In practice, however, especially in shops where a variety of metals must be ground with as little wheel changing as possible, this rarely happens. Hence, silicon carbide and aluminum oxide wheels must be dressed, that is, sharpened. Diamond and cubic boron nitride wheels are not dressed as often, and never as severely, because of the expense of the abrasive grain and partly because they can be more accurately mounted. In addition, both kinds of grains are so much harder than aluminum oxide and silicon carbide that they do not wear as fast.

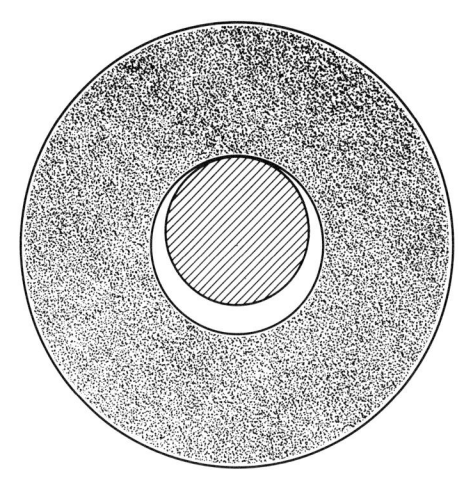

Figure 1. Sketch, exaggerated for effect, showing a grinding wheel on a spindle. Actually, the gap under the spindle may be only a few thousandths of an inch, but still enough to cause problems if the wheel is not trued to the center of the spindle.

Figure 2. Since the hole in the core of a diamond or CBN wheel is machined, it can be fitted to much closer tolerances (K & M Industrial Tool, Inc.).

Figure 3. Single point diamond dresser. The most important precaution in using such a dresser is to turn the diamond often to avoid grinding flats on it. This diamond is pointing to the right (Photograph courtesy of The Desmond-Stephan Manufacturing Co.).

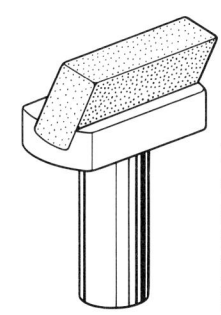

Figure 4. Cluster-type dressers have come into use mostly because several smaller diamonds are cheaper than one large diamond (Photograph by courtesy of The Desmond-Stephan Manufacturing Co.).

Trueing and balancing of silicon carbide and aluminum oxide wheels are essential in the first place, because it is difficult to mount common abrasive wheels true to the centerline of the spindle. Clearance at the bottom of the center hole of the wheel is always a little greater than at the top, where there is none. Figure 1 illustrates the point, although the clearance is, of course, greatly exaggerated. The machined hole of a diamond wheel (Figure 2) makes possible a much closer fit on the spindle. Balancing is usually a factor for wheels with a 12 in. diameter and larger, because of the weight involved. Below that size, an out-of-balance condition would probably cause problems only in supercritical finishing. Wheels are in balance when they leave the factory, but not necessarily after they are mounted on the spindle with mounting flanges, as is often the case with large wheels on several types of grinders. To provide good finish, wheels must be true and in balance.

It is important to realize that while a grinding wheel is removing metal from a workpiece, the workpiece is also removing material from the wheel, although, of course, at a much slower rate. This reverse step of the operation means that the wheel eventually will have to be sharpened or dressed by removing material (the dulled abrasive grain and some bond material) from the wheel. Dressing of precision, vitrified-bonded wheels is typically done with a diamond tool, usually a single diamond in a holder (Figure 3), that can be moved back and forth across the grinding face of the rotating wheel. Sometimes a number of smaller diamonds are imbedded in a metal matrix, a comparatively soft material with the diamond chips across its face (Figure 4). This second type of dresser, sometimes called a cluster dresser, may be wide enough to cover the grinding face of the wheel without being traversed. These are used for dressing a flat grinding surface on the wheel.

Even for a flat grinding face it is usually considered good practice to dress a little radius between the grinding face and the side of the wheel, since a sharp corner could leave undesirable marks on the surface of the workpiece. This is done by swinging the dresser in a small freehand arc around the corner or, to use the shop term, breaking the corner. Holding a dressing stick lightly against the corner for a couple of seconds will probably do the job even more easily.

As a point of information, dressing of resinoid bonded grinding wheels for snagging and other rough grinding operations, where the only purpose is the quick and cheap removal of metal, is done with a Huntington dresser (Figure 5). This type is made up of a group of spurlike discs, strung on a shaft in a holder with a handle. The discs can spin freely, while the dresser is held against the surface of the rotating wheel, and the points remove dull grains and embedded metal to renew the grinding face (Figure 6a). It is worth noting, also that grinding wheels can be used to dress other grinding wheels (Figure 6b).

Vitrified grinding wheels can also grind forms (slots, grooves, and contours) in workpieces. Mostly this is done by form-dressing the wheel to the *reverse* of the form desired in the workpiece. That is, where there is a ridge in the workpiece, there is a corresponding valley in the wheel. Sometimes this is done as a finishing operation on a form started by some other operation, such as milling. More often, however, the economical way is to grind the form from the solid, or flat — in short, doing the whole job by grinding.

For example, if you need a rounded groove in the workpiece, there are radius dressers available to swing the single point diamond dresser in an arc of the correct radius. This is done mechanically and more accurately than the job referred to earlier as breaking the corners of the grinding wheel (Figure 7). Refinements of this kind of mechanical linkage make it possible, for instance, to reproduce on the grinding face of a wheel some fairly complicated contoured forms. It should be clear by now that, for all its capabil-

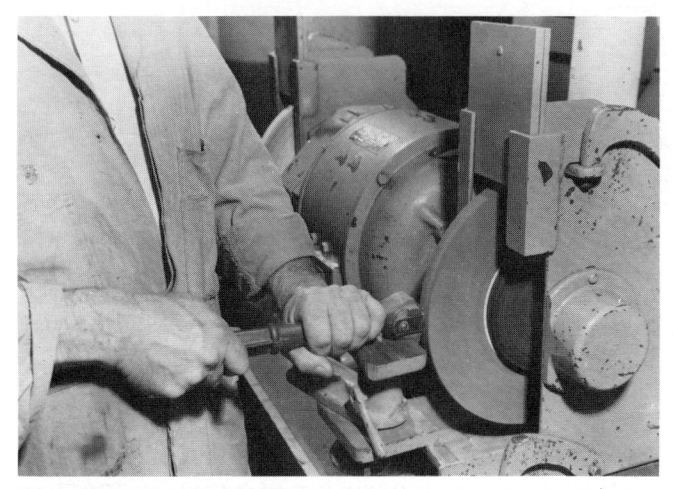

Figure 6a. Huntington dresser set up for dressing. The dresser has a hook or lug that fits over the workrest of the grinder (Photo courtesy of Norton Company).

Figure 6b. Sometimes it is practical to use one grinding wheel to dress another. The abrasive wheel in the dresser is being used with a metal-bonded diamond wheel on the grinding machine (Photo courtesy of Norton Company).

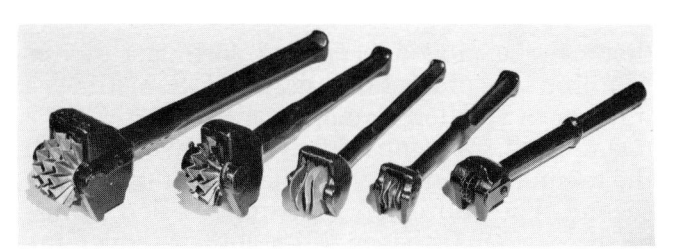

Figure 5. Huntington-type dresser (sometimes called a star) used for coarse wheels intended for rough grinding (Photo courtesy of Norton Company).

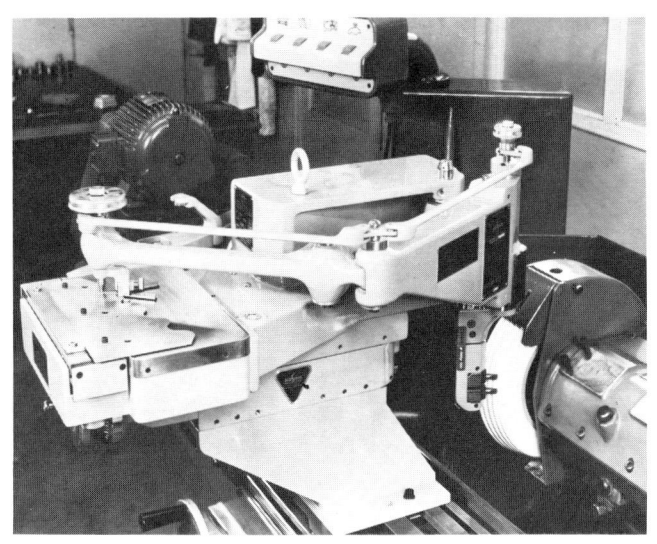

Figure 7. A dressing unit such as this can be set to dress practically any desired shape in a wheel. It is very versatile (Courtesy of Engis Co., Diamond Tool Division).

Figure 8. This is a crush roll, which literally crushes out the form in the wheel, instead of cutting it, as with diamond rolls or blocks (Courtesy of DoAll Company).

Figure 9. The crush roll must be mounted to provide support against considerable force, either as here on a center-type cylindrical grinder, or on a surface grinder (Courtesy of Bendix Automation and Measurement Division).

ity to remove material, the vitrified grinding wheel can be readily formed.

Of the several kinds of form dressers available, the crush roll and the formed diamond dresser deserve brief mention, although both are production tools. In crush forming with the crush roll, the rapidly rotating wheel is pressed with considerable force against a rotating crush roll (Figure 8), which has the form desired in the finished workpiece. Once the form is crushed into the grinding face of the wheel (Figure 9), the wheel can be used to grind the form into workpieces. This operation can be done on either flat or cylindrical surfaces with, in the same order, either a surface or a center-type cylindrical grinder. Crushing leaves the abrasive grain very sharp; this does not result in the best finished surface on the work, although it is adequate for most uses.

Diamond roll form dressers or diamond dressing blocks (Figure 10) can be made for many of the same applications as crush rolls and also for other applications. The action of this dresser involves less pressure and more cutting of the abrasive grain, so that the wheel produces a somewhat better surface finish. Both crush rolls and diamond form rolls are expensive and specialized tools that can only be justified by large production and substantial savings over alternate

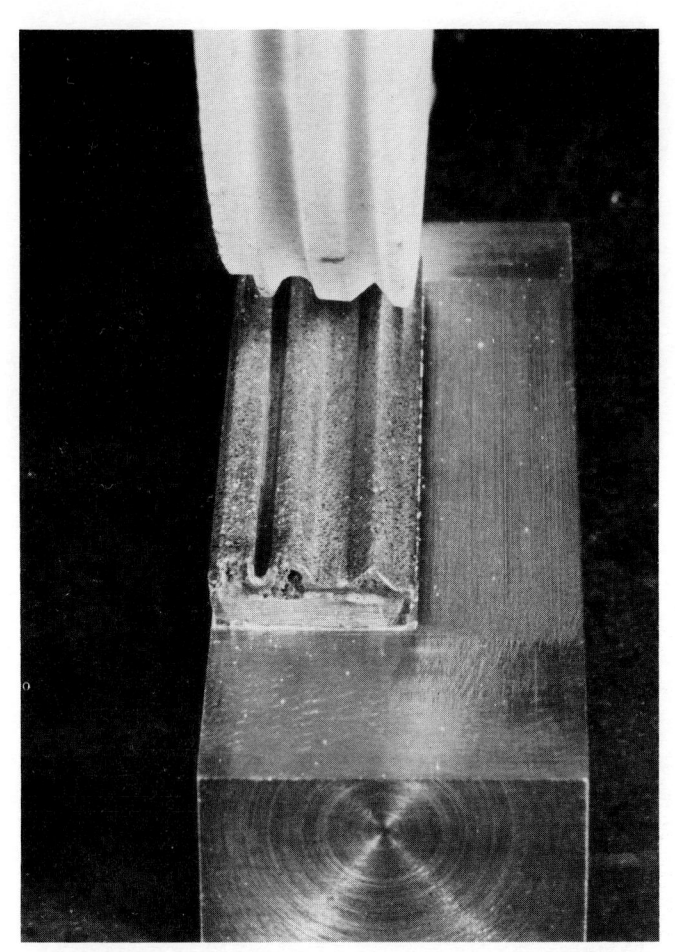

Figure 10. Diamond-plated dressing block, intended for use on a surface grinder. The block is held flat on the magnetic chuck, and the wheel is traversed back and forth along it. The block is formed to the shape of the finished workpiece (Courtesy of Engis Co., Diamond Tool Division).

Figure 11. Overhead-type dresser for a surface grinder. The calibrated micrometer dial controls the downfeed of the diamond. The handle at the left cross feeds the diamond back and forth. In operation, of course, the wheel and the dresser would be concealed by the front piece of the safety guard (Courtesy of DoAll Company).

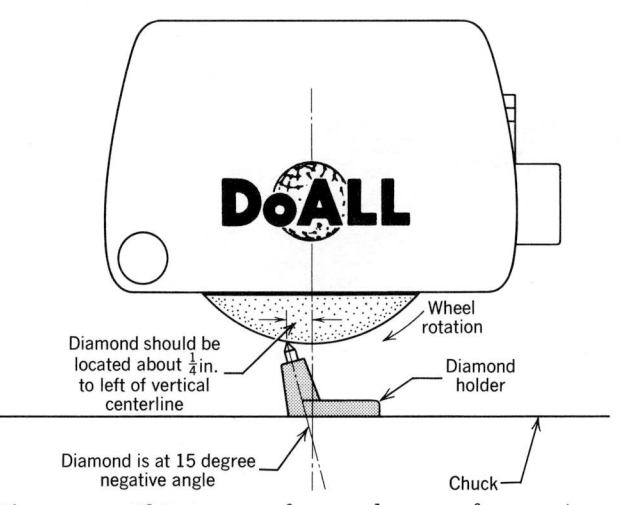

Figure 12. This is one of several ways of mounting a dresser on a surface grinder. The dresser with its diamond is simply spotted on the clean magnetic chuck. Note, however, that the diamond is slanted at a 15 degree angle and slightly past the vertical centerline of the wheel, as the wheel turns (Courtesy of DoAll Company).

methods. Each new form requires a new set of rolls.

Chances are that you will most often be dressing the wheel so that the peripheral surface is concentric with the center of the machine spindle and parallel to the centerline of the spindle, and with a single point diamond dresser. The dresser may be mounted above the wheel (Figure 11), on the magentic chuck that is the workholder of the machine (Figure 12), or to the side (Figure 13). In any of these positions, it is possible to move the dresser back and forth across the face of the wheel. This is called traversing the dresser.

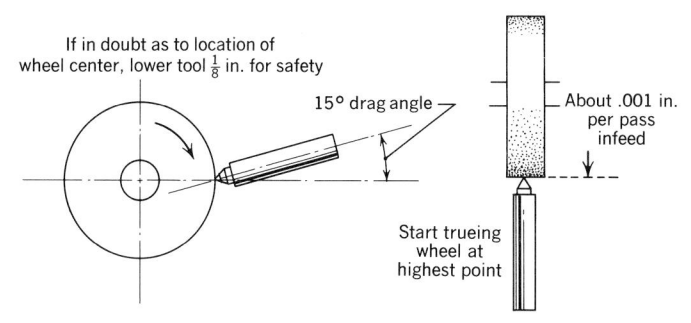

Figure 13. This illustrates the idea of wheel dressing instead of any specific setup. Dressing is rarely done free hand. Note that the diamond is always a little past the centerline and on an angle. (Photograph by courtesy of The Desmond-Stephan Manufacturing Co., Urbana, Ohio).

The diamond is mounted at an angle of 15 degrees, so that it contacts the wheel just after the low or high point of the wheel, or just below the centerline, depending on the location of the dresser (Figures 11, 12 and 13). Thus the wheel is cutting toward the point of the diamond. Remember that dressing with a diamond is always a two-way operation; that is, the diamond is kept sharp while it sharpens the wheel grinding face. Often there is an arrow or other indicator on the dresser to indicate its position. The diamond always points in the direction of the wheel's rotation.

Any new wheel, or a wheel that has just been reflanged, must first be trued. In trueing, the wheel and the dresser are brought together so that the dresser is touching the high point of the wheel. Otherwise, the traversing of the diamond might cause it to dig too deeply into the wheel, which could ruin the diamond. On the other hand, if you start at the high point, which is the point furthest from the center, the cross cut or traverse is short at first and gradually becomes longer until finally you are dressing the entire width of the wheel.

Infeed of the diamond into the wheel should be light, about .001 in. per pass; if dressing is being done dry, there should be frequent pauses, after every three or four passes, to allow the diamond to cool off. A hot diamond can be shattered if a drop of water or other liquid hits it. Turn the diamond frequently. This helps to keep it sharp.

Trueing the wheel is accomplished, then, by moving the dresser back and forth across the grinding face of the wheel while the wheel is rotating at operating speed. It is preferably done wet. If trueing is done wet, continuous coolant must be assured. If it is done dry, as it must be on many machine shop machines, dressing must be interrupted at intervals to allow the diamond to cool off.

The speed of traverse is probably the remaining point of concern. Generally, the faster the dresser traverses across the wheel's grinding face, the sharper the dress will be and the better suited the wheel will be for rough grinding. Slower traverse means that the diamond does more cutting on the abrasive, dulling it a little bit, so that the wheel is better suited to finishing than to roughing work.

With a new wheel or one that has just been mounted between a pair of flanges, the first step is to true the wheel. This is done as above, by traversing a diamond tool back and forth across the grinding face of the wheel until the dresser is contacting the wheel at all times.

After trueing, the wheel may or may not need to be dressed. It depends on the surface. If the wheel is to be used for roughing, then the surface should be open and the grains should be sharp. If the wheel is to be used for finishing, the grains should be a little duller. Sometimes when a wheel is intended for finishing, it is good to take a few passes across the wheel without any infeed. The point is that with a little experience you can tell the degree of sharpness that is needed in the wheel face and dress the wheel accordingly, depending on what you want to do with it afterward. As grain becomes dulled, it tends to polish more and cut less, and size becomes less of a factor in the action of the grain.

It was mentioned earlier that coarse grain is recommended for cutting and material removal and fine grain is recommended for finishing. This is true, provided both are in the same degree of sharpness. It can now be said that by proper dressing, it is possible to make a finish with comparatively coarse grain like that of a much finer grain. For example, a 46 grit wheel, dressed to dull the grain, could give the same finish as a 120 grit wheel. The reverse is not true, however.

The special situation of diamond and cubic boron nitride wheels has already been mentioned. Such wheels can be centered on the spin-

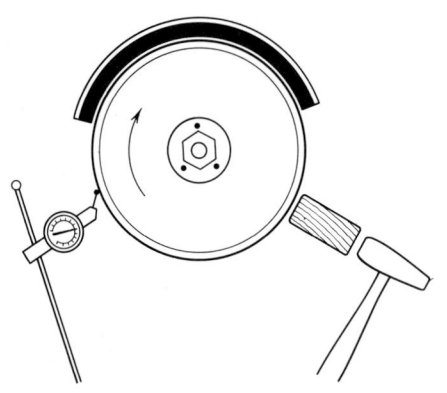

Figure 14a. Runout on a diamond wheel is checked with a dial indicator and must be within .0005 in. for resinoid wheels or .00025 in. (half as much) for metal bonded wheels. Tapping a wooden block held against the wheel to shift the wheel on the spindle is often enough to bring it within limits. Otherwise, it will have to be trued (Courtesy of Precision Diamond Tool Co.).

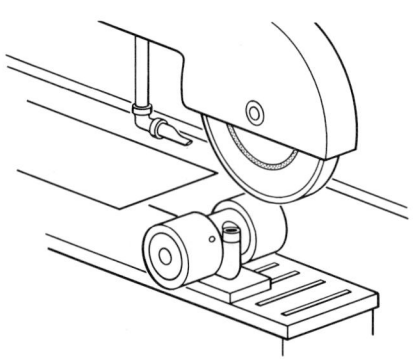

Figure 14b. Trueing a diamond wheel, when necessary, is best done with a brake-type trueing device, as shown. For resinoid wheels, the job may also be done by grinding a piece of low carbon steel (Courtesy of Precision Diamond Tool Co.).

dle with more accuracy. Furthermore, it is critical that everything else (spindle, flanges, spacers) be running true and be clean and free from burrs. If there is anything amiss, it should be corrected.

Diamond wheels are customarily trued to the bore in the manufacturing process. But if a wheel runs out more than .0005 in. (resinoid bond) or .00025 in. (metal bond), it can usually be trued by lightly tapping a wooden block held against the wheel (Figure 14a). If this does not work, then the wheel must be trued. This can be done with a brake-type trueing device (Figure 14b). Resinoid bond wheels can be trued by grinding a piece of low carbon steel.

After trueing, as with other types of wheels, it may be necessary to dress the wheel. For this, use either a dressing stick (usually provided with the wheel) or lump pumice.

BALANCING

Once the wheel is trued and dressed, it is ready to be balanced, although this step may be eliminated (except perhaps for particularly critical work) if the wheel is 12 in. in diameter or smaller. In such small wheels the weight of the wheel is not great enough to disturb the finish if the wheel is a little out of balance. For wheels 14 in. in diameter and larger, balancing is important. It might be noted that there are no 13 in. standard wheels manufactured.

Balancing of grinding wheels is done for the same reasons, on similar equipment, and by similar methods as the balancing of automobile tires. The job is done on either balancing ways with overlapping discs (Figure 15) or on parallel ways

Figure 15. This type of balancing device with overlapping wheels or discs is quite common. It has an advantage in that it need not be precisely leveled (Courtesy of Bay State Abrasives, Dresser Industries, Inc.).

Figure 16. Balancing a wheel on two knife edges, as on this unit, is very accurate, because there is minimum friction. Of course, the unit must be perfectly level and true. Otherwise, the wheel may roll from causes other than out-of-balance (Courtesy of Bay State Abrasives, Dresser Industries, Inc.).

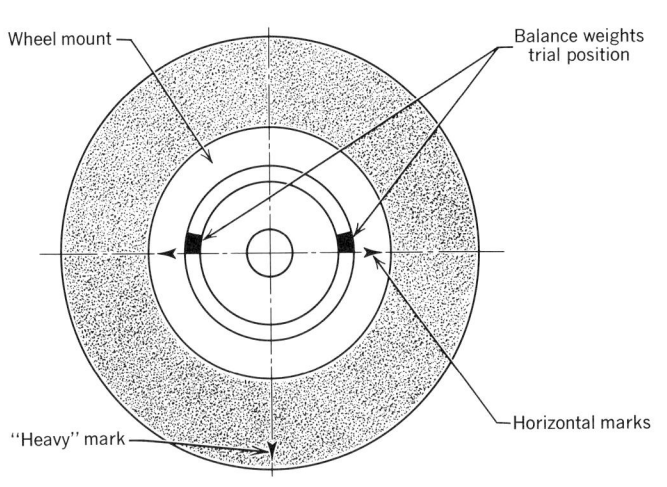

Figure 17. With the weights at some point between the vertical and the horizontal centerlines, the wheel should be in proper balance, stationary in any position. If not, one or two other balance weights should be used (Courtesy of Bay State Abrasives, Dresser Industries, Inc.).

(Figure 16), which must, of course, be perfectly level. The wheel, on a balancing arbor, is placed on the ways. The heavy point of the wheel is marked with chalk after the wheel has been allowed to come completely to rest. A horizontal line is also drawn through the center of the wheel. Then weights in the flanges (Figure 17) are shifted for a trial balance and the heavy point is turned to the top. This sequence of shifting weights and turning the wheel is continued until the wheel remains at rest in any position. This is called static balancing. Some production machines, usually some form of cylindrical grinders, are equipped so that the wheel can be automatically balanced while it is still in the machine.

self-evaluation

SELF-TEST

1. What is meant by trueing a grinding wheel, and why is it important?

2. All grinding wheels are balanced at the factory. Why and under what circumstances is it necessary to rebalance it? On what size wheels?

3. Explain the difference between dressing and trueing a wheel. How often should a wheel be dressed?

4. In your own words, define form dressing. Mention at least two methods of form dressing and one advantage and one disadvantage of each.

    a. _____

b. _____

5. Explain why it takes longer to get a diamond or cubic boron nitride wheel ready to grind than it does an aluminum oxide wheel. Be specific as to details.
   _____
   _____

6. Explain the essentials in the placement of a single point diamond dresser.
   _____

7. List the steps in trueing an aluminum oxide or silicon carbide wheel after mounting.
   _____
   _____

8. Explain the essential differences between dressing a grinding wheel for roughing as against finishing.
   _____

9. Under what circumstances might it be necessary to true a diamond wheel and how is it done?
   _____
   _____

10. Generally, what determines whether a wheel needs to be balanced? Describe the procedure.
    _____
    _____

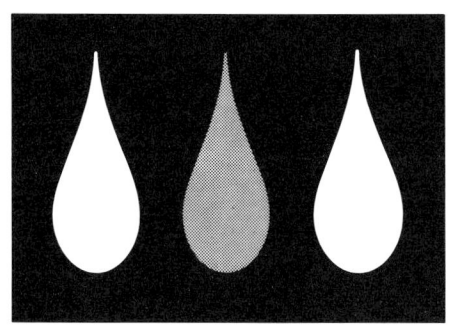

unit 4
grinding fluids

Grinding produces very high temperatures and is really a very hot cutting process. Temperatures at the interface (the small area where the abrasive grains are actually cutting metal) are reliably estimated to be over 2000° F, and that is enough to warp even fairly thick workpieces. Nor is it safe to assume that just because there is a lot of grinding fluid (sometimes called coolant in the shop) flowing around the grinding area, the interface is cooled. With the grinding wheel rotating at its usual 5000 SFPM plus rate, it creates enough of a fan effect to blow coolant away from the interface. This creates a condition sometimes referred to as "grinding dry with water." In other words, in spite of the amount of coolant close by, the actual cutting area may be dry and hot.

Furthermore, it is not just a matter of *one* coolant. There are at least three principal groupings, many subgroups, and dozens of brand names. So there is a concern about which coolant to choose; you should know which main group to pick and why.

objectives After completing this unit, you will be able to:
1. List three principal jobs of a grinding fluid.
2. List the three principal types of grinding fluids with their major advantages and disadvantages.
3. Sketch a design of an effective nozzle for flood coolant application, and explain briefly how it works
4. Explain the advantages and disadvantages of mist coolant and through-the-wheel application of coolant.
5. Explain the advantages and disadvantages of various coolant cleaning methods.

information

The terms grinding fluids and coolants will be used interchangeably; the first is the engineering term and the second is the shop term. Coolants are used because they make grinding a more efficient operation. They help produce a better product the first time with less touch-up work to be done. Some specific notations about coolants are as follows.

☐ They reduce the temperature in the workpiece, thus reducing warping, especially in thin workpieces, making a more accurate product.

☐ They lubricate the interface between the wheel and the workpiece, making it more difficult for abrasive particles and bits of metal to stick in the wheel's grinding surface. Coolant

also helps to "soften" the abrasive action and produce a better finish on the workpiece.
☐ They also flush away the bits of metal and abrasive (called swarf) into the coolant tank, where they can be filtered out and not recirculated to make random scratches on the finished surface of the workpiece.

Our discussion, then really begins from four generally accepted statements.

1. Grinding fluids or coolants are necessary in practically every type of grinding. But it would be simpler from the operator's viewpoint, for machine design, and for shop housekeeping if they were not needed.
2. Coolants vary greatly in qualities. Although there are some general rules (mostly worked out from experience) for selecting one over another, it is often puzzling as to why one works in a given situation while a similar one does not.
3. The design of the coolant and filtration system is highly important. These are not an operator's responsibility, but you should always be alert to signs that the filter is not working properly; "tramp" metal and abrasive recirculated over the grinding area can cause poor finish very quickly. Fortunately, these signs are easily recognized as random scratches without any pattern. The condition is also referred to as "fishtailing."
4. The method of application, and particularly the design and placement of the nozzle that delivers the coolant to the grinding area, are at least as important as the selection of the coolant itself. It is even possible that redesigning the nozzle and making sure that it is always located where it would do the most good is even more important than coolant selection.

Figure 1. This is a very common method of flood coolant application. For the photograph, the volume of coolant has been reduced (Lane Community College).

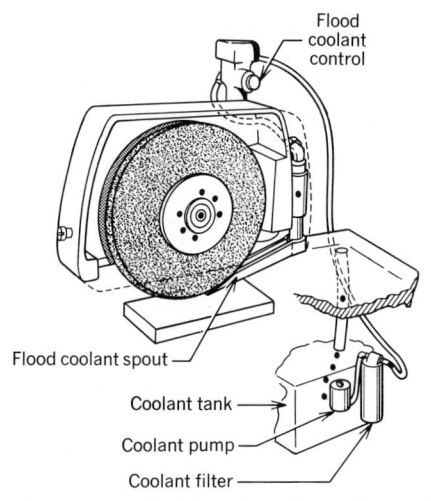

Figure 2. Fluid recirculates through the tank, piping, nozzle, and drains in flood grinding system (Courtesy of DoAll Company).

METHODS OF COOLANT APPLICATION

Most grinding fluids are applied in what is called flooding. A stream of coolant under pressure is directed from a pipe, sometimes shaped but often just round, in the general direction of the grinding area (Figures 1 and 4). The fluid collects beneath the grinding area, is piped back to a tank where it is allowed to settle, and is cleaned; once more it is pumped back around the wheel (Figure 2). There is always a little waste, and periodically either more coolant concentrate or water is added to the solution. Flooding is an effective method of applying coolant; provided the solution stays within the effective range, neither too rich nor too lean, it works very well. However, the fluid cannot just be in the vicinity of the grinding area; it must be right *in* that area if it is to do its job. For that purpose, particularly on high speed work, something like the nozzle shown in Figure 3 may be needed. In some cases, for example, on a small surface grinder processing small parts, the nozzle can almost be pointed at the grinding wheel-workpiece interface (Figure 4), although

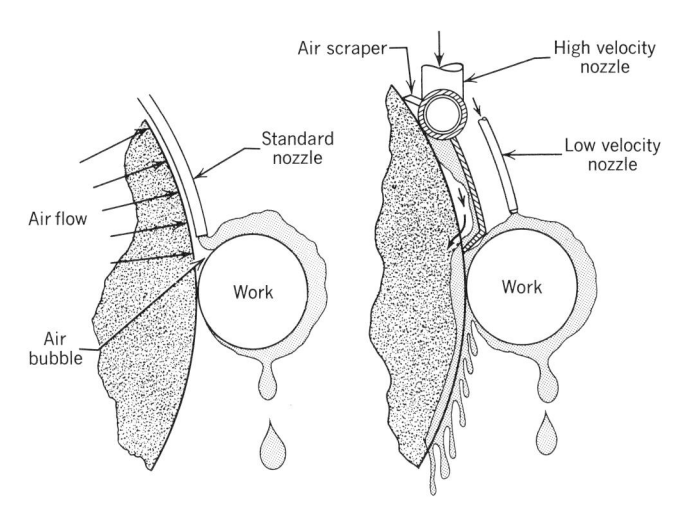

Figure 3. A specially designed nozzle like this helps to keep the fanlike effect of the rapidly rotating wheel from blowing coolant away from the wheel-work interface (Courtesy of Cincinnati Milacron, Products Division).

Figure 4. Here is another coolant application method. Under these circumstances, the pipe apparently supplies enough fluid (Courtesy of Diamond Abrasives Corporation).

here the flow of coolant may have been cut down a little when the picture was taken. Often the nozzle is lower down and closer to the wheel-work interface.

Mist lubrication is a second method of applying coolant, often used where there is an occasional need for wet grinding. In contrast to flood cooling, where fluid flow is measured in gallons per minute, mist coolant may be measured in ounces per day. The mist lubricator (Figure 5) is a small nozzle through which a mist of coolant is blown; the mist evaporates as it hits the work-wheel interface and does a remarkable job of cooling. The heat required to evaporate the coolant is considerable. The lubricating depends on the fluid used, and there is, of course, little or no removal of swarf. However, it provides an unobstructed view of the wheel-work interface; for that reason is popular with the operator who occasionally runs across the job that simply cannot be ground dry. Finally, it requires no return piping.

It was mentioned earlier that vitrified grinding wheels are porous, not solid, and this characteristic is put to work in through-the-wheel grinding fluid application. As noted in Figure 6, the wheel flange has holes leading toward the wheel. As the wheel revolves, grinding fluid is fed through the curving tube into the circular groove, through the holes into the wheel and, finally, out through the pores in the wheel to the

Figure 5. Mist grinding fluid application is sometimes used on a normally dry grinder. The cooling effect is excellent, lubrication practically nil, but no recirculating system is required (Courtesy of DoAll Company).

grinding area and to the inside of the safety guard. Centifugal force keeps the fluid moving outward. This approach has a lot of merit. It absolutely assures that the fluid will get to the wheel-work interface. It must be said, however, that at the same time it assures that fluid will be sprayed all over the inside of the guard. The fluid must be superclean. Otherwise it will clog the pores in the wheel. The guard must be very clean, or dirty fluid could drop onto the work surface.

Figure 6. Nozzle and flange design for through-the-wheel application. This allows coolant to filter through the wheel right to the cutting area. It also sprays coolant over the inside of the wheel guard. (Courtesy of DoAll Company).

As a point of interest, most machine shop-type machines have individual coolant tanks and recirculating systems that include one or more means of cleaning the fluid (Figures 7, 8, and 11). Many production grinding shops, however, have central systems that supply all the machines in the plant. Here, the coolant tanks may be larger than many swimming pools, and the cleaning system may be quite complex. However, in this discussion, we will be referring to the individual machine type, which is more in keeping with the subject matter. The principles are the same in either case.

TYPES OF FLUIDS

The three principal reasons for coolants (cooling, lubricating, and flushing away swarf) have already been mentioned. In the following discussion, each type of coolant covered will be rated in terms of these and perhaps other requirements.

Water, for example, is excellent for cooling. It carries away heat very well. However, water by itself has practically no lubricating ability, although it is good for cleaning away swarf. It also has the disadvantage of causing rust on ground surfaces. For these reasons, water by itself is not a

Figure 7. This is a settling tank for coolant and is used in connection with a single machine, in this case a cylindrical grinder. On some heavy stock removal operations, such as on a vertical spindle rotary surface grinder, the tank might have a conveyor to pull the chips off the bottom of the tank and out (Courtesy of Cincinnati Milacron).

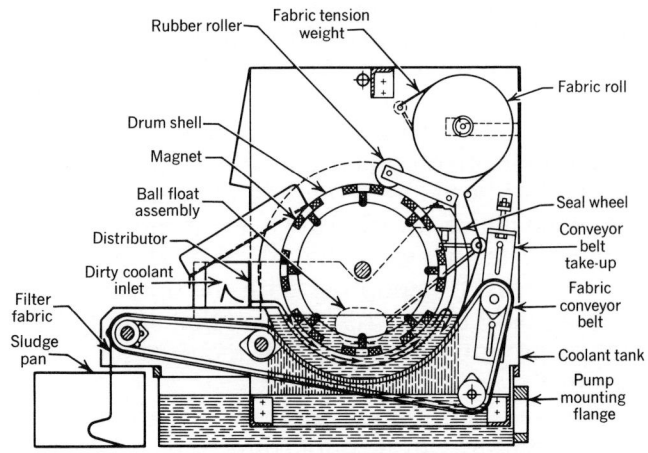

Figure 8. One of the possibilities of a combination filter for a single machine. This combines magnetic filtering to remove iron and steel with settling. Steel or iron chips picked up by the magnets are scraped off into the chute and fall down onto the dirty side of the filter fabric. Other swarf picked up by the filter goes along with the steel chips into the sludge pan (Courtesy of Barnes Drill Co., Rockford, Ill.).

good grinding fluid. However, the addition of water soluble chemicals or water soluble oils to plain water (which make two different types of water based coolants) makes up the larger share of coolants used in grinding.

Water soluble chemical grinding fluids (often called chemical fluids or coolants) are transparent and are the largest single group of grinding fluids used. Their cooling qualities are excellent and their lubricating ability is adequate in the medium to high stock removal applications where they are usually used. The additives include rust preventatives, detergents, water soluble polymers for better lubrication, and bactericides to control bacteria that cause "Monday morning odor," which is often a problem. During the workweek the agitation of the fluid helps control bacterial growth, but bacteria flourish over the weekend in the still water of the coolant tank.

Chemical fluids score high on cooling, good enough for all but the heaviest grinding jobs on lubrication and satisfactory on cleaning. A critical point for the operator on a machine with an individual system, if the additives come as a paste to be added to the water, is: Be sure that you maintain the solution at its proper richness, such as 20 parts of water to one part of paste (called a 20 to 1 solution).

"Oil and water don't mix" is an old saw that can be bent, in the case of water soluble oil grinding fluids, by the addition of emulsifying agents to the mixture. With these agents, even though the water and oil do not mix in the common sense of the word, they are dispersed within each other to an extent that provides the effect of a mixture. The resulting milky solution reduces visibility. Otherwise, its cooling and cleaning ability are about the same as the water soluble chemical type, and its lubrication is somewhat better. This type of fluid is often used on applications involving light to medium stock removal.

Straight oils have no equal for lubrication, but they are used mainly on jobs were lubrication is the prime consideration, such as thread grinding and heavy form grinding, usually on very hard materials. The reasons are not hard to find. For all their lubricating quality, oils are much less effective than the water based fluids for cooling, and they do an average job of cleaning swarf from the work area. Furthermore, the swarf does not settle out as quickly nor as thoroughly as it does in water based coolants. Oils have two disadvantages. They cost more than other fluids and, without or sometimes in spite of a mist-collecting system, they seem to form an objectionable oil mist over the machine and over the whole shop area. But there are some jobs where the high rating on lubrication cancels out all the minuses listed above.

Two related practices in dry grinding should be mentioned. A jet of air has both a cooling and a cleaning effect on a grinding operation, and it is occasionally used by itself. It should be noted, as was mentioned in the beginning of the unit, that the fast rotation of the wheel produces an air current that may have some cooling effect in dry grinding, but it is likely to be more of a hindrance than a help in wet grinding.

Another possibility of "dry" grinding, when a particularly good finish is needed, is to use a stick of a natural wax blend and hold it against the rotating wheel for a couple of seconds. As the wax heats up, it melts and forms a coating that acts as a lubricant. This will improve the finish for a few parts. This technique of "loading" the surface of a wheel with wax, incidentally, is much more common in coated abrasive belt grinding than it is in wheel grinding.

CLEANING THE GRINDING FLUID

It is clear that if grinding fluid is going to be recirculated, it must also have the swarf removed before it gets back to the grinding area. Clean coolant is a must in quality grinding.

There are a number of ways of doing this job. Obviously, if the tank is big enough for the coolant to become completely still, the larger bits will settle out. One way of doing this is shown in Figure 9. However, the smaller the particle, the longer it takes to settle, and many plants do not have that much time.

Next is a cloth filter or a filter of some other material. Although this loads up, it can be designed with a float switch that becomes activated when the filter is sufficiently loaded with swarf. The clean filter fabric is on a roll on one side of the tank, and there is a supporting roll on each side of the tank. When the switch is connected, a fresh section of filter is pulled into place, and the used section of filter dumps the swarf into a tote box arrangement for removal (Figure 10).

This type of filter is obviously limited in the minimum size of particle it will remove in relation to the volume of coolant flow. If the fabric will catch the smallest particles, it will at the same time slow the flow of coolant. On the other

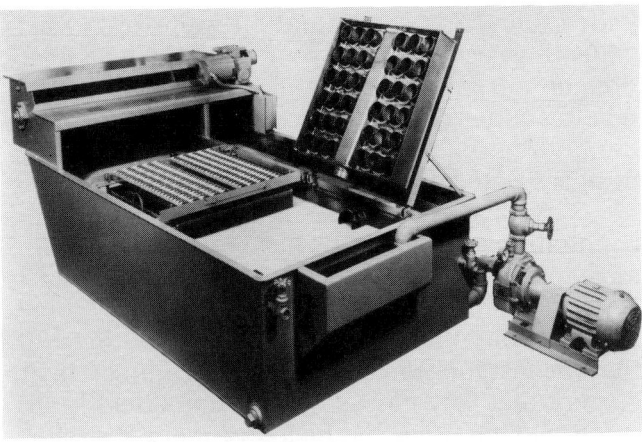

Figure 9. A settling tank. The two requirements for cleaning coolant by settling are that the tank be big enough to allow the coolant to remain still for enough time to allow the swarf to settle out, and that there always be enough coolant in the system. This also means that the tank must be kept clean; otherwise, dirty coolant is recirculated (Courtesy of The Carborundum Company).

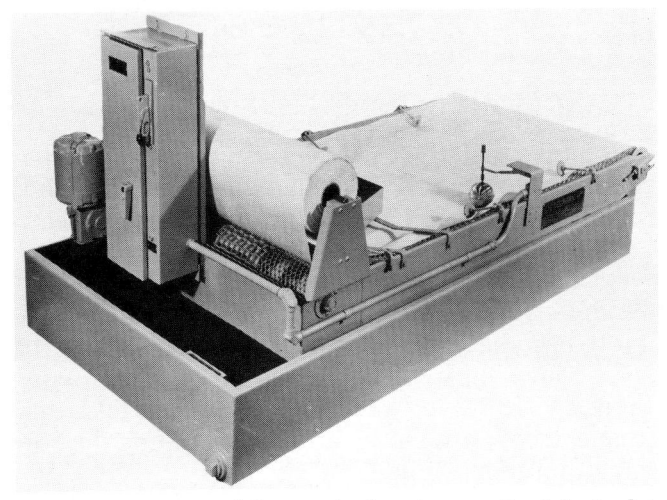

Figure 10. A widely used cleaning method is to let the fluid fall through filter fabric directly into a tank. Then, when the load on the fabric becomes heavy enough, it actuates a switch that pulls the loaded fabric into a waste container and also pulls a fresh section of fabric into place (Courtesy of The Carborundum Company).

Figure 11. The centrifugal unit on top spins at high speed to remove swarf from the fluid. Dirty fluid, already partly cleaned by settling, is pumped into the unit by the large hose from the tank and recirculated through the smaller hose and piping (Courtesy of Barrett Centrifugals).

hand, the openings must be small if the filter is to be effective. While the original cost of the unit is not excessive, the cost of the filter cloth can be a problem if there is any large amount of swarf. It is a simple method, and where the demands for clean coolant are not too tight and the volume of fluid to be handled is not too great, they are popular.

Three other types of filters, at least, are in use to varying degrees. These are the centrifugal separator, the cyclonic separator, and the magnetic separator. The cyclonic separator was adapted to metal working from a type long in use in paper mills.

The centrifugal separator, as the name implies, has an inner bowl that spins at high speed. The unit is designed so that the swarf spins to the outside of the bowl and the clean coolant stays toward the center. It is very efficient where the volume of coolant is not too great, and it is most economical in space, taking up perhaps the least space of any of the possibilities. Of course, the bowl must be cleaned from time to time, and most installations have a spare bowl so that one may be cleaned while the other is in use (Figure 11).

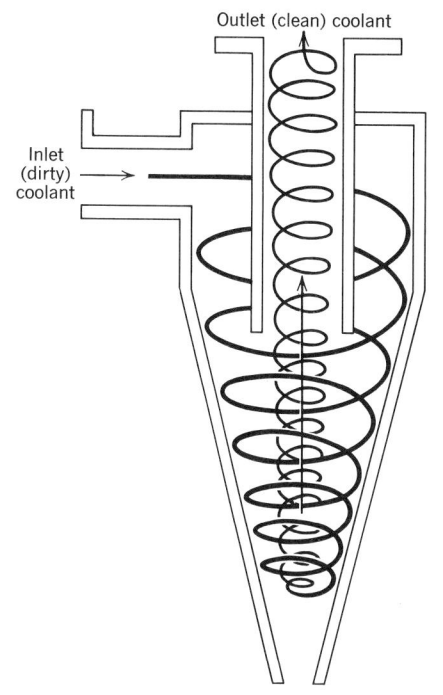

Figure 12. A cyclonic filter. The dirty coolant is fed in from the side, swirls around in the cone as the clean coolant goes out through the top, and the swarf through the bottom of the cone (Courtesy of Barnes Drill Co., Rockford, Ill.).

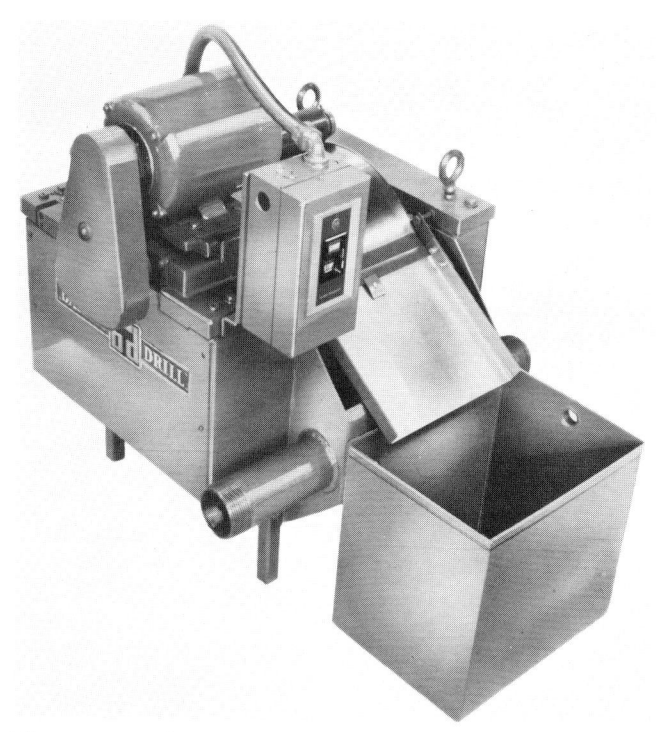

Figure 13. This magnetic separator removes iron and steel chips from water soluble oils and cutting oils. The magnets are in the roll visible just past the electrical control box. Chips picked up by the roll are scraped off onto the slide and go down into the sludge box (Courtesy of Barnes Drill Co., Rockford, Ill.).

The cyclonic separator also uses centrifugal force for separation of swarf from clean coolant. Figure 12 explains the principle of the operation. It must be admitted that the technical reasons for the separation may not be as clear as for the centrifugal type, although you will note that the results are similar. The swarf goes to the outside, while the clean coolant stays toward the center. Both light-duty and heavy-duty types are available, the latter with automatic sludge removal, a larger settling tank, and a larger dirt discharge opening.

The magnetic separator is, for obvious reasons, used where the grinding of iron and steel is the primary operation. Essentially, the unit is a magnetized drum that rotates into and out of the coolant flow and picks up chips of metal. As the drum rotates, a scraper, called a doctor blade, scrapes off the chips into a waste container (Figure 13).

The cleaning system or combination of systems used is not ordinarily a responsibility of an operator. However, once the system is in use, you should know in general how it works and what your responsibilities are in connection with it. These, of course, will vary with different shop situations, but the following are some examples.

1. Making sure that the coolant is maintained at the proper level, particularly on systems for individual machines. There is always a certain amount of loss from splashing, spillage, and evaporation, and the system is built to operate on a certain amount of fluid. If this amount becomes less, the fluid circulates faster, gets warmer, as does the workpiece, and the entire heating process continues to get worse, to say nothing of the probability of recirculation of more swarf and the harmful effects that follow.

2. Maintaining the proper solution of water based fluids. It will probably not be your concern to determine the relative percentages of water and concentrate, but once these are determined, you should add water or concentrate at the times and in the amounts set up. Too rich a mixture will probably have bad effects on surface efficiency and on the effectiveness

of the fluid. Besides, it boosts fluid costs. A mixture that is too lean may seem to cut coolant costs, but it actually raises them because of the scrap rate or other difficulties.

3. Keep on the lookout for signs that lubricating oil from the grinding machine is getting into the coolant. If you are using an oil coolant, this may not be too much of a problem but, in any case, it would be difficult to detect. The addition of a few drops of lubricating oil in a water based coolant could be very bad for the finish and the coolant's efficiency.

4. Finally, always make sure that the coolant nozzle is properly adjusted and aimed. What you are doing is directing the coolant at a very specific spot.

As you gain experience, you will come to appreciate some of the fine points of coolant effectiveness. If you become a good operator in a small shop, you may very well be in the position of making the decision as to which coolant is purchased. If you are in a large shop, particularly one that is part of a larger company, your part in coolant selection may be limited to expressing an opinion to an engineer who makes the selection particularly if the plant has a central coolant system.

However, when it comes to matters like checking the cleanliness of the coolant and keeping the coolant nozzle properly adjusted and directed, there is no question about who is responsible: you are.

self-evaluation

SELF-TEST

1. Why is a knowledge of grinding fluids important?

2. Explain the three major functions of grinding fluids.

   a.

   b.

   c.

3. Explain why the design and the placement of the grinding fluid flood nozzle are so important. Use sketches if helpful.

4. For one of the three major groups of grinding fluids, evaluate it in terms of the three major functions mentioned in Problem 2.

5. Repeat the same process for another of the three major fluids.

6. Do the same for the third.

Unit 4 Grinding Fluids 395

7. Explain how a mist coolant system works and under what circumstances one might be used.

   _____

8. List four methods of cleaning grinding fluids, and include advantages and disadvantages of each.

   a. _____

   b. _____

   c. _____

   d. _____

9. List reasons for one side or the other (it does not matter which) of the proposition that the operator should take part in the selection of grinding fluids for his machine.

   _____

10. List the four major responsibilities of an operator in *using* grinding fluids.

    a. _____

    b. _____

    c. _____

    d. _____

unit 5
horizontal spindle, reciprocating table surface grinders

The small horizontal spindle, reciprocating table surface grinder is considered to be basic to most general purpose machine shops. With accessories, they are extremly versatile and can do a remarkable variety of work. It is important that you learn the nomenclature and functions of this type of grinder and that you are aware of the many accessories that can be obtained to make this machine very useful in a wide variety of situations.

objectives After completing this unit, you will be able to:
1. Name the components of the horizontal spindle surface grinder.
2. Define the functions of the various component parts of this grinder.
3. Name and describe the functions of at least four accessory devices used to increase the versatility of the surface grinder.

information

By any measure, the small horizontal spindle, reciprocating table surface grinder (Figure 1) must be considered the basic grinder. There is at least one in nearly every machine shop, toolroom, school shop, and anywhere else that grinding is done. It is economical both in original cost and floor space, some requiring only a 3 foot square space for the machine itself. It can be used to demonstrate the principles of grinding that apply to all kinds of grinders, such as selection of abrasive wheels and selection and application of coolants, and the results of too much or too little infeed (which on a surface grinder is called downfeed).

There are probably more machines, more different makes and models, and more companies making them than any other single type of grinder. They come in a broad range of prices; in general, the more precise the machine, the more expensive it is; but principles can be taught on an inexpensive machine as well as on a more expensive one. Operating one on straightforward work can be relatively simple. In fact, it has been said that three fourths or more of all grinding work could be done on a small 6 by 12 in. capacity surface grinder.

In a sense, of course, all grinding is done on a surface, but a surface grinder is one for producing flat surfaces (or, with these horizontal spindle machines, formed surfaces) as long as the hills and valleys of the form run parallel with the path of the wheel as the magnetic chuck moves

Figure 1. Typical toolroom-type surface grinder (Courtesy of Boyar-Schultz Corp.).

Figure 2. Grinding a simple rounded form. Note that the wheel is dressed to the reverse of the form in the finished workpiece.

the work under it. Figure 2 shows a simple form. Vertical spindle machines, however, are usually limited to flat surfaces, because the grinding is done on the flat of the wheel with the abrasive "scratches" making an overlapping, circular path.

It is also possible on the horizontal spindle, reciprocating table grinder, with proper fixturing, to grind surfaces that are not parallel but are either flat or formed. Finally, with accessories, it is possible to do almost any kind of grinding: center-type, cylindrical, centerless, and internal. This is usually incidental work and small workpieces. But these show that the small, hand-operated surface grinder is a very versatile machine.

The best place to start learning the part names of a hand-operated horizontal spindle, reciprocating table grinder is the wheel-work interface, the point where the action is. In this unit, you should understand that the term "grinder" means this type of machine.

The cutting element is the periphery or OD of the grinding wheel. The wheel moves up and down under the control of handwheel A, as shown in Figure 3. The doubleheaded arrow A shows the direction of this movement. The wheel, together with its spindle, motor drive, and other necessary attachments, is often referred to as the wheelhead.

The work usually is held on a flat magnetic chuck, which is clamped to the table which, in turn, is supported on saddle ways. The magnetic chuck, which is the most common accessory for this grinder, holds iron and steel firmly enough to be ground. Using blocking of iron or steel, it will also hold nonmagnetic metals like aluminum, brass, or bronze. Only very rarely is it necessary to fasten a workpiece directly to the machine table.

So, for many workpieces, all that is needed is to clean off the chuck surface and the workpiece, place the part, turn on the current, check for firm workholding, and start grinding (assuming that the wheel on the grinder is suited to the workpiece). On all such pieces, it is imperative that

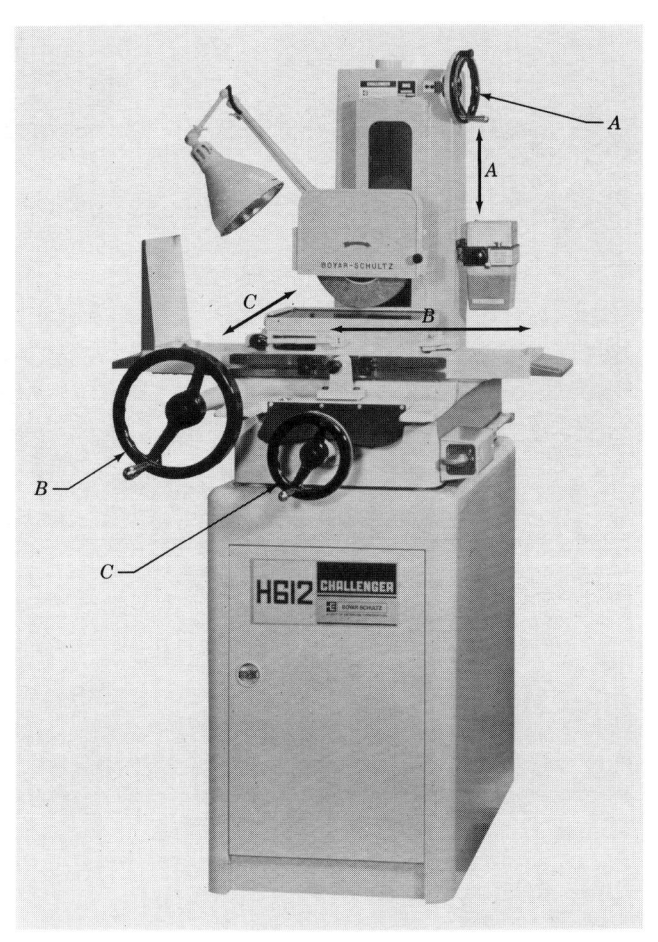

Figure 3. Surface grinder with direction and control of movements indicated by arrows. Wheel A controls downfeed A. Large wheel B controls table traverse B. Wheel C controls crossfeed C (Courtesy of Boyar-Schultz Corp.).

the surface to be ground is parallel with the surface resting on the magnetic chuck. The surface of the chuck and the "grinding line" of the wheel must always be parallel. For this reason part of the downfeed of the wheel is to compensate for wheel wear, and the chuck must be "ground in" when it is first mounted and periodically reground. This grinding and regrinding of the chuck will be discussed further in the next unit.

However, surfaces that are not parallel can be ground. As a simple illustration, consider a beveled edge to be ground around all four sides of the top of a rectangular block. You have two choices. One is to use a fixture that can be set to the angle you want, which is a magnetic sine chuck, as shown in Figure 13. The other is to dress the angle you want on the wheel with an attachment called a sine dresser and then grind in the usual fashion. The choice is up to the operator. These applications are covered in detail in Unit 7.

The chuck moves from left to right and back again (traverse), as shown in Figure 3, arrow B. On many grinders, traverse and crossfeed increments are controlled hydraulically with the table moving back and forth between preset stops. When the hydraulic control is actuated, the traverse wheel is automatically disengaged. This movement is controlled by large handwheel B. The chuck also moves toward and away from the operator (crossfeed), as shown by arrow C, with the motion controlled by handwheel C. This is the standard arrangement, since most machines are designed for right-handed operators. In at least one make (Figure 1) the traverse handwheel can be reinstalled on the right of the crossfeed wheel to accommodate a left-handed operator. The grinding wheel, incidentally, always runs clockwise.

On a completely hand-operated machine, the operator stands in front of the machine with his left hand on the traverse wheel and his right hand on the crossfeed wheel (if he is right-handed), swinging the table and the magnetic chuck back and forth with his left hand and cross feeding with his right hand at the end of each pass across the workpiece (Figure 4). The wheel clears the workpiece at both ends of the pass, and the crossfeed is always less than the width of the wheel, so that there is an overlap. At the end of each complete pass across the entire surface to be ground, he feeds the wheel down with the downfeed wheel on the column.

The traverse handwheel is not marked, because all that is necessary is to clear both ends of the surface. However, both of the other wheels have very accurately engraved markings, so that downfeed and crossfeed can be accurately measured. The downfeed handwheel (Figure 5) has 250 marks around it, and turning the wheel from one mark to the next lowers or raises the wheel .0001 in. This is one ten-thousandth of an inch, known in the shop usually as a "tenth." The crossfeed handwheel has 100 marks, and moving the wheel from one mark to the next moves the workpiece toward or away from the operator .001 in.

With the 250 marks on the downfeed handwheel and .0001 in. movement per mark, a complete revolution of the handwheel moves the

Figure 4. Operator in position at grinder. On many small surface grinders the crossfeed and traverse are hydraulic (Lane Community College).

grinding wheel .025 in. (250 × .0001 in.). A complete revolution of the crossfeed wheel, which has 100 marks, moves the table and the chuck .100 in. (100 × .001 in.).

The other feature needing mention is the zeroing slip ring on the downfeed handwheel or the crossfeed handwheel. You cannot predict in advance just where on the scale either wheel will be when the grinding wheel first contacts the workpiece, and where the control wheels ought to be when the surface is ground. These can be figured out mathematically, but with considerable chance of error. However, with the zeroing slip ring, the starting point on the scale on each wheel is simply set at zero, locked in place (Figure 6), and then ground until the required amount of stock has been removed. Allowance must be made for wheel wear, especially where considerable stock is removed by grinding.

A skilled operator develops a rhythm as he traverses the workpiece back and forth under the grinding wheel, cross feeding at the end of each pass and downfeeding when the whole surface has been covered. This is a knack you develop with experience. Using machines with hydraulic traverse and crossfeed takes out the need for physical coordination on the part of the operator,

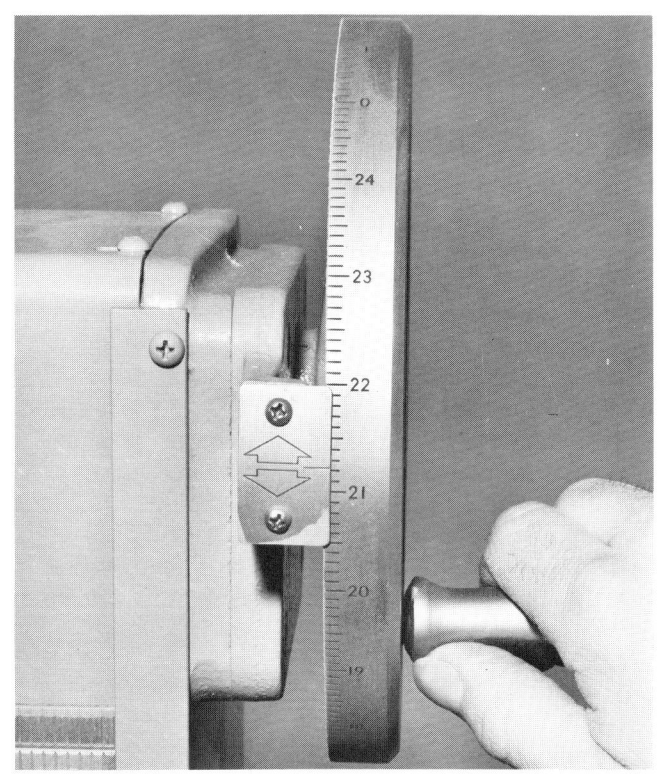

Figure 5. Closeup of downfeed handwheel. Moving from one mark to the next lowers or raises the grinding wheel .0001 in. (Courtesy of DoAll Company).

Figure 6. Same control wheel with slip ring set to zero. Now it is simpler for operator to down feed the grinding wheel as he grinds (Courtesy of DoAll Company).

but the skill necessary to make good choices on the traverse speed and on the amount of crossfeed for each pass is very important to first-class grinding results. Generally, combinations of large crossfeed movements on the order of one half the wheel width and relatively small amounts of downfeed are preferred because wheel wear is distributed better this way.

Actually, in the wheelhead and the magnetic chuck, with the three control wheels, you have the basic parts of a surface grinder. You do not need a dressing device, which may be just a holder with the diamond mounted at the proper angle that can be mounted on the magnetic chuck or a built-in dresser (Figure 7). But everything else on the machine is either for support, for instance, the table, saddle ways, base, and column that holds the wheelhead; or an accessory that makes it possible to do something you could not do otherwise or that makes the job easier to do.

ACCESSORIES

The list of accessories for a hand and hydraulically operated toolroom-type surface grinder may be quite extensive. As mentioned before, with the proper accessories, almost any type of grinding can be done on one of these little grinders within maximum size ranges. Finally, accessories are a major point of difference between toolroom and production machines. Toolroom grinders must handle a variety of work, so accessories are needed. Production machines are used mostly for one purpose; if that purpose changes, the machine is rebuilt or modified for its new use or it is removed from service.

ATTACHMENTS

For practical purposes it probably makes very little difference whether something is called an attachment or an accessory. Both make it possible to do something with the machine that could not otherwise be done or at least could not be done so easily or quickly.

Rotary Chuck
A rotary magnetic chuck (Figure 8) can be mounted on the regular rectangular magnetic chuck and then locked in place. The rotary chuck is independently magnetized and powered. This, in effect, converts the grinder into a horizontal spindle, rotary table grinder, which is useful for

Figure 7. Built-in wheel dresser. Lever traverses dresser across wheel (Courtesy of DoAll Company).

grinding work where a circular scratch pattern around the center is desirable in the workpiece. A good example would be the grinding of custom piston ring thicknesses for use in modified engines.

Swivel Table
With some modifications, the rotating table described above can index as well as rotate for work that needs to be ground at some angle other than parallel to the front edge of the basic chuck (Figure 8).

Centerless Grinding Attachment
For the surface grinding shop that has an occasional need to do some small centerless grinding (Figure 9), there is an attachment to do just that. The part is held between the grinding wheel and two rollers. The wheel supplies the power for rotation, and the rollers act as brakes so that the part is ground and not just spun. This is one of the largest of the attachments. It can grind parts up to 5 in. in diameter and can be mounted on any surface grinder 6 × 12 in. or larger.

Figure 8. Rotary magnetic chuck turns the grinder into a rotating table type, good for grinding parts that require a circular scratch pattern; for instance, like grinding a metal-to-metal seal (Courtesy of M & M Precision Systems, Inc., Roto Grand®).

Figure 9. Centerless grinding attachment mounted on surface grinder with an assortment of parts that can be ground on it (Courtesy of Unison Corporation).

Center-Type Cylindrical Attachment
This is basically a workholder with a headstock and a tailstock (Figure 10); it is mounted crosswise on the locked table so that the crossfeed of the grinder makes the wheel traverse end to end on the workpiece. If a flat is needed on an essentially cylindrical part, then all that is necessary is to stop the rotation of the work, reciprocate the table just a little, and cross feed as for any flat surface.

High-Speed Attachment
For incidental internal grinding, this attachment (Figure 11), driven by a belt from the grinder spindle, provides the high speed that is needed to make the small mounted wheels run at the high RPMs that are needed to make them grind efficiently. Of course, it is essential, usually, to provide an attachment for mounting the workpiece also.

Another form of the high speed spindle is the variable speed spindle. Most grinders are built to run at a constant speed in RPMs, but sometimes where a grinder is used for both regular and diamond wheels, it is worthwhile to have an attachment to vary that speed. Diamond wheels are most effective at speeds that are well below the 5000 FPM or more that is common for

Figure 10. Center-type cylindrical attachment mounted on surface grinder. Attachment can be tilted for grinding a taper, as shown here, or set level for grinding a straight cylinder (Photo of Harig Lectric-Centers, courtesy of Harig Mfg. Corp., Chicago, Ill.).

vitrified bonded aluminum oxide or silicon carbide wheels.

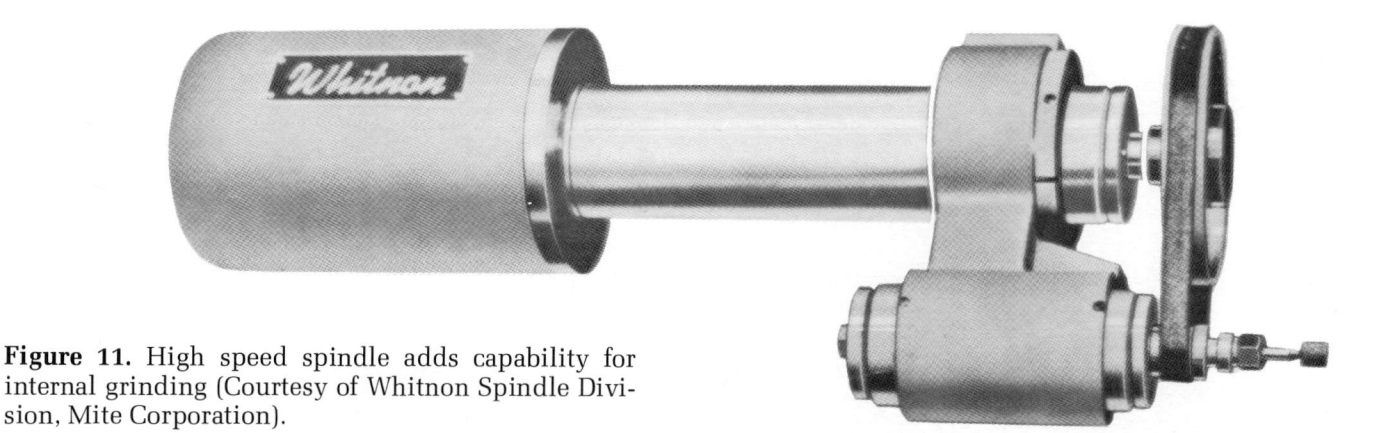

Figure 11. High speed spindle adds capability for internal grinding (Courtesy of Whitnon Spindle Division, Mite Corporation).

Vacuum Chuck

This replacement for an electromagnetic chuck (Figure 12) holds the work by exhausting the air from under it. Thus, it makes no difference whether the workpiece is magnetic or not. It is also recommended by some experts for holding pieces as thin as only a few thousandths.

Magnetic Sine Chucks

The surface of the magnetic chuck is always parallel with the line on which the wheel grinds. A sine chuck (Figure 13), usually electromagnetic, can be ajdusted so that a surface to be ground on angular work can be made parallel to the chuck. Some sine chucks are also designed so that compound angles can be ground. These devices are typically set with gage blocks in the same fashion as a sine bar of the same length.

Dressers

It makes some sense to consider all dressing devices as accessories, although a flat dresser, as discussed in Unit 3, is a necessity for any surface grinder, and when the dresser is built into the safety guard, as shown in Figure 7, then it is an integral part of the machine. Form dressing with the equipment needed will be discussed in a later unit.

The decision as to which of the attachments or accessories are justified depends mainly on production and economic factors. It is beneficial to know that there are such units available. It may be, especially in a small shop, that particular information that would widen the range of work that can be done, which is a major purpose of each attachment and accessory listed above.

Some shops, even some very large shops, find it to their advantage to have a surface grind-

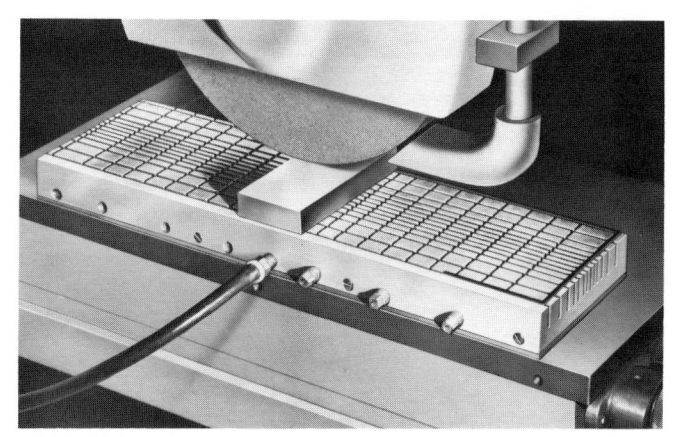

Figure 12. Vacuum chucks such as this one hold practically anything and are considered good for thin work (Courtesy of Thompson Vacuum Co., Inc.).

Figure 13. Magnetic sine chuck needed for grinding nonparallel surfaces. It is, of course, adjustable (Courtesy of Hitachi Magna-Lock Corp.).

er set up with a centerless grinding attachment, for example. It can be less expensive than a small centerless grinder; for certain work it can be adequately productive. If there comes a slack time in work for the attachment, the machine can readily be converted back for surface grinding.

self-evaluation

SELF-TEST

1. Give at least two reasons for the importance of studying the small hand or hydraulically operated horizontal spindle, reciprocating table surface grinder.

    a. _____

    b. _____

2. What three or more characteristics or qualities make the electromagnetic chuck such a valuable part of a grinder?

    a. _____

    b. _____

    c. _____

4. What is the principal result of these motions in terms of the wheel and the surface to be ground?

5. The handwheel that controls wheel motion (downfeed) is 10 times as precise as the one controlling crossfeed, and the traverse handwheel (if there is one) is not precise at all. Why is this?

6. What is the principal advantage of zeroing slip rings on the handwheels?

7. Why is form grinding possible only on a horizontal spindle surface grinder and not on a vertical spindle grinder? What limitations does this place on the form?

8. List at least two attachments or accessories that increase the kinds of grinding that can be done on a surface grinder. Include a sentence to tell what each unit does.

    a. _____

    b. _____

9. Make a similar list for other workholding attachments, including a sentence for each as to what it does.

a. _____

b. _____

10. What three reasons can you give for the popularity of these toolroom surface grinders?

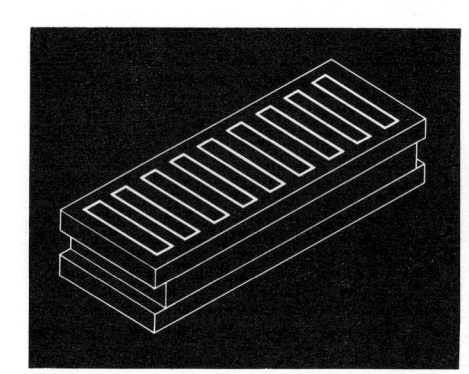

a. _____

b. _____

c. _____

unit 6
workholding on the surface grinder

The development of the surface grinder and of the electromagnetic chuck have been so closely tied together that it is almost impossible to consider one without the other. It is true, of course, that some work must be clamped onto the surface grinder, and that the magnetic chuck is used on some machine tools other than the surface grinder, but these are both minor. Essentially, in industrial shops the electromagnetic chuck is the basic workholder for the surface grindder, but the permanent magnetic chuck is often used for instruction. There is no basic difference in principle.

This is true of all surface grinders but, in this unit, the discussion will be related to horizontal spindle, reciprocating table surface grinders. Others are different only in detail, not in principle.

Permanent magnet chucks are used to some extent; they are made up of a series of alternating plates composed of powerful alnico magnets and a nonmagnetic material. Some improved types use ceramic magnets alternating with stainless steel plates. They exert a more concentrated holding force and can be used for milling as well as grinding. Permanent magnet chucks are used for small work and are not as widely used as electromagnetic chucks for industrial purposes. Therefore the term "chuck" will be considered as meaning an electromagnetic chuck for the remainder of this unit.

objectives After completing this unit, you will be able to:
1. Describe how a chuck is made and how it operates.
2. Detail the daily care and periodic care of a chuck.
3. Given a sketch or a description of a workpiece where the surface to be ground is not parallel to the side to be placed on the chuck (the chucking side), describe how to set it up for surface grinding, including accessories.
4. Detail the major steps in grinding a workpiece square on all six sides, including the accessories needed.
5. List precautions to be used in setting up thin workpieces for surface grinding.

information

On reciprocating grinders the chuck is rectangular (Figures 1 and 2). On rotary grinders it is round. On reciprocating grinders, the dimensions of the chuck describe the size of the machine. Most of the grinders you will work on will be referred to as 6 by 12 in. or 10 by 30 in. or some other combination. Rotary grinders use the diameter of the chuck to define the machine size. Some companies, in fact, use the diameter as a model number.

Chucks operate on direct current (dc) at 24, 110, or 220 V. They are made up of alternating strips or rings of steel and some nonmagnetic metal, usually lead or brass. The holding power depends on two factors: the area of the workpiece in contact with the chuck and the amount of current that is used. Thus, a workpiece with a flat and relatively large chucking side is easily held; one with a projection of the chucking presents other problems, particularly if the surface to be ground is larger than the surface contacting the chuck. A problem like that can be handled by supporting the shoulders of the workpiece on magnetic parallels that are a little thicker than the projection is high (Figure 3).

The percentage of total available power that is used is often critical in grinding thin parts, where the full power of the chuck could pull a warped piece down to the chuck, only to have the warping recur when the power is turned off and the workpiece is removed from the chuck.

Another unique feature about the chuck is

Figure 1. Common type of magnetic chuck for reciprocating surface grinder. The guards at the back and left side are usually adjustable and help keep work from sliding off the chuck (Courtesy of DoAll Company).

the special switch. If there were only a regular on-off switch, a large, heavy workpiece with a considerable area of chuck contact would be hard to remove after the power was turned off because of magnetism remaining in the chuck (called residual magnetism). The special switch gets rid of residual magnetism, so that any part is easy to remove. It is important to leave the part in place until the demagnetizing cycle is complete (about 20 sec), or the part will still have residual magnetism.

Mention should also be made about the

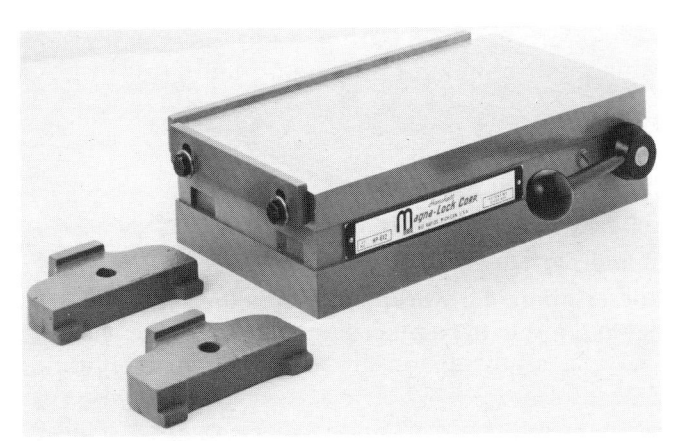

Figure 2. The permanent magentic chuck looks about the same as any other chuck of the same shape and often is used where wiring would be inconvenient (Courtesy of Hitachi Magna-Lock Corp.).

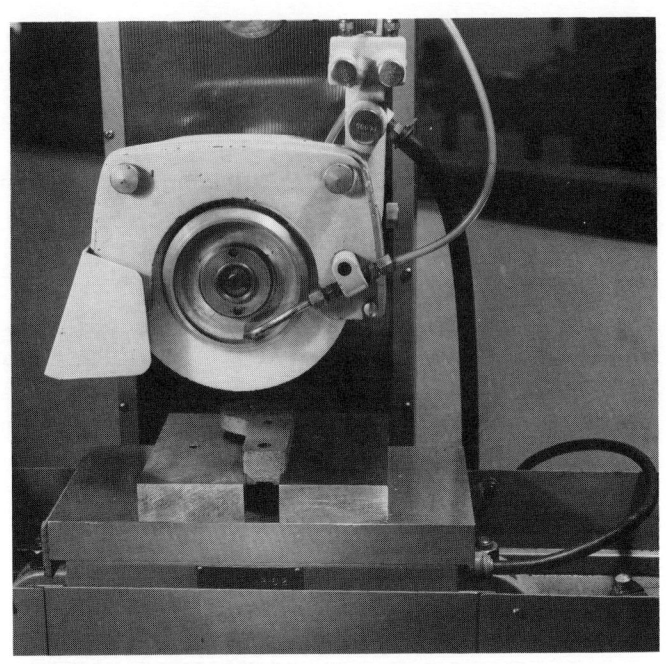

Figure 3. Chuck setup for workpiece with projection on chucking side. Work is supported on laminated magnetic parallels (Courtesy of DoAll Company).

Figure 4. Much machine shop work is chucked as simply as this. If the chuck is clean, the operator can spot the workpiece on the chuck, check to see that it is firmly in place, and proceed to grind (Courtesy of DoAll Company).

weight of the workpiece and the friction that this generates between the workpiece and the chuck. This is a factor, but one not generally of much concern in most toolrooms and machine shops where the parts tend to be small anyway (Figure 4). In fact, most such workpieces have the side to be ground parallel with the chucking side. About the only precautions that need be observed are to clean the chuck of chips or fluid before each piece is mounted and to spot the pieces on various parts of the chuck face so that you do not wear a depression in the center that would eventually require the regrinding of the entire chuck face. The quickness and simplicity of loading or chucking the majority of parts is one principal reason for the increasing use of the surface grinder. When producing many parts, you should load as many parts on the chuck as is practical, because for such parts you can grind several about as easily as you can grind one.

ROUTINE CARE

There are a few points for routine care of the chuck that ought to be second nature with you at any time you use a surface grinder. Some of these are:

1. Clean off the top of the chuck before you place any workpiece on it. This is sometimes done with a squeegee.
2. Place the workpieces carefully on the chuck to avoid nicking, burring, or scratching the chuck. Sometimes it is even worthwhile to place a thin piece of paper between the chuck and the workpiece. However, if you are grinding more than one part at a time, it is considered good practice to place the workpiece in some kind of regular pattern. Also, particularly in placing small parts, it is important to

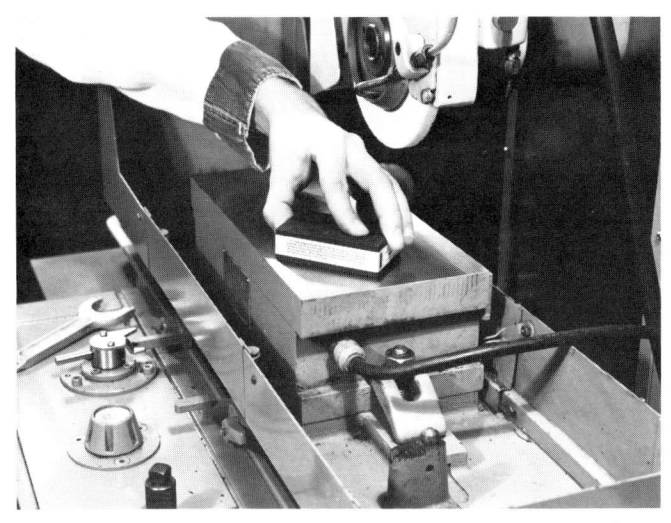

Figure 5. Periodic deburring of the chuck with a granite deburring stone like this, or with a fine grit oilstone, is a good practice (Courtesy of DoAll Company).

Figure 6. In regrinding the chuck, plenty of coolant is a necessity (Courtesy of DoAll Company).

have each part span as many of the magnetized strips in the chuck as possible to get maximum holding power. Use the same care when placing any accessory holding device on the chuck.

3. Rub the chuck occasionally with a deburring stone (Figure 5) or a medium to fine grit oilstone to remove the nicks or burrs that will come even with the best of care. When wear begins to affect work quality, you will have to regrind the chuck. (See point 6.)
4. A word of precaution: Never wear a watch while you are operating a surface grinder. It is unsafe practice and you may magnetize the watch mechanism and ruin its operation as well.
5. In case of chuck failure, a service representative of the manufacturer should be called. Any tinkering with the chuck by anyone else could void the warranty; furthermore, it could be dangerous. A chuck failure is a serious safety hazard. DO NOT use a chuck that gives any indication of weakness; parts slipping on the chuck can lead to broken wheels, bent spindles, and impact hazards to the operator and others in the same work area.
6. Because of nicking, burring, or ordinary wear, you will occasionally have to regrind the chuck, although it would be rare for you to have to install a new chuck. Also, a chuck should be indicated in every time it is clamped on the machine. In most cases, it must be reground lightly because clamping alone is not usually accurate enough for the parallelism you need. The chuck must be mounted and indicated parallel to the table and the saddle ways under it and, most important, to the grinding line of the wheel above it as the chuck moves the work under the wheel.

Use plenty of coolant if it is available (Figure 6). Otherwise, you may be limited to light cuts with time out at intervals for the chuck to cool off. The wheel specification recommended is friable (usually white) aluminum oxide, 46 grit, H grade, 8 structure, vitrified bond (friable A46-H8-V) dressed with rapid passes of a sharp diamond for "open" cutting. Use light downfeed, a fairly rapid table speed, and slow crossfeed. *Do not take off any more stock than you have to.* The following are some steps in regrinding the chuck.

1. Mark the entire top surface of the chuck with a thin application of layout dye or perhaps with Prussian blue. Any blue mark remaining on the chuck after the first full pass indicates a low spot.
2. Magnetize the chuck.
3. Downfeed should be about .0002 in.
4. Usually the flatness will be restored in less than .001 in. of metal removal. If more metal must be removed, you may want to redress the wheel rough before dressing it for the finish cuts.

5. Demagnetize the chuck.
6. Test the flatness of the chuck with a dial indicator fastened to the wheel guard. In appearance, the chuck surface should look dull, but almost polished.
7. If you have not done so, clean off the chuck carefully and remove any residual microburrs with a granite stone (Figure 5) or a fine India oilstone. The chuck should now be ready for use.

So far we have been discussing iron or steel workpieces on which the surface to be ground is approximately parallel to the surface that rests on the chuck. This may be 85 to 90 percent of all surface grinding work. There are other classes of work that should be covered: nonmagnetic metals such as brass and aluminum, thin work, odd-shaped work, and work to be ground square. These were mentioned briefly in the previous unit.

HOLDING NONMAGNETIC WORK

The critical point in holding nonmagnetic work on a magnetic chuck is to block it with steel bars, either with the vises that are shown separately in Figure 7 or that are shown in use in Figure 8. The surface to be ground must, of course, be higher than any of its retainers.

The other possibility for holding nonmagnetic work is the vacuum chuck, although this is more likely to be used in production grinding of sheet metal. It has a steel bottom plate and can be used directly on top of a magnetic chuck.

HOLDING THIN WORK

Surface grinding of thin steel that has been rolled is likely to release some of the stresses in the metal caused by the rolling process; the result is that the metal may warp or twist in some unpredictable fashion. The trick here is to use only the minimum power needed to hold the work. If more power is used, the work will be flat while you are grinding, but it will spring out of flat the moment you turn off the power. Grinding with partial chuck power requires great care to be done safely.

Blocking with thin precision ground stock is often done to prevent end movement of the part. Do not grind under partial power without instructor approval for your specific setup. Then, if you have nonmagnetic thin work to grind, you may have to resort to something like double-faced tape, which is sticky on both sides. Then stick the tape to the chuck, using one or more pieces according to the size of the work, peel off the second piece of backing paper, and lay the work firmly on top of the strips of tape. For this method, you need no power for the chuck and light cuts should be taken. Do not use fluids, since they usually cause the adhesive to become slippery.

HOLDING ODD-SHAPED WORK

There are two basic principles that apply to all surface grinding work, but particularly to odd-shaped work.

1. The surface to be ground must be parallel to

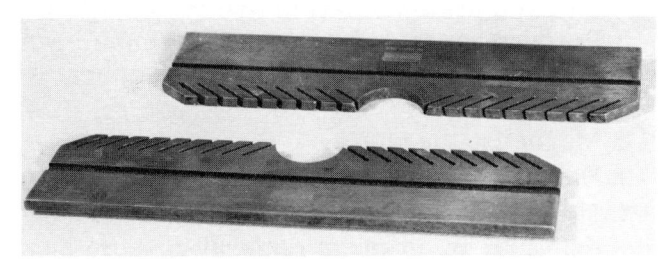

Figure 7. The comblike teeth in these magnetically activated clamps help hold nonferrous metal work (Courtesy of DoAll Company).

Figure 8. Tooth clamps in use. Note that the toothed clamps are lower than the surface to be ground (Courtesy of DoAll Company).

Figure 9. A set of magnetic parallels and V-blocks can be very useful (Courtesy of Hitachi Magna-Lock Corp.).

the chuck surface and, hence, also parallel to the wheel's line of grinding.
2. There must be a parallel and opposite surface of sufficient area to hold in place the surface to be ground, either on the chuck itself, or on something that in turn is held on the chuck, such as magnetic bars or parallels (Figure 9) or the magnetic sine chuck (Figure 10). Sometimes you may have to grind a chucking surface on the workpiece, but keep the setup as simple as possible.

A steel vee block (Figure 11), such as the project suggested for you to finish in Unit 7, could be very useful. You must bear in mind that only specially constructed laminated vee blocks will hold a workpiece magnetically; if you use a conventional vee block, you must hold the workpiece into the vee by mechanical means while the block itself is held magnetically to the chuck surface. A steel plate with a number of drilled and tapped holes in it might also be another handy thing to have around. With a few little bars of steel with similar bolt holes drilled and some miscellaneous blocking, this could be used as a base for strapping down nonmagnetic work.

Another difficult part to hold is one in which the ends to be ground are small in area compared to its length, such as a part 4 in. long by 1 in. by 1 in. Probably the best setup here is to clamp the part to an angle plate so that the end to be ground projects above the plate. Blocking is preferable under the workpiece, but it is optional because the pressures involved are typically quite small.

Figure 10. Adjustable magnetic sine chuck is useful for holding a variety of work where the chucking side is not parallel to the side to be ground (Courtesy of Hitachi Magna-Lock Corp.).

Figure 11. A steel V-block such as this has many uses in a shop (Lane Community College).

There are a great many accessories to help you in holding work on the surface grinder. Probably the most important point about these for most machine shops is that the fixtures be versatile, so that they can be used on a wide variety of work.

However, as far as you are concerned, the most important need is that you have a clear idea of what you want to accomplish on the part. With that and a good knowledge of what is at hand, you can probably determine what you need to do to hold the work for finishing the job. Remember, the chuck and all the accessories that go on it are precision equipment and should be treated with great care.

self-evaluation

SELF-TEST
1. Describe the construction of an electromagnetic chuck, including the type of controls and power used, and how the construction affects the use of the chuck.

2. Explain how the chucking of a workpiece is affected by its size, shape, and material.

3. Discuss three frequent routines for caring for the magnetic chuck.

   a. _____

   b. _____

   c. _____

4. List at least four major points in the regrinding of a chuck.

   a. _____

   b. _____

   c. _____

   d. _____

5. List or write a paragraph about the advantages and disadvantages of electromagnetic chucks.

6. Detail at least three methods of holding metals like aluminum, brass, or bronze on a magnetic chuck.

   a. _____

   b. _____

   c. _____

7. Thin workpieces present special problems. Discuss the special precautions to be taken and why they are necessary.

   _____

8. Describe the holding fixtures that you might use on workpieces like those shown in end views (Figure 12).

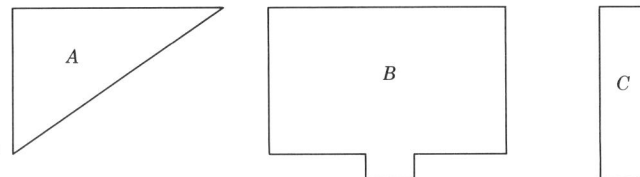

   _____

Figure 12. In each instance the top surface is to be ground.

9. Describe two or more basic principles of workholding on the surface grinder. A brief comment as to the importance of each would be desirable.

   a. _____

   b. _____

10. Where parts have a small cross section to present to the magnetic chuck, what should you do to overcome the problem?

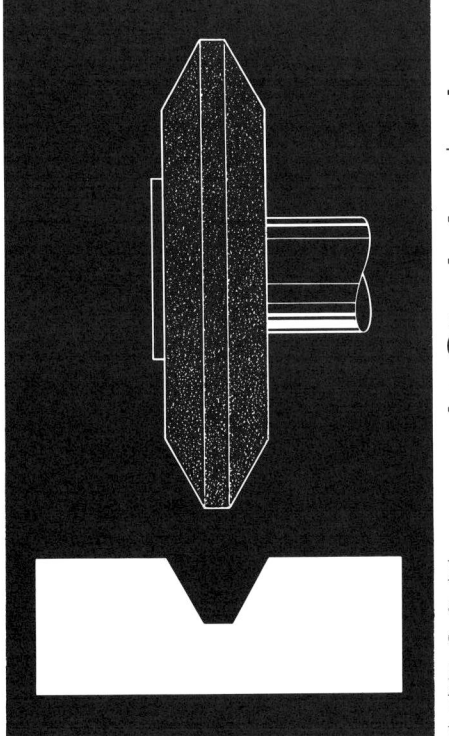

unit 7
form grinding and wheel dressing

Form grinding on surface grinders is becoming more widely accepted because it is the only effective way of producing angled or contoured forms on flat surfaces of hardened metals. Most form grinding requires dressing the grinding face of the wheel to something other than its usual straight surface. So form dressing of wheels is an important part of any surface grinder's skills.

Section H Grinding Machines

objectives After completing this unit, you will be able to:
1. Explain the wheel dressing and workholding requirement for producing simple forms such as radii, angles, and slots.
2. List the differences and similarities between normal surface wheel dressing and form dressing.

information

Surface grinding may be defined as the development of flat surfaces by grinding. *Form* surface grinding can be best described as any change in a major surface that is essentially flat, a change such as a bevel, a slot, a surface other than at a 90 degree angle to the adjoining sides and, of course, any kind of cross-sectional contour, as viewed from the end of the workpiece.

As we have said, the wheel of any surface grinder is limited to an up and down, straight line motion. The table of a reciprocating grinder is limited to a straight line, back and forth motion. Thus, any angular change from this level, parallel motion must be accomplished either by dressing the wheel or using a magnetic sine chuck (Figure 1), as described in Unit 6. The hills and valleys or the angle of the form must run parallel to the line of chuck or table travel.

WHEEL FORMS

Standard shapes available from wheel manufacturers are shown in Figure 2. The number of workpieces usually ground in one lot in a machine shop probably does not justify the purchase of these wheels, which can all be dressed by a skilled operator, but it is a standard and quick way of referring to most different shapes of grinding wheel faces throughout this unit.

Shape A is simply an ordinary wheel, like the majority of wheels that come into any shop. You can consider this as the shape from which any form dressing will start. Furthermore, whenever any workpiece on which a straight bevel or similar form must be ground is too big for the magnetic sine chuck, this is the method that must be used.

ANGLED FORMS

One of the simplest forms to grind is a 45 degree bevel on four sides of a block (Figure 3). There are two side bevels and the bevels on the ends.

Assuming that the workpiece is of a size suited to your magnetic sine chuck, you have two choices. You can either mount the workpiece on the sine chuck and grind it with a standard A-face wheel, or you can dress the wheel to a 45 degree angle (Figure 2, shape C) and mount the part on the machine chuck in the usual fashion.

A third example of angle grinding, vee-shaped grooves in a flat surface, is shown in Figure 4. In most machine shops, this probably would be rough ground flat, then each groove would be ground with a 45 degree E-face wheel (Figure 2, shape E) to produce a 90 degree included angle in the groove. Then the flats or lands remaining could be finish ground. In production grinding, however, this could be ground in one setup with either a crush-dressed wheel (Figure 5), or a wheel dressed to form with a diamond-plated form block, depending on equipment available and on the number of pieces to be ground.

Another possibility is the slot shown in Fig-

Figure 1. Workpiece to be ground mounted on a magnetic sine chuck for grinding one side of the large v-shaped notch at the top (Courtesy of Hitachi Magna-Lock Corp.).

Unit 7 Form Grinding and Wheel Dressing 413

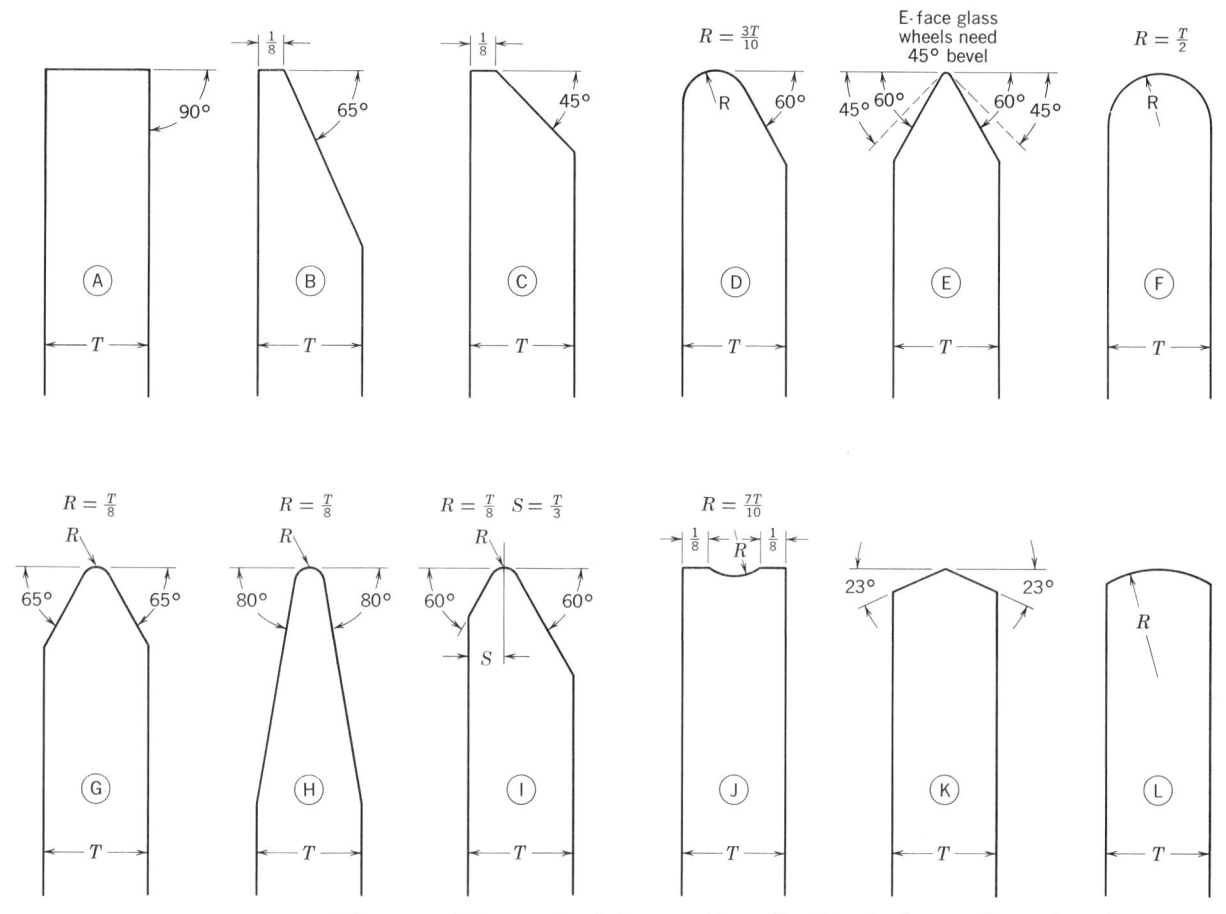

Figure 2. These wheel shapes, identified by the letters from this chart, can be supplied by wheel manufacturers. While they are used mostly for production grinding, they are also an easy way of identifying wheel faces for various purposes.

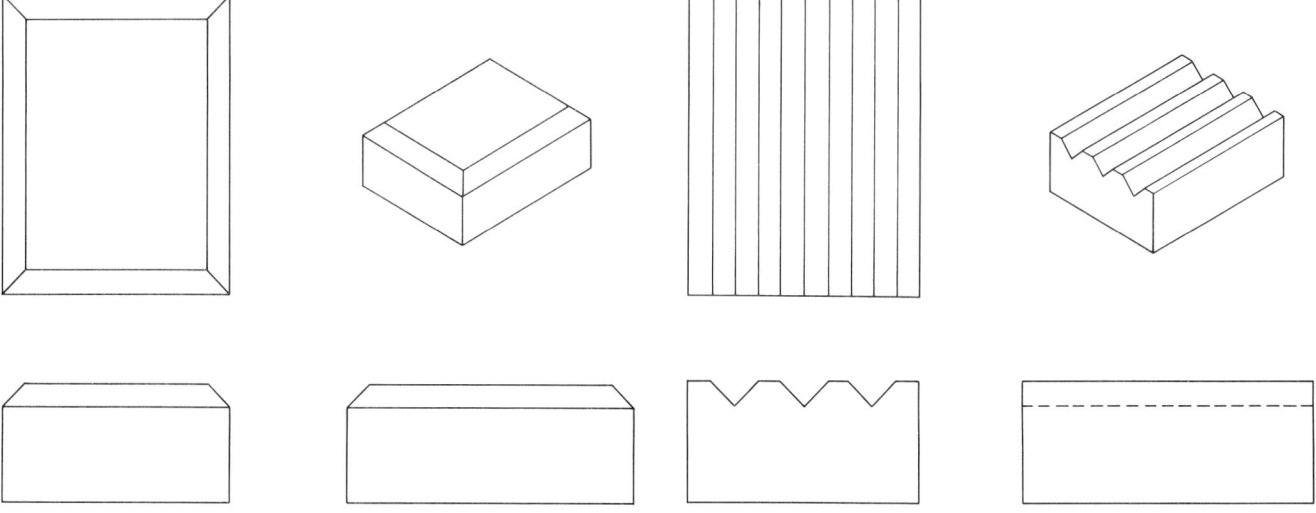

Figure 3. This workpiece is a simple rectangle or square, with four bevels on the four sides to be ground.

Figure 4. The forms to be ground are the three grooves running the length of the workpiece and the flats adjoining them.

Figure 5. Surface grinder set up for renewing the form on a crush dresser, the roll below the wheel. For a surface like that of Figure 4, the roll would be shaped like the top of the finished part. A diamond-plated block shaped like the finished top is another method for production grinding. Crush rolls and plated form blocks are generally too expensive and too specialized for general machine shop work (Courtesy of DoAll Company).

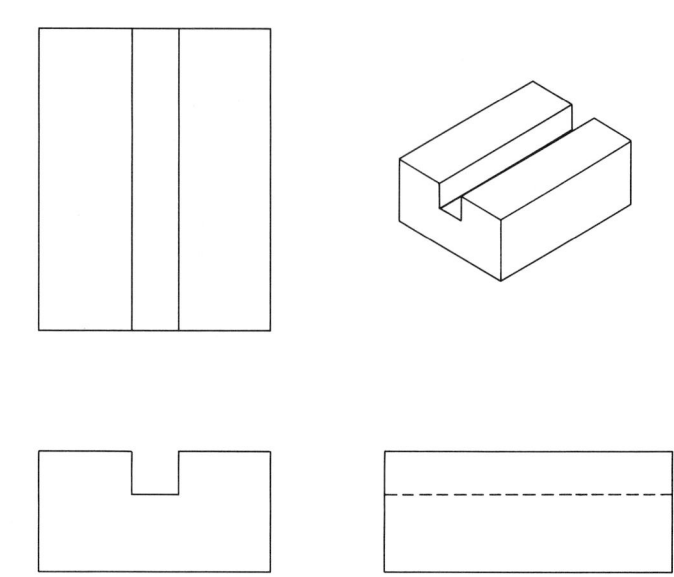

Figure 6. In this sketch the form to be ground is the square slot.

ure 6. For this example, the slot is $\frac{1}{2}$ in. wide and $\frac{1}{4}$ in. deep. The length is of little concern, as long as it is not longer than the chuck. If a $\frac{1}{2}$ in. thick wheel is available, then there is no problem. However, if one is not available, then it is not too difficult to improvise a dresser that will dress the side of the wheel to $\frac{1}{2}$ in. and a little further in from the periphery than the slot is deep. One such device consists of a steel block with a hole drilled at a 10 to 15 degree angle off the vertical, with a drilled and tapped hole for a setscrew to hold a diamond on its standard $\frac{7}{16}$ in. shank. The diamond must be at an angle large enough for the dressed wheel to clear the shank. Then, using the suggested dressing procedure, downfeed the wheel to be dressed, using the crossfeed to control the amount of abrasive removed from the side of the wheel. In this case, it does not matter too much whether the abrasive is all removed from one side or whether equal cuts are made on both sides. When the wheel is $\frac{1}{2}$ in. thick, set the workpiece on the chuck and grind the slot.

ROUNDED OR CONTOURED FORMS

You may have noted that in form grinding the form dressed in the wheel is exactly opposite to the form required in the finished workpiece. This shows up even more clearly in Figure 7a in a concave slot, which is ground with a wheel with an F-face (Figure 2); that is, of course, if you have a wheel of the correct thickness which, according to the note in Figure 2, is twice the radius ($R = \frac{T}{2}$). If not, you will have to thin down a slightly thicker wheel and then dress the radius for producing convex forms on the work (Figure 7b), although the J-face (Figure 2) bears a resemblance.

FORM DRESSING

The actual operation of form dressing (Figure 8) with a single point diamond tool is very similar to flat wheel dressing, and you will probably not be doing any form dressing until you have become proficient in flat dressing. The downfeeds, use of coolant (if possible), and care of the diamond are all identical. Probably the two major differences are that you will be removing more abrasive from the wheel when you dress it to a form and that the form must be quite accurate. A couple of extra passes in flat dressing probably mean at most that you have wasted a little abrasive but, in form dressing, it could result in an inaccurate form. For instance, the arc produced by radius dressing (Figures 9 and 10) *must* be accurately centered on the grinding face of the

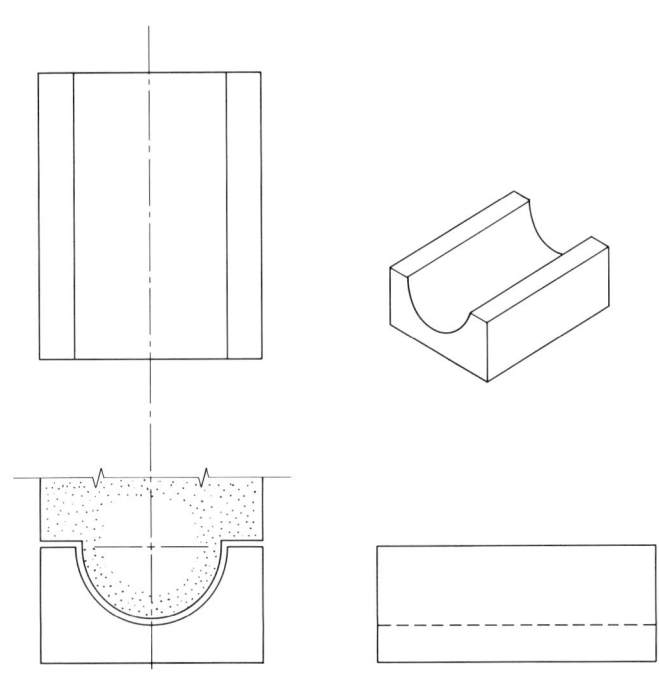

Figure 7a. Sketch exaggerated for clarity. The wheel fits very closely in the slot, as it does in all the other similar illustrations except Figure 3. The wheel, in fact, makes the slot. This workpiece would probably be ground flat first.

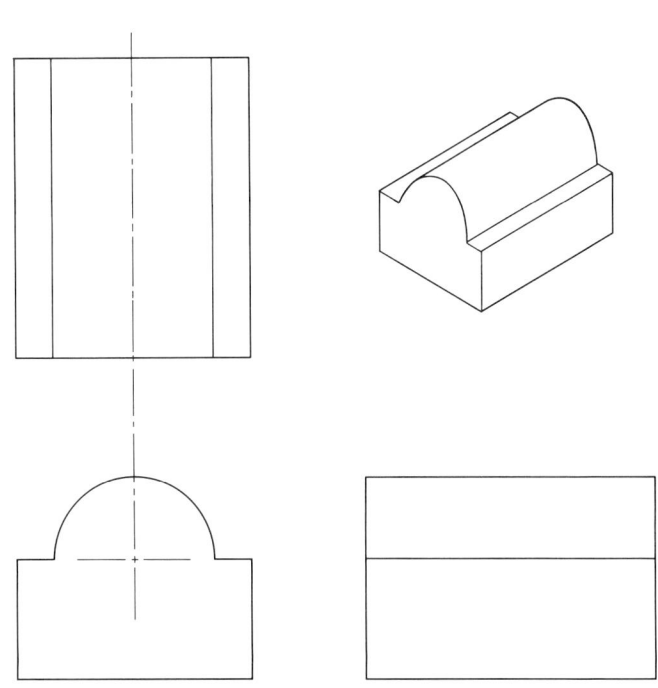

Figure 7b. This is the reverse of Figure 7a. A similar form is illustrated in Figure 2, Unit 5. Assuming that the wheel is thick enough, with a concave radius centered on its face, the top of this workpiece could be ground in one setup.

Figure 8. Relationship of finished workpiece and the wheel that ground it shows clearly here (Courtesy of Engis Corporation).

Figure 9. Radius dresser set up for grinding a convex radius on the wheel. Dresser can be adjusted for various radii. For a concave radius, the diamond would be reversed in the holder and the arm would be up instead of down (Courtesy of Harig Manufacturing Corp.).

Figure 10. Head-on view of a similar dresser set up for a convex radius wheel face. In fact, the face has been nearly formed (Courtesy of DoAll Company).

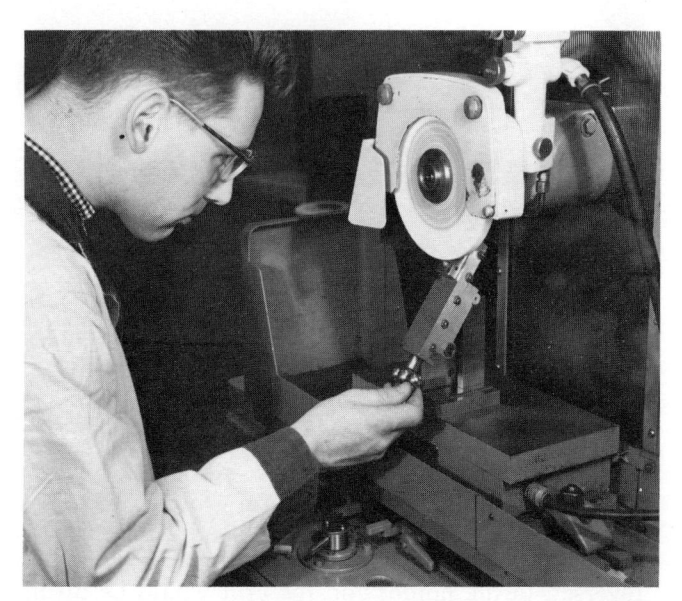

Figure 11. Sine dresser set up to dress an angled face on the wheel. The finished face could look like the B or C-faces of Figure 2 (Courtesy of DoAll Company).

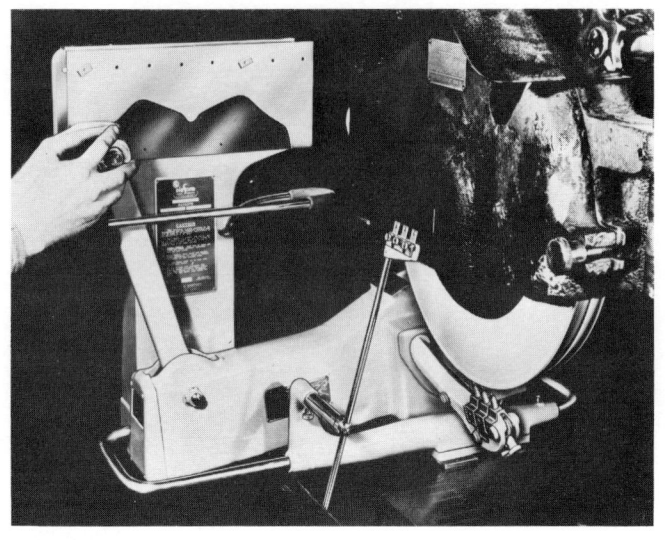

Figure 12. Template-type dresser mounted on surface grinder. This portable unit can be mounted on the chuck, as shown here or, more commonly, permanently mounted on the grinder itself (Courtesy of Engis Corporation).

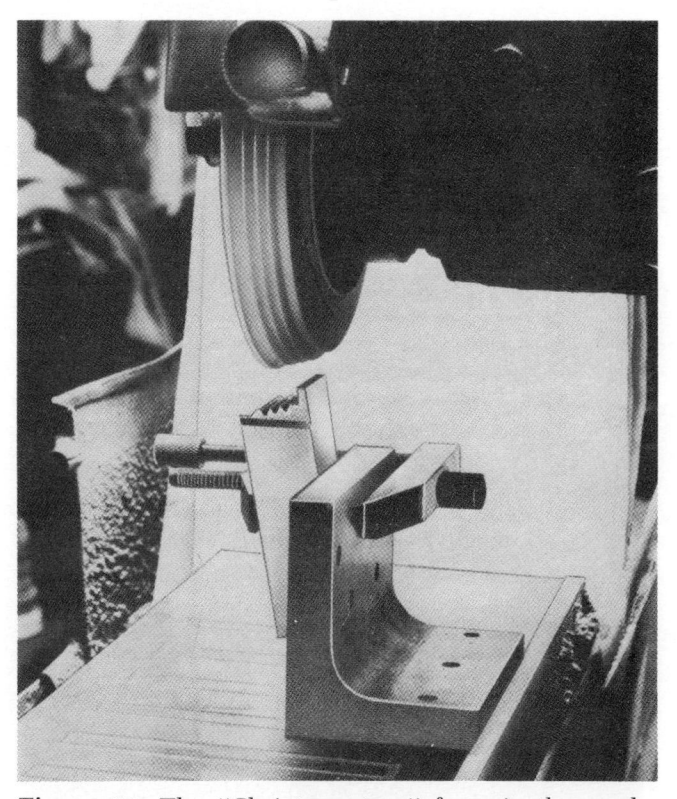

Figure 13. The "Christmas tree" form in the workpiece clamped to the angle plate is only one of many forms that can be produced on a template dresser. Note here the relationship between the formed wheel and the formed workpiece (Courtesy of Engis Corporation).

wheel. And the angle produced on a sine dresser (Figure 11) *must* be accurate.

POTENTIALITY OF FORM GRINDING

In the beginning, of course, your experience will probably be limited to simple forms like the ones described earlier: radii, angles, slots, and similar shapes. However, you should be aware that the possibilities of form grinding and the equipment for doing such work are increasing rapidly. Note the template-type dresser shown in Figure 12. In using it to dress the wheel, the operator traces a stylus over an easily prepared template to produce forms like the one in Figure 13. The template is several times larger than the form, so that any errors in tracing are reduced accordingly in the finished workpiece. This is only one of several such devices available.

self-evaluation

SELF-TEST

1. In what ways does form surface grinding differ from normal surface grinding?

2. What are the limitations in the design and construction of a reciprocating grinder for form grinding? How do these limit the forms that can be ground?

3. In what two ways can you compensate for these design limitations?

   a. _____

   b. _____

4. In grinding an angled (other than 90 degrees) surface, on a workpiece, what choices do you have as to method?

5. How does the form dressed face of the wheel differ from the form of surface it is to grind? Sketch a simple form, a V-groove in a block or a U-shaped slot, and then sketch in the face and a little more of the wheel, to grind it.

6. What would you do if you had to grind a ¾ in. wide slot, ¼ in. deep, and had only a 1 in. thick wheel for the job?

7. What are the basic differences between normal flat wheel dressing and form dressing?

8. What differences are there between dressing a wheel for an angular form and dressing it for a concave form?

This unit has no post-test.

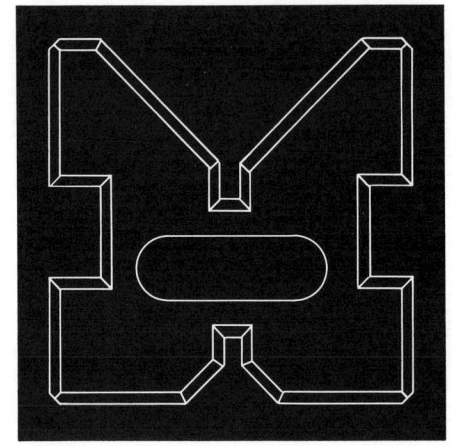

unit 8
using the surface grinder

Demonstrating any skill is, of course, the final test of whether you have learned it. The workpieces selected for this unit on using the surface grinder are two matching vee blocks of SAE 4140 or a similar alloy steel. A series of grinding steps are given in the text to detail the process for making these vee blocks. Some of these steps will include grinding operations on both blocks in the same setup at the same time. The specifications for surface grinding are often much closer than those required on other machine tools. These grinding machines, however, are capable of holding tolerances within a few tenths of a thousandth of an inch.

objectives After completing this unit, you will be able to:
1. Surface grind to specification two vee blocks of SAE 4140 or similar alloy steel.
2. Prepare a worksheet for and grind to specification a similar part involving surface grinding.

information

A SAE 4140 steel in the Rockwell C hardness range of 48 to 52 is regarded as being not too difficult to grind. It can be ground satisfactorily with a number of wheel specifications that are likely to be on hand in practically any grinding shop. As stated earlier, wheels for machine shop use have to be able to grind a wide range of materials, because the number of parts to be ground in any single lot is not likely to be large enough to warrant trials to find *exactly* the best wheel specification. Here, however, are some guidelines in the selection.

The abrasive will be aluminum oxide because this is steel, and the bond will be vitrified because this is precision grinding. Given the requirements for this particular workpiece, below are the recommendations of several specialists in the grinding field giving a range of four possible selections of wheels.

1. 9A46-H8V.
2. 9A60-K8V.
3. 32A46-I8V.
4. DA46-J9V.

All of these may be regarded as general purpose specifications for a part of this configuration, material, and hardness.

Looking at these recommendations in order, the abrasive recommendation of the first two is a white aluminum oxide, which is the most friable (brittle). The third and fourth are for a mixture of white and regular (gray) aluminum oxide, which is slightly tougher and perhaps wears a little longer. In wet grinding, as this is, probably the tougher abrasive would be preferred. However, in dry grinding, there would probably be no question about the use of white abrasive; it does not tend to burn the work as much as a tougher abrasive.

A grit size of 46 would be indicated for grinding efficiency, but 60 would provide a slightly better finish. Wheel grade provides the widest range (H to K), which could be interpreted to mean that any grade within this range would do the job. However, the wheel with the finest grit (60) also has the hardest grade (K). Such a wheel would wear a little longer than the others. Grit size and abrasive types are both likely to be fairly constant from various manufacturers.

The range of structure is only from 8 to 9, which is hardly significant. Grade and structure, however, result from manufacturing procedures, and hence are at least comparable from one wheel supplier to another.

Vitrified bond is indicated. Vitrified bonds are somewhat varied, although all wheel manufacturers have one or two general purpose or standard bonds for machine shop work.

Most general purpose grinding shops have neither the time nor the need to work out detailed comparisons and will no doubt depend on the experience of their better machinists in making such determinations. Good machinists develop opinions about which wheels perform best for them on given materials, but it is likely to be more of a process of instinct or feeling than of factual proof. The final choice of combinations of abrasive factors for particular types of workpieces is quite individual. The more aggressive machinist often uses a relatively harder bond because his procedures lead to more rapid wheel breakdown. Selection of the wheel to match jobs and machinist characteristics can still be considered something of an art rather than a specific science.

GRINDING THE VEE BLOCKS

The part selected for this precision grinding project is illustrated in Figure 1. Figure 2 is a sketch with the necessary dimensions and other data. These particular workpieces, a matching pair of vee blocks, follow through on a project started in a previous section on milling and shaper work. The purpose of this project is to show that you can surface grind a part flat and square to common tolerances and surface finish.

At this point, you have completed the previous steps of preparing the workpiece on other machines and completed the heat treatment of the vee blocks to RC 48-52. It is assumed that you have allowed appropriate stock for grinding .015 in. on all surfaces to be ground, except the grooves at the bottom of the vees and the ½ by ¼ in. side grooves.

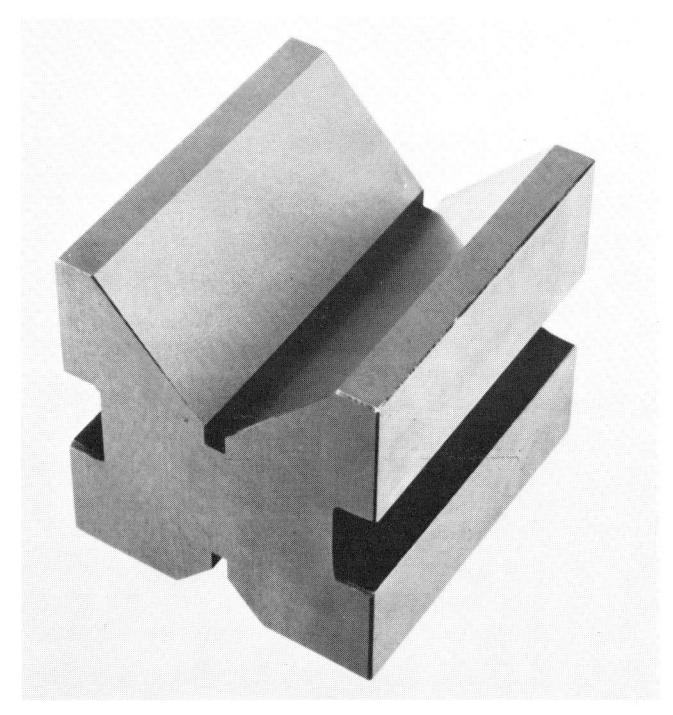

Figure 1. Finished, hardened, and ground precision vee block.

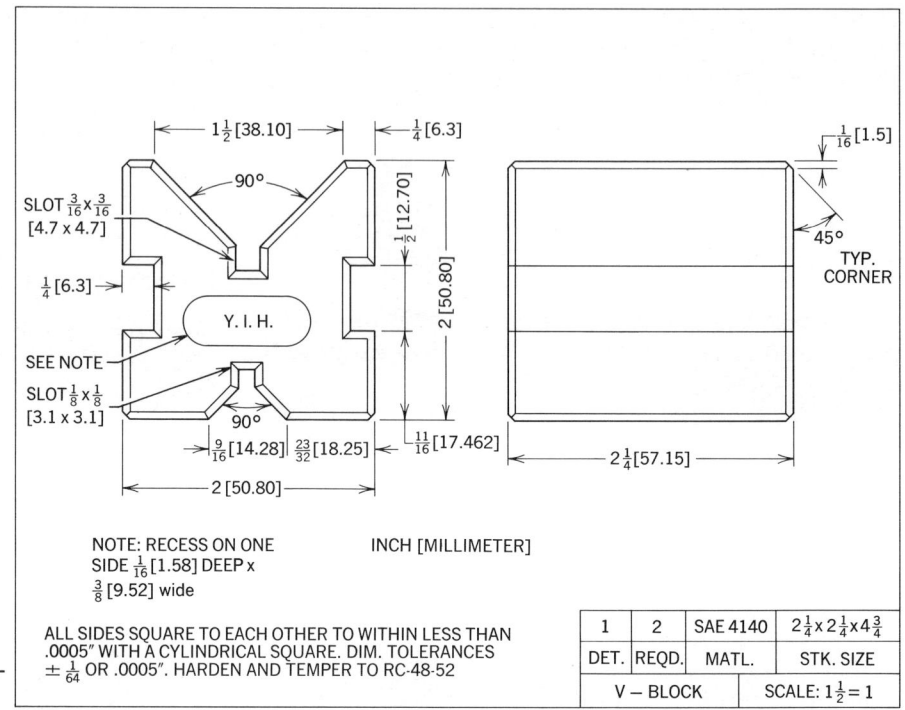

Figure 2. Dimensions and information for grinding the vee block.

STEPS IN GRINDING

1. *Select the wheel.* The previous information would help you do this.
2. *Clean the spindle (Figure 3).* Use a soft cloth to remove any grit or dirt from the spindle. Note that the chuck is protected by a cloth to prevent nicks or burrs from tools laid out on it.
3. *Ring test the wheel (Figure 4).* As indicated in Unit 2, this is a safety precaution any time a wheel is mounted.
4. *Mount the wheel (Figure 5).* The wheel should fit snugly on the spindle, and the outside flange must be of the same size as the inner flange, as shown in Figure 4. This wheel has blotters attached, but if they were damaged or there were none attached, it would be necessary to get some new blotters. Flanges should also be checked occasionally for burrs or nicks and flatness. The flange is held by a nut.
5. *Tighten the nut (Figure 6).* The nut should be tightened snugly. The blotters will take up a little extra force, but if either flange is warped or otherwise out of flat, it is possible to crack a wheel. Overtightening can also crack a wheel.
6. *Replace the safety guard (Figure 7).* This is a necessary precaution for safety.
7. *Place the diamond for dressing the wheel (Figure 8).* The wheel, as noted by the arrow on the safety guard, revolves clockwise so that the correct placement of the diamond is a little past the bottom point of the wheel, at about 6:30. This type of dresser has the correct "drag angle" of 10 to 15 degrees built into it.
8. *Dress the wheel (Figure 9).* This operation is to dress the wheel for stock removal so that the diamond can be cross fed rapidly back and forth across the wheel. Coolant is used,

Figure 3. Cleaning the wheel spindle with a soft cloth (Lane Community College).

Unit 8 Using the Surface Grinder 421

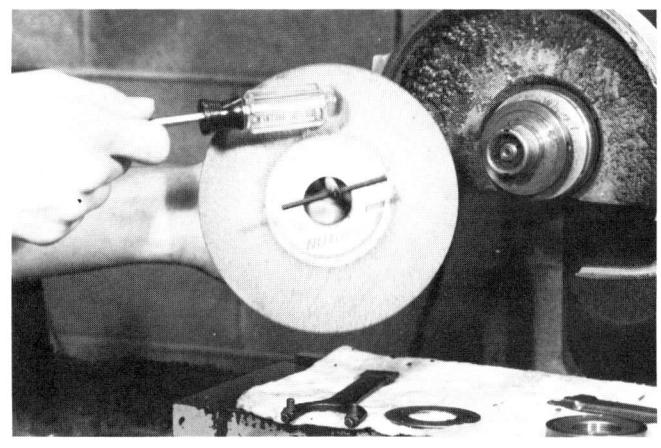

Figure 4. Ring testing the wheel (Lane Community College).

Figure 5. Mounting the wheel, flange, and nut (Lane Community College).

Figure 6. Tightening the nut with spanner wrenches (Lane Community College).

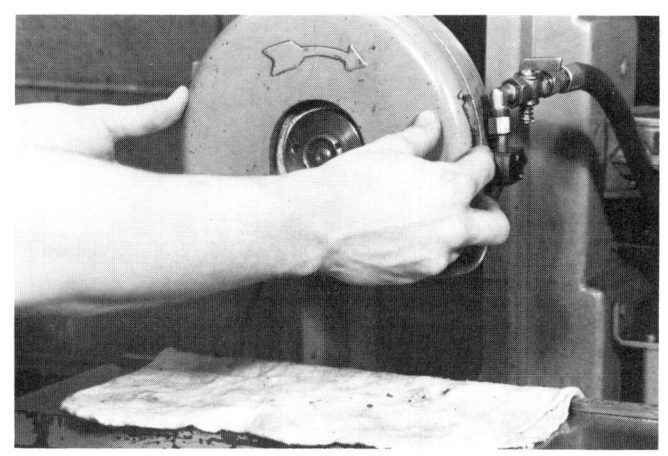

Figure 7. The safety guard being replaced (Lane Community College).

Figure 8. Diamond dresser in the correct position. The camera angle does not show the location of the diamond clearly (Lane Community College).

Figure 9. Dressing the wheel using coolant (Lane Community College).

since the grinding is done wet, but at greater volume than is shown in the illustration. The coolant was reduced for better visibility but, in actual dressing, the dresser should be either completely drenched with coolant or no coolant should be used at all.

9. *Wipe off the top of the chuck (Figure 10).* This is to remove any chips or bits of abrasive that may be on the chuck face. It is done with the wheel at a full stop and with the magnetic chuck turned off. A squeegee is often used before finishing the job with a shop cloth.

10. *Check the chuck for nicks and burrs (Figure 11).* The quickest and best way to do this is to run your hand over the face of the chuck. If the previous step has been done adequately, there should be no slivers or bits of metal to stick in your hand.

11. *Place the two blocks for the first grind (Figure 12).* Note that the sides with the large vees are up and that the blocks are placed near the center of the chuck. The paper protects the chuck and the workpieces from each other. For a single job like this, most operators tend to place the workpieces at the center of the chuck. However, if the grinder is in regular use, the paper would probably not be used and each group of parts would be spotted at a different place on the chuck, to equalize wear.

12. *Turn on the chuck (Figure 13).* Magnetic flux is applied on this chuck by moving a handle from left to right. On other chucks, such as the electromagnetic type, magnetism is activated by an electrical switch.

13. *Downfeed the wheel close to the top of the vee block (Figure 14).* This is simply a matter of turning the downfeed (sometimes called the infeed) handwheel at the top of the column to lower the wheel to a point an inch or so above the workpieces.

14. *Position the workpieces for grinding (Figure*

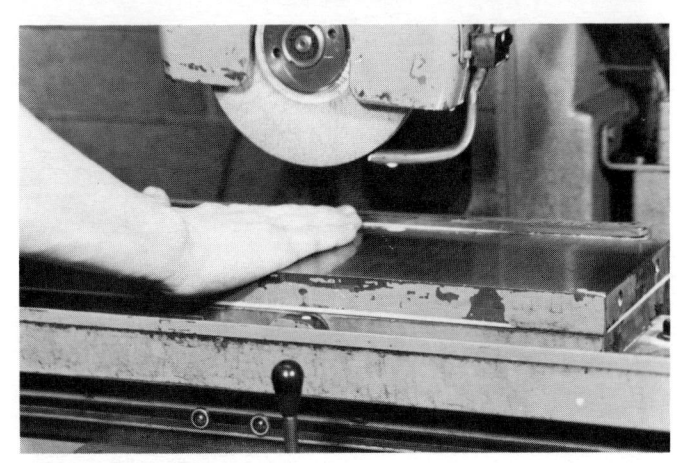

Figure 11. Checking for nicks or burrs on the magnetic chuck (Lane Community College).

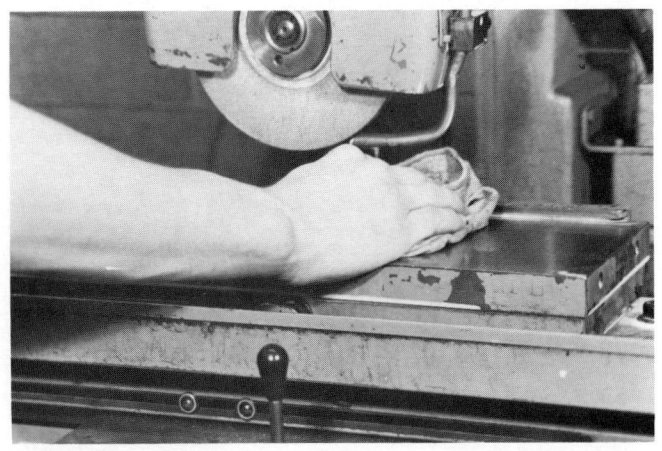

Figure 10. Cleaning the magnetic chuck with a cloth. The wheel must be completely stopped when this is done (Lane Community College).

Figure 12. The rough blocks in place with the large vee side up ready to be ground (Lane Community College).

Figure 13. Activating the magnetic chuck (Lane Community College).

Figure 14. Downfeeding the wheel (Lane Community College).

15). The large wheel, held by the operator's left hand, controls the left and right (traverse) movement of the table and chuck. The smaller wheel, held in the operator's right hand, cross feeds the table and chuck toward or away from the operator. During grinding the table is usually traversed hydraulically with the large handwheel disengaged. The large handwheel is mainly used for positioning work.

15. *Set the table stops (Figure 16).* Table stops are sometimes called trip dogs. Their purpose is to set the limits within which the table can travel, and these are usually about an inch off either end of the workpiece. All that is needed is to make sure that the wheel clears the end of the workpiece and allows time for cross feeding between traverse passes.
16. *Turn on the grinding wheel spindle.* Also turn on the hydraulic pump motion if your

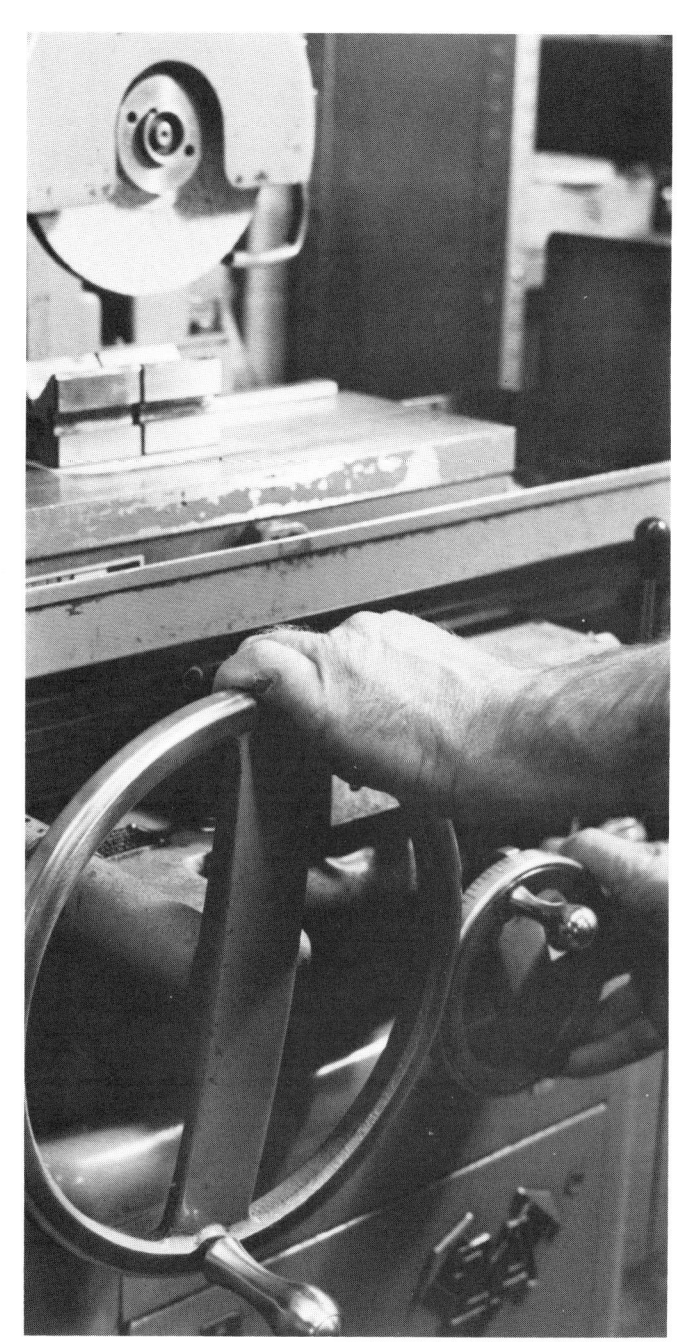

Figure 15. Positioning the blocks to set the stroke length (Lane Community College).

grinder is so equipped. As a safety precaution, let the wheel run for a minute, taking care that you do not place yourself, or allow anyone else, to stand in line with the plane of rotation. The grinding wheel is then brought down close to the work (Figure 17). The high point on the work is sought by manually

Figure 16. Setting the table stops (Lane Community College).

Figure 17. The wheel is turned on and, after running a minute for safety reasons, the rotating wheel is brought close to the workpiece (Lane Community College).

cross feeding during the table traverse to find the high spot. When it is found, set the downfeed dial to zero as a reference. Now turn on the traverse (Figure 18), if your machine is equipped with one.

17. *Turn on the coolant (Figure 19).* With the coolant flowing, downfeed about .002 in. With about .015 in. stock to be removed from each dimension and leaving about .003 to .005 in. for the finish cuts, remove about .010 to .012 in. from each of the sides and from the two ends. How much of this total comes off each side or end is usually left up to the operator's judgment, unless it is covered in a job sheet. In this case, perhaps removing a few thousandths of an inch for cleanup would be best until the squaring procedure is finished.

18. *Turn on the table crossfeed (Figure 20).* With power for both traverse and crossfeed, watch the wheel and listen for unusual sounds indicating overloading as it grinds. Downfeed .001 to .002 in. at the end of each complete pass across the two flat surfaces on either side of the large vees on both blocks.

19. *Clean up the opposite sides (Figure 21).* This step repeats steps 17 and 18 with the large vees turned down. Grinding should continue until the block is just cleaned up. (Steps 20 to 22 are done on each block singly.)

20. *Clamp either finished side of one block to a precision angle plate laid on its side on a surface plate (Figure 22).* Using a dial indicator and either a height or surface gage, clamp the vee block parallel to the surface plate. One end of the block should project

Figure 18. Starting the table traverse. The wheel is carefully adjusted to just touch the high point of the work (Lane Community College).

slightly beyond the angle plate so it may be ground. This is probably the most critical step in the procedure.

21. *Set up for grinding the end (Figure 23).* Return the angle plate and block to the grinder and place them on the magnetic chuck so that one end of the block is in position to be ground. Turn on the magnetic chuck and grind the end to clean up. The ground end should now be relatively square to the two ground sides.

22. *Grind one of the remaining sides square (Figure 24).* Use the same procedure as that explained in step 21, except that the end you have just ground on each block is clamped to

Unit 8 Using the Surface Grinder 425

Figure 19. The coolant is turned on and an additional downfeed of .002 in. is made (Lane Community College).

Figure 20. Turning on the table crossfeed; both surfaces are ground on the large vee side (Lane Community College).

Figure 21. The blocks are turned over and the opposite sides (small vee) are ground (Lane Community College).

Figure 22. A ground side of the vee block is clamped to a precision angle plate that is turned on its side and the top side of the vee block is being leveled with a dial indicator that is mounted on a height gage (Lane Community College).

Figure 23. The precision angle plate and vee block setup is turned with the vee block end up on the magnetic chuck. The end of the vee block is ground square to a ground side (Lane Community College).

Figure 24. The other two sides being ground square to the vee block end that was previously ground (Lane Community College).

the angle plate. The side to be ground must be made parallel to the surface plate. The side to be ground must project above the end of the angle plate so there will be no interference.

23. *Grind the remaining ends at the same time (Figure 25).* Both blocks can be set on the magnetic chuck without additional support and ground separately. Grind to within .003 to .005 in. of finished dimension.

24. *Grind the remaining sides and the two cleaned up sides to .003 to .005 in. oversize (Figure 26).* This is essentially the same setup as for step 23. These steps can be done without the angle plate after two ajoining sides and one end are square with each other, because surfaces parallel to square surfaces are also square to each other.

25. *Redress the wheel for finish grinding.* This is essentially the same as step 8, except that the dresser is cross fed more slowly. Downfeed for the finish grinding should be about .0002 in., about one tenth of the downfeed used for rough grinding.

26. *Check all sides and ends for square (Figure 27).* This requires a precision cylindrical square and a .0001 in. reading dial indicator supported by a height gage on a surface plate. Correct any errors of squareness by touch-up grinding, using tissue paper shims under one side to make the faulty surface parallel with the magnetic chuck. Remember to put the tissue paper under the thick side.

27. *Check dimensions.* This can be done by using a ten-thousandths reading micrometer (Figure 28) or by using gage blocks.

28. *Grind sides and ends to finish dimensions.* If your work to this point has been done carefully, both blocks can be done at the same time, as in steps 23 and 24. Just be sure that you are grinding corresponding sides on each setup. No tissue paper is used in finish grinding.

29. *Grind one side of the large Vee (Figure 29).* Set up both blocks in a magnetic vee block, making sure that the magnetic vee block is aligned parallel with the magnetic chuck. Turn on the magnetic chuck. Turn on the

Figure 26. The sides that have not been ground are also being ground in one setup to make them parallel to the other sides (Lane Community College).

Figure 25. Grinding the opposite ends of both blocks in one setup to make them parallel (Lane Community College).

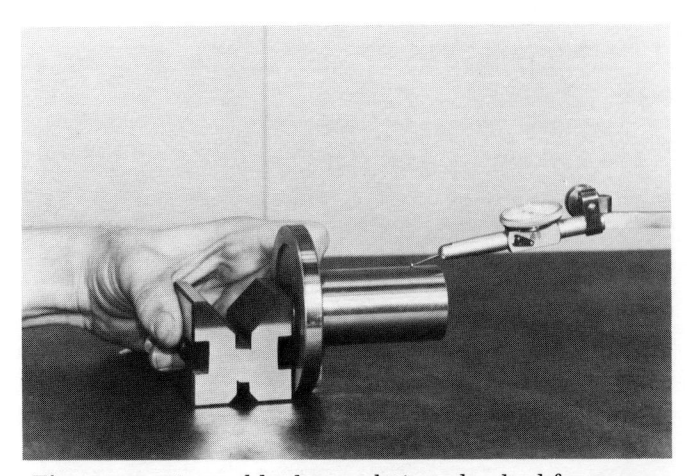

Figure 27. A vee block now being checked for square on all sides using a precision cylindrical square, a dial indicator, and a height gage on a surface plate (Lane Community College).

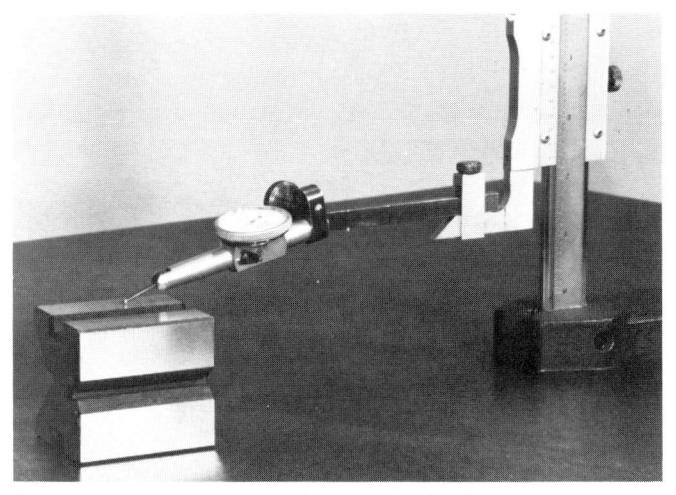

Figure 28. Dimensions being checked using a .0001 in. reading dial indicator on a height gage. Precision gage blocks may be used for comparison measurement for this operation (Lane Community College).

Figure 29. Setting up the magnetic vee block to grind the angular surfaces on the large vees (Lane Community College).

coolant and grind the side for cleanup. Use light downfeed.

30. *Reverse the vee blocks and repeat Step 29 for the other side of the large vee with the same setting (Figure 30).* This will center the vee. Check the large vee for square. The angle between the two sides of the large vee should be exactly 90 degrees. Repeat this procedure on the small vees on the opposite side.

31. *Clean up your machine.* Run the wheel for 5 min to remove coolant, which helps to maintain balance when the wheel is next used. Return the tooling to its proper place. Measure and record the finished dimensions of the vee block. Submit the vee block to your instructor for evaluation.

Figure 30. The vee blocks are repositioned and the other side of the large vees are being ground (Lane Community College).

self-evaluation

SELF-TEST Go to the surface grinder that you will be using and familiarize yourself with the controls and tooling. If someone is using the machine, observe the operation for a time. Check the supply of grinding wheels and determine their possible uses by observing their color and by reading the printed symbol on the blotter.

WORKSHEET 1 Given a surface grinder and accessories and a set of vee blocks previously machined and heat treated, you will:

1. Grind all sides square to each other within plus or minus .0005 in.

2. Hold the decimal dimensions to within .0005 in. and the fractional dimensions within plus or minus $\frac{1}{64}$ in.
3. Obtain a fine finish (about 16 to 20 microinch) on all surfaces with the final grinds. Check with your instructor on measurement, squareness, and finish from time to time as you progress with this project.
4. Turn in the completed vee blocks to your instructor for evaluation and grading.

WORKSHEET 2 Given a surface grinder and accessories, do an alternate project assigned by your instructor. It will be a grinding project similar to the vee blocks you have just studied, which will involve setting up the machine, testing and mounting the wheel, starting the grinder, and finishing the pieces.

Completing the project satisfactorily will involve your grinding the piece to specifications and writing a worksheet indicating the steps needed to complete the job. Your worksheet should state the steps that you will follow. Turn in your grinding project to your instructor for evaluation when you have completed it.

unit 9
problems and solutions in surface grinding

Producing quality work on a surface grinder bears some resemblance to driving a car. It is not too difficult when everything is going right. It is knowing what to do when something is not going right that separates the skilled from the unskilled. This unit is a discussion of some of the common problems of grinding, how to recognize them, and what to do about them. It should also give you some insight into whether you should try to do something about the problem yourself or whether you should report it to your superior or call a serviceman.

Unit 9 Problems and Solutions in Surface Grinding

objectives After completing this unit, you will be able to:
1. Recognize common surface defects resulting from surface grinding, and suggest ways of correcting them.
2. Suggest ways of correcting situations that show up in postinspection such as work that is not flat, parallel, or to form.

information

Conditions causing problems in surface grinding can conveniently be divided into two groups.

1. *Mechanical.* Those problems having to do with the condition of the machine, for example, worn bearings; or those problems having to do with the electrical or hydraulic systems of the machine.
2. *Operational.* Those problems having to do with the operation of the machine; for example, the selection and condition of the grinding wheel, selection of coolant, wheel dressing, and other similar responsibilities.

Obviously, there is not a clear line that can be drawn between these two. The division of responsibilities varies between shops; in general, the mechanical condition of the machine, aside from routine daily lubrication, is a shop maintenance responsibility, while the operation of the machine is your responsibility as its operator. Still, there are some close decisions to be made; they are usually decided on the basis of shop policy or practice. For example, if the machine ways are not lubricated often enough, which causes the table to "stick-slip" as it traverses, shop policy or practice determines who is responsible.

Two other general observations are in order. One is that any surface grinder is limited in the degree of precision and the quality of surface finish that it will produce. A lightly built, inexpensive grinder simply will not produce the quality of finish nor the precision of a more heavily built and more expensive machine.

The second is that any machined surface, even the finest, most mirrorlike surface, is a series of scratches. It is true that on the finer finishes the scratches are finer, closer together, and follow a definite pattern, but there are scratches nonetheless, and they will show up if the surface is sufficiently magnified.

Although you have not studied surface finish in any detail in this section, it is easy to understand that on a reciprocating table surface grinder, the scratches will be parallel and running in the direction of table travel. Anything that differs significantly from this pattern can be considered a surface defect and a problem.

Dirt, heat, faulty wheel dressing, and vibration can cause problems, as indicated earlier. The condition of the wheel's grinding surface can also cause problems. A wheel whose surface is either loaded or glazed will not cut well or produce a good flat surface. *Loading* means that bits of the work material have become embedded in the wheel's face. This usually means that the wheel's structure should be more open. *Glazing* means that the wheel face has worn too smooth to cut. It may result from using a wheel with a grit that is too fine, a structure that is too dense, or a grade that is too hard.

OPERATOR'S RESPONSIBILITIES

In general, it can be said that you, as the operator, are responsible for the daily checking and running of the machine. This includes things such as selecting and dressing the wheel, selecting the coolant, checking to see that the coolant tank is full and the filters are working as they should, and that the coolant is flowing in sufficient volume. You should observe the lubrication of the machine to see whether there is too little or too much. You should be alert for signs that the wheel is not secure or if the bearings are beginning to wear too much. You should not only check the work as instructed, but you should also observe it for surface irregularities. Of course, you will not be working entirely on your own; much of this will be done with the advice and agreement of your instructor or supervisor, especially when you first begin to use the machine.

GOOD MACHINE CONDITION

Its an old axiom of grinding that you can not produce quality work on a machine in poor con-

Table 1
Summary of Surface Grinding Defects and Possible Causes. Source: *Precision Surface Grinding*. Data courtesy of DoAll Company, 1964.

	Problems								
Causes	Burning or Checking	Burnishing of Work	Chatter Marks	Scratches on Work	Wheel Glazing	Wheel Loading	Work Not Flat	Work Out of Parallel	Work Sliding on Chuck
Machine Operation									
Dirty coolant				x		x			
Insufficient coolant	x						x	x	
Wrong coolant					x	x			
Dirty or burred chuck				x			x	x	
Inadequate blocking									x
Poor chuck loading							x	x	x
Sliding work off chuck				x					
Dull diamond					x				
Too fine dress	x				x	x	x		
Too long a grinding stroke								x	
Loose dirt under guard				x					
Grinding Wheel									
Too fine grain size	x				x	x			
Too dense structure					x	x			
Too hard grade	x	x	x		x	x	x		
Too soft grade			x	x					
Machine Adjustment									
Chuck out of line								x	
Loose or cracked diamond				x			x	x	
No magnetism									x
Vibration			x						
Condition of Work									
Heat treat stresses							x		
Thin							x		

dition. It is true that you can compensate for worn bearings to some degree by substituting a softer wheel, but this is a trap to avoid if at all possible. Not that machine condition is the only factor to be considered, but it is probably the most important single factor in preventing problems.

One frequently hears of unusual causes of trouble, such as the story about a machine that had been performing satisfactorily for years on a given wheel specification. Then suddenly, with the same specification, the operator has nothing but problems. When the manufacturer checked, he found that the customer had moved the machine from the ground floor to the sixth floor, and the added vibration required a new specification. With a wheel that suited the new conditions, there were no further problems. Many problems are not so straightforward. Indeed, the ones that are most difficult are those where there is more than one condition causing the problem. However, in the discussion that follows, the best that can be done is to indicate that there is more than one condition that may cause a particular problem.

Before discussing specific problems, many of which are shown in Table 1, it is worth noting that good dressing practice is probably the best single way an operator can avoid problems. For example, if you cross feed the dresser too fast, you run the risk of causing distinctive spiral scratches on the surface of the workpiece. If you cross feed the diamond too slowly, the wheel is dressed too fine and the effect is similar to that resulting from grit size that is too fine or a wheel that is too hard. A wheel that is too hard also causes burning or burnishing of the work, or hollow spots because an area has become too hot, expanded, been ground off, and then contracted below the surface as it cooled.

Surface finish problems can occur if you have not dressed a small radius on the corner of the wheel, as recommended in Unit 3. The first corrective measure you should take, if you have surface finish problems, is to redress the wheel.

SURFACE FINISH DEFECTS

Surface defects usually show up as unwanted scratches in the finish. Figure 1 shows four slightly oversize (1.25X) pieces that have been selected to illustrate common problems of this type. Not all of those discussed in the following paragraphs lend themselves to illustration.

Chatter Marks

These are sometimes referred to as "vibration marks" (Figure 2) because vibration is usually the cause. The reason may be from some outside source such as a punch press operating nearby and transmitting vibration through the machine to the wheel, so that the wheel "slips" instead of cutting for a moment. It may also come from within the machine, a wheel that is not balanced, or a wheel with one side soaked with coolant. It may be from worn wheel bearings, or even from a wheel that loads and/or glazes, and alternately slips, drops its load, and resumes cutting. This slipping and cutting alternation usually produces chatter marks that are close together; the more irregular and wider spaced marks are likely to come from some other source. The remedy may be just redressing, but it could also require a change of grinding wheel, a check on the wheel bearings, or even, if none of these work, a relocation of the machine.

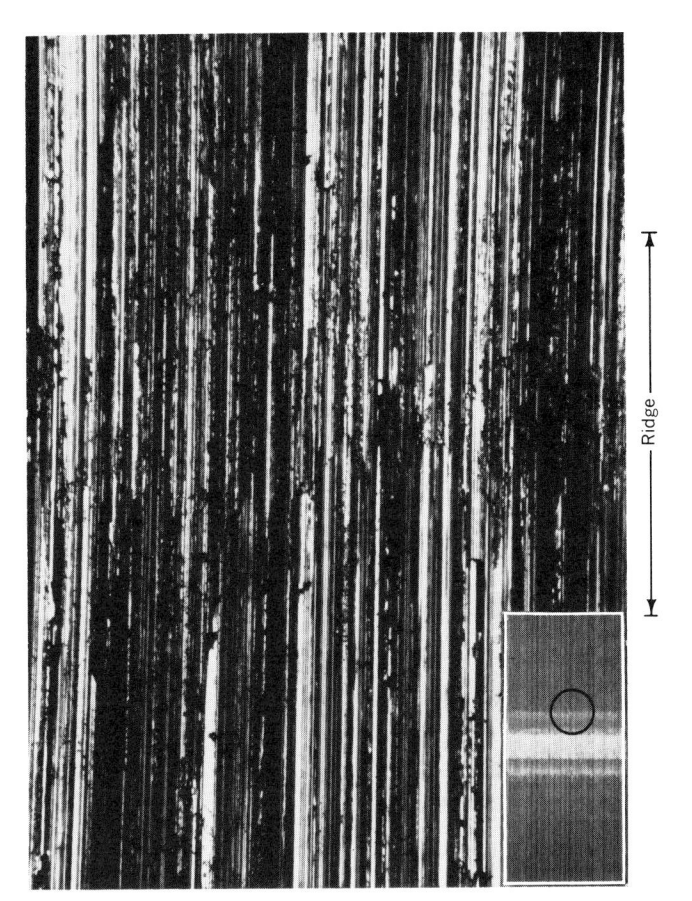

Figure 2. Chatter marks enlarged 80 ×. Marked inset (1.25 ×) shows area enlarged (Courtesy of Mark Drzewiecki, Surface Finishes, Inc.).

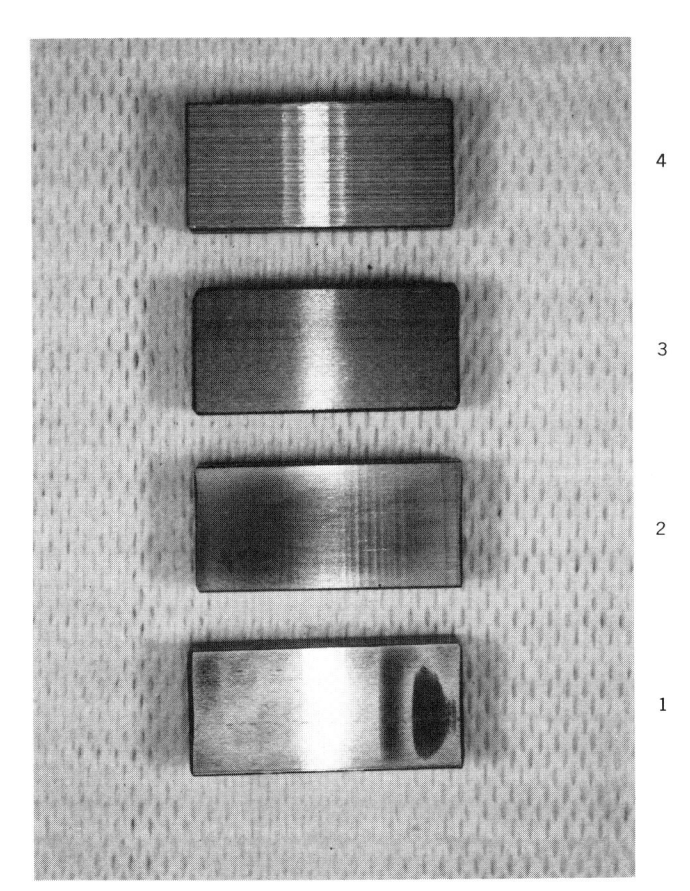

Figure 1. Four specimens of oil hardening steel, approximately 60 RC, with specific surface defects (Courtesy of Mark Drzewiecki, Surface Finishes, Inc.).

Irregular Scratches

These are random scratches (Figure 3), often called, for obvious reasons, "fishtails." Usually the problem is the recirculation of dirt, bits of abrasive or "tramp" metal in the coolant, or dirt falling from under the wheel guard. If there is not enough coolant, for instance, it may be recirculating too fast for the settling of swarf (grinding particles) to take place. A similar result can occur if you slide a workpiece off of a dirty chuck instead of lifting it off, which may be more difficult.

Figure 3. Grinding marks or "fishtails" also enlarged 80 ×. Inset (enlarged 1.25 ×) shows the damaged area (Courtesy of Mark Drzewiecki, Surface Finishes, Inc.).

The first thing to do to remedy these problems is to clean out the inside of the wheel guard; then replace or clean the filters in the coolant tank and make sure there is enough coolant in the tank. It should be worthwhile to take some extra time to ensure that the chuck is clean before you load it again.

Discoloration
Discoloration, which is also known as burning or checking, results when a workpiece becomes overheated. It can be caused by insufficient coolant, improperly applied coolant, a wheel that is too hard or too fine, or by removing too much metal too rapidly from a small area (Figure 4). If carried on too long, it can result in expansion of the metal, probably in the middle of the surface being ground. Then, if the hump is ground off, there will be a low spot when the workpiece cools to normal temperature. However, the burning may be considered in the first place as a surface defect that may or may not be a problem, depending on the final use of the workpiece.

Probably the first remedy is to try to get the wheel to *act* softer by speeding up table traverse, redressing the wheel rougher, or taking lighter cuts and, of course, by checking the supply and the application of the coolant. Whenever there is not enough coolant, it is recirculated before it has had a chance to cool off as it should, and so it simply becomes warmer and warmer.

Burnishing
A burnished surface (Figure 5) is one that is smoothed by abrasive rubbing instead of by cutting. It often looks good, and it may indeed not be a problem if the surface is not subject to wear. Technically, what happens when a surface is burnished is that the hills of the surface are heated enough so that they can be pushed into

Figure 4. Discoloration or burning also enlarged 80 ×, with damaged area marked on inset (Courtesy of Mark Drzewiecki, Surface Finishes, Inc.).

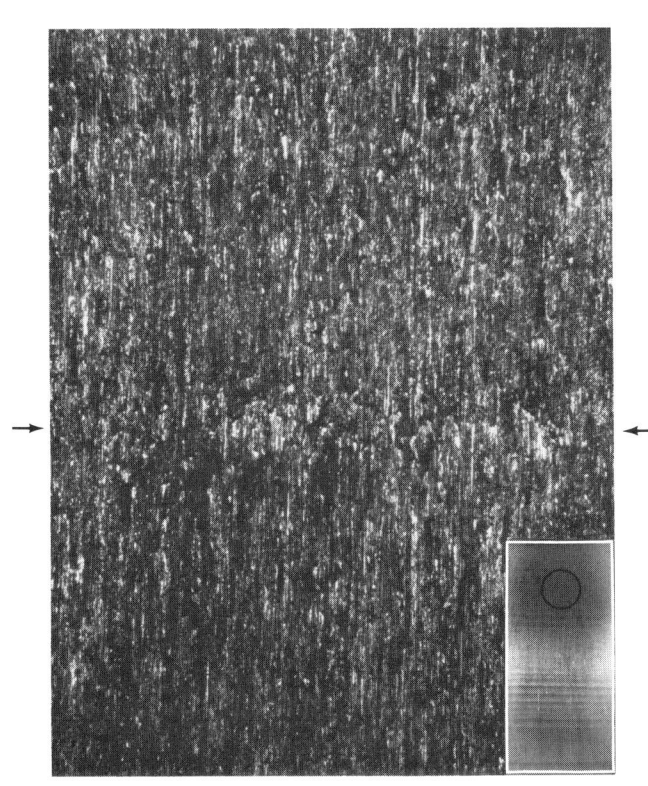

Figure 5. Burnished area enlarged 80 ×. Inset shows damaged area. (Courtesy of Mark Drzewiecki, Surface Finishes, Inc.).

the valleys of the scratches. But with any wear at all, the displaced metal breaks loose, and the surface suddenly becomes much rougher. It usually results from using a wheel that is too hard. For that reason, the usual remedy is to get a softer wheel or to change grinding conditions to make the wheel act softer by increasing work speed, redressing the wheel, or taking lighter cuts.

Miscellaneous Surface Defects
As mentioned earlier in this unit, any ground surface is a planned series of scratches, preferably of uniform depth and direction. On the grinder you have been studying, the scratches are parallel and at right angles to the direction of work traverse. Any other pattern, scratches without a pattern, or any discoloration of the work surface can be considered a defect. Some of the causes of these defects are that the wheel may be loose, the bearings may be worn, there may be vibration from some unsuspected source, the wheel may have been dressed too fast or too slow, or the wheel may be too rough or too fine.

Sometimes the causes are so remote or obscure as to puzzle even experienced troubleshooters.

Work Not Parallel
It has been said repeatedly that the grinding line of the wheel is parallel to the top of the chuck, and this is true if the grinder is in good condition. However, if the chuck is out of line, dirty, or burred, this parallelism may no longer exist, and any workpiece may be out of parallel. Of the three conditions mentioned above, dirt and burring of the chuck are definitely your responsibility, but the chuck alignment may or may not be. The remedies are usually obvious, as are those of other possible causes, as shown in Table 1. As you progress, you will develop a sort of routine to be followed for a given machine with this condition.

Work Out of Flat
Lack of flatness in a thick workpiece is likely to be the result of some local overheating, which causes an area of the surface to bulge. Then when the bulge is ground off and the work cools, there is a low spot.

Most flatness problems arise with thin work, and for very obvious reasons. Thin work does not have the bulk to absorb grinding heat without distortion. If it has been rolled, then stresses caused by the passage of the metal through the rolls may have been created, and the grinding may release these stresses on one side, causing the metal to warp or bow out of flat.

Correcting warpage in a thin workpiece is a matter of patience, a right start, and the minimum chuck power required to hold the workpiece in place. The procedure is as follows.

1. Using the least practical amount of chuck power, place the work on the chuck with the bowed side up so that the work rests on the ends.
2. Take a light cut with minimum downfeed. This should grind only the high spots on the work. Cutting should begin near the center.
3. Turn over the workpiece, shim the ends with paper, and take another light cut. This time cutting begins near the ends.
4. Repeat these steps, reducing the shims gradually, until the part is flat within specifications.

Finally, it should only have to be mentioned, as a reminder, that it is impossible to grind work flat if the chuck surface is not flat. When the chuck surface is between .0001 and .0002 in. out of flat, it is time to regrind it.

Perhaps the last word on solving the problems that arise in surface grinding is that care in avoiding the causes of trouble, such as careful and thorough cleaning of the chuck, frequent checking of the coolant level, the condition of the filters, and care in placing and removing workpieces from the chuck, can prevent many of the problems before they happen. It might be called preventive operation of the grinder.

self-evaluation

SELF-TEST
1. List at least three actions of an operator that will help prevent problems in surface grinding.

   a. _____

   b. _____

   c. _____

2. Define the term "surface defect" as it applies to surface grinding.

   _____

3. Name at least three general causes of surface grinding problems.

   a. _____

   b. _____

   c. _____

4. What is the principal cause of chatter marks? What are some possible remedies?

   _____

5. "Fishtails" are another common problem. What are they, what causes them, and what should you do to get rid of them?

   _____

6. Name two of the problems that can result from overheating work. What two or three things might you do if you suspect that a workpiece is getting too hot?

   a. _____

   b. _____

   c. _____

   d. _____

   e. _____

7. What is a burnished surface? Why is it objectionable? What are some remedies?

8. List at least two conditions that could produce out-of-parallel work.

   a. _____

   b. _____

9. List at least two conditions that could cause out-of-flat work.

   a. _____

   b. _____

10. Why is it difficult to grind thin work flat? How do you correct the condition?

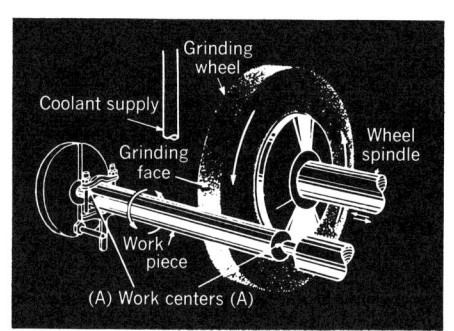

unit 10
cylindrical grinders

Cylindrical grinding, along with surface grinding, which you have already studied, can be considered as one of the basic forms of grinding. Indeed, practically all forms of grinding with wheels or belts are some kind of variation of these two.

It was pointed out in the introduction to this section that there are several types of grinders for grinding the outside diameters of workpieces, and the internal grinder for grinding inside diameters.

The grinder that is the subject of this unit is called a universal grinder, more accurately called a universal center-type cylindrical grinder. Within its size range, it will do everything that any other center-type cylindrical grinder will do. Center type means that the workpiece is held on centers, like a lathe. The universal grinder is a toolroom-type machine, which means that it is very flexible and can do many jobs, although it does not do any one of them as efficiently as a more specialized machine. Hence, it is a very valuable machine to use to get acquainted with all the variations of center-type cylindrical grinding.

Section H Grinding Machines

objectives After completing this unit, you will be able to:
1. Identify and explain the purpose of the major operating components of the center-type cylindrical grinder.
2. Describe the movements of the principal machine components.
3. Explain the major differences between a universal grinder, a plain grinder, an angle-head grinder, and a plunge grinder.
4. Explain when steadyrests are necessary, and how they are used.
5. List some of the major differences between surface grinding and center-type cylindrical grinding.

information

The center-type cylindrical grinder is the oldest type of grinder, and it is generally considered that the plain grinder, which has the grinding wheel moving toward or away from the operator on a fixed slide at right angles to the axis of the work, was the oldest. However, very shortly there were many more of the universal grinders with both the wheelhead and the table able to swivel, because there was more of a need for a finishing machine than for anything else. Furthermore, it is difficult to conceive a small 3 or 4 in. wheel, less than ½ in. thick, doing very much in the way of metal removal.

The earlier development of cylindrical grinding can probably be credited to two conditions. One was that the means for holding such work had already been pretty well developed in the lathe. In fact, the first cylindrical grinders were simply lathes with a grinding wheel mounted on the toolpost in place of the cutting tool; they were called grinding lathes. Another condition was probably the fact that the magnetic chuck had not yet been developed for the surface grinder, which made workholding more difficult.

DIFFERENCES BETWEEN SURFACE AND CYLINDRICAL GRINDING

Figure 1 shows a universal center-type cylindrical grinder. Figure 2 is a sketch of the general motions of cylindrical grinding. These provide sufficient background for some comparisons between surface grinding and cylindrical grinding.

For one thing, on a surface grinder the pressure of the wheel on the workpiece is always downward with the magnetic chuck providing the backup. The chuck obviously provides a much more stable platform for grinding than the two centers at either end of the cylindrical work-

Figure 1. Modern universal cylindrical grinder (Courtesy of Landis Tool Co., Division of Litton Industries).

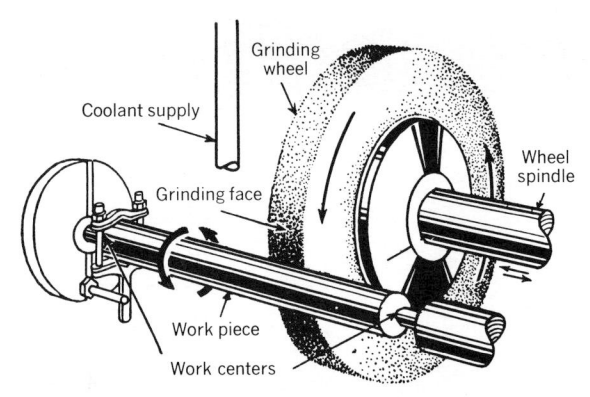

Figure 2. Sketch of center-type cylindrical grinder set up for traverse grinding. Note particularly the direction of travel of the grinding wheel and the workpiece, and the method of rotating the workpiece (Courtesy of Bay State Abrasives, Dresser Industries, Inc.).

piece, in which case the pressure is outward (toward the operator) and to some degree downward. The workpiece provides its own backup, except in the case of long, thin work when additional supports called steadyrests must be provided.

Another difference on a cylindrical grinder (Figure 2) is that the wheel-work contact area is a line at the intersection of the arc of the wheel and the smaller arc of the workpiece. Since the contact area in surface grinding is that of the arc of a wheel and a flat surface, it is slightly greater in area than that of the cylindrical grinder. The practical effect of this is that cylindrical grinding wheels tend to be just a little harder in grade than surface grinding wheels.

You will also notice in Figure 2 that the wheel cuts in only one direction (down) on the workpiece; in surface grinding the workpiece goes either with or against the direction of the abrasive cuts as it traverses.

Another significant point in terms of dimensions in cylindrical grinding is that the reduction in diameter is always approximately twice the infeed, or the reduction of the radius roughly equals the infeed. In surface grinding the reduction in the thickness of the workpiece always approximates the downfeed.

Cylindrical grinding is almost exclusively a wet operation, which aids in flushing away the swarf and dirt of grinding. This is helped by the force of gravity on the coolant. Thus, it is quite possible that the coolant on a cylindrical grinder does a better job of flushing away the grinding swarf than the same coolant would do on a surface grinder chuck and table.

Cylindrical grinding is always done with the periphery of the wheel, either a Type 1 straight wheel or a similar wheel with a shallow recess on one or both sides, so that with a given spindle you can secure a wider wheel face if that is needed.

GROUPING OF CENTER-TYPE CYLINDRICAL GRINDERS

As you will note from Figures 1 and 3, cylindrical grinders look pretty much alike from the outside. They vary in flexibility all the way from the universal (Figures 1 and 4) to the straight plunge grinder (Figure 6). On the universal, everything can swivel; on the straight plunge type, which is a specialized production grinder, the wheel moves only forward and back and the table may not even swivel. For these reasons, the plunge grinder is the most rigid, and the universal grinder is the least rigid.

The type of grinding that can be done on

Figure 3. Head-on view of a plain grinder (Courtesy of Landis Tool Co., Division of Litton Industries).

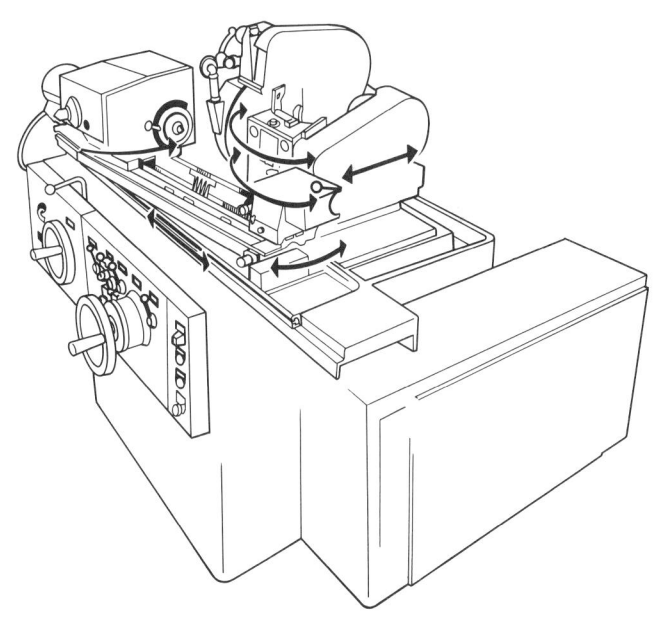

Figure 4. View of universal cylindrical grinder with arrows indicating the swiveling capabilities of the various major components (Courtesy of Cincinnati Milacron).

various types of these grinders is shown in Figures 5 to 8. Traverse grinding (Figure 5) can be done on any grinder that has the ability to traverse the workpiece back and forth across the wheel face. It does not matter whether the surface is interrupted, as it is in the illustration, or continuous, as it usually is.

Straight plunge grinding (Figure 6) is done

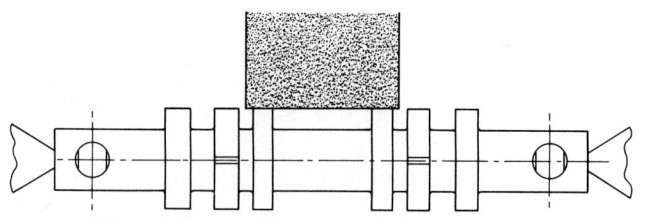

Figure 5. Sketch of traverse grinding with interrupted surfaces. Wheel should always be thick enough to span two surfaces or more at once (Courtesy of Cincinnati Milacron).

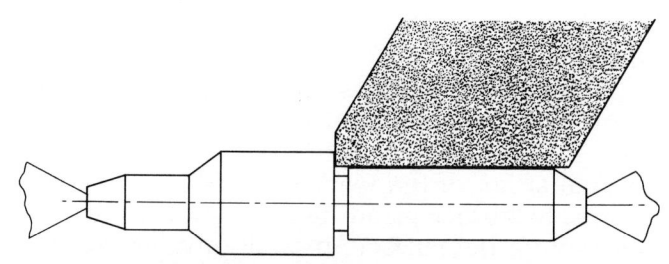

Figure 7. Angular plunge grinding with shoulder grinding. Note the dressing of the grinding wheel (Courtesy of Cincinnati Milacron).

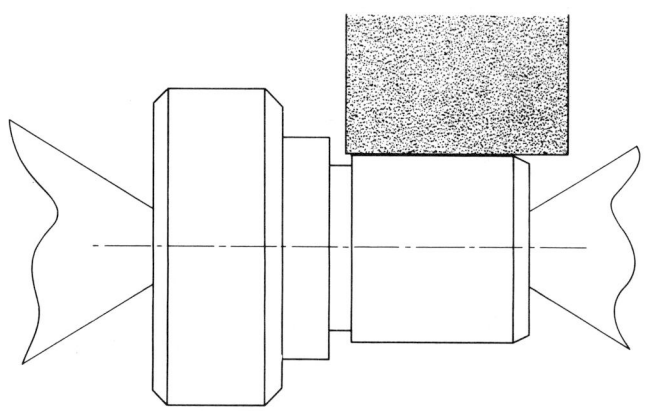

Figure 6. Straight plunge grinding, where the wheel is usually thicker than the length of the workpiece, except where the intent is to take a series of overlapping plunge cuts across a longer piece and finish with several traverses along the entire length (Courtesy of Cincinnati Milacron).

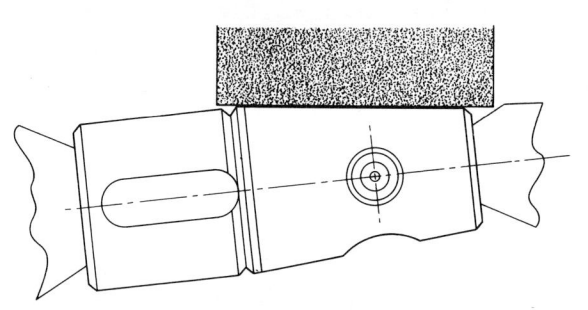

Figure 8. Taper grinding with the workpiece swiveled to the desired angle. For a steeper taper, the wheel might also have to be dressed at an angle less than 90 degrees (Courtesy of Cincinnati Milacron).

MAJOR PARTS OF THE GRINDER

Machine Bed

This is the base of the machine. It must be strong and heavy enough to support the moving elements of the grinder and rigid enough to reduce vibration to a minimum. A universal grinder has a swivel trunnion to support the wheelhead slide; on other types the slides are cast as part of the base. The base also contains ways for traversing the sliding table.

Sliding and Swivel Tables

The sliding table goes on top of the base to provide traverse motion of the work. The swivel table is mounted above the sliding table and supports both the headstock and the footstock, which actually hold the work. At one end of the table is a calibrated plate to indicate the taper both in degrees and in inches of taper per foot. Occasionally, it is desirable to make an independent check of the settings to make sure you are getting a true reading.

Headstock

The headstock (Figure 9) is generally at the left end as you face the grinder. It rotates the work

with a straight wheel; the only requirement, with one exception, is that the wheel be wider than the work is long. The exception is on a nominal traverse grinding job. The procedure is to make a succession of overlapping plunge cuts to produce a workpiece that is slightly oversize, and then finish it with a series of light traverse cuts. In many cases this is a faster way of doing what would normally be a traverse grinding job.

Angular plunge grinding (Figure 7) is similar, except that it uses a wheelhead at a 45 degree angle with the face dressed to the same angle. As the illustration suggests, and as you learned in Unit 7, this is an efficient and safe way of grinding a diameter and an adjoining shoulder.

Swiveling the table, as shown in Figure 8, is one way of grinding a taper or of straightening a tapered piece. This will be discussed at greater length under workholding.

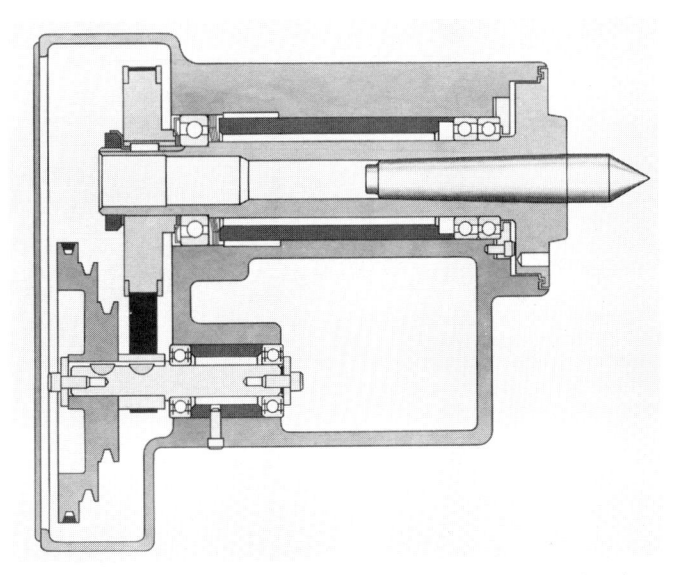

Figure 9. Typical headstock of a center-type cylindrical grinder with cutaway sketch. Note the size and complexity of the headstock in comparison with the footstock in the following figure (Courtesy of Landis Tool Co., Division of Litton Industries).

There is usually a means of selecting the proper work speed.

Footstock

The footstock (Figure 10a and 10b) essentially supports and aligns the other end of the workpiece. Besides support, it is usually equipped with a lever, as shown, to retract the footstock center for loading or unloading. It is also smaller and easier to move than the headstock, making it convenient to move the footstock to adjust for different length workpieces. However, it is also desirable to have the center of the workpiece length approximately in line with the center point of the swivel table. If only the footstock is moved over a long enough period of time, there is a possibility of extra wear and consequent vibration on the footstock end of the swivel table.

and supports one end of the work, ensuring precise alignment. Universal grinder headstocks usually have a combination live (revolving) and dead (not revolving) spindle nose. There is usually a selector lever to change from one to the other. If you select the live spindle drive, the entire spindle revolves and either a chuck or a faceplate can be mounted for work not held between centers. However, the bulk of work done on any center-type grinder is done with a dead spindle with a center that does not revolve and a workdriver plate that does. The plate has a rod or hook projection that pushes against a work driving dog clamped to the end of the workpiece.

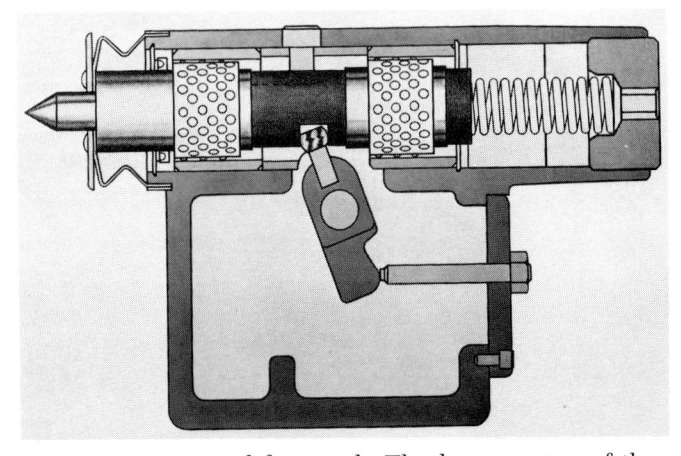

Figure 10. Typical footstock. The lever on top of the footstock retracts the work center so that the workpiece can be mounted on the grinder. The spring (right end of sketch) provides tension to hold the workpiece in place (Courtesy of Landis Tool Co., Division of Litton Industries).

Both of these conditions make it desirable to move the headstock occasionally instead of, or in addition to, moving the footstock.

Wheelhead
The wheelhead contains the grinding wheel and its driving mechanism. Both solid construction and perfect balance are essential, because any vibration from neglect in either case will be directly transmitted to the work as chatter marks.

Grinding Machine Controls
The controls for a universal center-type cylindrical grinder are quite similar in function, although they may be placed somewhat differently on different makes of machines. The controls on one make of machine will be discussed and illustrated in Unit 11.

WORKHOLDING

The basic principle of workholding on a center-type cylindrical grinder is simple. The workpiece is supported on two conical points (work centers) inserted into two matching conical center holes in either end of the blank cylinder. The work centers are supported by the headstock, which supplies the power to rotate the work, and by the footstock, which simply supports the other end. The headstock and the footstock are supported on the swivel table, which can be swung either one way or the other in order to create a tapered product or to straighten out the taper of a finished piece that is supposed to be straight. If the taper is steeper than can be produced by swiveling the table, then (on a universal grinder) the wheelhead can be swiveled. When it becomes too steep to be produced in that fashion, the face of the wheel can be dressed to whatever angle is needed.

The importance of the correct location and grinding of center holes (Figures 11, 12, and 13) cannot be too strongly emphasized. Together with painstaking care and lubrication of the work centers, this is the key to precise cylindrical grinding. No workpiece can be any better than the location and the proper angling of its center holes.

WORK POSSIBLE ON A UNIVERSAL GRINDER

Figures 14 to 21 represent a series of illustrations of the kinds of work that can be done on a universal grinder. The series starts with simple traverse

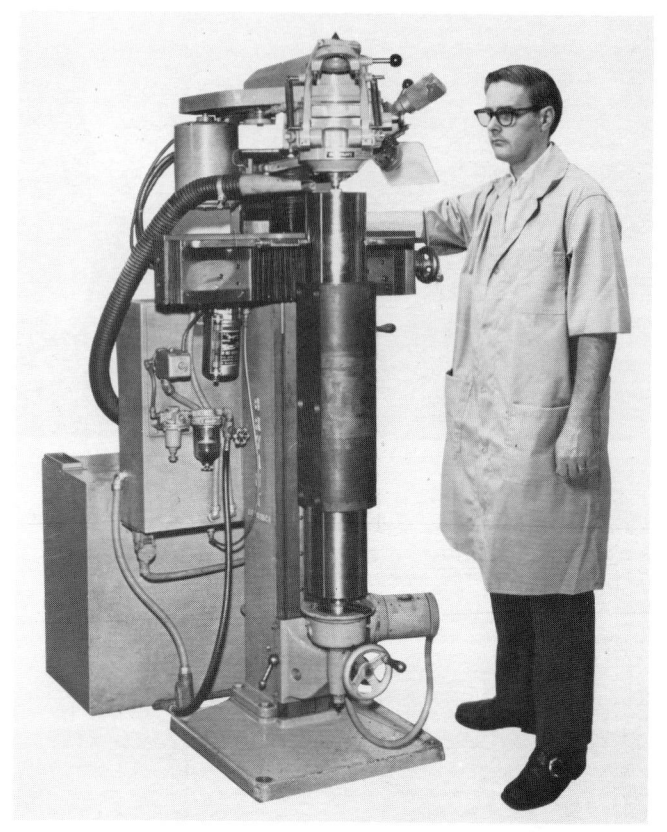

Figure 11. Center hole grinding machine (Courtesy of Bryant Grinder Corp.).

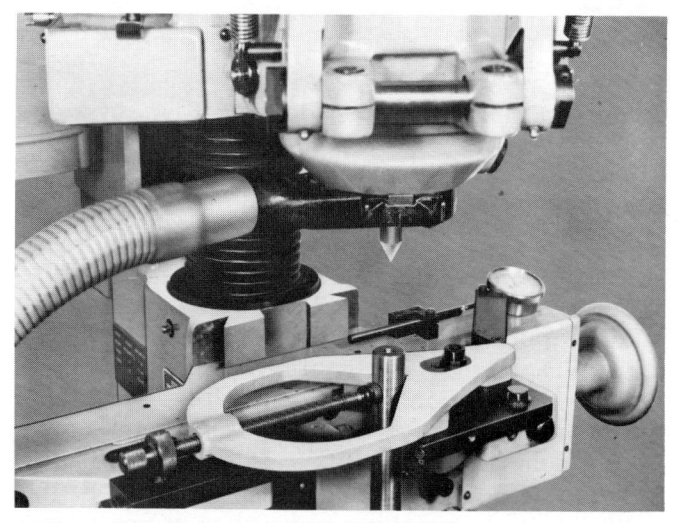

Figure 12. Closeup of center hole locating setup. Exact location of the center is a most critical step in the operation (Courtesy of Bryant Grinder Corp.).

grinding (Figure 14), progresses through OD taper grinding (Figure 15), and traverse grinding with a backrest or steadyrest (Figure 16). You will note that the workpiece in Figure 16 is long

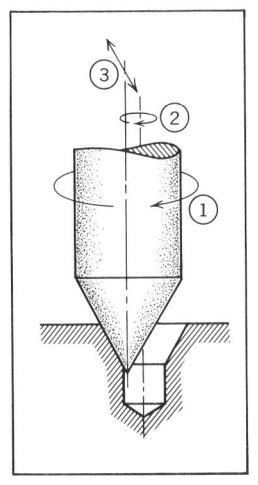

Figure 13. Sketch showing the motions of the grinding wheel in center hole grinding (Courtesy of Bryant Grinder Corp.).

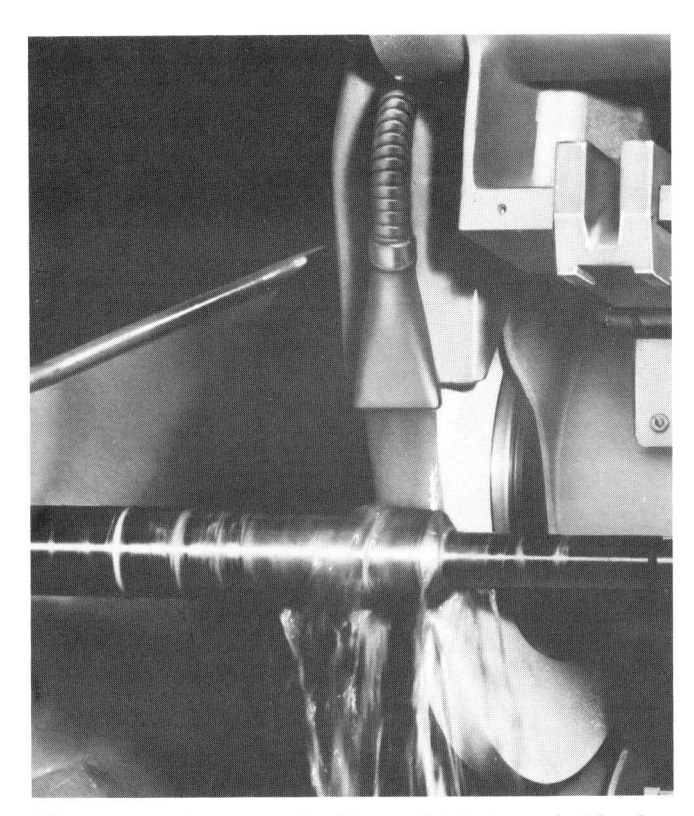

Figure 14. Traverse grinding, which is probably the most common type of cylindrical grinding (Courtesy of Cincinnati Milacron).

Figure 15. OD taper grinding, which may be done to produce a tapered finished workpiece or to straighten up a rough piece that was previously tapered (Courtesy of Cincinnati Milacron).

Figure 16. Backrest or steadyrest used to support a thin piece for grinding. If only one is used, as here, it is in the center of the workpiece. If more are used, it is always an odd number, and they are equally spaced along the workpiece (Courtesy of Cincinnati Milacron).

and thin in comparison with most of the others shown.

Steep taper grinding (Figure 17) involves swiveling the wheelhead as well as the table. Multiple diameter grinding, which can be done by dressing the wheel in procedures like those discussed in Unit 7, is illustrated in Figure 18. Figure 19 shows simultaneous grinding of a

Figure 17. Steep taper grinding. Here the wheelhead has been swiveled to grind the workpiece taper (Courtesy of Cincinnati Milacron).

Figure 18. Multiple-diameter or form grinding. This is usually a plunge grinding operation with the form dressed in the wheel face in the same manner as for reciprocal surface grinding (Courtesy of Cincinnati Milacron).

Figure 19. Angular-shoulder grinding. This is very often a production-type operation, but it is shown here on a universal grinder (Courtesy of Cincinnati Milacron).

Figure 20. Internal grinding, which requires a special high speed attachment that is mounted on the wheelhead so that it can be swung up out of the way when not in use. It also requires either a chuck or a face plate on the headstock (Courtesy of Cincinnati Milacron).

diameter and a shoulder. Figures 20 and 21 illustrate internal grinding.

These are not necessarily the most efficient ways of doing these jobs. The sequence is only

Figure 21. ID taper grinding. This is the same sort of operation as shown in Figure 20, but with the workpiece swiveled at the required angle (Courtesy of Cincinnati Milacron).

designed to illustrate the range of the machine. On the other hand, it does emphasize that the universal grinder is very flexible, and that a shop equipped with a universal grinder and a small surface grinder can handle a great variety of work.

ATTACHMENTS AND ACCESSORIES

Accessories for the universal grinder can be grouped as those for headstocks, those for table mounting, and all others designed for a variety of uses. For the headstock, the units are generally three- or four-jawed chucks for mounting the small percentage of parts for which drilling and grinding a center hole would be impractical or impossible.

For the table-mounted group, there are diamond dressers, either for straight or contour dressing, that have the same purpose but a different mounting as the form dressers described in Unit 7. Others, like the center rest, can supply extra support for very thin work, and the dial indicator for table swivel adjustment simply provides greater accuracy in tapering or removing taper than the standard machine.

self-evaluation

SELF-TEST

1. Explain the basic differences in construction between the universal center-type grinder and other types of center-type cylindrical grinders.

2. Name the most important reason that the center-type cylindrical grinder was the first to be developed.

3. List at least three differences between center-type cylindrical grinding and surface grinding.

   a. _____

   b. _____

   c. _____

4. List and discuss in a sentence or two the different setups for grinding cylindrical tapers.

5. What would be the major factors to consider in setups to be used?

6. How do the functions of the headstock differ from those of the tailstock? Which is moved more often? Why?

7. Describe the basic method by which a workpiece is held on a center-type cylindrical grinder.

8. What is a steadyrest and when is it used? List some of the factors that determine how many will be used on a given workpiece.

9. What is the relationship between the variety of work possible on a center-type cylindrical grinder and the rigidity of the grinder?

10. Why are the center holes so important in this type of cylindrical grinding?

unit 11
using the cylindrical grinder

The workpiece selected for the cylindrical grinder in this unit is a lathe mandrel, slightly tapered, which requires the application of many of the techniques that you studied in the previous unit. The material is SAE 4140 or a similar alloy steel, which is a widely used metal. The unit includes a step-by-step procedure for grinding a lathe mandrel. Some of the steps are not discussed because they are sufficiently clear from the illustrations.

objectives After completing this unit, you will be able to:
1. Grind the tapered mandrel of SAE 4140 or similar alloy steel on a universal center-type cylindrical grinder to your instructor's satisfaction.
2. Explain any part of the operation to your instructor on request.
3. Prepare a worksheet for and actually grind a similar tapered part involving center-type cylindrical grinding to your instructor's satisfaction.

information

The part selected for this demonstration project is sketched in Figure 1 and shows the necessary dimensions. In this unit it is also assumed that you have mastered the machining skills necessary to machine the part to allow about .025 to .030 in. oversize for grinding.

The selection of the grinding wheel, which was discussed in Unit 8, will only be reviewed here, since the only difference is that this is OD grinding instead of flat surface grinding. Hence, it will be a general purpose wheel of friable or semifriable aluminum oxide (because it is for grinding steel), a little finer (60 or 54 instead of 46) and definitely harder grit (K, L, or M grade instead of H), a shade more dense in structure (if this is mentioned, as it is frequently not mentioned in such wheels), and with a general purpose vitrified bond.

The testing and mounting of the wheel (steps 2 to 6 of Unit 8) are very similar, and the work centers on the headstock and the footstock should be wiped off with a clean, soft cloth.

STEPS IN GRINDING

1. *Set up and adjust the diamond for dressing (Figure 2).* The footstock work center in Figure 2 provides a reference point, but there are times in actual practice when it would be desirable to move the footstock out of the way after the diamond has been located in relation to the wheel.
2. *Turn on the coolant and dress the wheel (Figure 3).* The first dressing of a wheel in a situation like this is usually for rough grinding for stock removal, and the diamond is traversed back and forth across the wheel rather rapidly.

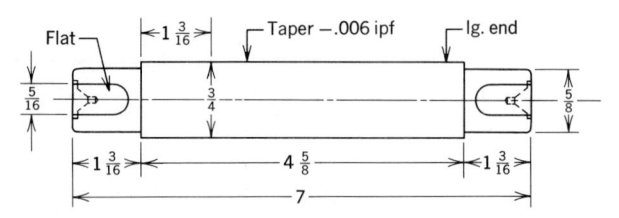

Figure 1. Sketch of tapered lathe mandrel.

Figure 2. Setting up the diamond for wheel dressing (Lane Community College).

Figure 3. Dressing the wheel with the coolant on (Lane Community College).

Figure 4. Setting up the parallel test bar (Lane Community College).

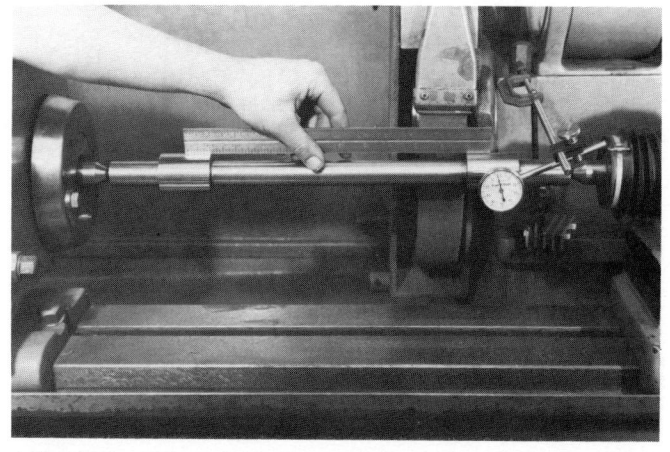

Figure 5. Second reading of the dial indicator (Lane Community College).

3. *Set up the parallel test bar (Figure 4).* The test bar is mounted between centers and a dial indicator with a magnetic base is mounted at a convenient spot on the wheelhead. The indicator is zeroed at the right side of the test bar. (This is at the operator's left.)
4. *Read the indicator on the other end, 12 inches from the original position (Figure 5).* For the required taper of .006 IPF, the indicator should read .003 in. counterclockwise on the radius. You will remember that in one revolution of the workpiece the wheel removes about *twice* the amount of infeed of the wheel.
5. *Adjust the swivel table (Figure 6).* In a sense this is a sort of interim step with steps 3 and 4. This is the adjustment that has to be made to obtain the two readings called for in those steps. The scale gives an approximation, but you should follow the readings from the dial indicator.
6. *Lubricate the mandrel ends (Figures 7 a and b).* The center hole at the ends of each workpiece must be lubricated with center lube, since both the headstock and the footstock centers are dead; that is, they do not rotate. The work dog should also be mounted on the mandrel at this point.

Unit 11 Using the Cylindrical Grinder 447

Figure 6. Adjusting the swivel table (Lane Community College).

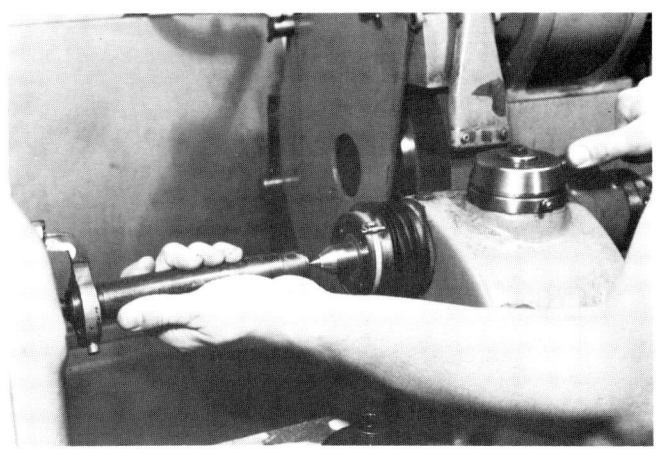

Figure 8. Clamping the footstock (Lane Community College).

Figure 9. Checking the work dog clearance and drive pin contact (Lane Community College).

Figure 7. Lubricating the center holes in the two ends of the workpiece (Lane Community College).

7. *Clamp the footstock to the table (Figure 8).* The distance between the headstock and the footstock should be slightly less than the workpiece length, so that when the workpiece is mounted on centers there will be a little tension on it from the slightly compressed spring in the footstock.
8. *Check the work dog (Figure 9).* The work dog on the workpiece must have clearance to turn freely without binding. Also, it must contact the drive pin. This combination is what rotates the workpiece.
9. *Set stops for wheel overrun (Figure 10).* To make sure that the ends of the workpiece are ground, the wheel must definitely travel a little farther than the end of the work. About one third of the wheel thickness is recommended.

Figure 10. Adjusting the wheel overrun (Lane Community College).

Figure 11. Checking wheel-work dog clearance (Lane Community College).

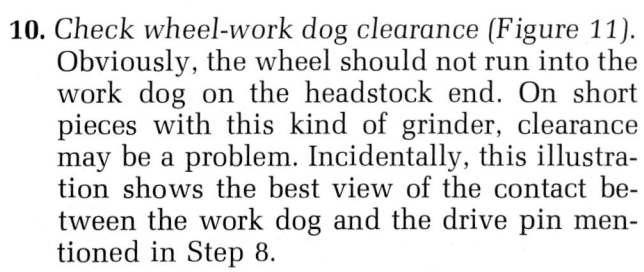

Figure 12. Adjusting the wheel to the workpiece (Lane Community College).

Figure 13. Turning on the wheel (Lane Community College).

10. *Check wheel-work dog clearance (Figure 11).* Obviously, the wheel should not run into the work dog on the headstock end. On short pieces with this kind of grinder, clearance may be a problem. Incidentally, this illustration shows the best view of the contact between the work dog and the drive pin mentioned in Step 8.
11. *Adjust the wheel to the work (Figure 12).* Remove the workpiece and advance the wheel, backing it off if necessary to assure that it will clear the workpiece. Replace the workpiece. The wheel is at what will be the smaller end of the mandrel.
12. *Turn on the wheel (Figure 13).* For safety's sake, stand out of line with the wheel for a minute or so. After starting the wheel, feed it in close to the workpiece.
13. *Turn on the spindle (Figure 14).* This starts the workpiece rotating. Set the work speed. Infeed the wheel slowly toward the work until sparks begin to show.
14. *Turn on the coolant (Figure 15).*
15. *Turn on the table feed and adjust the speed (Figure 16).* For roughing, the traverse rate will be about one-half to two-thirds of the wheel width for each revolution of the work. Wheel infeed should be about .001 to .002 in. Take a roughing cut or two.
16. *Check taper (Figures 17 a and b).* Retract the wheel and make two marks 4 in. apart. Measure the differences in *diameter* with a micrometer to determine whether the taper per foot is correct. The difference in a 4 in. length should be .002 in. Adjust the swivel table as necessary.

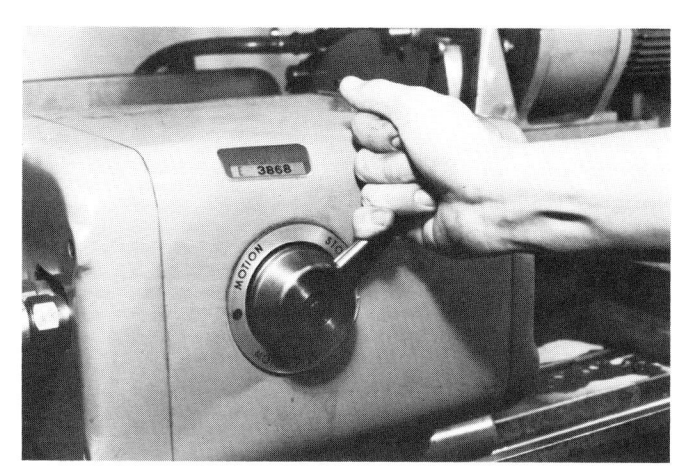

Figure 14. Turning on the spindle (Lane Community College).

Figure 15. Turning on the coolant (Lane Community College).

Figure 16. Adjusting the table speed (Lane Community College).

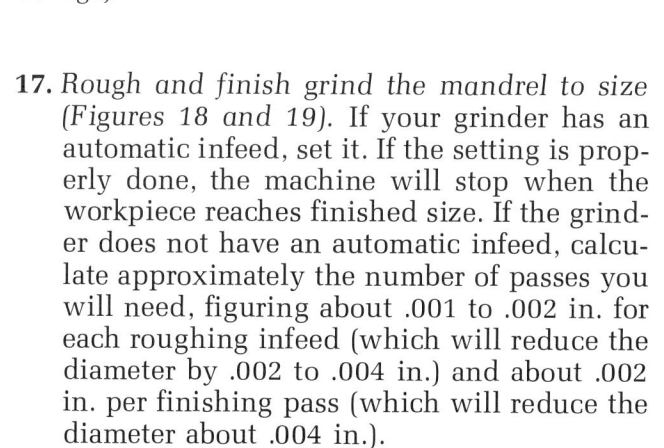

Figure 17. Checking the taper (Lane Community College).

17. *Rough and finish grind the mandrel to size (Figures 18 and 19).* If your grinder has an automatic infeed, set it. If the setting is properly done, the machine will stop when the workpiece reaches finished size. If the grinder does not have an automatic infeed, calculate approximately the number of passes you will need, figuring about .001 to .002 in. for each roughing infeed (which will reduce the diameter by .002 to .004 in.) and about .002 in. per finishing pass (which will reduce the diameter about .004 in.).

Stop and check diameter and taper periodically. The main point to avoid is grinding undersize, because then you have only scrap.

Traverse rate in finish grinding is no more than ½ in. per work revolution, and the work speed ranges up to 100 SFPM. On a center-type cylindrical grinder, the wheel, at 12, 14, or 16 in. in diameter, may wear enough that wheel speed

Figure 18. Grinding the mandrel to size (Lane Community College).

Figure 19. The finished mandrel (Lane Community College).

(SFPM) will be affected. If this happens, it may be advisable to reduce the work speed, also. This will be further discussed in Unit 12.

Clean up the machine and return the tooling to proper storage. Show the finished mandrel to your instructor for his evaluation.

self-evaluation

SELF-TEST Your instructor will assign you a center-type grinding project similar to the one you have just studied, which will involve setting up the machine from scratch, perhaps setting and checking a taper, starting up the grinder, and roughing and finishing the workpiece.

Completing the test satisfactorily will involve writing a worksheet indicating the steps needed to complete the job and grinding the workpiece to specifications. Your worksheet will follow the general pattern of what has been listed for each step; there may be some other steps to include or some of these to leave out. What follows the step and figure numbers is explanatory information that you will not have to include in your worksheet.

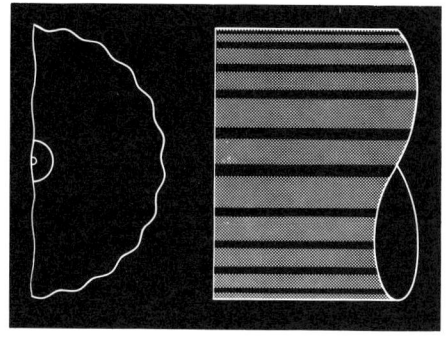

unit 12
problems and solutions in cylindrical grinding

Since both surface grinders and cylindrical grinders are grinding machines, often grinding the same kinds of materials, it follows that they have many problems in common with similar solutions. Hence, this unit is similar to Unit 9, although it will concentrate on the situations that are limited to center-type cylindrical grinding. However, there will be as little repetition as possible; when it occurs, it will be primarily to avoid referring back in your book.

objectives

After completing this unit, you will be able to:
1. Define and recognize surface defects resulting from cylindrical grinding and suggest ways of correcting them.
2. Suggest ways of correcting faults that show up in postinspection of work that is not round, not tapered correctly or not straight, or not to size.

information

You have read already that cylindrical workpieces are possibly, in most cases, less solidly supported than flat work done on the surface grinder. The area contact between wheel and work is less in cylindrical grinding resulting in the need to use slightly harder wheels. All cylindrical grinding work is done on the periphery of the wheel instead of on either the periphery or the side of the wheel as is true of surface grinding. All of these make for a different set of problems, or sometimes, different solutions to the same problems. At the same time, both are still grinding operations, and problems such as burning or cracking of surfaces and discoloration are common to both.

As with surface grinding, cylindrical grinding defects could be subdivided into those caused by the machine or the shop situation, such as worn bearings and misalignment, and those caused in the operation of the grinder, such as wheel or grinding fluid selection, dressing, traverse, and work speed rates. In general, the first group is not under the control of the operator, while the second group most likely is. It is also true that different shops have different policies as to the division of work between the mechanics or other maintenance men and the operators; and you will, of course, follow whatever policy is set down in the shop where you are working.

It is also possible to classify grinding faults as those having to do with finish and those having to do with sizing or dimensions. Table 1 is a quick survey of center-type cylindrical faults; one group is concerned with finish, but the other three involve sizing. Causes for the condition are given and, in most cases, the remedy is clear. For instance, if a diamond is dull, you rotate it in the holder to engage a sharp point, or you replace it. If it is cracked, it is replaced.

CHATTER MARKS

Not all of the parallel-line faults in a ground workpiece are necessarily chatter marks (Figure 1), but it is a good assumption to make because most of them are. Chatter marks are caused by vibration, which produces variations in the contact between the wheel and the workpiece. Some of the possible causes and some hints on finding and correcting them are discussed below.

Vibration is the most likely cause and the orderly approach to locating it is not too difficult. Eliminating vibration after you find it, particularly if it is outside the machine, may be another problem.

To begin, you shut down the machine entirely. Then you place a pan of water on the wheelhead. If the water ripples, the vibration is outside the machine; then you either have to insulate the grinder in some way, such as extra padding under the base, or move the grinder, if this is feasible. You may need to search further for the vibration. The step you take depends on

Table 1
Causes of Problems in Cylindrical Grinding. Source. Reprinted from American Machinist, August 17, 1953, by McGraw-Hill Publishing Co., 330 W. 42nd St., New York, N. Y., 1953.

POOR FINISH

1. Dull or cracked diamond
2. Loose diamond or diamond holder
3. Diamond overhangs too far
4. Poor grade of diamond
5. Wheel out of balance or roundness
6. Oil on portions of wheel
7. Dull grinding wheel
8. Incorrect wheel grade
9. Sharp corners on wheel
10. Improper speeds and feeds
11. Excessive stock removal
12. Headstock belts slipping
13. Unmatched belts on wheelhead
14. Dirty, incorrect or insufficient coolant
15. Insufficient coolant when dressing
16. Dirty workpieces
17. Centers too long or loose
18. Loose wheel or wheel mount
19. Loose sheave on grinding-wheel spindle
20. Vibration from outside source

SIZING TROUBLES

1. Loose or incorrect taper on centers
2. Insufficient dwell
3. Insufficient coolant when grinding
4. Improper wheel
5. Bad centers in workpiece
6. Incorrect lubricant or hydraulic oil
7. Inconsistent operating cycle

OUT OF ROUNDNESS

1. Loose or incorrect taper on centers
2. Workpiece not properly supported
3. Too much footstock tension on slender work
4. Not enough tension on footstock
5. Headstock or footstock not clamped properly
6. Condition of workpiece center holes
7. Lack of lubrication on centers
8. Overheating due to lack of coolant
9. Stock removal too rapid
10. Insufficient dwell
11. Out-of-balance workpiece

TAPER

1. Loose or incorrect taper on centers
2. Loose footstock spindle
3. Footstock or headstock not clamped tightly
4. Machine out-of-level
5. Condition of center holes in workpiece
6. Too much or too little dwell at one end
7. Over-travel of wheel at end of work
8. Dirt under swivel table
9. Insufficient coolant when dressing wheel

Figure 1. Chatter marks result from irregular pressure of the wheel on the work, usually causing lack of contact in some regular pattern (Photo courtesy of Norton Company).

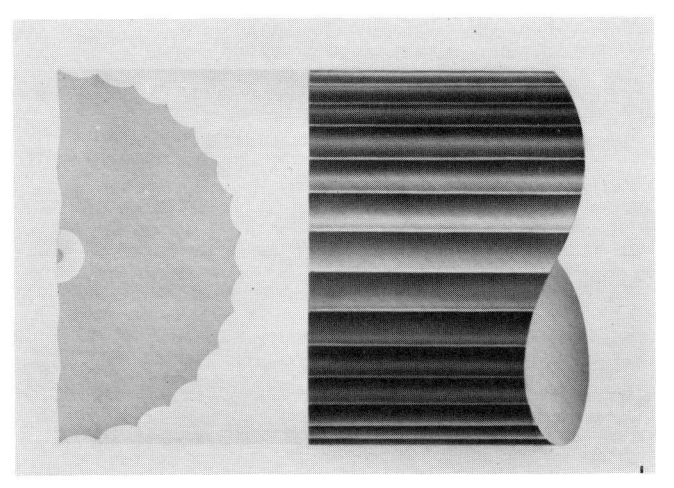

Figure 2. Artist's conception of short wave marks with an end view at the left and a side view at the right. Wave marks can occur without breaking contact between the wheel and the workpiece (Photo courtesy of Norton Company).

whether the vibration can be located and stopped or whether it is a necessary part of shop operation.

On the other hand, if the water remains still, the vibration is in the machine. The cause may be discovered by the process of elimination by operating singly as many of the machine units as you can, the grinding wheel, table traverse, and so on, by placing the pan of water on the operating unit that you are testing.

If this checks out, then you might look for trouble in the workpiece support, not enough or no steadyrests, too much or not enough tension on the workpiece, and so on. Finally, look at the wheel, which may be too hard. The less rigid the machine, the harder the wheel that is needed and vice versa. If you can not change wheels, maybe you can make the wheel *act* softer by speeding up the work rotation by dressing the wheel a little coarser. You should never forget the possibility that the wheel may be out of balance.

WAVE MARKS

Wave marks look like chatter, as shown in Figures 2 and 3, except that contact between the wheel and the workpiece is not broken. Whether they are short wave marks (Figure 2) or long wave marks (Figure 3), they also differ from chatter in that they are regularly spaced around the workpiece. You will not go far wrong if you treat them as chatter.

Figure 3. Long wave marks may be as much as $\frac{1}{4}$ to 1 in. apart, depending on the speed of the workpiece and the wheel, but they result from the same causes as short wave marks (Photo courtesy of Norton Company).

"BURLAP FINISH"

This is a fault named because of its appearance (Figure 4). It is almost always an indication that something is loose in the diamond dresser, either in the tool or the holder.

SPIRAL LINES

These are sometimes called trueing lines (Figure 5), feed lines (Figure 6), or "barber pole" finish.

Figure 4. "Burlap finish." Typical pattern caused by a wheel dressed with a diamond that was not held firmly in the shank (Photo courtesy of Norton Company).

Figure 5. Trueing lines. Coarse finish with threadlike lines resulting from dressing with too sharp a diamond and too much infeed on the final passes of the dressing cycle. Use of a cluster diamond dressing tool or making the final dressing passes without infeed will help avoid this condition (Photo courtesy of Norton Company).

Figure 6. Spiral lines like these usually occur during traverse grinding, and the lines have a lead similar to the traverse rate. Most often they occur because of misalignment between the headstock and the footstock; but, like most surface defects, they can be caused by poor dressing technique — in this case, neglecting to round off the wheel corners (Photo courtesy of Norton Company).

They have a great deal in common in appearance. If you are plunge grinding and you spot spiral lines, they are probably trueing lines. However, if the lines have the same lead as the rate of traverse, they are feed lines. If you are traverse grinding, you may have one or another of these conditions.

Trueing lines generally come from using too sharp a diamond, or from not rounding the edges of the wheel. They can usually be avoided by using a cluster-type tool or a worn diamond instead of a sharp single diamond and by radiusing the corners of the wheel. The cluster type is becoming more popular as time goes on because it is usually less expensive, easier to use and probably, on the whole, does a better job than a single diamond.

Feed lines generally occur because of something out of alignment or because the wheel is not suitable. If the headstock and the footstock are out of alignment, if steadyrests are being used and are not in line, or sometimes even if work speed is not accurate, feed lines can show up. If the wheel is the cause, it is probably worthwhile to have a knowledgeable abrasive engineer or perhaps an experienced grinding wheel salesman check the wheel specification. Otherwise, of course, it is a matter of checking out the alignment of the work and the machine elements involved.

COMBINATION PROBLEMS

Perhaps the most puzzling of all problems are those that arise out of some combination of conditions. It should be said that poor quality work very often does result from a combination of problems; situations are rarely as clean-cut as you would like them to be. In fact, sometimes it is necessary to try two or three remedies before you find one that works. You may find that the situation has been corrected without knowing just what did the job. For that reason, it may be well to conclude with a few general observations that should have been made clear by now.

☐ If in doubt, go for the softer of two wheels, in most cases. Many shops prefer hard wheels on the basis that they last longer (which they probably do), that they perform better on machines that are not well maintained (which they do), and that they are less expensive (which is debatable). Actually, in any grinding outside of the very high volume, high production operations like those in automobile production, the cost of the abrasive used is one of the minor expenses, especially when you look at the number of problems whose cause is too-hard wheels.

☐ In case of problems, look first at the work, then at the grinder, and then at the wheels and the coolant. You should not get caught in the trap of trying to compensate for worn wheel bearings or table misalignment by trying to adjust the wheels or the coolant.

☐ Make sure that your machine, including the coolant tank, is kept clean. If swarf accumulates in the bottom of the coolant tank, for example, it is the same as cutting down the capacity of the system, and fishtails (Figure 7) may develop on the work.

Figure 7. When you see little random scratches like these in an otherwise good finish, you had better check quickly to see that the coolant is being filtered and cleaned and that the coolant tank is full. These are the direct result of recirculation of dirt or bits of metal in the coolant, or possibly falling from the wheel guard, although this is not as likely in cylindrical grinding (Photo courtesy of Norton Company).

☐ If your coolant is a concentrate to which water is added, make sure that the additions are measured in the right proportions. Coolant that is not in the right concentration can cause problems and be wasteful.

☐ As a temporary measure, using a smaller wheel of the same specification (which has the effect of cutting down wheel speed) or increasing work speed will make the wheel act softer. It has the same effect as using a softer wheel.

self-evaluation

SELF-TEST 1. List two or more factors that make the problems of cylindrical grinding different from those of surface grinding.

a. _____

b. _____

2. How are the problems of the two methods alike?

   _____

   _____

3. Briefly describe the pan-of-water method of locating vibration both outside and within the cylindrical grinder.

   _____

4. List at least three general causes of problems in cylindrical grinding.

   a. _____

   b. _____

   c. _____

5. How do wave marks resemble chatter? How are they different?

   _____

6. Name the principal causes of "burlap finish." How are they corrected?

   _____

7. How do feed lines and trueing lines resemble each other? How are they different?

   _____

8. What is the relationship of wheel grade and machine condition? Of wheel grades in surface and cylindrical grinding?

   _____

   _____

9. Name a general rule for deciding between two grades of wheels. Name one or more, if possible, temporary adjustments that are possible for wheels that are too hard.

   _____

   _____

10. Name at least two responsibilities of an operator for coolant.

    a. _____

    b. _____

unit 13
cutter and tool grinder features and components

After you have had experience with other types of grinding machines, such as surface grinders and center-type grinders, and after you have had practice with selecting grinding wheels to match grinding requirements, you will probably be ready for work on the cutter and tool grinder. This versatile machine tool can be used for a great number of grinding operations.

objectives After completing this unit, you will be able to:
1. Identify and name the features and components of the cutter and tool grinder.
2. Identify and name at least three accessory attachments for the cutter and tool grinder.
3. Name two types of workholding devices used for holding arbor mounted cutting tools.
4. Name two types of tooth rest supports used with cutter and tool grinders.

information

The cutter and tool grinder may well be the most versatile grinding machine made. It is principally used for the sharpening of rotary cutting tools such as reamers and milling cutters but, with various accessories, it can do a remarkable number of grinding operations. These include internal and external cylindrical and taper grinding, surface grinding, cutting off operations and single point tool grinding. The standard machine (Figure 1) comes with a variety of components as standard equipment, including a workhead that may be swiveled on two axes and tailstock centers for mounting tools that have centers in each end. Also supplied are wheel guards, tooth rests, supports, and mounting equipment for positioning tooth rests.

The cutter and tool grinder may also be equipped with power table traverse and with a motor drive for the workhead, called a cylindrical grinding attachment (Figure 2). This combination makes internal and external cylindrical grinding easy to do. The motor driven workhead

458   Section H Grinding Machines

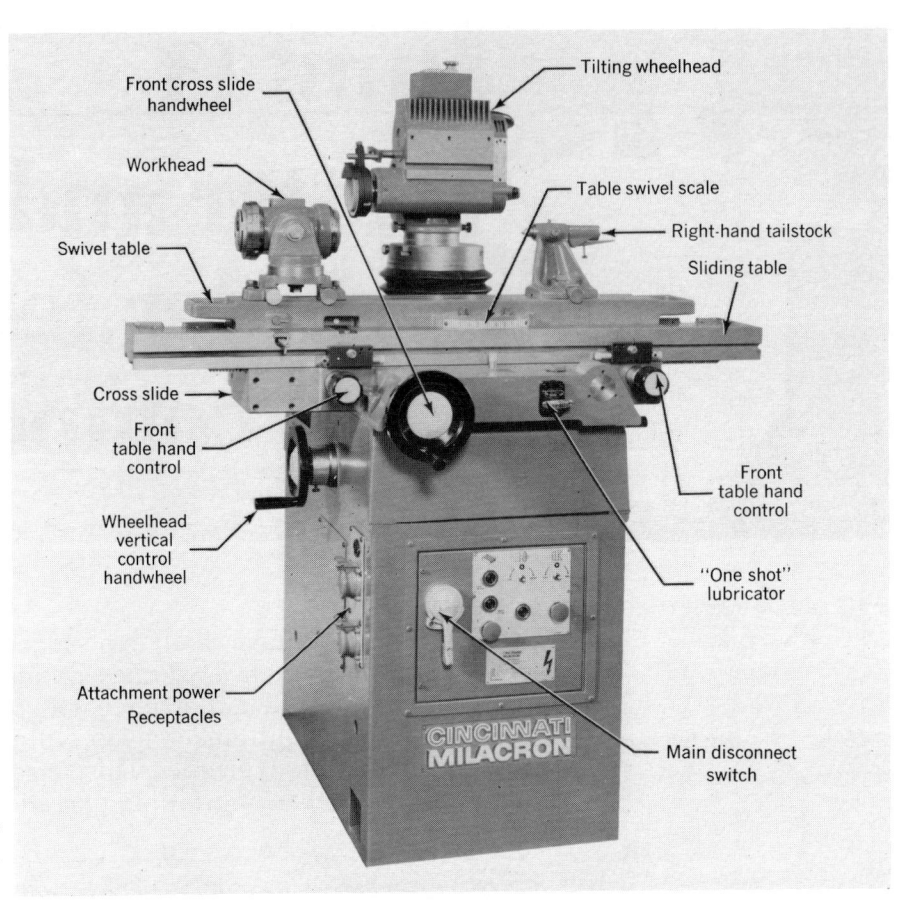

Figure 1. Components of the cutter and tool grinder (Courtesy of Cincinnati Milacron).

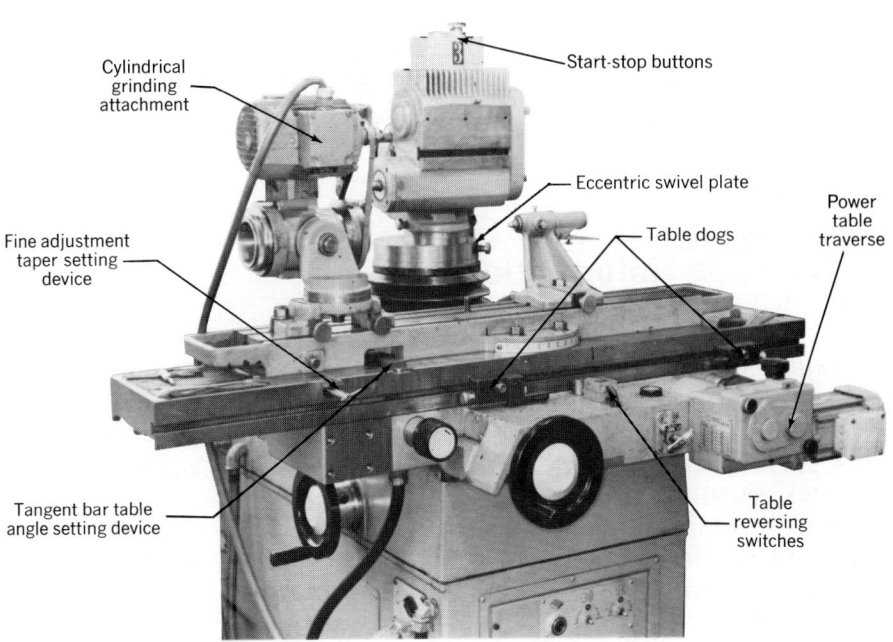

Figure 2. Cutter and tool grinder equipped with power table traverse and with a cylindrical grinding attachment mounted on the workhead (Industrial Plastics Products, Inc.).

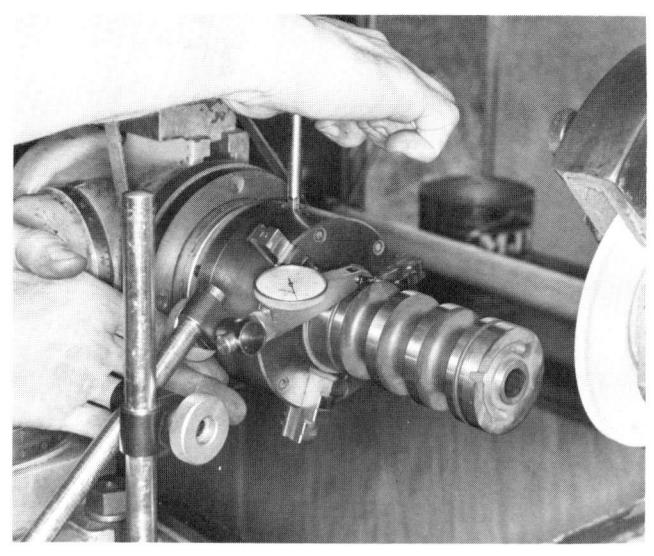

Figure 3. Cylindrical grinding attachment being used with adjustable scroll chuck to set up part for cylindrical grinding (Industrial Plastics Products, Inc.).

Figure 4. A permanent magnet chuck is also very useful for cylindrical grinding (Industrial Plastics Products, Inc.).

Figure 5. Internal grinding attachment set up for use with a mounted grinding wheel. Only the forward portion of the wheel is used; the rest is dressed away for clearance (Industrial Plastics Products, Inc.).

Figure 6. Another use for the cylindrical grinding attachment is the reconditioning of centers (Industrial Plastics Products, Inc.).

can be fitted with a scroll chuck with an adjustable backing plate that can be used for setting up work for cylindrical grinding (Figure 3).

A magnetic chuck (Figure 4) is also very useful as a cylindrical grinding attachment accessory, especially when external cylindrical grinding is preceded or followed by internal grinding (Figure 5). The cylindrical grinding attachment also makes it easy to recondition centers (Figure 6). This is done to restore centers for other machine tools such as lathes, but it is also done, when necessary, before mounting work between centers to ensure accuracy when parts are cylindrically ground on centers (Figure 7). The workhead is also equipped with a lock so that the center in the workhead may be held stationary (as a "dead" center) while the workpiece is driven, in the same way that it is done on regular universal and plain cylindrical grinders.

Figure 7. Resizing the pilot portion of a counterbore by cylindrical grinding (Industrial Plastics Products, Inc.).

Figure 8. Tailstock centers are basic to a large portion of cutter grinding (Industrial Plastics Products, Inc.).

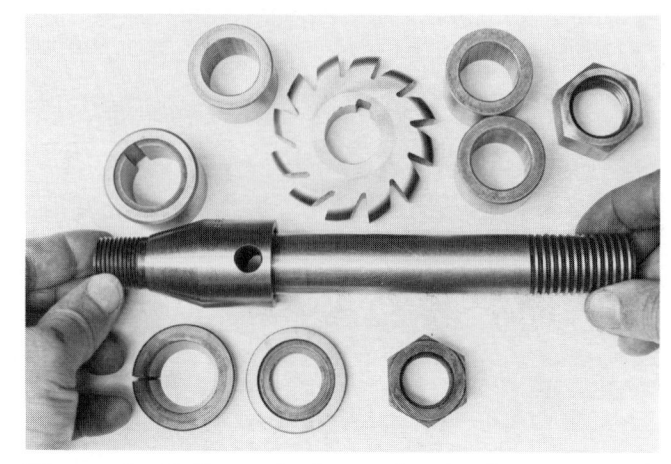

Figure 9. Cutter grinding arbor and components (Lane Community College).

Figure 10. Slitting saw being mounted on grinding arbor (Lane Community College).

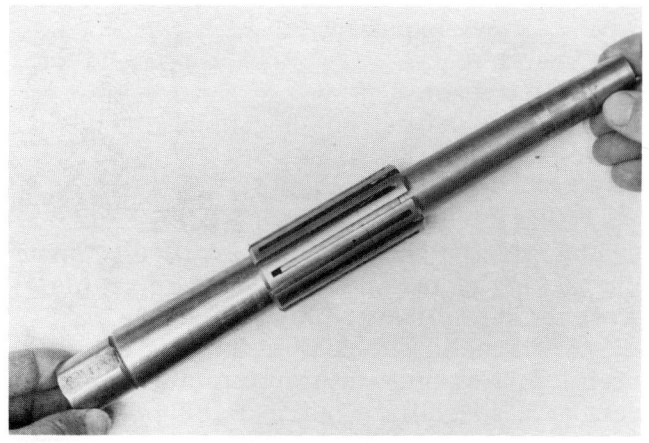

Figure 11. Adjustable grinding mandrel. The slotted bushing is moved along the mandrel to adjust for the ID of the tool to be mounted (Lane Community College).

The greatest amount of work on the cutter and tool grinder is done between tailstock centers (Figure 8) or, as will be seen later, from a manually operated workhead. Cutters such as formed cutters and plain milling cutters are usually mounted between centers on accurately made hardened and ground cutter grinding arbors (Figure 9) with precise centers in each end. This type of arbor can also be used for mounting slitting saws for resharpening (Figure 10).

Another type of holding device that can accommodate variations in internal diameter is the adjustable mandrel (Figure 11). This type of

Figure 12. Various designs of tooth rest blades (Lane Community College).

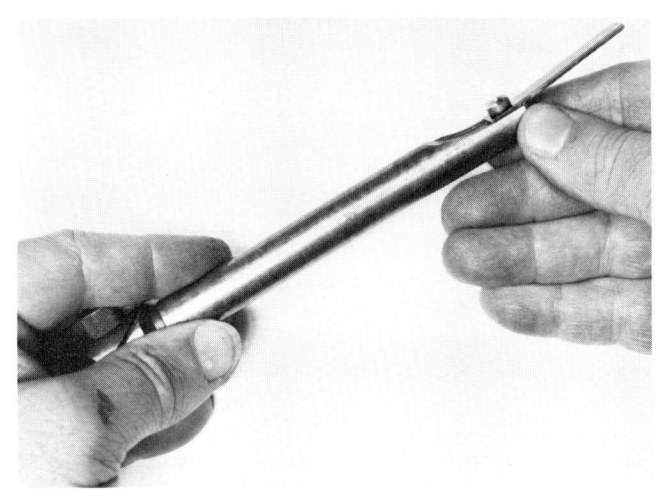

Figure 13. Plain-type tooth rest support (Lane Community College).

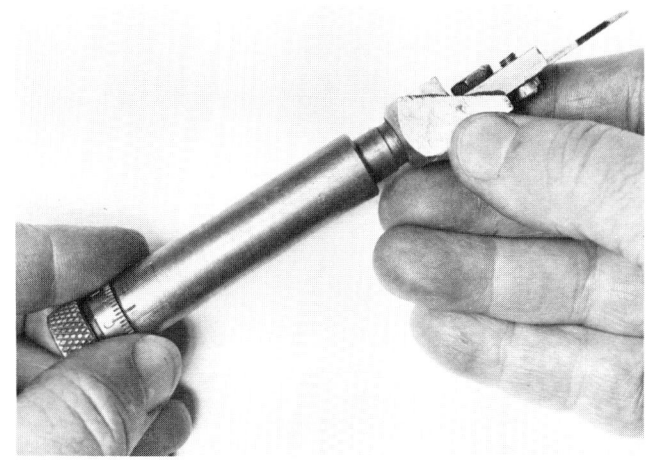

Figure 14. Micrometer-type tooth rest support. This example also has provision for spring loading of the finger to permit ratcheting of the cutter tooth, called a "flicker finger."

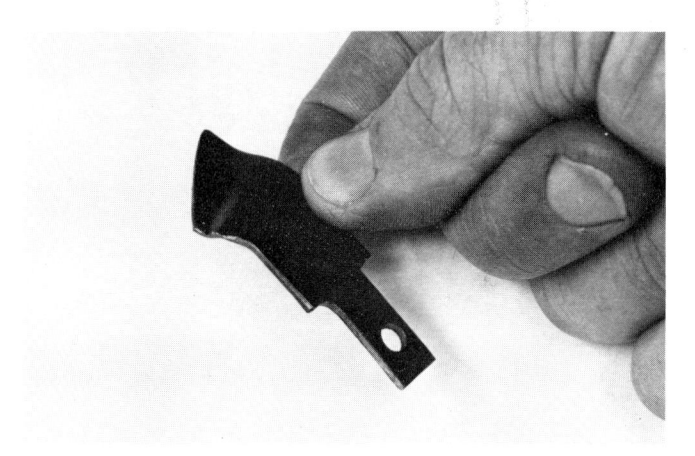

Figure 15. Offset tooth rest blade for use in grinding helical milling cutters.

holding device requires special care to get accurate results.

Most of the cutter grinding operations require a support for the tooth being ground. This is usually done by mounting specially shaped tooth rest blades (Figure 12) in either plain tooth rest supports (Figure 13) or in a support with a micrometer adjustment (Figure 14). The micrometer adjustment is especially useful for making small adjustments in setup and, as will be seen later, as a feeding device for finish grinding of form relieved tools.

For grinding helical milling cutters, an offset tooth rest (Figure 15) is obtained or made up to match the helix direction and angle of the cutter to be sharpened. Other shapes of tooth rests will be seen in action later.

462   Section H Grinding Machines

self-evaluation

SELF-TEST 1. Name the features and components indicated in Figure 16.

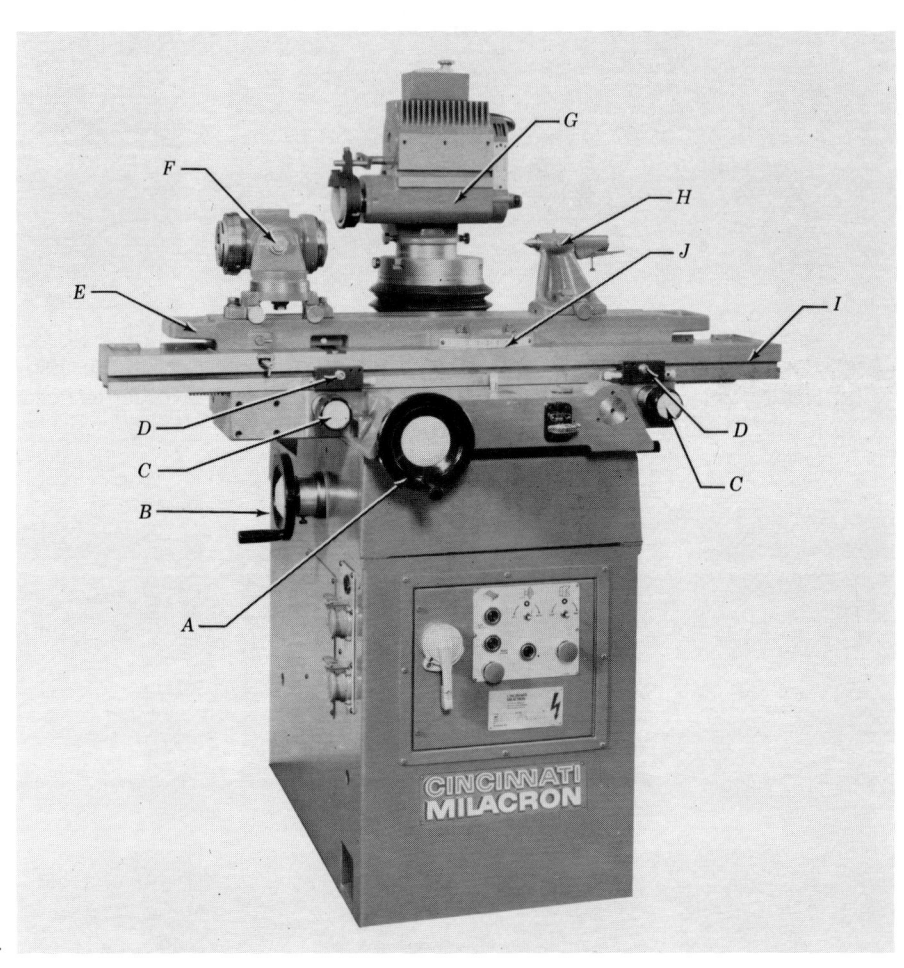

Figure 16. Features and components of the cutter and tool grinder.

a. _____

b. _____

c. _____

d. _____

e. _____

f. _____

g. _____

h. _____

i. _____

Unit 13 Cutter and Tool Grinder Features and Components   463

    j. _____

2. Name two major types of grinding work made possible by the cylindrical grinding attachment.

    a. _____

    b. _____

3. Name at least two ways the grinding arbor can be used.

    a. _____

    b. _____

4. Name two types of supports that are commonly used for supporting the teeth of cutters for sharpening.

    a. _____

    b. _____

5. What is the purpose of an offset tooth rest?

    _____

6. Why is the cutter and tool grinder considered to be one of the most versatile machine tools?

    _____

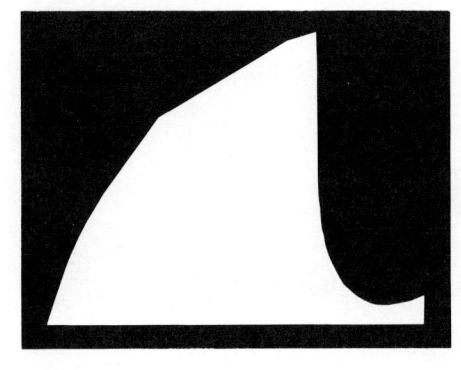

unit 14
cutter and tool grinder safety and general setup procedures

The cutter and tool grinder presents special problems that arise out of the versatility of the machine. The variety of grinding wheels, shapes, and sizes that can be mounted, and the varieties of ways that the wheel can be presented to the work sometimes makes the guarding of the wheel difficult. The variations in available wheel diameters also make it necessary for the operator to pay strict attention to spindle RPM and speed restrictions on the grinding wheel. Methods of machine alignment and methods of establishing and measuring cutter geometry are also important information to have before setting up to sharpen cutters.

objectives After completing this unit, you will be able to:
1. Match spindle speed to maximum permissible grinding wheel RPM.
2. Determine the wheel shape appropriate to the grinding application.
3. State at least four important safety precautions.
4. Make an alignment of the swivel table.
5. Establish and check cutter clearances by two methods.
6. Calculate, establish, and evaluate cutter clearances.
7. Differentiate between flat and hollow sharpening methods.

information

Although grinding wheel safety and selection factors have already been covered, some of these precautions should be repeated. First, as a matter of personal safety, always wear safety glasses in the shop; when working around abrasives, side shields on the safety glasses are indicated. Always use wheel guards, even though it is a temptation not to use them when working space is restricted. Before you mount a wheel, give it a ring test, as described in Unit 2. Once the wheel is mounted on a wheel collet and trued, it should be handled with special care, because it is inconvenient to demount the wheel to make a ring test. These mounted wheels should be stored on pegs and not permitted to lay around on table tops or be placed in drawers without proper separators where they can be bumped together.

GRINDING WHEEL SPEED

Cutter and tool grinding wheels are clearly marked with the maximum permissible operating RPM. It is very important, particularly with the larger diameter wheels, that you ensure that the wheel spindle RPM is less than the maximum RPM listed on the wheel. In some cases, special pulley belts and guards are required to match the wheel requirements.

GENERAL GRINDING PROCEDURE PRECAUTIONS

The shape of the wheel must be appropriate to the application. Side grinding must **not** be done on straight (Type 1) wheels, intended for peripheral grinding. For cutter and tool grinding work, only cup wheels, flaring cup wheels, and dish wheels are suited to side grinding stresses. After you mount the wheel, you should stand out of line of the plane of rotation, and start the wheel and permit it to run for at least a full minute before attempting to true, dress, or grind with it. If the wheel is damaged, most likely it will break during the first minute.

GENERAL SETUP PROCEDURES FOR MOST CUTTER SHARPENING

After centers are mounted and secured at the correct distance for the arbor to be mounted, it is important to check grinding arbors both for run-out on centers and, in the case of cylindrical cutters like plain milling cutters, the table alignment should also be checked (Figures 1a and 1b). If the table is not aligned, it should be adjusted until the indicator hand remains stationary as the table is tracked past the indicator tip. The table should then be locked in alignment (Figure 2) and reindicated to insure that it did not shift during the securing process. When using tailstock centers, it is desirable to position a T-bolt between the centers suitable for mounting either a tooth rest plate or a dressing diamond.

Figure 1. After checking the grinding arbor for run-out on the centers at each end, the bezel should be zeroed and table alignment checked (Lane Community College).

Figure 2. Adjusting and locking the swivel table in alignment (Lane Community College).

ESTABLISHING AND CHECKING CUTTER CLEARANCES

For tools that are sharpened on the periphery (profile ground tools) such as plain milling cutters, stagger tooth cutters, and saws, it is necessary to establish the needed relief and clearance angles to match the cutting requirements (Figure 3). For sharpening form relieved tools, like gear milling cutters, where form and relief are manufactured into the tool, the sharpening takes place on the face of the tool, and special procedures are used that will be covered later.

Correct reliefs and clearances are of critical importance to efficient cutting, satisfactory tool life, and to avoid cutter breakage (Figure 4).

For profile ground cutters, the following angles are recommended (Table 1) for the outside diameter (peripheral clearance).

MEASURING OF CUTTER CLEARANCE ANGLES

One method of checking peripheral relief and clearance angles may be easily done right on the machine while the tool is still between centers. This is called the "indicator drop" method (Figure 5). By using a table that takes into account radial relief angle, land width, and cutter diameter, the amount of drop can easily be determined (Table 2).

The cutter illustrated in Figure 6 is 3 in. OD. The radial movement is approximately .028 in. and the indicator drop approximately .003 in. The table shows that this tool has a primary radial relief angle midway between 5 and 8 degrees, or about $6\frac{1}{2}$ degrees.

This type of checking is done mainly for demonstration purposes, since it would not be economical to do this kind of checking where high productivity is critical. A much faster and more convenient method for checking clearances than the indicator drop method is the use of a cutter clearance gage (Figure 7).

Most experienced cutter grinding specialists use a chart (based on the following formula) for guidance in setting relief and clearances. For flat

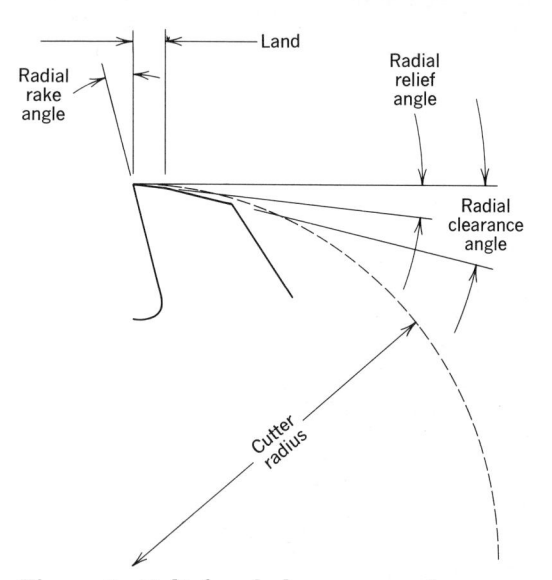

Figure 3. Relief and clearance angles.

Figure 4. This cutter was weakened by incorrect excessive clearance.

Table 1
Peripheral (OD) Relief and Clearance Angles Suggested for High Speed Steel Cutters

Material	Primary Relief, Degrees	Secondary Clearance, Degrees
Carbon steels	3—5	8—10
Gray cast iron	4—7	9—12
Bronze	4—7	7—12
Brasses and other copper alloys	5—8	10—13
Stainless steels	5—7	11—15
Titanium	8—12	14—18
Aluminum and magnesium alloys	10—12	15—17

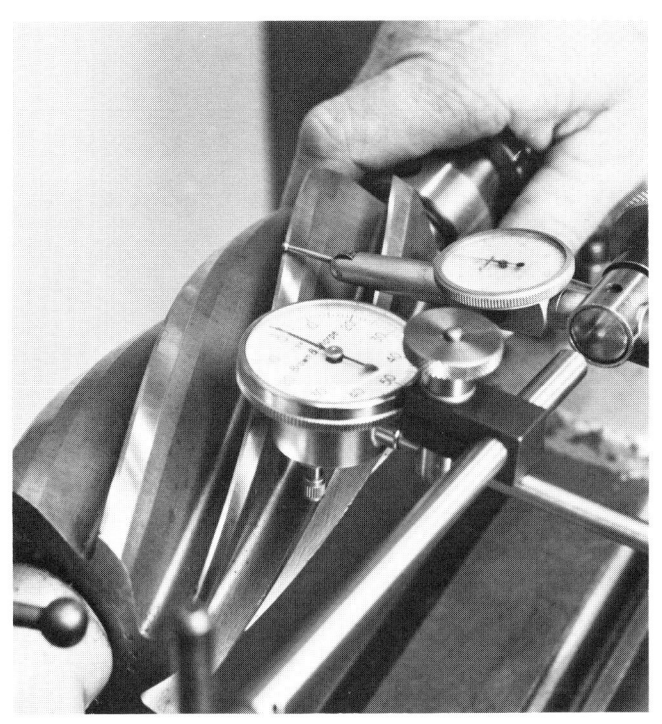

Figure 5. Setting up indicator for the indicator drop method of checking clearances (Lane Community College).

Figure 6. Checking the primary radial relief by the indicator drop method (Lane Community College).

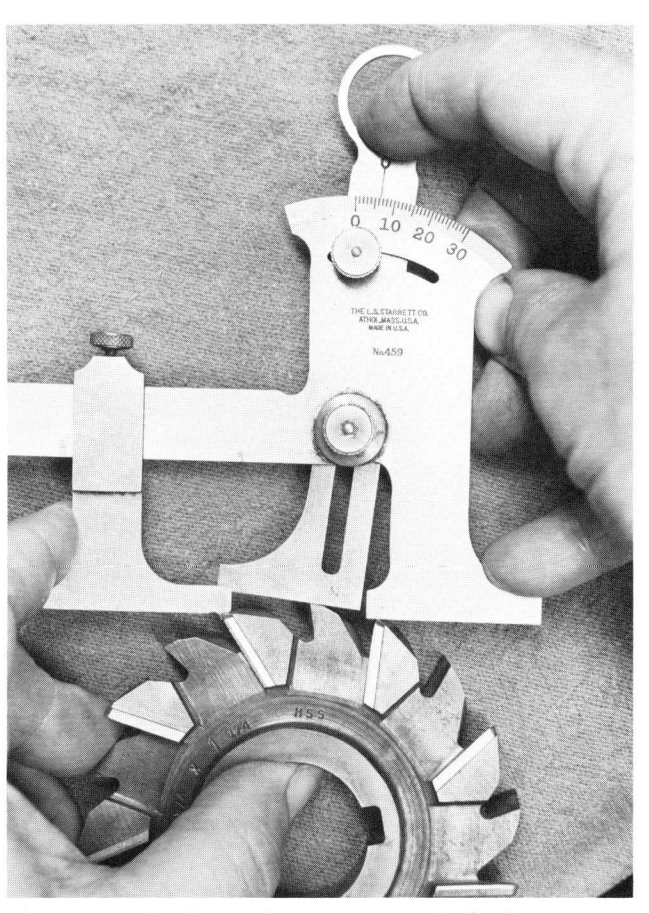

Figure 7. Checking the primary relief of a stagger tooth milling cutter with a Starrett Cutter clearance gage (K & M Tool, Inc.).

Table 2
Indicator Drop Method of Determining Radial Relief Angles

Diameter of Cutter, Inches	Average Range of Radial Relief, Degrees	Indicator Drop for Range of Radial Relief Shown Minimum	Indicator Drop for Range of Radial Relief Shown Maximum	Radial Movement for Checking
$\frac{1}{16}$	20–25	.0018	.0027	.010
$\frac{1}{8}$	15–19	.0021	.0032	.015
$\frac{3}{16}$	12–16	.0020	.0034	.020
$\frac{1}{4}$	10–14	.0019	.0033	.020
$\frac{5}{16}$	10–13	.0020	.0033	.020
$\frac{7}{16}$	9–12	.0025	.0038	.025
$\frac{1}{2}$	9–12	.0027	.0040	.025
$\frac{5}{8}$	8–11	.0028	.0045	$\frac{1}{32}$
$\frac{7}{8}$	8–11	.0033	.0049	$\frac{1}{32}$
1	7–10	.0028	.0045	$\frac{1}{32}$
$1\frac{1}{4}$	6–9	.0025	.0042	$\frac{1}{32}$
$1\frac{1}{2}$	6–9	.0026	.0043	$\frac{1}{32}$
$1\frac{3}{4}$	6–9	.0027	.0044	$\frac{1}{32}$
2	6–9	.0028	.0045	$\frac{1}{32}$
$2\frac{1}{2}$	5–8	.0024	.0040	$\frac{1}{32}$
3	5–8	.0024	.0041	$\frac{1}{32}$
4	5–8	.0025	.0042	$\frac{1}{32}$
5	4–7	.0020	.0037	$\frac{1}{32}$
6	4–7	.0021	.0037	$\frac{1}{32}$
8	4–7	.0021	.0037	$\frac{1}{32}$

grinding of relief or clearance with a flaring cup wheel, the amount to lower the wheel with the tooth rest attached to the wheelhead to obtain a specified primary relief angle is:

Sine of the desired angle × radius of the cutter

For example, if you wish a 5 degree primary relief angle on a 3 in. OD cutter, then:

Sine of 5 degrees =
.08715 × 1.5 (cutter radius) = .131 in.

This is the distance to **lower** the wheelhead to obtain this angle.

The additional drop for the secondary clearance is done in the same way. If you are seeking a secondary clearance angle of 9 degrees, then lower the wheelhead and its attached tooth rest an additional 4 degrees.

Sine of 4 degrees = .06976 × cutter radius = .06976 × 1.5 = .105 in.

This is the additional distance for the secondary clearance.

If you have a grinding machine with a tilting head, all that is necessary is to tilt the spindle for the required primary relief and secondary clearance angles when using a flaring cup wheel. In this case, the cutter tooth tip remains level instead of being lowered as in the previous flaring cup or flat grinding method.

Another method of sharpening cutters on the periphery is by using a relatively large diameter straight (Type 1) grinding wheel (Figure 8). Using this type of wheel produces a slight hollow grind (Figure 9) behind the cutting edge. The larger the wheel, of course, the less the amount of hollow. This method is popular with cutter sharpening job shops; with the larger wheel there is more available abrasive for cutting without losing dimensions from tooth to tooth, and more cutters can be ground between wheel dressings. This method is quite similar to the flat grind method and involves either lowering the wheelhead with finger attached to produce the relief or clearance angles, or raising the wheelhead for clearance when the tooth rest is attached to the table, the cutter tooth remains level and the wheelhead is raised to produce the relief or clearances. The formula when the wheel is raised is similar to the flat grind method, except that the **grinding wheel radius** instead of the cutter diameter is used in the computation.

Amount to raise the wheelhead **with the cutter tooth maintained level with work centers** =

Figure 8. A plain grinding wheel can be used for cutter sharpening (K & M Tool, Inc.).

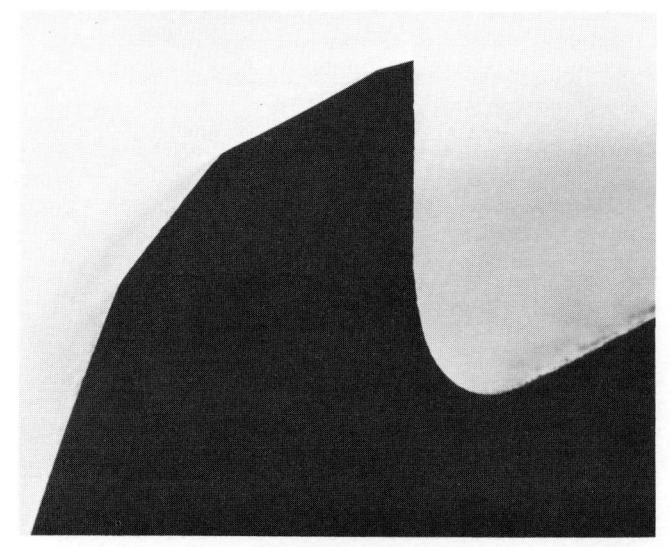

Figure 9. Tooth form produced by "hollow grinding" with a relatively large grinding wheel.

Sine of the angle desired × radius of the grinding wheel

For example, if a 6 degree primary relief was desired and the machine is fitted with a 7 in. diameter grinding wheel:

Sine 6 degrees = 0.1453 × 3.5 (wheel radius)
= .366 in.

the amount to raise the wheelhead.

If the finger is attached to the wheelhead, then the problem is treated in the same way as in using a flaring cup wheel.

Sine of desired angle × *cutter* radius = *drop* required

self-evaluation

SELF-TEST

1. Name at least four safety precautions that must be observed in the use of the cutter and tool grinding machine.

   a. _____

   b. _____

   c. _____

   d. _____

2. Why should flaring cup wheels and dish wheels be used for side grinding and *not* straight wheels?

3. Why should you stand out of line with the wheel during the first minute of operation?

4. Why is alignment of the swivel table so important in most cutter sharpening?

5. Why are correct reliefs and clearances so important in cutter and tool grinding?

6. Describe two methods of checking cutter clearance angles.

   a. _____

   b. _____

7. If you are flat grinding with a flaring cup wheel with a finger attached and you require a 7 degree primary relief angle on a 4 in. diameter cutter, will you raise or lower the wheelhead, and by what amount?

8. If the machine is being used with a flaring cup wheel and has a tilting spindle arrangement, is it necessary to lower the head?

9. If the outside diameter of a 6 in. diameter straight wheel is being used to grind a 5 degree relief on a 3 in. diameter plain milling cutter and the tooth rest is attached to the table, will the wheelhead be raised or lowered to make the grind, and how far?

10. If in Problem 9 the tooth rest is attached to the wheelhead, how will the problem be worked out?

unit 15
sharpening plain milling cutters

The sharpening of plain milling cutters is one of the most common operations on the cutter and tool grinder. It is really important to learn how to do this operation with smooth coordination before attempting to sharpen other types of cutters. This unit will show two methods of sharpening that apply to cutter and tool grinders without spindle tilting provisions.

objectives After completing this unit, you will be able to:
1. Set up a cutter and tool grinding machine for producing primary relief and secondary clearance with a flaring cup wheel (flat grinding method).
2. Determine if the swivel table is correctly aligned, and make corrections as needed.
3. Set up a cutter and tool grinding machine for producing primary relief and secondary clearance with a straight grinding wheel (hollow grinding method).

information

SHARPENING OF PLAIN MILLING CUTTERS USING THE CUP WHEEL (FLAT GRINDING METHOD)

The following steps demonstrate the basic method of using a flaring cup wheel to sharpen a plain milling cutter. Look over the machine you have available while you are studying this unit, and determine what differences you must allow for when you do the same operation in your own shop.

1. Mount the tailstock on the table to match the length of the correct size grinding mandrel.
2. Lubricate the centers of the grinding mandrel and fit it between centers so that the mandrel is free to rotate, but completely free of end play.
3. Indicate the mandrel for runout on centers and then also indicate to check swivel table alignment. Adjust table angle to zero indicator travel as necessary.
4. Set the wheelhead square to table and install a flaring cup wheel (Figure 1). Install wheel guard.
5. Start the wheel, true the face of the wheel with a diamond (Figure 2), and dress the wheel to a narrow land (about $\frac{1}{8}$ in. wide) with a dressing stick. Stop the wheel.
6. Install offset tooth rest, and set tooth rest, centers, and wheel axis in the same plane with the center gage (Figure 3).

Figure 2. True the wheel with a diamond, then dress to a narrow land using a dressing stick (Lane Community College).

Figure 3. Use the center gage to set wheel and tooth rest blade height (Lane Community College).

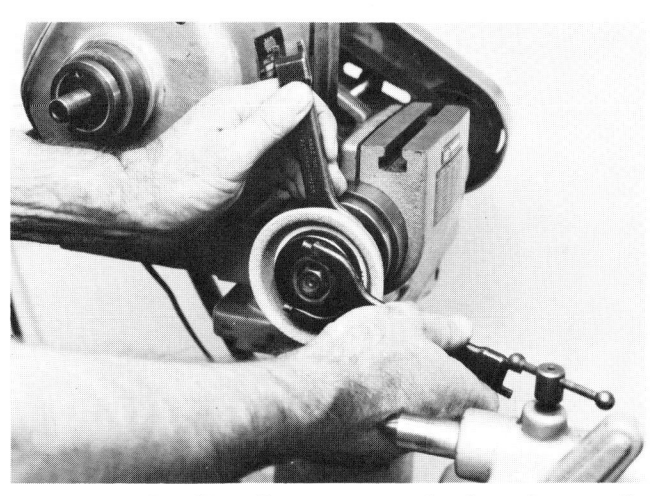

Figure 1. Install a flaring cup wheel with spindle square to the table (Lane Community College).

7. Unlock and rotate the wheelhead counterclockwise about 1 degree so that the trailing edge of the cup wheel will clear the cutter on the right side (Figure 4).
8. With cutter mounted on the grinding arbor and with the tooth to be ground resting on

Figure 4. Rotate wheelhead at least 1 degree counter-clockwise (Lane Community College).

the tooth rest blade, use the center gage to check the level of the tooth. Readjust the tooth rest height if necessary to level the tooth to the center height (Figure 5).

9. Slip the calibrated ring on the wheelhead vertical control handwheel to zero, without changing the elevation (Figure 6).
10. Calculate the necessary drop to produce the required primary clearance. (Drop = sine of clearance angle × radius of cutter.)
11. Adjust the wheelhead vertical control handwheel to lower the grinding head the required value.
12. With the cutter just clear of the wheel, start the wheel spindle and let the machine run for at least a minute before starting to grind. This is not only a safety measure, but it allows the bearings to warm up and can result in better finish.
13. Traverse the cutter past the grinding wheel while guiding the cutter gently along the top of the tooth rest blade (Figure 7); add small increments of cross feed until a very narrow land is produced. Note the cross slide handwheel reading.
14. With the cutter clear of the tooth rest blade and of the wheel, rotate the cutter to the opposing tooth (180 degrees).
15. Grind the diametrically opposite tooth to the same cross slide handwheel reading.
16. Stop the spindle and bring the cutter clear of the wheel in order to check for taper with a micrometer.

Figure 5. Recheck tooth level and adjust tooth rest blade to level if necessary (Lane Community College).

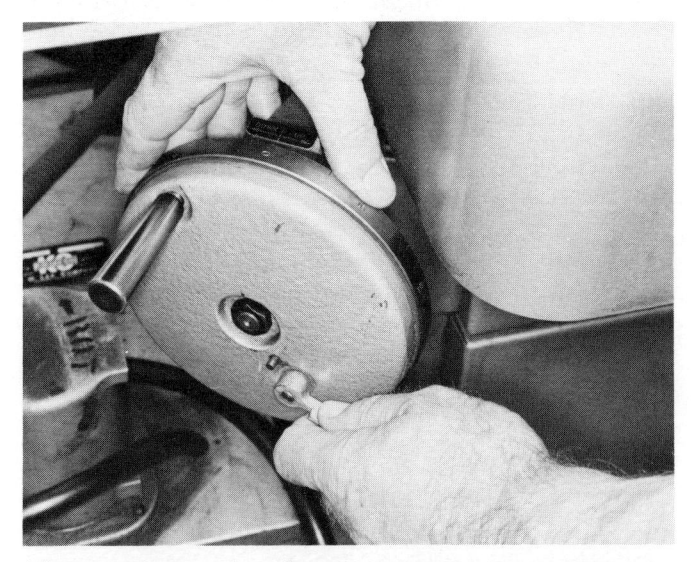

Figure 6. Zero the elevation setting (Lane Community College).

17. Check the cutter for taper. If an unacceptable amount of taper exists (over .001 in. would be considered poor practice), remove the cutter from the grinding arbor and recheck the alignment procedure (step 3).
18. If the taper check is correct, grind the remaining teeth with additional infeed and make the

Unit 15 Sharpening Plain Milling Cutters

Figure 7. Make the grinding passes (Lane Community College).

Figure 8. Set the wheelhead and tooth rest blade height to center (K & M Tool, Inc.).

final sparking-out pass on all the teeth with no additional infeed.

19. Set the additional clearance angle required for the secondary clearance.
20. Grind the secondary clearance until a primary land of .030 to .040 in. remains.
21. Check the clearances with a cutter clearance gage or by the indicator drop method, if you do not have a cutter clearance gage.

SHARPENING OF PLAIN MILLING CUTTERS USING THE STRAIGHT WHEEL (HOLLOW GRINDING METHOD)

This type of grind will result in slight hollow grinding. To reduce the amount of hollow, use the largest straight wheel consistent with the spindle RPM and with an available wheel guard. A grinding wheel of 6 in. diameter is usual. If your grinder has wheels larger than 6 in., a collet or wheel flanges having a minimum diameter of 3 in. must be used to meet safety standards (ANSI Safety Code 5.8.1-1964). It is also of *critical importance* that the spindle speed be checked and adjusted (usually by changing the spindle drive belts or pulley positions) to reduce the RPM to or below the speed marked on the grinding wheel blotter.

1. Mount the grinding wheel. If the wheel is a collet mount type, trueing may not be necessary.
2. Mount the tailstock to match the length of the grinding arbor for the cutter to be ground.
3. Indicate the arbor for runout and for table alignment.
4. Mount a tooth rest and blade consistent with the helix angle of the cutter.
5. Using the center gage, adjust the wheelhead to center height and position the height of the tooth rest blade to center (Figure 8).
6. Mount the wheel guard.
7. Mount the cutter on the arbor and place between centers, free to rotate but without end play.
8. Recheck the center height of the tooth to be ground and adjust the tooth rest blade height to level the tooth tip, if required (Figure 9).
9. Set the elevation handwheel collar to zero.
10. Calculate the amount of wheelhead movement necessary to produce a clearance of 6 degrees. Will the wheelhead be raised or lowered in the situation shown in Figures 8 and 9?

Figure 9. Establish the tooth tip level to the center (K & M Tool, Inc.).

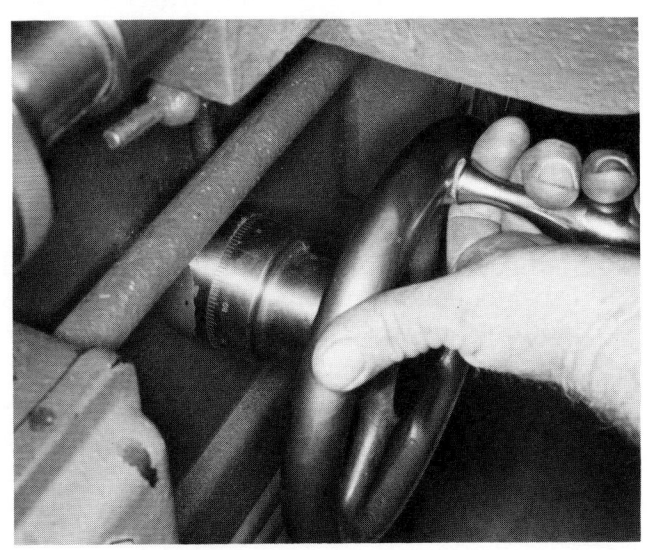

Figure 10. Position the head elevation to provide clearance (K & M Tool, Inc.).

Figure 11. Traverse the table while maintaining cutter contact with the support finger (K & M Tool, Inc.).

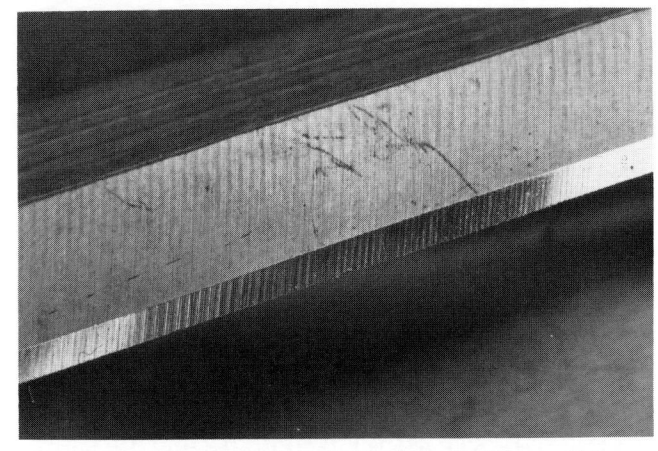

Figure 12. The completed grind should have this appearance (K & M Tool, Inc.).

11. Since the tooth rest is attached to the wheelhead, the wheelhead will be lowered by:

   Sine of clearance angle × radius of the cutter

   Lower the wheelhead. If the tooth rest had been attached to the table, you would raise the wheelhead by:

   Sine of clearance × radius of the grinding wheel (Figure 10)

12. With the cutter clear of the grinding wheel, you are ready to start the spindle. For safety, let the spindle run for approximately 1 min. before making the grind.

If your instructor has approved your use of the machine, prepare Worksheet 2 and then start the wheel and follow the additional steps 13 to 21 of the procedure established for grinding a plain milling cutter with a flaring cup wheel, described earlier in this unit (Figure 11).

The completed appearance of the cutter ground surfaces should be like that shown in Figure 12.

self-evaluation

WORKSHEET 1
1. Without additional reference to the steps listed in the information, generate a process sheet for grinding a plain milling cutter utilizing a flaring cup wheel (*without* using the tilting head of the grinder, if it is so equipped).
2. Follow your process sheet and practice the steps required, but without starting the spindle, trueing, dressing, or making a grind.
3. Correct any step left out.
4. Show your instructor that you have done the required steps and that you have developed coordinated movement. Request permission to make a "live" grind.

WORKSHEET 2
1. Generate a process sheet for the hollow grinding method of cutter grinding, specific to the cutter and tool grinder and equipment you have available. Try to do this without referring to the steps listed in the information.
2. After you have finished your sheet, compare it with these steps. Do you have more or fewer steps? What changes were necessary to match your available machine and equipment?
3. Perform the grinding sequence.
4. Show your sharpened cutter to your instructor.

This unit has no post-test.

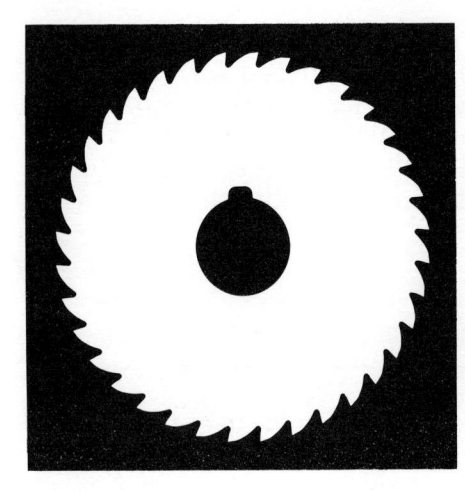

unit 16
sharpening slitting saws and stagger tooth milling cutters

Slitting saws and stagger tooth milling cutters are both used to produce slots on the milling machine. A properly sharpened slitting saw or stagger tooth milling cutter can produce an excellent finish on the side of the slot. In the case of the stagger tooth milling cutter, side grinding is done as little as possible, because these cutters are intended to produce slots of a specific width; however, it is important to be able to grind both the peripheral teeth and side teeth as required.

objectives After completing this unit, you will be able to:
1. Set up a cutter and tool grinder and sharpen the peripheral teeth on a slitting saw using a table mounted tooth support.
2. Calculate and enter the required relief and clearance angles for both slitting saws and stagger tooth milling cutters.
3. Set up a cutter and tool grinder to sharpen the peripheral teeth of a stagger tooth milling cutter and make corrections as required.
4. Set up and sharpen the side teeth of stagger tooth milling cutters as needed.

information

SHARPENING SLITTING SAWS

You can use the same basic procedure used to setup a plain milling cutter with a straight (Type 1) wheel, to grind a slitting saw. The difference in procedure shown will illustrate the use of a *table mounted* tooth support (Figure 1).

1. Position the tooth rest blade so that the tip of the tooth is level to the work centers, then use the centering gage to set the wheelhead at the same level. [It should be noted that on this machine the distance from the top of the spindle assembly is designed to be the same as that from the table to the work centers — which is quite a convenience (Figure 2).].

2. Now set the micrometer collar on the

Unit 16 Sharpening Slitting Saws and Stagger Tooth Milling Cutters   477

Figure 1. Setting up the tooth rest on the table to grind a slitting saw (K & M Tool, Inc.).

Figure 2. Setting the wheelhead level to the tip of the tooth with the center gage (K & M Tool, Inc.).

Figure 3. The micrometer collar is zeroed (K & M Tool, Inc.).

wheelhead handwheel to zero (Figure 3).

3. Determine the amount of distance that the wheelhead must be raised in order to produce the required clearance. In this case, the cutter remains leveled on the tooth rest blade, so we need to know the diameter of the grinding wheel to work out the problem. The formula is the same as before.

   Sine of clearance angle desired × radius of the grinding wheel

   Let us assume that the grinding wheel is 7 in. in diameter and that we desire a radial relief angle of 4 degrees for use in low carbon steel.*

   Sine 4 degrees × 3½ = .06976 × 3.5 = .244 in.

4. For the primary grind, the wheelhead will then be *raised* .244 in. (Figure 4).
5. Grind by passing the table and cutter past the grinding wheel, using the back controls for table traverse (Figure 5).

   This application is one of the best to illustrate both the use of the hook- or L-type tooth rest blade and the flicker-type tooth rest support (Figure 6). The next tooth is quickly ratcheted into place for grinding.

---

*A very important source for accurate recommendation for tool geometry derived from experimentation for use with specific materials is *Machining Data Handbook*, 1972, published by Metcut Research Associates, Cincinnati, Ohio.

Figure 4. The toolhead is raised for the primary relief angle (K & M Tool, Inc.).

The final grinding should be done with no additional cross slide or saddle infeed to ensure that all of the teeth are the same distance from center.

6. Calculate the additional height necessary for a secondary clearance of 10 degrees.
7. Set the secondary clearance by additional raising of the wheelhead. Care must be taken that in grinding the secondary clearance that the primary relief of the above tooth is clear; otherwise, a smaller diameter grinding wheel must be used. In many instances, only primary sharpening needs to be done. Side grinding is not usually done on slitting saws.

SHARPENING STAGGER TOOTH MILLING CUTTERS

The stagger tooth cutter is one of the more interesting milling cutters to sharpen, because you are sharpening on a differing helix on each successive tooth. For this type of grinding, a straight (Type 1) wheel is used.

Figure 5. Grinding the primary relief (K & M Tool, Inc.).

Figure 6. The hook- or L-type finger and the flicker-type tooth rest support are ideal for this application (K & M Tool, Inc.).

Figure 7. Preparing the wheel for sharpening a stagger tooth milling cutter with a boron nitride dressing stick (K & M Tool, Inc.).

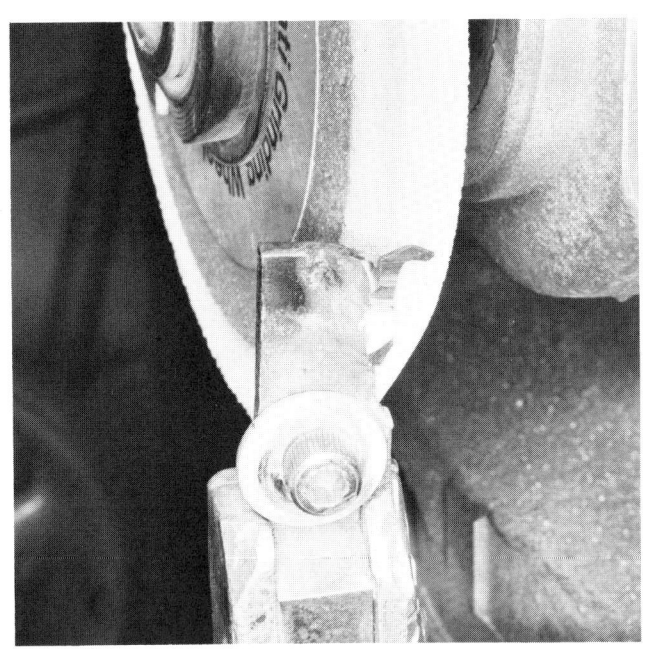

Figure 8. An inverted V tool rest is used and must be centered on the grinding land (K & M Tool, Inc.).

1. Mount, true, and narrow the face of the wheel to almost a sharp V (Figures 7a and 7b).
2. Install a tooth rest in the form of an inverted V (Figure 8). This must be accurately centered to the grinding land on the wheel face; if it is not centered, every other tooth will be higher or lower than the adjacent tooth. This difference must not exceed .0003 in. For this operation, the tooth rest support is mounted to the wheelhead assembly.
3. Set the wheelhead and tooth rest blade height to the level of the work centers (Figure 9). Since the tooth rest is attached to the wheelhead, the wheelhead is lowered (Figure 10) to provide the primary relief by the same formula used before.

    Sine of the desired relief × radius of the cutter

    The cutter shown is 4 in. in diameter and a 6 degree radial relief is desired, so the wheelhead would be *lowered* by

    Sine of 6 degrees × 2   or   .0145 × 2 = .209 in.

4. Lower the wheelhead, start the spindle, and grind two successive teeth lightly (Figure 11).
5. Check for concentricity (Figure 12) on both helixes of the cutter. If the difference exceeds .0003 in., then the tooth rest blade must be adjusted. The adjustment is to move the apex of the V in the direction of the helix with the high tooth. This lifts that tooth slightly higher to remove additional material.
6. Then grind the cutter all the way around,

Figure 9. Wheelhead and tooth rest height are set to center and the micrometer collar is zeroed (K & M Tool, Inc.).

Figure 10. The wheelhead is lowered to provide the primary relief (K & M Tool, Inc.).

Figure 11. Grinding two successive teeth to check for centering of the toothrest (K & M Tool, Inc.).

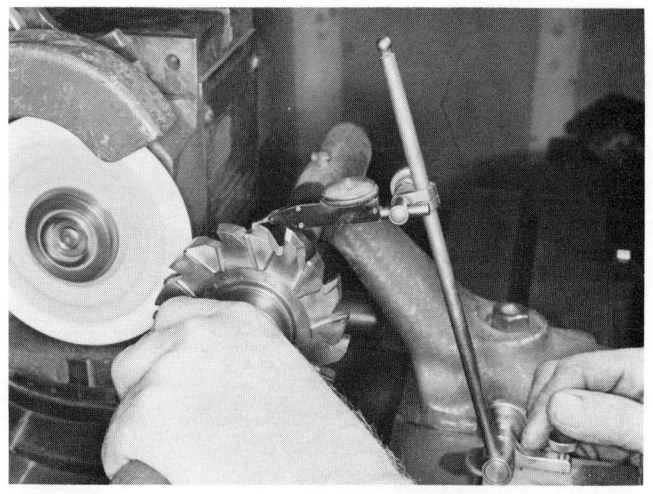

Figure 12. Checking for concentricity on the two teeth of opposite helix (K & M Tool, Inc.).

again finishing up all the teeth on the last pass with no additional infeed of the cross slide handwheel.

For the secondary clearance, there are some options. The most straightforward way is to lower the wheelhead for the additional clearance. Another is to use a flicker-type tooth rest with a straight blade and a micrometer adjustment to lower the tooth. With this combination, it is necessary to swivel the table for each hand of helix in order to obtain a parallel land width on the primary relief (Figure 13). The important

Figure 13. Completing the primary relief (K & M Tool, Inc.).

Figure 15. Setting the side tooth horizontal with a level (K & M Tool, Inc.).

Figure 14. Correct appearance of the sharpened tooth (K & M Tool, Inc.).

thing is that whatever procedure is followed, the result should be a parallel primary relief (Figure 14).

SHARPENING THE SIDE TEETH OF A STAGGER TOOTH MILLING CUTTER

For this operation, the milling cutter is mounted on a stub arbor and installed in a universal workhead.

1. Set a tooth horizontal and lock the workhead spindle temporarily in place (Figure 15).
2. Install a flicker-type tooth rest support with a micrometer adjustment on the workhead with

Figure 16. Installing a tooth rest assembly on the universal workhead (K & M Tool, Inc.).

a hook- or L-type tooth rest blade (Figure 16). Set the required relief angles into the workhead (Figure 17). Stagger tooth cutters have minimal relief on the land to maintain the original cutter width. The side teeth do essentially no cutting. The secondary clearance is used to reduce the primary land width to the point that very little heat is generated by land

Figure 17. The relief angles are set into the universal workhead (K & M Tool, Inc.).

Figure 18. The cutter is held upward against the tooth rest and is traversed across the wheel (K & M Tool, Inc.).

Figure 19. The workhead is given additional tilt to provide for secondary clearance (K & M Tool, Inc.).

Figure 20. Appearance of the completed sharpening of the side cutting teeth of a stagger tooth milling cutter (K & M Tool, Inc.).

contact, thus preserving the cutter width. The cutter is also relieved just slightly toward the center (Figure 17) (about ½ degree) to reduce contact further. This also results in an improved side finish in the deep slots that these cutters are designed to produce.

3. Turn the wheelhead square to the table and mount a flaring cup wheel.
4. True and narrow the wheel with a dressing stick to a nearly sharp edge.
5. Traverse the cutter past the wheel while the tooth of the cutter is held *up* against the tooth rest (Figure 18). After all of the primary reliefs have been cleaned up, make a final sparking-out pass on all teeth with no additional infeed of the cross slide.
6. Withdraw the cutter and swivel the toolhead upward an additional amount to provide secondary clearance.
7. Grind the secondary clearance (Figure 19). The completed side tooth sharpening should have the appearance shown in Figure 20. The sharpening of a stagger tooth milling cutter is usually to a specific width. After the first side is sharpened, the cutter should be carefully measured for width to assess the amount to be removed from the opposite side.

The procedure for producing the primary relief and secondary clearance for a stagger tooth milling cutter may also be applied to sharpening the end teeth on a shell milling cutter as well.

self-evaluation

WORKSHEET 1
1. Prepare a process sheet for sharpening a slitting saw.
2. With your instructor's permission, set up and sharpen an example of this class of tool. Check carefully for interference before attempting to make a secondary clearance.
3. Submit the completed tool to your instructor for approval.

WORKSHEET 2
1. Examine the tooth rest blades that you have available for the cutter and tool grinder in your shop. If you have an inverted V tooth rest blade, mount it on the machine and practice the movements for grinding a stagger tooth cutter.
2. Make up a process sheet to match your situation.
3. With your instructor's approval, sharpen the peripheral teeth of a stagger tooth milling cutter. Be sure that the difference between suceeding teeth does not exceed a runout of .0003 in.

WORKSHEET 3
1. Set up a workhead for grinding the side teeth on a stagger tooth milling cutter.
2. Make up a process sheet for the operation consistent with the cutter and tool grinder and components that you have available.
3. With your instructor's approval, sharpen the side teeth of a stagger tooth milling cutter on both sides to a specified width. If your inventory of cutters does not permit this exercise, use another cutter, such as the end teeth of a shell milling cutter, to obtain this practice.
4. Show the sharpened cutter to your instructor.

This unit has no post-test.

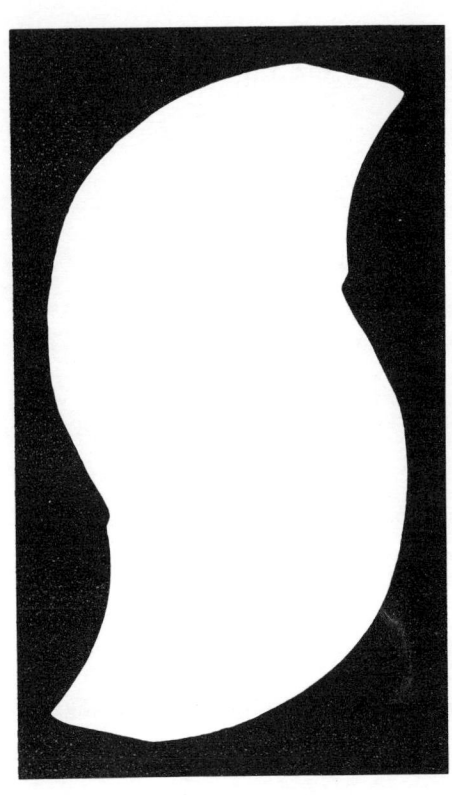

unit 17
sharpening of end mills

The sharpening of end mills is probably one of the most frequently done cutter grinding operations. You should have experience with sharpening plain milling cutters and gain experience in making smooth, coordinated movements before attempting to sharpen end mills. We will cover both conventional primary and secondary grinding methods and eccentric relief methods in this unit. We will also cover the grinding of end teeth.

objectives After completing this unit, you will be able to:
1. Set up a cutter and tool grinder with an accessory grinding fixture for end mills.
2. Make conventional primary relief and secondary clearance sharpening grinds on the sides of end mills.
3. Sharpen the end of an end mill using the universal workhead.
4. Sharpen the side of an end mill by the eccentric relief method.

information

CONVENTIONAL SHARPENING OF THE SIDE OF AN END MILL

For the grinding of end mills, an accessory spindle is usually used that has a very freely moving spindle that both rotates and slides axially with ease. In some cases, this spindle is floating in an air bearing, as will be seen later in this unit. These fixtures also have the capability of being rocked sideways to clear the end mill from contact with the grinding wheel. The fundamental practices for producing primary relief and secondary clearances are the same as for sharpening plain milling cutters but, as will be seen, there are some substantial detail differences. See Table 1 for recommended radial relief and clearance angles for end mills.

Table 1
Radial Relief Angles Suggested for High Speed Steel End Mills

Workpiece Material	End Mill Diameter, inches
	1/8 1/4 3/8 1/2 3/4 1 1½ 2
Carbon steels*	16° 12° 11° 10° 9° 8° 7° 6°
Nonferrous metals	19° 15° 13° 13° 12° 10° 8° 7°

*For tool steels, decrease indicated values about 20 percent. For secondary clearance angle, increase value by about ⅓.

Since the spindle that carries the milling cutter provides its own alignment, the table of the cutter and tool grinder can be swung at an angle for

Figure 1. The height of the wheelhead and the tooth rest blade are adjusted to the center of the end mill mounted in the special grinding fixture (K & M Tool, Inc.).

Figure 2. The wheelhead is lowered with the tooth rest blade to establish the primary relief (K & M Tool, Inc.).

the operator's convenience, as seen in Figure 3.

The steps to sharpen end mills with this type of fixture are as follows:

1. Mount a straight wheel, true if necessary, and narrow the wheel with a dressing stick to a land of about $\frac{1}{16}$ in. width and then stop the spindle.
2. Mount the accessory grinding spindle on the table and install a tooth rest on the wheelhead with a narrow blade shaped to fit the helix of the end mill.
3. Install the end mill and adjust the wheelhead and tooth rest blade to the height of the centerline of the mounted end mill (Figure 1). The tooth tip must be horizontal.
4. Zero set the elevation handwheel micrometer collar.
5. Calculate the required wheelhead drop.

   Sine of the primary relief angle × the radius of the cutter

6. Lower the wheelhead with tooth rest for the primary relief angle (Figure 2).
7. Start the spindle and allow it to run for a minute for safety reasons.
8. Rock the fixture away from the grinding wheel to clear the cutter (Figure 3) and move the cutter forward along the tooth rest blade.

Figure 3. The fixture is rocked away from the wheel and the cutter is moved forward (K & M Tool, Inc.).

9. Release the fixture gently when the tooth adjacent to the shank end is on the tooth rest blade (Figure 4). The reason for starting at the shank end is to prevent accidental contacts between the wheel and the end of the cutter. The end may not need sharpening or it may have been sharpened before the side. It is important for beginners to start this way until coordination is developed. With experience, the operation can start from either end of the tooth.
10. Make additional passes with increasing in-

Figure 4. The grinding pass is started at the shank end of the tooth (K & M Tool, Inc.).

Figure 5. Passes are made with additional infeed until the primary relief is complete (K & M Tool, Inc.).

feed increments (Figure 5) until all evidence of cutter damage has been ground away and the primary land is complete.

11. Lower the head an additional amount, using the same formula as before, to provide for secondary clearance.
12. Grind the secondary clearance (Figure 6) with added infeed until a primary land of correct width remains (about $\frac{1}{32}$ in. for the example shown) (Figure 7).
13. Inspect the cutter and remove.

SHARPENING THE END OF AN END MILL USING THE UNIVERSAL WORKHEAD

1. Install a relatively small diameter flaring cup wheel. Start, true if necessary, and dress to a narrow edge ($\frac{1}{8}$ to $\frac{3}{16}$ in.) (Figure 8) and then stop the spindle.
2. Set the workhead square to the table, then turn it slightly counterclockwise (2 degrees is recommended, 3 degrees at the most) to make the center a little lower than the outside edge (Figure 9).
3. Set in the amount of axial relief angle required by tilting up the workhead spindle and securing it in place (Figure 10). The amount of axial relief for an end mill should be about 5 degrees for cutting most steels and about 10 degrees for cutting most nonferrous metals.
4. Attach a flicker-type micrometer tooth rest support to the workhead to provide for

Figure 6. The secondary clearance is ground (K & M Tool, Inc.).

Figure 7. Correct appearance of the completed side grinding of the end mill (K & M Tool, Inc.).

Figure 8. A small flaring cup wheel is installed and prepared (K & M Tool, Inc.).

Figure 9. Workhead is squared to table allowing for center relief (K & M Tool, Inc.).

Figure 10. The primary relief angle is set in (K & M Tool, Inc.).

Figure 11. Flicker tooth rest support with a micrometer base is used for ratchet indexing (K & M Tool, Inc.).

ratchet indexing (Figure 11). A regular indexing attachment may be used, but it tends to be slower both in use and in setup.

5. Install the end mill with adapters and level the tooth using a spirit level and the micrometer base of the tooth rest support (Figure 12).
6. Traverse the end mill in the workhead so that the center of the cutter is just across from the edge of the wheel and set the table stop (Figure 13).
7. Start the spindle and allow it to run for a minute.
8. Infeed the cutter about .003 in. and traverse across the face of the wheel (Figure 14), then ratchet the spindle to the next tooth.
9. Continue the infeed and grinding routine until all damage has been ground away and the cutter is sharp. The last grind around the teeth is made as a spark-out with no additional infeed.
10. Raise the workhead spindle an additional amount for the secondary clearance. This should be about twice the value of the axial

Figure 12. The tooth is leveled (K & M Tool, Inc.).

Figure 13. A stop is set to limit cross travel to the center of the cutter (K & M Tool, Inc.).

Figure 14. The primary relief is ground on all the teeth (K & M Tool, Inc.).

Figure 15. Grinding the secondary clearance (K & M Tool, Inc.).

relief angle that you used. Adjust the spindle head height to compensate and grind the secondary clearance (Figure 15).

11. Examine the cutter and remove (Figure 16).

SHARPENING THE SIDE OF AN END MILL BY THE ECCENTRIC RELIEF METHOD

Eccentric relief grinding is commonly done by manufacturers of end mills made to a specific size for use in situations like numerically controlled machining, where the tool is discarded (for that use) after it wears. This type of grind may not require a secondary clearance since, in effect, both primary relief and secondary clearance are produced at the same time. It is especially useful where you have a large number of

Figure 16. Appearance of the completed end grinding of the end mill.

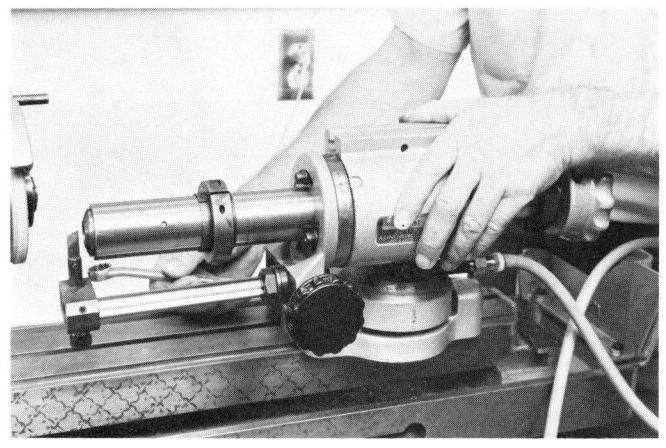

Figure 17. The air bearing fixture is mounted (Lane Community College).

Figure 18. The wheel is trued parallel to the table (Lane Community College).

cutters of the same hand, helix angle, and diameter to grind, because the setups are specific to these factors. Because of these factors, this is not a popular grind for cutter sharpening job shops; it is somewhat inflexible. A grind of this type works especially well in materials of low tensile strength, and where surface finish requirements are high.

For this grinding procedure, the machine will be fitted with an air spindle grinding accessory with the support finger integral with the accessory device (or equivalent of having the support attached to the table). The steps to take are as follows.

1. Install the fixture on the machine table and connect the air line so that the spindle both will rotate and move freely in the axial direction (Figure 17). The air source should have a moisture filter and a lubricator in the line.
2. Since this fixture is capable of being tilted up and down, indicate along the top surface of the spindle.
3. Set the spindle of the machine parallel to the table. Start the spindle and true the wheel square on the face with a diamond dresser (Figure 18). Then narrow the face of the wheel to a width equal to the width of the cutter land measured along the axis (Figure 19) and stop the wheel.
4. Rotate the wheelhead counterclockwise several degrees and secure (Figure 20). The amount is relative to clearance, helix angle, and cutter diameter (Table 2).
5. Set the cutter tooth to center height (horizontal) and the wheelhead to the same height (Figure 21).
6. Set the tooth rest to make contact at the right

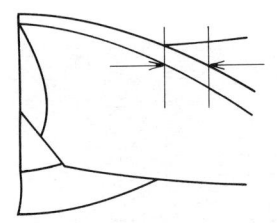

Figure 19. The wheel is narrowed to this width (Courtesy of National Twist Drill, Division of Lear Siegler, Inc.).

Figure 20. The wheelhead is rotated several degrees counterclockwise and secured (Lane Community College).

Figure 21. The air spindle, cutter tooth, and wheelhead are set in the same plane (Lane Community College).

Figure 22. The tooth rest must be correctly positioned and of the correct shape (Lane Community College).

Table 2
Grinding Wheel Angles to Produce Radial Relief Angles for Eccentric Relief Sharpening

Cutter Diameter	Radial Relief Angle, Degrees	Cutter Helix Angle, Degrees		
		20	30	40
		\multicolumn{3}{c}{Wheelhead Angle Required, Degrees}		
$\frac{1}{8}$	13	$4\frac{1}{2}$	7	11
$\frac{1}{4}$	$10\frac{1}{2}$	$3\frac{3}{4}$	6	10
$\frac{1}{2}$	10	$3\frac{1}{4}$	$5\frac{1}{4}$	$9\frac{1}{4}$
$\frac{3}{4}$	9	3	$4\frac{2}{3}$	$8\frac{1}{2}$
$1\frac{1}{4}$	8	$2\frac{1}{2}$	4	$7\frac{1}{2}$

edge of the wheel, angled to support the tooth fully across the wheel face (Figure 22).

7. Start the grinding wheel, allow to run for a minute, and start the grinding (Figure 23). In this case, the grinding is started from the end of the cutter instead of cutting from the shank end, as suggested for conventional grinding. Because of the possibility of hitting the end of the tooth by grinding in this direction, it is preferable to leave the end grinding of the teeth until last, if required. The finished appearance is substantially different than conventional grinding (Figure 24).

A view of the end of an eccentrically relieved cutter shows the way in which the relief or clearance increases from the cutting edge (Figure

Figure 23. Making the eccentric relief grind (Lane Community College).

Figure 24. Close-up appearance of an eccentric relief grind.

25). The greater clearance is generated by the relatively larger left side of the wheel as the cutter is rotated past on the tooth rest. The cutting edge is generated by the relatively smaller right side of the wheel. For this grind to result in a highly finished tool, good coordination is essential. Keep in mind that the wheelhead is not raised or lowered for relief and clearance, but that the relief and clearance is *generated* as the cutter passes the wheel.

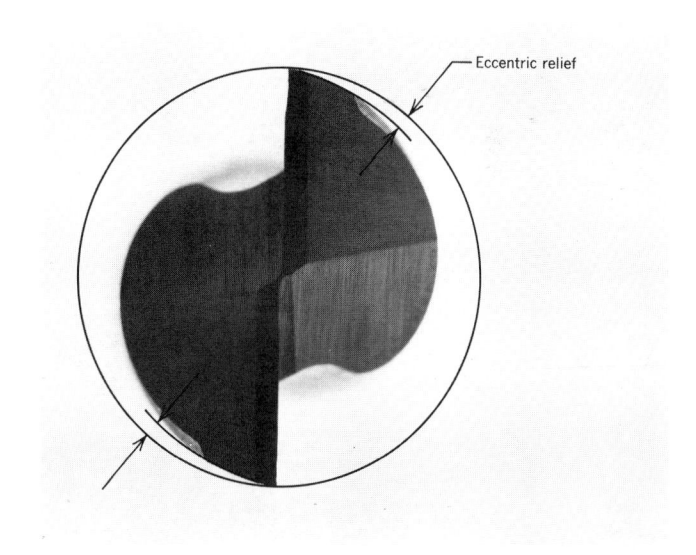

Figure 25. Form of the eccentric relief grind as viewed from the end.

self-evaluation

WORKSHEET 1
1. Develop a process sheet for grinding the side of an end mill with the cutter and tool grinder and accessories you have available.
2. Make a series of practice "dry runs" with the cutter clear of the grinding wheel to develop coordination.
3. With your instructor's permission, follow your process sheet and grind both primary relief and secondary clearance.
4. Show the completed job to your instructor.

WORKSHEET 2
1. Develop a process sheet for grinding the end of an end mill with the cutter and tool grinder and accessories that you have available.
2. With your instructor's permission, follow your process sheet and grind both primary relief and secondary clearance.
3. Show the result to your instructor.

WORKSHEET 3
1. Determine if you have the necessary tooling to perform an eccentric relief grind on the cutter and tool grinder you have available.
2. If you have the means to perform this grind, develop a process sheet.
3. Make a number of "dry runs" with the spindle stopped and the cutter clear of the wheel to develop coordination. Make sure that the action is smooth and that the flutes of the cutter are free of any burrs that could cause lack of smoothness.
4. With your instructor's permission, make an eccentric relief grind. Do not be greatly concerned if the appearance is strange on the first passes, especially if the tool has been conventionally sharpened before; continue infeeding until the land is completely ground.
5. Show the completed grind to your instructor.

unit 18
sharpening form relieved cutters

Form relieved cutters are a very interesting class of milling cutters. Their design dates back to the 1850s when Brown and Sharpe began cutting gear teeth with this type of cutter. Form relieved cutters keep the same profile along the width of the tooth and are sharpened only on the face of the tooth. These cutters are quite expensive and should not be allowed to become excessively dull before sharpening. You will find that they are easy to sharpen if you follow the procedure.

objectives After completing this unit, you will be able to:
1. Determine if a form relieved cutter requires a radial (or axial) rake.

2. Determine if the cutter must be prepared by "back grinding" before being sharpened.
3. Set up a form relieved cutter on centers.
4. Prepare the grinding wheel for both roughing and finishing cuts.
5. Rough grind the cutter with downfeed.
6. Finish grind the cutter with traverse, using micrometer feeding.

information

Form relieved cutters are those in which the reverse of the shape to be machined is permanently generated into the cutter, and the relief is also permanently a part of the design. Examples of this type of cutter are found in gear tooth milling cutters, corner rounding cutters, convex milling cutters, and concave milling cutters, which will be discussed here. This type of cutter is sharpened on the face. These cutters are frequently made without rake (radial) on the face of the tool, but those in which the radial rake is figured into the final form are usually marked with the required rake angles to guide the sharpening technician (Figure 1). Sometimes axial rake on the face of the tooth is indicated, making it necessary to swivel the table also, but this is quite rare and will not be considered further here.

If the cutter has not previously been sharpened, it is necessary to use a micrometer to check the thickness of the tooth from face to back to determine if the teeth are uniform. A flange or disc micrometer works especially well. If the variation exceeds .002 in., it is necessary to grind the backs of the teeth before grinding the face of the tooth to ensure uniformity on the completed cutter. This cutter (Figure 2), has had this operation done. If grinding of the tooth back is required, the same procedure is followed as for sharpening the front of the tooth, except that only a reference spot is required on the back of each tooth instead of a full grind to the bottom of the tooth gullet. The following steps are used to set up and grind a zero radial rake milling cutter. The example is a concave milling cutter with considerable damage resulting from not taking it out of service for sharpening soon enough.

Figure 1. Form relieved cutter marked for radial rake angle (K & M Tool, Inc.).

Figure 2. This cutter has been ground for a reference surface on the back of the tooth (K & M Tool, Inc.).

The steps to sharpen this type of cutter are as follows:

1. Mount the cutter on an arbor between centers with a flicker-type micrometer tooth rest support positioned so that the face of the top tooth is accurately vertical (Figure 3). (Notice that an extension has been added to the spindle to position the wheel forward for clearance.)
2. Mount a dish wheel (Type 12) with a guard. Start and form the wheel to the shape of the tooth gullet with a dressing stick (Figure 4). Make the outer side of the wheel somewhat concave to limit wheel contact.
3. Position the rotating wheel clear of the cutter face in the top of the tooth gullet (Figure 5), and adjust the cross slide saddle until just grazing contact is made. Set the micrometer dial on the cross slide hand wheel to zero (Figure 6). This continues as the zero point for roughing, or until another dressing needs to be made on the wheel face.
4. Add infeed with the micrometer screw on the base of the tooth rest support (see Figure 10).
5. Traverse the cutter past the wheel with wheelhead *downfeed* on each pass for roughing (Figure 7), which gives this appearance (Figure 8).

   Where substantial metal removal is required to recondition the cutter, it may be necessary to redress the wheel and make several complete passes around the cutter to full depth, until uniformity is obtained.
6. For finishing, dress the wheel again and relieve the side of the wheel to reduce contact (Figure 9).
7. Reestablish the vertical position accurately and zero the cross slide. Make infeed increments by rotating the cutter on its own axis with the micrometer tooth rest support to ensure that the finished grind remains radial on the tooth

Figure 3. Cutter is mounted with the face of the top tooth accurately vertical (K & M Tool, Inc.).

Figure 4. The wheel is shaped with a dressing stick (K & M Tool, Inc.).

Figure 5. The wheel is positioned relative to the face of the tooth (K & M Tool, Inc.).

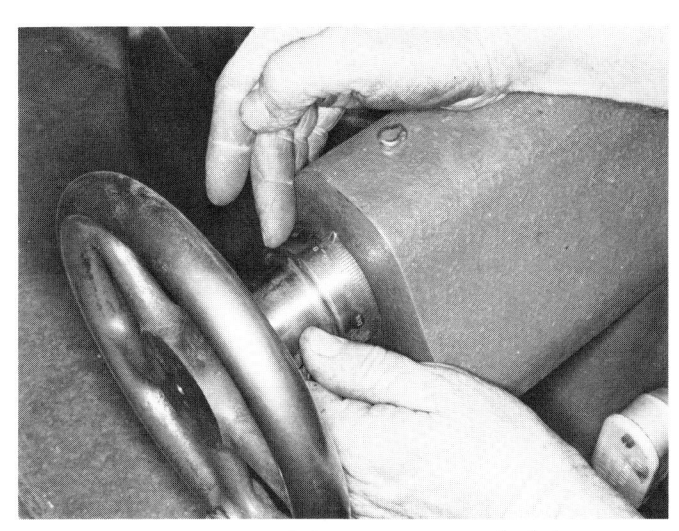

Figure 6. The cross slide micrometer dial is zeroed (K & M Tool, Inc.).

Figure 7. Cutter is traversed with increments of downfeed for roughing (K & M Tool, Inc.).

Figure 8. Close-up appearance of the roughing grind (K & M Tool, Inc.).

Figure 9. Dressing the wheel for the finishing cut (K & M Tool, Inc.).

face (Figure 10). Mark the initial tooth for reference (Figure 11).

8. If additional circular infeed with the micrometer tooth rest is necessary, do it on the side away from the initial tooth, and mark the tooth "2" (Figure 12).

9. Spark-out the cutter all the way around. Indicate for uniformity of height on the cutting surfaces and remove.

Figure 10. Infeeding for finishing is done with the micrometer tooth rest support (K & M Tool, Inc.).

Figure 11. The initial tooth is marked "1" before finish grinding (K & M Tool, Inc.).

Figure 12. If more circular infeed is required for "spark-out," it is added on the side opposite and marked "2" (K & M Tool, Inc.).

self-evaluation

WORKSHEET
1. Obtain a form relieved cutter that requires sharpening.
2. Determine if you have the necessary equipment to do the sharpening operation on the cutter and tool grinder you have available. You can employ alternative ways of tooth support (Figure 13), but direct support behind the specific tooth to be ground is recommended.

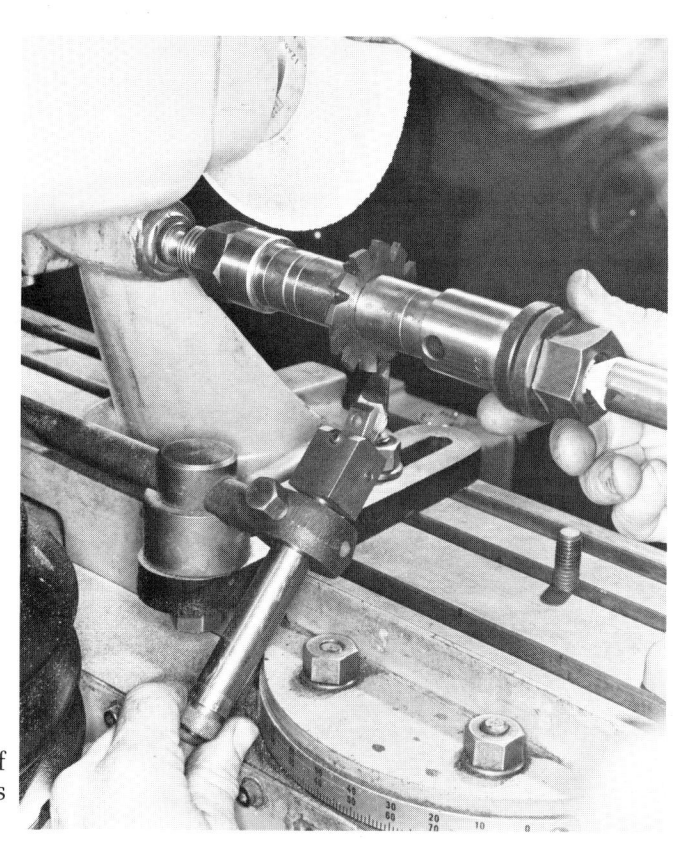

Figure 13. Alternative method of tooth support (Industrial Plastics Products, Inc.).

3. Make up a process sheet consistent with the cutter and tool grinder and components that you have available.
4. With your instructor's approval, rough grind and finish sharpen a form relieved cutter.
5. Indicate the radial height of the teeth (difference should not exceed .0005 in.), and show the sharpened cutter to your instructor.

This unit has no post-test.

unit 19
sharpening machine reamers

The sharpening of machine reamers is a very frequent cutter and tool grinding operation, possibly only exceeded by end milling cutters and plain milling cutters in the volume of sharpening. For a great majority of machine reamer sharpening, perhaps more than 95 percent, the sharpening operation is limited to work on the chamfer. Two methods of sharpening the chamfer will be considered, one using the tilting spindle method, the other using the horizontal spindle method. The grinder with the tilting spindle also has a double eccentric wheelhead mount that can be used for additional clearance when needed.

Even though resizing of reamers is a relatively infrequent operation, a method of resizing and margin preparation called circle grinding will be shown with a method of reestablishing secondary clearance.

objectives After completing this unit, you will be able to:
1. Set up either a tilting spindle or horizontal spindle cutter and tool grinder and sharpen the chamfer of a machine reamer.
2. Set up a cylindrical grinding attachment and correctly perform a circle grinding operation to produce a margin on a machine reamer.
3. Adjust the swivel table to provide longitudinal relief or back taper on a machine reamer.
4. Set up, calculate, and grind a secondary clearance on a machine reamer.

information

SHARPENING THE CHAMFER OF A MACHINE REAMER ON CENTERS

The sharpening of machine reamers is rarely done one at a time, except in emergencies. Normally, dozens of dull reamers are accumulated and classified first for length; then the chamfers are sharpened in batches, with the right tailstock of the cutter and tool grinder positioned to adjust for each group of reamers.

After the chamfer is sharpened, the reamers that are to be resized on the diameter are clustered by diameter and by length. For each diameter and length, the circle grinding is done and the table is swivelled for a longitudinal relief; again, they are ground in batch lots.

The procedure for secondary clearances is similar, with the reamers grouped in the interest of economics. The procedure used in this unit

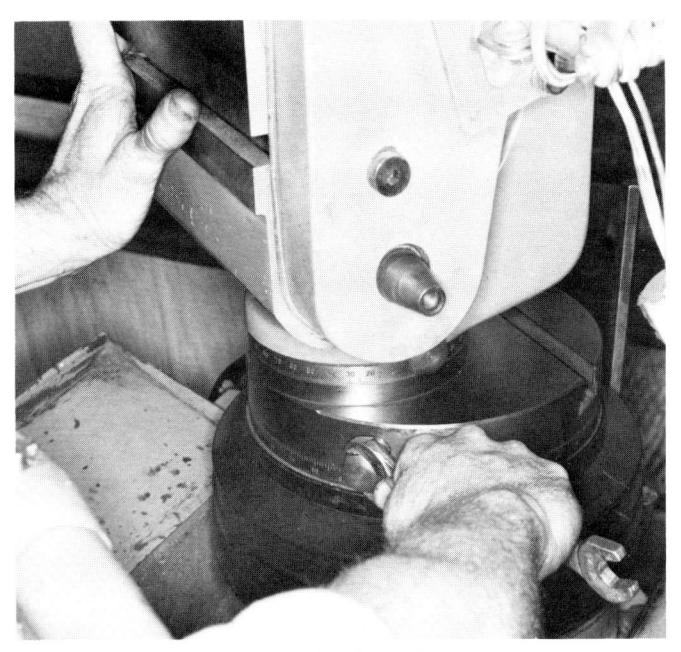

Figure 1. Using the wheelhead eccentric to gain clearance (Industrial Plastics Products, Inc.).

Figure 2. The table is swiveled to 45 degrees (Industrial Plastics Products, Inc.).

carries you through the complete operation for a single reamer, but please realize that doing it this way for a single reamer would *not* be considered normal practice.

Tilt Head Method

The following steps will show a procedure for sharpening the chamfer of a machine reamer between centers with a tilt head cutter and tool grinder.

1. Loosen the wheelhead eccentric screws and swivel the head to the rear to obtain maximum clearance (Figure 1).
2. Loosen the table clamping nuts and swivel the table to 45 degrees and retighten the clamping nuts (Figure 2).
3. Adjust the distance between centers until the reamer placed on centers rotates freely, but does not have any axial movement (Figure 3).
4. Mount a micrometer-type tooth rest support to the table with a flicker finger, and level the cutting edge of the reamer to center height (Figure 4).
5. Tip the spindle assembly down approximately 7 degrees, which is about average for use in most machine steels. Install a dish wheel and guard. Start the wheel and dress to shape with

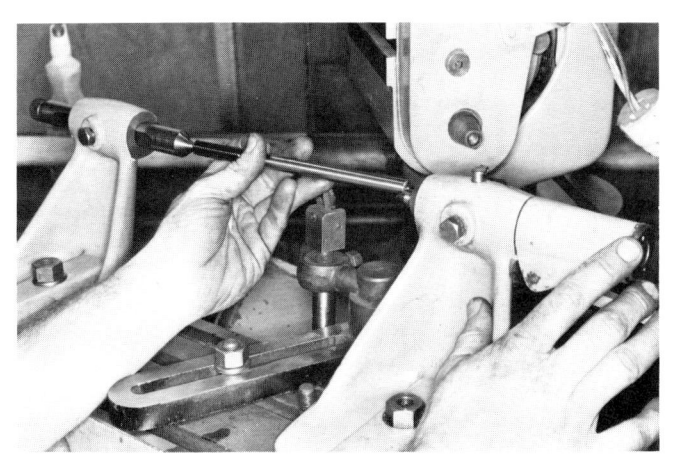

Figure 3. The center distance is set correctly (Industrial Plastics Products, Inc.).

a dressing stick (Figure 5). Then stop the spindle.
6. Set a table stop to prevent damage to the centers or to the wheel (Figure 6). **This is a very important step.**
7. Restart the wheel and bring the reamer into contact by feeding the cross slide (saddle) toward the wheelhead with the *back* cross slide handwheel (Figure 7).
8. Sharpen by making coordinated movement of

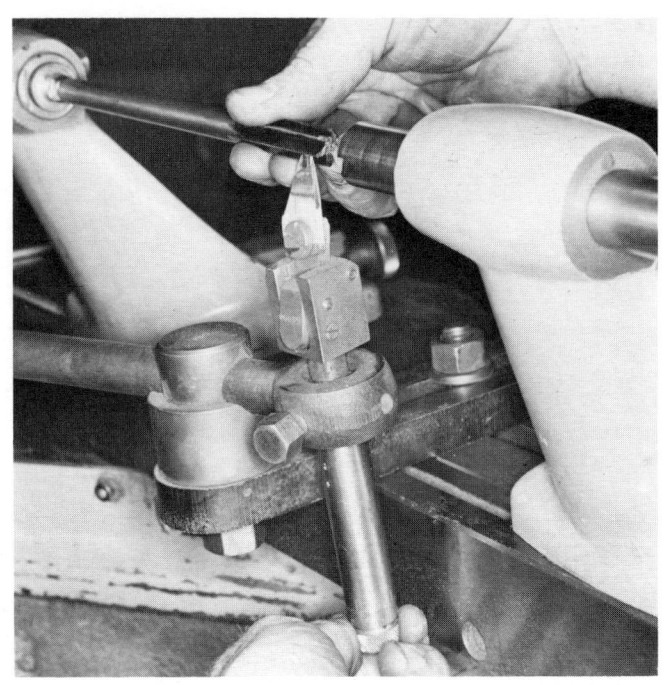

Figure 4. Center height is established on the reamer (Industrial Plastics Products, Inc.).

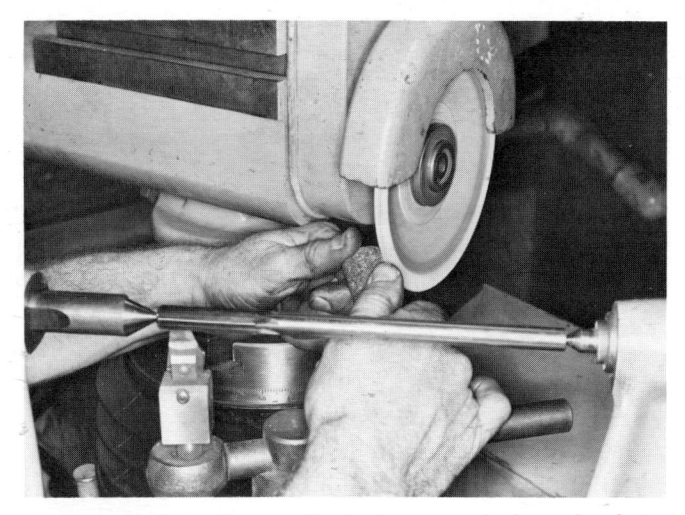

Figure 5. Spindle is tilted down and the wheel is dressed to shape (Industrial Plastics Products, Inc.).

Figure 6. The table stop is set for safety reasons (Industrial Plastics Products, Inc.).

Horizontal Spindle Method

For the sharpening of reamer chamfers on cutter and tool grinders with fixed horizontal spindles, take the following steps:

1. Set the wheelhead of the machine with the spindle parallel to the table travel direction.
2. Mount a 6 in. diameter dish wheel and guard. Start the wheel and true the OD of the wheel (if necessary).
3. Dress the wheel, using a dressing stick, so that there is a narrow margin on the "outboard" side, and hollow the outer side of the wheel as well. Stop the wheel spindle. Set the wheelhead to center height.
4. Loosen and swing the swivel table to 45 degrees and retighten the clamping nuts (Figure 2).
5. Adjust the distance between centers so that the reamer placed on centers rotates freely but does not have axial movement (Figure 3).
6. Mount a micrometer-type tooth rest support

the table with the left hand and by ratcheting the reamer over the flicker finger for each individual chamfer (Figure 8). As in other cutter sharpening, make the final spark-out passes without additional infeed to ensure even chamfer length.

The angle of the sharpening can easily be seen by placing the end of a steel rule against the chamfer (Figure 9).

Figure 7. Feeding the reamer to the wheel (Industrial Plastics Products, Inc.).

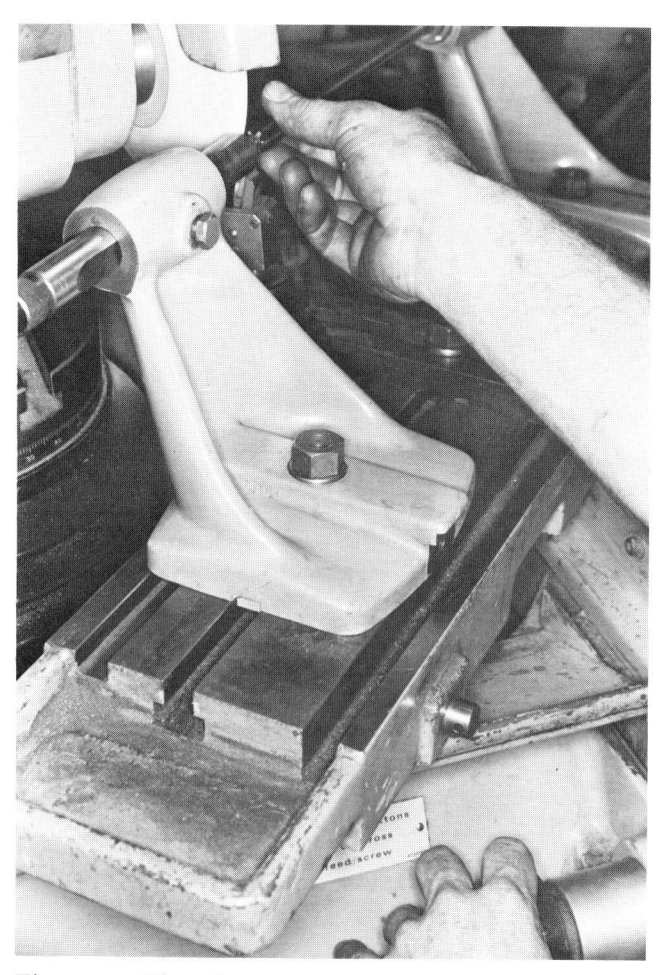

Figure 8. The sharpening is done by coordinated movements (Industrial Plastics Products, Inc.).

Figure 9. Appearance of the finished chamfer (Industrial Plastics Products, Inc.).

to the table with a flicker finger and level the cutting edge to center height (Figure 4).

7. Bring the chamfer of the reamer close to the outer edge of the wheel and position so that the grinding margin is on a line just slightly on the tailstock center side of the reamer chamfer and set the table stop to arrest any further movement toward the tailstock center.
8. Calculate the rise of the wheelhead needed for a 7 degree (average) clearance angle.

Sine of angle × wheel radius

For the example given, sine of 7 degrees = .12187. .12187 × 3 = .366, the necessary rise for this situation. Raise the wheelhead.

9. Restart the wheel spindle and bring the reamer to the grinding wheel by feeding the cross slide to the wheel with the back cross slide handwheel.
10. Sharpen by making coordinated movements of the table with the left hand and by ratcheting the reamer over the flicker finger for each individual chamfer. As in other cutter sharpening, make the final spark-out passes without additional infeed to ensure even chamfer length.

GRINDING THE BODY DIAMETER OF A REAMER TO BE USED IN STEEL (CIRCLE GRINDING)

This operation is not frequently done, except for resizing a reamer to a smaller size, because so little wear usually occurs on the margins of a machine reamer and because nearly all of the cutting is done on the chamfer. The margins are essentially a device for guidance instead of for cutting. This operation requires a special type of cylindrical grinding procedure to give a very small amount of clearance to the margin of the reamer which, after the secondary clearance is made, will only be .007 to .009 in. wide. For grinding the margin, the following steps are used.

1. Set the spindle parallel to the travel of the table and install a straight (Type 1) grinding wheel (Figure 10).
2. Install the cylindrical grinding attachment with the center spindle locked so the cylindrical grinding will be done on dead centers (Figure 11). Since this spindle can be tilted up and down, check the center height by either indicating along the top of a test bar mounted between centers or, as is more commonly done, by indicating on a test bar held in the spindle by a collet. The table should also be indicated in to zero taper.
3. True the face of the wheel and slowly dress with the diamond to obtain a good finish on the margins of the reamer. It is also a good idea to break the edges of the wheel very slightly with a fine boron carbide dressing stick to avoid feeding lines as the reamer is ground.
4. Color the lands of the reamer with layout dye so that you can see the effects of the grinding action to follow, and then lubricate the centers. Mount the reamer on the centers with a

Figure 10. A straight wheel is installed (Industrial Plastics Products, Inc.).

Figure 11. Cylindrical grinding attachment is installed (with a dead spindle) (Industrial Plastics Products, Inc.).

Figure 12. After indicating in the machine, the reamer is mounted for cylindrical grinding (Industrial Plastics Products, Inc.).

Figure 13. Setting up the traverse rate (Industrial Plastics Products, Inc.).

Figure 14. Table traverse microswitches (Industrial Plastics Products, Inc.).

driving dog to couple with the workhead driving ring (Figure 12).

The table traverse should be considered next. On many machines, hand traverse is all that is available but, typically, the hand traverse has two speeds. Select the lowest speed for smooth control.

If your machine has a power traverse, it may be one of two types, variable speed control or a change gear type (Figure 13).

5. For either manual or power feed table traverse, set the table stops or table traverse microswitches (Figure 14) so that the grinding wheel comes off a trifle more than half the wheel width at each end of the fluted portion of the reamer. A desirable traverse is about 4 to 5 sec. to move across the length of the flutes, with a somewhat slower series of passes for the last .001 in. to size.

6. Set the direction of rotation of the reamer opposite to its normal rotation when cutting. This causes the back of the margin to contact the wheel before the cutting edge does. **This direction of rotation is critical for success in margin grinding.** If the direction is incorrect, there will be **no** clearance behind the cutting edge; in fact, the clearance effect will be reversed. The controls for both spindle direction and workhead rotation are located on the front of this machine (Figure 15).

7. Determine the grinding wheel direction of rotation. The grinding wheel should *oppose* the direction of the reamer at the point of contact. Newer machines, and some older machines, have grinding spindle motor reverse. *As a point of caution,* if the spindle is to be reversed, you must be certain that the wheel will remain on the spindle and not be spun off and become a missile when the spindle is reversed. Newer machines with collets typically have a keyed washer to make spindle reversals safe. **Make sure that reversing the spindle is safe by checking out the holding method so that the wheel will stay on regardless of spindle direction. This is important.**

8. Start the spindle (Figure 16) and run for a

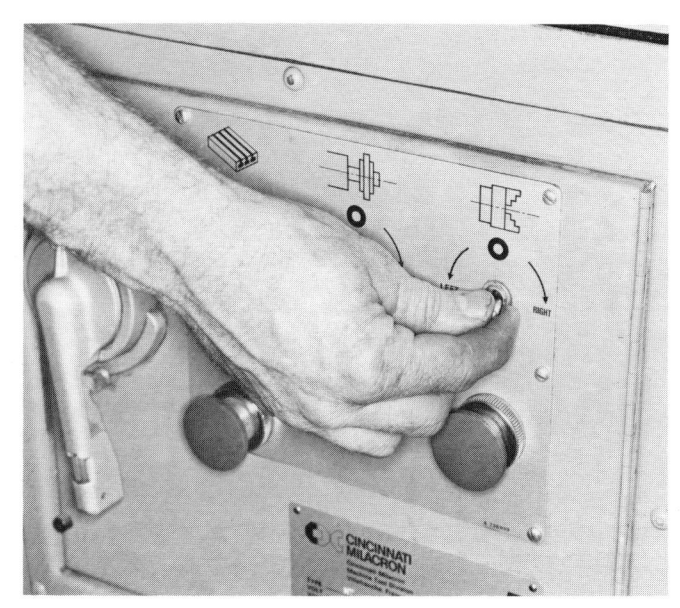

Figure 15. The workhead must rotate the reamer in the opposite direction of its normal cutting direction (Industrial Plastics Products, Inc.).

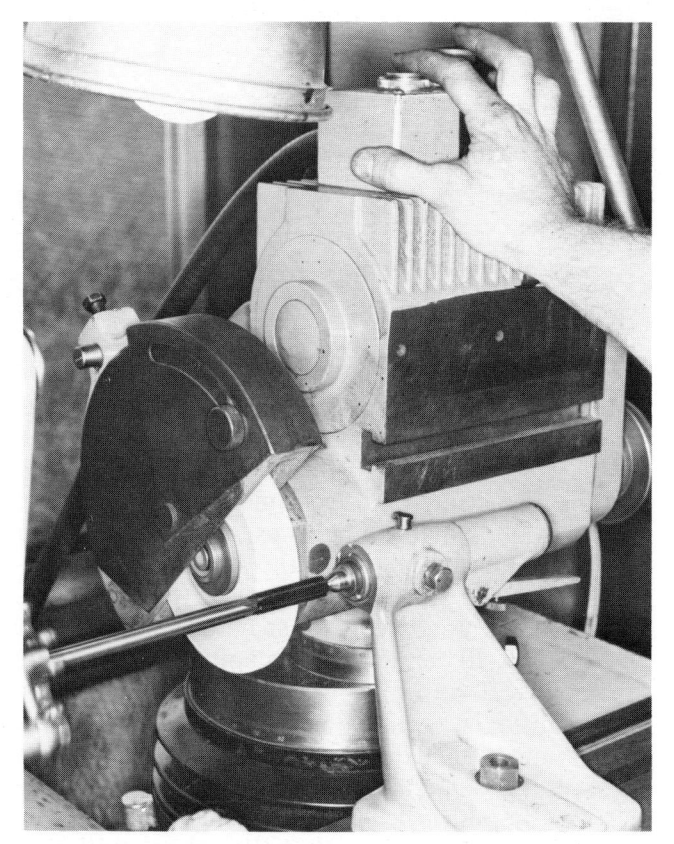

Figure 16. Starting the spindle after making the setup (Industrial Plastics Products, Inc.).

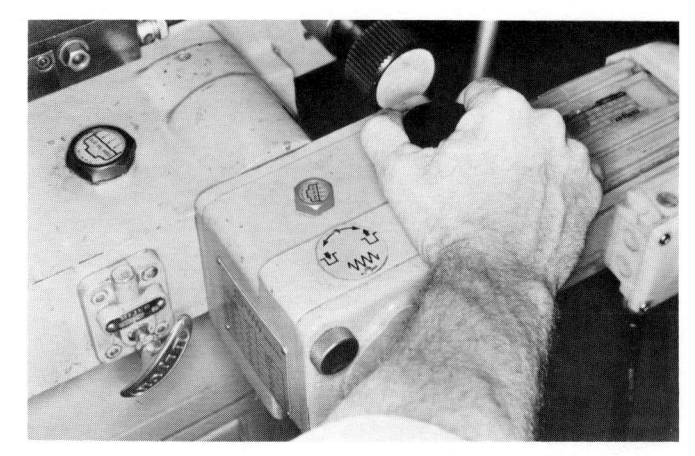

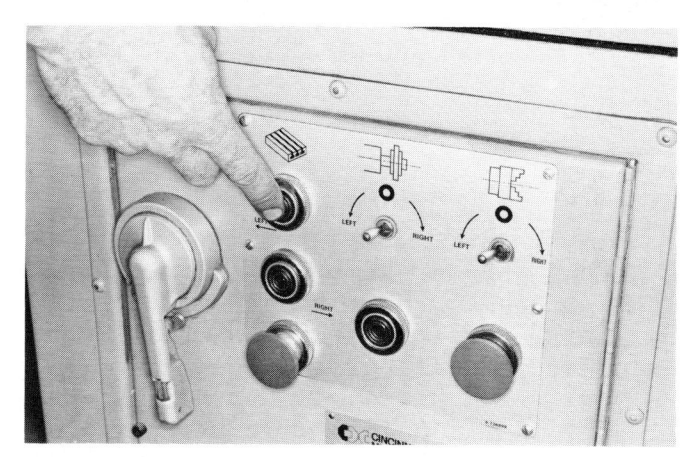

Figure 17. Engaging and starting the table traverse (Industrial Plastics Products, Inc.).

minute. Determine that the workhead and the spindle are rotating in a correct relationship, then engage the power table traverse and start the table drive motor (Figures 17a and 17b), if your machine is so equipped. If you are using a manual traverse, apply it with the next step.

9. Use the front cross slide (saddle) handwheel to infeed the reamer to the grinding wheel to make the body grind of the reamer margin to the required size. Then draw the saddle back and stop the machine.

10. Loosen the swivel table clamping nuts slightly and adjust the swivel table very slightly for longitudinal relief or what is commonly called back taper (Figure 18). The result should be about .0002 IPI. The adjustment should be made by traversing the table

Figure 18. Setting the longitudinal relief or "back taper" (Industrial Plastics Products, Inc.).

past a .0001 graduation test indicator while adjusting the taper.

11. Give the lands a thin coating of layout dye and restart the machine. Gently bring in the cross slide movement and grind the back taper until it covers two thirds of the margin length on the shank end of the reamer. Draw back the cross slide and stop the spindle.
12. Reset the swivel table to zero, indicate in, and secure the swivel locking nuts.
13. Remove the reamer from between centers and remove the driving dog.
14. Remove the cylindrical grinding attachment and replace it with the left tailstock center.

GRINDING THE SECONDARY CLEARANCE

The procedure used here is basically the same as for grinding the secondary clearance on a plain milling cutter or on an end mill, using a straight wheel.

If you already have a straight wheel of about a 6 in. diameter with a narrow land mounted in a collet, change to that wheel; otherwise:

1. Dress the face of a straight wheel to a land on the circumference no wider than $\frac{1}{16}$ in.
2. Mount the reamer on centers so that the initial tooth to be given clearance is facing downward on the side toward the grinding wheel.
3. Mount a tooth rest assembly with a flicker finger under the tooth and adjust height until the tip of the tooth is level with the centers.
4. Check the wheelhead height for center, correct to center if needed.
5. Compute the required secondary clearance. For steel, 12 degrees is about right.

Sine of 12 × radius of the wheel (assuming a 6 in. diameter wheel) = .20791 × 3 = .624 in.

6. Raise the wheelhead by the amount calculated for your wheel diameter.
7. Start the spindle and infeed the cross slide. Carefully grind the secondary clearance until a cylindrically ground margin of .007 to .009 in. remains. This is an especially careful job and *very* easy to overdo.

If you were careful and used good coordination, you now have a reamer ready for service.

self-evaluation

WORKSHEET 1
1. Depending on the type of cutter and tool grinder that you have available, make up a process sheet for grinding the chamfer on a machine reamer between centers.
2. With your instructor's permission, set up and sharpen the chamfer on a machine reamer. Look up recommendations for clearance angles for the materials on which the reamer is to be used. (It should be between 7 and 12 degrees.)
3. Show the finished chamfer sharpening to your instructor.

WORKSHEET 2
1. If the cutter and tool grinder that you have available is equipped with a cylindrical grinding attachment, prepare a process sheet for cylindrically grinding the margin and the longitudinal relief on a machine reamer.
2. Prepare a process sheet for grinding of secondary clearance on a machine reamer using a straight grinding wheel.
3. With your instructor's permission, circle grind the margin and perform a secondary clearance grind on a machine reamer.
4. Show the completed job to your instructor.

This unit has no post-test.

unit 20
miscellaneous cutter and tool grinding operations

The cutter and tool grinder is extremely versatile, as you have seen in the preceding units. The tool grinder, as a person, must also be extremely skillful in developing methods and procedures for sharpening and restoring cutting tools to make the rest of the chip-making process accurate and efficient. This unit will cover an assortment of additional cutter and tool grinding processes, including further end work on end mills, hand reamers, step tools, combined drill-countersinks, and relief sharpening of taps. It will also show the use of the surface grinding attachment.

objectives After completing this unit, you will be able to:
1. **Cut off damaged end mills and rough shape the end for sharpening.**
2. **Gash the sharpened end of a milling cutter to permit cutting to center.**
3. **Adjust, mount, and sharpen an adjustable hand reamer and other hand reamers.**
4. **Set up a tilt head cutter and tool grinder for sharpening step tools.**
5. **Describe a type of form relief device for sharpening combined drill-countersinks.**

6. Describe the motions of a form relief attachment for sharpening taps and other similar tools.
7. Describe the use of a surface grinding attachment on a cutter and tool grinder.

information

ADDITIONAL END MILL PREPARATION

For reconditioning of end mills that have been damaged excessively on the end by breakage or by excessive speed resulting in burning of the corners, it is necessary to do a cutting off operation, followed by a reshaping of the end of the tool. The tool is then resharpened and gashed carefully to center to recondition. To do these operations, take the following steps:

1. Mount the damaged end mill in a workhead set square to the table. Mount a narrow (about $\frac{1}{16}$ in.) reinforced cutting off wheel and a wheel guard.
2. Start the spindle and dress the edge of the wheel with a dressing stick (Figure 1).
3. Set the required distance to remove the damaged portion and lock the cross slide, if your machine is so equipped.
4. Rotate the workhead by hand to do the cutting off. Hand rotation is recommended because the usual helix on the flutes of an end mill tend to deflect the wheel initially during the cutting off process and, with hand turning, you can be sensitive to crowding of the wheel. If you should employ a cylindrical grinding attachment for cutting off (Figure 2), use a reinforced wheel of $\frac{1}{8}$ in. thickness.
5. Using the same reinforced wheel, gash across the face of each tooth (Figure 3) and then remove the excess material behind each tooth (Figure 4). This procedure is followed for multiple fluted end mills as well as the two flute type shown.
6. Mount the end mill for regular primary and secondary grinding as described in the unit on sharpening end mills (Figure 5).
7. Mount a large diameter straight wheel and dress to a sharp edged bevel (Figure 6).
8. Grind across the face of each tooth to center, as shown in Figure 7.

Figure 1. Dressing a cutting off wheel set up for cutting off damaged end mills (K & M Tool, Inc.).

Figure 2. Cutting off the damaged end (K & M Tool, Inc.).

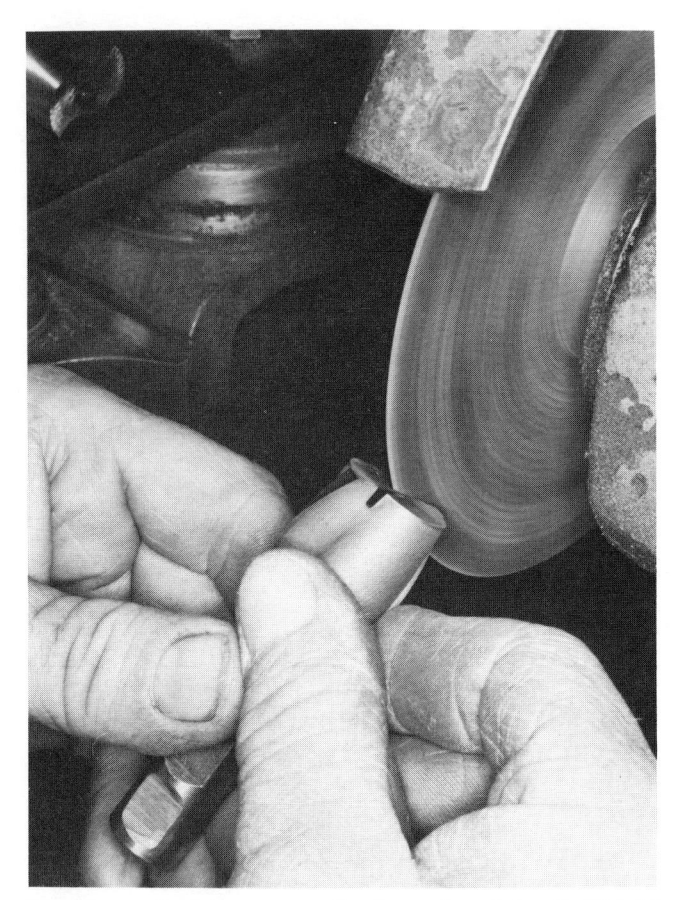

Figure 3. Gashing across the face of each tooth (K & M Tool, Inc.).

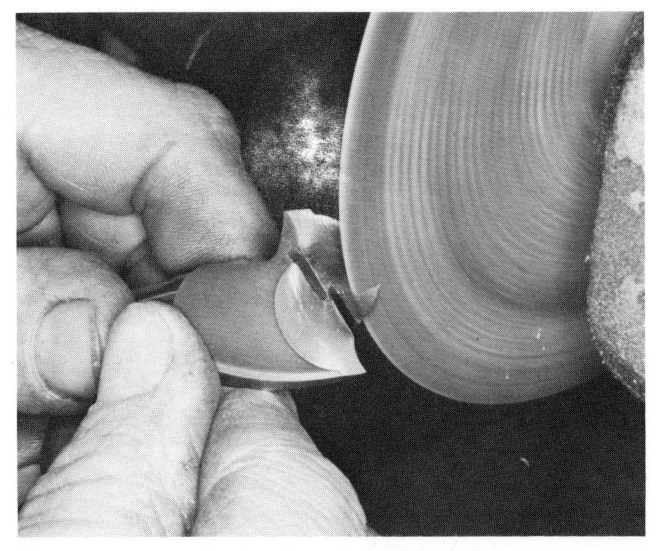

Figure 4. Rough shaping the end of the tool (K & M Tool, Inc.).

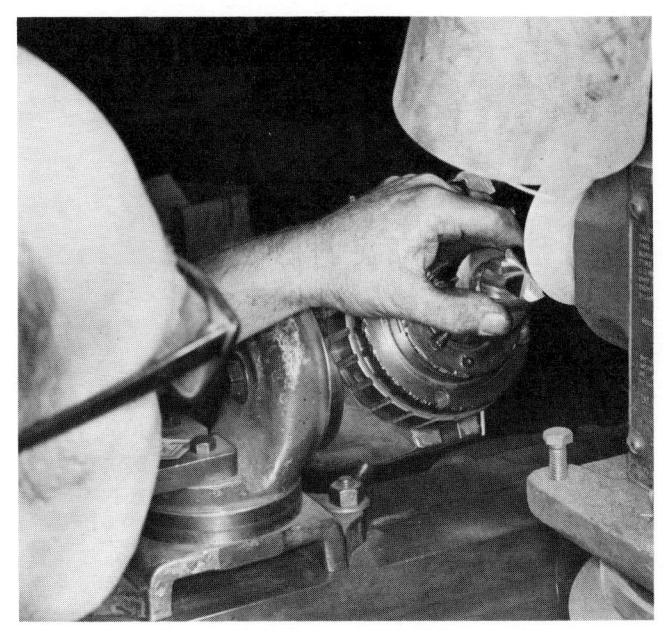

Figure 5. Sharpening the end mill in a conventional way (K & M Tool, Inc.).

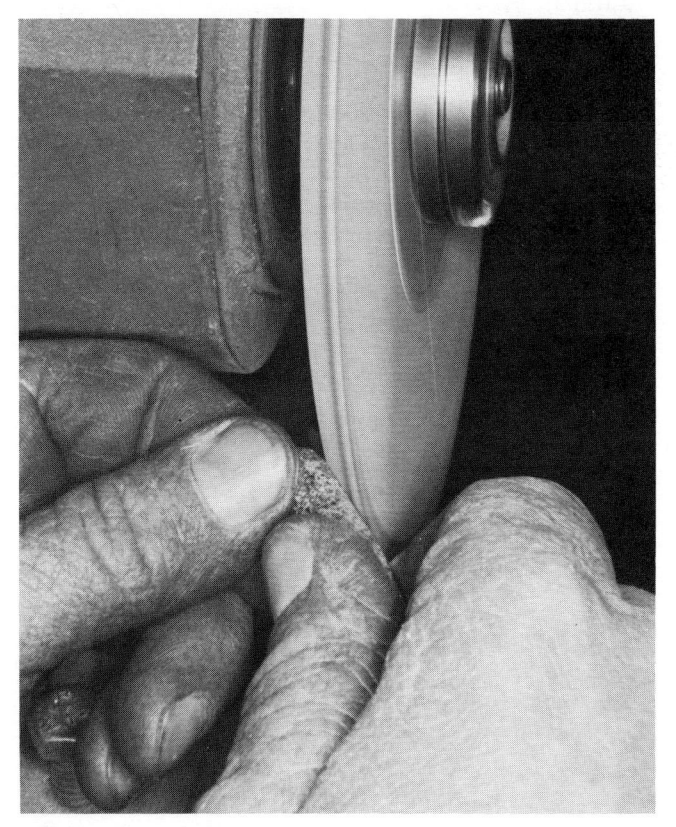

Figure 6. Dressing a straight wheel to a sharp bevel (K & M Tool, Inc.).

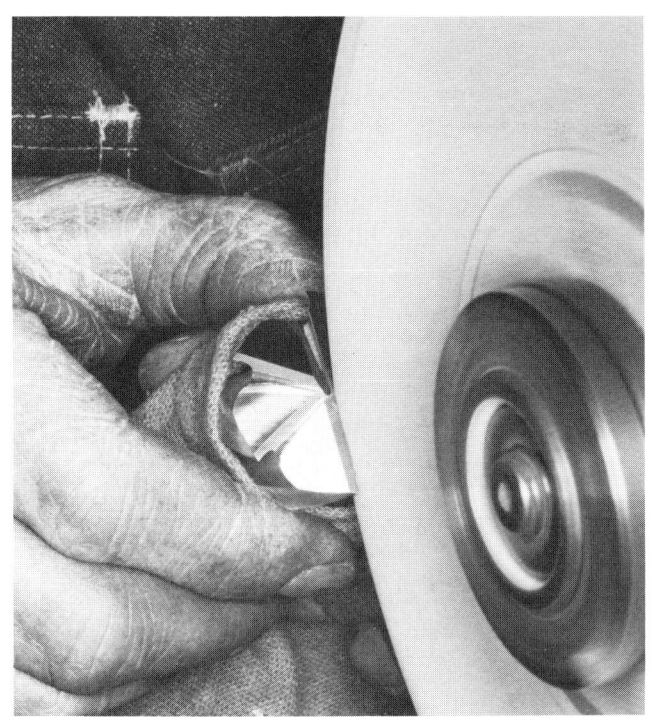

Figure 7. Grinding across the face of the tooth to center (K & M Tool, Inc.).

Figure 8. Remove the burrs — note the completed appearance (K & M Tool, Inc.).

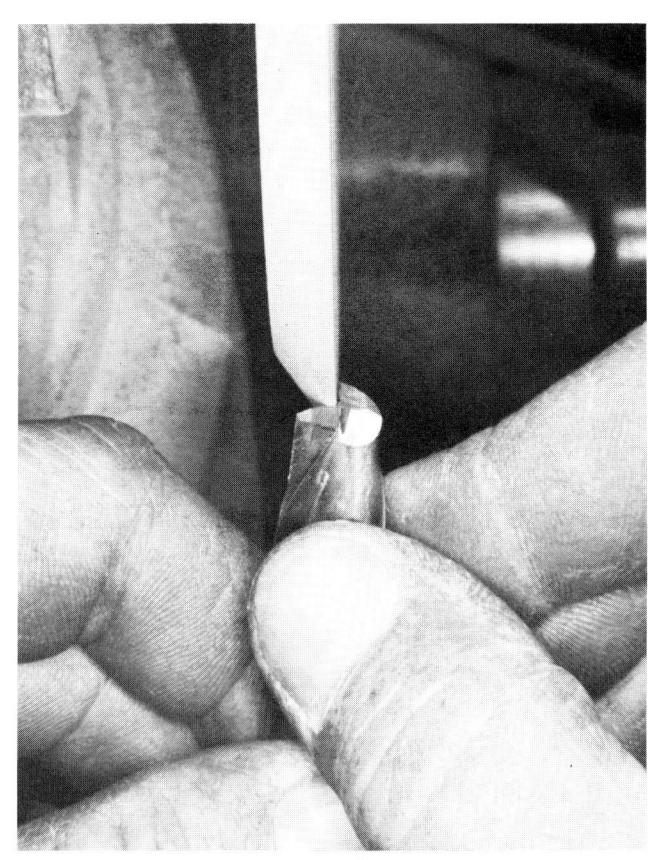

Figure 9. The same method is used for two and four fluted end mills (K & M Tool, Inc.).

9. Examine the result and remove any remaining burrs with a hone or a fine dressing stick (Figure 8). Note the completed appearance.

The same procedure is followed for two and four fluted end mills (Figure 9). This procedure requires great care and skillful coordination to obtain satisfactory results. You should become skilled on end mills that are shortened by many grindings before doing it on a cutter in which you have a good deal of sharpening time invested.

The hand gashing method is described because it is the most widely used method in tool sharpening job shops. It requires substantial skill to do this operation well. The workhead can also be used for this operation in conjunction with an indexing attachment. Figure 15 shows an indexing attachment being used for another application. Using the indexing attachment is much slower than hand gashing. If you find it necessary to use the workhead to get good results, be

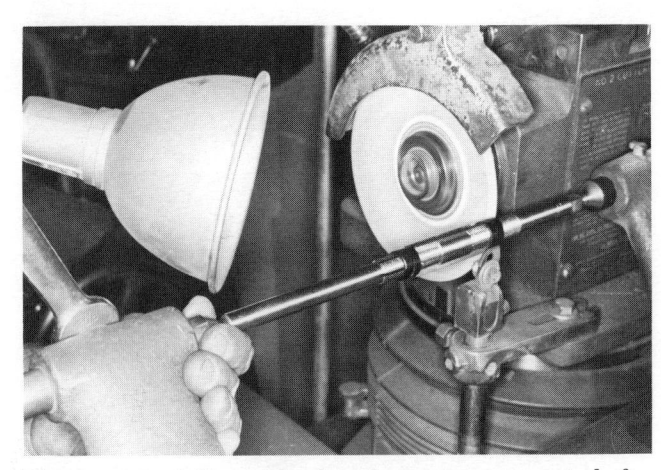

Figure 10. Adjustable hand reamer mounted for sharpening (K & M Tool, Inc.).

sure to set it around the vertical axis so a positive axial rake angle is produced on the face of the tooth being gashed.

SHARPENING AN ADJUSTABLE HAND REAMER

The adjustable hand reamer (Figure 10) is used a good deal in maintenance work, particularly in situations where bushings must be reamed to accommodate odd sized mating parts that are not of a standard hand or machine reamer size. The adjustable hand reamer is a difficult tool to use at best, because the very act of adjustment often leaves the blades in an out-of-parallel condition. For that reason, it is recommended that for producing an accurate hole, you adjust the reamer to a slight oversize and sharpen it to the finished size, just as you would an expansion reamer. To sharpen this type of tool, take the following steps.

1. Mount a straight or dish wheel and dress the OD to a narrow (about $\frac{1}{16}$ in. width) land.
2. Mount a tooth rest support with a narrow blade to the wheelhead, with the blade directly in line with the narrow land on the wheel. The tooth rest is mounted to the wheelhead instead of to the table, because the blades are sometimes not perfectly straight, so it should be treated like a helically fluted tool.
3. Set the wheelhead and tooth rest height to center.
4. Check the fit of the blades in their slots, adjust the reamer about .010 in. oversize, mount on centers, and start the wheel.
5. Compute the required reliefs and clearances (Table 1).
6. With the reamer just clear of the wheel, lower the wheelhead for the primary relief and record your setting.

Sine of desired relief angle × radius of the tool

7. Grind a narrow margin on opposing teeth. Stop the spindle and check for both taper and size with a micrometer (Figure 11).
8. Complete the primary sharpening to finished

Table 1
Hand Reamer Relief and Clearance Recommendations

Reamer Diameter	Hand Reamer for Bronze and Cast Iron (Margin .025-.030 in.)		Hand Reamer for Steel (Margin .005-.007 in.)	
	Primary Relief Angle, Degrees	Secondary Clearance Angle, Degrees	Primary* Relief Angle, Degrees	Secondary Clearance Angle Degrees
$\frac{1}{2}$	$7\frac{1}{2}$	17	3	12
1	$4\frac{1}{2}$	14	$1\frac{1}{2}$	$10\frac{1}{2}$
$1\frac{1}{2}$	$3\frac{1}{2}$	13	1	10
2	$3\frac{1}{4}$	$12\frac{1}{2}$	$\frac{3}{4}$	10

*Note the indicated margin width. If a margin wider than shown remains, the relief may be insufficient to prevent land interference with the hole. The cylindrical grinding attachment may also be used, as described for sharpening machine reamers, to produce the primary relief on a hand reamer.

Figure 11. Checking for size and taper (K & M Tool, Inc.).

Figure 12. Grinding the secondary clearance (K & M Tool, Inc.).

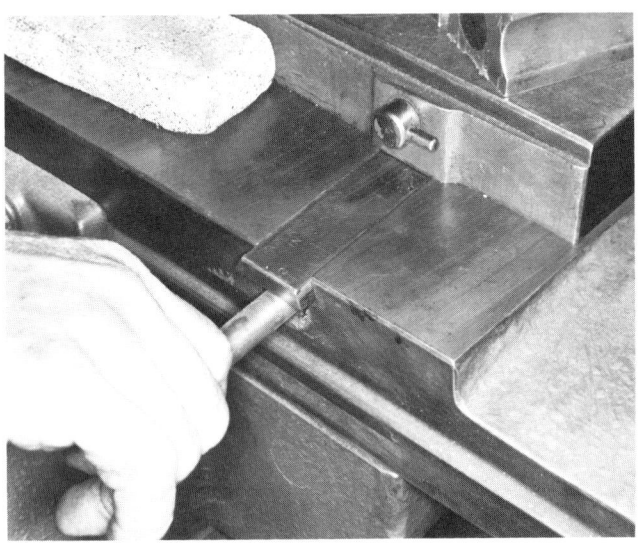

Figure 13. Setting the starting taper angle (K & M Tool, Inc.).

size, making a sparking-out final pass all the way around without additional infeed.

9. Set in the secondary clearance angle and record your setting.
10. Grind the secondary clearance (Figure 12) and withdraw the work when the correct margin has been obtained — in this case, about .025 in. Stop the wheel spindle.
11. Reset the workhead height to center, loosen the swivel table locking nuts, and swivel the table to $\frac{1}{4}$ IPF to produce a starting taper on the tool (Figure 13).
12. Adjust the wheelhead height to the same elevation used for the primary relief. Start the spindle and infeed and traverse until the primary starting taper is about one fourth of the flute length (Figure 14), all the way around the tool. Spark out for accurate concentricity. Clear the tool from the wheel. In a few cases, it is desirable to produce a secondary clearance on the starting taper. This would typically be for tools to be used in steel where the primary relief is so small. In this case, adjust the wheelhead height to the same value as the secondary clearance in the OD of the reamer, and then grind the second-

Figure 14. Grinding the starting taper (K & M Tool, Inc.).

Figure 15. Sharpening a three fluted step drill with tilt head machine and indexing attachment on the workhead (Industrial Plastics Products, Inc.).

ary clearance until the correct land width is obtained.

All that has been said about sharpening the adjustable hand reamer is equally applicable to solid hand reamers except, of course, for the preliminary sizing of the tool. A straight fluted reamer can also be done with the tooth rest attached to the table.

SHARPENING OTHER AXIAL CUTTING TOOLS AND SINGLE POINT TOOLS

Here we will see some additional setups for grinding specific classes of tools. This is for information instead of providing step-by-step sequences.

Step Drills

Sharpening a step drill requires careful attention to both concentricity and uniform step length in order to have a tool that cuts both to size on the diameter and to length. In some cases, particularly where two or more steps are related for length, it is necessary to set the length between the steps accurately. In Figure 15, a single step, three fluted step drill is being sharpened with a tilting head cutter and tool grinder, utilizing an indexing attachment mounted on a universal workhead. The required clearance angle is set directly as the wheelhead tilt angle.

Combined Drill-Countersinks

It is important that the drill portion and the countersink portion of a combined drill-countersink are concentric, and that the relief or clearances behind the countersink cutting edges are uniform. This can be taken care of by using a device called a form relief attachment to advance and withdraw the tool twice per revolution of the hand wheel (Figure 16). With this arrangement, the grinding wheel is formed (Figure 17) to provide the required angle (60 degrees) to make holes suitable for mounting work on centers. The drill point is sharpened in a secondary operation.

Grinding of Taps

The fluted portion of a tap may be sharpened in the same way as a form relieved cutter, but it results in the tap cutting a smaller pitch diameter after sharpening. Sharpening is best done on the chamfer below the cutting edges. For this type of sharpening, a special type of form relieving attachment is needed (Figure 18). This device both rocks the tool toward and away from the grinding wheel, and moves the work across the wheel face at the same time. The number of times this is done per work spindle revolution can be set by changing the gears at the left end of the head.

Figure 16. Form relief fixture for sharpening combined drill-countersinks (K & M Tool, Inc.).

Figure 17. The wheel is formed to produce the required cutting angle (K & M Tool, Inc.).

Figure 18. Form relief attachment for taps and similar tools (K & M Tool, Inc.).

Figure 19. Adjustment for the amount of clearance (K & M Tool, Inc.).

The amount of relief or clearance is adjusted by rotating a sine bar-like device (Figure 19) on the side of the attachment. The greater the angle, the more the attachment rocks the tool into the grinding wheel for each flute, resulting in increasing clearance. Before the grinding is begun, the chamfer is coated with layout dye (Figure 20) to show the extent of grinding and to be able to match or pick up the grind where the needed chamfer is wider than the available wheel face. The grinding is done by infeeding the cross feed saddle while keeping the tool consistently against the driving pin that is rotating it (Figure 21). To pick up on the chamfer to continue the cut the cross slide is backed away, the table moved to the right, and then the saddle is infed until the new surface matches the first ground surface. The chamfer is then complete and the tool is dipped in plastic to prevent damage to its restored surfaces (Figure 22). This should be done to all newly sharpened tools that are not to be put into immediate use. As a caution, this material is usually kept about 350° F. Do not

Figure 20. Coloring the chamfer (K & M Tool, Inc.).

Figure 21. Grinding the chamfer (K & M Tool, Inc.).

Figure 22. Protect sharpened tools with a plastic dip (K & M Tool, Inc.).

Figure 23. Using a surface grinding attachment (K & M Tool, Inc.).

make the mistake of reaching your hand into the pot if you should accidentally drop a tool into the molten plastic! It can result in third-degree burns!

Figure 24. Diamond wheel being used to grind the top surface of a carbide turning tool (K & M Tool, Inc.).

Grinding Single Point Tools

The cutter and tool grinder is also useful for grinding tools like carbide tipped lathe tools (Figure 23). For this type of work, a surface grinding attachment is used; it consists of a swivel vise and intermediate support that allows movement in two planes. This attachment can be used for sharpening flat forming tools, flat thread chasers, and similar other work, in addition to lathe and planer tools. For sharpening carbide tools, the diamond wheel is the most economical cutting tool to use (Figure 24). Silicon carbide grinding wheels will work, but the rate of breakdown necessitates nearly constant dressing attention.

self-evaluation

SELF-TEST

1. When cutting off and rough shaping an end mill, why is the reinforced wheel necessary?

2. End mills are frequently "gashed" after the ends have been sharpened. Why is this done?

3. Why is it desirable to sharpen adjustable reamers to size when possible instead of simply adjusting them to size as needed?

4. Why is secondary clearance sometimes done on the starting taper of hand reamers?

5. Why is concentricity important in the sharpening of step tools?

6. Why is the countersink portion of a combined drill-countersink set at 60 degrees?

7. Taps are sometimes sharpened by grinding the flutes, like a form relieved cutter. What is the chief limitation to this kind of sharpening?

8. Why must a form relief grinding attachment have provision for setting the number of relief movements per work spindle revolution?

9. Why should the amount of relief be controlled in tap and countersink grinding?

10. Single point tools are also sharpened on the cutter and tool grinder. Why are carbide tipped cutting tools usually ground with diamond grinding wheels?

WORKSHEET 1
1. Check your available cutter and tool grinder for a reinforced cutting off wheel.
2. Obtain a damaged end mill that requires a cutting off procedure.
3. Make up a process sheet to include the specific equipment you have available.
4. With your instructor's permission, cut off, gash, and shape the end mill for sharpening.
5. Sharpen the end of the end mill.
6. Finish gash the teeth to center, using a correctly shaped straight wheel.
7. Deburr and show the completed job to your instructor.

WORKSHEET 2
1. Obtain either an adjustable or solid hand reamer that needs sharpening.
2. Make up a process sheet to match your available machine and accessories.
3. With your instructor's permission, follow your process sheet and sharpen the tool.
4. Try the tool; make a hole finished to your expected size (*by hand*, of course).
5. Show the finished tool and the holes that you have reamed with it to your instructor.

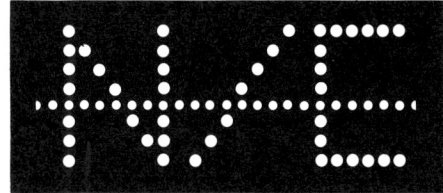

section i
numerical control of machine tools

Numerical control of machine tools is of major importance in modern machining technology. With numerical control (N/C), machining productivity is increased. This is an important consideration in an age where manufacturing is faced with rising costs. The N/C machine tool is helping to keep production costs down. Manufacturers have seen the value and versatility of these machines. They are appearing in machine shops of every size. Numerical control can be productively applied to almost any machining task. An individual beginning machinist training today does so in a numerical control environment.

As you begin a career in machining, N/C will play an ever increasing part. The field of N/C machining technology is constantly expanding. What was new in numerical control only a short time ago is becoming rapidly obsolete. In this section you will have an opportunity to investigate N/C technology in a general way. As you proceed into further training and possibly to a career in machining, you may find the area of N/C to be of great interest. In any case, the field is sure to be expanding and changing. For this reason, the numerical control branch of machining technology presents a number of opportunities for specialized study.

INTRODUCTION TO NUMERICAL CONTROL

Machine Control by Numerical Instructions

Numerical control is a method and a system of controlling a machine or process by instructions in the form of numbers. On a manually operated machine tool, the operator turns cranks in order to move milling machine tables or lathe cross slides and carriages. On the N/C machine tool, cranks are replaced by drive motors or hydraulic mechanisms. These are controlled from an external machine control unit (MCU).

Machine control functions previously provided by the operator are translated into numeric instructions that can be understood by the machine control unit. These control functions include positioning tables and spindles, setting milling feedrates, setting spindle speeds, cycling drill press quills, changing cutting tools, and turning coolant on and off. A simple on/off control function provided by an electric switch might be used to start and stop an electric motor or open and close a hydraulic valve. This might be expressed as a simple numeric instruction of "one" to open a valve or start a motor. A numeric instruction of "zero" might close a valve or stop a motor. If the motor or hydraulic mechanism is connected to the leadscrew of a lathe or the table screw of a milling machine, the simple on/off, one/zero numerical control function can be used to position the component of the machine tool. The distance traveled and the direction can be controlled as well. In actual practice, numerical control is somewhat more complicated than a simple on/off function. However, this example will give you some idea of control by a numeric instruction.

Historical Development of Numerical Control The development of true numerical control dates from about 1945. However, research and development in automatic control had been done as early as 1725 (Figure 1).

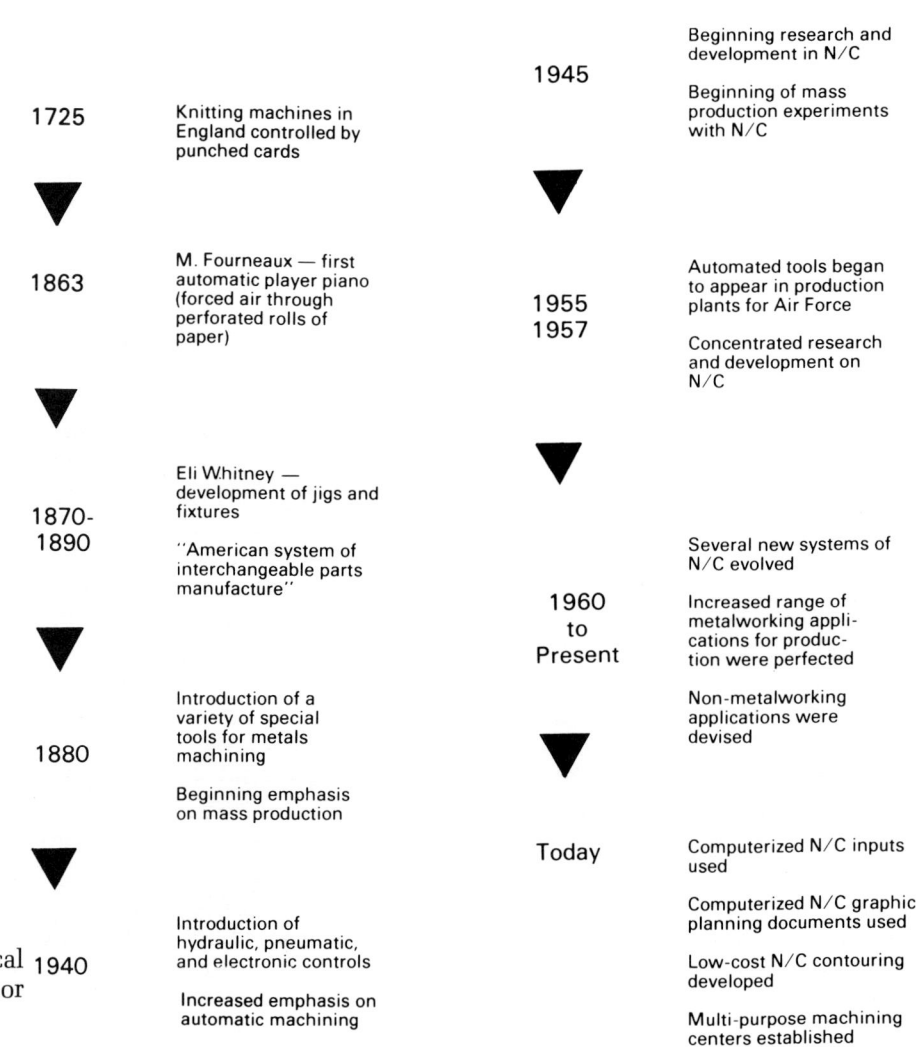

Year	Event	Year	Event
1725	Knitting machines in England controlled by punched cards	1945	Beginning research and development in N/C. Beginning of mass production experiments with N/C
1863	M. Fourneaux — first automatic player piano (forced air through perforated rolls of paper)	1955 1957	Automated tools began to appear in production plants for Air Force. Concentrated research and development on N/C
1870-1890	Eli Whitney — development of jigs and fixtures. "American system of interchangeable parts manufacture"	1960 to Present	Several new systems of N/C evolved. Increased range of metalworking applications for production were perfected. Non-metalworking applications were devised
1880	Introduction of a variety of special tools for metals machining. Beginning emphasis on mass production	Today	Computerized N/C inputs used. Computerized N/C graphic planning documents used. Low-cost N/C contouring developed. Multi-purpose machining centers established
1940	Introduction of hydraulic, pneumatic, and electronic controls. Increased emphasis on automatic machining		

Figure 1. Time line for numerical control (Courtesy of the Superior Electric Company.)

Construction Features of N/C Machine Tools

A numerical control system can be added to an existing machine tool (Figure 2). This is called a retrofit and presents a less expensive way to gain N/C capability in the machine shop. Retrofit numerical controls are frequently found in school shops.

The many advantages of N/C have brought about extensive development of machine tools designed specifically for N/C operations. In conventional machining, a complex workpiece may have to be set up on several machine tools in order to complete all required machining tasks. The development of numerical control technology has brought about the concept of a "machining center" on which a wide variety of machining tasks can be accomplished on the same machine tool. The N/C machines used in machining centers are patterned after their conventional counterparts. However, the construction of an N/C machine tool is often quite different.

N/C machine tools are designed for long hours of continuous production. This necessitates that they be built so that accuracy will be retained as long as possible. Wear is a problem associated with

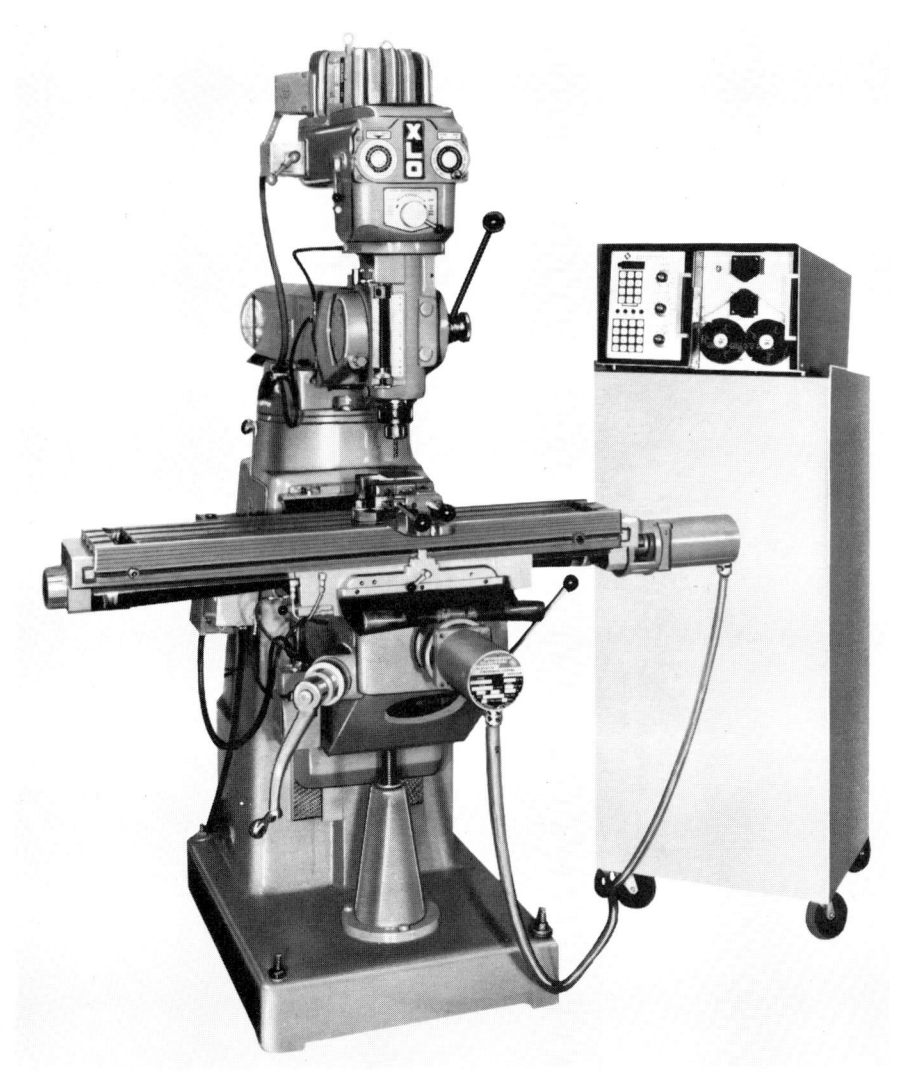

Figure 2. A numerical control system added to an existing milling machine (Courtesy of the Superior Electric Company).

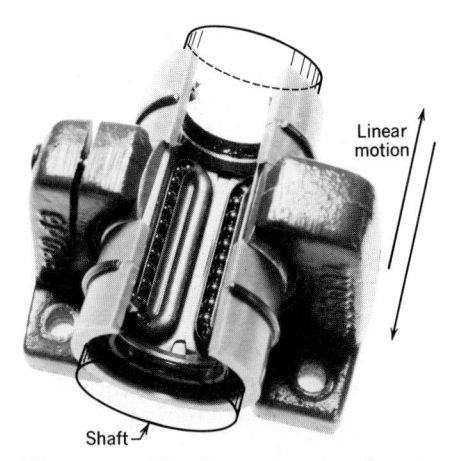

Figure 3. The linear motion bearing permits a close fit along with low wear and nearly friction free movement (California State University at Fresno).

any mechanical device. Wear in a machine tool is particularly significant in that the accuracy of the machine is directly affected. A possible advantage of a manually operated machine tool over an N/C machine is that a good machinist can "know" his machine and compensate for wear. However, the increased productivity and versatility of the N/C machine tool far outweighs this factor. N/C machines, like all machines, are subject to wear. Because of this, manufacturers have introduced many features designed to prolong the machine's accuracy holding capabilities.

Sliding components have given way to rolling components in order to provide nearly friction-free movement and still retain the alignment and close fit of a slide. Several types of linear motion bearings are used on N/C machines. These bearings permit the rolling advantages of the ball bearing to be used in a linear motion application (Figure 3).

Conventional machine tools such as milling machines use a screw turning in a nut to operate the table saddle and knee. Threads are subject to backlash, and this places certain limitations on machining operations. You will remember from your study of milling machines that because of backlash, climb milling is not recommended under most conditions. You will also remember that backlash must be taken into consideration when positioning a mill table or lathe cross slide by hand. The N/C machine tool cannot be limited by this factor if maximum productivity is to be realized.

One of the features designed to overcome the problem of backlash is the recirculating ball screw. The principle of this device has been used for many years in automobile steering gears. Ball screws are used extensively on N/C machine tools. Machines equipped with them are not limited by a backlash problem. Where numerical controls have been added to existing machine tools, electromechanical backlash compensation has been built into the machine control unit.

Hydraulics are employed to overcome the limitations of threads. Some N/C machine tools use full hydraulic positioning systems, thus eliminating the use of any thread mechanism for moving machine components.

N/C Machine Tools and Machining Centers

VERTICAL SPINDLE N/C MACHINING CENTERS. Vertical spindle N/C machining centers patterned after the vertical milling machine are very popular. The versatility of the vertical mill is mated to the advantages of numerical control. Vertical spindle machines may be equipped with an eight-position turret tool holder (Figure 4) or a drum or carousel tool holder containing many tools (Figure 5). Tool changing is numerically controlled. The vertical spindle design also includes large capacity multispindle milling machines such as the bridge-type profiler (Figure 6) and the gantry-type profiler (Figure 7). These machine tools permit several workpieces to be machined at the same time.

HORIZONTAL SPINDLE N/C MACHINE TOOLS. Horizontal spindle N/C machining centers, patterned after the horizontal mill and horizontal boring machine, are also very popular and versatile. These machines may be equipped with side mounted tool drums (Figure 8)

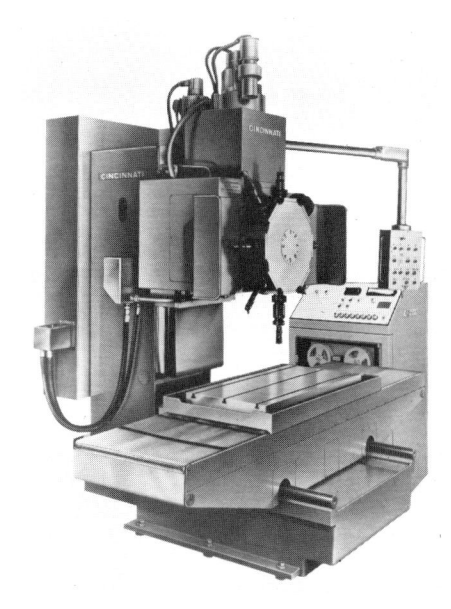

Figure 4. Vertical spindle N/C milling machine with eight position turret tool holder (Courtesy of Cincinnati Milacron).

Figure 5. Vertical N/C milling machine with side mounted tool drum (Courtesy of Hydra-Point Division, Moog Inc.).

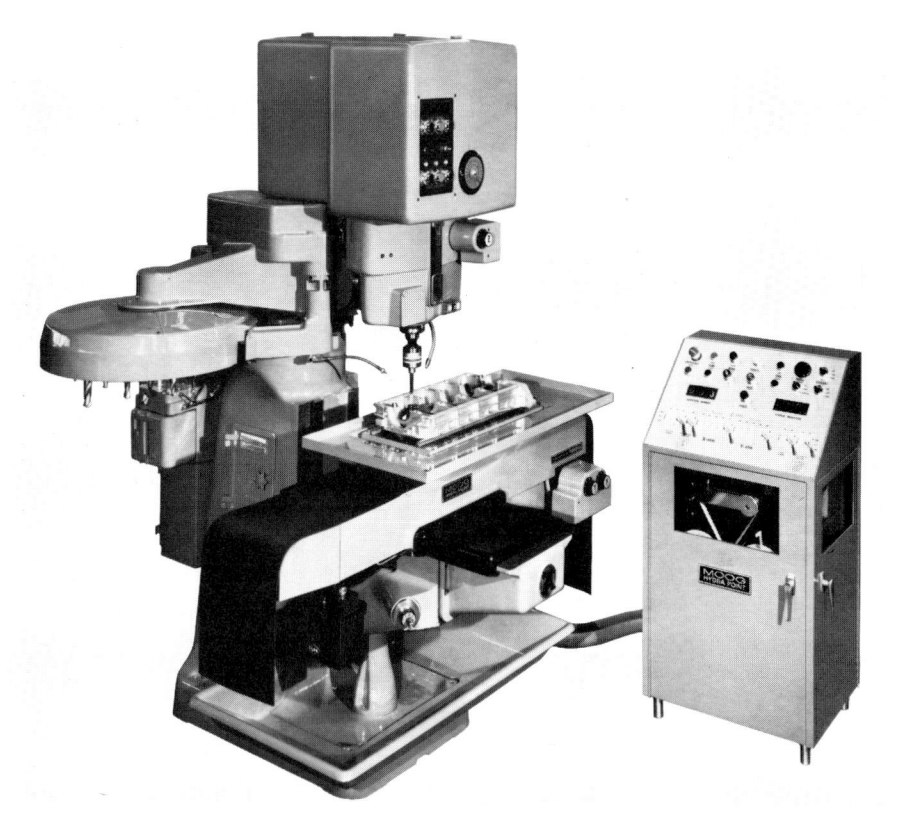

Figure 6. Bridge-type multiple spindle N/C profiler (Courtesy of Cincinnati Milacron).

or top mounted drums (Figure 9). Tool changing is numerically controlled. The workpiece may be mounted on a rotary table, enabling both sides to be machined (Figure 10).

Large capacity horizontal spindle machines include the traveling column profiler (Figure 11). The machine tool shown is equipped with a conveyor system to remove chips. Horizontal multiple spindle profilers may be used to machine several workpieces at the same time (Figure 12).

Figure 7. Gantry-type multiple spindle N/C profiler (Courtesy of Cincinnati Milacron).

Figure 8. Horizontal spindle N/C machining center with side mounted tool drum (Courtesy of Cincinnati Milacron).

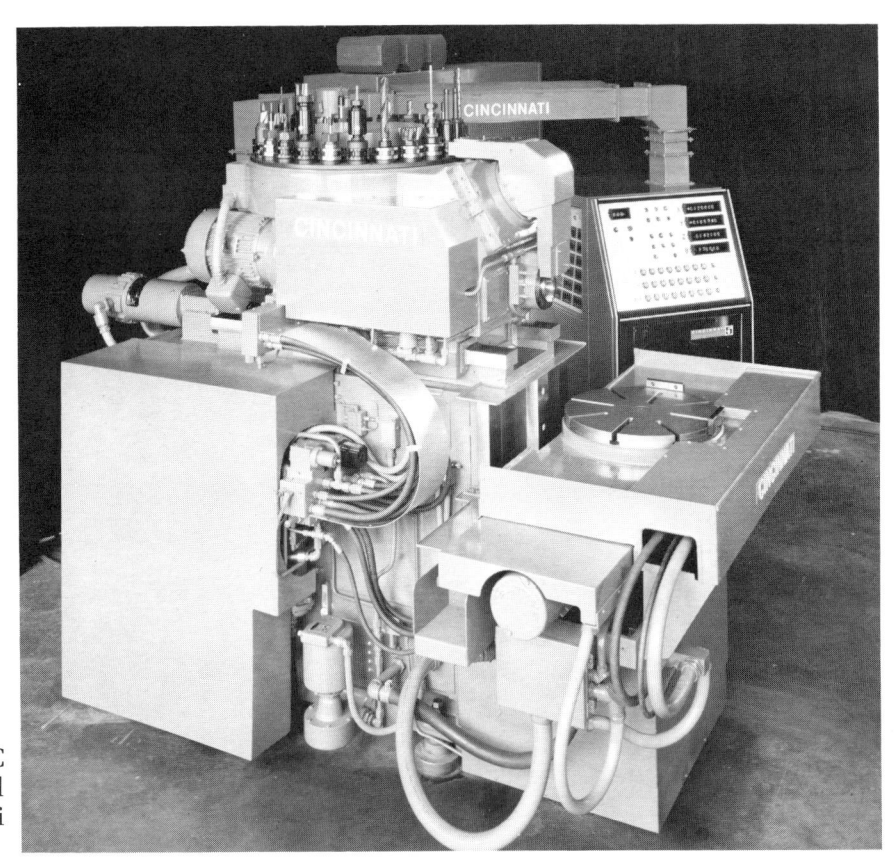

Figure 9. Horizontal spindle N/C machining center with top mounted tool drum (Courtesy of Cincinnati Milacron).

Figure 10. Rotary table on a horizontal spindle machining center (Courtesy of Cincinnati Milacron).

N/C LATHES. A sophisticated N/C lathe or turning center (Figure 13) may have numerically controlled turret tool holders for inside diameter turning (Figure 14). With the workpiece held in a chuck, a

Figure 11. Traveling column N/C profiler with chip conveyor (Courtesy of Cincinnati Milacron).

Figure 12. Horizontal N/C profiler with three spindles (Courtesy of Cincinnati Milacron).

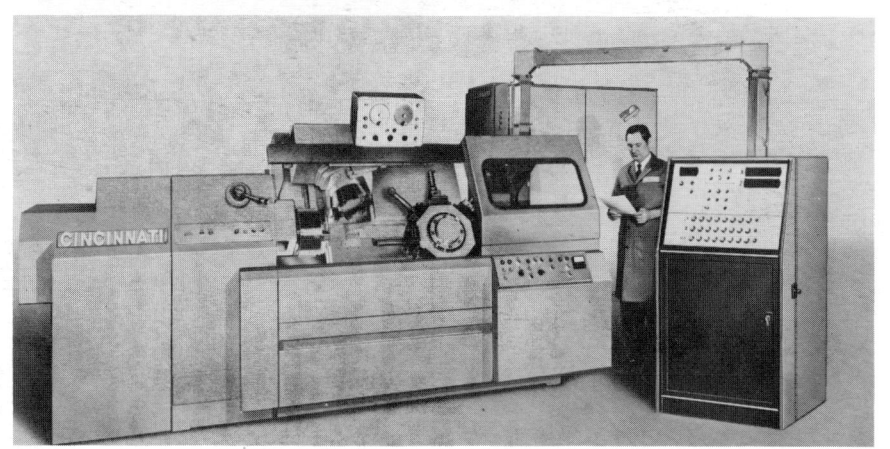

Figure 13. Dual turret N/C turning center (Courtesy of Cincinnati Milacron).

Figure 14. Inside and outside diameter N/C turning center turrets (Courtesy of Cincinnati Milacron).

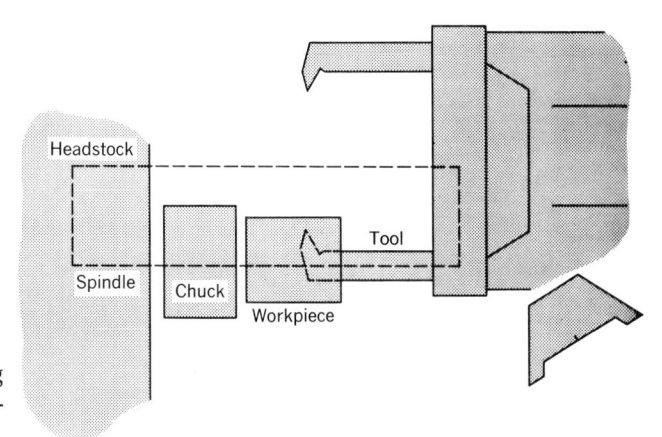

Figure 15. Inside diameter turning loop (Courtesy of Cincinnati Milacron).

variety of cutting tools can be applied in the inside diameter turning loop (Figure 15). The outside diameter turret (Figure 16) operates in the outside diameter turning loop (Figure 17). Numerically controlled lathes are also used in shaft turning operations (Figure 18). The shaft turning loop requires use of a tailstock center (Figure 19).

The Input Medium of Numerical Control Once a control function has been translated into numeric form, the information must be provided to the machine tool. This can be done through a control medium such as punched cards, magnetic tape, or

Figure 16. Outside diameter turret on the N/C turning center (Courtesy of Cincinnati Milacron).

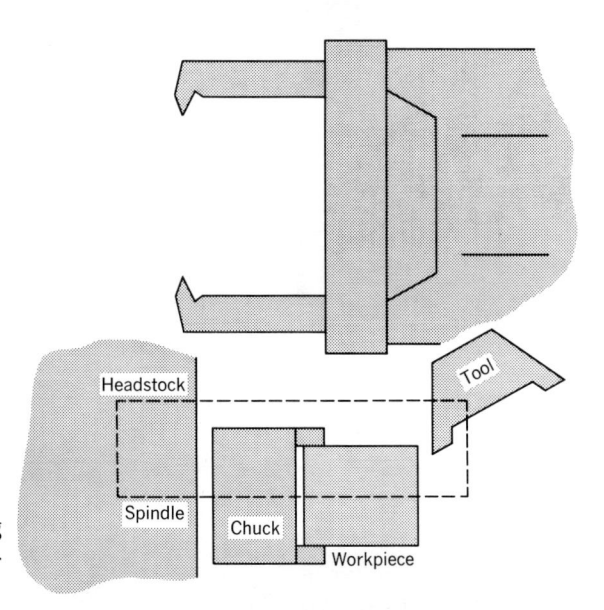

Figure 17. Outside diameter turning loop (Courtesy of Cincinnati Milacron).

Figure 18. Shaft turning on the N/C turning center (Courtesy of Cincinnati Milacron).

Section I Numerical Control of Machine Tools 527

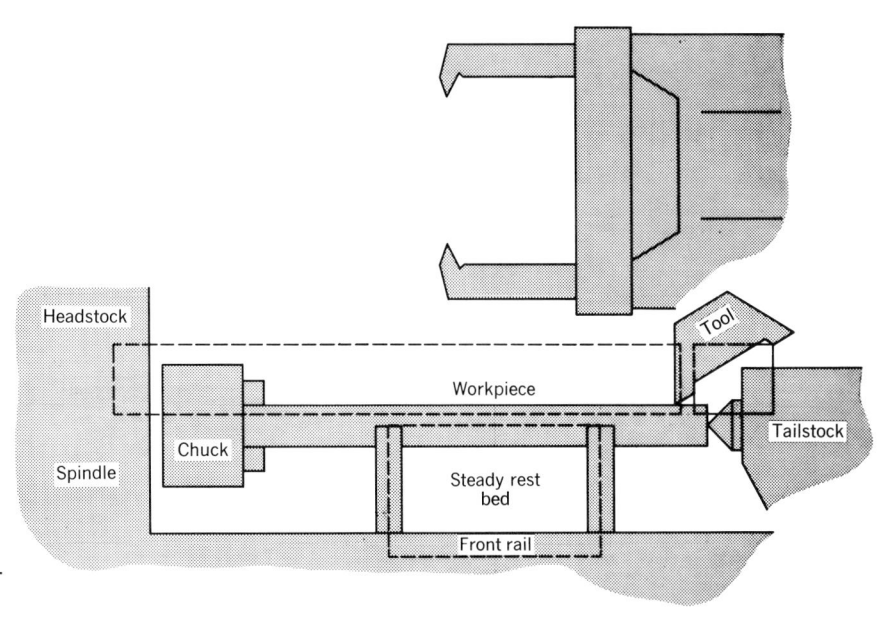

Figure 19. Shaft turning loop (Courtesy of Cincinnati Milacron).

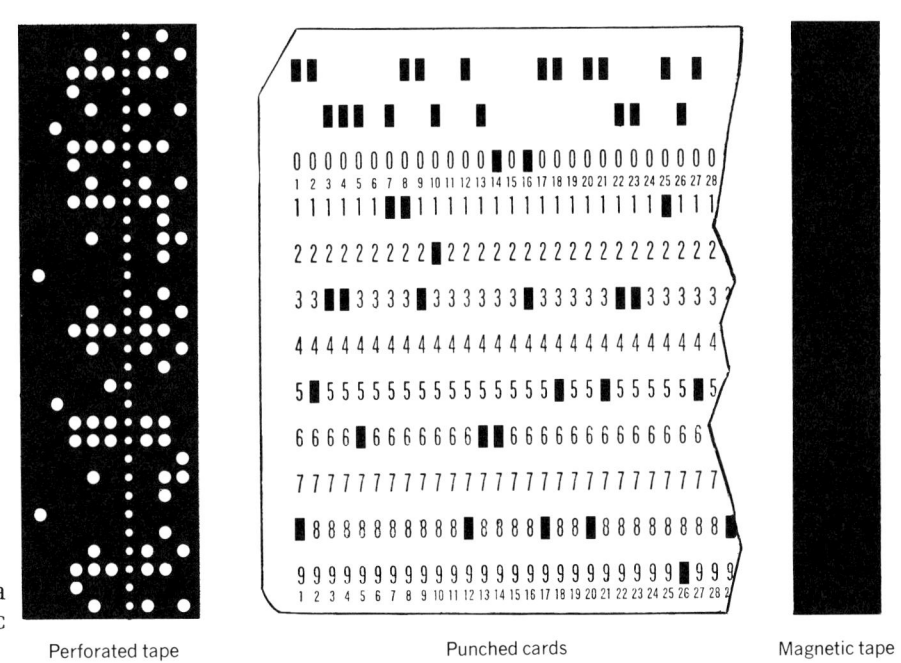

Figure 20. N/C input program media (Courtesy of the Superior Electric Company).

Perforated tape   Punched cards   Magnetic tape

punched tape (Figure 20). The punched tape medium is very popular at the present stage of the technology. Instructions are translated into numeric form and appear as specific patterns of holes punched into a tape. The tape is read by the tape reader in the machine control unit. The MCU then interprets the numerical information on the control tape and, through electric and mechanical means, controls components of the machine tool. As you study numerical control, you may hear terms such as tape machine and tape control. These are references to machine control by a tape medium.

Advantages of Numerical Control Machining

REPEATABILITY. The N/C machine tool can produce 1, 10, or 10,000 parts with unvarying accuracy. This is important to the pro-

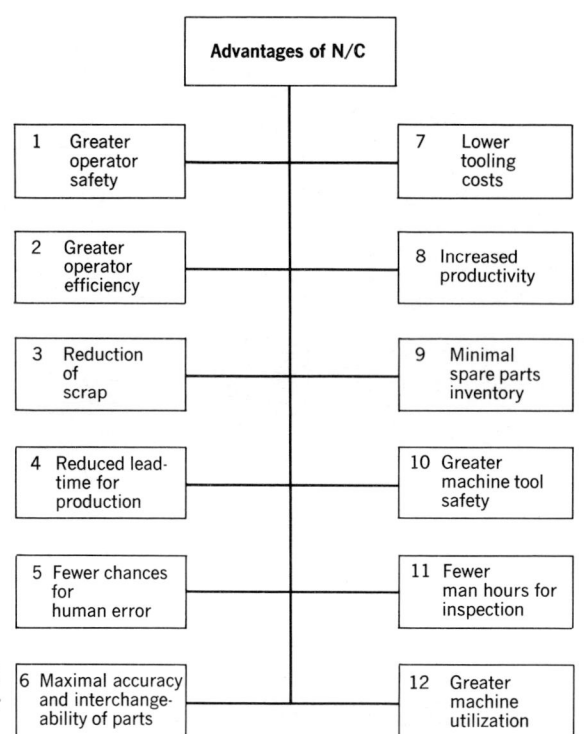

Figure 21. Advantages of numerical control (Courtesy of the Superior Electric Company).

```
                    Advantages of N/C
                           |
   ┌───────────────────────┼───────────────────────┐
 1 Greater                                       7 Lower
   operator                                        tooling
   safety                                          costs

 2 Greater                                       8 Increased
   operator                                        productivity
   efficiency

 3 Reduction                                     9 Minimal
   of                                              spare parts
   scrap                                           inventory

 4 Reduced lead-                                10 Greater
   time for                                        machine tool
   production                                      safety

 5 Fewer chances                                11 Fewer
   for                                             man hours for
   human error                                     inspection

 6 Maximal accuracy                             12 Greater
   and interchange-                                machine
   ability of parts                                utilization
```

duction of duplicate parts that are within tolerance. A set of tape instructions does not vary. The machine tool does not become fatigued or bored, as does the operator of its manually operated counterpart. The machine will repeat precisely during each machining cycle. Workpieces need only be removed and replaced with new stock. Except for downtime due to resharpening of cutting tools or routine maintenance, the N/C machine tool can function 24 hours a day, year round, if necessary. In this respect, it has a decided advantage over its manually controlled counterpart (Figure 21).

VERSATILITY. The machining function of the N/C tool can easily be changed by inserting a new tape. With computer control, N/C programs can be stored in computer memory and the machine tool can be operated directly, thus bypassing the tape. Tape N/C programs are permanent and can be stored for future use. A long production run can be stopped and a short run can be inserted. The machine tool can then be returned to its long run in a very short time.

MACHINING CAPABILITY. With the numerically controlled tool changer, the N/C machine tool or machining center can accomplish a wide variety of machining tasks. Tool changing may be a two-step operation where the drum is mounted on the side of the machine (Figure 22). The drum rotates the required tool into position, where it is pivoted down and grasped by the changing arm. The tool is then placed in the spindle (Figure 23). At the same time, the tool already in the spindle is returned to the pivot arm and replaced in the drum. Where the drum is on the top of the machine, the tool in the drum and the tool in the spindle are removed at the same time and exchanged (Figures 24a to 24d). Tools are secured in a self-releasing

Figure 22. Pivoting presenter arm on a side mounted tool changer (Courtesy of Heald Machine Division/Cincinnati Milacron).

Figure 23. Tool changing arm (Courtesy of Heald Machine Division/Cincinnati Milacron).

taper shank tool holder. The holder is secured in the machine by a locking mechanism contained within the spindle (Figure 25).

Drilling (Figure 26) is a common N/C machining capability. Spindle speed, feedrate, and depth can be controlled from tape instructions. A "peck drilling" cycle can be used for deep hole or small hole drilling. When the peck drill cycle is initiated, the drill feeds partway into the workpiece, where it dwells for a short time. The drill is then automatically withdrawn in order to clear chips. The cycle is automatically repeated and the drill is permitted to feed further into the workpiece. The cycle continues until the final depth is reached. Dwell time during each "peck" can be varied manually.

Milling an enclosed feature or "pocket milling" (Figure 27) is another very common and useful machining capability. Spindle speed, milling feedrates, and direction and distance of the cuts are controlled from tape instructions.

Close tolerance boring (Figure 28) is frequently done on the N/C machine tool. Spindle speeds and bore feedrates are tape controlled. The boring bar may be withdrawn from the bore at the same feedrate used during boring. This reduces tool marks in the workpiece. Progressive boring bars may be used where close tolerances must be maintained. The hole is first bored with a roughing bar and finished with a finishing bar. Boring bars must be preset for correct diameter before they are used.

Tapping is another machining operation that is well suited to N/C operations. One method is leadscrew tapping (Figure 29). The leadscrew causes the machine spindle to feed or lead the same amount as the tap lead. Leadscrews are changed to correspond to different tap leads.

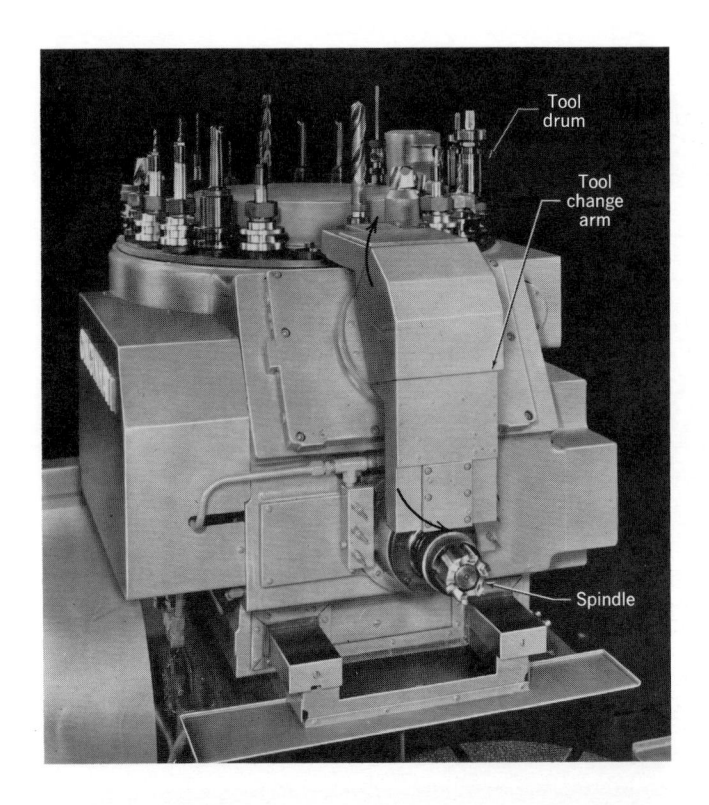

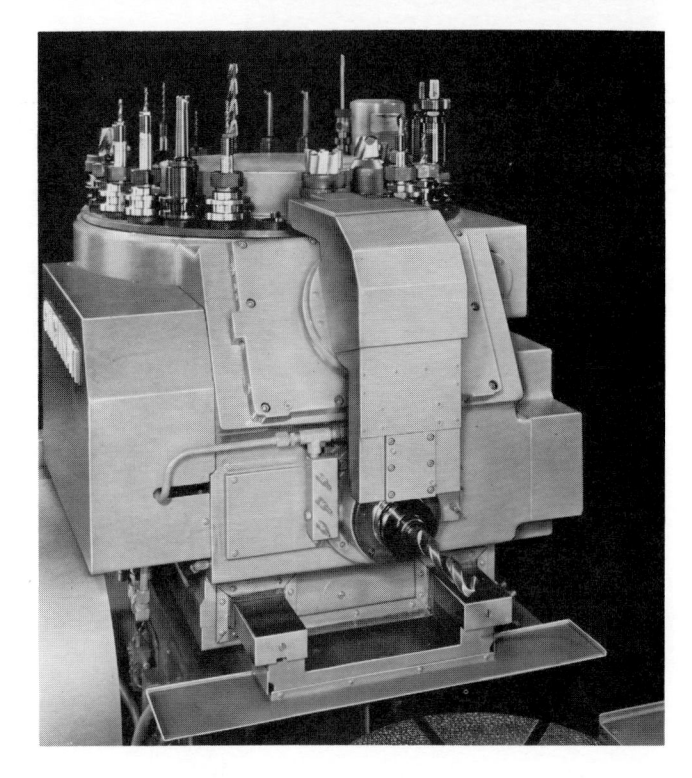

Figure 24. Tool changing sequence from a top mounted tool drum (Courtesy of Cincinnati Milacron).

Probably the greatest machining capability of the N/C machine tool is that of contouring or continuous path machining (Figure 30). This includes circles, angles, and radius cuts, as well as irregular shapes in two or three dimensions. In fact, the N/C machine tool can

Figure 25. Spindle tool holding mechanism (Courtesy of Hydra-Point Division, Moog Inc.).

Figure 26. N/C drilling (Courtesy of Hydra-Point Division, Moog Inc.).

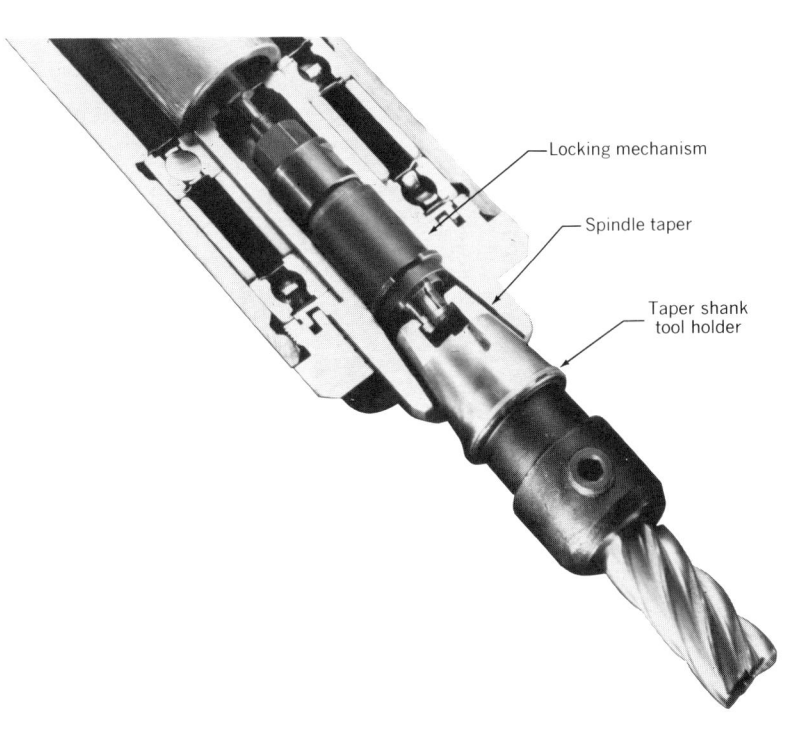

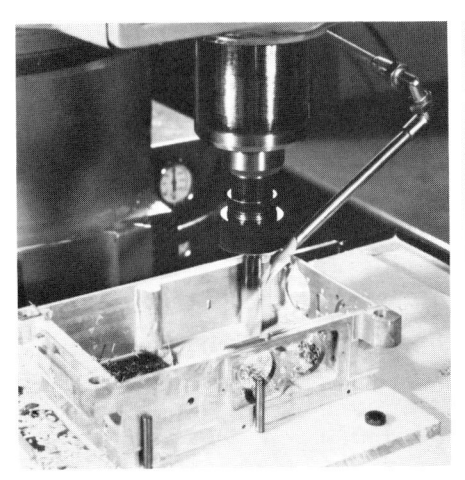

Figure 27. N/C pocket milling (Courtesy of Hydra-Point Division, Moog Inc.).

Figure 28. N/C boring (Courtesy of Hydra-Point Division, Moog Inc.).

Figure 29. N/C leadscrew tapping (Courtesy of Hydra-Point Division, Moog Inc.).

produce shapes that would be quite impossible to machine by manual means. This capability has opened new avenues of study for the designer of mechanical hardware.

Numerical Control and the Machinist The N/C machine tool, with its amazing capabilities, seldom requires a fully qualified machinist as an operator. In fact, the talent of the machinist would be wasted if time were spent changing workpieces on an N/C machine. Most N/C machine tools can be operated

Figure 30. Continuous path N/C machining (Yuba College).

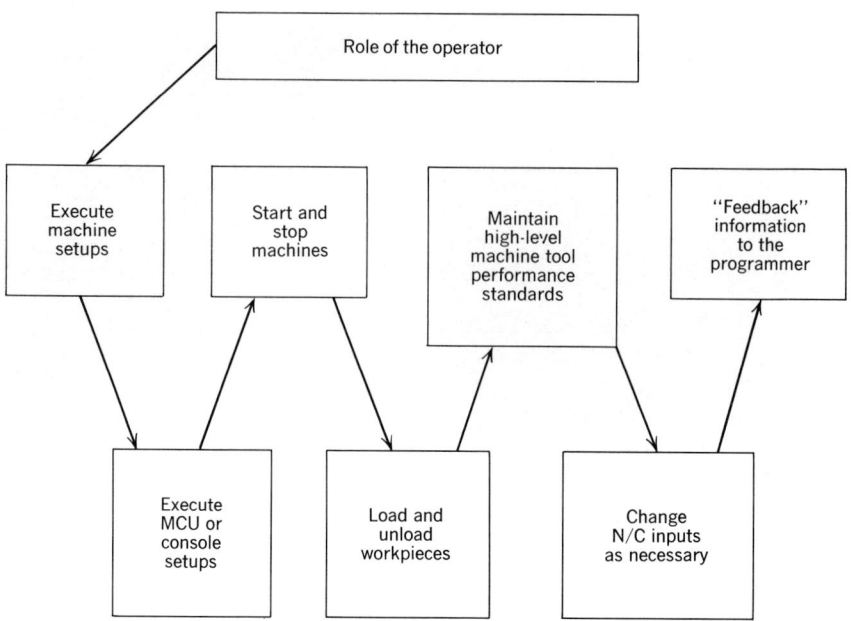

Figure 31. Role of the N/C machine tool operator (Courtesy of the Superior Electric Company).

by a competent machine operator. However, the N/C operator has several important functions to perform (Figure 31). He must know the machine and its operating characteristics.

Numerical Control Systems

CLOSED LOOP SYSTEMS. In a closed loop N/C system, a signal is fed back to the machine control unit confirming the specific instruction (Figure 32). For example, if the MCU instructs a milling machine table to move 10 in., a signal would be fed back to the MCU from a sensor on the drive motor indicating that the table has moved the instructed distance. Closed loop provides the MCU with a check on the accuracy of machine movement.

OPEN LOOP SYSTEMS. No feedback signal is used in an open loop N/C system. The open loop system may use an electric stepping motor to control the movement of machine components. The stepping motor is used on many numerical controls that are added to existing machine tools.

A stepping motor operates on a pulse of electric current supplied by the MCU. Each current pulse causes the motor rotor to turn or "step" a fraction of a revolution. When the motor is coupled to a mill table, lathe cross slide, or lathe leadscrew, it can act to move the screw specific amounts according to the number of pulses received from the MCU. Stepping motors are often designed to move machine tool tables a distance of .001 in. per pulse. They are reversible and can move a component in either direction.

Numerical Control and the Computer

The computer is an important and valuable component of the modern numerical control system. The computer can do mathematics with great speed and accuracy. This has made it a valuable tool in

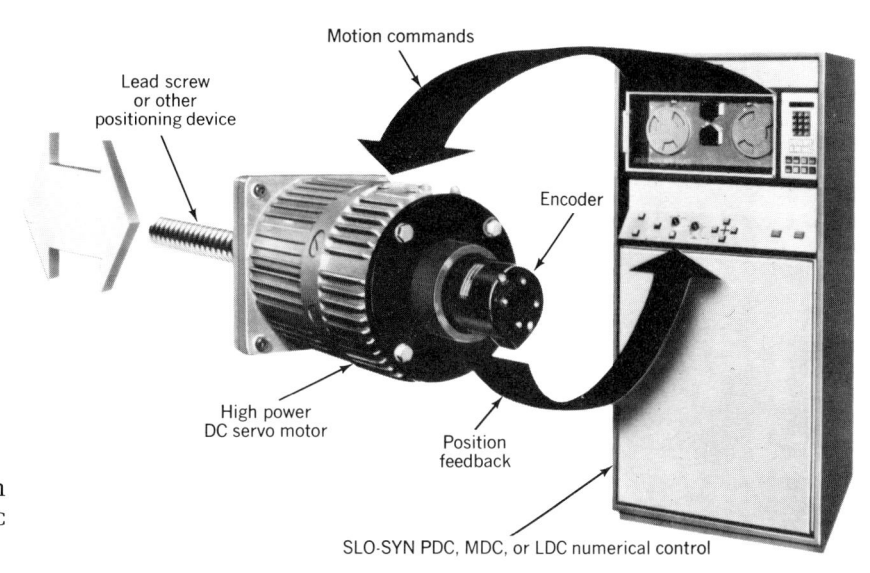

Figure 32. Closed loop N/C system (Courtesy of the Superior Electric Company).

the N/C programming of complex machining operations. Mathematical computation necessary for continuous path machining would be difficult and time consuming if it had to be done by hand. The computer can accommodate this kind of calculation easily.

A computer can understand direct descriptions of workpiece geometry, machine control functions, and machining operations. This permits the N/C programmer to program much in the same way that he would verbally describe the machining task to be done. The computer can understand direct statements such as GO TO, MILL, DRILL, or BORE. These direct descriptions are translated by the computer into appropriate instructions for a specific N/C machine tool.

COMPILER POST-PROCESSOR SYSTEM. Large capacity computers that can accommodate complex N/C programming are expensive. For this reason, they must be put to productive use. A manufacturing concern may use its computer facility for many purposes in addition to N/C programming. Where N/C programming is only a portion of a computer's total work, the "compiler post-processor" system may be used (Figure 33).

A computer programmer first enters a special set of instructions called a "compiler." The compiler is used to direct computer operations when N/C program instructions are entered (Figure 33). N/C instructions to the computer can now be entered in the form of an N/C computer language consisting of abbreviated "words" and punctuation. Workpiece geometry, machine control functions, and machining operations can be described directly. The computer acts on the N/C program instructions and produces an output in the form of magnetic tape or punched computer cards. This information must then be translated into a format suitable for a specific N/C machine tool.

Different types of N/C machine tools use different program formats. An additional set of instructions called a "post processor" is used to translate the computer output into specific instructions in the format required by a specific machine tool. Manufacturers of N/C

Figure 33. N/C complier postprocessor computer system.

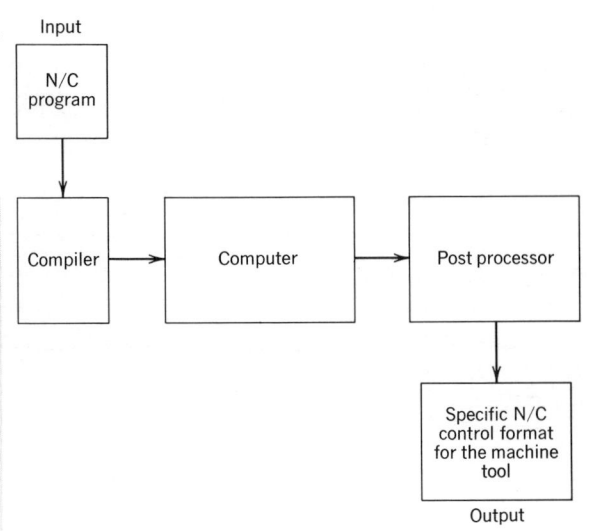

machine tools and numerical controls often provide post processors for their specific machines. Punched N/C tape may be produced directly by the computer and post processor. However, as this is a time consuming operation as compared to the computer time required for the programming operation, actual tape preparation may be done on a separate tape preparation unit.

DIRECT COMPUTER NUMERICAL CONTROL. Modern numerical controls are incorporating the computer into the machine control unit (Figure 34). This permits direct computer control of one or more N/C machine tools. The advantages of direct computer N/C, or C N/C, include machine tool operation directly from programs stored in the computer memory. The tape and tape reader are bypassed. Computer numerical controls also permit tapes to be verified at the machine tool location.

One versatile feature of C N/C is tape editing. With the tape editor, a tape error can be corrected or a change can be made in machining operations or the location of a part feature.

Figure 34. MCU computer numerical control (Courtesy of Cincinnati Milacron).

review questions

1. What is N/C?
2. What are the advantages of N/C?
3. What is a popular N/C program input medium?
4. What are the duties of the N/C machine tool operator?
5. What advantage is the computer in N/C?

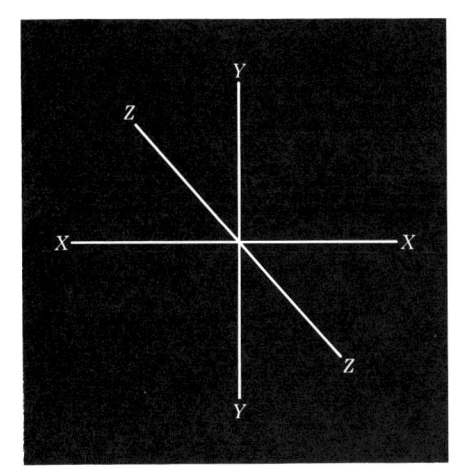

unit 1
numerical control dimensioning

The numerical control programmer studies a drawing of the workpiece and determines the direction and distance that the cutting tool must travel. The programmer then directs the machine movements along these paths by indicating the appropriate numerical instructions. In order to do this, the programmer must be able to define and identify the travel direction of a specific machine component. He must also be able to differentiate travel directions of different machine components. This is the purpose of machine tool axis identifications and N/C dimensioning.

objectives After completing this unit, you will be able to:
1. Describe the application of rectangular coordinates to machine tools.
2. Describe the purpose of rotational axes.
3. Describe the direction of machine spindle movements in a coordinate system.

information

MACHINE TOOL AXES

Basic Axes
The rectangular coordinate system consists of the two perpendicular axes of X and Y (Figure 1). The X and Y axes lie in the same plane and are known as coordinate axes. With the addition of a third axis, Z, that is perpendicular to the X-Y plane, a three-dimensional volume of space can be described and identified (Figure 2). The point at which the axes intersect is called the origin and has a numeric value of zero.

These notations are applied to N/C machine tools in order to identify the basic machine axes. The Z axis is always the spindle axis, even though the machine spindle may be horizontal or vertical. On a typical vertical spindle machine tool, such as a vertical mill, the spindle axis is Z. The X axis is the table and the Y axis is the saddle (Figure 3). The knee is in the Z axis.

On a horizontal spindle machine tool, Z remains the spindle axis while Y becomes vertical and X remains horizontal (Figure 4). The N/C lathe is also a horizontal spindle machine tool. The spindle axis is Z and the cross slide is X, or the horizontal axis perpendicular to Z (Figure 5). Since the lathe tool holder is not moved vertically, the Y axis is not used.

Rotational Axes
Rotational axes define numerically controlled motion around the X, Y, and Z basic axes. An N/C machine tool may have a rotary table or part indexer (Figure 6). These accessories may operate from tape instructions, rotationally around the basic axes. The direction of rotation, as well

535

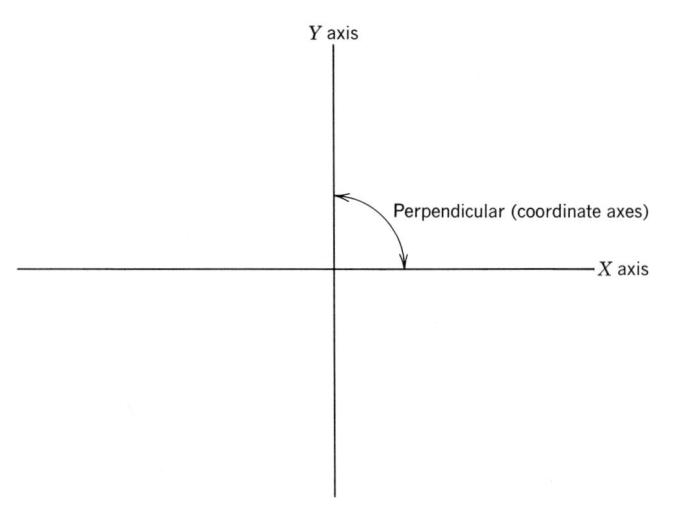

Figure 1. Coordinate axes X and Y.

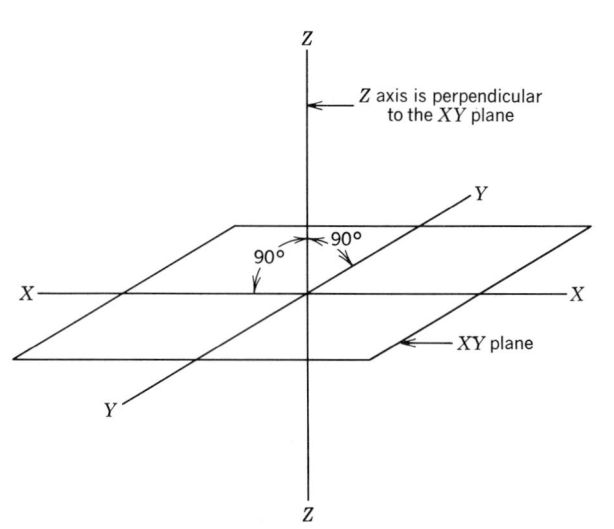

Figure 2. The Z axis is perpendicular to the XY plane.

as the basic axes around which rotation occurs, must be identified. Rotational axes are identified as a, b, and c (Figure 7). Discussions about four and five axis N/C machine tools refer to the basic axes of X, Y, and Z and the rotational axes of a, b, and c.

QUADRANTS

The perpendicular coordinate axes X and Y form four quadrants (Figure 8). Quadrants are numbered in a counterclockwise direction beginning at the upper right. The point of axial intersection or origin has a numeric value of zero. All points to the right of zero along the X axis have positive

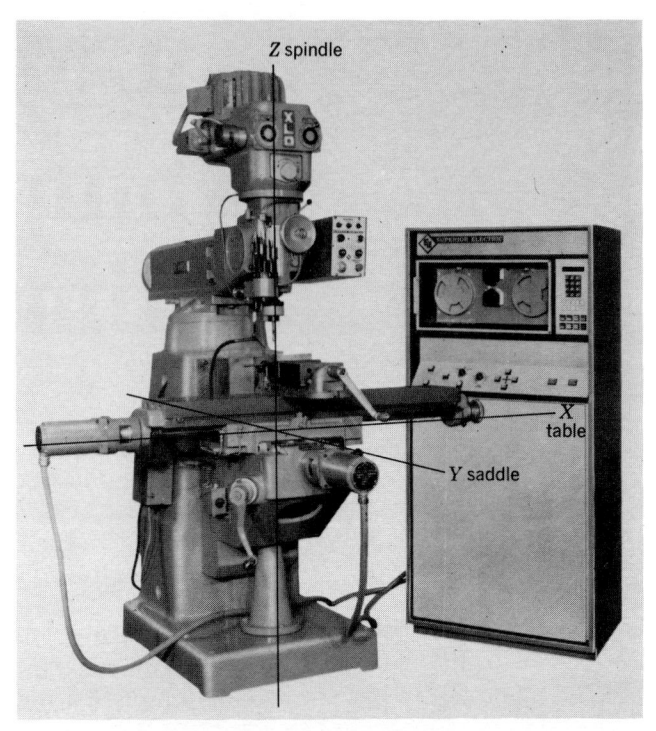

Figure 3. Basic axes of a vertical spindle N/C machine tool (Courtesy of the Superior Electric Company).

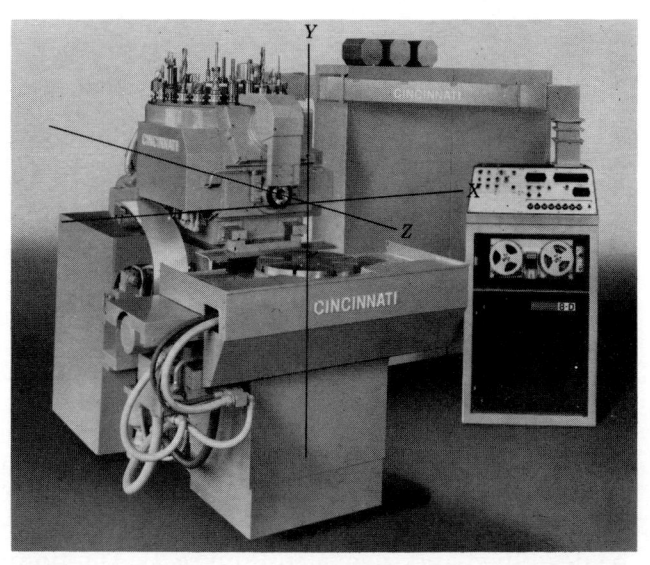

Figure 4. Basic axes of a horizontal spindle N/C machine tool (Courtesy of Cincinnati Milacron).

value. All points to the left of zero have negative values. Points on the Y axis above zero are positive and points below zero are negative.

If the X-Y plane is horizontal, as it is on all

Unit 1 Numerical Control Dimensioning 537

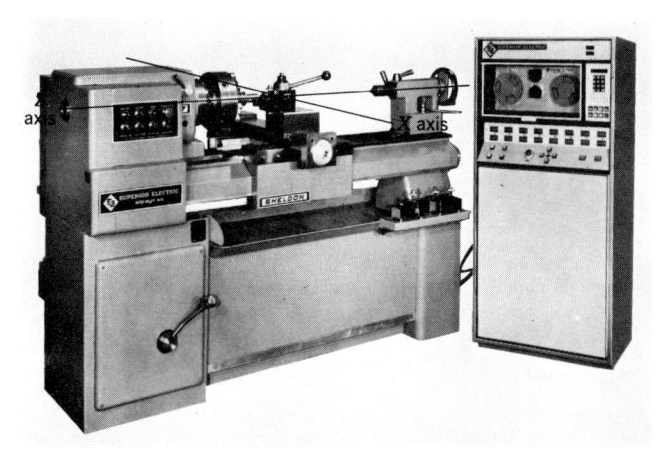

Figure 5. Basic axes of an N/C lathe (Courtesy of the Superior Electric Company).

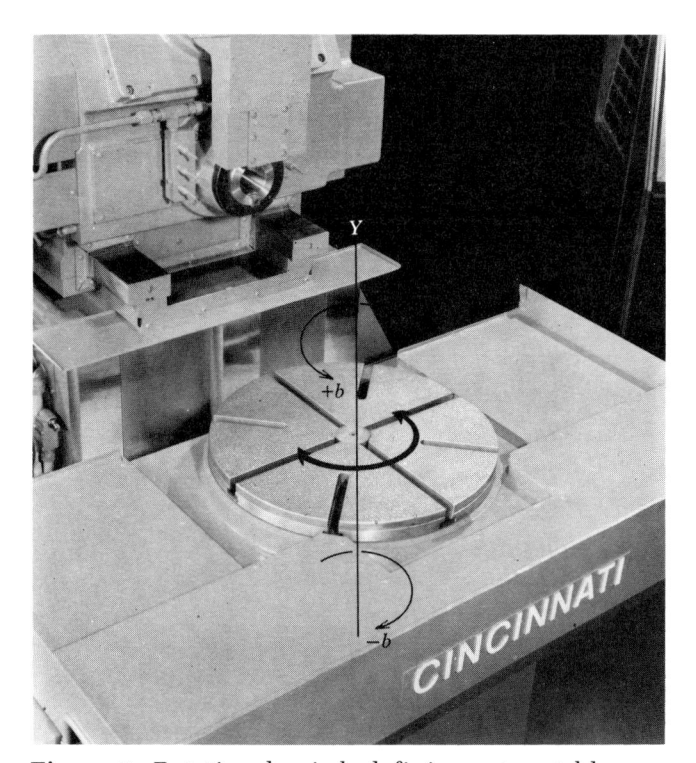

Figure 6. Rotational axis b defining rotary table motion (Courtesy of Cincinnati Milacron).

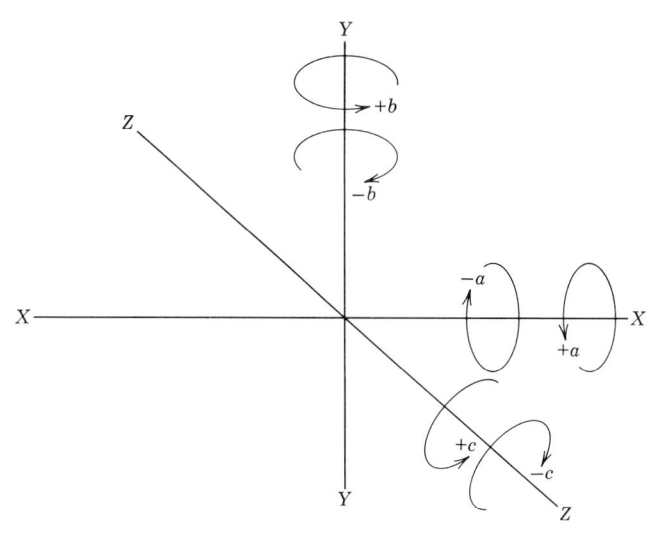

Figure 7. Rotational axes.

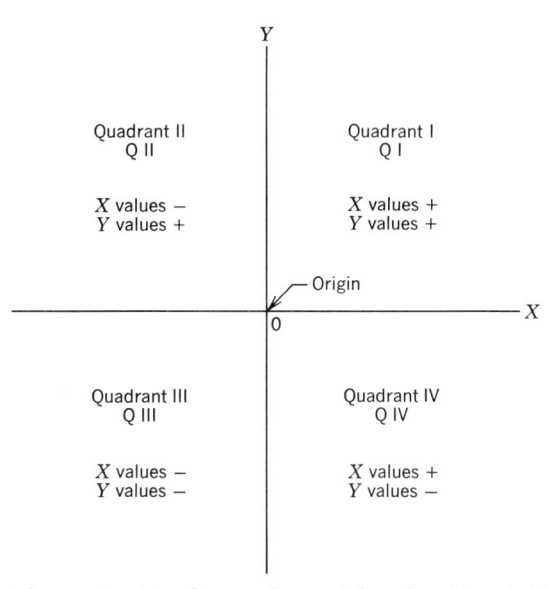

Figure 8. Quadrants formed by the X and Y coordinate axes.

vertical spindle machine tools, there is no geographic location of above and below zero. In this case, points from the origin away from you are positive, while points toward you from zero are negative.

Quadrant Point Values
Point values in the four quadrants are as follows.

Quadrant I: X positive, Y positive.
Quadrant II: X negative, Y positive.
Quadrant III: X negative, Y negative.
Quadrant IV: X positive, Y negative.

DIRECTIONS OF MACHINE TOOL SPINDLE TRAVEL

Understanding quadrant point values is important to the preparation of certain types of N/C tape instructions. The MCU must tell the machine tool spindle or worktable to move in a

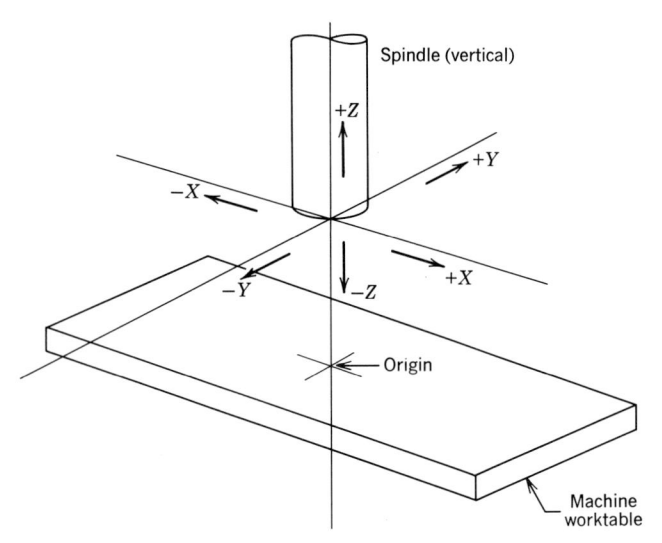

Figure 9. Directions of vertical spindle movement.

certain direction or to a specified location. This may be done by providing tape instructions that indicate positive and negative movement directions.

N/C programming is always done as if the tool were moving, even though the worktable is the moving component on a machine tool with a fixed spindle. The direction of spindle movement is expressed by noting its direction of travel along a specified axis (Figure 9). Spindle movement in the Z axis is also defined in terms of positive and negative directions. If the distance between the worktable and spindle is decreasing, the spindle is moving in a minus (−) Z direction. If the distance between worktable and spindle is increasing, the spindle is moving in a positive (+) Z direction.

SPINDLE POSITIONING BY INCREMENTAL MEASUREMENT

Certain numerical control programs instruct the machine tool to position the spindle by incremental measurement. This means that the spindle measures the distance to its next location from the position at which it was last located. Incremental positioning requires positive and negative travel directions.

EXAMPLE

A certain workpiece is to be drilled on an N/C machine tool positioning by incremental measurement. The workpiece is set up so that the spindle start point is over one corner and the edges of the part are parallel with the coordinate axes. The N/C program instructs the machine spindle to move in a +X direction a distance of 1 in. and in a +Y direction a distance of 4 in. (Figure 10). This will position the spindle over drilling location 1.

To reach the second location, the spindle moves in the +X direction an additional distance of 2 in. However, it must move in a −Y direction a distance of 1 in. to reach location 2. The machine uses location 1 as a new origin from which to measure its movement to location 2. After drilling is completed, it is desired to return the spindle to the start point. This requires a −X move of 3 in. and a −Y move of 3 in. Once again, location 2 becomes a new origin from which to measure the distance back to the start point.

SPINDLE POSITIONING BY ABSOLUTE OR COORDINATE MEASUREMENT

Other types of N/C machine tools position the spindle by absolute or coordinate measurement. Two systems are used.

With fixed zero absolute positioning, all measurements are taken from the same zero reference point located at the lower left corner of the workpiece or worktable. With fixed zero, the need for positive and negative moves is eliminated. For practical purposes, the spindle is always operating in Quadrant I, where all points have positive value. All points are specified as coordinate locations in terms of the distance from the coordinate axes.

EXAMPLE

The same workpiece is to be drilled on a machine positioning by absolute measurement from fixed zero. Each drilling location is expressed as a dimension from the absolute zero point (Figure 11). Coordinate dimensions are measured parallel to the coordinate axes.

Drilling location 1 is at point (1X, 4Y) from zero. This is a coordinate location. Location 2 is at point (3X, 3Y) from zero. The N/C program instructs the spindle to position to coordinate location (1X, 4Y). After drilling the first hole, the spindle is instructed to position to location 2 at coordinate (3X, 3Y). Since this second location is measured from zero and not from the previous location, the machine positions to the new location without the need of movement in a negative direction. Return to the start point is ac-

Unit 1 Numerical Control Dimensioning 539

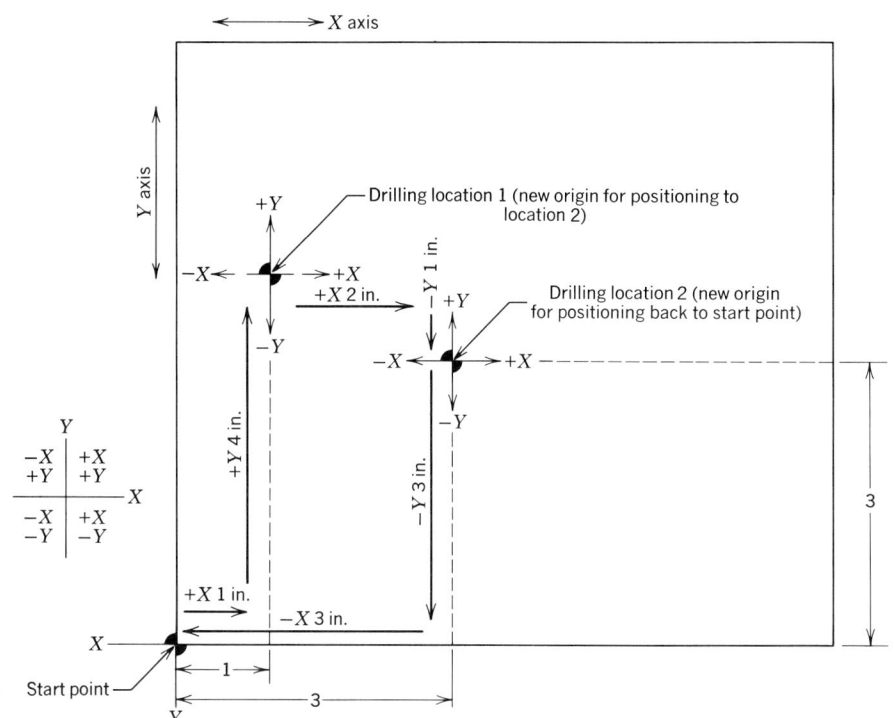

Figure 10. Positioning by incremental measurement.

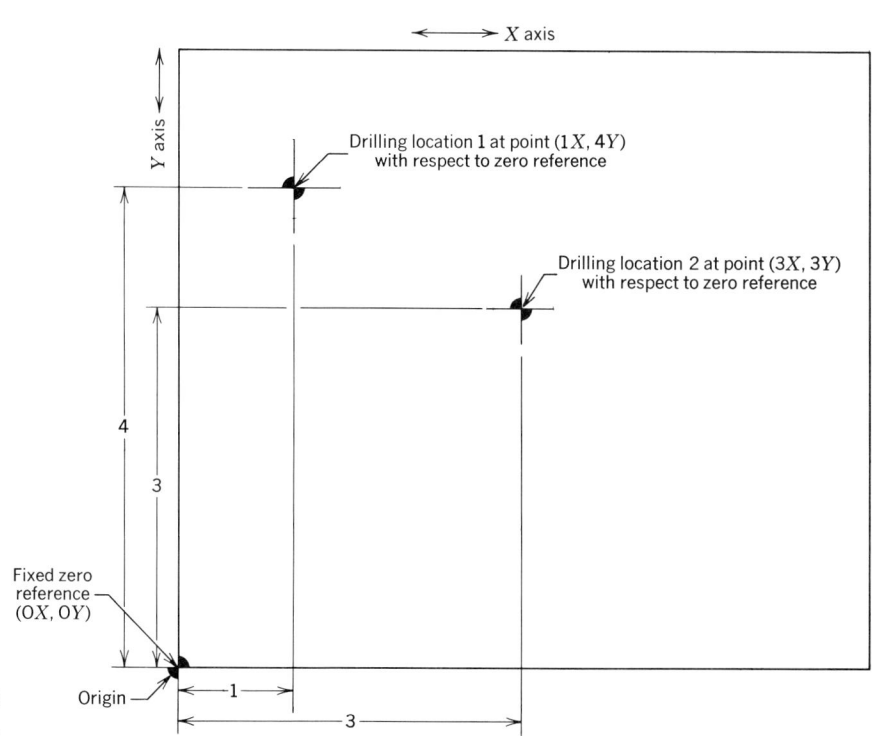

Figure 11. Positioning by fixed zero absolute or coordinate measurement.

complished by instructing the spindle to position to the coordinate location (0X, 0Y).

Some N/C machines permit any point to be established as absolute zero. This is known as floating zero and can be used to make certain programming easier.

EXAMPLE
When drilling a symmetric pattern, it might be

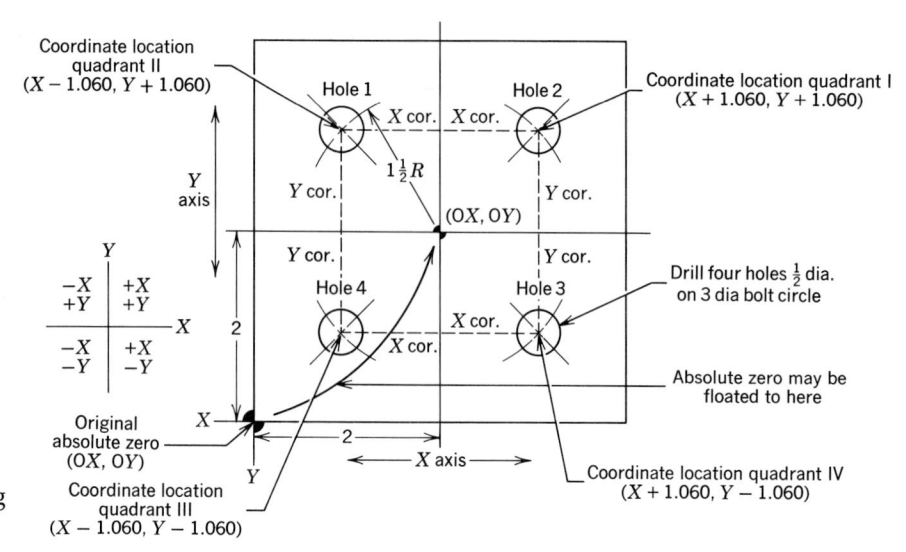

Figure 12. Positioning by floating zero absolute measurement.

more convenient to start the program from a central location. Instead of starting the program over the corner of the part (Figure 12), the absolute zero point may be "floated" to a central location. Tool positioning to the hole locations is still done by specifying coordinate locations. However, holes are to be drilled in all four quadrants. This requires that positive and negative coordinate locations be specified.

When programming for an N/C machine using absolute positioning, drawing dimensions must be expressed in terms of absolute measure from an appropriate zero reference point.

Absolute positioning may be used on N/C machine tools that are designed to position with extreme accuracy. Many N/C machines can position their spindles or worktables within a few ten thousandths of an inch. Small errors that might be cumulative with incremental positioning, do not pose a problem in absolute positioning.

self-evaluation

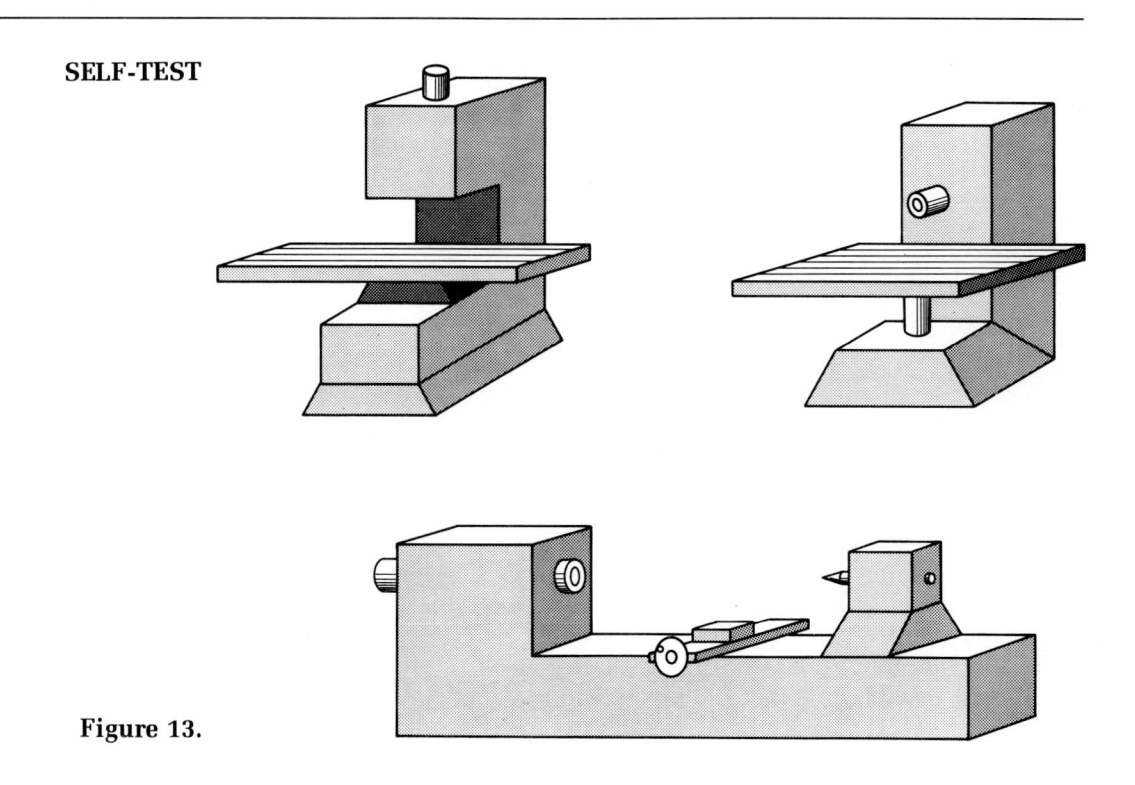

SELF-TEST

Figure 13.

Unit 1 Numerical Control Dimensioning 541

1. Sketch the X, Y, and Z axes on the machine tool outlines (Figure 13).

2. Where might rotational axes be used?

3. An N/C instruction on a drill press instructs the spindle to move toward the workpiece. What is the axis and direction of travel?

4. The worktable on a vertical mill is moving the workpiece to the right of the operator. The saddle is stationary. What is the direction and axis of travel with reference to the machine spindle?

5. The tailstock of an N/C lathe is located in the _____ axis.

This unit has no post-test.

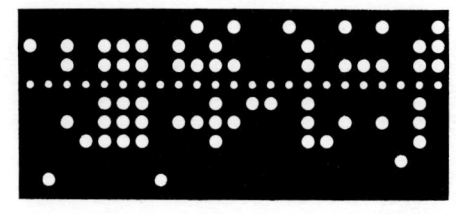

unit 2

numerical control tape, tape preparation, tape code systems, and tape readers

Before advancing to actual N/C programming, you should be familiar with N/C tape materials, tape preparation, tape code systems, and tape readers. N/C programming is sure to involve you in operating a typewriter tape punch unit to produce new tapes, verify programs, duplicate tapes, and correct tape errors. An understanding of tape readers will be useful to your overall understanding of the entire numerical control system.

objectives After completing this unit, you will be able to:
1. Describe and cite advantages of tape materials.
2. Describe the various functions of the tape preparation unit.
3. Name standard tape codes.
4. Recognize types of tape readers appearing on N/C machine tools.

information

NUMERICAL CONTROL TAPES

Tape Materials

Eight channel punched tape (Figure 1) is a very common and popular means by which control instructions are given to an N/C machine tool. N/C tape materials include paper or paper-plastic and aluminum-plastic laminates. Blank tape may be supplied in 1000 or 2000 ft. rolls, depending on thickness. The plastic and aluminum laminate materials are generally more durable than paper. They are more expensive, but they will withstand many trips through the tape reader without wear or damage. The plastic materials are also less subject to damage by oils and grease found around the machine and shop. Manufacturing tolerances are held closely (Figure 2) so that tape may be standardized between different manufacturers.

Paper tape is available in several different

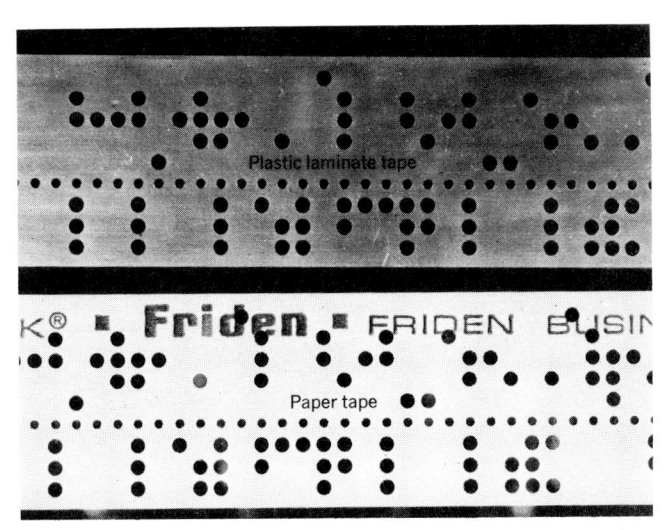

Figure 1. Paper and plastic laminate tape materials.

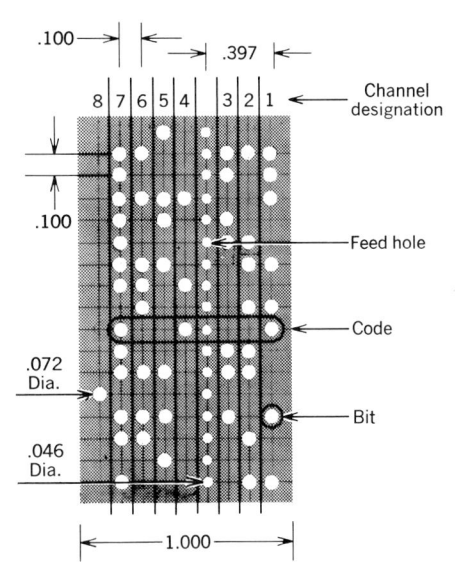

Figure 2. Dimensions of N/C tape features (Courtesy of the Superior Electric Company).

colors, including pink, yellow, and black. The color has no significance except that certain colors are more suitable for use in photoelectric tape readers. In these readers, light passes through the tape punches and activates photoelectric cells. A photoelectric tape reader using paper tape may require a black or dark color so that light will not pass through the tape material and activate improper photocells.

Standard N/C tape is 1 in. wide and contains eight channels for information (Figure 2). A row of sprocket or feed holes also appears on the tape. These are necessary for transport through certain types of tape readers. The line of feed holes is punched off center to eliminate confusion as to how the tape is to be placed in the tape reader.

TAPE PREPARATION

Control information is placed on the N/C tape by punching a specific pattern of holes. This is accomplished on a special typewriter tape punch machine (Figure 3). The typewriter keyboard operates in a similar manner as a standard typewriter. The tape punch typewriter has the same letters and numbers found on a standard typewriter. In addition, several extra symbols are included, as well as control keys for the tape punch.

The tape punch is activated as each typewriter key is depressed. This produces a pattern of holes in the tape that is unique to that typewriter symbol. As the tape is punched, a printed record is typed on paper in the typewriter car-

Figure 3. Typewriter tape punch (California State University at Fresno).

riage. The tape feed key (Figure 4) causes blank tape to feed through the punch. Feed holes are produced during this operation (Figure 5). Blank tape is run out to provide a leader that can be wound on the machine tool tape reader reels.

The typewriter tape punch also has a tape reading head. The reading head is not unlike the tape reader on the N/C machine tool. The function of the typewriter reader is to operate the typewriter from the punched tape. The reader is used to obtain a printed record of a punched tape. This is useful for verifying tape accuracy.

Figure 4. Tape feed key (California State University at Fresno).

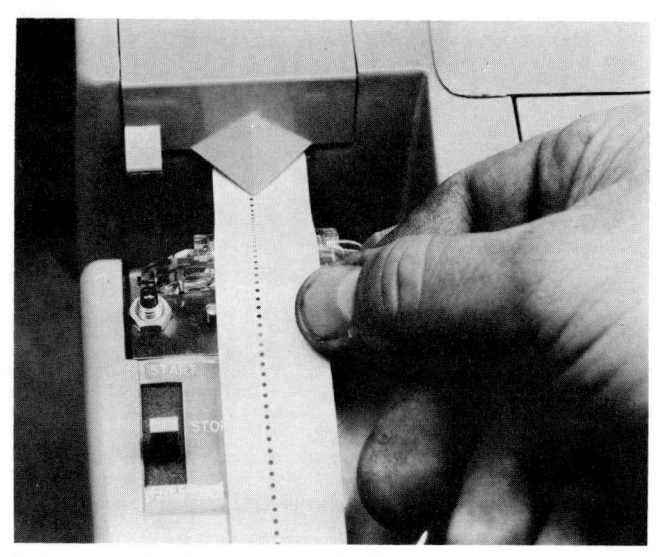

Figure 5. Blank tape with feed holes (California State University at Fresno).

After a tape has been prepared, it may be inserted in the typewriter reader (Figure 6). The tape will now activate the typewriter and a record of the information will be typed out. If there is an error in the tape information, it can be detected and corrected. The typewriter reader is also used to duplicate tapes.

Correcting Tape Errors

If an error in typing is made and detected at that time by the tape punch operator, a correction may be made by pressing the "delete" key. The delete key causes all seven rows on the tape to be punched out. The operator may then retype the correct information and continue with the remainder of the program.

If an error is not detected until the tape is completed and read back using the typewriter reader, a new tape will have to be produced. This can be done by inserting the incorrect tape in the reader and duplicating a new tape to the point of the error. The operator will, of course, have to watch the typed printout and stop the duplicating process when it reaches the last correct entry. The correct information is then typed from the keyboard. The incorrect information on the original tape is advanced through the reader by hand and the duplicating process is resumed.

Punched tape may also be corrected by inserting a splice at the appropriate point. This requires that the feed holes and tape perforations be precisely aligned.

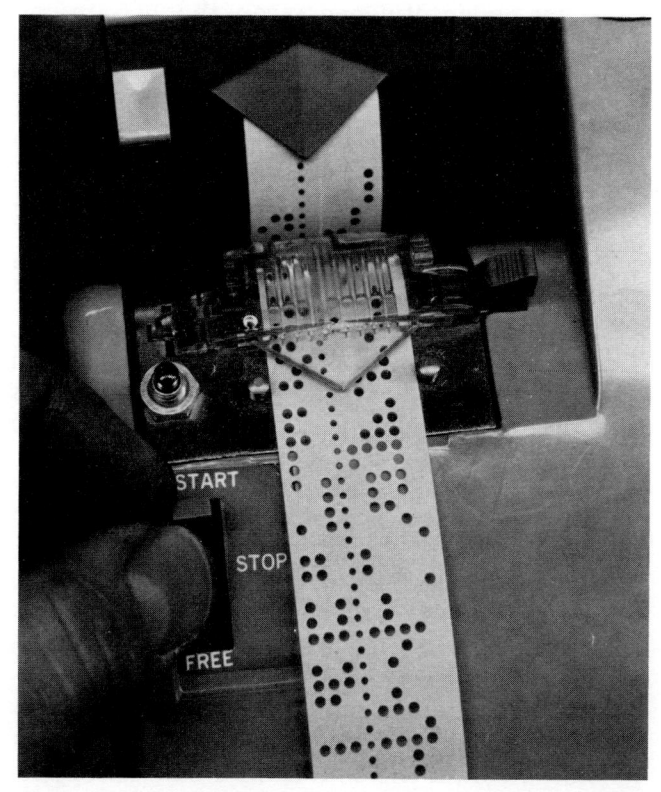

Figure 6. Typewriter tape reader (California State University at Fresno).

TAPE CODE SYSTEMS

Standard systems of tape codes are used throughout industry. One system is the Elec-

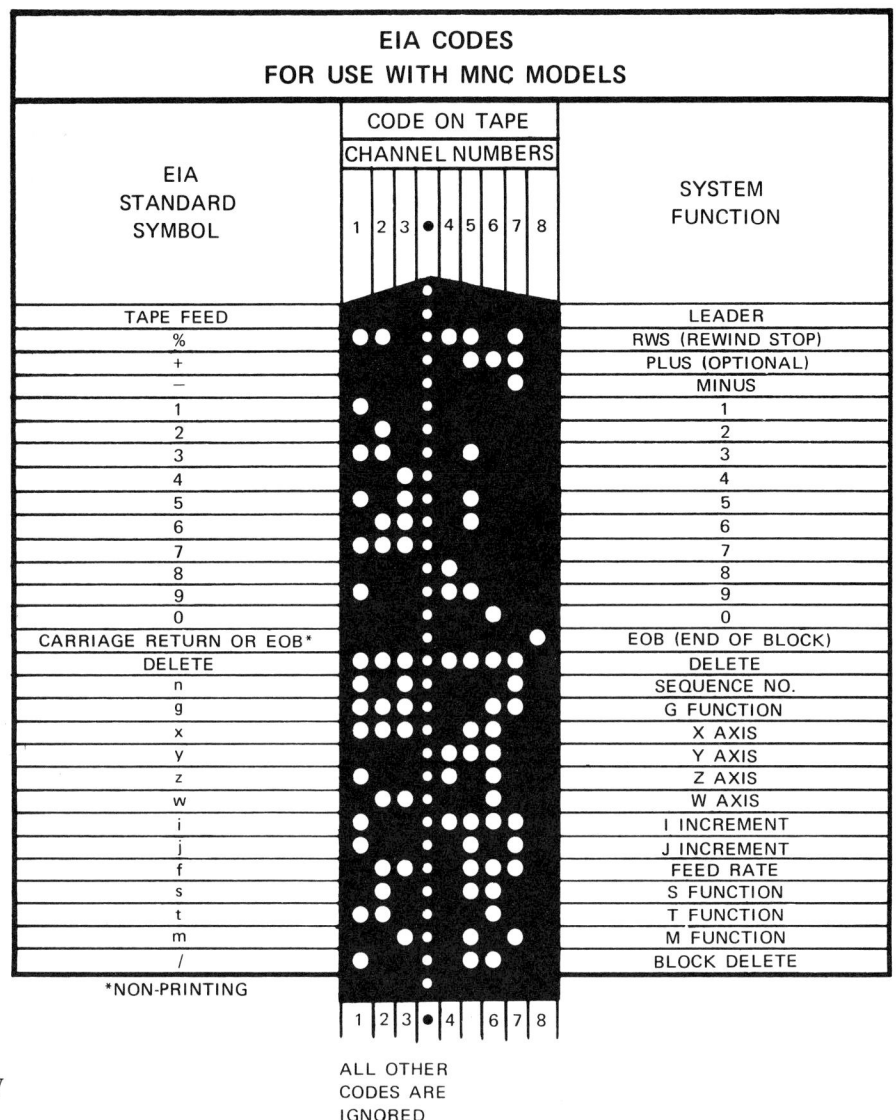

EIA CODES FOR USE WITH MNC MODELS		
EIA STANDARD SYMBOL	CODE ON TAPE CHANNEL NUMBERS 1 2 3 • 4 5 6 7 8	SYSTEM FUNCTION
TAPE FEED		LEADER
%		RWS (REWIND STOP)
+		PLUS (OPTIONAL)
−		MINUS
1		1
2		2
3		3
4		4
5		5
6		6
7		7
8		8
9		9
0		0
CARRIAGE RETURN OR EOB*		EOB (END OF BLOCK)
DELETE		DELETE
n		SEQUENCE NO.
g		G FUNCTION
x		X AXIS
y		Y AXIS
z		Z AXIS
w		W AXIS
i		I INCREMENT
j		J INCREMENT
f		FEED RATE
s		S FUNCTION
t		T FUNCTION
m		M FUNCTION
/		BLOCK DELETE

*NON-PRINTING

ALL OTHER CODES ARE IGNORED

Figure 7. EIA tape codes (Courtesy of the Superior Electric Company).

tronics Industries Association code, known as EIA (Figure 7). Another system is the U.S.A. Standard Code for Information Interchange, known as ASCII (Figure 8). Note that each typewriter symbol has a specific pattern of holes in the tape. Typewriter keys such as "tab" and "carriage return" also produce a specific tape punch code. However, these codes do not produce a printed symbol.

TAPE READERS

The tape reader is usually found in the machine control unit. The MCU may be attached to the machine tool, or it may be freestanding and connected to the machine by appropriate wiring. Tape readers may be electromechanical, photoelectric, or pneumatic. A tape reader consists of the tape reading head and tape transport system. The transport system includes the drive sprocket, tension arms, and tape reels. The tension arms maintain a taut tape as it passes by the reading head. In photoelectric tape readers, the tape may be transported through the reader by a pinch roll instead of by a drive sprocket.

Electromechanical Readers

The electromechanical reader uses electrical contacts that operate through the tape punches. Electromechanical readers are quite fast reading.

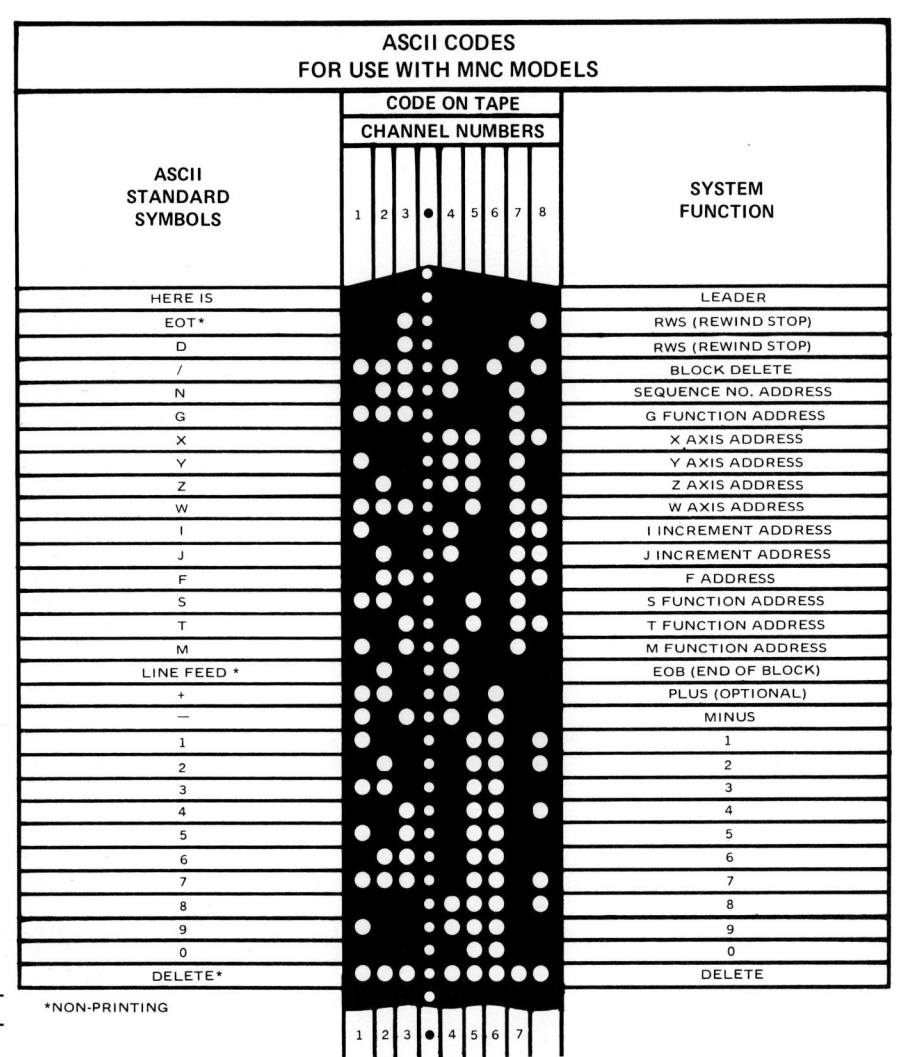

ASCII STANDARD SYMBOLS	CODE ON TAPE (Channel Numbers 1 2 3 • 4 5 6 7 8)	SYSTEM FUNCTION
HERE IS		LEADER
EOT*		RWS (REWIND STOP)
D		RWS (REWIND STOP)
/		BLOCK DELETE
N		SEQUENCE NO. ADDRESS
G		G FUNCTION ADDRESS
X		X AXIS ADDRESS
Y		Y AXIS ADDRESS
Z		Z AXIS ADDRESS
W		W AXIS ADDRESS
I		I INCREMENT ADDRESS
J		J INCREMENT ADDRESS
F		F ADDRESS
S		S FUNCTION ADDRESS
T		T FUNCTION ADDRESS
M		M FUNCTION ADDRESS
LINE FEED *		EOB (END OF BLOCK)
+		PLUS (OPTIONAL)
–		MINUS
1		1
2		2
3		3
4		4
5		5
6		6
7		7
8		8
9		9
0		0
DELETE*		DELETE

*NON-PRINTING

Figure 8. ASCII tape codes (Courtesy of the Superior Electric Company).

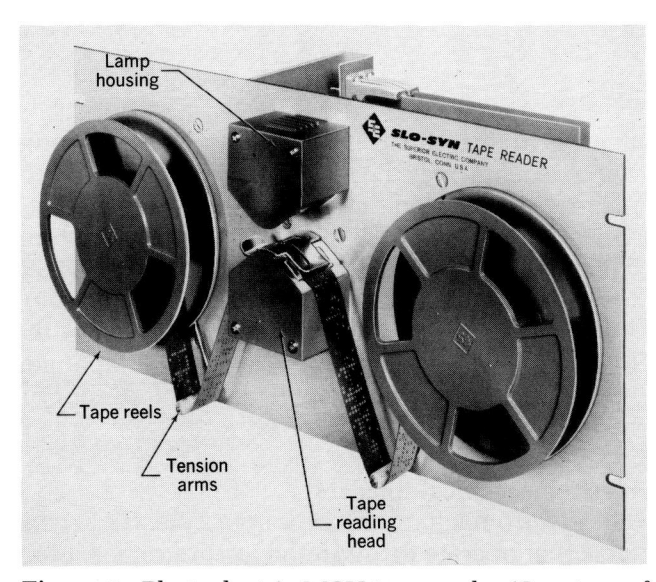

Figure 9. Photoelectric MCU tape reader (Courtesy of the Superior Electric Company).

Photoelectric Readers

The photoelectric reader (Figure 9) uses a concentrated light source that beams light through the tape punches. The light beam activates photoelectric cells. Photoelectric readers are fast reading and are used on numerical controls designed for continuous path machine tools. Continuous path machining may involve many small cuts to approximate a radius, angle, or irregular shape. Instructions for each move must be provided from tape instructions. If the machine tool has to wait for the tape reader to read a tape instruction, time will be lost, resulting in lowered productivity.

To reduce waiting time while instructions are read from the tape, the tape reader reads and stores one instruction ahead. The MCU may be equipped with a memory for this purpose. This is called a buffer storage. While the machine is performing its cut, the tape reader is reading and storing the next instruction. When the cut is completed, the next instruction is instantly available from buffer storage. In this way, the machine does not have to wait for the tape reader to read the next instruction from the tape.

Pneumatic Readers
The pneumatic tape reader uses air flowing through the tape punches to activate electromechanical switches. Pneumatic readers are slower reading than the photoelectric or electromechanical types. They also depend on a precise alignment of the tape over the reader air jets.

self-evaluation

SELF-TEST 1. Name two common tape materials.

a. _____

b. _____

2. What are the advantages of the laminate materials?

_____

3. What are the various functions of the typewriter tape punch?

_____

4. Name two standard tape code systems.

a. _____

b. _____

5. Name three types of tape readers.

a. _____

b. _____

c. _____

EXERCISE Familiarize yourself with the operation of the typewriter tape punch unit in your shop.

This unit has no post-test.

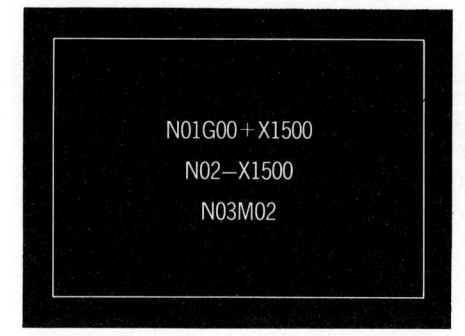

unit 3

basic format numerical control programs

```
N01G00+X1500
N02-X1500
N03M02
```

When you begin to program for N/C, you will need to study the part drawing to determine the control functions and machine positioning information normally provided by the operator. You will then translate this information into a format that can be understood by the machine control unit. N/C machine tools vary widely in capability. A drill press, for example, may only have table positioning in the X and Y axes. The quill may have the capability of cycling from tape instructions. Other functions such as spindle speeds and drilling depth control may be set and varied by hand.

With a multipurpose machining center, a great many more functions may be controlled from tape instructions. These include setting spindle speeds and milling feedrates, indexing drill stop turrets, workpiece indexing, positioning of machine components in three, four or five axes, and instructions to perform continuous path cuts in three dimensions. N/C programming for this type of machine tool is necessarily more complex and may require the use of a computer and special vocabulary to describe workpiece geometry.

objective After completing this unit, you will be able to:
1. Recognize basic format N/C programs.

information

The first step in N/C programming is to study the workpiece drawing and prepare a handwritten manuscript in the program format for your specific machine tool. In order to do this, you will have to have a knowledge of tooling and machining practice so that you can specify workholding fixtures, cutting tools, feeds, and speeds.

The manufacturers of numerical controls and N/C machine tools provide programming manuals detailing exact programming procedures. You should obtain a programmer's manual for your machine if it is available. Printed program forms are also available in the specific formats. These are very helpful in constructing N/C programs.

MISCELLANEOUS FUNCTIONS

Basically, the N/C program must position the spindle or worktable and then instruct the machine to accomplish a desired machining task.

If, for example, the task were milling, the next logical step after positioning would be to turn on coolant and lower the cutting tool in order to begin the cut. Lowering the spindle and turning on coolant are miscellaneous functions that would normally be provided by the operator on a conventional machine tool.

With numerical control, it is desired to control as many miscellaneous functions as possible by tape instructions. Therefore, specific tape codes are used to indicate specific miscellaneous or "m" functions. N/C machine tools, depending on capability, will make use of only a few or many m functions. An m function is noted by the letter m followed by a numeric code. Some of the common m functions are:

m01—index.
m02—end of program (tape rewind).
m06—tool change.
m08—coolant on.
m09—coolant off.
m52—quill down.
m53—quill up.
m55—override milling feedrate.

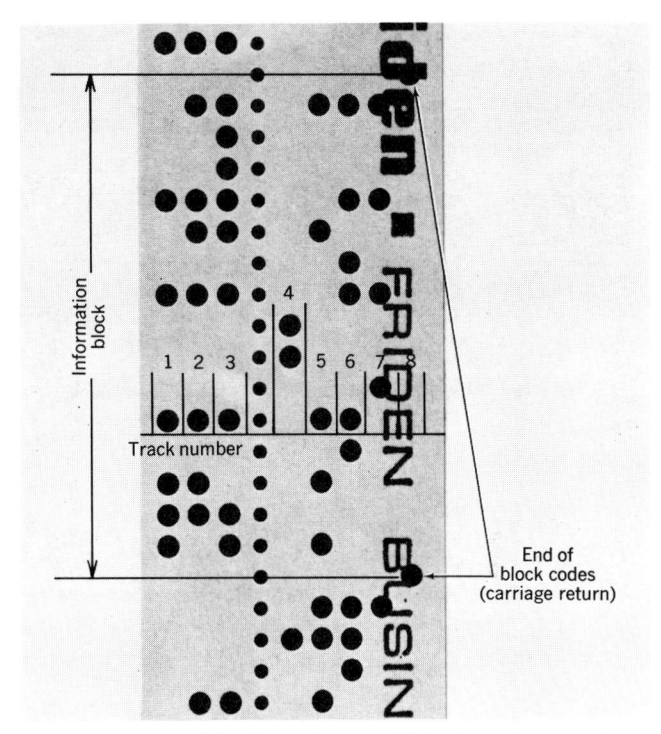

Figure 1. Tape block and end of block code punches.

TAPE BLOCKS AND END OF BLOCK CODE

Information is placed on the N/C tape in a block. The block may contain several lines of information or only a few lines. The MCU tape reader reads a block and then acts on the information contained therein. Blocks are separated by the end-of-block code, or EOB code. This is the code that is punched in the tape when the carriage return key is pressed on the typewriter tape punch. The EOB code is a single punch in the number eight tape channel (Figure 1). In some types of N/C program formats, the length of the block may vary. In other formats, each block is the same length.

TAPE REWIND STOP CODE

The tape rewind stop code is usually the first code on the tape. When the end of a tape run is reached, the MCU reader will read an m02 miscellaneous function instruction, indicating end of program. This will rewind the tape in preparation to begin a new run. It is desirable to stop the rewind before the tape winds completely off the reader reels. When the stop rewind code reaches the reading head, rewind is stopped.

TAB SEQUENTIAL FORMAT PROGRAMS

Tab sequential is one of the common basic format N/C programs. It is popular for point-to-point N/C applications such as drilling. Tab sequential is also useful for milling in axis or out of axis at 45 degree angles. With an interpolation feature, tab sequential can be used in a continuous path contouring program.

In tab sequential the code produced on the tape when the typewriter is tabbed is used to separate axial positioning and m function information. This permits the electrical sections of the MCU to tell the difference between X axis positioning, Y axis positioning, Z axis positioning, and m functions. Information in tab sequential is set in a specific sequence. The positioning or machining operation step or sequence number appears first, followed by a tab code separating it from X axis positioning information, which appears next. A second tab code separates X positioning from Y positioning, and a third tab separates Y positioning from m functions. If Z posi-

Figure 2. Machine control unit console and tape reader (Courtesy of the Superior Electric Company).

tioning or rotational axis positioning is required, these are also separated by tab codes.

EXAMPLE 1. TAB SEQUENTIAL PROGRAM FOR DRILLING. Drilling is a typical point to point operation. The drill positions to a point, drills, and then positions to another point. The start point for a given program may be established at any appropriate point. This is usually done off the workpiece to facilitate inserting new parts. The machine table or spindle may be positioned over the start point by manual control from the MCU console (Figure 2).

In the example shown (Figure 3) a simple single axis drilling operation is accomplished with a tab sequential program. As you can see, tab sequential involves positive and negative moves. The positive and negative travel in any axis must add to zero if the tool is to return to the starting point after each tape·run. In the program, you will note that the total travel in the +X axis amounted to 4 in. In the program sequence four, the tool moves back to the start point a distance of 4 in. along the −X axis. The sum of positive and negative travel adds to zero, indicating that the tool has returned to the start point. Sequences are explained in the program notes.

EXAMPLE 2. TAB SEQUENTIAL PROGRAM FOR DRILLING INVOLVING A TOOL CHANGE. In this example (Figure 4) a tool change is required. The m06 instruction is indicated in the m function column and is separated by tab codes from the axial positioning information. When the block containing the m06 code is read, the program will stop and a tool change indicator light will appear on the MCU. The operator will then change the tool and restart the tape run. The program sequences are explained in the program notes.

EXAMPLE 3. TAB SEQUENTIAL PROGRAM USING A THIRD AXIS. An m54 code indicates that the information after the second tab is instructions for a third axis control function. A third axis function would normally control some machine component in the spindle or Z axis. On a vertical milling machine, for example, a Z axis control function might raise or lower the knee or quill. On some numerical controls, the third axis control can be used to operate auxiliary equipment. In this example (Figure 5) the third axis control function is used to operate a rotary table. The rotary table is driven by a stepping motor. Table motion and other program sequences are explained in the program notes.

EXAMPLE 4. TAB SEQUENTIAL PROGRAM FOR MILLING. Milling involves positioning of the tool or workpiece and the entry of the cutter into the work, followed by further positioning at appropriate feedrates (Figure 6). The milling feedrate may be set manually or controlled from the tape, if the N/C machine tool has this capability. In this example, the milling feedrate is preset from the MCU console. In order to save time during the initial positioning, the m55 code is used to override the milling feedrate in one block only. This permits the tool to position rapidly to the location where milling will begin. The rate of tool entry into the work can also be regulated. A N/C milling machine will have features that permit the cutter to approach the workpiece at a rapid traverse rate and then enter at an appropriate feedrate.

MIRROR IMAGES

A mirror image is a reversed duplicate of the workpiece (Figure 7). Some numerical controls permit the directions of travel to be reversed electrically from the MCU. When mirror image is activated, all positive direction movement becomes negative and all negative becomes positive. Right-handed and left-handed parts can be produced by using the mirror image feature.

LINEAR AND CIRCULAR INTERPOLATION

For a cutting tool to move along an angular, radius, or other continuous path, the axis drive motors must be able to operate at different rates of speed. Linear interpolation is the control of a travel rate in two directions that is proportional to the distance traveled. Linear interpolation permits an angle to be closely approximated when using stepping motors. Circular interpolation permits machining of circles of radii.

Tab sequential format can be used with a numerical control having the interpolation capability. Two additional tab codes are used (Figure 8). X and Y positioning information appears in the same manner as in point-to-point programs. The i and j coordinates are next, with m functions following the fifth tab code. Information for the continuous path is provided by the i and j coordinates. Sequences are explained in the program notes.

WORD ADDRESS FORMAT PROGRAMS

The work address format uses a single letter code A to Z to denote an address. Word address differs from tab sequential in that the differentiation of control information is determined by the single letter address instead of a specific sequence separated by tab codes.

Address Words

Generally N/C machine tools making use of word address will have added capabilities. Common letter words include:

X for X axis distance and direction (+ and −).
Y for Y axis distance and direction (+ and −).
Z for Z axis distance and direction (+ and −).
m for miscellaneous functions.
g for preparatory functions.
n for program sequence numbers.
f for feedrates.

Specific instructions are indicated by adding a numeric code to the letter address word. For example, m02 is an end-of-program instruction, just as it was in tab sequential. However, in word address, the letter is typed with the numeric code.

Preparatory Functions

Preparatory or "g" functions call up specific machine cycles. An example of this is a g81 instruction. This would cause a milling machine or drill press quill to cycle for a drilling operation. A g85 instruction is a bore cycle that will withdraw the boring bar at the boring feedrate, thus preventing tool marks on the workpiece.

EXAMPLE 1. WORD ADDRESS PROGRAM FOR DRILLING. In word address, the sequence number may be indicated by the letter n followed by the appropriate digits. The stop rewind entry can be given a sequence number if desired. This is not necessary, but it is sometimes indicated by the sequence n00. Be sure to type zeros and not the letter o in place of a zero.

In the sample drilling operation (Figure 9) it is desired to move the tool over the drilling location as quickly as possible. Therefore, a rapid traverse feedrate is programmed into the f address. The rate in the example is f1800, meaning a feedrate of 180 IPM. The positioning feedrate can be the maximum rapid traverse rate that the machine tool has available. The remaining sequences are explained in the program notes. When the tool returns to the start point, the g81 drill cycle established in the first sequence will be canceled by the m02 end of program instruction. This will prevent unnecessary tool actuation over the start point.

EXAMPLE 2. WORD ADDRESS PROGRAM FOR MILLING. In this example (Figure 10) the tool is positioned and lowered into the workpiece. A milling feedrate of 6 IPM is programmed in sequence n2 by the f6 address. The tool is lowered into the workpiece by the m52 code. Sequences eight through twelve are the final cuts at an increased feedrate programmed by f120 in sequence n8.

EXAMPLE 3. WORD ADDRESS PROGRAM USING A THIRD AXIS. When the third axis is used, information is programmed into the z address (Figure 11). In this case, the third axis is rotary table control. Program sequences are explained in the program notes.

EXAMPLE 4. WORD ADDRESS PROGRAM FOR A CONTINUOUS PATH. The contouring program (Figure 12) uses the i and j information to indicate circular interpolation. Program sequences are explained in the program notes.

Word Address Continuous Path Machining by Incremental Steps

The preparatory function g01 can be used to call

FOCUS
A. Identification of a starting point.
B. Basic drilling procedure.
C. Automatic tool function procedure.
D. Tool movement — x axis only.

Note:
1. Machine operation emphasis is drilling.
2. The starting point is located to the left of the part. This point has been selected to allow for loading or unloading the part with minimum interference from the tool and holding fixtures.
3. Basic drilling procedure prescribes the center-drilling of each hole prior to drilling with 1/2" drill.

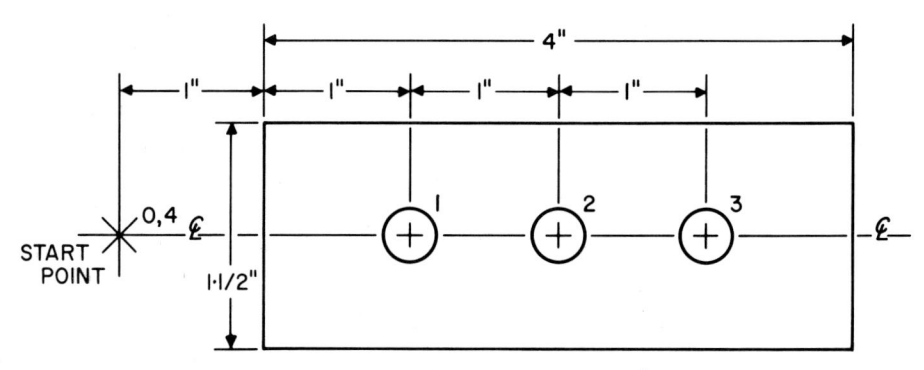

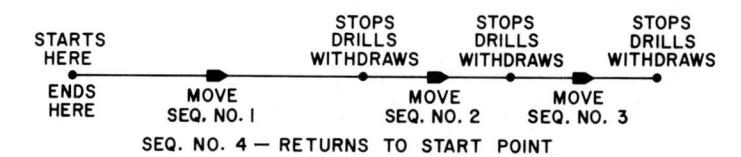

Figure 3. Example 1. Tab sequential program for drilling (Courtesy of the Superior Electric Company).

Program Notes:

1. Console presets
   a. Tool switch—auto position: Tool will drill and retract automatically at each position.
   b. Feed Rate—Hi: Tool will position at "Hi" rate of speed.
   c. Backlash—#2 or #3: Promotes precise positioning by compensating for lead-screw backlash.
2. Program begins with E. O. B. code.
3. Program statements are preceded by RWS code. Rewind stop (RWS) is an instruction to the tape reader to stop at this point after rewinding, thus enabling recycling of the program.
4. Sequence No. 1 statement—prescribes the tool movement on the x axis from the starting point to hole (1). The hole will be center-drilled automatically.
5. Sequence No. 2 statement—prescribes the tool movement on the x axis from hole (1) to hole (2). The second hole will be center-drilled automatically.
6. Sequence No. 3 statement—prescribes the tool movement on the x axis from hole (2) to hole (3). The third hole will be center-drilled automatically.
7. Sequence No. 4 statement—prescribes the tool movement on the x axis from hole (3) back to the starting point and instructs the reader to rewind the tape. Machine will not "drill" at starting point.
8. Center-drill must be replaced with 1/2" drill.
9. Sequences No. 1 through 4 will be repeated. Replace 1/2" drill with center-drill for the next cycle.
10. For this sample, all tool positioning data are recorded in the "x" increment column only.

SLO-SYN™ NUMERICAL TAPE CONTROL PROGRAM

COMPANY NAME		ADDRESS	
PREPARED BY / DATE	PART NAME	PART NO. Sample Drilling	OPER. NO.
CK'D BY / DATE	Plate	Program No. 1	
SHEET OF	REMARKS: Tool Switch—Auto Tools: Center Drill		
DEPT	Feed Rate—Hi 1/2 Dia. Drill		
TAPE NO.	Backlash—No. 2 or No. 3 Run Program For Each Tool		

SEQ. NO.	TAB OR EOB	+ OR −	"x" INCREMENT	TAB OR EOB	+ OR −	"y" INCREMENT	TAB OR EOB	"m" FUNCT	EOB	INSTRUCTIONS
									EOB	
0	RWS								EOB	Change Tool, Load, Start
1	TAB		2000						EOB	
2	TAB		1000						EOB	
3	TAB		1000						EOB	
4	TAB	−	4000	TAB			TAB	02	EOB	

Section I Numerical Control of Machine Tools

FOCUS
   A. Programming for parts requiring several size holes

Notes:
1. Machine operation is drilling.
2. Program will include two tool changes. All holes will first be drilled to 3/16" diameter. The 5/16" diameter holes will then be finish drilled, followed by the 1/4" diameter holes.

Program Notes:
1. Console presets
   a. Tool switch at Auto.
   b. Feed Rate switch at Hi.
   c. Backlash switch at #2.
2. Sequence 1 through 12 statements move the tool to each location where a hole is required. A 3/16" diameter hole is drilled at each point.
3. Sequence 13 statement returns the tool to the starting point. The 06 (tool change) code will stop the control and light the Tool Change lamp. The instruction column tells the operator to change to a 5/16" diameter drill at this point.

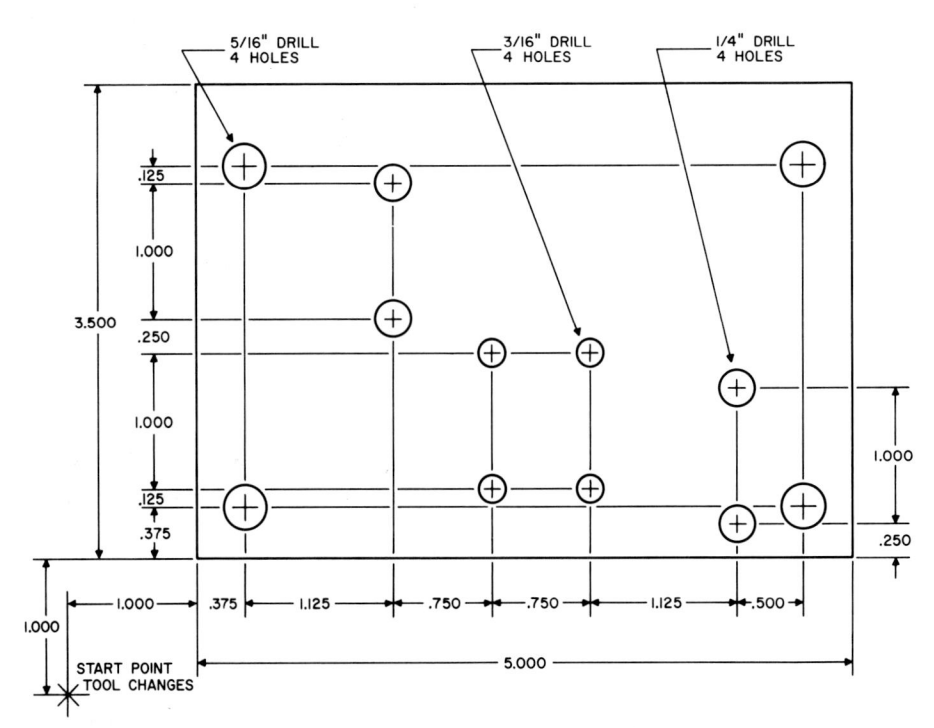

Figure 4. Example 2. Tab sequential program for drilling involving a tool change (Courtesy of the Superior Electric Company).

Unit 3 Basic Format Numerical Control Programs

4. Sequence 14 through 17 statements contain the information needed to position the tool and drill each of the four 5/16" diameter holes. Sequence 18 statement returns the tool to the starting point and calls for another tool change — this time to a 1/4" diameter drill.
5. Sequence 19 through 22 statements position the tool to the locations requiring 1/4" diameter holes where the holes are drilled automatically. Sequence 23 statement returns the tool to the starting point and initiates the tape rewind mode.

SLO-SYN™ NUMERICAL TAPE CONTROL PROGRAM

COMPANY NAME _____ ADDRESS _____

PREPARED BY DATE	PART NAME	PART NO. Sample Drilling Program No. 5	OPER. NO.
CK'D BY DATE			
SHEET OF	REMARKS: Tool Switch—Auto	Tools: 3/16" Drill	
DEPT.	Feed Rate—Hi	5/16" Drill	
TAPE NO.	Backlash—No. 2	1/4" Drill	

SEQ. NO.	TAB OR EOB	+ OR −	"x" INCREMENT	TAB OR EOB	+ OR −	"y" INCREMENT	TAB OR EOB	"m" FUNCT	EOB	INSTRUCTIONS
									EOB	Load Part
0	RWS								EOB	3/16" Stub Drill
1	TAB		1375	TAB		1375			EOB	Start Program
2	TAB		1875	TAB		125			EOB	
3	TAB		750						EOB	
4	TAB		1125	TAB	−	250			EOB	
5	TAB		500	TAB		125			EOB	
6	TAB	−	500	TAB		875			EOB	
7	TAB		500	TAB		1625			EOB	
8	TAB	−	1625	TAB	−	1375			EOB	
9	TAB	−	750						EOB	
10	TAB	−	750	TAB		250			EOB	
11	TAB			TAB		1000			EOB	
12	TAB	−	1125	TAB		125			EOB	
13	TAB	−	1375	TAB	−	3875	TAB	06	EOB	Tool Change, 5/16" Stub Drill
14	TAB		1375	TAB		1375			EOB	
15	TAB		4250						EOB	
16	TAB			TAB		2500			EOB	
17	TAB	−	4250						EOB	
18	TAB	−	1375	TAB	−	3875	TAB	06	EOB	Tool Change 1/4" Stub Drill
19	TAB		2500	TAB		2750			EOB	
20	TAB			TAB		1000			EOB	
21	TAB		2625	TAB	−	1500			EOB	
22	TAB			TAB	−	1000			EOB	
23	TAB	−	5125	TAB	−	1250	TAB	02	EOB	

FOCUS
  A. Programming for a three-axis control

Notes:
  1. Machine operation is drilling.
  2. Program will be run using a three-axis control with a rotary table for the third axis.

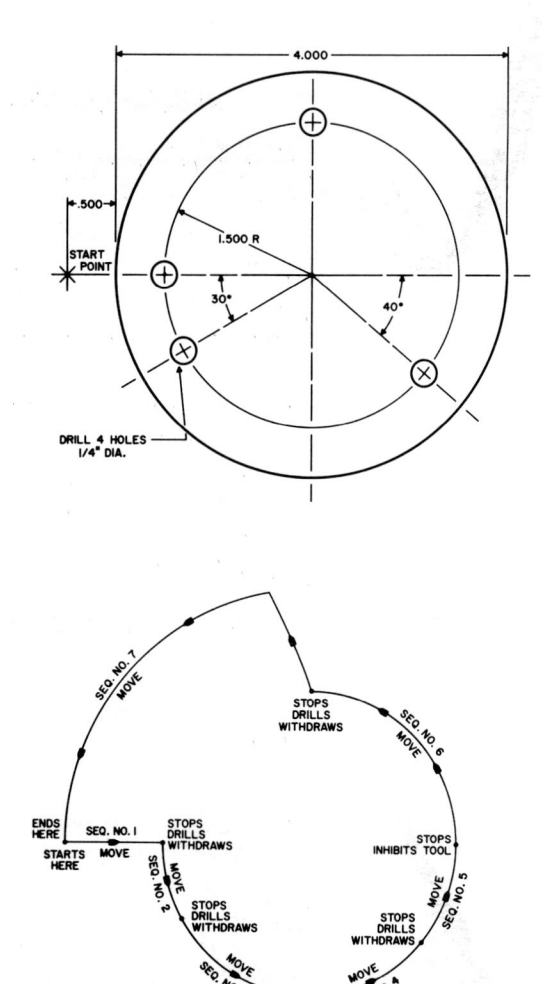

Figure 5. Example 3. Tab sequential program using a third axis (Courtesy of the Superior Electric Company).

Program Notes:

1. Console presets
   a. Tool switch at Auto
   b. Feed Rate switch at Hi
   c. Backlash switch at #2
2. Sequence 1 statement moves the tool from the starting point to the first hole where the hole is drilled automatically. Tool motion is along the x axis only.
3. Sequence 2 statement calls for a tool movement along the third (rotary) axis of 30 degrees. A 54 code is entered in the M function column to inform the control that the numerical information following the second tab code is for the third axis. The hole will be drilled automatically.
4. Each motor step moves the rotary table .01 degrees. Therefore, the maximum rotary movement that can be done in one block is 99.99 degrees. Since the angular increment to the next hole is 110 degrees, the movement must be done in two blocks of information. Sequence 3 statement accomplishes the first portion of this tool motion (60 degree) and contains a 54 (third axis) code and 56 (tool inhibit) code. Since the leading 5 on these codes need only be entered once, the codes are written as 546.
5. Sequence 4 through 6 statements move the tool to the remaining hole locations where the holes are drilled automatically. A tool inhibit code is used again in sequence 5 where the rotary tool motion must be done in two steps.
6. Sequence 7 statement returns the tool to the starting point and rewinds the tape. Please note the "x" axis increments add to zero whereas the rotary axis increments add to 36,000, equivalent to a complete revolution.

SLO-SYN™ NUMERICAL TAPE CONTROL PROGRAM

COMPANY NAME _____ ADDRESS _____

PREPARED BY DATE		PART NAME		PART NO. Sample Drilling Program No. 7	OPER. NO.
CK'D BY DATE					
SHEET OF		REMARKS: Tool Switch—Auto Feed Rate—Hi Backlash—No. 2		Tool: 1/4" Drill	
DEPT.					
TAPE NO.					

SEQ. NO.	TAB OR EOB	+ OR −	"x" INCREMENT	TAB OR EOB	+ OR −	"y" INCREMENT	TAB OR EOB	"m" FUNCT.	EOB	INSTRUCTIONS
									EOB	
0	RWS								EOB	
1	TAB		1000						EOB	
2	TAB			TAB		3000	TAB	54	EOB	
3	TAB			TAB		6000	TAB	546	EOB	
4	TAB			TAB		5000	TAB	54	EOB	
5	TAB			TAB		4000	TAB	546	EOB	
6	TAB			TAB		9000	TAB	54	EOB	
7	TAB	−	1000	TAB		9000	TAB	0254	EOB	

FOCUS
   A. Basic procedure for milling — same basic part as in Sample Milling Program #1, but part will be located differently on table.

Notes:
1. Machine operation is milling.
2. Tool will be advanced and retracted from tape.
3. Hi feed will be called from tape.
4. Backlash circuit will be inoperative.

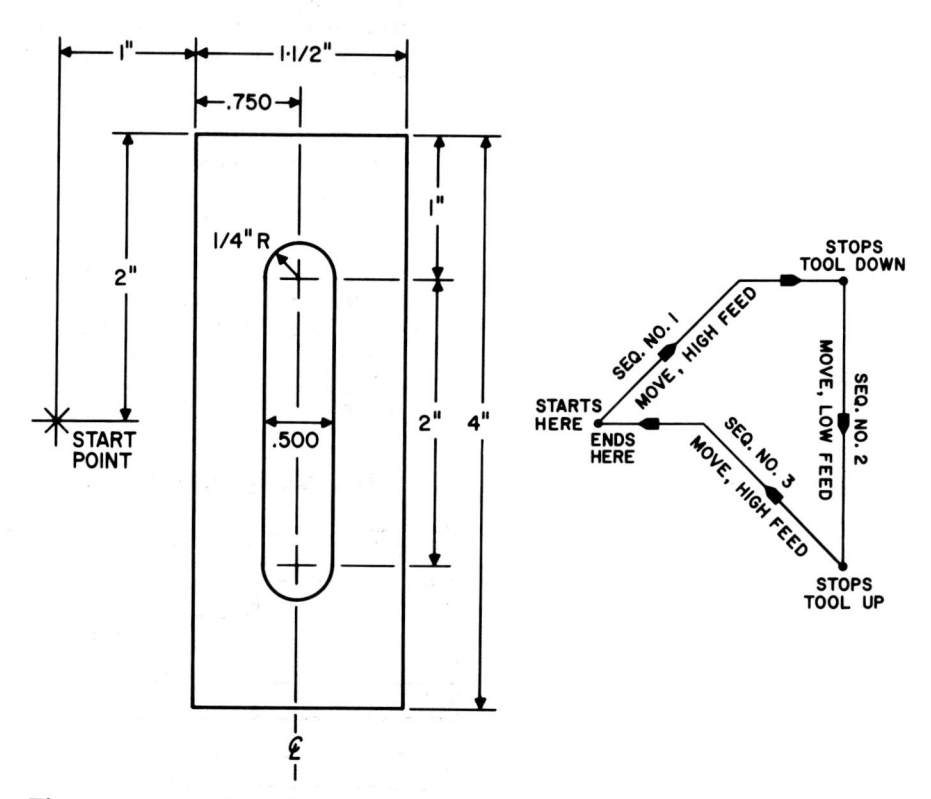

Figure 6. Example 4. Tab sequential program for milling (Courtesy of the Superior Electric Company).

Unit 3 Basic Format Numerical Control Programs

Program Notes:
1. Console presets.
   a. Tool switch at Off.
   b. Feed Rate switch at 6 ipm.
2. Sequence 1 statement prescribes the x and y axes tool motion from the starting point to the point where milling will begin and calls for the move to be made at the Hi feed rate.
3. Sequence 2 statement calls for the tool to be fed down into the workpiece and prescribes the y axis motion needed to mill the slot.
4. Sequence 3 statement contains the information needed to withdraw the tool from the work, position back to the starting point at Hi feed and rewind the tape.

SLO-SYN™ NUMERICAL TAPE CONTROL PROGRAM

COMPANY NAME _____ ADDRESS _____

PREPARED BY DATE	PART NAME	PART NO. Sample Milling Program No. 2	OPER. NO
CK'D BY DATE			
SHEET OF	REMARKS:		
DEPT	Tool Switch—Off	Tools: 1/2" End Mill	
TAPE NO.	Feed Rate—6 ipm		
	Backlash—No. 1 or No. 2		

SEQ. NO.	TAB OR EOB	+ OR −	"x" INCREMENT	TAB OR EOB	+ OR −	"y" INCREMENT	TAB OR EOB	"m" FUNCT.	EOB	INSTRUCTIONS
									EOB	
0	RWS								EOB	
1	TAB		1750	TAB		1000	TAB	55	EOB	
2	TAB			TAB	−	2000	TAB	52	EOB	
3	TAB	−	1750	TAB		1000	TAB	02535	EOB	

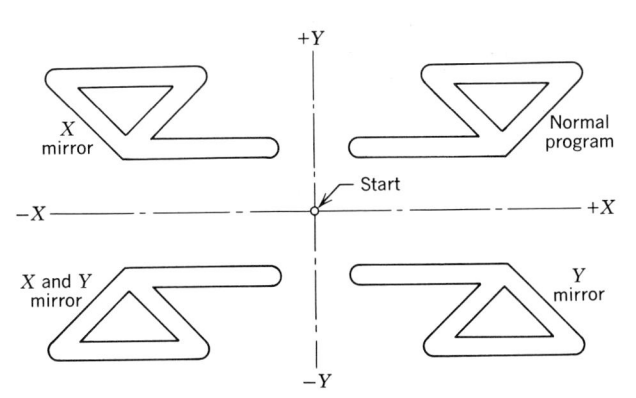

Figure 7. Mirror images (Courtesy of the Superior Electric Company).

Focus:
1. Milling a full circle using circular interpolation

Notes:
1. Machine operation is milling.
2. Tool feed and retract will be controlled from tape.
3. Hi speed will be called from tape.

Program Notes:
1. Console presets
   a. Tool Switch at Off
   b. Speed switch at Lo.
   c. Speed at 10 ipm.
2. Sequence 1 statement moves the tool to the nearest part of the circle at Hi speed.
3. Sequence 2 statement lowers the tool into the work and moves the tool through the first quadrant of the arc. The tool will stay down until the complete circle has been milled.
4. Sequence 3 through 5 statements provide the information needed to mill the remainder of the circle, one quadrant at a time.
5. Sequence 6 statement brings the tool back to the starting point and rewinds the tape.

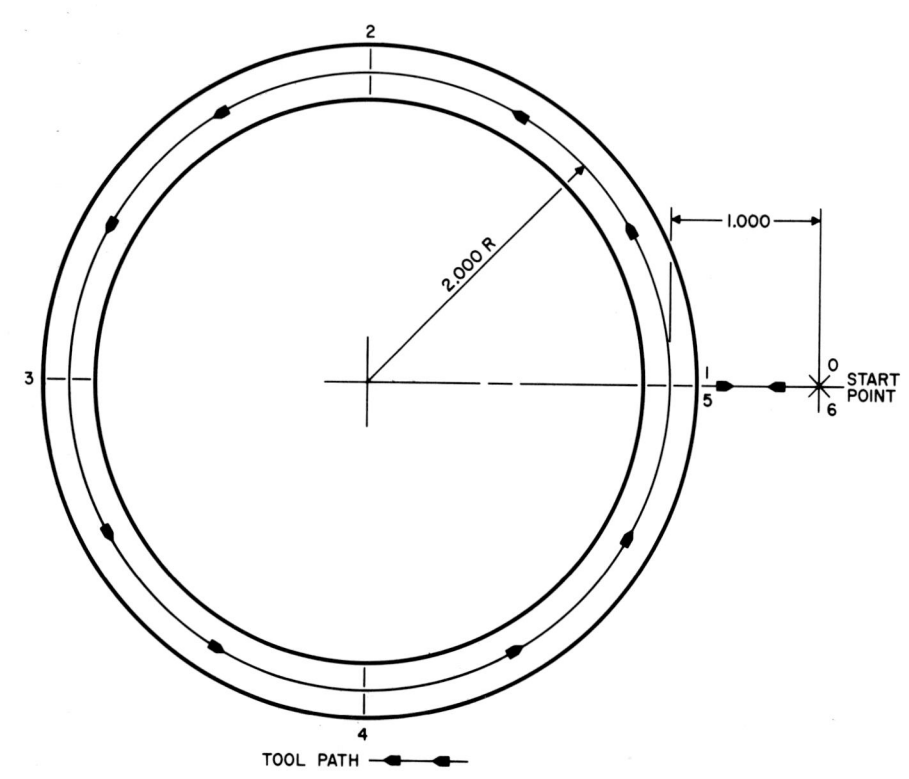

Figure 8. Tab sequential program for a continuous path involving circular interpolation (Courtesy of the Superior Electric Company).

Unit 3 Basic Format Numerical Control Programs

SLO-SYN™ CONTINUOUS PATH CONTOURING CONTROL PROGRAM

PREPARED BY ___ DATE ___
CHECKED BY ___ DATE ___
SHEET ___ OF ___
TAPE NO. ___ DEPT. ___

PART NAME: Contouring Sample No. 5
PART NO. ___
OPER. NO. ___

REMARKS: Tool Switch—Off / Speed Switch—Lo / Lo Speed Control—10 ipm

COMPANY NAME ___
ADDRESS ___

SEQ NO.	TAB	+ OR −	"x" INCREMENT	TAB	+ OR −	"y" INCREMENT	TAB	+ OR −	INCREMENT	TAB	"m" FUNCT.	EOB
0	RWS											EOB
1	TAB	−	1000	TAB			TAB					EOB
2	TAB	−	2000	TAB		2000	TAB	−	2000	TAB	55	EOB
3	TAB	−	2000	TAB		2000	TAB				52	EOB
4	TAB		2000	TAB		2000	TAB					EOB
5	TAB		2000	TAB			TAB		2000	TAB		EOB
6	TAB		1000	TAB			TAB			TAB	02535	EOB

Drilling Sample No. 1

FOCUS

A. Identification of a starting point.
B. Basic drilling procedure
C. Automatic tool function procedure.
D. Tool movement — X axis only.

NOTE:

1. Machine operation emphasis is drilling
2. The starting point is located to the left of the part. This point has been selected to allow for loading or unloading the part with minimum interference from the tool and holding fixtures.
3. Basic drilling procedure prescribes the center-drilling of each hole prior to drilling with 1/2" drill.

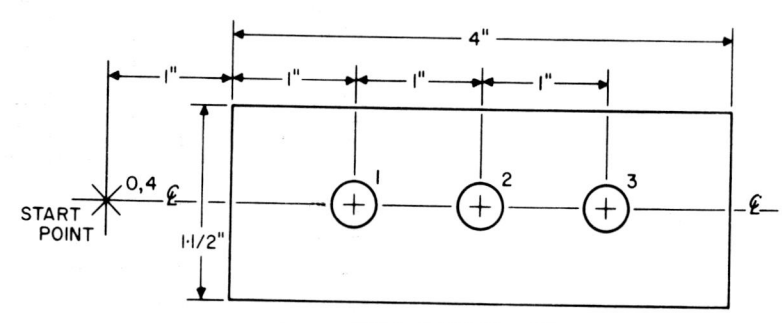

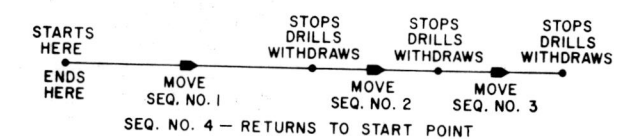

Program Notes

1. Console presets
    Tool Mode Switch — AUTO. Tool will drill and retract automatically when a g81 code is in effect.

2. Program begins with an EOB code.

3. Program statements are preceded by an RWS (Rewind Stop) code.

4. Sequence No. 1 statement — prescribes the g81 (Automatic Tool Cycle); tool movement on the X axis from the starting point to hole (1) — hole will be drilled automatically; maximum feed rate for units with M113-FJ40 motors. See table preceding the program samples for maximum feed rates for other motors.

5. Sequence No. 2 statement—prescribes the tool movement on the X axis from hole (1) to hole (2). The second hole will be center-drilled automatically.

6. Sequence No. 3 statement — prescribes the tool movement on the X axis from hole (2) to hole (3). The third hole will be center-drilled automatically.

7. Sequence No. 4 statement — prescribes the tool movement on the X axis from hole (3) back to the starting point and instructs the reader to rewind the tape. Machine will not "drill" at starting point and the g81 (Automatic Tool Cycle) will be cancelled.

8. Center-drill must be replaced with 1/2" drill.

9. Sequence Nos. 1 through 4 will be repeated. Replace 1/2" drill with center-drill for the next cycle.

10. For this sample, all tool positioning data are recorded in the "X" increment column only.

Figure 9. Example 1. Word address program for drilling (Courtesy of the Superior Electric Company).

Unit 3 Basic Format Numerical Control Programs

SLO-SYN® N/C WORD ADDRESS PROGRAM

PART NO. SAMPLE DRILLING PROGRAM #1
PART NAME PLATE
REMARKS TOOL MODE SWITCH -- AUTO   RUN PROGRAM FOR EACH TOOL
TOOLS: CENTER DRILL
1/2" dia. drill

n ADDRESS	SEQ. NO.	g ADDRESS	g DATA	x ADDRESS	x DATA (SIGN IF REQUIRED)	f ADDRESS	f DATA	m ADDRESS	m DATA	EOB	INSTRUCTIONS
	% (RWS)										
n	1	g	81	x	2000					EOB	
										EOB	CHANGE TOOL, LOAD, START
n	2			x	1000	f	1800			EOB	
n	3			x	1000					EOB	
n	4			x	-4000			m	02	EOB	

Section I Numerical Control of Machine Tools

FOCUS

1. Programming low feed rate changes.

NOTES:

1. Machine operation is milling.

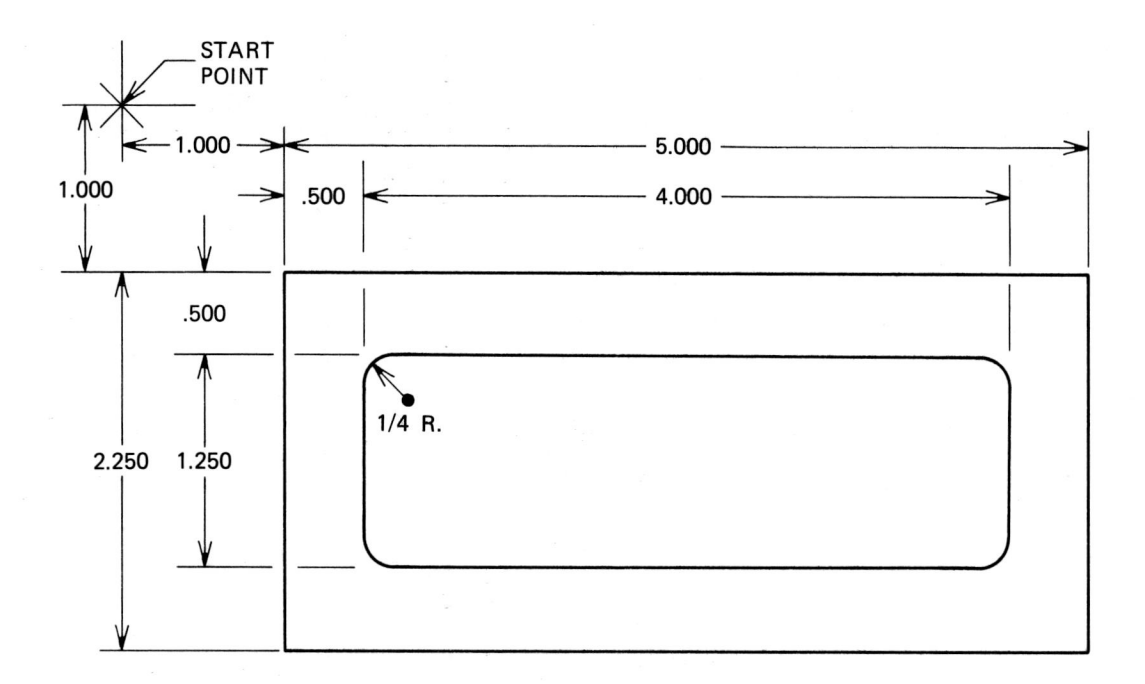

Program Notes

1. Console presets: Tool Mode Switch at Off.

2. Sequence No. 1 prescribes a g80 (Cancel Automatic Tool Cycle); X and Y axis data to move the tool into position to begin milling the pocket; and a maximum feed rate for the M113-FJ40 motor. For maximum feed rates of units with other motors see chart preceding the sample program.

3. Sequence No. 2 lowers the tool (m52) and begins milling the pocket at the programmed feed rate of 6 ipm. Sequence Nos. 3 through 7 complete the rough milling operation. The pocket will be milled .020" undersize. Sequence Nos. 2 through 5 operate at 6 ipm. A feed rate of 8 ipm is programmed in sequence No. 6 and held through sequence No. 7.

4. Sequence Nos. 8 through 12 mill the pocket to the final size. A feed rate of 12 ipm is programmed for these sequences.

5. Sequence No. 13 raises the tool (m53) and returns it to the start point at a maximum feed rate.

6. Sequence No. 14 calls for rewinding the tape (m02) back to the Rewind Stop Code.

Figure 10. Example 2. Word address program for milling (Courtesy of the Superior Electric Company).

Unit 3 Basic Format Numerical Control Programs 565

SLO-SYN® N/C WORD ADDRESS PROGRAM

PART NO. SAMPLE MILLING PROGRAM #4
TOOL: 1/2" END MILL
REMARKS: TOOL MODE SWITCH -- OFF

seq. no.	g addr	g data	x addr	x data	y addr	y data	z addr	z data	i addr	i data	j addr	j data	f addr	f data	s addr	s data	t addr	t data	m addr	m data	EOB
% (RWS)																					
n 1	g	80	x	1760	y	-1760															EOB
n 2			x	3480									f	1800							EOB
n 3					y	-730							f	60					m	52	EOB
n 4			x	-3480																	EOB
n 5					y	730															EOB
n 6					y	-365															EOB
n 7			x	3490									f	80							EOB
n 8					y	-375															EOB
n 9			x	-3500									f	120							EOB
n 10					y	750															EOB
n 11			x	3500																	EOB
n 12					y	-375															EOB
n 13			x	-5250	y	2125							f	1800					m	53	EOB
n 14																			m	02	EOB

Section I Numerical Control of Machine Tools

FOCUS

1. Programming for a three-axis N/C

NOTES:

1. Machine operation is drilling.
2. Program will be run using a three-axis control with a rotary table for the third axis.

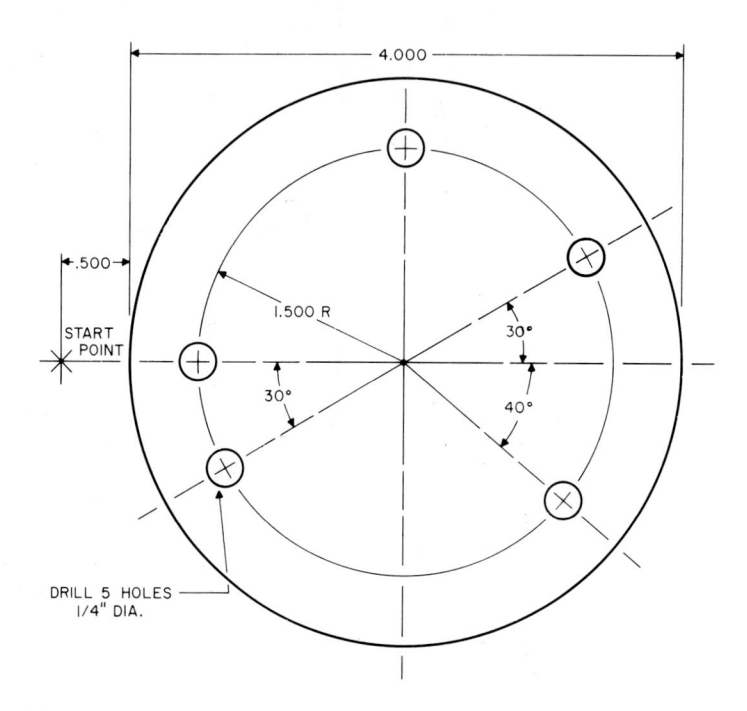

Program Notes

1. Console presets: Tool Mode Switch at Auto.

2. Sequence No. 1 prescribes a g81 (Automatic Tool Cycle); X axis data that moves the tool from the start point to the position where the first hole will be drilled automatically; and a feed rate that is maximum for the M113-FJ40 motor. The maximum feed rate for units with different motors is listed in the table preceding the sample programs.

3. Sequence No. 2 calls for a tool movement along the Z or third (rotary) axis of 30°. With the SLO-SYN rotary table, each motor step equals .01° of rotation. On standard units the third axis is time shared with the Y axis. Whichever axis is addressed (Y or Z) is the one that is active for that block. The hole will be automatically drilled when movement stops.

4. Sequence Nos. 3, 4 and 5 move the tool along the third axis to drill the remaining holes.

5. Sequence No. 6 returns the tool to the starting point and rewinds the tape. The machine does not drill at the start point and the g81 (Automatic Tool Cycle) is cancelled.

Figure 11. Example 3. Word address program using a third axis (Courtesy of the Superior Electric Company).

Unit 3 Basic Format Numerical Control Programs

SLO-SYN® N/C WORD ADDRESS PROGRAM

PART NO. SAMPLE DRILLING PROGRAM #6
REMARKS: TOOL MODE SWITCH -- AUTO TOOL: 1/4" DIA. DRILL
THIRD AXIS IS ROTARY TABLE

seq. no.	address g	g data	address x	x data	address y	y data	address z	z data	address i	i data	address j	j data	address f	f data	address s	s data	address t	t data	address m	m data	EOB
% (RWS)																					EOB
n 1	g	81	x	1000																	EOB
n 2							z	3000													EOB
n 3							z	11000													EOB
n 4							z	7000													EOB
n 5							z	6000					f	1800							EOB
n 6			x	-1000			z	9000											m	02	EOB

Section I Numerical Control of Machine Tools

FOCUS

1. Milling a full circle using circular interpolation.

NOTES:

1. Machine operation is milling.
2. Tool lowering and raising will be controlled from tape.

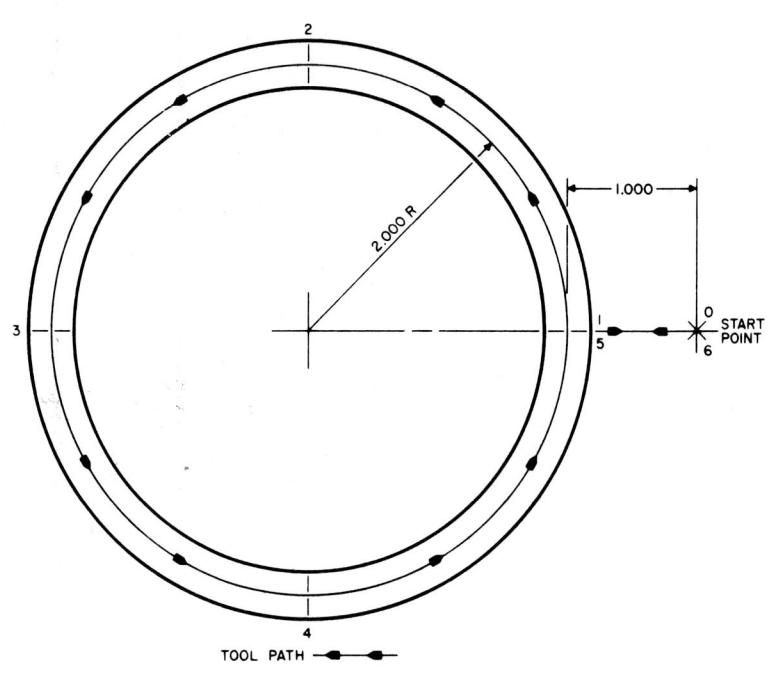

Program Notes

1. Console presets: Tool Mode Switch at Off.

2. Sequence No. 1 statement moves the tool to the nearest part of the circle at a maximum feed rate for the M113-FJ40 motor. For maximum feed rates for units with other motors see chart preceding the sample programs.

3. Sequence No. 2 statement lowers the tool (m52) into the work and moves the tool through the milling of the first quadrant of the arc at a feed rate of 28 ipm. The tool will stay down until the complete circle has been milled when a command (m53) raises it again. (NOTE: Whenever the value of I or J is zero the entire word address must be omitted as is done in this and succeeding blocks.)

4. Sequence Nos. 3 through 5 statements provide the information needed to mill the remainder of the circle, one quadrant at a time.

5. Sequence No. 6 statement raises the tool and brings it back to the starting point at a maximum feed rate.

6. Sequence No. 7 rewinds the tape (m02) back to the Rewind Stop (RWS) code.

Figure 12. Example 4. Word address program for contouring (Courtesy of the Superior Electric Company).

Unit 3 Basic Format Numerical Control Programs

SLO-SYN® N/C WORD ADDRESS PROGRAM

PART NO. CONTOURING SAMPLE PROGRAM #5

REMARKS: TOOL MODE SWITCH -- OFF

n SEQ. NO.	g ADDR	g DATA	x ADDR	x DATA	y ADDR	y DATA	z ADDR	z DATA	i ADDR	i DATA	j ADDR	j DATA	f ADDR	f DATA	s ADDR	s DATA	t ADDR	t DATA	m ADDR	m DATA	EOB	INSTRUCTIONS
% (RWS)																					EOB	
1	g	80	x	-1000																	EOB	
2			x	-2000	y	2000															EOB	
3			x	-2000	y	-2000			i	-2000			f	1800					m	52	EOB	
4			x	2000	y	-2000			i	2000			f	280							EOB	
5			x	2000	y	2000					j	-2000									EOB	
6			x	1000							j	2000	f	1800					m	53	EOB	
7																			m	02	EOB	

569

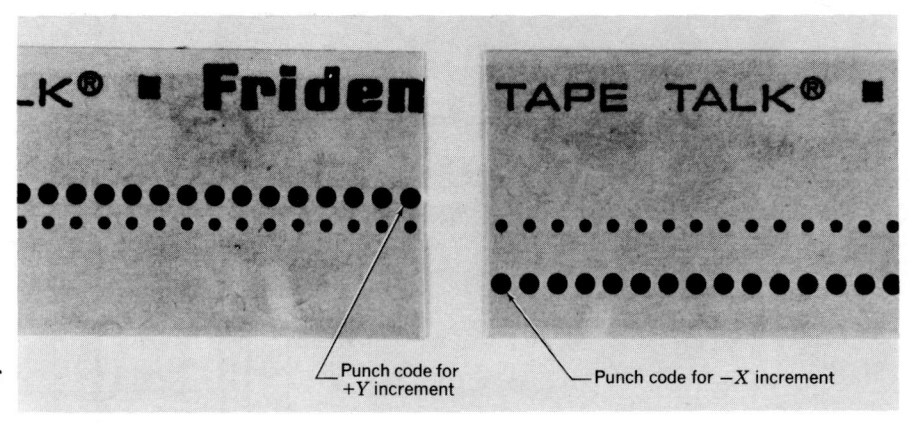

Figure 13. Tape punch patterns for direct control of stepping motors.

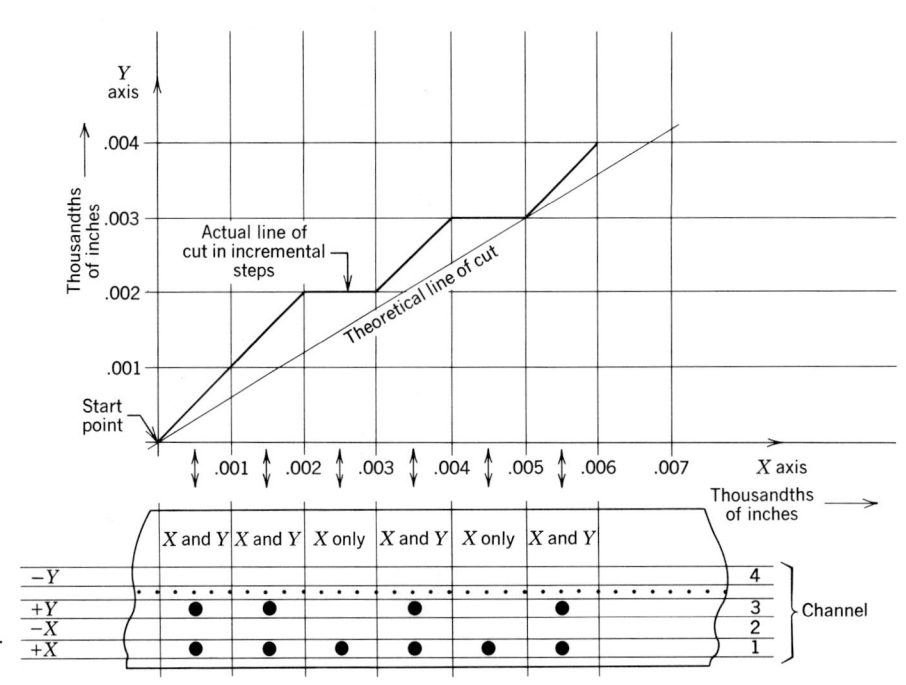

Figure 14. Approximating a continuous path by incremental steps.

up an interpolation cycle. On some N/C machine tools, this is done by sending pulses to the stepping motors in accordance with the pattern of tape punches passing the tape reader. After a g01 code, tape punches appear in lines corresponding to positive and negative travel along specified axes (Figure 13). As each tape punch passes the reader, either the X or Y stepping motor will step. Motors may be stepped separately or both at the same time. Stepping both motors causes the tool to move at a 45 degree angle to the basic axes. In this manner, an angle, radius, or other continuous path may be approximated. If the stepping motor advances the cutting tool or worktable a distance of .001 in. per step, a continuous path may be quite closely approximated. Note the relationship of the theoretical line of cut (Figure 14) to the actual line of cut. The corresponding tape codes are illustrated. The deviation of the cutter from the theoretical path must be taken into consideration as far as part tolerance is concenred. Contouring by this method is not a true continuous path, but a series of short straight line paths that approximate a continuous path.

FIXED BLOCK FORMAT PROGRAMS

In the fixed block format, each tape block is the same length. A popular system is the 20 digit block. When programming in a fixed block format, all spaces must be filled with a symbol. The 20 digit block is constructed as follows (Figure 15).

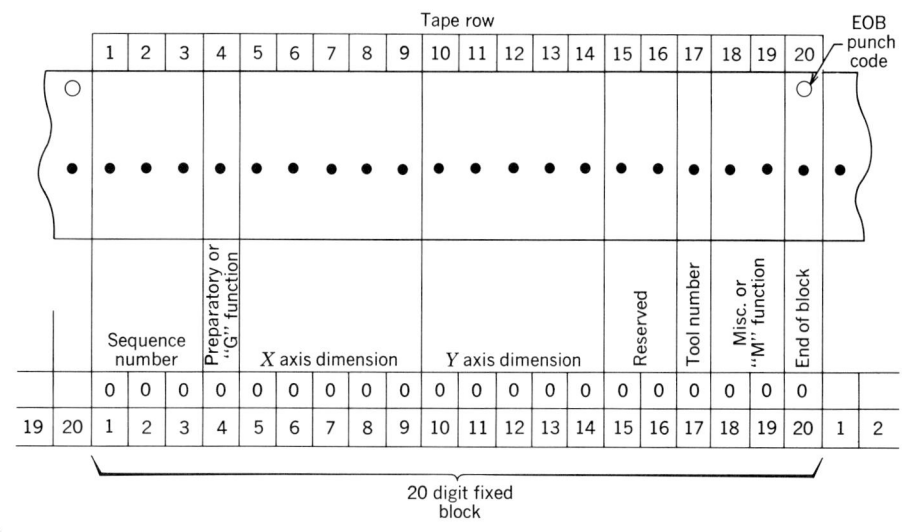

Figure 15. Construction of a fixed block.

Tape Row	
1, 2, 3	Sequence number in three digits (000 to 999).
4	One digit preparatory function. This is the same as the g function in word address. However, since only one row is allowed for this in fixed block, a single digit is used. A drill cycle would be indicated by the digit 5 as opposed to a g81 in word address. If you are programming in fixed block, consult the manufacturer's programming handbook for specific preparatory function code digits.
5, 6, 7, 8, 9	The X axis positioning dimension appears in five digits. The decimal point is understood to be after the first digit (X.XXXX). Decimal points are not typed when preparing the tape.
10, 11, 12, 13, 14	The Y axis dimension appears in five digits (Y.YYYY).
15, 16	These rows are reserved and must be filled with zeros.
17	A tool number appears as one digit. This can operate an automatic tool changer or activate an indicator light on a tool rack near the machine. The operator can then change the tool.
18, 19	Miscellaneous functions appear as two digits. The letter m is not typed. Only the numeric portion is required.
20	The last row is the EOB (carriage return) code.

self-evaluation

SELF-TEST

1. What basic format N/C program uses tab codes to separate axial positioning from m functions?

2. What basic format uses a single letter code to indicate positioning, preparatory, and miscellaneous functions?

3. In what basic format are block lengths the same?

4. Give an example of a preparatory function.

5. Give an example of a miscellaneous function.

EXERCISE Prepare N/C programs for the machine tool in your shop.

This unit has no post-test.

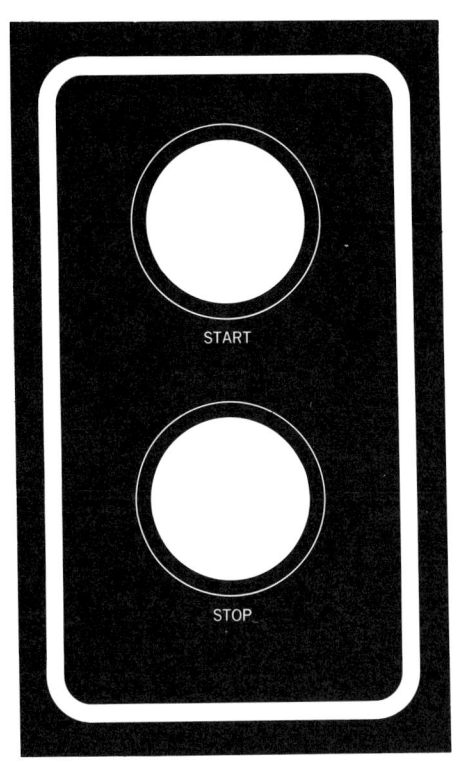

unit 4
operating the n/c machine tool

Before operating the N/C machine tool from tape instructions, you must first become totally familiar with its manual operation. The N/C machine is somewhat different than its conventional counterpart in that control functions are accomplished from the MCU for both manual and tape operation. To protect yourself from injury and to protect an expensive machine tool from damage, it is most important to know ahead of time exactly what will happen when an MCU console control is actuated.

objectives After completing this unit, you will be able to:
1. Describe safe operating procedure for an N/C machine tool.
2. Describe the purpose of tool length gaging equipment.

information

N/C MACHINE SAFETY

The same safety rules apply to N/C machine tools as apply to all machine tools. Always wear appropriate eye protection and short sleeves. Remove rings and watches before operating the machine. See that the workpiece is properly secured in a vise or by suitable clamps. Be sure that the cutter is clamped securely in its arbor or collet. Use proper speeds and feeds as you would for any machining task.

MANUAL OPERATION FROM THE MCU CONSOLE

Many N/C machine tools, especially those that have been retrofitted with a numerical control system, may have some controls on the MCU console and other controls on the machine itself. Because of this, you must exercise additional caution when operating the machine manually.

The MCU console may have an emergency stop control that will stop worktable or spindle positioning in case a problem should develop during a cut. However, the emergency stop control may not stop spindle rotation. This may be a separate control attached to the machine tool and may be quite far removed from the MCU console.

When positioning the worktable or spindle manually during the setup of the workpiece, be sure that the cutter and machine spindle are clear of vises and clamps before entering positioning data on the MCU console (Figure 1). Check clearances before lowering the quill on a vertical mill or drill press. It is good practice to set up the workpiece on a milling machine so that the quill must be lowered a small amount before beginning a milling cut. The quill should not be extended excessively as this will affect rigidity. If a problem should develop, the cutter can be quickly raised clear of the workpiece by actuating the quill up control on the MCU. However, if

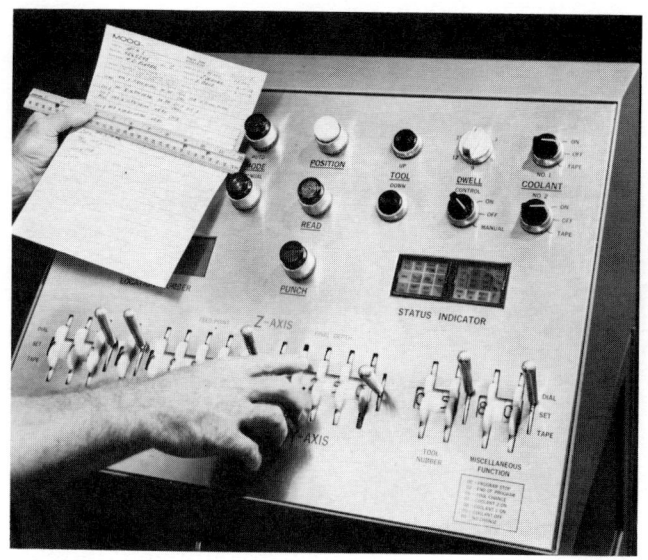

Figure 1. Entering positioning data on the MCU console (Courtesy of Hydra-Point Division, Moog Inc.).

Figure 2. Manually operated turret depth stop on an N/C milling machine (California State University at Fresno).

the setup has been made such that the quill is already in the full up position during milling, it cannot be raised clear of the workpiece. In case of a problem, the cutter, machine, or workpiece may be damaged.

If the N/C machine has a manually indexed turret to control quill travel (Figure 2), do not forget to index the turret at the appropriate points during the machining cycle. Be precise in adjusting the rapid traverse of the quill as it approaches the workpiece. If the cutter should run into the work during rapid traverse, damage can result to the machine or cutter in addition to the hazard of flying metal.

USING THE N/C MACHINE AS A CONVENTIONAL TOOL

The N/C machine tool may be used as a substitute for its conventional counterpart. This will generally not be done in industry, since the reason for having the N/C machine is to realize its increased productivity and other advantages. However, in the school shop, the N/C machine tool may be used to supplement conventional machines. In fact, the accurate positioning of an N/C machine tool often makes it very effective for a routine machining task.

N/C TAPE OPERATION

After a tape has been prepared, it should be read back at the typewriter tape punch to determine if any errors are present (Figure 3). The printout can be checked against the program manuscript. If the tape is correct, it should be verified by a "dry run" on the machine tool.

Be sure that the spindle and cutter are clear of all obstructions. On a milling machine this can be insured by moving the knee below the maximum extension of the spindle. Insert the tape in the MCU reader and observe the machine as it completes all programmed instructions. N/C machine tools will have a feature that permits tape blocks to be read one at a time. Each sequence can be initiated from the MCU. The machine will read and execute one tape block and stop. Positioning and miscellaneous func-

Figure 3. After punching a tape, obtain a printout using the typewriter reader (Courtesy of Hydra-Point Division, Moog Inc.).

Figure 4. Micrometer tool length gage (Courtesy of Hydra-Point Division, Moog Inc.).

tions can be observed and checked for accuracy. A dry tape run will safely verify the tape, thus preventing possible damage to the machine, cutter, or workpiece.

GAGING TOOL LENGTHS FOR N/C

An N/C machining task often requires a number of different tools. For example, a drilling operation may require center drilling, drilling, and reaming. If the holes are through, the drill and reamer will have to extend from the tool holder the appropriate distance. Since the center drill is probably shorter, the machine spindle will have to extend further or the worktable raised accordingly.

In an industrial setting, an entire set of tools required for a specific job may be preset for length and stored for future use. When a machining job comes to the shop, the N/C machine operator obtains the complete set of preadjusted tools and places them in the machine tool changer or in a tool rack.

The tool length gage is used to measure the projection of cutting tools from their tool holders. Common length gages consist of a series of accurately spaced rings mounted on a column (Figure 4). The instrument is not unlike a precision height gage. Ring spacing is usually 1 in. A

Figure 5. Dial indicator gage for adjusting length and diameter (Courtesy of Cincinnati Milacron).

Figure 6. Electronic digital tool length gage (Courtesy of Cincinnati Milacron).

Figure 7. Setting a tool length with the electronic gage (Courtesy of Cincinnati Milacron).

Figure 8. Setting an insert tooth milling cutter for diameter using the electronic gage (Courtesy of Cincinnati Milacron).

micrometer head with 1 in. travel spans the distance between the rings and can be placed at any desired height within the range of the gage. The cutting tool to be set is placed in its holder and the amount of projection is adjusted according to the job requirements. Tool length gages may also use dial indicators (Figure 5). This type of gage can be used to set boring bars and insert tooth cutters for specific diameters.

High discrimination electronic tool length gages are also used. These instruments are equipped with digital readouts (Figure 6). They can be used for tool length adjustments (Figure 7) and cutter diameter adjustments (Figure 8). One advantage of the electronic gage is its ability to read in inch and metric dimensions.

self-evaluation

SELF-TEST 1. After you have punched a tape, what should you do before placing the tape in the machine tool for a dry run?

2. What is the purpose of a dry tape run?

3. What is an important precaution that must be considered when operating the machine tool from the MCU console?

4. What is the function of the tool length gage?

5. Name two types of tool length gages.

   a.

   b.

EXERCISE Familiarize yourself with the manual and tape operation of the N/C machine tool in your shop.

This unit has no post-test.

appendix I self-test answers

section a unit 1 vertical band machine safety

SELF-TEST ANSWERS

1. Fingers and hands in close proximity to the blade.
2. Changing blades, friction, and high speed sawing.
3. A pusher is a piece of material used to push the workpiece into the blade.
4. Round stock may turn if handheld. This can cause an injury and damage the saw blade.
5. Eye protection is a primary piece of safety equipment.

section a unit 2 preparing to use the vertical band machine

SELF-TEST ANSWERS

1. The ends of the blade should be ground with the teeth opposed. This will insure that the ends of the blade are square.
2. The blade ends are placed in the welder with the teeth pointed in. The ends must contact squarely in the gap between the jaws. The welder must be adjusted for the band width to be welded. You should wear eye protection and stand to one side during the welding operation. The weld will occur when the weld lever is depressed.
3. The weld is ground on the grinding wheel attached to the welder. Grind the weld on both sides of the band until the band fits the thickness gage. Be careful not to grind the saw teeth.
4. The guides support the band. This is essential to straight cutting.
5. Band guides must fully support the band except for the teeth. A wide guide used on a narrow band will destroy the saw set as soon as the machine is started.
6. The guide setting gage is used to adjust the band guides.
7. Annealing is the process of softening the band weld in order to improve strength qualities.
8. The band should be clamped in the annealing jaws with the teeth pointed out. A small amount of compression should be placed on the movable welder jaw prior to clamping the band. The correct annealing color is dull red. As soon as this color is reached, the anneal switch should be released and then operated briefly several times to slow the cooling rate of the weld.
9. Band tracking is the position of the band as it runs on the idler wheels.
10. Band tracking is adjusted by tilting the idler wheels until the band just touches the backup bearing.

section a unit 3 using the vertical band machine

SELF-TEST ANSWERS

1. The three sets are straight, wave, and raker. Straight set may be used for thin material, wave for material with a variable cross section, and raker for general purpose sawing.
2. Scalloped and wavy edged bands might be used on nonmetallic material where blade teeth would tear the material being cut.
3. Band velocity is measured in feet per minute.
4. The variable speed pulley is designed so that the pulley flanges may be moved toward and away from each other. This permits the belt position to be varied, resulting in speed changes.
5. The job selector provides information about recommended saw velocity, saw pitch, power feed, saw set, and temper. Band filing information is also indicated.
6. Speed range is selected by shifting the transmission.
7. Speed range shift must be done with the band speed set at the lowest setting.
8. The upper guidepost must be adjusted so that it is as close to the workpiece as possible.
9. Band pitch must be correct for the thickness of material to be cut. Generally, a fine pitch will be used on thin material. Cutting a thick workpiece with a fine pitch band will clog saw teeth and reduce cutting efficiency.
10. Band set must be adequate for the thickness of the blade used in a contour cut. If set is insufficient, the blade may not be able to cut the desired radius.

section b unit 1 vertical milling machine safety

SELF-TEST ANSWERS

1. When sick or tired, one's reflexes are slow and one can not think a problem through. A safe machine operator is alert and uses his eyes, ears, sense of smell, and often touch while observing a machine operation. He needs to be able to make a split second decision when something unexpected happens.
2. Safe dress includes short sleeves or snug sleeves, eye protection, covering for long hair, and shoes that provide protection from falling objects. Safe dress excludes rings, bracelets, and wristwatches.
3. Chips should never be handled with bare hands because they are sharp, possibly hot, and often contaminated from cutting fluids.
4. Eye protection is always worn in a machining facility.
5. Machine guards are there for safety. Guards not in place leave operators or bystanders exposed to dangerous moving parts of machines.
6. A clean work area prevents one from stumbling over parts on the floor and slipping on spilled oil or cutting fluids.
7. Heavy objects should be lifted with a hoist or with the help of another person. Any lifting should be done with the leg muscles, keeping the back straight and vertical.
8. The cutting edges of tools are sharp and can cause cuts. Use a rag when handling tools.
9. The motor on top of the work head is heavy and will flip down if the clamping bolts are loosened completely. The clamping bolts should be kept snug to provide enough drag so that the head only swivels when pushed against.
10. Measurements should only be made when the machine spindle has stopped.

section b unit 2 the vertical spindle milling machine

SELF-TEST ANSWERS

1. The column, knee, saddle, table, ram, and toolhead.
2. The table traverse handwheel and the table power feed.
3. The cross traverse handwheel.
4. The quill feed hand lever and handwheel.

5. The table clamp locks the table rigidly and keeps it from moving while other table axes are in movement.
6. The spindle brake locks the spindle while tool changes are being made.
7. The spindle has to stop before speed changes from high to low are made.
8. The ram movement increases the working capacity of the toolhead.
9. Loose machine movements are adjusted with the big adjustment screws.
10. The quill clamp is tightened to lock the quill rigidly while milling.

section b unit 3 vertical milling machine operations

SELF-TEST ANSWERS

1. Accurate centering of a cutter over a shaft is done with the machine dials.
2. The feed direction against the cutter rotation assures positive dimensional movement. It also prevents the workpiece from being pulled into the cutter because of any backlash in the machine.
3. End mills can work themselves out of a split collet if the cut is too heavy or when the cutter gets dull.
4. Angular cuts can be made by tilting the workpiece or by tilting the workhead.
5. Circular slots can be milled by using a rotary table or an index head.
6. Squares, hexagons, or other shapes that require surfaces at precise angles to each other are made by using a dividing head.
7. Square holes or other internal hole shapes can be made by using a vertical shaping attachment.
8. With a right angle milling attachment, milling cuts can be made in very inaccessible places.
9. Layout lines are used as guides to indicate where machining should take place. Layouts should be made prior to machining any reference surfaces away.
10. A T-slot cutter only enlarges the bottom part of a groove. The groove has to be made before a T-slot cutter can be used.

section b unit 4 cutting tools for the vertical milling machine

SELF-TEST ANSWERS

1. When viewed from the cutting end, a right-hand cut end mill will rotate counterclockwise.
2. An end mill has to have center cutting teeth to be used for plunge cutting.
3. End mills for aluminum usually have a fast helix angle and also highly polished flutes and cutting edges.
4. Carbide end mills are very effective when milling abrasive or hard materials.
5. Roughing mills are used to remove large amounts of material.
6. Tapered end mills are mostly used in mold or die making to obtain precisely tapered sides on workpieces.
7. Carbide insert tools are used because new cutting edges are easily exposed. They are available in grades to cut most materials and they are very efficient cutting tools.
8. Straight shank mills are held in collets or adapters.
9. Shell end mills are mounted on shell mill arbors.
10. Quick change tool holders make presetting of a number of tools possible and tools can be changed with a minimum loss of time.

section b unit 5 setups on the vertical milling machine

SELF-TEST ANSWERS

1. Workpieces can be aligned on a machine table by measuring their distance from the edge of the table, by locating against stops in the T-slots, or by indicating the workpiece side.

2. To align a vise on a machine table, the solid vise jaw needs to be indicated.
3. Tool head alignment is checked when it is important that machining takes place square to the machine table.
4. When the knee clamping bolts are loose, the weight of the knee makes it sag. But when the knee clamps are tightened, the knee is pulled into its normal position in relation to the column.
5. When the toolhead clamping bolts are tightened, it usually produces a small change in the toolhead position.
6. A machine spindle can be located over the edge of a workpiece with an edge finder or with the aid of dial indicator.
7. The spindle axis is one-half of the tip diameter away from the workpiece edge when the tip walks off sideways.
8. An offset edge finder works best at 600 to 800 RPM.
9. To eliminate the effect of backlash, always position from the same direction.
10. The center of a hole is located with a dial indicator mounted in the machine spindle.

section b unit 6 using end mills

SELF-TEST ANSWERS

1. Lower cutting speeds are used to machine hard materials, tough materials, abrasive material, on heavy cuts, and to get maximum tool life.
2. Higher cutting speeds are used to machine softer materials, to obtain good surface finishes, with small diameter cutters, for light cuts, on frail workpieces and on frail setups.
3. The calculated RPM is a starting point and may change depending on conditions illustrated in the answers to Problems 1 and 2.
4. Cutting fluids are used with HSS cutters except on materials such as cast iron, brass, and many nonmetallic materials.
5. Cutting with carbide cutters is performed without a cutting fluid, unless a steady stream of fluid can be maintained at the cutting edge of the tool.
6. The thickness of the chips affects the tool life of the cutter. Very thin chips dull a cutting edge quickly. Too thick chips cause tool breakage or the chipping of the cutting edge.
7. The depth of cut for an end mill should not exceed one-half of the diameter of the cutter.
8. Limitations on the depth of cut for an end mill are the amount of material to be removed, the power available, and the rigidity of the tool and setup.
9. The RPM for a $\frac{3}{4}$ in. diameter end mill to cut brass is:

$$\text{RPM} = \frac{\text{CS } 4}{\text{D}} = \frac{200 \times 4}{\frac{3}{4}} = 1067 \text{ RPM}$$

10. The feedrate for a two flute, $\frac{1}{4}$ in. diameter carbide end mill in medium alloy steel is:

$$\text{feedrate} = f \times \text{RPM} \times n$$
$$f = .0005$$
$$\text{RPM} = \frac{\text{CS} \times 4}{\text{D}} = \frac{150 \times 4}{\frac{1}{4}} = 2400$$
$$n = 2$$
$$\text{Feedrate} = .0005 \times 2400 \times 2 = 2.4 \text{ IPM}$$

section b unit 7 using the offset boring head

SELF-TEST ANSWERS

1. An offset boring head is used to produce standard and nonstandard size holes at precisely controllable hole locations.
2. Parallels raise the workpiece off the table or other workholding device to allow through holes to be bored.

3. Unless the locking screw is tightened after toolslide adjustments are made, the toolslide will move during the cutting operation, resulting in a tapered or odd-sized hole.
4. The toolslide has a number of holes so that the boring tool can be held in different positions in relation to the spindle axis for different size bores.
5. It is important that you know if one graduation is one-thousandths of an inch or two-thousandths of an inch in hole size change.
6. The best boring tool to use is the one with the largest diameter that can be used and the one with the shortest shank.
7. It is very important that the cutting edge of the boring tool is on the centerline of the axis of the toolslide. Only in this position are the rake and clearance angles correct as ground on the tool.
8. The hole size obtained for given amount of depth of cut can change depending on the sharpness of the tool, the amount of tool overhang (boring bar length), and the amount of feed per revolution.
9. Boring tool deflection changes when the tool gets dull, the depth of cut increases or decreases, or the feed is changed.
10. The cutting speed is determined by the kind of tool material and the kind of work material, but boring vibrations set up through an unbalanced cutting tool or a very long boring bar may require a smaller than calculated RPM.

section c unit 1 horizontal milling machine safety

SELF-TEST ANSWERS

1. Loose clothing and long hair can easily be caught in revolving machine parts, pulling the operator into the machine and causing serious injury. Rings, bracelets, and necklaces can catch on revolving cutters and draw the operator into the machine.
2. Gloves can be worn while the machine is stopped to handle sharp edged materials. They should be removed before the machine is started.
3. Eye protection has to be worn in a shop at all times because of the danger from flying particles.
4. Do not reach over a revolving cutter. Do not try to remove chips while the spindle is turning.
5. Do not go out of reach of the machine controls while the spindle is revolving. Operate all machine controls by yourself; let no one turn the machine on or off for you. Do not lean on a running machine.
6. If possible, measurements should be made while the workpiece is still clamped in the machine, but only after the machine is turned off.
7. Compressed air can make chips into missiles that are dangerous to you and anyone around you. Compressed air also forces small chips into the slide ways of the machine.
8. Loads should be lifted with the legs by keeping the back straight and in a vertical position.
9. Oversize or worn-out wrenches slip when force is applied, resulting in bruises or cuts. They also damage the parts they are used on by rounding the corners of nuts and bolts or other fastening devices.
10. Keep the floor around the machine clean and dry. Remove chips with brushes, never with bare hands.

section c unit 2 plain and universal horizontal milling machines

There are no self-test answers.

section c unit 3 types of spindles, arbors, and adaptors

SELF-TEST ANSWERS

1. Face mills over 6 in. in diameter.
2. The two classes of taper are self-holding, with a small included angle, and self-releasing, with a steep taper.

3. 3½ IPF.
4. Any small nick or chip between shank and socket or between spacers will cause the cutter to run out and will mar these contact surfaces.
5. Where small diameter cutters are used, on light cuts, and where little clearance is available.
6. As close to the cutter as the workpiece and workholding device permit.
7. A Style C arbor is a shell end mill arbor.
8. Tightening or loosening an arbor nut without the arbor support in place will bend or spring the arbor.
9. To increase the range of cutters that can be used on a milling machine with a given size spindle socket.
10. Arbor extensions beyond the outer arbor support are often the cause of vibration and chatter.

section c unit 4 arbor-driven milling cutters

SELF-TEST ANSWERS

1. Profile sharpened cutters and form relieved cutters.
2. Light duty plain milling cutters have many teeth. They are used for finishing operations. Heavy duty plain mills have few but coarse teeth, designed for heavy cuts.
3. Plain milling cutters do not have side cutting teeth. This would cause extreme rubbing if used to mill steps or grooves. Plain milling cutters should be wider than the flat surface they are machining.
4. Side milling cutters, having side cutting teeth, are used when grooves are machined.
5. Straight tooth side mills are used only to mill shallow grooves because of their limited chip space between the teeth and their tendency to chatter. Stagger tooth mills have a smoother cutting action because of the alternate helical teeth; more chip clearance allows deeper cuts.
6. Half side milling cutters are efficiently used when straddle milling.
7. Metal slitting saws are used in slotting or cut off operations.
8. Gear tooth cutters and corner rounding cutters.
9. To mill V-notches, dovetails, or chamfers.
10. A right-hand cutter rotates counterclockwise when viewed from the outside end.

section c unit 5 setting speeds and feeds on the horizontal milling machine

SELF-TEST ANSWERS

1. Cutting speed is the distance a cutting edge of a tool travels in one minute. It is expressed in feet per minute (FPM).
2. Starting a cut at the low end of the speed range will save the cutter from overheating.
3. Carbide tools are operated at two to six times the speed of HSS tools. If a HSS tool is 100 FPM, the carbide tool would be 200 to 600 FPM.
4. Too low a cutting speed is inefficient because a cutter can do more work in a given time period.
5. The feedrate on a milling machine is given in inches per minute (IPM).
6. The feedrate is the product of RPM times the number of teeth of the cutter times the feed per tooth.
7. Feed per revolution does not consider the number of teeth on different cutters.
8. Too low a feedrate causes the cutter to rub and scrape the surface of the work instead of cut. Because of the high friction, the tool will dull quickly.
9. $\text{RPM} = \dfrac{\text{CS} \times 4}{\text{D}} = \dfrac{60 \times 4}{3} = 80 \text{ RPM}$.
10. $\text{RPM} = \dfrac{\text{CS} \times 4}{\text{D}} = \dfrac{150 \times 4}{4} = 150 \text{ RPM}$.
    Feed per tooth = .003 in.
    Number of teeth = 5
    Feedrate = RPM × feed per tooth × number of teeth = 150 × .003 × 5 = 2.25 = 2¼ IPM.

section c unit 6 workholding and locating devices on the milling machine

SELF-TEST ANSWERS

1. A clamping bolt should be close to the workpiece and a greater distance away from the support block.
2. Finished surfaces should be protected with shims from being marked by clamps or rough vise jaws.
3. Screwjacks are used to support workpieces or to support the end of a clamp.
4. A stop block prevents a workpiece from being moved by cutting pressure.
5. Quick action jaws are two independent jaws mounted anywhere on a machine table to form a custom vise.
6. Swivel vise has one movement in a horizontal plane, where a universal vise swivels both horizontally and vertically.
7. All-steel vises are used to hold rough workpieces such as castings or forgings.
8. A rotary table is used to mill gears, circular grooves, or for angular indexing.
9. A dividing head is used to divide accurately the circumference of a workpiece into any number of equal divisions.
10. A fixture is used when a great number of pieces have to be machined in exactly the same way, and when the cost of making the fixture can be justified in savings resulting from its use.

section c unit 7 plain milling on the horizontal milling machine

SELF-TEST ANSWERS

1. Because the movable jaw is not solidly held. It can move and swivel slightly to align itself with the work to some extent.
2. The dial indicator is more accurate. It will show you the amount of misalignment when you make adjustments, you can see when alignment is achieved.
3. Keys are designed to align a vise or other attachments with the T-slots on a machine table.
4. No, indicating the table from the column only measures the table sliding in its ways, but it does not show if it travels parallel to the column.
5. Yes, if possible the cutting tool pressure should be against the solid jaw. This makes the most rigid setup.
6. The smallest diameter cutter that will do the job will be the most efficient, because it requires a shorter movement to have the workpiece move clear of the cutter.
7. In conventional milling, the cutting pressure is against the feed direction and also up, where in climb milling the cutting pressure is down and the cutter tends to pull the workpiece under itself.
8. A good depth for a finish cut is .015 to .030 in. Less than .010 in. makes the cutter rub, which causes rapid wear.
9. Cutting vibrations and cutting pressures may make the table move when it should be rigidly clamped.
10. When a revolving cutter is moved over a just machined surface, it will leave tool marks.

section c unit 8 using side milling cutters on the horizontal milling machine

SELF-TEST ANSWERS

1. Full side milling cutters are used to cut slots and grooves and where contact on both sides of the cutter is made.
2. Half side milling cutters make contact on one side only, as in straddle milling where a left-hand and a right-hand cutter are combined to cut a workpiece to length.
3. The best cutter is the one with the smallest diameter that will work, considering the clearance needed under the arbor support.
4. Usually a groove is wider than the cutter.
5. A layout shows the machinist where the machining is to take place. It helps in preventing errors.

6. Accurate positioning is done with the help of a paper feeler strip. Often adequate accuracy is achieved by using a steel rule or by aligning by sight with the layout lines.
7. If a workpiece is measured while it is clamped in the machine, additional cuts can be made without additional setups being made.
8. Shims and spacers control workpiece width in straddle milling operations.
9. The diameters of the individual cutters determine the relationship of the depth of the steps in gang milling.
10. Interlocking side mills are used to cut slots over 1 in. wide and also when precise slot width is to be produced.

section c unit 9 using face milling cutters on the horizontal milling machine

SELF-TEST ANSWERS

1. Face mills are mounted on the spindle nose, driven by two keys, and held with four capscrews.
2. A shell end mill is a face mill 6 in. in diameter and smaller.
3. Light duty face mills have a great number of teeth and are used in finishing operations. Heavy duty face mills have fewer, stronger teeth and are used in roughing operations.
4. Positive rake angles on cutters use less power, have less cutting pressures, generate less heat, and work well on soft and ductile materials.
5. Negative rake angles make cutters very strong. They will withstand heavy interrupted cuts.
6. A large lead angle on a cutter produces a thinner but wider chip under equal depth of cut and feed conditions as a small lead angle would. A large lead angle helps in reducing initial cutting pressure on the tool because it gradually eases the cutter into the work.
7. Cuts as wide as the diameter of the face mill result in excessive rubbing and friction of the cutting edge, which causes rapid tool wear.
8. Cutting fluids are used with high speed steel cutters. With carbide tools, it is better not to use coolant than to apply it intermittently to the cutting edge.
9. A concave surface results when the cutter is tilted so that the trailing edge is not in contact with the workpiece surface.
10. Effective face milling requires:
    a. A rigid setup.
    b. A sharp cutting tool.
    c. The right cutting speed.
    d. The correct feed.

section d unit 1 indexing devices

SELF-TEST ANSWERS

1. When accurate spacings are made as with gears, splines, keyways, or precise angular spacings.
2. The use of a worm and worm wheel unit.
3. To make the spindle freewheeling and when direct indexing.
4. For direct indexing.
5. The most common ratio is 40:1.
6. Different hole circles are used to make precise partial revolutions with the index crank.
7. The sector arms are set to the number of holes that the index crank is to move. They eliminate the counting of these holes for each spacing.
8. The spindle lock is tightened after each indexing operation and before a cut is made to prevent any rotary movement of the spindle.
9. With the use of high number index plates or with a wide range divider.
10. When the rotation of the index crank is reversed, the backlash between the worm and worm wheel affects the accuracy of the spacing.

section d unit 2 direct and simple indexing

SELF-TEST ANSWERS

1. Direct indexing is performed from index holes on the spindle nose. Simple indexing uses the worm and worm wheel drive and the side index plate.
2. Use a marking pen or layout dye to mark the holes to be used in direct indexing.
3. Twenty-four index holes let you make equal divisions of 2, 3, 4, 6, 8, 12, and 24 parts.
4. The sector arms can be adjusted so that the number of holes to be indexed plus one are between the beveled edges.
5. For the highest degree of accuracy use the largest possible hole circle.
6. $\frac{40}{6} = \frac{20}{3} = 6\frac{2}{3}$ turns. Use the 57 hole circle for the $\frac{2}{3}$ turn. $\frac{2}{3} \times \frac{19}{19} = \frac{38}{57}$. The fraction is 38 holes in the 57 hole circle.
7. $\frac{40}{15} = \frac{8}{3} = 2\frac{2}{3}$ turns. Use the same hole circle as for Problem 6 above. Two turns and $\frac{38}{57}$ turn.
8. $\frac{40}{25} = \frac{8}{5} = 1\frac{3}{5}$ turns. $\frac{3}{5} \times \frac{6}{6} = \frac{18}{30}$. One turn and 18 holes in the 30 hole circle.
9. $\frac{40}{47}$. There is a 47 hole circle, so use 40 holes in the 47 hole circle.
10. $\frac{40}{64} = \frac{5}{8}$ turns. The only hole circle divisible by 8 is 24. So, $\frac{5}{8} \times \frac{3}{3} = \frac{15}{24}.$ = 15 holes in the 24 hole circle.

section d unit 3 angular indexing

SELF-TEST ANSWERS

1. 15 degrees.
2. Three holes.
3. Nine degrees.
4. All those hole circles are divisible by 9.
5. $\frac{17}{9} = 1\frac{8}{9}$ turn of the index crank.
6. 30 min or $\frac{1}{2}$ degrees.
7. 15 min or $\frac{1}{4}$ degrees.
8. 10 min or $\frac{1}{6}$ degrees.
9. 540 min.
10. Converting 54°30′ into minutes = 3270 min.

$$\frac{\text{Required minutes}}{540} = \frac{3270}{540} = 6\frac{30}{540} = 6\frac{3}{54} = 6 \text{ turns}$$

and 3 holes in the 54 hole circle.

section e unit 1 introduction to gears

SELF-TEST ANSWERS

1. Spur gears and helical gears.
2. Helical gears run more smoothly than spur gears because more than one tooth is in mesh at all times. Helical gears, however, generate axial thrust that has to be offset with thrust bearings.
3. The pinion in mesh with an internal gear rotates in the same direction as the gear.
4. The shafts will be at 90 degrees to each other.
5. 50:1.
6. No, a single start worm can only be replaced with a single start worm. To change the gear ratio, the number of teeth in the gear need to be changed.
7. Hardened steel.
8. Nonmetallic materials.
9. By making the pinion harder than the gear.
10. Nonferrous materials.

section e unit 2 spur gear terms and calculations

SELF-TEST ANSWERS

1. Pressure angles on gears vary from 14½ to 20 to 25 degrees.
2. Larger pressure angles make stronger teeth. They also allow gears to be made with fewer teeth.
3. $C = \dfrac{N_1 + N_2}{2P} = \dfrac{20 + 30}{2 \times 10} = \dfrac{50}{20} = 2.500$ in.
4. $C = \dfrac{D_1 + D_2}{2} = \dfrac{3.500 + 2.500}{2} = \dfrac{6.000}{2} = 3.000$ in.
5. The whole depth of a tooth is how deep a tooth is cut. The working depth gives the distance the teeth from one gear enter the opposing gear in meshing.
6. The addendum is above the pitch diameter and the dedendum is below the pitch diameter of a gear.
7. $D_o = \dfrac{N + 2}{P} = \dfrac{50 + 2}{5} = \dfrac{52}{5} = 10.400$ in.

   $t = \dfrac{1.5708}{P} = \dfrac{1.5708}{5} = .314$ in.
8. $P = \dfrac{N}{D} = \dfrac{36}{3} = 12$
9. $D_o = \dfrac{N + 2}{P} = \dfrac{40 + 2}{8} = \dfrac{42}{8} = 5.250$ in.

   $h_t = \dfrac{2.250}{P} = \dfrac{2.250}{8} = .2812$ in.

   $D = \dfrac{N}{P} = \dfrac{40}{8} = 5.000$ in.

   $b = \dfrac{1.250}{P} = \dfrac{1.250}{8} = .1562$ in.
10. $D_o = \dfrac{N + 2}{P} = \dfrac{48 + 2}{6} = \dfrac{50}{6} = 8.3333$ in.

    $c = \dfrac{.157}{P} = \dfrac{.157}{6} = .0261$ in.

    $h_t = \dfrac{2.157}{P} = \dfrac{2.157}{6} = .3595$ in.

    $t = \dfrac{1.5708}{P} = \dfrac{1.5708}{8} = .2618$ in.

    $D = \dfrac{N}{P} = \dfrac{48}{6} = 8.000$ in.

section e unit 3 cutting a spur gear

SELF-TEST ANSWERS

1. Gears cut on a milling machine lack a high degree of accuracy and are expensive to make.
2. Eight.
3. No. 3.

Appendix I Self-test Answers 589

4. No, the tooth profile differs with different pressure angles.
5. For a 17 tooth gear.
6. The number of the cutter, the diametral pitch, the number of teeth the cutter can cut, the pressure angle, and the whole depth of tooth.
7. To get a setup as rigid as possible.
8. The number of holes to be indexed plus one.
9. To check the correctness of the number of spaces required.
10. A center rest is used under gear blanks, mandrels, or shafts to help prevent chatter and deflection.

section e unit 4 gear inspection and measurement

SELF-TEST ANSWERS

1. Gear measurements with a gear tooth vernier caliper and with a micrometer over two wires or pins.
2. The chordal tooth thickness.
3. The chordal addendum.
4. The chordal addendum.
5. The circular thickness.
6. By calculation or from *Machinery's Handbook* tables.
7. No, the pin sizes are specifically calculated for each differing diametral pitch.
8. By consulting a *Machinery's Handbook* table.
9. The dimension given is divided by the diametral pitch used; in this question, the divisor is 12.
10. The optical comparator magnifies the gear tooth profile and projects it on a screen. On the screen there is also a transparent drawing with the tooth profile; by aligning the shadow with the drawn profile, any variation between the two can be seen and measured.

section f unit 1 features and tooling on the horizontal shaper

SELF-TEST ANSWERS

1. A. Tilt table
   B. Apron
   C. Cross rail
   D. Crossfeed engagement lever
   E. Rail elevating crank
   F. Stroke adjusting shaft
   G. Ram
   H. Ram adjusting shaft
   I. Tool (swivel) head
   J. Tool lifter
2. It permits the cutting tool to tilt up on the return (feeding) stroke of the ram without damaging the cutting edge.
3. The crank shaper uses a crank gear with a movable pivot that drives a rocker arm connected to the ram. The return stroke is faster than the cutting stroke by about 3:2. The hydraulic shaper uses high pressure oil to drive a piston connected to the ram. The cutting and return rates are independently set by the operator.
4. The feeding of the table is done by a ratchet mechanism that is driven by a cam or adjustable eccentric. On many larger shapers, vertical feed can be selected as an alternative to table crossfeed.
5. The size of a shaper is the maximum tool movement that can be obtained. Shapers up to about 16 in. can be expected to hold a cubic workpiece of the shaper's specified size. For larger sizes, vertical height is essentially limited to 16 in.
6. The single screw vise, usually with a swivel base and the two or multiple screw vises for out-of-parallel parts, with or without a swivel base.
7. The use of a stop to arrest end motion and poppets or bunters and toe dogs to secure the part down to the table.
8. The fixture permits more rapid location and holding of the workpiece than direct attachment to the machine table.
9. Shaper and planer tools prepared for cast iron usually have less side and back rake angles than those prepared for steel.
10. Placing the tool on the ram side is better for roughing cuts because it permits the tool to deflect without increasing the load. Where visibility is more important and cuts are light, it is acceptable to place the tool on the front side.

section f unit 2 cutting factors on shapers and planers

SELF-TEST ANSWERS

1. Negative back and side rake give better support beneath the cutting edge, so that at the instant of tool contact with the workpiece there is a lessened tendency for tool breakage.
2. M-2 and M-3 high speed steels are typically recommended. For carbides, the C-2 grade is recommended for cast iron and nonferrous materials and the C-6 grade for steel cutting.
3. The large amount of tool contact on planer finishing tools can lead to chatter at higher speeds.
4. *Machining Data Handbook* is the chief source of useful data for speeds and feeds on planer (and shaper) applications.
5. The depth of cut and the feed rate should be as much as can be tolerated by the weakest link of cutting tool, workholding, part strength, or machine tool rigidity and power, consistent with the required workpiece surface characteristics.
7. $\dfrac{CS \times 7}{L} = \dfrac{12 \times 7}{15 + 1} = \dfrac{84}{16} = 5.25$ strokes per minute.
8. $\dfrac{CS \times 7}{L} = \dfrac{60 \times 7}{4\frac{1}{4} + \frac{3}{4}} = \dfrac{420}{5} = 84$ strokes per minute.
9. Deep roughing = depth $\times$ 10 percent = .450 $\times$ .10 = .045 in.; shallow roughing = depth $\times \frac{1}{3}$ to $\frac{1}{2}$ = .100 $\times$ .33 to .5 = .033 to .050 in.
10. Finishing feed with square nose finishing tool = approximately three quarter tool width. $\frac{1}{2} \times \frac{3}{4}$ = .50 $\times$ .75 = .375 in. desired feedrate.

section f unit 3 shaper safety and using the shaper

WORKSHEET EXAMPLES

Worksheet 1 Example
1. Toolhead vertical and retracted.
2. Toolholder removed from toolpost.
3. Set lowest number of strokes per minute.
4. Set ram travel.
5. Set ram position.
6. Attach an indicator.
7. Indicate the table in two directions.
8. "Indicate in" vise and lock it in position.

Worksheet 2 Example
1. Secure the workpiece.
2. Install the cutting tool.
3. Set stroke length; allow about 1 in. overtravel.
4. Set the ram position.
5. Set the apron angle.
6. Reset the cutting tool.
7. Bring up the table and clamp it in place.
8. Cross feed to the workpiece.
9. Position the table support.
10. Calculate the stroke rate.
11. Set the stroke rate into the machine or "dial in" the cutting speed if the shaper is hydraulic.
12. Set the feedrate and direction of feed.
13. Position the toolslide and lock it in place.
14. Operate the machine.

Worksheet 3 Example
1. Turn the deburred flat surface to the solid jaw.
2. Obtain a cylindrical piece and place it on movable jaw side.
3. Machine the top surface.
4. Remove and deburr the part. Clean vise and parallels.
5. Check with a square.
6. Turn the part over end for end so that the last surface is toward the parallels on the bottom of the vise and the previous flat surface (1) is in contact with the vise solid jaw. Use round bar against movable jaw.
7. Machine the top. Bring the part to the required dimensions.
8. Remove and deburr the part; clean parallels and vise.
9. Replace the workpiece in the vise without the round bar and tap it down on the parallels as you tighten the vise.
10. Machine the top surface of the part to the required dimensions.
11. Remove the part, deburr, and clean parallels and vise.
12. Check the part with a square, the dimensions with a micrometer or with a height gage and indicator.

Worksheet 4 Example
1. Set the vise parallel to the table cross travel.
2. Indicate the vise for parallelism to crossfeed, on the top and on the face of the fixed jaw.
3. Use a square to set the part vertically.
4. Recalculate the stroke rate and reset the transmission accordingly.
5. Reposition the ram.
6. Machine the end of the part.
7. Remove, deburr, and turn the part over to machine the opposite end.

Worksheet 5 Example
1. Position the vise and "indicate in" as for surfacing.
2. Calculate and set the stroke rate.
3. Adjust the ram position.
4. Set the toolhead to vertical.
5. Set the clapper box apron to relieve the tool on the return stroke.
6. Install the roughing tool.
7. "Rough out" to near layout line by hand contouring.
8. Set the toolhead to 45 degree angle.
9. Reposition the apron on the clapper box to relieve the tool on the return stroke.
10. Install the finishing tool to clear the part throughout the cut.
11. Make the cut by infeeding at 45 degrees with the toolhead feedscrew.
12. Retract the tool and maintain the table position.
13. Turn the part 180 degrees to place the unfinished part of the "V" under the tool and tighten the vise.
14. Machine the second "V": surface to the identical machine settings to assure symmetry.
15. Clean the machine completely and put the tooling away.
16. Leave the machine with minimum ram travel and in the lowest speed range with the clutch disengaged.

SELF-TEST ANSWERS
1. The greatest hazard arises out of the shaper ram travel. The ram movement is a hazard to people both in front of and behind the shaper.
2. The ram must be completely stopped before measuring or making adjustments. Do not allow other people to operate the machine controls. Use a chip screen. Check all workholding with special care. Remove all loose handles before operating. "Turn down" vise handle. When securing, remove cutting tools and tool holder; set toolhead to vertical and retract the toolslide; set stroke length to minimum; set stroke rate to minimum; check to see that the clutch is disengaged.
3. Positive pressure lubrication is usually supplied by a lubricating pump, or from the main hydraulic system.
4. Have your instructor inspect it.
5. At least $\frac{1}{2}$ in. before the cut and $\frac{1}{4}$ in. after (unless you are working to a shoulder).
6. The table should be indicated in the crossfeed direction. The tilt plate should be indicated (if that side is up).

7. $\dfrac{CS \times 7}{L} = \dfrac{35 \times 7}{L + 1} = \dfrac{245}{6.5} = 37.6$ strokes per minute.

8. The line contact provided by the round bar forces the reference surface into full contact with the solid jaw of the vise.
9. With vertical squaring the vise could be out of square considerably without affecting part squareness. When squaring from the end, the squareness of the part is completely dependent on the squareness of the vise.
10. Reversing the work while maintaining the table position assures that both angles will be exactly the same and will intersect at the center of the workpiece.

section f unit 4 features and tooling of the planer

SELF-TEST ANSWERS

1. The rail provides mounting and guidance for the railhead.
2. Yes, the clapper box is capable of being swiveled to relieve the tool as on a shaper.
3. Yes.
4. Rack and pinion drives are usually associated with direct current drive motors to obtain quick reverse and return. The hydraulic drive is used also; it used hydraulic oil under pressure to drive pistons connected to the machine table.
5. Plain slot, gooseneck, single end finger, double end finger, adjustable step pattern "U" strap, universal adjustable.
6. These are used mainly for stops and poppets.
7. Yes, either together or separately.
8. First, place a stop at the end of the part to arrest the end thrust of the cut. Then use rows of poppets and toe dogs to secure the part down to the table.
9. Planer jacks, riser blocks, step blocks, and straps.
10. Because they must be able to stand up to loads that can reach eighteen tons or more on large planers.
11. A. Railhead.
    B. Table (platen).
    C. Bed.
    D. Table start-stop control.
    E. Cutting speed control.
    F. Feed control.
    G. Sidehead.
    H. Housing (column).
    I. Feedbox.
    J. Rail.

section f unit 5 planer safety and using the planer

SELF-TEST ANSWERS

1. The most basic hazards in planer usage are operator entrapment and the dangers arising from the handling of heavy workpieces.
2. A slightly compressible laminated plastic material is often used. Other firm compressible materials are also used. Shims to remove "rock" from irregular forgings, castings, and weldments.
3. If the eye of a cable sling is not free to swivel, it can sometimes exert enough leverage to wrench the clamp from the workpiece and allow the part to drop.
4. Poppets and planer pins are especially useful to apply force to the workpiece.
5. When the cross rail is raised slightly after having been lowered, it takes up backlash and reduces the possibility of having the head creep downward after tightening.
6. The feeding function of the planer must be allowed for in this distance.
7. The broad nose finishing tool should be fed from two-thirds to three-quarters of tool width on each stroke.

8. The part must be supported by some external means until the stops are set and clamping is done to offset the out of balance. Then the external support should be removed before jacks or other supports are brought into contact.
9. If a part is being leveled to a surface generated from the surface to be cut, a height gage and indicator can be used to check part level. The same device can be zeroed to the table surface and used to establish the actual workpiece height.
10. Sometimes planer jacks have too much leverage applied, which distorts the part level. Careful readjustment is needed to obtain part level without loss of support.

section g unit 1 mechanical and physical properties of metals

SELF-TEST ANSWERS

1. Creep occurs within the elastic range of metals.
2. Creep failures occur over a period of time; the rate of creep increasing as the temperature increases.
3. A metal that is ductile above its transition temperature behaves as a brittle metal below the transition temperature and loses much of its toughness.
4. Hardness, strength, and modulus of elasticity are increased with a decrease in temperature.
5. Nickel will lower the transition temperature when alloyed with steel.
6. Resistance to penetration may be measured by the Rockwell or Brinell testers; elastic hardness may be measured with a scleroscope, and abrasive resistance may be tested in the shop to some degree with a file.
7. Tensile, compressive, and shear.
8. Unit stress $= \dfrac{\text{Load}}{\text{Area}} = \dfrac{40{,}000}{4} = 10{,}000$ PSI
9. Ductility is the ability of a metal to permanently deform under a tensile load.
10. Malleability is the ability of a metal to deform permanently under a compressive load.
11. Fatigue strength can be improved by eliminating sharp undercuts, deep tool marks, and other forms of stress concentration.
12. Electrical and thermal conductivity are related. A metal that is a good conductor of electricity is also a good conductor of heat.
13. A pure metal (unalloyed) will conduct best.
14. The rate of thermal expansion is expressed in inches (of dimensional change) per inch (of material length or diameter) per degree Fahrenheit. The value for each material is called the coefficient of thermal expansion.
15. A machinist works with dimensional tolerances of a few thousandths or ten-thousandths of an inch and the temperature rise caused by machining friction often exceeds these tolerances. Warping of parts and surface cracking when grinding, can also cause damage and scrapping of parts.

section g unit 2 the crystal structure of metals

SELF-TEST ANSWERS

1. Atoms are composed of a nucleus consisting of positively charged protons and neutrons. Electrons are negative in charge and revolve in paths called shells. The outer shell or valence electrons are important in determining chemical and physical properties.
2. The electron cloud of free valence electrons creates a mutual attraction for the metal atoms. This free movement accounts for their high electrical and thermal conductivity, plasticity, and elasticity.
3. a. Body-centered cubic.
   b. Face-centered cubic.
   c. Close-packed hexagonal.
   d. Cubic.
   e. Body-centered tetragonal.
   f. Rhombohedral.

4. Dendrite.
5. No. Each grain grows from its own nucleus in independent orientation and so the lattice structure of adjacent grains jam together in a misfit pattern.
6. Austenitized carbon steel is a solid solution of carbon in the interstices of the iron. When the steel is quenched, the FCC attempts to transform into BCC, but BCC can contain almost no carbon in its interstices. The result is a hard, elongated body-centered tetragonal cubic structure.
7. A material that changes its crystal lattice structure is called allotropic. Iron changes from body-centered cubic to face-centered cubic as the temperature rises. It is therefore an allotropic element.
8. Fine grained steels are preferred over coarse grain for almost every application. Coarse grain has increased hardenability; that is, it will carburize faster than fine grain steels.
9. No. Some metals such as lead and zinc will separate when molten, as oil and water do.
10. **a.** Substitutional and interstitial.
    b. Continuous solid solution.
    c. Terminal solid solution.
    d. Combined solution.
    Mixtures are also often found in the solid state.

section g unit 3 phase diagrams for steels

SELF-TEST ANSWERS

1. Check Figure 2 in this unit for the correct answers.
2. It is a short horizontal section on the line.
3. Lowest transition temperature. Eutectic is the lowest melting temperature. Eutectoid is the lowest transition temperature from one solid phase to another.
4. A_3 shows the beginning of transition from austenite to ferrite. A_1 shows the completion of austenite transition to ferrite and pearlite. A_{cm} shows the limit of carbon solubility in austenite.
5. Cementite. It appears dark when in the form of lamellar pearlite, but white when massive such as in grain boundaries or in white cast iron.
6. It is FCC. Iron with a high solubility for carbon. Austenite normally exists at temperatures between 2700°F (1482°C) and 1330°F (721°C), depending on carbon content.
7. Ferrite. It appears light under the microscope. It dissolves very little carbon, about .008 percent at room temperature and .025 percent at 1330°F (721°C).
8. Pearlite. It looks like a fingerprint. It is alternating layers of ferrite and cementite.
9. Ferrite. 1330°F (721°C).
10. The A_1 and $A_{3,1}$. Ferrite forms austenite at the A_3 line.
11. No.
12. It moves it to the left or decreases the carbon content of the eutectoid.
13. Chromium, titanium, or molybdenum.
14. Nickel or manganese.
15. Knowing the transformations that take place during heat treatments and understanding something about carbon and alloy additions and their effect on transformation, the heat treater will be better equipped to predict the outcome of the specific part.

section g unit 4 I-T diagrams and cooling curves

SELF-TEST ANSWERS

1. About 50° F (10° C) above the A_3 or $A_{3,1}$ lines.
2. Martensite is produced by rapid quenching from the austenitizing temperature to the Mf or near room temperature. Time is the major consideration.

Appendix I Self-test Answers 595

3. The Ms temperature is the point at which martensite begins to form and Mf is the point where it is at 100 percent transformation.
4. The critical cooling rate is the time necessary to undercool austenite below the Ms temperature to avoid any transformation occurring at the nose of the S-curve.
5. The microstructure should be partly fine pearlite and partly martensite.
6. An increase in carbon content moves the nose of the S-curve to the right.
7. Hardening and tempering.
8. This could be caused by the difference in internal and external cooling rates. Changing to an oil or air hardening steel may correct the problem.
9. It can be observed that a cooling curve will always cut through the nose of the S-curve no matter how fast the part is quenched. It is therefore evident that little or no martensite may be produced.
10. A steel that is hardenable and is deep hardening shows an S-curve that is moved to the right.

section g unit 5 hardenability of steels and tempered martensite

SELF-TEST ANSWERS

1. The Jominy End-Quench Hardenability Test.
2. A 1 in. diameter specimen about 4 in. long is heated to the quenching temperature and placed in a jet of water so that only one end is cooled without getting the side wet. When the whole specimen has cooled, flat surfaces are ground on the sides and Rockwell C scale readings are taken at $\frac{1}{16}$ in. intervals.
3. The rate of cooling and hence the microstructure as shown on the I-T diagram are related to the depth of hardening as shown by the Jominy end-quench graph.
4. Circulation of the cooling medium in all cases increases depth of hardening.
5. RC67 is about as hard as any carbon steel will get. Steels with less than .83 percent carbon will not get this hard.
6. Coarse pearlite.
7. Austempering is isothermal quenching in the lower bainite region of the S-curve. Since no martensite is formed, a tougher, more ductile microstructure is the result, and it is superior to a quenched and tempered product of the same hardness.
8. The best time to temper is immediately after quenching, as soon as the piece is cool enough to be hand held.
9. The blue brittle tempering range is found between 400° F (204° C) and 800° F (427° C). Only some alloy steels show loss of notch toughness at higher tempering ranges. This is called temper brittleness. It can be avoided by quenching from the tempering temperature.
10. You can predict the hardness that a tempered part will be by consulting a mechanical properties chart for that particular grade of steel.

section g unit 6 heat treating steels

SELF-TEST ANSWERS

1. Electric, gas, oil fired, and pot furnaces.
2. The surface decarburizes or loses surface carbon to the atmosphere as it combines with oxygen to form carbon dioxide.
3. Dispersion of carbon atoms in the solid solution of austenite may be incomplete, and little or no hardening in the quench takes place as a result. Also, the center of a thick section takes more time to come to the austenitizing temperature.
4. Circulation or agitation breaks down the vapor barrier. This action allows the quench to proceed at a more rapid rate.
5. By furnace.
6. They run from the surface toward the center of the piece. The fractured surfaces usually appear blackened. The surfaces have a fine crystalline structure.

7. a. Overheating.
   b. Wrong quench.
   c. Wrong selection of steel.
   d. Poor design.
   e. Time delays between quench and tempering.
   f. Wrong angle into the quench.
   g. Not enough material to grind off decarburization.
8. a. Controlled atmosphere furnace.
   b. Wrapping the piece in stainless steel foil.
   c. Covering with cast iron chips.
9. Changes in hardness of the surface area and the development of high internal stresses during grinding.
10. An air hardening tool steel should be used when distortion must be kept to a minimum.

section h unit 1 selection and identification of grinding wheels

SELF-TEST ANSWERS

1. Use silicon carbide for the bronze, aluminum oxide for the steel, either manufactured or natural diamond for the tungsten carbide inserts, and cubic boron nitride or aluminum oxide for the high speed steel. Cubic boron nitride could be eliminated.
2. For a straight wheel the third dimension is the hole size. On a cylinder wheel the third dimension is the wall thickness.
3. Side grinding: cylindrical (Type 2), straight cup (Type 6), flaring cup (Type 11), and dish (Type 12). Peripheral grinding: straight (Type 1) and dish (Type 12). The saucer or dish wheel is the only shape on which both peripheral and side grinding are rated safe.
4. These cutoff wheels would be rubber-bonded, because they are the only ones that can be made so thin.
5. Ball grinding requires the hardest wheels, then cylindrical grinding (cylinder ground with a peripheral wheel), then flat surface grinding with a peripheral wheel, then internal grinding and, finally, the softest, flat surfacing with a side grinding wheel.
6. Wheel 1 is a resinoid-bonded (B) aluminum oxide (A) wheel, very coarse grit (14), very hard (Z) and dense (3) structure. It is a typical specification for grinding castings. Wheel 2 is a vitrified-bonded (V) silicon carbide (C) wheel, same grit size but with medium (J) grade and structure (6), for grinding some soft metal like copper.
7. Wheel 1 is the peripheral grinding wheel, because it is a harder (K) and denser (8) wheel. Both wheels have the same abrasive, silicon carbide, and the same bond, vitrified. The abrasive in wheel 2 is coarser, 24 grit as against 36, and may be used for side grinding.
8. This is a vitrified aluminum oxide wheel, H grade, 8 structure, in a 46 grit size. The 32 indicates a particular kind of aluminum oxide, and the "BE" a particular vitrified bond.
9. In a straight wheel for flat grinding, you would use a softer (J or I) and more open structure (7 or 8) wheel. With a cup or segmental wheel, you would use a still softer grade, like H. For flat grinding, probably a size coarser grit, 46, would be used. The important thing is to go softer.
10. The material to be ground usually determines the abrasive to be used, although it is not practical to stock cubic boron nitride or diamond wheels unless you have a good deal of work to be done with them.

    Hard materials require fine grit sizes and soft grades. With soft, ductile materials, coarser grit sizes and harder, longer-wearing wheels are needed.

    Except for very hard materials, the rule is that grit sizes are coarser as the amount of stock to be removed becomes greater.

    Generally, the finer the grit, the better the finish.

section h unit 2 grinding wheel safety

SELF-TEST ANSWERS

1. A 12 in. wheel can be safe on a 2000 RPM grinder. The calculation is either: 12 × 3.1416 ÷ 12 × 2000 = 6283 SFPM or

Appendix I Self-test Answers 597

$$\frac{6500}{12 \times 3.1416 \div 12} = 2069 \text{ RPM}$$

2. a. Tap about 45 degrees either side of the vertical centerline.
   b. Use a wooden tapper or mallet.
   c. Suspend the wheel, or place it on a clean, hard floor. Make sure there is nothing to muffle the sound.
3. a. On a receipt from supplier, to insure getting sound wheels.
   b. Before first mounting, to catch cracks developed in toolroom storage.
   c. Before each remounting, to catch cracks developed in storage at machine.
4. a. Do not bump or drop wheels.
   b. Do not roll wheel like a hoop.
   c. If you can not carry a wheel, use a hand truck, but support and guard the wheel to prevent cracking.
   d. Never stack anything, particularly metal tools, on top of wheels.
   e. In storage, separate wheels from each other with cardboard or other protection.
5. Flanges must be equal in diameter, clean and flat on the mounting rim, but hollow inside the rim so there is no pressure at the wheel hole. Flanges must be checked to see that there are no burrs and no nicks. They must be at least one third the wheel diameter.
6. a. Ring test the wheel.
   b. Check machine RPM against RPM of wheel. Wheel must be equal or higher.
   c. See that wheel fits snugly on spindle or mounting flange. It should not be so tight that it must be forced on, nor should it be loose and sloppy.
   d. Check flanges as above. Blotters should be larger than flanges.
   e. Stand to one side while starting the wheel.
   f. Run the wheel at least a minute without load before starting to grind.
7. a. Always wear approved safety glasses or other face protection.
   b. Grind on the periphery, except for Type 2 cylinder wheels, cup wheels, and other wheels designed for side grinding.
   c. Never jam work into the wheel.
   d. Always make sure safety guard is in place and properly adjusted.
   e. Never use a wheel that has been dropped.
   f. If using coolant, turn coolant off a minute or so before stopping the wheel. This keeps the wheel from becoming unbalanced.
8. If RPM remains the same, then wheel speed (SFPM) increases as RPM increases. Or to put it another and more practical way, as wheel diameter decreases through wear and dressing, the RPM must be increased to maintain the same speed in surface feet.
9. The fact that a wheel will run at 150 percent of safe speed in a factory test room, under extremely safe conditions with extra protection, and *without any load*, is no guarantee that it would be equally safe *with load under operating* conditions. The 50 percent overspeed has been determined by research to be a reasonable assurance of safe operation at the listed safe speed. It is common manufacturing practice to test products under conditions much more severe than the suppliers ever expect the product to meet in actual use.
10. Long sleeves, neckties, or anything that could be caught in a machine is not acceptable in a grinding shop or in any other shop. This includes finger rings and wrist watches. In fact, you should not wear any watch at all around a surface grinder, where the magnetic chuck could magnetize it and ruin it.

section h unit 3 care of abrasive wheels: trueing, dressing, and balancing

SELF-TEST ANSWERS

1. Trueing a grinding wheel means making the OD of the wheel concentric with the center of the spindle after the wheel is mounted. This means that the wheel will be a little out-of-round with its own center. On aluminum oxide wheels and silicon carbide wheels this usually means dressing abrasive grain off the periphery of the wheel. On diamond wheels the adjustment is made as close as possible by tapping the wheel into place. Trueing is needed to insure a good finish on the workpiece.
2. Wheels, like automobile tires, must be balanced when they are mounted, particularly on mounting flanges, as is true for many large wheels. In fact, the reasons and the procedure are very similar to the ones for tires.

Balancing is required on all wheels over 12 in. in diameter; it is useful on smaller wheels when finish is critical.
3. Trueing is a dimensional job, making the wheel's OD concentric with the machine's spindle. Dressing is resharpening the wheel by removing dull grain and bits of work material. Wheels should be dressed when the grain becomes dulled or loaded (filled with bits of metal), and also when you change from roughing to finishing or back to roughing.
4. Form dressing is simply shaping the grinding surface of a wheel to produce something other than a flat surface on the workpiece. It can be done by crush roll, diamond-plated form block or roll, or single point diamond tools mounted so that it can accurately generate a shape in the wheel face by following a pattern.
5. Diamond and cubic boron nitride wheels have to be mounted to much closer tolerances than other types of wheels. With a common abrasive wheel, after the first mounting and trueing, it is often enough to remount the wheel and dress it for whatever type of work you want to do. With diamond wheels, even though the core is machined to a much better fit than is possible with any abrasive, the maximum runout is about .0005 in., and that takes time to achieve. It is just a fussier job.
6. The essential point in placing the diamond for wheel dressing is that it contact the wheel just a little past the vertical or horizontal center line of the wheel, depending on the placement of the dresser, and that the diamond be pointing in the direction of wheel travel at an angle of about 15 degrees. On most dressers, the angle is taken care of. If in doubt, however, move it about $\frac{1}{8}$ in. ahead. If the wheel is rotating clockwise, then the diamond should be at about 12:05 p.m., 3:20 p.m., or 6:35 p.m.
7. The steps in trueing a wheel are:
   a. Assuming the diamond is properly placed, adjust the wheel so that the diamond is touching the high point of the wheel.
   b. Turn on the coolant (if possible) and start the wheel, allowing it to run for about a minute before starting to true the wheel.
   c. Use light infeed, .001 in. or less per pass across the wheel. For trueing, the speed of traverse does not matter.
   d. If dressing is done dry, stop traversing after every three or four passes to allow the diamond to cool.
   e. Continue trueing only until the diamond is contacting the complete grinding surface of the wheel. Going further wastes abrasive, and stopping short of that point means there is still a low point on the grinding surface of the wheel.
8. For roughing, you want the grinding surface to be as sharp and open as possible, so you take quick passes across the wheel. For finish grinding, the passes are slower, not so deep, and you probably will finish with two or three passes without any downfeed or infeed at all.
9. Assuming that bearings and other machine components are running true, if diamond wheel runout is still too much, the wheel must be trued. This can be done with a brake trueing device or, in the case of a resinoid-bonded diamond wheel, by grinding a piece of low carbon steel.
10. Generally, any wheel 12 in. or over in diameter should be balanced the first time it is mounted, particularly when it is mounted on flanges, and any other time it is remounted. If there is a pattern of recurring marks on the finish of the workpiece, then rebalancing may be in order also.

section h unit 4 grinding fluids

SELF-TEST ANSWERS

1. The direct answer to this question is that most grinding is done wet, and the grinder operator is in the best position to see whether the workpieces from the grinder are coming from the grinder in the condition they should. Fluids are necessary to reduce heat and he can observe the grinding operation; with a little experience he can begin to "sense" whether or not the coolant is doing its job. Finally, there are many brands of grinding fluids, and the operator who knows something of their uses can understand why some of the instructions are given. He will not be following them blindly.
2. The three major functions of the grinding fluid are to cool, lubricate, and clean the surface being ground. Cooling is essential to keep the heat from distorting and perhaps warping the workpiece, particularly thin work. It also prevents "burning" or heat discoloration of the ground surface. Lubrication affects the grinding action of the wheel. The quality of the finished surface is directly affected by the lubricating qualities of the fluid. Furthermore, lubrication probably tends to make the wheel act a little softer, so it can have some effect on wheel selection. The third action is cleaning, and the effectiveness here depends on both the volume and the force of the coolant stream. Water based fluids are probably superior in this function.

Appendix I Self-test Answers

3. The design and placement of the flood coolant nozzle are important because together they ensure that the coolant penetrates the curtain of air that surrounds the revolving wheel and actually gets to the grinding or cutting point. It makes no difference how much fluid is splashing about. It is the amount that actually gets to the grinding line that counts.
4. Water soluble chemical fluids are the fluids used most with excellent cooling ability, additives that make them satisfactory in lubrication on most grinding jobs, and quite satisfactory in removing grinding swarf. An added benefit is that they are transparent, probably giving the best view of the work of all the fluids. Swarf settles out easily.
5. Water soluble oil grinding fluids are probably most often used on light to medium stock removal jobs. They are on a par with chemical fluids in cooling and cleaning, are slightly better in lubrication, and the swarf settles out easily. The addition of emulsifying agents causes the closest thing to an actual oil-water mixture. The result is a milky solution that cuts down on visibility.
6. Straight oils rate well in lubrication, but low in practically every other comparison. They are used only on the heaviest stock removal jobs, where lubrication is the top requirement that makes up for below average cooling ability, tendency to retain chips instead of allowing them to settle out, and other shortcomings. Oil does a pretty good job of cleaning but, as mentioned above, it is used mainly where its lubricating ability is a must.
7. A mist coolant system is simply one with a small container, a hose with a nozzle equipped so that a fine mist of the coolant is directed against the grinding area. The heat evaporates the mist, so there is nothing to recirculate; and consumption is closer to pints per day than gallons per hour. These systems are used mainly on dry grinders that need an occasional job done with a grinding fluid. These are usually portable and need only an air connection after they are mounted on the machine to be ready to go to work.
8. a. Settling

    Advantages. Easy to maintain. Needs only occasional cleaning out of accumulated swarf in the bottom, but usually comes equipped with a "drag-out" cleaner that makes this easy. Virtually no maintenance cost. With water based coolants, it does a good job if the tank is large enough and some recirculation of fines is acceptable.

    Disadvantages. Needs a larger total volume of coolant to allow swarf time to settle and more floor space than almost any other method. Does not work well if coolant volume is down. Needs time when fluid is quite low for the swarf to settle out.

    b. Filtering

    Advantages. Effective if mesh size of filter provides removal of enough swarf to satisfy finish requirements. Floor space and coolant volume requirements moderate. Simple piece of equipment that rarely breaks down, and easy to fix if it does.

    Disadvantages. Filter cloth with small enough mesh to remove very fine particles would also slow down fluid flow too much. Filter cloth can not be reused. On large volumes of swarf, the cost of cloth can become prohibitive.

    c. Magnetic separator

    Advantages. Probably the most positive protection against the recirculation of "tramp iron" in the fluid. Properly installed, it should remove all iron and steel from the fluid.

    Disadvantages. Almost requires use in combination with another separator to remove nonmagnetic materials such as brass or abrasive particles. Not efficient in shops where general grinding, including a large percentage of nonmagnetic metals, are ground.

    d. Centrifugal

    Advantages. Both the cyclonic and the bowl-type centrifugal do a good, fast job of separating all the swarf from the fluid. Removes both ferrous and nonferrous metal and abrasive. Floor space requirement is least for the bowl type, and perhaps moderate for the cyclonic type.

    Disadvantages. For the bowl type, the cleaning of bowls may present a problem. There must be periodic shutdown, either to change bowls (if there is a spare) or, otherwise, to clean bowls. These are probably the most complicated and expensive of the cleaners. Furthermore, maintenance cost of the bowl types, because of the high speed or rotation, is likely to be high.
9. The principal argument for having the operator involved in the selection of the coolant is probably that he is in the best position to judge how it is working. In a small shop with perhaps six or less operators, the owner is likely to feel that there is not that much difference between water based coolants or oil coolants. Thus, he may not be inclined to buy something that he thinks his operators do not like or feel is effective. In a large shop, it is not usually practical to take a vote on what the operators feel is best. Usually in such operations, which are most often production rather than toolroom-type jobs, there is a person who has specialized in coolants and the plant's products and, in such cases, would make the decision. A good engineer always asks the opinions of operators, however. If the shop has a central coolant system, then someone in management must make the decision.

10. a. Make sure that there is always enough fluid in the system.
    b. Add water or concentrate as directed to keep the solution in proper balance.
    c. Check frequently for lubricating oil in the coolant and report it to the proper person.
    d. Check frequently to see that the nozzle is always in proper adjustment and accurately aimed.

section h unit 5 horizontal spindle, reciprocating table surface grinders

SELF-TEST ANSWERS

1. There are several reasons for studying this grinder:
   a. It is a very popular machine in machine shops.
   b. Ability to operate one is expected of most machinists.
   c. The same principles of abrasive and coolant selection and wheel dressing apply to it as much as they do to other machines.
   d. Workholding for many parts is usually simple.
2. These are characteristics of the electromagnetic chuck:
   a. It virtually eliminates clamping of parts.
   b. Ordinarily the entire top surface of the workpiece is available to be ground.
   c. It is probably the fastest way to set up a workpiece for grinding.
   d. It holds iron and steel readily and, with simple blocking, will hold nonferrous work as well.
3. The wheelhead moves up and down in the column and always grinds in a plane parallel to the top of the magnetic chuck and parallel to the front edge of the chuck. The magnetic chuck moves from right to left and left to right (reciprocates), and both toward and away from the operator (crossfeed).
4. It means that the plane of abrasive action or grinding action of the wheel is always parallel to the top of the chuck. Even when the wheel is formed in some way with several hills and valleys, each point across the wheel is always grinding parallel to the top of the chuck.
5. The downfeed is the most critical motion of all, because the amount of material taken off is often measured in "tenths" (ten-thousandths of an inch), sometimes even less. The crossfeed is much less critical; a couple of thousandths one way or the other are not usually critical. The traverse is not critical at all. The wheel must definitely clear the edge of the workpiece, but whether by half an inch or an inch is not critical. More than an inch, however, and you are wasting too much time.
6. Zeroing slip rings make calculating downfeed and crossfeed much easier and less likely to be in error. It eliminates all calculation in determining the total amount of stock to be removed and the amount of crossfeed per pass.
7. Vertical spindle machines that grind on the flat side of the wheel have tiny scratches caused by the abrasive which are usually seen as overlapping circles. The form on a horizontal spindle must run parallel to the path of the wheel as the workpiece passes under it. Of course, if you want to make a form with grooves crossing at right angles, all you have to do is to grind one set, turn the workpiece 90 degrees, and grind the other set.
8. a. Rotary chuck. Turns the grinder into a horizontal spindle, rotating table type.
   b. Centerless grinding attachment. Makes possible centerless grinding of relatively small parts.
   c. Center-type grinding attachment. Similar to centerless grinder.
   d. High speed attachment. Increases spindle speed to a point where internal grinding or any other type requiring a small, very high speed wheel can be used.
9. a. Vacuum chuck. Holds all kinds of work, both magnetic and nonmagnetic. Particularly good on thin work.
   b. Magnetic sine chucks. Provides workholding for surfaces of many different angles that are not parallel.
10. These grinders are very economical of floor space, probably requiring less than any other type. They are probably the most inexpensive grinders although, of course, the price varies with the quality of the grinder. Within the size capacity of the grinder, they can handle, with attachments, a wide variety of both flat and cylindrical work. Finally, they are probably the most suited of all grinders for teaching the principles of grinding.

section h unit 6 workholding on the surface grinder

SELF-TEST ANSWERS

1. An electromagnetic chuck is made up of alternating strips (rectangular) or rings (rotary) of steel and some nonmagnetic metal like brass or lead. Chucks operate on direct current, 24, 110, or 220 V, depending on the size of the machine. This type of construction has proved itself adequate for the various sizes of grinders in use today. There is a special control that demagnetizes the workpiece quickly; otherwise magnetism remaining in large parts could hold them on the machine, possibly for a long time.
2. While it is said that about three quarters of all work done on surface grinders could be done on a 6 × 12 in. grinder, there is a size limit that is imposed by the dimensions of the chuck. The shape of the workpiece determines how much extra fixturing must be done. If there is a side for chucking parallel to the side to be ground, there is no problem. However, if there is no such side for chucking, then fixtures must be used to hold the surface to be ground parallel to the top face of the chuck. As to weight, the heavier the workpiece, the easier it is to hold. Nonmagnetic materials cannot be held on the chuck without other blocking or fixturing.
3. a. Wipe the top of the chuck clean before each use, preferably with a squeegee. Cloth can be used, but there is a possibility of picking up chips of metal that could later cut your hand.
   b. Run your hand lightly and carefully over the chuck to detect burrs or scratches. If any of these are noticed, remove them with either a black granite deburring stone or a fine grit abrasive stone.
   c. The deburring operation should probably be done at the beginning of each shift, also.
   d. Always be careful in placing workpieces or accessories on the chuck. It is a precision instrument, and its top surface should always be treated with care.
4. a. Use plenty of coolant.
   b. Use a white (friable) wheel, aluminum oxide, medium grit, medium structure, vitrified bond. (Friable A46-HV was the specification listed in the text.) Dress the wheel "open."
   c. Thinly cover the top of the chuck with layout dye, Prussian blue, or something similar for quick detection of low spots. Continue grinding until all the marking is gone.
   d. Downfeed about two "tenths" per pass. A few passes should ordinarily make the chuck flat again.
   e. Test flatness with a dial indicator mounted on the wheelguard.
   f. Deburr the chuck.
5. The electromagnetic chuck is one of the easiest workholding devices to use in the machine tool field. It is fast; most of the work, that with parallel sides for chucking and grinding, can simply be laid on the chuck top. After the current is turned on, you are often ready to grind. It is much faster than clamping or holding work between centers, for example. It will handle the bulk of the work (any iron or steel) without any added fixturing, if the geometry is right. It cuts setup time considerably.
6. Metals like aluminum and brass are nonmagnetic; for that reason they need to be held in place with either
   a. Steel blocking.
   b. Magnetic vises backed up by steel blocking.
   c. Clamping, preferably to a steel plate.
   d. If thin pieces, can sometimes be held by double-faced tape stuck to the chuck. No power is needed for this.
7. The basic precautions for grinding thin pieces are to use only the power that will hold the work in place and to take light cuts. Excessive power may flatten the work on the chuck, only to let it spring out of flat if rolling stresses are relieved. Too heavy cuts could knock the work off the chuck.
8. a. Magnetic sine chuck.
   b. Magnetic parallels or steel blocking of equal height, probably placed crosswise of the chuck.
   c. Requires at least an angle plate and a clamp. Depending on the squareness tolerance and the parallelism of the opposite end, might need a second angle plate and a steel ball under the part and on the chuck.
9. The surface to be ground must be parallel with the top of the chuck and the saddle ways; hence, parallel with the cutting or grinding line of the wheel as the work is traversed beneath it. The surface grinder is essentially a parallel grinder.

    There must be a parallel opposite face for chucking and, if not, the workpiece must be fixtured to make the surface to be ground parallel with the chuck top.

    The fixturing should be kept as simple as it can be to do the job to specifications.
10. Use an auxiliary holding device such as a precision vise or an angle plate. These devices present sufficient area to the magnetic chuck to provide a basic surface for good workholding.

section h unit 7 form grinding and wheel dressing

SELF-TEST ANSWERS

1. Normally, the grinding face of a wheel is dressed at right angles to the sides of the wheel. Form dressing requires changing that face in a way to suit the form to be ground. Form dressing usually requires more abrasive removal and must usually be more accurate.
2. The wheel travels only up and down, and the table or chuck travels only left and right. Neither can be swiveled or tilted. So the contour of the form must run parallel to the line of table travel.
3. Usually, by form-dressing the wheel. Sometimes, by using a sine chuck.
4. For bevel or similar surface, if the part can be held on a magnetic sine chuck, you can use the chuck and grind with a regular wheel face, or you can dress the wheel to the correct angle and grind the part held flat on the machine chuck. If it is too big for a sine chuck, you have to dress the wheel face.
5. The form in the wheel is always the opposite of the form to be ground.
6. You would have to dress a section of the wheel to $\frac{3}{4}$ in. thickness, either taking $\frac{1}{4}$ in. from one side of the wheel or $\frac{1}{8}$ in. equally from both sides. The thickness can be set accurately by using the crossfeed.
7. Form dressing usually involves taking off more abrasive than flat dressing. In flat dressing it does not matter if you take an extra pass or so — only a little wasted abrasive. Form dressing, however, has to be quite accurate.
8. For the angled form you would use a sine dresser adjusted and checked to the angle desired. For a concave form you would use a dresser with the diamond mounted on a swinging arm that can be set for the radius desired. The radius must be centered on the face of the wheel.

section h unit 8 using the surface grinder

There are no self-test answers.

section h unit 9 problems and solutions in surface grinding

SELF-TEST ANSWERS

1. Careful dressing of the wheel, keeping coolant tank full, checking coolant filters, thorough cleaning of chuck before each loading, checking tightness of wheel.
2. A surface defect is anything that is out of pattern; any kind of unwanted scratches or discoloration.
3. Vibration, heat, and dirt, plus poor quality wheel dressing.
4. Vibration is the principal cause. Off-grade wheels and wheels principally too hard.
5. Fishtails are random scratches caused most often by dirt and grit in the coolant. Sliding the work off the chuck might also be considered a cause; and less often, a wheel that is very much too soft. Usually, the starting point in getting rid of them is to check the coolant level and filters.
6. Burning, checking, or discoloration is one. Low spots (out of flatness) is the other. Probably you would check first to see that there was enough coolant and that it was actually getting to the cutting area. Speeding up the table speed, if this is possible, could also help.
7. A burnished surface is one in which the hills of the scratches have been rubbed over into the valleys. It is bad because the surface will not resist wear. Probably the wheel is too hard and you should change it.
8. Poor chuck alignment is probably the first cause. You might also look for dirt or swarf on the chuck or insufficient coolant.
9. Overheating, particularly of thin work, is one. Dressing a wheel too fine could also be a cause. A wheel that is too hard is another cause. In fact, although work can be parallel and not flat, or flat and not parallel, the two conditions usually go together.
10. Mostly because it can not absorb the heat, but also because the grinding may release stresses rolled into the workpieces.

section h unit 10 cylindrical grinders

SELF-TEST ANSWERS

1. The basic difference between the universal grinder and other center-type grinders is that on the universal grinder all major components swivel; certainly the wheelhead and the swivel table do. However, the swivel table is a feature of most grinders of this type.
2. The center-type cylindrical grinder was probably the first developed because it was essentially a matter of replacing the cutting tool on a lathe with a grinding wheel. In fact, the first cylindrical grinders were called grinding lathes.
3. Aside from the fact that surface grinding is flat grinding and cylindrical grinding is the grinding of shafts and similar parts, there are these differences, at least.
   a. Work is more solidly supported on the surface grinder because it is backed up by the magnetic chuck.
   b. On a surface grinder, the reduction in thickness is about the same as the downfeed (infeed) of the wheel; in cylindrical grinding, the reduction in diameter of a part is roughly twice the infeed.
   c. The wheel and the work always move in the same direction with relation to each other in cylindrical grinding. In surface grinding, the work moves alternately with and against the direction of wheel travel.
4. The first step in grinding a taper (or, for that matter, in correcting one you do not want) is to swivel the table to the desired taper. This is assuming that a square or A face is on the wheel. With a universal grinder, when you reach the limit of table swivel, you can also swivel the wheel without redressing the face of the wheel. Otherwise, you would have to dress the wheel to something other than a right angle or A face. Of course, even with a universal grinder you might reach a taper steep enough to require angle dressing of the wheel.
5. Obviously, as indicated in number 4, the steepness or degree of taper must be considered. Most tapers are usually ground by just swiveling the table.
6. In addition to supporting one end of the workpiece, the headstock also supplies the power to rotate it. The headstock can also be modified with accessories to hold work that cannot be held by centers. This might be a part that could be held in a chuck or on a faceplate.
7. Basically, on a center-type cylindrical grinder, the work is supported on two conical work centers that project into two matching conical holes in the ends of the workpiece.
8. A steadyrest is a support to back up a long, thin workpiece against the outward and downward pressure of the grinding wheel. They keep the workpiece from bending. With proper use of steadyrests in odd numbers (one, three, five, etc.) it is possible to support a very thin workpiece so that it will be ground straight. Steadyrests are always used in odd numbers, so that as each pair is added, after the first one, they are placed in the center of the remaining length to be supported.
9. Generally, the greater the variety of work that can be done on a cylindrical grinder, the less its rigidity. The universal grinder, with a swiveling wheelhead, is the least rigid, and the straight plunge grinder, with nothing swiveling, is the most rigid. Also, most workpieces that are plunge-ground are shorter than the thickness of the wheel, which provides a great deal of stiffness.
10. If the center holes are not very accurate and round, it is going to be impossible to make the workpiece they help support accurate and round. In other words, any lack of precision in the center holes will be reflected in the quality of the finished workpiece.

section h unit 11 using the cylindrical grinder

There are no self-test answers.

section h unit 12 problems and solutions in cylindrical grinding

SELF-TEST ANSWERS

1. Factors that make cylindrical grinding problems different from those of surface grinding include:
   a. Cylindrical grinding has a smaller area of contact between the work and the wheel, really only a line contact. Thus a harder wheel is required.

604 Machine Tools and Machining Practices

    b. Cylindrical grinding has a different method of supporting the work.
    c. Cylindrical grinding is always done with the periphery of the wheel, while a lot of surface grinding is done with the wall or flat side of a cylindrical, cup, or segmental wheel.
2. Problems like burning or discoloration of the surface and chatter marks are the same for all types of grinding.
3. To find whether vibration is within the grinder or outside, shut down the grinder and place a pan of water on the wheelhead, for instance. If the water makes ripples, then there is vibration from somewhere outside affecting the machine. If there are no ripples, the vibration is somewhere in the machine. To find vibration inside the machine, operate one element at a time. Again, if there are ripples, then the element that is running is causing the vibration.
4. In cylindrical grinding, as in most other forms of grinding, wheels that are too hard and machine vibrations are two of the broad causes of problems. Coolants, either the wrong kind or mixtures too rich or too lean, can also cause many problems. Sometimes there is no logical explanation of why a certain combination of wheel specification and coolant works on a particular material.
5. The two resemble each other in that both are parallel marks across the workpiece surface, and both are often the result of vibration. However, wave marks are the result of vibration while the workpiece and the wheel are in contact, while chatter is the result of periodic loss of contact between the two.
6. A "burlap finish" is the direct result of something not tightened in the dresser, either the diamond itself or its support. It is the direct result of a dressing error.
7. Both are lines spiraling around the work, and mostly in traverse grinding. Feed lines usually result from something out of line, the headstock and footstock, for example, or a steadyrest exerting too much pressure on a thin workpiece. Trueing lines are the direct result of dressing with a diamond that is too sharp or of failing to round off the corners of the wheel.
8. The general rule of the relationship of wheel grade and machine condition is that, other things being equal, the more rigid the machine, the softer the wheel that can be used on it. Conversely, machines that are in poor condition require harder wheels. In fact, the same relationships hold for vibration. The less vibration, the softer the wheel that can be used.
9. The general rule is: If in doubt, take the softer wheel, at least to start. The kind of grinding done in most machine shops if not of a size or character where wheel wear will be a big factor, although a very soft wheel, for instance, could cause some taper on a very long shaft through wheel wear. The savings resulting from using hard wheels are rarely as much as most managers imagine.
10. If as an operator you would like to avoid trouble caused by coolant:
    a. Keep it clean. Check filters and clean them where necessary.
    b. Maintain the proper and recommended dilution, which would be 20 to 1 to 50 to 1. Measure additions, do not guess at the amounts.
    c. Observe what happens with different coolants.

section h unit 13 cutter and tool grinder features and components

SELF-TEST ANSWERS

1. *A.* Front cross slide hand wheel.
   B. Wheelhead vertical control hand wheel.
   C. Front table hand control.
   D. Table dogs.
   E. Swivel table.
   F. Workhead.
   G. Tilting workhead.
   H. Tailstock.
   I. Sliding table.
   J. Table swivel scale.
2. External cylindrical grinding, grinding of centers, and internal grinding.
3. Grinding arbors can be used for mounting plain and formed milling cutters and also can be used for mounting slitting saws.
4. The plain tooth rest support and the micrometer-type tooth rest support are the two types.
5. The offset tooth rest provides support for milling cutters with helical teeth.

Appendix I Self-test Answers

6. The cutter and tool grinder can be used for external and internal cylindrical grinding, surface grinding, cutting off operations, and single point tool grinding in addition to the sharpening of rotary cutting tools.

section h unit 14 cutter and tool grinder safety and general setup procedures

SELF-TEST ANSWERS

1. a. Always wear safety glasses.
   b. Always use wheel guards.
   c. Ring test grinding wheels before mounting.
   d. Store wheels with special care to prevent impact damage that can result in safety hazards.
   e. Be sure that the spindle RPM either matches or is lower than the marked maximum on the grinding wheel.
2. Cup and dish wheels are designed to accept side grinding stress because of their shape. Straight wheels are not designed for side grinding stresses and can fracture and explode if used that way.
3. If the wheel is cracked, it will probably break during the first minute.
4. It is very difficult to mill flat surfaces with tapered cutters.
5. Incorrect cutting geometry can result in inefficient cutter performance and in cutter breakage from excessive clearance.
6. The indicator drop method uses two dial indicators; one to measure radial movement and the other to show the corresponding drop in the relief or clearance behind the cutting edge. The other is the use of a cutter clearance gage to check the angle of relief or clearance directly.
7. You will *lower* the wheelhead by sine of 7 degrees × 2 in. cutter radius = .12187 × 2 = .244.
8. No, the cutter tooth may be kept level and the angles set in directly on the head.
9. Under this condition, the radius of the grinding wheel is the controlling factor in calculation:
   Sine of 5 degrees × radius of the grinding wheel = .08715 × 3 in. radius = .261 in.
   The wheelhead will be *raised* .261 in.
10. Sine of 5 degrees × radius of the cutter = .08715 × 1.5 in. radius = .131 in.
    In this case, the wheelhead is *lowered,* as it would be in flat grinding with the flaring cup wheel.

section h units 15-19

There are no self-test answers.

section h unit 20 miscellaneous cutter and tool grinding operations

SELF-TEST ANSWERS

1. A thin wheel without reinforcement would shatter in this kind of service.
2. Gashing permits the tool to cut to center and also provides chip space.
3. When they are adjusted, the blades frequently get somewhat out of parallel, resulting in inaccurate, poorly finished holes.
4. With reamers having minimal primary relief and a narrow margin, land interference can result, even with the ¼ IPF starting taper.
5. If the tool is not concentric, it will produce an oversize hole.
6. Because these tools are used mainly for producing center holes in parts to be done on centers in other machines.
7. It changes the pitch diameter of the tap (unless the tap is completely concentric).
8. Taps and other tools (like countersinks) done on this device have differing numbers of cutting edges.
9. Differing materials require differing relief in the cutting tool for cutting efficiency.

section i unit 1 numerical control dimensioning

SELF-TEST ANSWERS

1. See Figure 14.
2. Rotational axes might be used to define the motion of a rotary table or part indexer that operates rotationally around one of the basic axes.
3. -Z axis.
4. -X axis, with respect to the spindle. As the worktable moves right, the apparent spindle motion is left or in a -X direction.
5. Z.

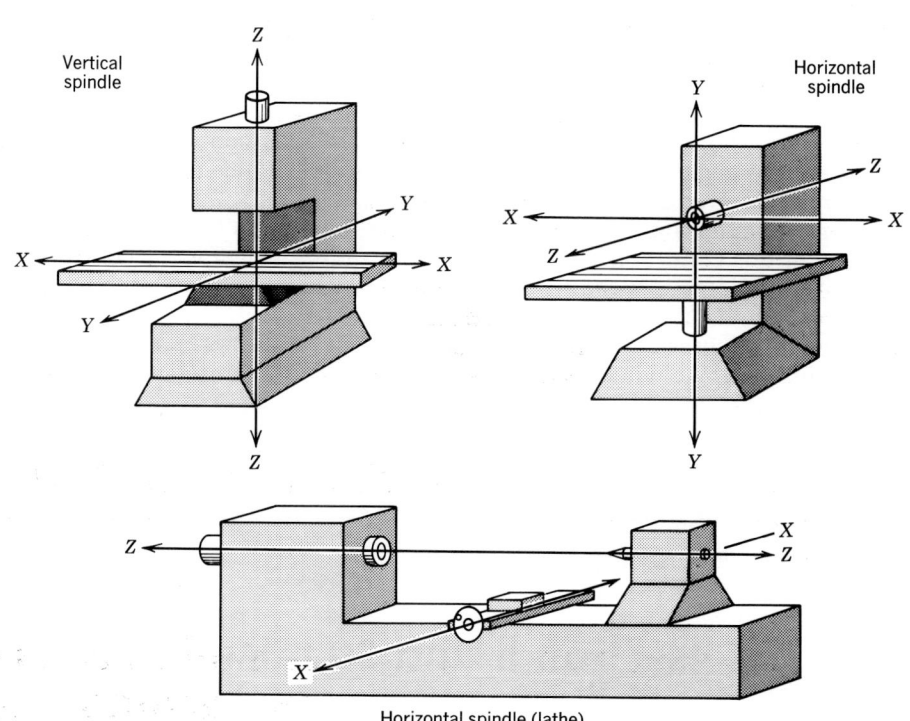

Figure 14.

section i unit 2 numerical control tape, tape preparation, tape code systems, and tape readers

SELF-TEST ANSWERS

1. Paper and plastic laminates.
2. More durable and less subject to damage and wear in the tape reader.
3. Preparation of new tapes, tape duplication, error correction, and typed records of tape information.
4. EIA and ASCII.
5. Electromechanical, photoelectric, and pneumatic.

section i unit 3 basic format numerical control programs

SELF-TEST ANSWERS

1. Tab sequential.
2. Word address.
3. Fixed block.
4. g81 drill or g85 bore are two examples.
5. m02 end of program or m06 tool change are two examples.

section i unit 4 operating the N/C machine tool

SELF-TEST ANSWERS

1. After the tape has been punched, a printout should be obtained and checked against the program manuscript.
2. The dry tape run will verify the accuracy of the tape and prevent possible damage to the machine tool.
3. Be sure that the spindle and cutter are clear of all obstructions before positioning the worktable.
4. The tool length gage is used to adjust tool lengths as well as diameters on boring bars and insert cutters.
5. Micrometers and dial indicator types, electronic digital types.

appendix II

Precision Vise: Drawings I, II, III, and IV
V-block: Drawing I

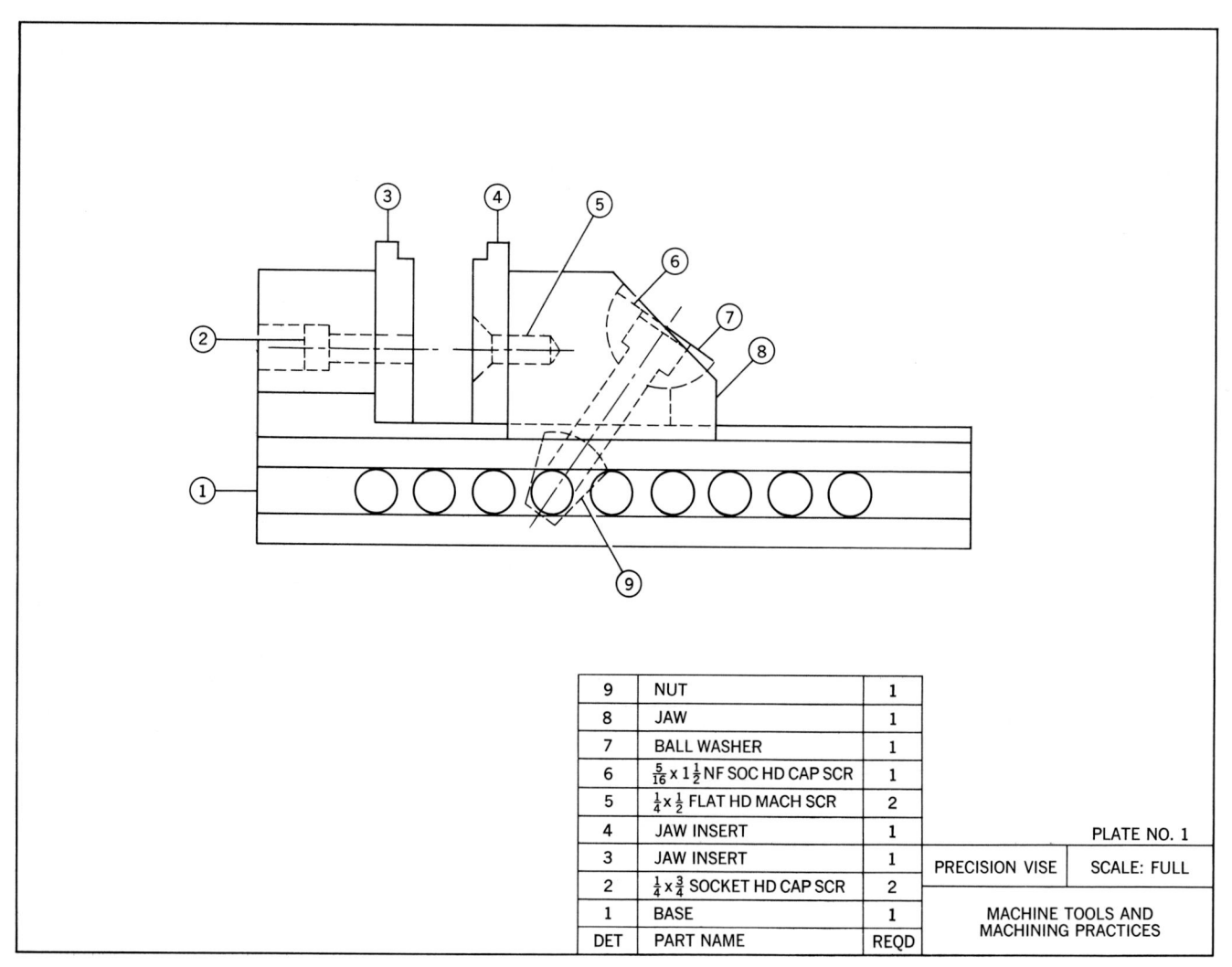

9	NUT	1
8	JAW	1
7	BALL WASHER	1
6	$\frac{5}{16}$ x $1\frac{1}{2}$ NF SOC HD CAP SCR	1
5	$\frac{1}{4}$ x $\frac{1}{2}$ FLAT HD MACH SCR	2
4	JAW INSERT	1
3	JAW INSERT	1
2	$\frac{1}{4}$ x $\frac{3}{4}$ SOCKET HD CAP SCR	2
1	BASE	1
DET	PART NAME	REQD

PLATE NO. 1
PRECISION VISE   SCALE: FULL
MACHINE TOOLS AND MACHINING PRACTICES

Drawing I.

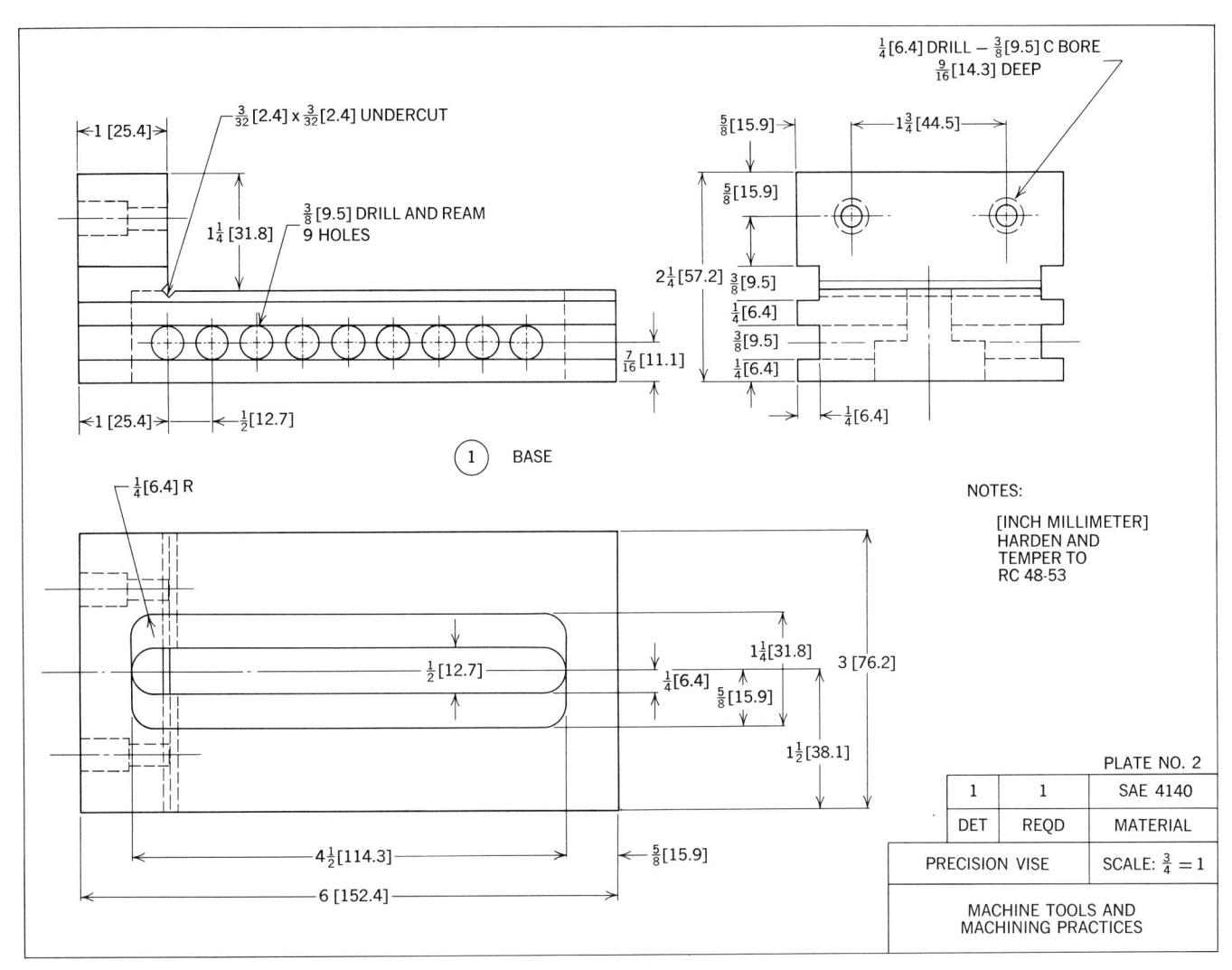

Drawing II.

Machine Tools and Machining Practices

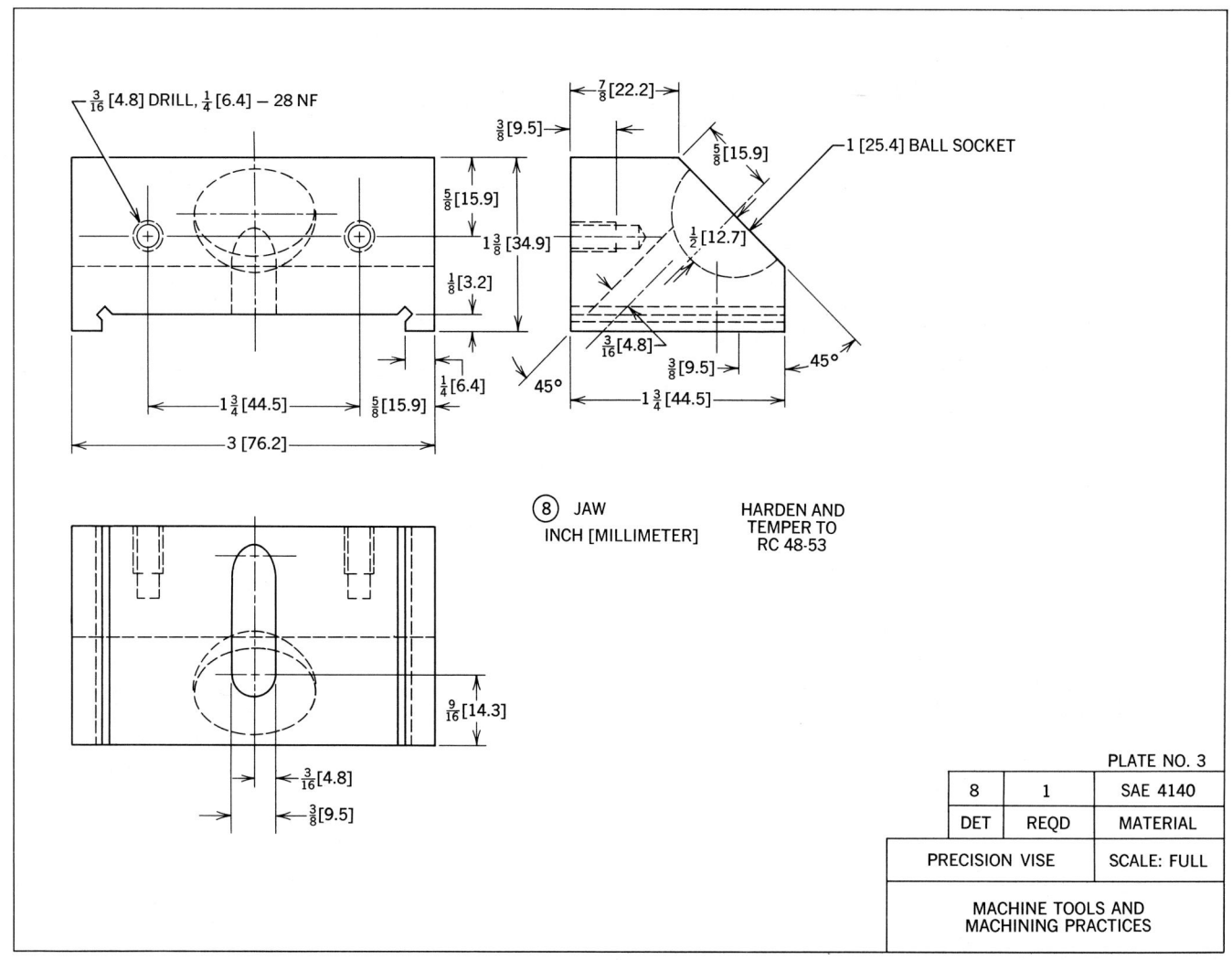

Drawing III.

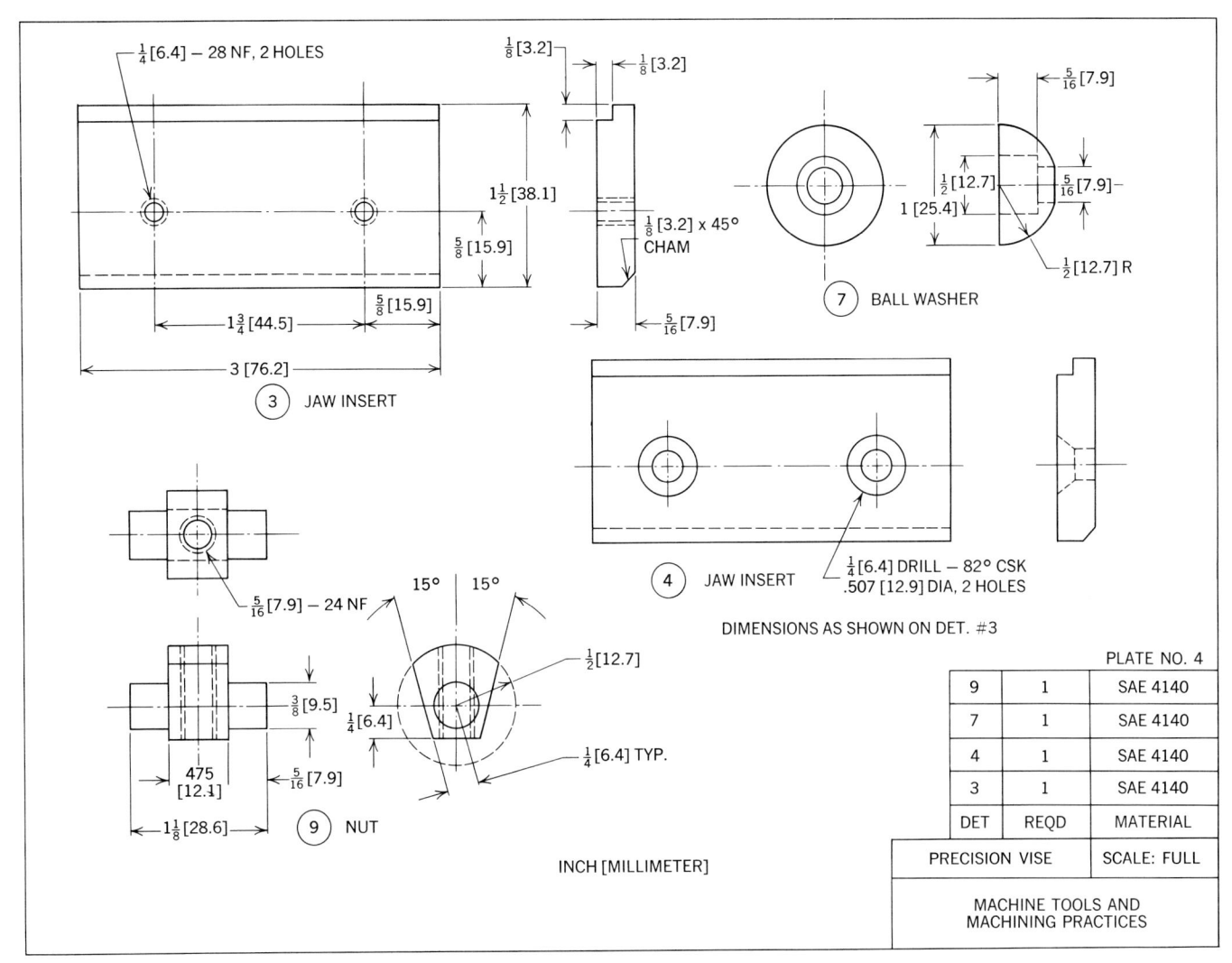

Drawing IV.

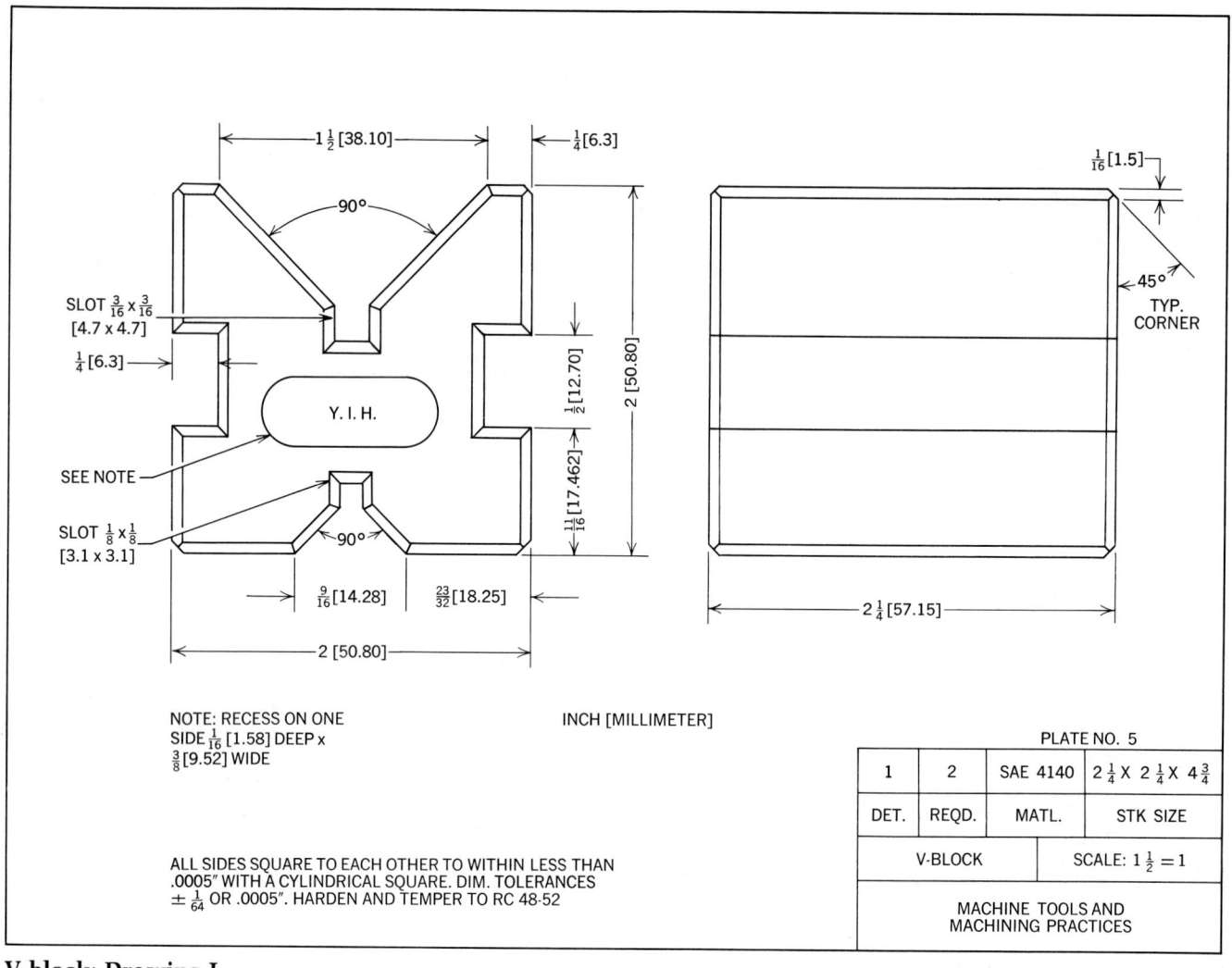

V-block: Drawing I.

index

Abrasives, grinding, 360–361
Adaptors, 94–98
  mounting, 98
  removal procedure, 97
Allotropic element, 293, 294
Allotropy, 293
Arbors, 94–98
  collets, 95
  milling cutters, 95
  mounting, 98, 99–105
  removal procedure, 97
ASCII, N/C tape code system, 545–546
Atom, 282
  bonding, 283
  electrons, 282
    valence, 282
  molecule, 283
  neutrons, 282
  oxygen, 283
  protons, 282
  shells, 282
Austempering, 313
Austenitizing, 304

Band blades, vertical band machine, 15–25, 27–33
Band machines, vertical, see Vertical band machines
Band machining, 1, 2
Band saws, vertical band machines, 15–25, 27–33
  annealing welds, 18–19
Band tools, 2, 9–10
Bevel gears, 172
Bonding of metals, covalent, 183
  ionic, 283
  metallic, 283
  Van der Waals, 283
Brittleness, 271

Collets, 55, 56, 95
Contour sawing, vertical band machine, 1, 2, 32–33
  blade width, selection of, 32
  layout, 33
  purpose, 1
Corrosion resistance, 275
Covalent bonding, 283
Crank shaper, 211, 226–227
  cutting speed formula, 226
  feedrates, 225
  stroke requirements, 226
Creep strength, table, 275
Crystalline lattice structures, 284–286
  body-centered cubic, 285
  close-packed hexagonal, 286
  face-centered cubic, 285

  grain boundary, 284
  heating changes, 286
Crystal structures of metals, 282–292
  atom, 282
  bonding, 283
Cutter and tool grinder, 457–463, 464–469
  cutter angles, 467
  grinding wheel speed, 465
  miscellaneous operations, 507–516
  parts, 458
  radial relief angles, 466–467
  safety, 464–469
  set up procedures, 465
Cutting fluids, milling machine, 108
  vertical band machine, 15, 22, 23, 31, 32
    adjusting coolant nozzle, 22, 23
    purpose, 31
    safety, 14
Cutting tools, sharpening, end mills, 484–491
  form relieved cutters, 492–497
  hand reamers, 510–512
  machine reamers, 498–506
  milling cutters, 470–475
  single point tools, 512–515
  slitting saws, 476–478
  stagger tooth milling cutters, 478–483
  step drills, 512
  taps, 512
Cylindrical grinder, 341–349, 435–443, 445–455
  center-type, 342–344, 437
    universal, 342–343
  centerless, 345–346
  internal, 346–347
  cutter and tool grinder, 347–349
  parts, 438–440
  problems and solutions, 451–456
  procedure, 445–449
  use of, 445–450
  workholding, 440

Dial indicator, 60
Dividing head, 114, 153
Dovetail, 47
Ductility, 272

EIA, N/C tape code system, 544–545
Elasticity, 268–272
  mechanical properties, 268
  modulus of, 272
  stiffness, 272
  stress-strain curve, 269, 270
  yield point, 269
  yield strength, 269

Electrons, atom, 282
  valence, 282
End mills, 45, 65–67, 484–491, 507
  cutting fluid, 65
    formula, 66
    speed, 65
    table, 66
  depth of cut, 66
  feeds, table, 66
  radial relief angles, 484
  sharpening, 484–491
    end, 486–488
    side, 488–490
Eutectic point, 293–294
Eutectoid point, 294–295, 312
Expansion, of metals, 276

Face milling cutters, 136–141
  cutting fluid, 139
  cutting inserts, 137
  lead angle, 137
  parts, 136
  surface finish, 138
Fatigue strength, 274
Flaring cup grinding wheel, for tools, 471–473
Form dressing, grinding, 414–417
Form relieved cutters, 492-497
  sharpening procedure, 494–497
Furnaces, heat treating, 319–320

Gears, 163–168
  bevel, 172
  helical, 173
  materials, 174
  spur, 171, 176–184. See also Spur gear
  worm, 173
Grain size, 286–287, 298
  classification, 287
Grinding, coolants, methods of application, 388–390
  fluids, 387–395
    cleaning, 391–394
    types, 390–391
  form, 411–417
    dressing, 414–417
  grit size, 419
  problems and solutions, 428–434
  procedure, 420–427
  warpage, 433
  wheel dressing, 411–417
  wheel forms, 412
Grinding machines, 331–516
  centerless, 345–346
  cutter and tool, 347, 457–463. See also Cutter and tool grinder
  cylindrical, 341–349, 435–455. See also Cylindrical grinder
  disc, 349
  gear, 349

  parts, 438, 458
  roll, 344
  surface, 337, 396–403, 404–410. See also Surface grinders
Grinding wheels, abrasive, 360–361
  balancing, 384–385
  blotter, 366
  bond, 363–364
  care, 376–385
  dressing, 379–384
  forms, 412
  grade, 361–362
  grit, 361
  hardness, 361–362
  identification, 357–368
  marking system, 360
  mounting, 374
  radius, 468–469
  relief angle, 466–469
  ring test, 371
  safety, 369–377
  selection, 357–368
  shapes, 358–359, 375, 413
  speed, 366, 370
  storage, 373
  structure, 362–363
  symbols, 360–364
  trueing, 379–384
  types, 360–361

Hand reamer, adjustable, sharpening, 510–512
Hardenability test, 311
  jominy end-quench, 311
Hardness, metals, 266
  elastic, 266
Hardness testers, Brinell, 266
  Izod-Charpy impact, 273, 276, 314
  microhardness, 266
  Rockwell, 266
  tensile, 269
  Tukon, 266
Heat treating, steels, 319–329
  double tempering, 323
  furnaces, 319
  grinding, 325
  problems, 323
  quenching media, 320
  soaking, 320, 322
  stress relieve, 325
  tempering, 322
  thermocouple, 319
Helical gears, 173
Horizontal milling machine, 75–141
  adaptors, 95–98
  arbors, 95–98
  control levers, 93
  cutter selection, 119

cutting fluid, 108
face milling cutters, 136–141
feed, 90, 108
indexing devices, 153–155
locating devices, 110–115
lubrication, 92
milling cutters, 99–105
parts, 91
plain milling, 116–123
rotary table, 114
safety, 87–88
set up, 120–121
side milling cutters, 101, 127–135
size, 92
speeds, formula, 107
  table, 106
spindle, 94
table alignment, 117
universal, 92
vise alignment, 117–119
workholding devices, 110–115
Horizontal shaper, 190, 209, 229–241
  care, 237
  controls, 231
  cutting characteristics, 225
  cutting tools, 217
  drives, 211
  feeds, 212, 226
  lubrication, 230
  operation, 230–241
  parts, 210
  speed, formula, 232
  workholding, 214–217
Hydraulic shaper, 212
Hydraulic tracing attachment, vertical band machine, 3–5

Indexing, milling machine, 157–161
  angular, 159
    by degrees, 160
  direct, 157
  simple, 157
Indexing devices, 143–161
  dividing head, 114, 153
  footstock, 154
  index head, 153
    wide range, 153
  rotary table, 155
  use, in making divisions, 157–158
Ionic bonding, 283
Iron allotropic, 294
  cooling curve, 294
  phases, 294
Iron carbon diagram, 294–296, 298
  alloying elements, 298
  delta iron region, 294
  steel portion, 294
  transformation line, 298

Isothermal transformation, I-T diagrams, 303–310
  cooling curve, 305
  cooling rate, 306
  hardening temperature, 304
  oil hardening steels, 307
  quenching rates, 307
products, 304, 312–314
temperature, 304, 312
Izod-Charpy impact tester, 273, 276, 314

Job selector, vertical band machine, 28–29
Jominy end-quench, 311

Keyways, 45

Machine reamers, 498–506
  grinding, body diameter, 502–505
    secondary clearance, 505
  sharpening, 498–506
    horizontal spindle method, 500–502
    tilt head method, 499–500
Malleability, 272
Martempering, 314
Martensite, tempered steel, 315–316
    procedures, 314
    temperatures, 315
Mechanical properties of metals, 265–281
  brittleness, 271
  chart, 270, 271
  corrosion resistance, 275
  creep strength, 275
  ductility, 272
  elasticity, 268
  fatigue, 274
  hardness, 266
  malleability, 272
  notch toughness, 272
  plasticity, 271
  scaling, 275
  stiffness, 272
  strength, 266
Metal, at low temperatures, 275
  bonding, 283
  conductivity, 276–278
  contraction, 276, 278
  crystal structure, 282–292
  expansion, 276
    thermal, 276
  grains, 286–287
  heat treating, 319–329
  mechanical properties, 265–281
  oil hardening steel, 307
  physical properties, 276–278
  properties, table, 284
  solutions, types of, 287
Metalloids, 284
Metallurgical microscope, 295
  parts, 302

Index

Metcalf experiment, 291
Microstructures, austenite, 295, 304
   bainite, 306
   cementite, 295, 304
   ferrite, 295
   iron carbide, 295
   martensite, 304, 312
   pearlite, 293, 295
Milling, plain, 116–123
   table alignment, 117
Milling cutters, 99–105, 470–483
   angular, 102
   convex, 103
   double angle, 102
   involute, 103
   plain, 100
   sharpening plain, 470–475
     cup wheel, 471–473
     straight wheel, 473–475
   sharpening stagger tooth, 478–483
     side teeth, 481–483
   side, 101
   teeth, 105
Milling machine, horizontal, 75–141
   indexing devices, 143–161
   safety, 39–40, 87–88
   universal, 92
   vertical, 35–74
Modulus of elasticity, formula, 272
   low temperatures, 275
   Young's modulus, 272
Molecule, atom, 283

Neutrons, atom, 282
Normalizing, 298, 313
Notch toughness, 272
   Izod-Charpy tests, 273
Numerical control, of machine tools, 517–534
   advantages, 527–530
   and computers, 532–534
   construction features, 519, 520
   control systems, 532
   historical development, 518
   input media, 525, 527
   and machining centers, 520, 521–524, 525
   and the machinist, 531–532
   by numeric instructions, 517, 518
Numerical control (N/C) and computers, 532–534
   compiler-post processor, 533, 534
   direct computer control, 534
Numerical control (N/C) dimensioning, 534–541
   basic machine tool axes, 535, 536, 537
   directions of machine spindle travel, 537–538
   quadrants, 536, 537
   rotational axes, 535, 537
   spindle positioning, by coordinate measurement, 538, 539
     by incremental measurement, 538, 539

Numerical control (N/C) input media, 525, 527
Numerical control machine tools, axis identification, 535, 538
   basic, 535–537
   rotational 535, 537
   boring, 529
   operation, 573–577
     manual, 573–574
     safety, 573
     tape, 574–575
   tapping, 529, 531
   tool changers, 520–522, 526, 529–531
Numerical control (N/C) machining, 529–531
   continuous path, 530, 531
   contouring, 530, 531
   drilling, 529, 531
   peck drilling, 529
   pocket milling, 529
   tapping, 529, 531
Numerical control programs, basic format, 548–572
   blocks and EOB code, 549
   examples, 552–569
   fixed block format, 570, 571
   linear and circular interpolation, 551
   mirror images, 550, 559
   miscellaneous or "m" functions, 548, 549
   preparatory or "g" functions, 551
   rewind stop code, 549
   tab sequential format, 549, 550, 552–571
   word address format, 551, 562–569, 570
Numerical control systems, 532–533
   closed loop, 532, 533
   open loop, 532
Numerical control (N/C) tape, 542–547
   ASCII, 545–546
   code systems, 544, 546
   correction of errors on, 544
   EIA, 544–545
   materials, 542–543
   preparation, 543–544
   readers, 544, 546–547

Offset boring head, 69–74
   depth of cut, 71–72
   parts, 70
   project, 73–74
   quill feed, 71
   tool movement, 70
Oxygen atom, 283

Phase diagrams, 292–301
   eutectic point, 293, 294
   eutectoid point, 294
   iron carbon, 294, 295
Physical properties of metals, 276–278
   conductivity, 276, 277
   contraction, 278

thermal expansion, 276–278
    formula, 278
Planer, 196, 242–259
    controls, 243
    cutting characteristics, 225, 254
        speeds, 225, 253–254
        tools, 246
    drive systems, 243–244
    feed, 225
    feedrate, 226, 254
    parts, 243
    safety, 249–253
    set up, 251, 255
    workholding, 244–246
Plasticity, 271
Precision vise, 62–64, 68–69, 122–126, 132–135, 607–610
Proton, atom, 282

Quenching, 307, 320–327
    austempering, 313
    cracks, 323
    isothermal, 322
    jominy end-quench, 311
    liquid, 320
    martempering, 314
    multiple, 322
    step, 322

Reamers, hand, adjustable, 510–512
    machine, 498–506. See also Machine reamers

Safety, band machines, 11–15
        checklist, 14–15
    cutter and tool grinder, 464–469
    grinding wheel, 369–377
    horizontal milling machine, 87–88
    numerical control machine tools, 573
    planer, 249–253
    shaper, 229–230
    vertical milling machine, 39–40
Saw blades, vertical band machines, 15–25, 27–33
        adjustment, of tracking position, 23, 24
            of tension, 23, 24
        coolant, 22, 31
        grinding, 19
        guides, 20
        gullet, 17
        installation, 20, 22, 23
        length calculations, formula, 16
        materials, 27
        problems, 20
        selection, 27
        sets, 27
        special types, 27, 28
        tooth forms, 27
        welds, 16–20
            annealing, 18

grinding, 17, 19, 20
            procedure, 16–18
Scaling, 275
Scleroscope, 266, 267
Screw jack, 112, 115
Shaper, 190–241
    crank, 211, 226
    cutting characteristics, 225
    cutting tools, 217
    drives, 211
    features, 210
    feeds, 212, 226
    fixtures, 217
    horizontal, 190, 209, 229–241
    hydraulic, 212
    parts, 210
    safety, 229–230
    size, 214
    speeds, 225
    tool holder, 222
    workholding devices, 214–217
Shells, atom, 282
Side milling cutters, 101, 127–135
    horizontal milling machine, 127–135
    layout, 128
Slitting saw, 101, 476–478
    sharpening, 476–478
        formula, 477
Soaking, 320, 322
Solid solutions, 287–290
    continuous, 287
    insoluble, 289
    interstitial, 289–290
    substitutional, 289–290
    terminal, 287
Spindles, milling machine, 94
Spur gear, 171, 176–184
    calculation, 177–179
    cutting, table, 181
        procedure, 181–183
    dimensions, formula, 178
    gear tooth vernier caliper, 185
    inspection, 185–187
    measurement, 185–187
    optical comparator, 187
    symbols, 177
    terms, definitions, 176
    tooth forms, 177
Stagger tooth milling cutters, sharpening, 478–483
        side teeth, 481–483
Steels, hardenability, 311
    heat treating, 319–329
    jominey end-quench, 311
    phase diagrams, 292–303
        eutectic point, 293–294
        heating, 297
        slow cooling, 295, 296
Stiffness, 272

Strap clamps, 246
Strength, compressive, 266
    fatigue, 274
    material, table, 268
    shear, 266
    stress-strain diagram, 269, 270
    tensile, 266
        tester, 267
    unit stress, 266, 269
    yield, 269
Stress-strain, curve, diagram, 269–270
Surface grinders, 337, 396–410
    attachments, 400–403
    care, 406–408
    chucks, 400, 405
        magnetic sine, 402
        regrinding, 407
        rotary, 400
        vacuum, 402
    dressers, 402
    finish defects, 431
    horizontal spindle, 396–403
    using, 418–427
    workholding, 404–410
        nonmagnetic work, 408
        odd-shaped work, 408
        thin work, 408

Tables, causes of problems, in cylindrical grinding, 452
    creep strengths, for several alloys, 275
    cutting speed, for milling, 106
        and starting values, for materials, 66
    feed, in inches per tooth, 108
        for end mills, 66
    gear cutter numbers, 105
    grinding wheel angles, to produce radial relief
        angles for eccentric relief sharpening, 490
        grinding wheel safety rules, 372
        hand reamer, relief and clearance
            recommendations, 510
        indicator drop method, of determining radial relief
            angles, 467
        mass effect data, SAE 4140 steel, 325
        material strength, 268
        mechanical properties, SAE 1015 steel, 270
            SAE 1095 steel, 271
        peripheral (OD) relief and clearance angles, 466
        properties of some common metals, 284
        radial relief angles, for high speed steel end mills, 484
        spur gear cutting, 181
        spur gear formulas, 178
        summary of surface grinding defects, and possible causes, 430
    Tapered lathe mandrel, 445–446
        procedure, 445–450
    Tempering, 314

        furnace, 322
        martensite, 315
        procedures, 314
        steels, 322
Tensile testing, 269
Thermal expansion, 276
    formula, 278
Thermal contraction, 278
Thermocouple, 319
Transformation products, 304, 312–314
    austempering, 313
    isothermal quenching, 313
    martempering, 314
    normalizing, 313
T-slot, 47

Unit stress, 266
Universal grinder, 440
    attachments, 443
Universal milling machine, 75–141
    set up, 120–121
    table alignment, 117
    vise alignment to table travel, 118–119

Van der Waals bonding, 283
Vee block, 409, 419, 420, 611
Vertical band machines, 1–33
    adjustments, 20, 21–25
    backup bearing, 20
    blades, 15, 27–28
    contour sawing, 1, 2, 7, 32–33
    coolant, 31–32
    cutting, 14
    cutting fluids, 15, 22–23, 31–32
    feeding mechanism, 3, 4
    filling of bands, 9
    friction sawing, 7, 8
    grinding of bands, 2
    guards, 12–14
    guidepost, 30
    hydraulic tracing attachments, 3–5
    installing bands, 22, 23
    job selector, 28
    lubrication, 24
    machining, 1, 2
    polishing of bands, 9
    pusher, 14
    safety, 11–15
        checklist, 14–15
    speed, 28–30
    straight cutting, 30, 31
    tool velocity, 5–6
    type, 2–6, 7
    use of, 15–20, 26–34
    variable speed drive, 28–30
    worktable, 2–5
Vertical milling machine, 35–74
    collets, 55, 56

Index

cutting tools, 40, 52–57
end mills, 45
feeds, 42
guards, 39
indexing devices, 143–161
machining, 45, 52
lubrication, 43
offset boring head, 69–72
operations, 45
parts, identification, 40
quill feed, 43
safety, 39–40
set up, 58, 59–60
speed, 42
toolhead alignment, 42
workholding, 58
Vises, 110–115, 118–119, 214
alignment, 118–119
all steel, 113
horizontal milling, 110–115
plain, 113
shaper, 214
swivel, 113
universal, 113

Welding of band saw blades, 16–20, 21
annealing of welds, 18, 19
grinding of welds, 19, 20
procedure, 16, 17, 18
Wheel dressing, 411–417
Worktables, vertical band machines, 2–5
power fed, 4, 5
tilting capacity, of fixed types, 2–3
Worm gears, 173

Yield strength, 274, 275
Young's modulus, 272